PERGAMON INTERNATIONAL LIBRARY
of Science, Technology, Engineering and Social Studies
The 1000-volume original paperback library in aid of education, industrial training and the enjoyment of leisure

Publisher: Robert Maxwell, M.C.

L. D. LANDAU & E. M. LIFSHITZ
COURSE OF THEORETICAL PHYSICS

Volume 4

Second Edition

QUANTUM ELECTRODYNAMICS

Other Titles in the Series

Vol. 1. *Mechanics*, 3rd Ed.

Vol. 2. *The Classical Theory of Fields*, 4th Ed.

Vol. 3. *Quantum Mechanics—Non-relativistic Theory*, 3rd Ed.

Vol. 5. *Statistical Physics*, Part 1, 3rd Ed.

Vol. 6. *Fluid Mechanics*

Vol. 7. *Theory of Elasticity*, 2nd Ed.

Vol. 8. *Electrodynamics of Continuous Media*

Vol. 9. *Statistical Physics*, Part 2

Vol. 10. *Physical Kinetics*

A SHORTER COURSE OF THEORETICAL PHYSICS

(Based on the Course of Theoretical Physics)

Vol. 1. *Mechanics and Electrodynamics*

Vol. 2. *Quantum Mechanics*

QUANTUM ELECTRODYNAMICS

by

V. B. BERESTETSKII, E. M. LIFSHITZ

and

L. P. PITAEVSKII

Institute of Physical Problems, U.S.S.R. Academy of Sciences

Volume 4 of *Course of Theoretical Physics*
Second edition

Translated from the Russian by

J. B. SYKES and J. S. BELL

PERGAMON PRESS

OXFORD · NEW YORK · TORONTO · SYDNEY · PARIS · FRANKFURT

U.K.	Pergamon Press Ltd., Headington Hill Hall, Oxford OX3 0BW, England
U.S.A.	Pergamon Press Inc., Maxwell House, Fairview Park, Elmsford, New York 10523, U.S.A.
CANADA	Pergamon Press Canada Ltd., Suite 104, 150 Consumers Rd., Willowdale, Ontario M2J 1P9, Canada
AUSTRALIA	Pergamon Press (Aust.) Pty. Ltd., P.O. Box 544, Potts Point, N.S.W. 2011, Australia
FRANCE	Pergamon Press SARL, 24 rue des Ecoles, 75240 Paris, Cedex 05, France
FEDERAL REPUBLIC OF GERMANY	Pergamon Press GmbH, 6242 Kronberg-Taunus, Hammerweg 6, Federal Republic of Germany

First edition 1971, 1974 (as *Relativistic Quantum Theory*)

Second edition 1982

British Library Cataloguing in Publication Data

Berestetskiĭ, V. B.
Quantum electrodynamics.—2nd ed.
—(Course of theoretical physics; V.4)
1. Quantum field theory
I. TITLE II. Lifshitz, E. M.
III. Pitaevskii, L.P. IV. Berestetskiĭ, V. B.
Relativistic quantum theory V. Series
530.1'2 QC174.45

ISBN 0-08-026503-0 (Hardcover)
ISBN 0-08-026504-9 (Flexicover)

Translated from **Kvantovaya elektrodinamika** published by Izdatel'stvo "Nauka", Moscow, 1980

Printed in Great Britain by A. Wheaton & Co. Exeter

PREFACE TO THE SECOND EDITION

THE first edition of this volume of the *Course of Theoretical Physics* was published in two parts (1971 and 1974) under the title "Relativistic Quantum Theory". It contained not only the basic material on quantum electrodynamics but also chapters on weak interactions and certain topics in the theory of strong interactions. The inclusion of those chapters now seems to us inopportune. The theory of strong and weak interactions is undergoing a vigorous development founded on new physical ideas, and the situation in this field is changing very rapidly, so that the time for a consistent exposition of the theory has not yet arrived. In the present edition, therefore, we have retained only quantum electrodynamics, and accordingly changed the title of the volume.

As well as a considerable number of corrections and minor changes, we have made in this edition several more significant additions, including the operator method of calculating the bremsstrahlung cross-section, the calculation of the probabilities of photon-induced pair production and photon decay in a magnetic field, the asymptotic form of the scattering amplitudes at high energies, inelastic scattering of electrons by hadrons, and the transformation of electron–positron pairs into hadrons.

A word regarding notation. We have reverted to the use of circumflexed letters for operators, in line with the other volumes in the *Course*. No special notation is used for the product of a 4-vector and a matrix vector γ^{μ}, previously denoted by a circumflexed letter; such products are now shown explicitly.

We have, alas, had to prepare this edition without the aid of Vladimir Berestetskiĭ, who died in 1977; but some of the added material mentioned above had been put together previously, by the three authors jointly.

Our sincere thanks are offered to all readers who have given us their comments on the first edition of the book, and in particular to J. S. Bell, V. P. Kraĭnov, L. B. Okun', V. I. Ritus, M. I. Ryazanov and I. S. Shapiro.

July 1979

E. M. LIFSHITZ
L. P. PITAEVSKIĬ

FROM THE PREFACE TO THE FIRST EDITION

IN ACCORDANCE with the general plan of this *Course of Theoretical Physics*, the present volume deals with relativistic quantum theory in the broad sense: the theory of all phenomena which depend upon the finite velocity of light, including the whole of the theory of radiation.

This branch of theoretical physics is still far from completion, even as regards its basic physical principles, and this is particularly true of the theory of strong and weak interactions. But even quantum electrodynamics, despite the remarkable achievements of the last twenty years, still lacks a satisfactory logical structure.

In the choice of material for this book we have considered only results which appear to be reasonably firmly established. In consequence, of course, the greater part of the book is devoted to quantum electrodynamics. We have tried to give a realistic exposition, with emphasis on the physical hypotheses used in the theory, but without going into details of justifications, which in the present state of the theory are in any case purely formal.

In the discussion of specific applications of the theory, our aim has been not to include the whole vast range of effects but to select only the most fundamental of them, adding some references to original papers which contain more detailed studies. We have often omitted some of the intermediate steps in the calculations, which in this subject are usually very lengthy, but we have always sought to indicate any non-trivial point of technique.

The discussion in this book demands a higher degree of previous knowledge on the part of the reader than do the other volumes in the *Course.* Our assumption has been that a reader whose study of theoretical physics has extended as far as the quantum theory of fields has no further need of predigested material.

This book has been written without the direct assistance of our teacher, L. D. Landau. Yet we have striven to be guided by the spirit and the approach to theoretical physics which characterized his teaching of us and which he embodied in the other volumes. We have often asked ourselves what would be the attitude of Dau to this or that topic, and sought the answer prompted by our many years' association with him.

Our thanks are due to V. N. Baĭer, who gave great help in compiling §§90 and 97, and to V. I. Ritus for great help in writing §101. We are grateful to B. É. Meĭerovich for assistance with calculations, and also to A. S. Kompaneets, who made available his notes of L. D. Landau's lectures on quantum electrodynamics, given at Moscow State University in the academic year 1959–60.

June 1967 V. B. BERESTETSKIĬ, E. M. LIFSHITZ, L. P. PITAEVSKIĬ

CONTENTS

IV. PARTICLES IN AN EXTERNAL FIELD

V. RADIATION

VI. SCATTERING OF RADIATION

VII. THE SCATTERING MATRIX

VIII. INVARIANT PERTURBATION THEORY

NOTATION

Four-dimensional

Four-dimensional tensor indices are denoted by Greek letters $\lambda, \mu, \nu, \ldots$, taking the values 0, 1, 2, 3.

A 4-metric with signature $(+---)$ is used. The metric tensor is

$$g_{\mu\nu}(g_{00} = 1,\ g_{11} = g_{22} = g_{33} = -1).$$

Components of a 4-vector are stated in the form $a^{\mu} = (a^0, \mathbf{a})$.

To simplify the formulae, the index is often omitted in writing the components of a 4-vector.† The scalar products of 4-vectors are written simply as (ab) or ab; $ab \equiv a_{\mu}b^{\mu} = a_0b_0 - \mathbf{a}\cdot\mathbf{b}$.

The 4-position-vector is $x^{\mu} = (t, \mathbf{r})$. The 4-volume element is d^4x.

The operator of differentiation with respect to the 4-coordinates is $\partial_{\mu} = \partial/\partial x^{\mu}$.

The antisymmetric unit 4-tensor is $e^{\lambda\mu\nu\rho}$, with $e^{0123} = -e_{0123} = +1$.

The four-dimensional delta function $\delta^{(4)}(a) = \delta(a_0)\delta(\mathbf{a})$.

Three-dimensional

Three-dimensional tensor indices are denoted by Latin letters $i, k, l, \ldots$, taking the values x, y, z.

Three-dimensional vectors are denoted by letters in bold type.

The three-dimensional volume element is d^3x.

Operators

Operators are denoted by italic letters with circumflex.‡

Commutators or anticommutators of two operators are written $\{\hat{f}, \hat{g}\}_{\pm} = \hat{f}\hat{g} \pm \hat{g}\hat{f}$.

The transposed operator is $\tilde{\hat{f}}$.

The Hermitian conjugate operator is $\hat{f}^{+}$.

Matrix elements

The matrix element of the operator $\hat{F}$ for a transition from initial state i to final state f is F_{fi} or $\langle f|F|i\rangle$.

† This way of writing the components is often used in recent literature. It is a compromise between the limited resources of the alphabet and the demands of physics, and means, of course, that the reader must be particularly attentive.

‡ However, to simplify the formulae, the circumflex is not written over spin matrices, and it is also omitted when operators are shown in matrix elements.

The notation $|i\rangle$ is used as an abstract symbol for a state independently of any specific representation in which its wave function may be expressed. The notation $\langle f|$ denotes a final ("complex conjugate") state.†

Correspondingly, $\langle s|r\rangle$ denotes the coefficients in the expression of a set of states with quantum numbers r as superpositions of states with quantum numbers s: $|r\rangle = \sum_s |s\rangle\langle s|r\rangle$.

The reduced matrix elements of spherical tensors are $\langle f\|F\|i\rangle$.

Dirac's equation

The Dirac matrices are γ^μ, with $(\gamma^0)^2 = 1$, $(\gamma^1)^2 = (\gamma^2)^2 = (\gamma^3)^2 = -1$. The matrix $\boldsymbol{\alpha} = \gamma^0\boldsymbol{\gamma}$, $\beta = \gamma^0$. The expressions in the spinor and standard representations are (21.3), (21.16) and (21.20).

$\gamma^5 = -i\gamma^0\gamma^1\gamma^2\gamma^3$, $(\gamma^5)^2 = 1$; see (22.18).

$\sigma^{\mu\nu} = \frac{1}{2}(\gamma^\mu\gamma^\nu - \gamma^\nu\gamma^\mu)$; see (28.2).

Dirac conjugation is expressed by $\bar{\psi} = \psi^*\gamma^0$.

The Pauli matrices are $\boldsymbol{\sigma} = (\sigma_x, \sigma_y, \sigma_z)$, defined in §20.

The 4-spinor indices are $\alpha, \beta, \ldots$ and $\dot{\alpha}, \dot{\beta}, \ldots$, taking the values 1, 2 and $\dot{1}$, $\dot{2}$.

The bispinor indices are $i, k, l, \ldots$, taking the values 1, 2, 3, 4.

Fourier expansion

Three-dimensional:

$$f(\mathbf{r}) = \int f(\mathbf{k})\, e^{i\mathbf{k}\cdot\mathbf{r}} \frac{d^3k}{(2\pi)^3}, \qquad f(\mathbf{k}) = \int f(\mathbf{r})\, e^{-i\mathbf{k}\cdot\mathbf{r}}\, d^3x,$$

and similarly for the four-dimensional expansion.

Units

Except where otherwise specified, *relativistic units* are used, with $\hbar = 1$, $c = 1$. In these units, the square of the unit charge is $e^2 = 1/137$.

Atomic units have $e = 1$, $\hbar = 1$, $m = 1$. In these units, $c = 137$. The atomic units of length, time and energy are $\hbar^2/me^2$, $\hbar^3/me^4$ and $me^4/\hbar^2$; the quantity $Ry = me^4/2\hbar^2$ is called a *rydberg*.

Ordinary units are given in the absolute (Gaussian) system.

Constants

Velocity of light $c = 2.998 \times 10^{10}$ cm/sec.

Unit charge‡ $|e| = 4.803 \times 10^{-10}$ CGS electrostatic units.

† This notation is due to Dirac.

‡ Throughout the book (except in Chapter XIV), e denotes the charge with the appropriate sign, so that $e = -|e|$ for an electron.

Electron mass $m = 9.11 \times 10^{-28}$ g.
Planck's constant $\hbar = 1.055 \times 10^{-27}$ erg. sec.
Fine-structure constant $\alpha = e^2/\hbar c$; $1/\alpha = 137.04$.
Bohr radius $\hbar^2/me^2 = 5.292 \times 10^{-9}$ cm.
Classical electron radius $r_e = e^2/mc^2 = 2.818 \times 10^{-13}$ cm.
Compton wavelength of the electron $\hbar/mc = 3.862 \times 10^{-11}$ cm.
Electron rest energy $mc^2 = 0.511 \times 10^6$ eV.
Atomic energy unit $me^4/\hbar^2 = 4.360 \times 10^{-11}$ erg $= 27.21$ eV.
Bohr magneton $|e|\hbar/2mc = 9.274 \times 10^{-21}$ erg/G.
Proton mass $m_p = 1.673 \times 10^{-24}$ g.
Compton wavelength of the proton $\hbar/m_p c = 2.103 \times 10^{-14}$ cm.
Nuclear magneton $|e|\hbar/2m_p c = 5.051 \times 10^{-24}$ erg/G.
Mass ratio of muon and electron $m_\mu/m = 2.068 \times 10^2$.

References to volumes in the *Course of Theoretical Physics*:

Mechanics = Vol. 1 (*Mechanics*, third English edition, 1976).
Fields = Vol. 2 (*The Classical Theory of Fields*, fourth English edition, 1975).
QM or *Quantum Mechanics* = Vol. 3 (*Quantum Mechanics*, third English edition, 1977).
ECM = Vol. 8 (*Electrodynamics of Continuous Media*, English edition, 1960).
PK = Vol. 10 (*Physical Kinetics*, English edition, 1981).

All are published by Pergamon Press.

INTRODUCTION

§ 1. The uncertainty principle in the relativistic case

THE quantum theory described in Volume 3 (*Quantum Mechanics*) is essentially non-relativistic throughout, and is not applicable to phenomena involving motion at velocities comparable with that of light. At first sight, one might expect that the change to a relativistic theory is possible by a fairly direct generalization of the formalism of non-relativistic quantum mechanics. But further consideration shows that a logically complete relativistic theory cannot be constructed without invoking new physical principles.

Let us recall some of the physical concepts forming the basis of non-relativistic quantum mechanics (*QM*, §1). We saw that one fundamental concept is that of *measurement*, by which is meant the process of interaction between a quantum system and a classical object or *apparatus*, causing the quantum system to acquire definite values of some particular dynamical variables (coordinates, velocities, etc.). We saw also that quantum mechanics greatly restricts the possibility that an electron† simultaneously possesses values of different dynamical variables. For example, the uncertainties Δq and Δp in simultaneously existing values of the coordinate and the momentum are related by the expression‡ $\Delta q \Delta p \sim \hbar$; the greater the accuracy with which one of these quantities is measured, the less the accuracy with which the other can be measured at the same time.

It is important to note, however, that any of the dynamical variables of the electron can individually be measured with arbitrarily high accuracy, and in an arbitrarily short period of time. This fact is of fundamental importance throughout non-relativistic quantum mechanics. It is the only justification for using the concept of the wave function, which is a basic part of the formalism. The physical significance of the wave function $\psi(q)$ is that the square of its modulus gives the probability of finding a particular value of the electron coordinate as the result of a measurement made at a given instant. The concept of such a probability clearly requires that the coordinate can in principle be measured with any specified accuracy and rapidity, since otherwise this concept would be purposeless and devoid of physical significance.

The existence of a limiting velocity (the velocity of light, denoted by c) leads to new fundamental limitations on the possible measurements of various physical quantities (L. D. Landau and R. E. Peierls, 1930).

† As in *QM*, §1, we shall, for brevity, speak of an "electron", meaning any quantum system.

‡ In this section, ordinary units are used.

In *QM*, §44, the following relationship has been derived:

$$(v'-v)\Delta p\,\Delta t \sim \hbar, \tag{1.1}$$

relating the uncertainty Δp in the measurement of the electron momentum and the duration Δt of the measurement process itself; v and v' are the velocities of the electron before and after the measurement. From this relationship it follows that a momentum measurement of high accuracy made during a short time (i.e. with Δp and Δt both small) can occur only if there is a large change in the velocity as a result of the measurement process itself. In the non-relativistic theory, this showed that the measurement of momentum cannot be repeated at short intervals of time, but it did not at all diminish the possibility, in principle, of making a single measurement of the momentum with arbitrarily high accuracy, since the difference $v'-v$ could take any value, no matter how large.

The existence of a limiting velocity, however, radically alters the situation. The difference $v'-v$, like the velocities themselves, cannot now exceed c (or rather $2c$). Replacing $v'-v$ in (1.1) by c, we obtain

$$\Delta p\,\Delta t \sim \hbar/c, \tag{1.2}$$

which determines the highest accuracy theoretically attainable when the momentum is measured by a process occupying a given time Δt. In the relativistic theory, therefore, it is in principle impossible to make an arbitrarily accurate and rapid measurement of the momentum. An exact measurement ($\Delta p \to 0$) is possible only in the limit as the duration of the measurement tends to infinity.

There is reason to suppose that the concept of measurability of the electron coordinate itself must also undergo modification. In the mathematical formalism of the theory, this situation is shown by the fact that an accurate measurement of the coordinate is incompatible with the assertion that the energy of a free particle is positive. It will be seen later that the complete set of eigenfunctions of the relativistic wave equation of a free particle includes, as well as solutions having the "correct" time dependence, also solutions having a "negative frequency". These functions will in general appear in the expansion of the wave packet corresponding to an electron localized in a small region of space.

It will be shown that the wave functions having a "negative frequency" correspond to the existence of antiparticles (positrons). The appearance of these functions in the expansion of the wave packet expresses the (in general) inevitable production of electron–positron pairs in the process of measuring the coordinates of an electron. This formation of new particles in a way which cannot be detected by the process itself renders meaningless the measurement of the electron coordinates.

In the rest frame of the electron, the least possible error in the measurement of its coordinates is

$$\Delta q \sim \hbar/mc. \tag{1.3}$$

This value (which purely dimensional arguments show to be the only possible one) corresponds to a momentum uncertainty $\Delta p \sim mc$, which in turn corresponds to the threshold energy for pair production.

In a frame of reference in which the electron is moving with energy ε, (1.3) becomes

$$\Delta q \sim c\hbar/\varepsilon. \tag{1.4}$$

In particular, in the limiting ultra-relativistic case the energy is related to the momentum by $\varepsilon \approx cp$, and

$$\Delta q \sim \hbar/p, \tag{1.5}$$

i.e. the error Δq is the same as the de Broglie wavelength of the particle.†

For photons, the ultra-relativistic case always applies, and the expression (1.5) is therefore valid. This means that the coordinates of a photon are meaningful only in cases where the characteristic dimensions of the problem are large in comparison with the wavelength. This is just the "classical" limit, corresponding to geometrical optics, in which radiation can be said to be propagated along definite paths or rays. In the quantum case, however, where the wavelength cannot be regarded as small, the concept of coordinates of the photon has no meaning. We shall see later (§4) that, in the mathematical formalism of the theory, the fact that the photon coordinates cannot be measured is evident because the photon wave function cannot be used to construct a quantity which might serve as a probability density satisfying the necessary conditions of relativistic invariance.

The foregoing discussion suggests that the theory will not consider the time dependence of particle interaction processes. It will show that in these processes there are no characteristics precisely definable (even within the usual limitations of quantum mechanics); the description of such a process as occurring in the course of time is therefore just as unreal as the classical paths are in non-relativistic quantum mechanics. The only observable quantities are the properties (momenta, polarizations) of free particles: the initial particles which come into interaction, and the final particles which result from the process (L. D. Landau and R. E. Peierls, 1930).

A typical problem as formulated in relativistic quantum theory is to determine the probability amplitudes of transitions between specified initial and final states ($t \to \mp\infty$) of a system of particles. The set of such amplitudes between all possible states constitutes the *scattering matrix* or *S-matrix*. This matrix will embody all the information about particle interaction processes that has an observable physical meaning (W. Heisenberg, 1938).

There is as yet no logically consistent and complete relativistic quantum theory. We shall see that the existing theory introduces new physical features into the nature of the description of particle states, which acquires some of the features of

† The measurements in question are those for which any experimental result yields a conclusion about the state of the electron; that is, we are not considering coordinate measurements by means of collisions, when the result does not occur with probability unity during the time of observation. Although the deflection of a measuring-particle in such cases may indicate the position of an electron, the absence of a deflection tells us nothing.

field theory (see §10). The theory is, however, largely constructed on the pattern of ordinary quantum mechanics. This structure of the theory has yielded good results in quantum electrodynamics. The lack of complete logical consistency in this theory is shown by the occurrence of divergent expressions when the mathematical formalism is directly applied, although there are quite well-defined ways of eliminating these divergences. Nevertheless, such methods remain, to a considerable extent, semiempirical rules, and our confidence in the correctness of the results is ultimately based only on their excellent agreement with experiment, not on the internal consistency or logical ordering of the fundamental principles of the theory.

CHAPTER I

PHOTONS

§2. Quantization of the free electromagnetic field

WITH the purpose of treating the electromagnetic field as a quantum object, it is convenient to begin from a classical description of the field in which it is represented by an infinite but discrete set of variables. This description permits the immediate application of the customary formalism of quantum mechanics. The representation of the field by means of potentials specified at every point in space is essentially a description by means of a continuous set of variables.

Let $\mathbf{A}(\mathbf{r}, t)$ be the vector potential of the free electromagnetic field, which satisfies the "transversality condition"

$$\operatorname{div} \mathbf{A} = 0. \tag{2.1}$$

The scalar potential $\Phi = 0$, and the fields $\mathbf{E}$ and $\mathbf{H}$ are

$$\mathbf{E} = -\dot{\mathbf{A}}, \qquad \mathbf{H} = \operatorname{curl} \mathbf{A}. \tag{2.2}$$

Maxwell's equations reduce to the wave equation for $\mathbf{A}$:

$$\triangle \mathbf{A} - \partial^2 \mathbf{A}/\partial t^2 = 0. \tag{2.3}$$

In classical electrodynamics (see *Fields*, §52) the change to the description by means of a discrete set of variables is brought about by considering the field in a large but finite volume V.† The following is a brief résumé of the argument.

The field in a finite volume can be expanded in terms of travelling plane waves, and its potential is then represented by a series

$$\mathbf{A} = \sum_{\mathbf{k}} (\mathbf{a}_{\mathbf{k}}\, e^{i\mathbf{k}\cdot\mathbf{r}} + \mathbf{a}_{\mathbf{k}}^*\, e^{-i\mathbf{k}\cdot\mathbf{r}}), \tag{2.4}$$

where the coefficients $\mathbf{a}_{\mathbf{k}}$ are functions of the time such that

$$\mathbf{a}_{\mathbf{k}} \sim e^{-i\omega t}, \qquad \omega = |\mathbf{k}|. \tag{2.5}$$

The condition (2.1) shows that the complex vectors $\mathbf{a}_{\mathbf{k}}$ are orthogonal to the corresponding wave vectors: $\mathbf{a}_{\mathbf{k}} \cdot \mathbf{k} = 0$.

The summation in (2.4) is taken over an infinite discrete set of values of the

† We shall take $V = 1$, in order to reduce the number of factors in the formulae.

wave vector (i.e. of its components k_x, k_y, k_z). The change to an integral over a continuous distribution may be made by means of the expression $d^3k/(2\pi)^3$ for the number of possible values of **k** belonging to the volume element $d^3k = dk_x dk_y dk_z$ in **k**-space.

If the vectors $\mathbf{a_k}$ are specified, the field in the volume considered is completely determined. Thus these quantities may be regarded as a discrete set of classical "field variables". In order to explain the transition to the quantum theory, however, a further transformation of these variables is needed, whereby the field equations take a form analogous to the canonical equations (Hamilton's equations) of classical mechanics. The canonical field variables are defined by

$$\left.\begin{aligned} \mathbf{Q_k} &= \frac{1}{\sqrt{(4\pi)}}(\mathbf{a_k} + \mathbf{a_k^*}), \\ \mathbf{P_k} &= \frac{-i\omega}{\sqrt{(4\pi)}}(\mathbf{a_k} - \mathbf{a_k^*}) = \dot{\mathbf{Q}}_\mathbf{k}, \end{aligned}\right\} \tag{2.6}$$

and are evidently real. The vector potential is expressed in terms of the canonical variables by

$$\mathbf{A} = \sqrt{(4\pi)} \sum_\mathbf{k} \left(\mathbf{Q_k} \cos \mathbf{k}\cdot\mathbf{r} - \frac{1}{\omega}\mathbf{P_k} \sin \mathbf{k}\cdot\mathbf{r}\right). \tag{2.7}$$

To find the Hamiltonian H, we must calculate the total energy of the field,

$$\frac{1}{8\pi}\int (\mathbf{E}^2 + \mathbf{H}^2)\, d^3x,$$

and express it in terms of the $\mathbf{Q_k}$ and $\mathbf{P_k}$. When **A** is written as the expansion (2.7), and **E** and **H** are found from (2.2), the result of the integration is

$$H = \tfrac{1}{2}\sum_\mathbf{k} (\mathbf{P_k^2} + \omega^2\mathbf{Q_k^2}).$$

Each of the vectors $\mathbf{P_k}$ and $\mathbf{Q_k}$ is perpendicular to the wave vector **k**, and therefore has two independent components. The direction of these vectors determines the direction of polarization of the corresponding wave. Denoting the two components of the vectors $\mathbf{Q_k}$ and $\mathbf{P_k}$ (in the plane perpendicular to **k**) by $Q_{\mathbf{k}\alpha}$, $P_{\mathbf{k}\alpha}$ ($\alpha = 1, 2$), we can write the Hamiltonian as

$$H = \sum_{\mathbf{k},\alpha} \tfrac{1}{2}(P_{\mathbf{k}\alpha}^2 + \omega^2 Q_{\mathbf{k}\alpha}^2). \tag{2.8}$$

Thus the Hamiltonian is the sum of independent terms, each of which contains only one pair of quantities $Q_{\mathbf{k}\alpha}$, $P_{\mathbf{k}\alpha}$. Each such term corresponds to a travelling wave with a definite wave vector and polarization, and has the form of the Hamiltonian for a one-dimensional harmonic oscillator. This expansion is therefore often referred to as an *oscillator expansion* of the field.

Let us now consider the quantization of the free electromagnetic field. The classical description of the field given above makes the manner of transition to the quantum theory obvious. We have now to use canonical variables (generalized coordinates $Q_{\mathbf{k}\alpha}$ and generalized momenta $P_{\mathbf{k}\alpha}$) as operators, with the commutation rule

$$\hat{P}_{\mathbf{k}\alpha}\hat{Q}_{\mathbf{k}\alpha} - \hat{Q}_{\mathbf{k}\alpha}\hat{P}_{\mathbf{k}\alpha} = -i; \tag{2.9}$$

operators with different values of $\mathbf{k}$ and α always commute. The potential $\mathbf{A}$ and, according to (2.2), the fields $\mathbf{E}$ and $\mathbf{H}$ likewise become (Hermitian) operators.

The consistent determination of the Hamiltonian requires the calculation of the integral

$$\hat{H} = \frac{1}{8\pi}\int (\hat{\mathbf{E}}^2 + \hat{\mathbf{H}}^2)\, d^3x, \tag{2.10}$$

in which $\hat{\mathbf{E}}$ and $\hat{\mathbf{H}}$ are expressed in terms of $\hat{P}_{\mathbf{k}\alpha}$ and $\hat{Q}_{\mathbf{k}\alpha}$. However, the fact that the latter do not commute is actually unimportant, since the products $\hat{Q}_{\mathbf{k}\alpha}\hat{P}_{\mathbf{k}\alpha}$ appear multiplied by $\cos \mathbf{k}\cdot\mathbf{r} \sin \mathbf{k}\cdot\mathbf{r}$, which becomes zero on integration over the whole volume. The resulting expression for the Hamiltonian is therefore

$$\hat{H} = \sum_{\mathbf{k},\alpha} \tfrac{1}{2}(\hat{P}^2_{\mathbf{k}\alpha} + \omega^2\hat{Q}^2_{\mathbf{k}\alpha}), \tag{2.11}$$

which is, as we might have expected, exactly the same in form as the classical Hamiltonian.

The determination of the eigenvalues of this Hamiltonian involves no further calculation, since it is equivalent to the familiar problem of the energy levels of linear oscillators (*QM*, §23). We can therefore immediately write down the field energy levels:

$$E = \sum_{\mathbf{k},\alpha} (N_{\mathbf{k}\alpha} + \tfrac{1}{2})\omega, \tag{2.12}$$

where the $N_{\mathbf{k}\alpha}$ are integers.

The further discussion of this formula will be left until §3; here we shall write out the matrix elements of the quantities $Q_{\mathbf{k}\alpha}$, which can be done at once by means of the known formulae for the matrix elements of the coordinates of an oscillator (see *QM*, §23). The non-zero matrix elements are

$$\begin{aligned}\langle N_{\mathbf{k}\alpha}|Q_{\mathbf{k}\alpha}|N_{\mathbf{k}\alpha}-1\rangle &= \langle N_{\mathbf{k}\alpha}-1|Q_{\mathbf{k}\alpha}|N_{\mathbf{k}\alpha}\rangle \\ &= \sqrt{(N_{\mathbf{k}\alpha}/2\omega)}.\end{aligned} \tag{2.13}$$

The matrix elements of the quantities $P_{\mathbf{k}\alpha} = \dot{Q}_{\mathbf{k}\alpha}$ differ from those of $Q_{\mathbf{k}\alpha}$ only by a factor $\pm i\omega$.

In subsequent calculations, however, it will be more convenient to replace the quantities $Q_{\mathbf{k}\alpha}$ and $P_{\mathbf{k}\alpha}$ by the linear combinations $\omega Q_{\mathbf{k}\alpha} \pm iP_{\mathbf{k}\alpha}$, which have non-zero matrix elements only for transitions $N_{\mathbf{k}\alpha} \to N_{\mathbf{k}\alpha} \pm 1$. We therefore define the operators

$$\left.\begin{aligned} \hat{c}_{\mathbf{k}\alpha} &= \frac{1}{\sqrt{(2\omega)}}(\omega\hat{Q}_{\mathbf{k}\alpha} + i\hat{P}_{\mathbf{k}\alpha}), \\ \hat{c}^{+}_{\mathbf{k}\alpha} &= \frac{1}{\sqrt{(2\omega)}}(\omega\hat{Q}_{\mathbf{k}\alpha} - i\hat{P}_{\mathbf{k}\alpha}); \end{aligned}\right\} \tag{2.14}$$

the classical quantities $c_{\mathbf{k}\alpha}$, $c^{*}_{\mathbf{k}\alpha}$ are the same, apart from a factor $\sqrt{(2\pi/\omega)}$, as the coefficients $a_{\mathbf{k}\alpha}$, $a^{*}_{\mathbf{k}\alpha}$ in the expansion (2.4). The matrix elements of these operators are

$$\begin{aligned} \langle N_{\mathbf{k}\alpha} - 1|\hat{c}_{\mathbf{k}\alpha}|N_{\mathbf{k}\alpha}\rangle &= \langle N_{\mathbf{k}\alpha}|\hat{c}^{+}_{\mathbf{k}\alpha}|N_{\mathbf{k}\alpha} - 1\rangle \\ &= \sqrt{N_{\mathbf{k}\alpha}}. \end{aligned} \tag{2.15}$$

The commutation rule for $\hat{c}_{\mathbf{k}\alpha}$ and $\hat{c}^{+}_{\mathbf{k}\alpha}$ is obtained by using the definitions (2.14) and the rule (2.9):

$$\hat{c}_{\mathbf{k}\alpha}\hat{c}^{+}_{\mathbf{k}\alpha} - \hat{c}^{+}_{\mathbf{k}\alpha}\hat{c}_{\mathbf{k}\alpha} = 1. \tag{2.16}$$

For the vector potential, we return to an expansion of the type (2.4), but with operator coefficients, writing it in the form

$$\hat{\mathbf{A}} = \sum_{\mathbf{k},\alpha} (\hat{c}_{\mathbf{k}\alpha}\mathbf{A}_{\mathbf{k}\alpha} + \hat{c}^{+}_{\mathbf{k}\alpha}\mathbf{A}^{*}_{\mathbf{k}\alpha}), \tag{2.17}$$

where

$$\mathbf{A}_{\mathbf{k}\alpha} = \sqrt{(4\pi)}\frac{\mathbf{e}^{(\alpha)}}{\sqrt{(2\omega)}}e^{i\mathbf{k}\cdot\mathbf{r}}. \tag{2.18}$$

The symbol $\mathbf{e}^{(\alpha)}$ denotes the unit vectors in the direction of polarization of the oscillators; these vectors are perpendicular to the wave vector $\mathbf{k}$, and for every $\mathbf{k}$ there are two independent polarizations.

Similarly, for the operators $\hat{\mathbf{E}}$ and $\hat{\mathbf{H}}$ we write

$$\left.\begin{aligned} \hat{\mathbf{E}} &= \sum_{\mathbf{k},\alpha} (\hat{c}_{\mathbf{k}\alpha}\mathbf{E}_{\mathbf{k}\alpha} + \hat{c}^{+}_{\mathbf{k}\alpha}\mathbf{E}^{*}_{\mathbf{k}\alpha}), \\ \hat{\mathbf{H}} &= \sum_{\mathbf{k},\alpha} (\hat{c}_{\mathbf{k}\alpha}\mathbf{H}_{\mathbf{k}\alpha} + \hat{c}^{+}_{\mathbf{k}\alpha}\mathbf{H}^{*}_{\mathbf{k}\alpha}), \end{aligned}\right\} \tag{2.19}$$

with

$$\mathbf{E}_{\mathbf{k}\alpha} = i\omega\mathbf{A}_{\mathbf{k}\alpha}, \qquad \mathbf{H}_{\mathbf{k}\alpha} = \mathbf{n}\times\mathbf{E}_{\mathbf{k}\alpha}, \qquad \mathbf{n} = \mathbf{k}/\omega. \tag{2.20}$$

The vectors $\mathbf{A}_{\mathbf{k}\alpha}$ are mutually orthogonal, in the sense that

$$\int \mathbf{A}_{\mathbf{k}\alpha}\cdot\mathbf{A}^{*}_{\mathbf{k}'\alpha'}\,d^3x = \frac{2\pi}{\omega}\delta_{\alpha\alpha'}\delta_{\mathbf{k}\mathbf{k}'}. \tag{2.21}$$

For, if $\mathbf{A}_{\mathbf{k}\alpha}$ and $\mathbf{A}^*_{\mathbf{k}'\alpha'}$ belong to different wave vectors, then their product contains a factor $e^{i(\mathbf{k}-\mathbf{k}')\cdot\mathbf{r}}$, which gives zero on integration over the volume; if they differ only in polarization, $\mathbf{e}^{(\alpha)}\cdot\mathbf{e}^{(\alpha')*}=0$, since the two independent directions of polarization are mutually orthogonal. Similar arguments apply to the vectors $\mathbf{E}_{\mathbf{k}\alpha}$ and $\mathbf{H}_{\mathbf{k}\alpha}$. They are conveniently normalized by imposing the condition

$$\frac{1}{4\pi}\int (\mathbf{E}_{\mathbf{k}\alpha}\cdot\mathbf{E}^*_{\mathbf{k}'\alpha'}+\mathbf{H}_{\mathbf{k}\alpha}\cdot\mathbf{H}^*_{\mathbf{k}'\alpha'})\,d^3x=\omega\delta_{\mathbf{k}\mathbf{k}'}\delta_{\alpha\alpha'}. \tag{2.22}$$

Substituting the operators (2.19) in (2.10), and carrying out the integration by means of (2.22), we obtain the field Hamiltonian expressed in terms of the operators $\hat{c}$, $\hat{c}^+$:

$$\hat{H}=\sum_{\mathbf{k},\alpha}\tfrac{1}{2}\omega(\hat{c}_{\mathbf{k}\alpha}\hat{c}^+_{\mathbf{k}\alpha}+\hat{c}^+_{\mathbf{k}\alpha}\hat{c}_{\mathbf{k}\alpha}). \tag{2.23}$$

This operator is diagonal in the representation considered (the matrix elements of the operators $\hat{c}$ and $\hat{c}^+$ being given by (2.15)), and its eigenvalues are of course (2.12).

In the classical theory, the field momentum is defined as the integral

$$\mathbf{P}=\frac{1}{4\pi}\int \mathbf{E}\times\mathbf{H}\,d^3x.$$

In changing to the quantum theory, we replace $\mathbf{E}$ and $\mathbf{H}$ by the operators (2.19), and thus easily find

$$\hat{\mathbf{P}}=\sum_{\mathbf{k},\alpha}\tfrac{1}{2}(\hat{P}^2_{\mathbf{k}\alpha}+\omega^2\hat{Q}^2_{\mathbf{k}\alpha})\mathbf{n}, \tag{2.24}$$

in agreement with the familiar classical relationship between the energy and momentum of plane waves. The eigenvalues of this operator are

$$\mathbf{P}=\sum_{\mathbf{k},\alpha}\mathbf{k}(N_{\mathbf{k}\alpha}+\tfrac{1}{2}). \tag{2.25}$$

The representation of operators by means of the matrix elements (2.15) is the "occupation number representation", corresponding to the description of the state of a system (the field) by specifying the quantum numbers $N_{\mathbf{k}\alpha}$ (the *occupation numbers*). In this representation the field operators (2.19), and therefore the Hamiltonian (2.11), act on the wave function of the system, expressed in terms of the numbers $N_{\mathbf{k}\alpha}$; let this be $\Phi(N_{\mathbf{k}\alpha}, t)$. The field operators (2.19) are not explicit functions of the time. This corresponds to the customary Schrödinger representation of operators in non-relativistic quantum mechanics. The state of the system, $\Phi(N_{\mathbf{k}\alpha}, t)$, does depend on the time, and this dependence is governed by Schrödinger's equation,

$$i\,\partial\Phi/\partial t=\hat{H}\Phi.$$

This description of the field is, by its nature, relativistically invariant, since it is based on the invariant Maxwell's equations. But this invariance is not explicitly shown, primarily because the space coordinates and the time appear in the description in a highly asymmetric manner.

In relativistic theory, it is convenient to put the description in a form which is more obviously invariant. To do so, we must use what is called the Heisenberg representation, in which the explicit time dependence is transferred to the operators themselves (see *QM*, §13). Then the time and the coordinates will appear on an equal footing in the expressions for the field operators, and the state of the system, Φ, will depend only on the occupation numbers.

For the operator $\hat{\mathbf{A}}$, the change to the Heisenberg representation amounts to replacing the factor $e^{i\mathbf{k}\cdot\mathbf{r}}$ in each term of the sum (2.17) by $e^{i(\mathbf{k}\cdot\mathbf{r}-\omega t)}$, i.e. to regarding the $\mathbf{A}_{\mathbf{k}\alpha}$ as the time-dependent functions

$$\mathbf{A}_{\mathbf{k}\alpha} = \sqrt{(4\pi)}\frac{\mathbf{e}^{(\alpha)}}{\sqrt{(2\omega)}}e^{-i(\omega t-\mathbf{k}\cdot\mathbf{r})}. \tag{2.26}$$

This is easily proved by noticing that the matrix element of the Heisenberg operator for the transition $i \to f$ must include a factor $\exp\{-i(E_i - E_f)t\}$, where E_i and E_f are the energies of the initial and final states (see *QM*, §13). For a transition in which $N_{\mathbf{k}}$ decreases or increases by 1, this factor becomes $e^{-i\omega t}$ or $e^{i\omega t}$ respectively, a condition which is satisfied by effecting the change mentioned above.

Henceforward, in discussing both the electromagnetic field and particle fields, we shall always assume that the Heisenberg representation of operators is used.

§3. Photons

We shall now further analyse the field quantization formulae obtained in §2.

First of all, formula (2.12) for the field energy raises the following difficulty. The lowest energy level of the field corresponds to the case where the quantum numbers $N_{\mathbf{k}\alpha}$ of all the oscillators are zero; this is called the *electromagnetic field vacuum* state. But, even in that state, each oscillator has a non-zero "zero-point energy" equal to $\frac{1}{2}\omega$. Summation over an infinite number of oscillators then gives an infinite result. Thus we meet with one of the "divergences" which are due to the fact that the present theory is not logically complete and consistent.

So long as only the field energy eigenvalues are under discussion, we can remove this difficulty by simply striking out the zero-point oscillation energy, i.e. by writing the field energy and momentum as†

$$E = \sum_{\mathbf{k},\alpha} N_{\mathbf{k}\alpha}\omega, \qquad \mathbf{P} = \sum_{\mathbf{k},\alpha} N_{\mathbf{k}\alpha}\mathbf{k}. \tag{3.1}$$

† This procedure can be formally carried out without contradiction if we agree to regard the products of operators in (2.10) as "normal" products, that is, as products in which the operators $\hat{c}^+$ are always placed to the left of the operators $\hat{c}$. Then formula (2.23) becomes

$$\hat{H} = \sum_{\mathbf{k},\alpha} \omega \hat{c}^+_{\mathbf{k}\alpha}\hat{c}_{\mathbf{k}\alpha}.$$

These formulae enable us to introduce the concept of *radiation quanta* or *photons*, which is fundamental throughout quantum electrodynamics.† We may regard the free electromagnetic field as an ensemble of particles each with energy ω ($=\hbar\omega$) and momentum $\mathbf{k}$ ($=\mathbf{n}\hbar\omega/c$). The relationship between the photon energy and momentum is as it should be in relativistic mechanics for particles having zero rest-mass and moving with the velocity of light. The occupation numbers $N_{\mathbf{k}\alpha}$ now represent the numbers of photons having given momentum $\mathbf{k}$ and polarization $\mathbf{e}^{(\alpha)}$. The polarization of the photon is analogous to the spin of other particles; the exact properties of the photon in this respect will be discussed in §6 below.

It is easily seen that the whole of the mathematical formalism developed in §2 is fully in accordance with the representation of the electromagnetic field as an ensemble of photons; it is just the second quantization formalism, applied to the system of photons.‡ In this treatment (see *QM*, §64), the independent variables are the occupation numbers of the states, and the operators act on functions of these numbers. The particle "annihilation" and "creation" operators are of basic importance; they respectively decrease and increase by one the occupation numbers. The $\hat{c}_{\mathbf{k}\alpha}$ and $\hat{c}^+_{\mathbf{k}\alpha}$ are operators of this kind: $\hat{c}_{\mathbf{k}\alpha}$ annihilates a photon in the state $\mathbf{k}$, α, and $\hat{c}^+_{\mathbf{k}\alpha}$ creates a photon in that state.

The commutation rule (2.16) corresponds to particles which obey Bose statistics. Photons, therefore, are bosons, as was to be expected, since the number of photons that can be in any one state must be unrestricted. The significance of this will be further discussed in §5.

The plane waves $\mathbf{A}_{\mathbf{k}\alpha}$ (2.26) which appear in the operator $\hat{\mathbf{A}}$ (2.17) as coefficients of the photon annihilation operators may be treated as the wave functions of photons having given momenta $\mathbf{k}$ and polarizations $\mathbf{e}^{(\alpha)}$. This corresponds to an expansion of the ψ-operator in terms of the wave functions of stationary states of a particle in the non-relativistic second quantization formalism; however, unlike the latter, the expansion (2.17) includes both particle annihilation and particle creation operators. The meaning of this difference is explained in §12.

The wave function (2.26) is normalized by the condition

$$\int \frac{1}{4\pi}(|\mathbf{E}_{\mathbf{k}\alpha}|^2 + |\mathbf{H}_{\mathbf{k}\alpha}|^2)\, d^3x = \omega. \tag{3.2}$$

This is the normalization to "one photon in the volume $V = 1$": the integral on the left is the quantum-mechanical mean value of the photon energy in the state having the given wave function. § The right-hand side of (3.2) is just the energy of a single photon.

The "Schrödinger's equation" for the photon is represented by Maxwell's equations. In the present case (when the potential $\mathbf{A}(\mathbf{r}, t)$ satisfies the condition

† This concept is originally due to A. Einstein (1905).

‡ The application of the second quantization method to the theory of radiation was first worked out by P. A. M. Dirac (1927).

§ It should be noted that the factor $1/4\pi$ in the integral (3.2) is twice the usual factor $1/8\pi$ (2.10). This is ultimately due to the fact that the vectors $\mathbf{E}_{\mathbf{k}\alpha}$, $\mathbf{H}_{\mathbf{k}\alpha}$ are complex, whereas the field operators $\hat{\mathbf{E}}$, $\hat{\mathbf{H}}$ are Hermitian.

(2.1)), this leads to the wave equation:

$$\partial^2\mathbf{A}/\partial t^2 - \triangle\mathbf{A} = 0.$$

The "wave functions" of the photon, in the general case of arbitrary stationary states, are complex solutions of this equation, whose time dependence is given by the factor $e^{-i\omega t}$.

In referring to the photon wave function, we must again emphasize that this can not be regarded as the probability amplitude of the spatial localization of the photon, in contrast to the fundamental significance of the wave function in non-relativistic quantum mechanics. This is because, as has been shown in §1, the concept of the coordinates of the photon has no physical meaning. The mathematical aspect of this situation will be further discussed at the end of §4.

The components of the Fourier expansion of the function $\mathbf{A}(\mathbf{r}, t)$ with respect to the coordinates form the wave function of the photon in the momentum representation; we denote this by $\mathbf{A}(\mathbf{k}, t) = \mathbf{A}(\mathbf{k})\, e^{-i\omega t}$. For example, in a state with a given momentum $\mathbf{k}$ and polarization $\mathbf{e}^{(\alpha)}$, the wave function in the momentum representation is given simply by the coefficient of the exponential factor in (2.26):

$$\mathbf{A}_{\mathbf{k}\alpha}(\mathbf{k}', \alpha') = \sqrt{(4\pi)}\frac{\mathbf{e}^{(\alpha)}}{\sqrt{(2\omega)}}\,\delta_{\mathbf{k}'\mathbf{k}}\delta_{\alpha'\alpha}\,. \tag{3.3}$$

Since the momentum of a free particle is measurable, the wave function in the momentum representation has a more profound physical significance than that in the coordinate representation: it enables us to calculate the probabilities $w_{\mathbf{k}\alpha}$ of various values of the momentum and polarization of a photon in a specified state. According to the general rules of quantum mechanics, $w_{\mathbf{k}\alpha}$ is given by the square of the modulus of the corresponding coefficient in the expansion of the function $\mathbf{A}(\mathbf{k}')$ in terms of the wave functions of states with given $\mathbf{k}$ and $\mathbf{e}^{(\alpha)}$:

$$w_{\mathbf{k}\alpha} \propto \left|\sum_{\mathbf{k}',\alpha'} \mathbf{A}^*_{\mathbf{k}\alpha}(\mathbf{k}', \alpha')\cdot\mathbf{A}(\mathbf{k}')\right|^2,$$

the proportionality coefficient depending on the way in which the functions are normalized. Substitution of (3.3) gives

$$w_{\mathbf{k}\alpha} \propto |\mathbf{e}^{(\alpha)}\cdot\mathbf{A}(\mathbf{k})|^2. \tag{3.4}$$

Summation over the two polarizations gives the probability that the photon momentum is $\mathbf{k}$:

$$w_{\mathbf{k}} \propto |\mathbf{A}(\mathbf{k})|^2. \tag{3.5}$$

§4. Gauge invariance

The field potential in classical electrodynamics is well known to be subject to an arbitrary choice: the components of the 4-potential A_μ can undergo any *gauge transformation* of the form

$$A_\mu \to A_\mu + \partial_\mu \chi, \tag{4.1}$$

where χ is any function of coordinates and time (see *Fields*, §18).

For a plane wave, if we consider only transformations which do not change the form of the potential (proportional to $\exp(-ik_\mu x^\mu)$), the freedom of choice reduces to the possibility of adding to the wave amplitude any 4-vector proportional to k^μ.

This arbitrariness in the potential persists in the quantum theory, of course, where it relates to the field operators or to the wave functions of photons. In order not to prejudice the choice of the potentials, we must replace (2.17) by the corresponding expansion for the operator 4-potential,

$$\hat{A}^\mu = \sum_{\mathbf{k},\alpha} (\hat{c}_{\mathbf{k}\alpha} A^\mu_{\mathbf{k}\alpha} + \hat{c}^+_{\mathbf{k}\alpha} A^{\mu *}_{\mathbf{k}\alpha}), \tag{4.2}$$

where the wave functions $A^\mu_{\mathbf{k}\alpha}$ are 4-vectors of the form

$$A^\mu_{\mathbf{k}} = \sqrt{(4\pi)} \frac{e^\mu}{\sqrt{(2\omega)}} e^{-ik_\nu x^\nu}, \qquad e_\mu e^{\mu *} = -1,$$

or more concisely, omitting the four-dimensional vector indices,

$$A_{\mathbf{k}} = \sqrt{(4\pi)} \frac{e}{\sqrt{(2\omega)}} e^{-ikx}, \qquad ee^* = -1. \tag{4.3}$$

Here the 4-momentum $k^\mu = (\omega, \mathbf{k})$ (and so $kx = \omega t - \mathbf{k}\cdot\mathbf{r}$), and e is the unit polarization 4-vector.†

If we consider only gauge transformations which do not alter the dependence of the function (4.3) on the coordinates and the time, the transformation must be

$$e_\mu \to e_\mu + \chi k_\mu, \tag{4.4}$$

where $\chi = \chi(k^\mu)$ is an arbitrary function. Since the polarization is transverse, it is always possible to choose a gauge such that the 4-vector e is

$$e^\mu = (0, \mathbf{e}), \qquad \mathbf{e}\cdot\mathbf{k} = 0; \tag{4.5}$$

this will be called the *three-dimensionally transverse* gauge. In invariant four-dimensional form, this condition becomes the condition of *four-dimensional transversality*

$$ek = 0. \tag{4.6}$$

† The expression (4.3) is not in a fully relativistic-covariant (4-vector) form; this is because the normalization to a finite volume $V = 1$, used here, is not invariant. This is, however, of no fundamental importance, and is entirely compensated by the advantages of the normalization used. We shall see later that it allows a simple and straightforward deduction of actual physical quantities in the necessary invariant form.

It should be noticed that this condition (like the normalization condition $ee^* = -1$) is preserved by the transformation (4.4), since $k^2 = 0$. If the square of the 4-momentum of a particle is zero, its mass must also be zero. This demonstrates the relationship between gauge invariance and the zero mass of the photon. Other aspects of the relationship will be discussed in §14.

There can be no change in any measurable physical quantities under a gauge transformation of the wave functions of photons concerned in a process. In quantum electrodynamics this requirement of *gauge invariance* is of even greater importance than in the classical theory. We shall see many examples of the fact that gauge invariance is here, like relativistic invariance, a valuable heuristic principle.

Gauge invariance is, in turn, closely related to the law of conservation of electric charge. This aspect will be discussed in §43.

It has already been mentioned in §3 that the coordinate wave function of the photon cannot be interpreted as the probability amplitude of its spatial localization. Mathematically, this is shown by the impossibility of constructing from the wave function any quantity which has even the formal properties of a probability density. Such a quantity would have to be expressed as a positive-definite bilinear combination of the wave function A_μ and its complex conjugate. Moreover, it would have to satisfy certain conditions of relativistic covariance by being the time component of a 4-vector. This is because the continuity equation, which expresses the conservation of the number of particles, is given in four-dimensional form by the vanishing of the divergence of the current 4-vector. The time component of the current is here the particle localization probability density; see *Fields*, §29. On the other hand, by the condition of gauge invariance, the 4-vector A_μ could appear in the current only as the antisymmetric tensor $F_{\mu\nu} = \partial_\mu A_\nu - \partial_\nu A_\mu = -i(k_\mu A_\nu - k_\nu A_\mu)$. Thus the current 4-vector would have to be a bilinear combination of $F_{\mu\nu}$ and $F^*_{\mu\nu}$ (and the components of the 4-vector k_μ). But such a 4-vector cannot be formed, since every expression (such as $k^\lambda F^*_{\mu\nu} F^\nu_\lambda$) which satisfies the conditions stated is zero by the transversality condition ($k^\lambda F_{\nu\lambda} = 0$), and in any case could not be positive-definite, since it contains odd powers of the components k_μ.

§ 5. The electromagnetic field in quantum theory

The description of the field as an ensemble of photons is the only description that fully accords with the physical significance of the electromagnetic field in quantum theory. It replaces the classical description in terms of field strengths. These appear in the mathematical formalism of the photon picture as second quantization operators.

The properties of a quantum system are known to be similar to the classical properties when the quantum numbers defining the stationary states of the system are large. For a free electromagnetic field (in a given volume) this means that the oscillator quantum numbers, i.e. the photon numbers $N_{\mathbf{k}\alpha}$, must be large. In this respect the fact that photons obey Bose statistics is of great importance. In the mathematical formalism of the theory, the relationship of the Bose statistics to the properties of the classical field is shown by the commutation rules for the operators

$\hat{c}_{\mathbf{k}\alpha}$, $\hat{c}^{+}_{\mathbf{k}\alpha}$. When the $N_{\mathbf{k}\alpha}$ are large, and the matrix elements of these operators are therefore large also, we may neglect unity on the right-hand side of the commutation rule (2.16), obtaining

$$\hat{c}_{\mathbf{k}\alpha}\hat{c}^{+}_{\mathbf{k}\alpha} \approx \hat{c}^{+}_{\mathbf{k}\alpha}\hat{c}_{\mathbf{k}\alpha};$$

these operators thus become the commuting classical quantities $c_{\mathbf{k}\alpha}$ and $c^{*}_{\mathbf{k}\alpha}$, which determine the classical field strengths.

The condition for the field to be quasi-classical needs to be made more precise, however, since, if all the numbers $N_{\mathbf{k}\alpha}$ are large, the energy of the field is certainly infinite on summation over all the states $\mathbf{k}$, α, and the condition then becomes meaningless.

A physically meaningful statement of the problem as to the conditions for a quasi-classical field can be based on a consideration of values of the field averaged over some short time interval Δt. If the classical electric field $\mathbf{E}$ (or magnetic field $\mathbf{H}$) is represented as a Fourier integral expansion with respect to the time, then, when it is averaged over the time interval Δt, only those Fourier components whose frequencies are such that $\omega\Delta t \lesssim 1$ will make a significant contribution to the mean value $\bar{\mathbf{E}}$, since otherwise the oscillating factor $e^{-i\omega t}$ almost vanishes on averaging. Thus, in determining the condition for the averaged field to be quasi-classical, we need consider only those quantum oscillators whose frequency $\omega \lesssim 1/\Delta t$. It is sufficient that the quantum numbers of these oscillators should be large.

The number of oscillators having frequencies between zero and $\omega \sim 1/\Delta t$ (for a volume $V = 1$) is, in order of magnitude,†

$$(\omega/c)^3 \sim 1/(c\Delta t)^3. \tag{5.1}$$

The total field energy per unit volume is proportional to $\overline{\mathbf{E}^2}$. Dividing this by the number of oscillators and by some mean value of the energy of a single photon ($\sim\hbar\omega$), we find as the order of magnitude of the numbers of photons

$$N_{\mathbf{k}} \sim \overline{\mathbf{E}^2}c^3/\hbar\omega^4.$$

With the condition that this number should be large, we obtain the inequality

$$|\mathbf{E}| \gg \sqrt{(\hbar c)}/(c\Delta t)^2. \tag{5.2}$$

This is the required condition, which allows the field averaged over time intervals Δt to be treated as classical. We see that the field must reach a certain strength, which increases as the averaging time Δt decreases. For variable fields, this time must not, of course, exceed the time during which the field changes appreciably. Thus variable fields, if sufficiently weak, can never be quasi-classical. Only for static (time-independent) fields can we make $\Delta t \to \infty$, so that the right-hand side of the inequality (5.2) tends to zero. Thus a static field is always classical.

† In this section, ordinary units are used.

It has already been mentioned that the classical expressions for the electromagnetic field as a superposition of plane waves must be regarded in quantum theory as operator expressions. These operators, however, have only a very limited physical meaning. A physically meaningful field operator would have to give zero field values in the photon vacuum state, whereas the mean value of the squared field operator $\hat{\mathbf{E}}^2$ in the ground state, which is the same as the zero-point energy of the field apart from a factor, is infinite; by the "mean value" is meant the quantum-mechanical mean value, i.e. the corresponding diagonal matrix element of the operator. This infinity cannot be avoided even by any formal cancelling operation (as was done for the field energy), since here this would have to be carried out by means of some appropriate modification of the operators $\hat{\mathbf{E}}$ and $\hat{\mathbf{H}}$ themselves (not their squares), which is impossible.

§6. The angular momentum and parity of the photon

The photon, like any other particle, can possess a certain angular momentum. In order to determine the properties of this quantity for the photon, let us first recall the relationship between the properties of the wave function of a particle and the angular momentum of the particle, in the mathematical formalism of quantum mechanics.

The angular momentum $\mathbf{j}$ of a particle consists of its orbital angular momentum $\mathbf{l}$ and its intrinsic angular momentum or spin $\mathbf{s}$. The wave function of a particle having spin s is a symmetrical spinor of rank $2s$, i.e. is a set of $2s+1$ components which are transformed into definite combinations of one another when the coordinate axes are rotated. The orbital angular momentum is related to the way in which the wave functions depend on the coordinates: states with orbital angular momentum l correspond to wave functions whose components are linear combinations of the spherical harmonic functions of order l.

The consistent distinguishability of the spin and the orbital angular momentum therefore requires that the "spin" and "coordinate" properties of the wave functions should be independent of each other: the dependence of the spinor components on the coordinates (at a given instant) must not be subject to any additional restrictions.

In the momentum representation of the wave functions, their dependence on the coordinates is replaced by their dependence on the momentum $\mathbf{k}$. The photon wave function (in the three-dimensionally transverse gauge) is the vector $\mathbf{A}(\mathbf{k})$. A vector is equivalent to a spinor of rank 2, and in this sense the photon might be said to have spin 1. But this vector wave function satisfies the transversality condition, $\mathbf{k}\cdot\mathbf{A}(\mathbf{k})=0$, which is a further condition imposed on the function $\mathbf{A}(\mathbf{k})$. Consequently, this function cannot be arbitrarily specified as regards every component of the vector at the same time, and therefore the orbital angular momentum and the spin cannot be strictly distinguished.

The definition of the spin as the angular momentum of a particle at rest is also inapplicable to the photon, because there is no rest frame for a photon, which moves with the velocity of light.

Thus only the total angular momentum of the photon has a meaning. It is, moreover, obvious that this total angular momentum must be integral, since the quantities describing the photon do not include any spinors of odd rank.

The state of a photon, like that of any particle, is also described by its parity, which refers to the behaviour of the wave function under inversion of the coordinates (see *QM*, §30). In the momentum representation, the change of sign of the coordinates is replaced by the change of sign of all the components of **k**. The effect of the inversion operator $\hat{P}$ on a scalar function $\phi(\mathbf{k})$ is simply to produce this change of sign: $\hat{P}\phi(\mathbf{k}) = \phi(-\mathbf{k})$. When it is applied to a vector function $\mathbf{A}(\mathbf{k})$, we must also take into account the fact that the reversal of the directions of the axes changes the sign of all the components of the vector; hence†

$$\hat{P}\,\mathbf{A}(\mathbf{k}) = -\mathbf{A}(-\mathbf{k}). \tag{6.1}$$

Although the separation of the angular momentum of the photon into the orbital angular momentum and the spin has no physical meaning, it is nevertheless convenient to define a "spin" s and an "orbital angular momentum" l as formal auxiliary quantities which express the transformation properties of the wave function under rotations: the value $s = 1$ corresponds to the fact that the wave function is a vector, and the value of l is the order of the spherical harmonics which occur in the wave function. Here we are considering the wave functions of states in which the photon angular momentum has a definite value; for a free particle, these are spherical waves. The number l, in particular, defines the parity of the photon state, which is

$$P = (-1)^{l+1}. \tag{6.2}$$

In the same way, the angular momentum operator $\hat{\mathbf{j}}$ may be represented as the sum $\hat{\mathbf{s}} + \hat{\mathbf{l}}$. The operator $\hat{\mathbf{j}}$ is related to the operator of an infinitesimal rotation of the coordinates, or, in the present case, to the action of this operator on a vector field. In the sum $\hat{\mathbf{s}} + \hat{\mathbf{l}}$, the operator $\hat{\mathbf{s}}$ acts on the vector index, transforming the components of the vector into combinations of one another. The operator $\hat{\mathbf{l}}$ acts on these components as functions of the momentum (or of the coordinates).

We may count the number of states (with a given energy) which are possible for a given value j of the photon angular momentum, ignoring the trivial $(2j+1)$-fold degeneracy with respect to the directions of the angular momentum.

When **l** and **s** are independent, this calculation is made by simply counting the number of ways in which the angular momenta **l** and **s** can be added, according to the rules of the vector model, so as to obtain the required value of j. For a particle

† We shall choose to define the parity of a state according to the effect of the inversion operator on a polar vector, such as **A** (or the corresponding electric vector $\mathbf{E} = i\omega\mathbf{A}$). This differs in sign from the effect on the axial vector $\mathbf{H} = i\mathbf{k}\times\mathbf{A}$, since the direction of such a vector is unaltered by inversion:

$$\hat{P}\mathbf{H}(\mathbf{k}) = \mathbf{H}(-\mathbf{k}).$$

with spin $s = 1$, and a given non-zero value of j, this would give three states, with the following values of l and the parity P:

$$l = j, \qquad P = (-1)^{l+1} = (-1)^{j+1};$$

$$l = j \pm 1, \qquad P = (-1)^{l+1} = (-1)^{j}.$$

If $j = 0$, however, only one state is obtained, with $l = 1$ and parity $P = +1$.

In this calculation the condition that the vector **A** is transverse has not been taken into account; all its three components have been assumed to be independent. We must therefore subtract, from the numbers of states found above, the numbers of states which correspond to a longitudinal vector. This vector may be written in the form $\mathbf{k}\phi(\mathbf{k})$, whence we see that its three components are equivalent, as regards their transformation properties (under rotations), to a single scalar ϕ.† We can therefore say that the extra state which is incompatible with the transversality condition would correspond to the state of a particle having a scalar wave function (spinor of rank 0), i.e. having "spin zero".‡ The angular momentum j of this state is therefore equal to the order of the spherical harmonics which occur in ϕ. The parity of the state as a state of the photon is determined by the action of the inversion operator on the vector function $\mathbf{k}\phi$:

$$\hat{P}(\mathbf{k}\phi) = -(-\mathbf{k})\phi(-\mathbf{k}) = (-1)^{j}\mathbf{k}\phi(\mathbf{k}),$$

and is therefore $(-1)^j$. Thus we must subtract one from the number of states found above which have the parity $(-1)^j$, i.e. two for $j \neq 0$ and one for $j = 0$.

The conclusion is, then, that when the photon angular momentum j is non-zero there is one even state and one odd state. When $j = 0$, no states exist. This means that a photon cannot have zero angular momentum; j therefore takes only the values $1, 2, 3, \ldots$. The impossibility of $j = 0$ is evident a priori, since the wave function of a state with zero angular momentum must be spherically symmetrical, and this cannot be true for a transverse wave.

The following terminology is customary to denote the various states of the photon. A photon with angular momentum j and parity $(-1)^j$ is called an *electric 2^j-pole* (or *Ej*) *photon*; one with parity $(-1)^{j+1}$ is called a *magnetic 2^j-pole* (or *Mj*) *photon*. For example, an odd state with $j = 1$ corresponds to an electric dipole photon, an even state with $j = 2$ to an electric quadrupole photon, and an even state with $j = 1$ to a magnetic dipole photon.§

† This is because the transformation of a quantity under rotation is a transformation at a given point, i.e. for a given value of **k**. Under such a transformation, $\mathbf{k}\phi(\mathbf{k})$ is unchanged, i.e. it behaves as a scalar.

‡ It should be again emphasized that this does not refer to a state of an actual particle. The calculation given here is a formal one, and amounts mathematically to a classification of the set of quantities which are transformed into combinations of one another, in terms of the irreducible representations of the rotation group.

§ This nomenclature corresponds to the terminology of classical radiation theory; we shall see later (§§46, 47) that the emission of electric and magnetic photons is governed by the electric and magnetic moments of a system of charges.

§ 7. Spherical waves of photons

Having ascertained the possible values of the photon angular momentum, we must now determine the corresponding wave functions.†

Let us first consider the formal problem of determining vector functions which are eigenfunctions of the operators $\hat{\mathbf{j}}^2$ and $\hat{j}_z$, without deciding as yet which of these functions will appear in the desired photon wave functions, and without taking account of the transversality condition.

We shall look for the functions in the momentum representation. In this representation, the coordinate operator is $\hat{\mathbf{r}} = i\partial/\partial\mathbf{k}$ (see *QM*, (15.12)). The orbital angular momentum operator is

$$\hat{\mathbf{l}} = \hat{\mathbf{r}} \times \mathbf{k} = -i\mathbf{k} \times \partial/\partial\mathbf{k},$$

and therefore differs from the angular momentum operator in the coordinate representation only in that $\mathbf{r}$ is replaced by $\mathbf{k}$. The solution of the problem is thus formally identical in the two representations.

Let the required eigenfunctions be denoted by $\mathbf{Y}_{jm}$ and referred to as *spherical harmonic vectors*. They must satisfy the conditions

$$\hat{\mathbf{j}}^2\mathbf{Y}_{jm} = j(j+1)\mathbf{Y}_{jm}, \qquad \hat{j}_z\mathbf{Y}_{jm} = m\mathbf{Y}_{jm}, \tag{7.1}$$

the z-axis being in a specified direction in space. We shall show that these conditions are satisfied by any function of the form $\mathbf{a}Y_{jm}$, where $\mathbf{a}$ is any vector formed from the unit vector $\mathbf{n} = \mathbf{k}/\omega$, and Y_{jm} are the ordinary (scalar) spherical harmonic functions. The latter will everywhere be defined as in *QM*, §28:

$$Y_{lm}(\mathbf{n}) = (-1)^{\frac{1}{2}(m+|m|)} i^l \sqrt{\frac{(2l+1)(l-|m|)!}{4\pi(l+|m|)!}}\, P_l^{|m|}(\cos\theta)\, e^{im\phi}, \tag{7.2}$$

where θ and ϕ are the spherical polar angles of the direction $\mathbf{n}$.‡

The proof is based on the commutation rule

$$\{\hat{l}_i, a_k\}_- = ie_{ikl}a_l$$

(*QM*, (29.4)). The right-hand side may be written as $-\hat{s}_i a_k$, where $\hat{\mathbf{s}}$ is the operator of spin 1; the effect of this operator on a vector function is in fact given by $\hat{s}_i a_k = -ie_{ikl}a_l$ (see *QM*, §57, Problem 2). Hence

$$\hat{l}_i a_k - a_k \hat{l}_i = -\hat{s}_i a_k,$$

† This problem was first discussed by W. Heitler (1936). The solution given here is due to V. B. Berestetskiĭ (1947).

‡ For future reference, the value of the function when $\theta = 0$ ($\mathbf{n}$ is along the z-axis) is

$$Y_{lm}(\mathbf{n}_z) = i^l\sqrt{\frac{2l+1}{4\pi}}\,\delta_{m0}. \tag{7.2a}$$

and therefore

$$\hat{j}_i a_k = (\hat{l}_i + \hat{s}_i) a_k = a_k \hat{l}_i.$$

Consequently

$$\hat{\mathbf{j}}^2(\mathbf{a} Y_{jm}) = \mathbf{a}\hat{\mathbf{l}}^2 Y_{jm}, \qquad \hat{j}_z(\mathbf{a} Y_{jm}) = \mathbf{a}\hat{l}_z Y_{jm}.$$

Since the spherical harmonic Y_{jm} is the eigenfunction of the operators $\hat{\mathbf{l}}^2$ and $\hat{l}_z$ which corresponds to the respective eigenvalues $j(j+1)$ and m, we arrive at equations (7.1).

The three essentially different types of spherical harmonic vectors are obtained by taking as the vector **a** the three following vectors:†

$$\frac{\nabla_{\mathbf{n}}}{\sqrt{[j(j+1)]}}, \qquad \frac{\mathbf{n}\times\nabla_{\mathbf{n}}}{\sqrt{[j(j+1)]}}, \qquad \mathbf{n}. \tag{7.3}$$

The spherical harmonic vectors are thus defined as

$$\left.\begin{aligned} \mathbf{Y}_{jm}^{(e)} &= \frac{1}{\sqrt{[j(j+1)]}}\nabla_{\mathbf{n}} Y_{jm}, & P &= (-1)^j; \\ \mathbf{Y}_{jm}^{(m)} &= \mathbf{n}\times\mathbf{Y}_{jm}^{(e)}, & P &= (-1)^{j+1}; \\ \mathbf{Y}_{jm}^{(l)} &= \mathbf{n} Y_{jm}, & P &= (-1)^j. \end{aligned}\right\} \tag{7.4}$$

The parity P is also shown for each vector. The three vectors are orthogonal, $\mathbf{Y}_{jm}^{(l)}$ being longitudinal and $\mathbf{Y}_{jm}^{(e)}$ and $\mathbf{Y}_{jm}^{(m)}$ transverse with respect to **n**.

The spherical harmonic vectors can be expressed in terms of the scalar spherical harmonics: $\mathbf{Y}_{jm}^{(m)}$ in terms of spherical harmonics of the order $l=j$ only, and $\mathbf{Y}_{jm}^{(e)}$ and $\mathbf{Y}_{jm}^{(l)}$ in terms of those of order $l=j\pm 1$. This is immediately evident on comparing the parities shown in (7.4) with the parity $(-1)^{l+1}$ of a vector field in terms of the order of the spherical harmonics concerned.

The spherical harmonic vectors of any one type are orthonormal:

$$\int \mathbf{Y}_{jm}\cdot\mathbf{Y}_{j'm'}^{*}\,do = \delta_{jj'}\delta_{mm'}. \tag{7.5}$$

For the vectors $\mathbf{Y}_{jm}^{(l)}$ this is obvious from the normalization condition for the

† The operator $\nabla_{\mathbf{n}} \equiv |\mathbf{k}|\nabla_{\mathbf{k}}$, and acts on functions which depend only on the direction of **n**. In spherical polar coordinates its two components are

$$\nabla_{\mathbf{n}} = \left(\frac{\partial}{\partial\theta}, \frac{1}{\sin\theta}\frac{\partial}{\partial\phi}\right).$$

The operator denoted below by $\Delta_{\mathbf{n}}$ is the angular part of the Laplacian operator:

$$\Delta_{\mathbf{n}} = \frac{1}{\sin\theta}\frac{\partial}{\partial\theta}\left(\sin\theta\frac{\partial}{\partial\theta}\right) + \frac{1}{\sin^2\theta}\frac{\partial^2}{\partial\phi^2}.$$

spherical harmonics Y_{jm}. For the vectors $\mathbf{Y}_{jm}^{(e)}$ the normalization integral is

$$\frac{1}{j(j+1)}\int \nabla_{\mathbf{n}} Y_{jm}\cdot\nabla_{\mathbf{n}} Y_{j'm'}^{*}\,do = -\frac{1}{j(j+1)}\int Y_{j'm'}^{*}\Delta_{\mathbf{n}} Y_{jm}\,do,$$

and, since $\Delta_{\mathbf{n}} Y_{jm} = -j(j+1)Y_{jm}$, equation (7.5) follows. The normalization for the vectors $\mathbf{Y}_{jm}^{(m)}$ leads to a similar integral.

The spherical harmonic vectors (7.4) could also be derived without the direct verification of equations (7.1) that has been carried out above, using only general arguments concerning the transformational properties of functions. In §6, these arguments were employed to show that a vector function $\mathbf{n}\phi$ corresponds to an angular momentum j which is the same as the order of the spherical harmonics occurring in ϕ. If we put simply $\phi = Y_{jm}$, the function $\mathbf{n}\phi$ will also correspond to a definite value m of the angular-momentum component. Thus we derive at once the spherical harmonic vectors $\mathbf{Y}_{jm}^{(l)}$. But the discussion of transformational properties in §6 is unaffected if the factor $\mathbf{n}$ in the product $\mathbf{n}\phi$ is replaced by the vector $\nabla_{\mathbf{n}}$ or by $\mathbf{n}\times\nabla_{\mathbf{n}}$. This leads to the other two types of spherical harmonic vectors.

Let us now consider the photon wave functions. For an electric photon of type Ej, the parity of the vector $\mathbf{A}(\mathbf{k})$ is $(-1)^j$. The spherical harmonic vectors $\mathbf{Y}_{jm}^{(e)}$ and $\mathbf{Y}_{jm}^{(l)}$ possess this parity, but only the former satisfies the transversality condition. For a magnetic photon of type Mj, the parity of the vector $\mathbf{A}(\mathbf{k})$ is $(-1)^{j+1}$; only $\mathbf{Y}_{jm}^{(m)}$ has this parity. The wave functions of a photon having a given angular momentum j, component thereof m, and energy ω, are therefore

$$\mathbf{A}_{\omega jm}(\mathbf{k}) = \frac{4\pi^2}{\omega^{3/2}}\,\delta(|\mathbf{k}|-\omega)\mathbf{Y}_{jm}(\mathbf{n}), \tag{7.6}$$

where $\mathbf{Y}_{jm}$ must be taken as $\mathbf{Y}_{jm}^{(e)}$ and $\mathbf{Y}_{jm}^{(m)}$ for electric and magnetic photons respectively. The given value of the energy is taken into account by the factor $\delta(|\mathbf{k}|-\omega)$.

The functions (7.6) are normalized by the condition

$$\frac{1}{(2\pi)^4}\int \omega\omega'\mathbf{A}_{\omega'j'm'}^{*}(\mathbf{k})\cdot\mathbf{A}_{\omega jm}(\mathbf{k})\,d^3k = \omega\delta(\omega'-\omega)\delta_{jj'}\delta_{mm'}. \tag{7.7}$$

For wave functions of the coordinate representation, the condition (7.7) is equivalent to the condition†

$$\frac{1}{4\pi}\int \{\mathbf{E}_{\omega'j'm'}^{*}(\mathbf{r})\cdot\mathbf{E}_{\omega jm}(\mathbf{r})+\mathbf{H}_{\omega'j'm'}^{*}(\mathbf{r})\cdot\mathbf{H}_{\omega jm}(\mathbf{r})\}\,d^3x = \omega\delta(\omega'-\omega)\delta_{jj'}\delta_{mm'}: \tag{7.8}$$

the integral on the left, when written in terms of the potentials, is

$$\frac{1}{2\pi}\int \mathbf{A}_{\omega'j'm'}^{*}(\mathbf{r})\cdot\mathbf{A}_{\omega jm}(\mathbf{r})\omega'\omega\,d^3x,$$

† This condition is of the same type as (2.22). The factor $\delta(\omega'-\omega)$ on the right-hand side appears because we are now considering a field (spherical wave) throughout infinite space instead of in the finite volume $V = 1$.

and with

$$\left.\begin{aligned} \mathbf{A}_{\omega jm}(\mathbf{r}) &= \int \mathbf{A}_{\omega jm}(\mathbf{k})\, e^{i\mathbf{k}\cdot\mathbf{r}} \frac{d^3k}{(2\pi)^3}, \\ \mathbf{A}^*_{\omega' j' m'}(\mathbf{r}) &= \int \mathbf{A}^*_{\omega' j' m'}(\mathbf{k}')\, e^{-i\mathbf{k}'\cdot\mathbf{r}} \frac{d^3k'}{(2\pi)^3}, \end{aligned}\right\} \tag{7.9}$$

the integral over d^3x gives the delta function $(2\pi)^3\delta(\mathbf{k}'-\mathbf{k})$. This is eliminated by integrating over d^3k, and the integral reduces to (7.7).

So far, we have assumed that the potentials are in the transverse gauge, for which the scalar potential $\Phi = 0$. In certain applications, however, other gauges of the spherical wave may be more convenient.

The transformation of the potentials that can be conducted in the momentum representation is

$$\mathbf{A} \rightarrow \mathbf{A} + \mathbf{n}f(\mathbf{k}), \qquad \Phi \rightarrow \Phi + f(\mathbf{k}),$$

where $f(\mathbf{k})$ is an arbitrary function. In the present case we shall choose it so that the new potentials are expressed in terms of the same spherical harmonics and again have a definite parity. For an electric photon, these conditions limit the choice of potentials to the following:

$$\left.\begin{aligned} \mathbf{A}^{(e)}_{\omega jm}(\mathbf{k}) &= \frac{4\pi^2}{\omega^{3/2}}\, \delta(|\mathbf{k}|-\omega)(\mathbf{Y}^{(e)}_{jm} + C\mathbf{n}Y_{jm}), \\ \Phi^{(e)}_{\omega jm}(\mathbf{k}) &= \frac{4\pi^2}{\omega^{3/2}}\, \delta(|\mathbf{k}|-\omega) C Y_{jm}, \end{aligned}\right\} \tag{7.10}$$

where C is an arbitrary constant. For a magnetic photon, this addition to $\mathbf{A}^{(m)}(\mathbf{k})$ would leave it without a definite parity, and (7.6) is therefore the only possible choice under these conditions.

The probability that a photon having a definite angular momentum and parity will be recorded as moving in a direction $\mathbf{n}$ which lies in the solid-angle element do is, according to (3.5) and (7.6),

$$w(\mathbf{n})\, do = |\mathbf{Y}^{(e)}_{jm}(\mathbf{n})|^2\, do. \tag{7.11}$$

This is the expression for an E photon, but, since $|\mathbf{Y}^{(m)}_{jm}|^2 = |\mathbf{Y}^{(e)}_{jm}|^2$, the probability distribution $w(\mathbf{n})$ is the same for both types of photon.

The squared modulus $|\mathbf{Y}^{(e)}_{jm}|^2$ is independent of the azimuthal angle ϕ, since the factors $e^{\pm im\phi}$ in the spherical harmonic functions cancel. The probability distribution $w(\mathbf{n})$ is therefore symmetrical about the z-axis. Moreover, since each of the spherical harmonic vectors has a definite parity, their squared moduli are unaffected by inversion, i.e. by the change of polar angle $\theta \rightarrow \pi - \theta$; this means that the expansion of the function $w(\theta)$ in Legendre polynomials will contain only those of even order. The determination of the expansion coefficients is equivalent to a calculation of the integrals of products of three spherical harmonic functions, followed by summation over components. These processes are effected by means of the formulae derived in *QM*, §§107, 108, and the result is

$$w(\theta) = (-1)^{m+1}\frac{(2j+1)^2}{4\pi}\sum_{n=0}^{\infty}(4n+1)\begin{pmatrix} j & j & 2n \\ 0 & 0 & 0 \end{pmatrix}\begin{pmatrix} j & j & 2n \\ m & -m & 0 \end{pmatrix}\begin{Bmatrix} j & j & 2n \\ j & j & 1 \end{Bmatrix}P_{2n}(\cos\theta). \quad (7.12)$$

Finally, we shall give the expressions for the components of the spherical harmonic vectors as expansions in terms of spherical harmonic functions. To do so, we shall use the "spherical components" of a vector, defined as in *QM*, §107. These components f_λ of a vector $\mathbf{f}$ are

$$f_0 = if_z, \qquad f_{+1} = -\frac{i}{\sqrt{2}}(f_x + if_y), \qquad f_{-1} = \frac{i}{\sqrt{2}}(f_x - if_y). \quad (7.13)$$

In terms of the "spherical unit vectors",

$$\mathbf{e}^{(0)} = i\mathbf{e}^{(z)}, \qquad \mathbf{e}^{(+1)} = -\frac{i}{\sqrt{2}}(\mathbf{e}^{(x)} + i\mathbf{e}^{(y)}), \qquad \mathbf{e}^{(-1)} = \frac{i}{\sqrt{2}}(\mathbf{e}^{(x)} - i\mathbf{e}^{(y)}), \quad (7.14)$$

where $\mathbf{e}^{(x,y,z)}$ are unit vectors in the direction of x, y and z, we have

$$\mathbf{f} = \sum_\lambda (-1)^{1-\lambda} f_{-\lambda}\mathbf{e}^{(\lambda)}, \qquad f_\lambda = (-1)^{1-\lambda}\mathbf{f}\cdot\mathbf{e}^{(-\lambda)*} = \mathbf{f}\cdot\mathbf{e}^{(\lambda)}. \quad (7.15)$$

The spherical components of the spherical harmonic vectors are expressed in terms of $3j$-symbols and spherical harmonic functions as follows:

$$\left.\begin{aligned}
(-1)^{j+m+\lambda+1}(\mathbf{Y}^{(e)}_{jm})_\lambda &= -\sqrt{j}\begin{pmatrix} j+1 & 1 & j \\ m+\lambda & -\lambda & -m \end{pmatrix}Y_{j+1,m+\lambda} + \\
&\quad + \sqrt{(j+1)}\begin{pmatrix} j-1 & 1 & j \\ m+\lambda & -\lambda & -m \end{pmatrix}Y_{j-1,m+\lambda}, \\
(-1)^{j+m+\lambda+1}(\mathbf{Y}^{(m)}_{jm})_\lambda &= -\sqrt{(2j+1)}\begin{pmatrix} j & 1 & j \\ m+\lambda & -\lambda & -m \end{pmatrix}Y_{j,m+\lambda}, \\
(-1)^{j+m+\lambda+1}(\mathbf{Y}^{(l)}_{jm})_\lambda &= \sqrt{(j+1)}\begin{pmatrix} j+1 & 1 & j \\ m+\lambda & -\lambda & -m \end{pmatrix}Y_{j+1,m+\lambda} + \\
&\quad + \sqrt{j}\begin{pmatrix} j-1 & 1 & j \\ m+\lambda & -\lambda & -m \end{pmatrix}Y_{j-1,m+\lambda}.
\end{aligned}\right\} \quad (7.16)$$

These formulae are derived in the following way. Each of the three spherical harmonic vectors is of the form $\mathbf{Y}_{jm} = \mathbf{a}Y_{jm}$, where $\mathbf{a}$ is one of the three vectors (7.3). Hence

$$\mathbf{Y}_{jm} = \sum_{l,m'}\langle lm'|\mathbf{a}|jm\rangle Y_{lm'},$$

and the problem is equivalent to that of finding the matrix elements of the vector **a** with respect to the eigenfunctions of the orbital angular momentum. According to *QM*, (107.6), we have

$$\langle lm'|a_\lambda|jm\rangle = i(-1)^{j_{\max}-m'}\begin{pmatrix} l & 1 & j \\ -m' & \lambda & m \end{pmatrix}\langle l\|a\|j\rangle,$$

where $j_{\max}$ is the larger of l and j. It is therefore sufficient to know the non-zero reduced matrix elements $\langle l\|a\|j\rangle$. These are given by the formulae

$$\left.\begin{aligned} \langle l-1\|n\|l\rangle &= \langle l\|n\|l-1\rangle^* = i\sqrt{l}, \\ \langle l\|\nabla_{\mathbf{n}}\|l-1\rangle &= i(l-1)\sqrt{l}, \\ \langle l-1\|\nabla_{\mathbf{n}}\|l\rangle &= i(l+1)\sqrt{l}, \\ \langle l\|\mathbf{n}\times\nabla_{\mathbf{n}}\|l\rangle &= i\sqrt{[l(l+1)(2l+1)]}. \end{aligned}\right\} \tag{7.17}$$

§ 8. The polarization of the photon

The polarization vector **e** acts for the photon as the "spin part" of the wave function (with the limitations stated in §6 in connection with the concept of photon spin).

The various cases which can occur with regard to the polarization of the photon are identical with the possible types of polarization of a classical electromagnetic wave (see *Fields*, §48).

Any polarization **e** can be represented as a superposition of two mutually orthogonal polarizations $\mathbf{e}^{(1)}$ and $\mathbf{e}^{(2)}$ ($\mathbf{e}^{(1)}\cdot\mathbf{e}^{(2)*}=0$), chosen in some specified manner. In the resolution

$$\mathbf{e} = e_1\mathbf{e}^{(1)} + e_2\mathbf{e}^{(2)}, \tag{8.1}$$

the squares of the moduli of the coefficients e_1 and e_2 determine the probabilities that the photon has polarization $\mathbf{e}^{(1)}$ and $\mathbf{e}^{(2)}$ respectively.

These polarizations may be taken to be two mutually perpendicular linear polarizations. We can also resolve any polarization into two circular polarizations having opposite directions of rotation. The vectors of the right-hand and left-hand circular polarizations will be denoted by $\mathbf{e}^{(+1)}$ and $\mathbf{e}^{(-1)}$ respectively; in coordinates ξ, η, ζ, with the ζ-axis in the direction of the photon $\mathbf{n}=\mathbf{k}/\omega$,

$$\mathbf{e}^{(+1)} = -\frac{i}{\sqrt{2}}(\mathbf{e}^{(\xi)} + i\mathbf{e}^{(\eta)}), \qquad \mathbf{e}^{(-1)} = \frac{i}{\sqrt{2}}(\mathbf{e}^{(\xi)} - i\mathbf{e}^{(\eta)}). \tag{8.2}$$

The possibility that the photon has two different polarizations (for a given momentum) is equivalent to the statement that each eigenvalue of the momentum is doubly degenerate. This property is closely related to the fact that the mass of the photon is zero.

A freely moving particle with non-zero mass always has a rest frame. The intrinsic symmetry properties of the particle, as such, will evidently appear in this particular frame of reference. Symmetry with respect to all possible rotations about the centre (i.e. with respect to the entire spherical symmetry group) must be considered. The property which describes the symmetry of the particle with respect to this group is its spin s; this determines the degree of degeneracy, the number of different wave functions which are transformed into linear combinations of one another being $2s+1$. In particular, a particle having a vector (three-component) wave function has spin 1.

If the mass of the particle is zero, however, there is no rest frame, since it moves with the velocity of light in every frame of reference. For such a particle, there is always a distinctive direction in space, the direction of the momentum vector $\mathbf{k}$ (the ζ-axis). In such a case there is clearly no symmetry with respect to the whole group of rotations in three dimensions, but only axial symmetry about the preferred axis.

When there is axial symmetry, only the *helicity* of the particle is conserved, i.e. the component of its angular momentum along the ζ-axis, which we denote by λ.† If we also impose the condition of symmetry under reflections in planes passing through the ζ-axis, the states differing in the sign of λ will be mutually degenerate, and when $\lambda \neq 0$ there is therefore twofold degeneracy.‡ The state of a photon having a definite momentum in fact corresponds to one type of these doubly degenerate states. It is described by a "spin" wave function which is a vector $\mathbf{e}$ in the $\xi\eta$-plane; the two components of this vector are transformed into combinations of each other by any rotation about the ζ-axis and by any reflection in a plane passing through that axis.

The various cases of the polarization of the photon are in a certain relationship to the possible values of its helicity. The relationship can be deduced from the formulae in *QM* (57.9), which connect the components of a vector wave function with those of the equivalent spinor of rank two.§ Vectors $\mathbf{e}$ with only the component $e_\xi - ie_\eta$ or $e_\xi + ie_\eta$ non-zero correspond to the components $\lambda = +1$ or -1 respectively; these are $\mathbf{e} = \mathbf{e}^{(+1)}$ and $\mathbf{e} = \mathbf{e}^{(-1)}$. In other words, the values $\lambda = +1$ and -1 correspond to right-hand and left-hand circular polarization of the photon. In §16 the same result will be derived by direct calculation of the eigenfunctions of the spin component operator.

Thus the component of the photon angular momentum along the direction of its motion can have only the two values ± 1; the value zero is not possible.

A state of the photon having a definite momentum and polarization is a pure state, in the sense defined in *QM*, §14; it is described by a wave function, and corresponds to a complete quantum-mechanical description of the state of the particle (the photon). "Mixed" photon states are also possible, which correspond to a less complete description by a density matrix only, not a wave function.

† This is to be distinguished from m, the component of the angular momentum in a specified direction in space (the z-axis), which was used in §7.

‡ This is the method of classifying the electron terms of the diatomic molecule (*QM*, §78).

§ It is the contravariant spinor components that correspond to the components of the wave function as the probability amplitudes of various values of the angular momentum of the particle (which are here considered).

Let us consider a state of the photon which is mixed as regards its polarization, but corresponds to a definite value of the momentum **k**. In such a state (called a state of *partial polarization*), a "coordinate" wave function exists.

The polarization density matrix of the photon is a tensor $\rho_{\alpha\beta}$ of rank two, in a plane perpendicular to the vector **n** (the $\xi\eta$-plane; the suffixes α, β take only two values). This tensor is Hermitian:

$$\rho_{\alpha\beta} = \rho^*_{\beta\alpha}, \tag{8.3}$$

and is normalized by the condition

$$\rho_{\alpha\alpha} \equiv \rho_{11} + \rho_{22} = 1. \tag{8.4}$$

From (8.3), the diagonal components ρ_{11} and ρ_{22} are real, and either is given in terms of the other by (8.4). The component ρ_{12} is complex, and $\rho_{21} = \rho^*_{12}$. The density matrix therefore involves three real parameters.

If the polarization density matrix is known, we can find the probability that the photon has any given polarization **e**. This probability is determined by the "projection" of the tensor $\rho_{\alpha\beta}$ on the direction of the vector **e**, i.e. by the quantity

$$\rho_{\alpha\beta} e^*_\alpha e_\beta. \tag{8.5}$$

For example, the components ρ_{11} and ρ_{22} are the probabilities of linear polarizations along the ξ and η axes. The probability of the two circular polarizations is given by taking the projections along the vectors (8.2):

$$\tfrac{1}{2}[1 \pm i(\rho_{12} - \rho_{21})]. \tag{8.6}$$

The properties of the tensor $\rho_{\alpha\beta}$ are essentially the same as those of the tensor $J_{\alpha\beta}$ which describes partially polarized light in the classical theory (see *Fields*, §50). Some of these properties are the following.

For a pure state with a definite polarization **e**, the tensor $\rho_{\alpha\beta}$ reduces to products of components of the vector **e**:

$$\rho_{\alpha\beta} = e_\alpha e^*_\beta, \tag{8.7}$$

and the determinant $|\rho_{\alpha\beta}| = 0$. In the opposite case of an unpolarized photon, all directions of polarization are equally probable, i.e.

$$\rho_{\alpha\beta} = \tfrac{1}{2}\delta_{\alpha\beta}, \tag{8.8}$$

and $|\rho_{\alpha\beta}| = \frac{1}{4}$.

In the general case, it is convenient to describe the partial polarization by means of three real Stokes parameters ξ_1, ξ_2, ξ_3,† in terms of which the density matrix can

† These are not to be confused with the ξ-axis.

be written

$$\rho_{\alpha\beta} = \frac{1}{2}\begin{pmatrix} 1+\xi_3 & \xi_1 - i\xi_2 \\ \xi_1 + i\xi_2 & 1-\xi_3 \end{pmatrix}. \tag{8.9}$$

All three parameters take values between -1 and $+1$. In the unpolarized state, $\xi_1 = \xi_2 = \xi_3 = 0$; for a completely polarized photon, $\xi_1^2 + \xi_2^2 + \xi_3^2 = 1$.

The parameter ξ_3 describes the linear polarization along the ξ or η axis; the probability that the photon is linearly polarized along these axes is respectively $\frac{1}{2}(1+\xi_3)$ and $\frac{1}{2}(1-\xi_3)$. The values $\xi_3 = +1$ and -1 therefore correspond to complete polarization in these directions.

The parameter ξ_1 describes the linear polarization along directions at angles $\phi = \pm\frac{1}{4}\pi$ to the ξ-axis. The probability that the photon is linearly polarized along these directions is respectively $\frac{1}{2}(1+\xi_1)$ and $\frac{1}{2}(1-\xi_1)$. This is easily shown by projecting the tensor $\rho_{\alpha\beta}$ on the directions $\mathbf{e} = (1, \pm 1)/\sqrt{2}$.

Finally, the parameter ξ_2 represents the degree of circular polarization: according to (8.6), the probability that the photon has right-hand or left-hand circular polarization is respectively $\frac{1}{2}(1+\xi_2)$ and $\frac{1}{2}(1-\xi_2)$. Since these two polarizations correspond to helicities $\lambda = \pm 1$, it is clear that ξ_2 is the mean value of the helicity of the photon. Moreover, for a pure state with polarization $\mathbf{e}$,

$$\xi_2 = i\mathbf{e} \times \mathbf{e}^* \cdot \mathbf{n}. \tag{8.10}$$

The quantities ξ_2 and $\sqrt{(\xi_1^2 + \xi_3^2)}$ are invariant under Lorentz transformations (see *Fields*, §50).

We shall later encounter the problem of the behaviour of the Stokes parameters under the operation of time reversal. It is easily seen that they are invariant. This property is evidently independent of the state of polarization, and therefore need be proved only for a pure state. In quantum mechanics, time reversal corresponds to replacing the wave function by its complex conjugate (*QM*, §18). For a plane-polarized wave, this implies the changes†

$$\mathbf{k} \to -\mathbf{k}, \qquad \mathbf{e} \to -\mathbf{e}^*. \tag{8.11}$$

Under this transformation, the symmetrical part

$$\tfrac{1}{2}(e_\alpha e_\beta^* + e_\beta e_\alpha^*)$$

of the density matrix is unchanged, and therefore so are ξ_1 and ξ_3. The fact that ξ_2 is unchanged by this transformation is seen from (8.10), and is also evident from the fact that ξ_2 is the mean value of the helicity: the helicity is the component of the angular momentum $\mathbf{j}$ in the direction of $\mathbf{n}$, i.e. the product $\mathbf{j}\cdot\mathbf{n}$, and both these vectors change sign under time reversal.

† The change in the sign of $\mathbf{e}$ is necessary because time reversal changes the sign of the vector potential of the electromagnetic field. The scalar potential, however, does not change sign, and the effect of time reversal on the 4-vector e is therefore as follows:

$$(e_0, \mathbf{e}) \to (e_0^*, -\mathbf{e}^*). \tag{8.11a}$$

In later calculations, we shall need the photon density matrix written in four-dimensional form, i.e. as a certain 4-tensor $\rho_{\mu\nu}$. For a polarized photon described by the 4-vector e_μ, this tensor can naturally be defined as

$$\rho_{\mu\nu} = e_\mu e_\nu^*. \tag{8.12}$$

In the three-dimensionally transverse gauge, $e = (0, \mathbf{e})$, and if one of the spatial coordinate axes is taken to be along $\mathbf{n}$ the non-zero components of this 4-tensor are the same as (8.7).

For an unpolarized photon the three-dimensionally transverse gauge corresponds to a tensor $\rho_{\mu\nu}$ having components

$$\rho_{ik} = \tfrac{1}{2}(\delta_{ik} - n_i n_k), \qquad \rho_{0i} = \rho_{i0} = \rho_{00} = 0; \tag{8.13}$$

if one of the axes is in the direction of $\mathbf{n}$, the result is again (8.8). It would, however, be inconvenient to use the tensor $\rho_{\mu\nu}$ in this three-dimensional form. But a gauge transformation can be applied, which for the density matrix is

$$\rho_{\mu\nu} \to \rho_{\mu\nu} + \chi_\mu k_\nu + \chi_\nu k_\mu, \tag{8.14}$$

where the χ_μ are arbitrary functions. Putting

$$\chi_0 = -1/4\omega, \qquad \chi_i = k_i/4\omega^2,$$

we obtain instead of (8.13) the simple four-dimensional expression

$$\rho_{\mu\nu} = -\tfrac{1}{2} g_{\mu\nu}. \tag{8.15}$$

The four-dimensional form of the density matrix for a partially polarized photon is easily found by first writing the two-dimensional tensor (8.9) in three-dimensional form:

$$\begin{aligned}\rho_{ik} = \tfrac{1}{2}(e_i^{(1)}e_k^{(1)} + e_i^{(2)}e_k^{(2)}) + \tfrac{1}{2}\xi_1(e_i^{(1)}e_k^{(2)} + e_i^{(2)}e_k^{(1)}) - \\ - \tfrac{1}{2}i\xi_2(e_i^{(1)}e_k^{(2)} - e_i^{(2)}e_k^{(1)}) + \tfrac{1}{2}\xi_3(e_i^{(1)}e_k^{(1)} - e_i^{(2)}e_k^{(2)}),\end{aligned}$$

where $\mathbf{e}^{(1)}$ and $\mathbf{e}^{(2)}$ are unit vectors along the ξ and η axes. The required generalization is obtained on replacing these 3-vectors by real space-like unit 4-vectors $e^{(1)}$, $e^{(2)}$ which are orthogonal to each other and to the photon 4-momentum k:

$$\left.\begin{aligned} e^{(1)2} = e^{(2)2} = -1, \\ e^{(1)}e^{(2)} = 0, \\ e^{(1)}k = e^{(2)}k = 0. \end{aligned}\right\} \tag{8.16}$$

In one particular frame of reference, $e^{(1)} = (0, \mathbf{e}^{(1)})$ and $e^{(2)} = (0, \mathbf{e}^{(2)})$. Thus the four-dimensional density matrix of the photon is

$$\begin{aligned}\rho_{\mu\nu} = \tfrac{1}{2}(e_\mu^{(1)}e_\nu^{(1)} + e_\mu^{(2)}e_\nu^{(2)}) + \tfrac{1}{2}\xi_1(e_\mu^{(1)}e_\nu^{(2)} + e_\mu^{(2)}e_\nu^{(1)}) - \\ - \tfrac{1}{2}i\xi_2(e_\mu^{(1)}e_\nu^{(2)} - e_\mu^{(2)}e_\nu^{(1)}) + \tfrac{1}{2}\xi_3(e_\mu^{(1)}e_\nu^{(1)} - e_\mu^{(2)}e_\nu^{(2)}).\end{aligned} \tag{8.17}$$

The convenience of any specific choice of the 4-vectors $e^{(1)}$, $e^{(2)}$ depends on the conditions of the problem concerned.

It must be noted that the conditions (8.16) do not uniquely define the choice of $e^{(1)}$ and $e^{(2)}$. If a 4-vector e_μ satisfies these conditions, then so does any 4-vector $e_\mu + \chi k_\mu$, since $k^2 = 0$. This non-uniqueness occurs because the density matrix is not invariant under gauge transformations.

The first term in (8.17) corresponds to the unpolarized state. According to (8.15), it can therefore be replaced by $-\frac{1}{2}g_{\mu\nu}$. This change is again equivalent to a certain gauge transformation.

The following formal device is useful in calculations with 4-tensors of the form (8.17) expressed in terms of two independent 4-vectors. We write the tensor (8.17) in the form

$$\rho_{\mu\nu} = \sum_{a,b=1}^{3} \rho^{(ab)} e_\mu^{(a)} e_\nu^{(b)},$$

and the coefficients $\rho^{(ab)}$ as a two-rowed matrix:

$$\rho = \begin{pmatrix} \rho^{(11)} & \rho^{(12)} \\ \rho^{(21)} & \rho^{(22)} \end{pmatrix}.$$

This, like any two-rowed Hermitian matrix, can be written in terms of four independent two-rowed matrices: the Pauli matrices σ_x, σ_y, σ_z and the unit matrix 1. The result is

$$\rho = \tfrac{1}{2}(1 + \boldsymbol{\xi} \cdot \boldsymbol{\sigma}), \qquad \boldsymbol{\xi} = (\xi_1, \xi_2, \xi_3), \tag{8.18}$$

as is easily seen by direct comparison with (8.17), using the expressions (18.5) for the Pauli matrices. The combination of the three quantities ξ_1, ξ_2, ξ_3 into a "vector" $\boldsymbol{\xi}$ is, of course, purely formal and is done only for convenience of notation.

PROBLEM

Write the photon density matrix for the case where the coordinate "axes" are the circular unit vectors (8.2).

SOLUTION. The components $\rho'_{\alpha\beta}$ of the tensor relative to the new axes ($\alpha, \beta = \pm 1$) are obtained by projecting the tensor (8.9) on the unit vectors (8.2):

$$\rho'_{11} = \rho_{\alpha\beta} e_\alpha^{(+1)*} e_\beta^{(+1)}, \qquad \rho'_{1,-1} = \rho_{\alpha\beta} e_\alpha^{(+1)*} e_\beta^{(-1)}, \ldots;$$

$$\rho' = \frac{1}{2}\begin{pmatrix} 1+\xi_2 & -\xi_3 + i\xi_1 \\ -\xi_3 - i\xi_1 & 1-\xi_2 \end{pmatrix}.$$

§9. A two-photon system

By arguments similar to those in §6, we can calculate the number of possible states in a more complicated case, that of a system of two photons (L. Landau, 1948).

We shall consider the photons in their centre-of-mass system; their momenta are $\mathbf{k}_1 = -\mathbf{k}_2 \equiv \mathbf{k}$.† The wave function of the two-photon system (in the momentum representation) can be written as a three-dimensional tensor of rank two $A_{ik}(\mathbf{n})$, formed by a bilinear combination of the vector wave functions of the two photons; each of the suffixes of this tensor corresponds to one of the photons ($\mathbf{n}$ being a unit vector in the direction of $\mathbf{k}$). The transversality of each photon is expressed by the orthogonality of the tensor A_{ik} to the vector $\mathbf{n}$:

$$A_{il}n_l = 0, \qquad A_{lk}n_l = 0. \tag{9.1}$$

An interchange of the photons corresponds to an interchange of the suffixes of the tensor A_{ik} and a simultaneous change in the sign of $\mathbf{n}$. Since photons obey Bose statistics, we have

$$A_{ik}(-\mathbf{n}) = A_{ki}(\mathbf{n}). \tag{9.2}$$

The tensor A_{ik} is not in general symmetrical with respect to its suffixes. It can be resolved into symmetric and antisymmetric parts: $A_{ik} = s_{ik} + a_{ik}$. The equation (9.2), and the orthogonality conditions (9.1), must evidently apply to each part separately. Hence we have

$$s_{ik}(-\mathbf{n}) = s_{ik}(\mathbf{n}), \tag{9.3}$$

$$a_{ik}(-\mathbf{n}) = -a_{ik}(\mathbf{n}). \tag{9.4}$$

Inversion of the coordinates does not affect the sign of the components of a tensor of rank two, but changes the sign of $\mathbf{n}$. From (9.3), therefore, the wave function s_{ik} is symmetrical under inversion, i.e. it corresponds to even states of the two-photon system, while the wave function a_{ik} corresponds to odd states.

An antisymmetric tensor of rank two is equivalent (dual) to a certain axial vector $\mathbf{a}$, whose components are given in terms of those of the tensor by

$$a_i = \tfrac{1}{2}e_{ikl}a_{kl},$$

e_{ikl} being the antisymmetric unit tensor; see *Fields*, §6. The orthogonality of the tensor a_{kl} and the vector $\mathbf{n}$ implies that the vectors $\mathbf{a}$ and $\mathbf{n}$ are parallel.‡ We can therefore write $\mathbf{a} = \mathbf{n}\phi(\mathbf{n})$, where ϕ is a scalar; according to (9.4), we must have $\mathbf{a}(-\mathbf{n}) = -\mathbf{a}(\mathbf{n})$, and therefore

$$\phi(-\mathbf{n}) = \phi(\mathbf{n}).$$

This equation signifies that the scalar ϕ can be formed linearly only from spherical harmonic functions of even order L (including order zero).

† This frame of reference always exists except in the case of two photons moving in the same direction. The total momentum $\mathbf{k}_1 + \mathbf{k}_2$ and the total energy $\omega_1 + \omega_2$ of such photons are related in the same way as those of a single photon, and there is therefore no frame of reference in which $\mathbf{k}_1 + \mathbf{k}_2 = 0$.

‡ For $a_{ik} = e_{ikl}a_l$, and the orthogonality condition gives $a_{ik}n_k = e_{ikl}a_l n_k = (\mathbf{n} \times \mathbf{a})_i = 0$.

We see that the transformation properties of the antisymmetric tensor a_{ik} under rotations are equivalent to those of a single scalar (cf. the second footnote to §6). When the latter is assigned a "spin" zero, the angular momentum of the state is found to be $J = L$. Thus the tensor a_{ik} corresponds to odd states of a photon system with even angular momentum J.

Let us now consider the symmetric tensor s_{ik}. Since this is unaltered when **n** changes sign, it corresponds to even states of the photon system. Hence all the components s_{ik} can be expressed in terms of spherical harmonic functions of even order L (including zero). It is well known that any symmetric tensor s_{ik} of rank two can be expressed as the sum of a scalar s_{ii} and a symmetric tensor s'_{ik} with zero trace ($s'_{ii} = 0$).

The scalar s_{ii} can be assigned a "spin" zero, and the angular momentum of the corresponding states is therefore $J = L$, i.e. is even. The tensor s'_{ik} has "spin" two (see *QM*, §57). Adding this "spin" to the even "orbital angular momentum" L by the law of addition of angular momenta, we find that for a given even $J \neq 0$ three states are possible (with $L = J \pm 2, J$), and for odd $J \neq 1$ two states (with $L = J \pm 1$). The exceptions are $J = 0$ with one state ($L = 2$) and $J = 1$ with one state ($L = 2$).

In these calculations, however, we have not yet included the condition that the tensor s_{ik} is orthogonal to the vector **n**. We must therefore subtract, from the numbers of states found above, the numbers of states corresponding to a symmetric tensor of rank 2 "parallel" to the vector **n**. Such a tensor, which we denote by s''_{ik}, can be written as

$$s''_{ik} = n_i b_k + n_k b_i,$$

where **b** is a certain vector. According to (9.3), this vector must be such that $\mathbf{b}(-\mathbf{n}) = -\mathbf{b}(\mathbf{n})$. Thus the tensor s''_{ik} which gives the "unwanted" states is equivalent to an odd vector. The latter must be expressible in terms only of spherical harmonics of odd order L. Moreover, the vector has a "spin" one, and therefore, for any even angular momentum $J \neq 0$, two states are possible (with $L = J \pm 1$), and for any odd J one state (with $L = J$); an exception is $J = 0$ with one state ($L = 1$).

Summarizing the results obtained, we obtain the following table giving the numbers of possible even and odd states of a two-photon system (with zero total momentum) for various values of the total angular momentum J:

J	even	odd
0	1	1
1	0	0
$2k$	2	1
$2k+1$	1	0

(9.5)

where k is any positive integer (not zero). We see that for odd J there are no odd states, and the value $J = 1$ cannot occur.†

The wave function A_{ik} of the two-photon system determines the correlation between the polarizations of the photons. The probability that both photons

† Another way of deriving these results is given in §70, Problem 1.

simultaneously have definite polarizations $\mathbf{e}_1$ and $\mathbf{e}_2$ is proportional to

$$A_{ik}e_{1i}^{*}e_{2k}^{*}.$$

Thus, if the polarization $\mathbf{e}_1$ of one photon is given, the polarization $\mathbf{e}_2$ of the other is

$$e_{2k} \propto A_{ik}e_{1i}^{*}. \tag{9.6}$$

In odd states of the system, A_{ik} is equal to the antisymmetric tensor a_{ik}, and

$$\mathbf{e}_2 \cdot \mathbf{e}_1^{*} \propto a_{ik}e_{1i}^{*}e_{1k}^{*} = 0,$$

so that the polarizations of the two photons are orthogonal. For linear polarization this means that the directions of polarization are perpendicular; for circular polarization, that the directions of rotation are opposite.

An even state with $J = 0$ corresponds to a symmetric tensor which reduces to a scalar,

$$s_{ik} = \text{constant} \times (\delta_{ik} - n_i n_k).$$

From (9.6), therefore, we have $\mathbf{e}_1 = \mathbf{e}_2^{*}$. For linear polarization this means that the directions of polarization are parallel; for circular polarization, that the directions of rotation are again opposite. The latter result is obvious, since when $J = 0$ the sum of the components of the photon angular momenta in the same direction $\mathbf{k}$ must always be zero, because the components in opposite directions $\mathbf{k}_1$ and $\mathbf{k}_2$, i.e. the helicities, are equal.

CHAPTER II

BOSONS

§ 10. The wave equation for particles with spin zero

IT HAS been shown in Chapter I how a quantum description of the free electromagnetic field can be constructed on the basis of the known properties of the field in the classical limit and the concepts of ordinary quantum mechanics. The resulting scheme for describing the field as a system of photons contains many features which occur also in the relativistic quantum theory description of particles.

The electromagnetic field is a system having an infinite number of degrees of freedom. For this system there is no law of conservation of number of particles (photons), and its possible states include states with an arbitrary number of particles.† In the relativistic theory, systems composed of any particles must in general share this property. The conservation of number of particles in the non-relativistic theory depends on the law of conservation of mass: the sum of the (rest) masses of the particles is unaffected by their interactions, and the constancy of the total mass in a system of electrons, say, implies that the number of electrons is also unchanged. In relativistic mechanics, however, there is no law of conservation of mass; only the total energy of the system is conserved, which includes the rest energy of the particles. The number of particles therefore need not be conserved, and consequently every relativistic theory of particles must be a theory of systems having an infinite number of degrees of freedom. That is to say, any such theory of particles must be a field theory.

The second quantization formalism (*QM*, §§64, 65) is a satisfactory means of describing systems with a variable number of particles. In the quantum description of the electromagnetic field, the second quantization operator is the 4-potential $\hat{A}$. This is expressed in terms of the (coordinate) wave functions of the individual particles (photons) and their creation and annihilation operators. The quantized wave function operator has a similar role in the description of a system of particles. To derive this operator, we must first know the form of the wave function of a single free particle and the equation satisfied by this function.

The concept of a field of free particles is, it must be emphasized, only an aid to the theory. Actual particles interact, and the task of the theory is to consider these interactions. But any interaction is equivalent to a collision, before and after which the system may be regarded as an ensemble of free particles. It has been remarked in §1 that the only measurable objects are of this kind. We therefore use the fields of free particles as a means of describing the initial and final states.

†In reality, of course, the number of photons changes only as a result of various interaction processes.

Let us first consider the relativistic description of free particles having spin zero. This case is mathematically simple, and illustrates most clearly the basic ideas and typical features of the description.

The state of a free particle (with spin zero) can be completely defined by specifying its momentum $\mathbf{p}$ only. The energy ε of the particle† is given by $\varepsilon^2 = \mathbf{p}^2 + m^2$ (where m is the mass of the particle) or, in four-dimensional form,

$$p^2 = m^2. \tag{10.1}$$

The laws of conservation of momentum and energy are well known to be related to the homogeneity of space and time, i.e. to the symmetry with respect to any parallel displacement of the 4-coordinate system. In the quantum description, this requirement of symmetry means that, under such a transformation of the 4-coordinates, the wave function of a particle having a given 4-momentum must be multiplied by a phase factor (of unit modulus). This can be true only for an exponential function, with the exponent linear in the 4-coordinates. Thus the wave function of the state of a free particle with a given 4-momentum $p^\mu = (\varepsilon, \mathbf{p})$ must be a plane wave:

$$\text{constant} \times e^{-ipx}, \quad px = \varepsilon t - \mathbf{p} \cdot \mathbf{r}; \tag{10.2}$$

the choice of sign of the exponent in the relativistic theory itself is arbitrary, and is here made in accordance with the non-relativistic case.

The wave equation must have the functions (10.2) as particular solutions for any 4-vector p which satisfies the condition (10.1). It must be linear, on account of the principle of superposition: any linear combination of the functions (10.2) also describes a possible state of the particle, and must therefore also be a solution. Finally, the equation must be of the lowest possible order; any higher order would bring in redundant solutions.

The spin is the angular momentum of the particle in a frame of reference in which the particle is at rest. If the spin of the particle is s, its wave function in the rest frame is a three-dimensional spinor of rank $2s$. To describe the particle in an arbitrary frame of reference, its wave function must be expressed in terms of four-dimensional quantities.

A particle with spin zero is described in the rest frame by a three-dimensional scalar. This scalar, however, can have more than one four-dimensional "origin": either a four-dimensional scalar ψ, or as the fourth component of a (time-like) 4-vector ψ_μ of which only the component ψ_0 is non-zero in the rest frame.‡

For a free particle, the only operator that can appear in the wave equation is the 4-momentum operator $\hat{p}$. Its components are the operators of differentiation with respect to coordinates and time:

$$\hat{p}^\mu = i\partial^\mu = \left(i\frac{\partial}{\partial t}, -i\nabla\right). \tag{10.3}$$

†We denote the energy of a single particle by ε, to distinguish it from the energy E of a system of particles.

‡Or, similarly, as the time component of a 4-tensor of higher order; but this would lead to higher-order equations.

The wave equation must be a differential relationship between the quantities ψ and ψ_μ through the operator $\hat{p}$. This relationship must, of course, be given by relativistically invariant expressions. Such expressions are

$$m\psi_\mu = \hat{p}_\mu\psi, \qquad \hat{p}^\mu\psi_\mu = m\psi, \tag{10.4}$$

where m is a dimensional constant characteristic of the particle.†

Substituting ψ_μ from the first equation (10.4) in the second equation, we obtain

$$(\hat{p}^2 - m^2)\psi = 0 \tag{10.5}$$

(O. Klein, and V. A. Fock, 1926; W. Gordon, 1927). The explicit form of this equation is

$$-\partial_\mu\partial^\mu\psi \equiv \left(-\frac{\partial^2}{\partial t^2} + \triangle\right)\psi = m^2\psi. \tag{10.6}$$

Substitution of ψ as the plane wave (10.2) gives $p^2 = m^2$, from which it is evident that m is the mass of the particle. We may note that the form of equation (10.5) is in any case obvious a priori, since $\hat{p}^2$ is the only scalar operator which can be derived from $\hat{p}$ (and, for the same reason, a similar equation is satisfied by every component of the wave function of a particle having any spin value, as will be seen on several occasions below).

Thus a particle with spin zero is essentially described by a single (four-dimensional) scalar ψ, which satisfies the second-order equation (10.5). In the first-order equations (10.4), the wave function is represented by the set of quantities ψ and ψ_μ, the 4-vector ψ_μ being the 4-gradient of the scalar ψ. In the rest frame, the wave function of the particle is independent of the (space) coordinates, and the space components of the 4-vector ψ_μ are therefore zero, as they should be.

In order to continue with the second quantization procedure, it is useful to express the energy and momentum of the particle as the space integrals of certain combinations bilinear in ψ and ψ^*, which represent a kind of space density of these quantities. We thus have to find an energy–momentum tensor $T_{\mu\nu}$ which corresponds to equation (10.5). In terms of this tensor the law of conservation of energy and momentum is expressed by the equation

$$\partial_\mu T^\mu_\nu = 0. \tag{10.7}$$

Following the general procedure of field theory (see *Fields*, §32), we write down a variational principle which would lead to equation (10.5). This principle must be that the "action integral"

$$S = \int L\, d^4x \tag{10.8}$$

† The constants m are shown in (10.4) so that ψ_μ and ψ shall have the same dimensions. There would be no point in using different constants m_1 and m_2 in the two equations, since they could always be made the same by redefining ψ or ψ_μ.

of some real 4-scalar L, the Lagrangian density of the field,† should take a minimum value. Using the scalar ψ (and the operator ∂^μ), we can construct a real bilinear scalar expression of the form

$$L = \partial_\mu\psi^* \cdot \partial^\mu\psi - m^2\psi^*\psi, \tag{10.9}$$

where m is a dimensional constant. Regarding ψ and ψ^* as independent variables describing the field ("generalized field coordinates" q), we easily see that Lagrange's equation

$$\frac{\partial}{\partial x^\mu}\frac{\partial L}{\partial q_{,\mu}} = \frac{\partial L}{\partial q} \tag{10.10}$$

(where $q_{,\mu} \equiv \partial_\mu q$) is in fact the same as the equation (10.5) for ψ and ψ^*, m being the mass of the particle. The sign of the expression (10.9) has been taken such that the square of the time derivative, $|\partial\psi/\partial t|^2$, appears in L with a positive sign; otherwise, the action could not take a minimum value (cf. *Fields*, §27). The choice of the numerical factor in L is arbitrary (and affects only the normalization factor in ψ).

The energy–momentum tensor can now be calculated from the formula

$$T_\mu^\nu = \sum q_{,\mu}\frac{\partial L}{\partial q_{,\nu}} - L\delta_\mu^\nu, \tag{10.11}$$

the summation being over all q. Substitution of (10.9) gives

$$T_{\mu\nu} = \partial_\mu\psi^* \cdot \partial_\nu\psi + \partial_\nu\psi^* \cdot \partial_\mu\psi - Lg_{\mu\nu}; \tag{10.12}$$

these quantities are real (as they should be), since L is real. In particular,

$$T_{00} = 2\frac{\partial\psi^*}{\partial t}\frac{\partial\psi}{\partial t} - L$$

$$= \frac{\partial\psi^*}{\partial t}\frac{\partial\psi}{\partial t} + \nabla\psi^* \cdot \nabla\psi + m^2\psi^*\psi, \tag{10.13}$$

$$T_{i0} = \frac{\partial\psi^*}{\partial t}\frac{\partial\psi}{\partial x^i} + \frac{\partial\psi^*}{\partial x^i}\frac{\partial\psi}{\partial t}. \tag{10.14}$$

The 4-momentum of the field is given by the integral

$$P_\mu = \int T_{\mu 0}\, d^3x, \tag{10.15}$$

i.e. T_{00} and T_{0i} act as the energy and momentum densities. The quantity T_{00} is essentially positive.

† The corresponding second-quantized operator $\hat{L}$ is called the *Lagrangian* of the field. To simplify the terminology, we shall use this term for either the "quantized" or the "non-quantized" Lagrangian density, as convenient.

Formula (10.13) can be used for the normalization of the wave function. A plane wave, normalized to "one particle in the volume $V = 1$", is

$$\psi_p = \frac{1}{\sqrt{(2\varepsilon)}} e^{-ipx}, \tag{10.16}$$

since for this function $T_{00} = \varepsilon$, and the total energy in the volume $V = 1$ is therefore equal to the energy of a single particle.

The angular momentum, whose conservation is due to the isotropy of space, can also be expressed as a space integral, but we shall not need this representation.

There is one further conservation law allowed by equations (10.4) in addition to those arising directly from space–time symmetry. It is easily seen that these equations and those for ψ^* lead to the equation

$$\partial_\mu j^\mu = 0, \tag{10.17}$$

where

$$\begin{aligned} j_\mu &= m(\psi^* \psi_\mu + \psi_\mu^* \psi) \\ &= i[\psi^* \partial_\mu \psi - (\partial_\mu \psi^*)\psi]. \end{aligned} \tag{10.18}$$

Thus j^μ acts as a current density 4-vector, and (10.17) is the equation of continuity expressing the law of conservation of the quantity

$$Q = \int j_0 \, d^3x, \tag{10.19}$$

where

$$j_0 = j^0 = i\left(\psi^* \frac{\partial \psi}{\partial t} - \frac{\partial \psi^*}{\partial t} \psi\right). \tag{10.20}$$

It should be noted that j_0 need not be positive. This shows that it cannot in general be interpreted as the probability density of spatial localization of the particle. The significance of the conservation law expressed by equation (10.17) will be shown in §11.

§ 11. Particles and antiparticles

In accordance with the general procedure of the second quantization method, we have to consider the expansion of an arbitrary wave function in terms of the eigenfunctions of a complete set of possible states of a free particle, for instance in plane waves ψ_p:

$$\psi = \sum_{\mathbf{p}} a_{\mathbf{p}} \psi_p, \qquad \psi^* = \sum_{\mathbf{p}} a_{\mathbf{p}}^* \psi_p^*.$$

The coefficients $a_{\mathbf{p}}$, $a^*_{\mathbf{p}}$ are then to be regarded as the annihilation and creation operators $\hat{a}_{\mathbf{p}}$, $\hat{a}^+_{\mathbf{p}}$ of particles in the corresponding states.†

Here, however, we immediately encounter a difference of principle as compared with the non-relativistic theory. In a plane wave which is a solution of equation (10.5), the energy ε need satsify (for a given momentum $\mathbf{p}$) only the condition $\varepsilon^2 = \mathbf{p}^2 + m^2$, i.e. it can have two values, $\pm\sqrt{(\mathbf{p}^2 + m^2)}$. Only positive values of ε can have the physical significance of the energy of a free particle. But the negative values cannot be simply omitted: the general solution of the wave equation can be obtained only by superposing all its independent particular solutions. This shows that the interpretation of the expansion coefficients for ψ and ψ^* in the second quantization method must be somewhat different.

We may write the expansion in the form

$$\psi = \sum_{\mathbf{p}} \frac{1}{\sqrt{(2\varepsilon)}} a^{(+)}_{\mathbf{p}} e^{i(\mathbf{p}\cdot\mathbf{r}-\varepsilon t)} + \sum_{\mathbf{p}} \frac{1}{\sqrt{(2\varepsilon)}} a^{(-)}_{\mathbf{p}} e^{i(\mathbf{p}\cdot\mathbf{r}+\varepsilon t)}, \tag{11.1}$$

where the first sum contains plane waves with positive "frequency", normalized according to (10.16), and the second sum contains those with negative "frequency", ε always denoting the positive quantity $+\sqrt{(\mathbf{p}^2 + m^2)}$. In the second quantization, the coefficients $a^{(+)}_{\mathbf{p}}$ in the first sum are replaced as usual by the particle annihilation operators $\hat{a}_{\mathbf{p}}$. In the second sum, we note that, in the subsequent derivation of the matrix elements, the time dependence of the terms will correspond to particle creation, not annihilation: the factor $e^{i\varepsilon t} = (e^{-i\varepsilon t})^*$ corresponds to one extra particle with energy ε in the final state (cf. the end of §2). Accordingly, the coefficients $a^{(-)}_{\mathbf{p}}$ are replaced by creation operators $\hat{b}^+_{-\mathbf{p}}$ relating to other particles. If the summation variable $\mathbf{p}$ in the second sum in (11.1) is replaced by $-\mathbf{p}$ in order to put the exponential factor in the form $e^{-i(\mathbf{p}\cdot\mathbf{r}-\varepsilon t)}$, the ψ-operators are obtained as

$$\left.\begin{aligned} \hat{\psi} &= \sum_{\mathbf{p}} \frac{1}{\sqrt{(2\varepsilon)}} (\hat{a}_{\mathbf{p}} e^{-ipx} + \hat{b}^+_{\mathbf{p}} e^{ipx}), \\ \hat{\psi}^+ &= \sum_{\mathbf{p}} \frac{1}{\sqrt{(2\varepsilon)}} (\hat{a}^+_{\mathbf{p}} e^{ipx} + \hat{b}_{\mathbf{p}} e^{-ipx}). \end{aligned}\right\} \tag{11.2}$$

Thus all the operators $\hat{a}_{\mathbf{p}}$, $\hat{b}_{\mathbf{p}}$ are multiplied by functions with the "correct" time dependence ($\sim e^{-i\varepsilon t}$), while the operators $\hat{a}^+_{\mathbf{p}}$, $\hat{b}^+_{\mathbf{p}}$ are multiplied by the complex conjugate functions. This makes it possible to interpret the former operators, in accordance with the general rules, as annihilation operators for particles with momentum $\mathbf{p}$ and energy ε, and the latter as creation operators for these particles.

In this way we arrive at the concept of particles of two types which occur simultaneously and on an equal footing. These are called *particles* and *antiparticles*; the significance of the names will be shown later. One type corresponds to the operators $\hat{a}_{\mathbf{p}}$, $\hat{a}^+_{\mathbf{p}}$ in the second quantization formalism, and the other type to $\hat{b}_{\mathbf{p}}$, $\hat{b}^+_{\mathbf{p}}$. The two types of particle have the same mass, since their operators appear in the same ψ-operator.

† The ψ function is given the 4-momentum p as suffix, since we intend to denote the functions with "negative frequency" by ψ_{-p}. The operators $\hat{a}$ and $\hat{a}^+$ are given the three-dimensional momentum $\mathbf{p}$ as suffix, since this entirely defines the state of an actual particle.

The reason for these results can also be examined from the point of view of the requirements of relativistic invariance.

The Lorentz transformations are, mathematically, rotations of the four-dimensional coordinate system which change the direction of the time axis; together with the purely spatial rotations which do not affect the time axis, they form the *Lorentz group* of transformations.† All the Lorentz transformations have the property that they leave the t axis within the corresponding light cone, and this expresses the physical principle that there exists a maximum possible velocity of propagation of signals.

In a purely mathematical sense, the simultaneous change of sign of all four coordinates (*four-dimensional inversion*) is also a rotation, since the determinant of this transformation is $+1$, like that of any rotational transformation. The time axis is thereby carried from one light cone to the other. Although this means that such a transformation is physically impossible (as a transformation of the frame of reference), the only difference mathematically is that, because the metric is pseudo-Euclidean, such a rotation cannot be effected continuously without allowing also a complex transformation of the coordinates.

It is reasonable to suppose that this difference is unimportant in relation to four-dimensional invariance. Then any expression which is invariant under the Lorentz transformations must be invariant under 4-inversion also. A precise statement of this condition as applied to the scalar ψ-operator will be given in §13, but here it may be noted that the condition will certainly make necessary the simultaneous presence in the ψ-operators of terms having both signs of ε in the exponents, since this sign is changed by the substitution $t \to -t$.

Let us return now to equations (11.2) and derive the commutation relations between the operators $\hat{a}_{\mathbf{p}}$, $\hat{a}_{\mathbf{p}}^{+}$ (and $\hat{b}_{\mathbf{p}}$, $\hat{b}_{\mathbf{p}}^{+}$). For photons (the operators $\hat{c}_{\mathbf{p}}$, $\hat{c}_{\mathbf{p}}^{+}$), this was done on the basis of the analogy with oscillators, that is, essentially from the properties of the electromagnetic field in the classical limit. Here there is no such analogy. In deriving the (Bose or Fermi) commutation rules between the operators, we can be guided only by the form of the Hamiltonian constructed from these operators.

This Hamiltonian is obtained (see *QM*, §64) by substituting $\hat{\psi}$ and $\hat{\psi}^{+}$ in place of ψ and ψ^{*} in the integral $\int T_{00}\, d^3x$.‡ We then find

$$\hat{H} = \sum_{\mathbf{p}} \varepsilon(\hat{a}_{\mathbf{p}}^{+}\hat{a}_{\mathbf{p}} + \hat{b}_{\mathbf{p}}\hat{b}_{\mathbf{p}}^{+}). \tag{11.3}$$

It is easily seen that a reasonable result is obtained for the eigenvalues of this Hamiltonian only if the operators satisfy the Bose commutation rules:

$$\{\hat{a}_{\mathbf{p}}, \hat{a}_{\mathbf{p}}^{+}\}_{-} = \{\hat{b}_{\mathbf{p}}, \hat{b}_{\mathbf{p}}^{+}\}_{-} = 1 \tag{11.4}$$

† The set of all three-dimensional (spatial) rotations is itself a group, which constitutes a subgroup of the Lorentz group. The set of the Lorentz transformations is not itself a group, since the result of successive Lorentz transformations may be a purely spatial rotation.

‡ In the non-relativistic theory, the conjugate operator $\hat{\psi}^{+}$ is by convention written to the left of $\hat{\psi}$. Here, the order is of no importance, since the interchange of $\hat{\psi}$ and $\hat{\psi}^{+}$ would cause only the interchange of the equivalent operators $\hat{a}_{\mathbf{p}}$ and $\hat{b}_{\mathbf{p}}$. However, once a particular order has been selected, the same order must be used throughout.

(all other pairs of operators commute, including each particle operator $\hat{a}_{\mathbf{p}}$, $\hat{a}_{\mathbf{p}}^{+}$ with each antiparticle operator $\hat{b}_{\mathbf{p}}$, $\hat{b}_{\mathbf{p}}^{+}$). For, in this case,

$$\hat{H} = \sum_{\mathbf{p}} \varepsilon(\hat{a}_{\mathbf{p}}^{+}\hat{a}_{\mathbf{p}} + \hat{b}_{\mathbf{p}}^{+}\hat{b}_{\mathbf{p}} + 1).$$

The eigenvalues of the products $\hat{a}_{\mathbf{p}}^{+}\hat{a}_{\mathbf{p}}$ and $\hat{b}_{\mathbf{p}}^{+}\hat{b}_{\mathbf{p}}$ are positive integers $N_{\mathbf{p}}$ and $\bar{N}_{\mathbf{p}}$, the numbers of particles and antiparticles. The infinite additive constant $\Sigma\, \varepsilon$ (the "energy of the vacuum") may again be simply omitted:

$$E = \sum_{\mathbf{p}} \varepsilon(N_{\mathbf{p}} + \bar{N}_{\mathbf{p}}); \tag{11.5}$$

cf. formula (3.1) and the footnote to it. This expression is essentially positive, and corresponds to the idea of two types of actual particles. Similarly, we have for the total momentum of the system

$$\mathbf{P} = \sum_{\mathbf{p}} \mathbf{p}(N_{\mathbf{p}} + \bar{N}_{\mathbf{p}}). \tag{11.6}$$

If, instead of (11.4), we used the Fermi commutation rules (anticommutators instead of commutators), we should obtain

$$\hat{H} = \sum_{\mathbf{p}} \varepsilon(\hat{a}_{\mathbf{p}}^{+}\hat{a}_{\mathbf{p}} - \hat{b}_{\mathbf{p}}^{+}\hat{b}_{\mathbf{p}} + 1),$$

and instead of (11.5) the physically meaningless expression $\Sigma\, \varepsilon(N_{\mathbf{p}} - \bar{N}_{\mathbf{p}})$, which is not positive-definite and hence cannot represent the energy of a system of free particles.

Particles with spin zero are therefore bosons.

Next, let us consider the integral Q (10.19). Replacing the functions ψ and ψ^* in j^0 by the operators $\hat{\psi}$ and $\hat{\psi}^+$, and carrying out the integration, we obtain

$$\hat{Q} = \sum_{\mathbf{p}} (\hat{a}_{\mathbf{p}}^{+}\hat{a}_{\mathbf{p}} - \hat{b}_{\mathbf{p}}\hat{b}_{\mathbf{p}}^{+}) = \sum_{\mathbf{p}} (\hat{a}_{\mathbf{p}}^{+}\hat{a}_{\mathbf{p}} - \hat{b}_{\mathbf{p}}^{+}\hat{b}_{\mathbf{p}} - 1). \tag{11.7}$$

The eigenvalues of this operator are (omitting the unimportant additive constant $\Sigma\, 1$)

$$Q = \sum_{\mathbf{p}} (N_{\mathbf{p}} - \bar{N}_{\mathbf{p}}), \tag{11.8}$$

and are therefore equal to the differences between the total numbers of particles and antiparticles.

So long as we are discussing free particles and ignoring any interaction between them, the law of conservation of the quantity Q is, of course, largely conventional (like those of total energy (11.5) and total momentum (11.6)): what is actually conserved is not only the sum Q but the numbers $N_{\mathbf{p}}$, $\bar{N}_{\mathbf{p}}$ individually. The nature of the interaction decides whether the quantity Q is conserved. If Q is conserved (i.e. if the operator $\hat{Q}$ commutes with the Hamiltonian of the interaction), the formula (11.8) shows the limitation imposed by the conservation law on the possible variation of the number of particles: only "particle–antiparticle" pairs can be formed or disappear.

If a particle is electrically charged, its antiparticle must have a charge of the opposite sign: if both had charges of the same sign, the creation or annihilation of the particle–antiparticle pair would contravene a rigorous law of nature, the conservation of total electric charge. We shall see later (§32) how the theory automatically leads to this oppositeness of the charges (for interactions of particles with an electromagnetic field).

The quantity Q is sometimes called the *charge* of the field of the particles concerned. For electrically charged particles Q gives, in particular, the total electric charge of the system in terms of the unit charge e. But particles and antiparticles may also be electrically neutral.

Thus we see that the nature of the relativistic relation between the energy and the momentum (the twofold root of the equation $\varepsilon^2 = \mathbf{p}^2 + m^2$), together with the requirements of relativistic invariance, leads in the quantum theory to a new principle of classification of particles: there can exist pairs of different particles (particle and antiparticle) which are interrelated in the way described above. This remarkable prediction was first made (for particles with spin $\frac{1}{2}$) by Dirac in 1930, before the discovery of the first antiparticle, the positron.†

§ 12. Strictly neutral particles

In the second quantization of the ψ-function (11.1), the coefficients $a_{\mathbf{p}}^{(+)}$ and $a_{\mathbf{p}}^{(-)}$ were treated as operators relating to different particles. This is not necessary, however: as a particular case, the annihilation and creation operators in $\hat{\psi}$ may relate to the same particles, as for photons (cf. (2.17)). Then, denoting these operators by $\hat{c}_{\mathbf{p}}$ and $\hat{c}_{\mathbf{p}}^{+}$, we write the ψ-operator as

$$\hat{\psi} = \sum_{\mathbf{p}} \frac{1}{\sqrt{(2\varepsilon)}} (\hat{c}_{\mathbf{p}} e^{-ipx} + \hat{c}_{\mathbf{p}}^{+} e^{ipx}). \tag{12.1}$$

The field described by this operator corresponds to a system of particles of one kind only, which may be said to be their own antiparticles.

The operator (12.1) is Hermitian ($\hat{\psi}^+ = \hat{\psi}$), and in this sense such a field has only half as many "degrees of freedom" as a complex field for which the operators $\hat{\psi}$ and $\hat{\psi}^+$ are not the same.

In consequence, the field Lagrangian, expressed in terms of the Hermitian operator $\hat{\psi}$, must contain a further factor $\frac{1}{2}$ in comparison with (10.9):‡

$$\hat{L} = \tfrac{1}{2}(\partial_\mu \hat{\psi} \cdot \partial^\mu \hat{\psi} - m^2 \hat{\psi}^2). \tag{12.2}$$

† The antiparticle concept was extended to bosons by V. Weisskopf and W. Pauli (1934).

‡ This resembles the extra factor $\frac{1}{2}$ in the operator (2.10) of the electromagnetic field energy density (when the field is expressed in terms of the Hermitian operators $\hat{\mathbf{E}}$ and $\hat{\mathbf{H}}$), in comparison with the photon energy density (3.2) expressed in terms of the complex wave function; cf. the last footnote to §3.

The corresponding energy–momentum tensor is

$$\hat{T}_{\mu\nu} = \partial_\mu \hat{\psi} \cdot \partial_\nu \hat{\psi} - \hat{L} g_{\mu\nu}, \tag{12.3}$$

and hence the energy density operator is

$$\begin{aligned} \hat{T}_{00} &= (\partial \hat{\psi}/\partial t)^2 - \hat{L} \\ &= \tfrac{1}{2}[(\partial \hat{\psi}/\partial t)^2 + (\nabla \hat{\psi})^2 + m^2 \hat{\psi}^2]. \end{aligned} \tag{12.4}$$

Substituting (12.1) in the integral $\int \hat{T}_{00}\, d^3x$, we obtain the field Hamiltonian:

$$\hat{H} = \tfrac{1}{2} \sum_{\mathbf{p}} \varepsilon (\hat{c}_{\mathbf{p}}^+ \hat{c}_{\mathbf{p}} + \hat{c}_{\mathbf{p}} \hat{c}_{\mathbf{p}}^+). \tag{12.5}$$

This again shows that Bose quantization is necessary:

$$\{\hat{c}_{\mathbf{p}}, \hat{c}_{\mathbf{p}}^+\}_- = 1, \tag{12.6}$$

and the energy eigenvalues (again without the additive constant) are

$$E = \sum_{\mathbf{p}} \varepsilon_{\mathbf{p}} N_{\mathbf{p}}. \tag{12.7}$$

Fermi quantization would lead to the absurd result that E is independent of $N_{\mathbf{p}}$.

The "charge" Q of this field is zero, as is evident from the fact that Q must change sign when particles are replaced by antiparticles, whereas in the present case there is no difference between the two. The current density 4-vector therefore does not exist, since the expression

$$\hat{j}_\mu = i[\hat{\psi}^+ \partial_\mu \hat{\psi} - (\partial_\mu \hat{\psi}^+) \hat{\psi}] \tag{12.8}$$

for the operator $\hat{j}$ of the conserved 4-vector is zero when $\hat{\psi} = \hat{\psi}^+$ (the vector $\hat{\psi} \partial_\mu \hat{\psi}$ is not itself conserved). This, in turn, means that there is no special conservation law restricting the possible changes in the number of particles. Such particles must clearly be electrically neutral.

Particles of this kind are said to be *strictly neutral*, as opposed to electrically neutral particles which are not their own antiparticles. Whereas the latter can be annihilated (transformed into photons) only as pairs, strictly neutral particles can be annihilated singly.

The structure of the ψ-operator (12.1) is similar to that of the electromagnetic field operators (2.17)–(2.20). In this sense we may say that photons are themselves strictly neutral particles. For the electromagnetic field, the operators are Hermitian because the fields are measurable physical quantities (in the classical limit) and are therefore real. For the ψ-operators of particles there is no such relation, since they do not correspond to any quantities that are directly measurable.

The absence of a conserved current 4-vector is a general property of strictly neutral particles, and does not require the spin to be zero; for instance, it occurs for photons also. Physically, it expresses the absence of the corresponding prohibitions on a change in the number of particles. There is a direct formal relation between the absence of a conserved current and the fact that the field is real (the operator $\hat{\psi}$ is Hermitian).

The Lagrangian of a complex field,

$$\hat{L} = \partial_\mu \hat{\psi}^+ \cdot \partial^\mu \hat{\psi} - m^2 \hat{\psi}^+ \hat{\psi}, \tag{12.9}$$

is invariant under multiplication of the ψ-operator by any phase factor, i.e. under the *gauge transformations*

$$\hat{\psi} \to e^{i\alpha} \hat{\psi}, \qquad \hat{\psi}^+ \to e^{-i\alpha} \hat{\psi}^+. \tag{12.10}$$

In particular, the Lagrangian is unchanged under the infinitesimal transformation

$$\hat{\psi} \to \hat{\psi} + i\,\delta\alpha \cdot \hat{\psi}, \quad \hat{\psi}^+ \to \hat{\psi}^+ - i\,\delta\alpha \cdot \hat{\psi}^+. \tag{12.11}$$

When the "generalized coordinates" q undergo an infinitesimal change, the change in the Lagrangian is

$$\begin{aligned} \delta\hat{L} &= \sum \left(\frac{\partial \hat{L}}{\partial q} \delta q + \frac{\partial \hat{L}}{\partial q_{,\mu}} \delta q_{,\mu} \right) \\ &= \sum \left(\frac{\partial \hat{L}}{\partial q} - \frac{\partial}{\partial x^\mu} \frac{\partial \hat{L}}{\partial q_{,\mu}} \right) \delta q + \sum \frac{\partial}{\partial x^\mu} \left(\frac{\partial \hat{L}}{\partial q_{,\mu}} \delta q \right) \end{aligned}$$

(with summation over all q). The first term is zero, from the "equations of motion" (Lagrange's equations). If the "coordinates" q are taken to be the operators $\hat{\psi}$ and $\hat{\psi}^+$, and with $\delta\hat{\psi} = i\,\delta\alpha \cdot \hat{\psi}$, $\delta\hat{\psi}^+ = -i\,\delta\alpha \cdot \hat{\psi}^+$, we obtain

$$\delta\hat{L} = i\,\delta\alpha \frac{\partial}{\partial x^\mu} \left(\hat{\psi} \frac{\partial \hat{L}}{\partial \hat{\psi}_{,\mu}} - \hat{\psi}^+ \frac{\partial \hat{L}}{\partial \hat{\psi}^+_{,\mu}} \right).$$

Hence we see that the condition for the Lagrangian to be invariant ($\delta\hat{L} = 0$) is equivalent to the equation of continuity ($\partial_\mu \hat{j}^\mu = 0$) for the 4-vector

$$\hat{j}^\mu = i \left(\hat{\psi}^+ \frac{\partial L}{\partial \hat{\psi}^+_{,\mu}} - \hat{\psi} \frac{\partial L}{\partial \hat{\psi}_{,\mu}} \right\} \tag{12.12}$$

It is easily shown that, with the Lagrangian (12.9), this formula yields the current (12.8).

Thus, in the mathematical formalism of the theory, the existence of a conserved current is related to the invariance of the Lagrangian under the gauge transformations (W. Pauli, 1941). The Lagrangian (12.2) of the strictly neutral field does not possess this symmetry.

§ 13. The transformations *C*, *P* and *T*

Unlike 4-inversion, three-dimensional (spatial) inversion is not reducible to any rotations of the 4-coordinate system; its determinant is -1, not $+1$. The symmetry properties of particles with respect to inversion (the *P* transformation) are therefore not determined already by considerations of relativistic invariance.†

The inversion operation, as applied to a scalar wave function, is the transformation

$$\hat{P}\psi(t, \mathbf{r}) = \pm\, \psi(t, -\mathbf{r}), \tag{13.1}$$

where the plus and minus signs on the right correspond to true scalars and pseudoscalars respectively.

Hence we see that two features of the behaviour of the wave function under inversion must be distinguished. One of these relates to the coordinate dependence of the wave function. In non-relativistic quantum mechanics, only this aspect was considered; it leads to the concept of the parity of the state (which we shall here call the *orbital parity*), describing the symmetry properties of the motion of the particle. If the state has a definite orbital parity ($+1$ or -1), this means that

$$\psi(t, -\mathbf{r}) = \pm\, \psi(t, \mathbf{r}).$$

The other feature is the behaviour of the wave function at a given point (which may conveniently be taken as the origin) under inversion of the coordinate axes. This leads to the concept of the *internal parity* of the particle. The two signs in (13.1) correspond to internal parity $+1$ and -1 (for a particle with spin zero). The total parity of a system of particles is given by the product of their internal parities and the orbital parity of their relative motion.

The "internal" symmetry properties of various particles appear, of course, only in their mutual transformation processes. In non-relativistic quantum mechanics, the analogue of the internal parity is the parity of a bound state of a composite system, such as a nucleus. In the relativistic theory, which makes no essential distinction between composite and elementary particles, this internal parity is no different from the internal parity of those particles which are regarded as elementary in the non-relativistic theory. In the non-relativistic case, where these particles are regarded as unalterable, their internal symmetry properties are not observable, and a discussion of these would therefore be devoid of physical significance.

In the second quantization formalism, the internal parity is expressed by the behaviour of the ψ-operators under inversion. Scalar and pseudoscalar fields correspond to the transformation laws

$$P:\ \hat{\psi}(t, \mathbf{r}) \to \pm\, \hat{\psi}(t, -\mathbf{r}). \tag{13.2}$$

The actual significance of the action of inversion on the ψ-operator must be

† The Lorentz group together with spatial inversion is called the *extended Lorentz group* (in contrast to the original group without *P*, which in this connection is called the *proper Lorentz group*). The extended group includes all transformations which leave the *t* axis within the corresponding light cone.

formulated as a particular transformation of the particle annihilation and creation operators, such as to lead to the result (13.2). It is easily seen that such transformations are

$$P: \hat{a}_{\mathbf{p}} \to \pm \hat{a}_{-\mathbf{p}}, \qquad \hat{b}_{\mathbf{p}} \to \pm \hat{b}_{-\mathbf{p}} \tag{13.3}$$

(and the same for the conjugate operators). For, on making these changes in the operator

$$\hat{\psi}(t, \mathbf{r}) = \sum_{\mathbf{p}} \frac{1}{\sqrt{(2\varepsilon)}} (\hat{a}_{\mathbf{p}}\, e^{-i\omega t + i\mathbf{p}\cdot\mathbf{r}} + \hat{b}_{\mathbf{p}}^{+}\, e^{i\omega t - i\mathbf{p}\cdot\mathbf{r}}) \tag{13.4}$$

and then changing the notation for the summation variable ($\mathbf{p} \to -\mathbf{p}$), we can bring it to the form $\pm\hat{\psi}(t, -\mathbf{r})$. Thus, if $\hat{\psi}^{P}(t, \mathbf{r})$ denotes the operator after the substitutions (13.3), we have

$$\hat{\psi}^{P}(t, \mathbf{r}) = \pm \hat{\psi}(t, -\mathbf{r}). \tag{13.5}$$

The transformation (13.3) is entirely reasonable, since inversion changes the sign of the polar vector $\mathbf{p}$, and particles with momentum $\mathbf{p}$ are therefore replaced by particles with momentum $-\mathbf{p}$.

In (13.3) the operators $\hat{a}_{\mathbf{p}}$ and $\hat{b}_{\mathbf{p}}$ are transformed either both with the upper sign or both with the lower sign. In the second quantization formalism, this expresses the fact that particles and antiparticles (with spin zero) have the same internal parity, a result which is evident because they are described by the same (scalar or pseudoscalar) wave functions.

The ψ-operator (13.4) is also symmetrical under a transformation which has no analogue in the non-relativistic theory, that of *charge conjugation* (the C transformation). If all the operators $\hat{a}_{\mathbf{p}}$ and $\hat{b}_{\mathbf{p}}$ are respectively interchanged:

$$C: \hat{a}_{\mathbf{p}} \to \hat{b}_{\mathbf{p}}, \qquad \hat{b}_{\mathbf{p}} \to \hat{a}_{\mathbf{p}} \tag{13.6}$$

(i.e. if particles and antiparticles are interchanged), then $\hat{\psi}$ becomes the charge-conjugate operator $\hat{\psi}^{C}$, where

$$\hat{\psi}^{C}(t, \mathbf{r}) = \hat{\psi}^{+}(t, \mathbf{r}). \tag{13.7}$$

This equation expresses the symmetry of the concepts of particles and antiparticles in the theory.

There is an unimportant formal arbitrariness in the definition of the charge-conjugation transformation. The significance of the transformation is unchanged if an arbitrary phase factor is included in the definition (13.6):

$$\hat{a}_{\mathbf{p}} \to e^{i\alpha}\hat{b}_{\mathbf{p}}, \qquad \hat{b}_{\mathbf{p}} \to e^{-i\alpha}\hat{a}_{\mathbf{p}}.$$

This would lead to

$$\hat{\psi} \to e^{i\alpha}\hat{\psi}^{+}, \qquad \hat{\psi}^{+} \to e^{-i\alpha}\hat{\psi},$$

and a twofold repetition of the transformation would again yield an identity ($\hat{\psi} \to \hat{\psi}$). All such definitions are equivalent, however. Since the properties of the ψ-operators are unchanged on multiplication by a phase factor (cf. the end of §12), we can simply write $\hat{\psi} e^{i\alpha/2}$ in place of $\hat{\psi}$, and thus again obtain the definition of charge conjugation (13.6), (13.7).

Since charge conjugation replaces a particle by its antiparticle, which is not identical with it, no new properties of a particle or a system of particles, as such, will in general arise.

An exception is formed by systems comprising equal numbers of particles and antiparticles. The operator $\hat{C}$ transforms such a system into itself, and so in this case the operator has eigenstates, corresponding to the eigenvalues $C = \pm 1$ (since $\hat{C}^2 = 1$). To describe the charge symmetry, we may regard the particle and the antiparticle as two different "charge states" of the same particle, differing in the value of the charge quantum number $Q = \pm 1$. The wave function of the system is the product of an orbital function and a "charge" function, and must be symmetrical with respect to simultaneous interchange of all the variables (coordinate and charge) of any pair of particles. The symmetry of the "charge" function determines the charge parity of the system (see the Problem at the end of this section).†

The concept of charge parity, which arises in a natural manner for "strictly neutral" systems, must apply also to strictly neutral "elementary" particles. In the second quantization formalism, this concept is represented by the equation

$$\hat{\psi}^C = \pm \hat{\psi}, \tag{13.8}$$

where the plus and minus signs correspond to charge-even and charge-odd particles respectively.

Relativistic invariance implies invariance under 4-inversion (see §11). For a scalar field operator (in the sense of 4-rotations) this means that 4-inversion must give

$$\hat{\psi}(t, \mathbf{r}) \to \hat{\psi}(-t, -\mathbf{r})$$

with the right-hand side always positive. In terms of transformations of the operators $\hat{a}_\mathbf{p}$, $\hat{b}_\mathbf{p}$, the transformation of $\hat{\psi}(t, \mathbf{r})$ into $\hat{\psi}(-t, -\mathbf{r})$ is obtained by interchanging the coefficients of e^{-ipx} and e^{ipx} in (13.4), i.e. by making

$$\hat{a}_\mathbf{p} \to \hat{b}_\mathbf{p}^+, \qquad \hat{b}_\mathbf{p} \to \hat{a}_\mathbf{p}^+. \tag{13.9}$$

Since a-operators are replaced by b-operators, this involves interchange of particles and antiparticles. We see that, in the relativistic theory, there is a natural requirement of invariance under a transformation in which spatial inversion (P) and time reversal (T) are accompanied by charge conjugation (C); this is called the *CPT theorem*.‡

† In this discussion we are considering a particle with spin zero. The treatment given here can be immediately generalized to other spin values; see, for instance, §27, Problem.

‡ This theorem was enunciated by G. Lüders (1954) and W. Pauli (1955).

Here, however, it must be emphasized that, although the arguments given in §§11 and 12 and the present section are a natural development of the ideas of ordinary quantum mechanics and classical relativity theory, the results thus obtained go beyond these both in form (ψ-operators including both particle creation and particle annihilation operators at the same time) and in content (particles and antiparticles). They cannot therefore be regarded as logically necessary, but embrace new physical principles whose correctness can be tested only by experiment.

If the operator (13.4) transformed by (13.9) is denoted by $\hat{\psi}^{CPT}(t, \mathbf{r})$, we can write

$$\hat{\psi}^{CPT}(t, \mathbf{r}) = \hat{\psi}(-t, -\mathbf{r}). \tag{13.10}$$

Thus, if 4-inversion is formulated as the transformation (13.9), we thereby establish also the formulation of the time-reversal transformation of the ψ-operator: together with the *combined inversion* transformation *CP*, it must give (13.9). Using the definitions (13.3) and (13.6), we therefore find

$$T: \hat{a}_{\mathbf{p}} \to \pm \hat{a}^{+}_{-\mathbf{p}}, \qquad \hat{b}_{\mathbf{p}} \to \pm \hat{b}^{+}_{-\mathbf{p}}, \tag{13.11}$$

where the signs $\pm$ correspond to those in (13.3). The significance of this transformation is obvious: time reversal not only changes motion with momentum $\mathbf{p}$ into motion with momentum $-\mathbf{p}$, but also interchanges initial and final states in the matrix elements. The annihilation operators for particles with momentum $\mathbf{p}$ are therefore replaced by creation operators for particles with momentum $-\mathbf{p}$. Making the substitutions (13.11) in (13.4) and changing the notation for the summation variable ($\mathbf{p} \to -\mathbf{p}$), we obtain†

$$\hat{\psi}^{T}(t, \mathbf{r}) = \pm \hat{\psi}^{+}(-t, \mathbf{r}). \tag{13.12}$$

This is similar to the general rule for time reversal in quantum mechanics: if a certain state is described by the wave function $\psi(t, \mathbf{r})$, then the "time-reversed" state is described by the function $\psi^*(-t, \mathbf{r})$. The change to the complex conjugate function is necessary because the "correct" time dependence must be restored, after being lost through the change in the sign of t (E. P. Wigner, 1932).

Since the transformation T (and therefore *CPT*) interchanges the initial and final states, there are no eigenstates and eigenvalues, and therefore no new properties of particles as such. The consequences as regards scattering processes will be discussed in §§69 and 71.

Let us see how the current 4-vector operator $\hat{j}^{\mu}$ (12.8) is affected by the transformations C, P and T. The transformation (13.2), together with $(\partial_0, \partial_i) \to (\partial_0, -\partial_i)$, gives

$$P : (\hat{j}^0, \hat{\mathbf{j}})_{t,\mathbf{r}} \to (\hat{j}^0, -\hat{\mathbf{j}})_{t,-\mathbf{r}}, \tag{13.13}$$

† If the operation T is defined without regard to the other transformations, there is the same arbitrariness in the choice of the phase factor as occurs for the operation C. The requirement of *CPT* symmetry implies that the phase factor can be chosen arbitrarily for only one of the transformations C and T.

as we should expect for a true 4-vector. The transformation (13.7) would give simply

$$C: (\hat{j}^0, \hat{\mathbf{j}})_{t,\mathbf{r}} \to (-\hat{j}^0, -\hat{\mathbf{j}})_{t,\mathbf{r}} \tag{13.14}$$

if the operators $\hat{\psi}$ and $\hat{\psi}^+$ commuted. However, the non-commutativity of these operators is due only to that of the operators $\hat{a}_\mathbf{p}$ and $\hat{a}_\mathbf{p}^+$ (or $\hat{b}_\mathbf{p}$ and $\hat{b}_\mathbf{p}^+$) with the same **p**, and from the commutation rules (11.4) the interchange of these operators produces only terms independent of the occupation numbers, i.e. independent of the state of the field. Omitting these terms as unimportant, as in (11.5), (11.6), we return to (13.14), whose significance is evident: charge conjugation replaces particles by antiparticles and thus changes the sign of every component of the 4-current.

Since the operation of time reversal involves transposing the initial and final states, it changes the order of the factors in a product of operators. For example,

$$(\hat{\psi}^+ \partial_\mu \hat{\psi})^T = (\partial_\mu \hat{\psi})^T (\hat{\psi}^+)^T.$$

Here, however, this is not important: since the ψ-operators commute (in the sense explained above), the result is unaffected by returning to the original order of factors. Since also $(\partial_0, \partial_i) \to (-\partial_0, \partial_i)$ under time reversal, the current transformation rule is

$$T: (\hat{j}^0, \hat{\mathbf{j}})_{t,\mathbf{r}} \to (\hat{j}^0, -\hat{\mathbf{j}})_{-t,\mathbf{r}}. \tag{13.15}$$

The three-dimensional vector $\hat{\mathbf{j}}$ changes sign, in accordance with its classical significance.

Finally, for the *CPT* transformation,

$$CPT: (\hat{j}^0, \hat{\mathbf{j}})_{t,\mathbf{r}} \to (-\hat{j}^0, -\hat{\mathbf{j}})_{-t,-\mathbf{r}}, \tag{13.16}$$

in accordance with the significance of this operation as 4-inversion. Here it must be emphasized that, since 4-inversion is a rotation of the 4-coordinate system, it does not correspond to two types (true and pseudo) of 4-tensors of any rank.

So far, we have assumed that the particles are free; but parity quantum numbers acquire real significance only when interacting particles are considered and definite selection rules are imposed which allow or forbid specified processes. Only conserved properties, however, can have this significance; that is, the eigenvalues of operators which commute with the Hamiltonian of the interacting particles.

Because of relativistic invariance, the *CPT* transformation operator always commutes with the Hamiltonian. For the *C* and *P* (and therefore *T*) transformations separately, experiment shows that the electromagnetic and strong interactions are invariant, and the corresponding parity quantum numbers are therefore conserved in these interactions. In a weak interaction, these conservation laws do not hold.†

† The idea that parity might not be conserved in weak interactions was first put forward by T. D. Lee and C. N. Yang (1956). The general notion that the laws of physics might not have *P* and *T* invariance had previously been suggested by Dirac (1949).

Anticipating a little, we may mention that the operator of the interaction between charged particles and the electromagnetic field is given by the product of the operator 4-vectors $\hat{A}$ and $\hat{j}$. Since charge conjugation changes the sign of $\hat{j}$, the invariance of the electromagnetic interaction under this transformation means that the sign of $\hat{A}$ must also be changed. Thus photons are charge-odd particles.

This behaviour of the operators $\hat{A}$ is in accordance with the properties of the 4-potential in the classical theory: from the transformations

$$\begin{aligned} C: &\quad (\hat{A}_0, \hat{\mathbf{A}}) \to (-\hat{A}_0, -\hat{\mathbf{A}})_{t,\mathbf{r}}, \\ P: &\quad (\hat{A}_0, \hat{\mathbf{A}}) \to (\hat{A}_0, -\hat{\mathbf{A}})_{t,-\mathbf{r}}, \\ CPT: &\quad (\hat{A}_0, \hat{\mathbf{A}}) \to (-\hat{A}_0, -\hat{\mathbf{A}})_{-t,-\mathbf{r}}, \end{aligned}$$

it follows that

$$T: \quad (\hat{A}_0, \hat{\mathbf{A}}) \to (\hat{A}_0, -\hat{\mathbf{A}})_{-t,\mathbf{r}},$$

in agreement with the classical rule for the transformation of the electromagnetic field potentials under time reversal.

The requirement of *CPT* invariance does not impose any limitations on the properties of the particles themselves, but it implies certain relations between those of particles and antiparticles. Firstly, their masses must be equal, as is evident from the relation described in §11 between 4-inversion and the basis of the concept of particles and antiparticles. Next, it follows from *CPT* invariance that there is only a difference of sign in the proportionality coefficients between the electric and magnetic moment vectors and the particle and antiparticle spin vector. The magnetic moment changes sign under the *C* and *T* transformations but (being an axial vector) is not affected by the *P* transformation. Hence the *CPT* transformation, which converts a particle into an antiparticle, does not change the sign of the magnetic moment; the spin vector does change sign. The same applies to the electric moment, which is unchanged by time reversal but changes sign under the *C* transformation and (being a polar vector) under spatial inversion.

The requirements of *P* and *T* invariance (if complied with) restrict the properties of each particle, prohibiting the existence of an electric dipole moment: the only vector that can be constructed from the ψ-operators of an elementary particle at rest is its spin operator vector, which is *P*-even and *T*-odd, and can therefore give rise to a magnetic moment but not an electric moment. We must emphasize that either *P* invariance or *T* invariance is sufficient to invoke this prohibition.

PROBLEM

Determine the charge and spatial parities of a system of two particles with spin zero (particle and antiparticle) and orbital angular momentum of relative motion l.

SOLUTION. Interchanging the coordinates of the particles is equivalent to inversion (about their centre of mass), and therefore multiplies the orbital function by $(-1)^l$; interchanging the charge variables is equivalent to charge conjugation, and multiplies the "charge" factor in the wave function by the required parity C. The condition $C(-1)^l = 1$ gives

$$C = (-1)^l.$$

The spatial parity P of the system is the product of the orbital parity and the internal parities of the two particles. Since the particle and the antiparticle have the same internal parity, in this case P is equal to the orbital parity:

$$P = (-1)^l.$$

§ 14. The wave equation for a particle with spin one

A particle with spin one is described in its rest frame by a three-component wave function, a three-dimensional vector; such a particle is often called a *vector particle*. The four-dimensional origin of this vector may be as the three spatial components of the space-like 4-vector ψ^μ or the mixed components of the antisymmetric 4-tensor $\psi^{\mu\nu}$ of rank two; the time component ψ^0 and the space components ψ^{ik} are zero in the rest frame.†

The wave equation is a differential relation between the quantities ψ^μ and $\psi^{\mu\nu}$, and will be written as the equations

$$i\psi_{\mu\nu} = \hat{p}_\mu\psi_\nu - \hat{p}_\nu\psi_\mu, \tag{14.1}$$

$$im^2\psi_\mu = \hat{p}^\nu\psi_{\mu\nu}, \tag{14.2}$$

with $\hat{p} = i\partial$ (A. Proca, 1936). Applying the operator $\hat{p}^\mu$ to both sides of equation (14.2), we have

$$\hat{p}^\mu\psi_\mu = 0, \tag{14.3}$$

since $\psi_{\mu\nu}$ is antisymmetric.

By substituting (14.1) in (14.2) to eliminate $\psi_{\mu\nu}$, and using (14.3), we obtain

$$(\hat{p}^2 - m^2)\psi_\mu = 0, \tag{14.4}$$

whence it is again evident (cf. §10) that m is the mass of the particle. Thus a free particle with spin one can be described by a single 4-vector ψ^μ, whose components satisfy the second-order equation (14.4), and also the further condition (14.3), which eliminates from ψ^μ the part pertaining to spin zero.

In the rest frame, where ψ_μ is independent of the spatial coordinates, we find that $\hat{p}^0\psi_0 = 0$. Since also $\hat{p}^0\psi_0 = m\psi_0$, it is seen that in the rest frame $\psi_0 = 0$, as it should be, and the ψ_{ik} are likewise zero.

A particle with spin one can have different internal parities, according as ψ^μ is a true vector or a pseudovector. In the former case

$$\hat{P}\psi^\mu = (\psi^0, -\psi^i),$$

and in the latter case

$$\hat{P}\psi^\mu = (-\psi^0, \psi^i).$$

† Anticipating, we may mention that the ensemble of the 4-vector ψ_μ and the 4-tensor $\psi^{\mu\nu}$ corresponds to that of the 4-dimensional spinors of rank two $\xi^{\alpha\beta}$, $\eta_{\dot\alpha\dot\beta}$, $\zeta^{\alpha\dot\beta}$, where $\xi^{\alpha\beta}$ and $\eta_{\dot\alpha\dot\beta}$ are symmetrical spinors changed into each other on inversion (§19).

Equations (14.1), (14.2) can be derived from the variational principle, using the Lagrangian

$$L = \tfrac{1}{2}\psi_{\mu\nu}\psi^{\mu\nu*} - \tfrac{1}{2}\psi^{\mu\nu*}(\partial_\mu\psi_\nu - \partial_\nu\psi_\mu) - \tfrac{1}{2}\psi^{\mu\nu}(\partial_\mu\psi_\nu^* - \partial_\nu\psi_\mu^*) + m^2\psi_\mu\psi^{\mu*}. \tag{14.5}$$

The independent generalized coordinates are here represented by ψ_μ, ψ_μ^*, $\psi_{\mu\nu}$, $\psi_{\mu\nu}^*$.†

To find the energy–momentum tensor, formula (10.11) is not entirely suitable here, since it would lead to an unsymmetrical tensor requiring further symmetrization. Instead, we can use the formula

$$\tfrac{1}{2}T_{\mu\nu}\sqrt{-g} = -\frac{\partial}{\partial x^\lambda}\frac{\partial(L\sqrt{-g})}{\partial g^{\mu\nu}{}_{,\lambda}} + \frac{\partial(L\sqrt{-g})}{\partial g^{\mu\nu}}, \tag{14.6}$$

in which L is assumed to be expressed in a form appropriate to any curvilinear coordinates (see *Fields*, §94). If L contains only the components of the metric tensor $g_{\mu\nu}$, and not their derivatives with respect to the coordinates, the formula becomes simply

$$T_{\mu\nu} = \frac{2}{\sqrt{-g}}\frac{\partial(L\sqrt{-g})}{\partial g^{\mu\nu}} = 2\frac{\partial L}{\partial g^{\mu\nu}} - g_{\mu\nu}L$$

(since $d\log g = -g_{\mu\nu}dg^{\mu\nu}$).

Since the differentiation in formula (14.6) is not with respect to the quantities ψ_μ, $\psi_{\mu\nu}$, these quantities need not be regarded as independent when applying the formula; we may immediately make use of the relationship (14.1) to rewrite the Lagrangian (14.5) as

$$L = -\tfrac{1}{2}\psi_{\mu\nu}\psi_{\lambda\rho}^* g^{\mu\lambda}g^{\nu\rho} + m^2\psi_\mu\psi_\nu^* g^{\mu\nu}. \tag{14.7}$$

Then

$$T_{\mu\nu} = -\psi_{\mu\lambda}\psi_\nu{}^{\lambda*} - \psi_{\mu\lambda}^*\psi_\nu{}^\lambda + m^2(\psi_\mu^*\psi_\nu + \psi_\nu^*\psi_\mu) + g_{\mu\nu}(\tfrac{1}{2}\psi_{\lambda\rho}\psi^{\lambda\rho*} - m^2\psi_\lambda^*\psi^\lambda). \tag{14.8}$$

In particular, the energy density is given by the essentially positive expression

$$T_{00} = \tfrac{1}{2}\psi_{ik}\psi_{ik}^* + \psi_{0i}\psi_{0i}^* + m^2(\psi_0\psi_0^* + \psi_i\psi_i^*). \tag{14.9}$$

The conserved current density 4-vector is given by

$$j^\mu = i(\psi^{\mu\nu*}\psi_\nu - \psi^{\mu\nu}\psi_\nu^*). \tag{14.10}$$

This can be obtained, in accordance with (12.12), by differentiating the Lagrangian

† If the variation were made with respect to ψ_μ only (assuming $\psi_{\mu\nu}$ already expressed in terms of ψ_μ by (14.1)), equation (14.3) would have to be imposed as an additional condition unrelated to the variational principle.

(14.5) with respect to the derivative $\partial_\mu \psi_\nu$. In particular

$$j^0 = i(\psi^{0k*}\psi_k - \psi^{0k}\psi_k^*) \tag{14.11}$$

and is not an essentially positive quantity.

A plane wave normalized to one particle in the volume $V = 1$ is

$$\psi_\mu = \frac{1}{\sqrt{(2\varepsilon)}} u_\mu e^{-ipx}, \qquad u_\mu u^{\mu *} = -1, \tag{14.12}$$

where u_μ is the unit polarization 4-vector, which, by (14.3), satisfies the condition of four-dimensional transversality,

$$u_\mu p^\mu = 0. \tag{14.13}$$

For, on substituting the function (14.12) in (14.9) and (14.11), we obtain

$$T_{00} = -2\varepsilon^2 \psi_\mu \psi^{\mu *} = \varepsilon, \qquad j^0 = 1.$$

Unlike the photon, a vector particle with non-zero mass has three independent directions of polarization. The corresponding amplitudes are given in (16.21).

The density matrix for partially polarized vector particles is defined so that in a pure state it reduces to the product

$$\rho_{\mu\nu} = u_\mu u_\nu^*$$

(similarly to (8.7) for photons). According to (14.12) and (14.13), it satisfies the conditions

$$p^\mu \rho_{\mu\nu} = 0, \qquad \rho_\mu^\mu = -1. \tag{14.14}$$

For unpolarized particles, $\rho_{\mu\nu}$ must have the form $ag_{\mu\nu} + bp_\mu p_\nu$. When the coefficients a and b are found from (14.14), the result is

$$\rho_{\mu\nu} = -\tfrac{1}{3}(g_{\mu\nu} - p_\mu p_\nu / m^2). \tag{14.15}$$

The quantization of the vector particle field is entirely analogous to the scalar case, and there is no need to repeat the arguments. The ψ-operators of the vector field are

$$\left.\begin{aligned} \hat{\psi}_\mu &= \sum_{\mathbf{p},\alpha} \frac{1}{\sqrt{(2\varepsilon)}} (\hat{a}_{\mathbf{p}\alpha} u_\mu^{(\alpha)} e^{-ipx} + \hat{b}_{\mathbf{p}\alpha}^{+} u_\mu^{(\alpha)*} e^{ipx}), \\ \hat{\psi}_\mu^{+} &= \sum_{\mathbf{p},\alpha} \frac{1}{\sqrt{(2\varepsilon)}} (\hat{a}_{\mathbf{p}\alpha}^{+} u_\mu^{(\alpha)*} e^{ipx} + \hat{b}_{\mathbf{p}\alpha} u_\mu^{(\alpha)} e^{-ipx}), \end{aligned}\right\} \tag{14.16}$$

where the suffix α labels the three independent polarizations.

As in the scalar case, Bose quantization is necessary because the expression (14.9) for T_{00} is positive definite and the expression (14.11) for j^0 is not.

There is a close connection between the properties of strictly neutral vector and electromagnetic fields. The neutral vector field is described by an Hermitian ψ-operator:

$$\hat{\psi}_\mu = \sum_{\mathbf{p},\alpha} \frac{1}{\sqrt{(2\varepsilon)}} (\hat{c}_{\mathbf{p}\alpha} u_\mu^{(\alpha)} e^{-ipx} + \hat{c}_{\mathbf{p}\alpha}^{+} u_\mu^{(\alpha)*} e^{ipx}). \tag{14.17}$$

The Lagrangian of this field is

$$\hat{L} = \tfrac{1}{4}\hat{\psi}_{\mu\nu}\hat{\psi}^{\mu\nu} - \tfrac{1}{2}\hat{\psi}^{\mu\nu}(\partial_\mu\hat{\psi}_\nu - \partial_\nu\hat{\psi}_\mu) + \tfrac{1}{2}m^2\hat{\psi}_\mu\hat{\psi}^\mu. \tag{14.18}$$

The electromagnetic field corresponds to $m = 0$. The 4-vector ψ^μ then becomes the 4-potential A^μ, and the 4-tensor $\psi^{\mu\nu}$ becomes the field tensor $F^{\mu\nu}$, which is related to the potential by the definition (14.1). Equation (14.2) becomes $\partial^\nu\psi_{\mu\nu} = 0$, corresponding to the second pair of Maxwell's equations. This does not imply the condition (14.3), which therefore is no longer obligatory. Since the extra condition has disappeared, there is no need to regard $\hat{\psi}_\mu$ and $\hat{\psi}_{\mu\nu}$ as independent "coordinates" in the Lagrangian, and (14.18) becomes

$$\hat{L} = -\tfrac{1}{4}\hat{\psi}_{\mu\nu}\hat{\psi}^{\mu\nu}, \tag{14.19}$$

in agreement with the familiar classical expression for the Lagrangian of the electromagnetic field. This Lagrangian, like the tensor $\hat{\psi}_{\mu\nu}$, is invariant under any gauge transformation of the "potentials" ψ_μ. There is an evident connection between this property and the zero mass: the Lagrangian (14.18) does not possess the property, because of the term $m^2\hat{\psi}_\mu\hat{\psi}^\mu$.

§ 15. **The wave equation for particles with higher integral spins**

Since the wave equations (14.3), (14.4) follow immediately when the particle mass and spin are given, the practical utilization of the Lagrangian involves not so much the derivation of these equations as the establishment of expressions for the field energy, momentum and charge.

To do so we can, as already mentioned, use in place of (14.5) the expression (14.7), and the latter can be further transformed as follows. From (14.1), it can be rewritten as

$$\begin{aligned} L &= -(\partial_\mu\psi_\nu^*)(\partial^\mu\psi^\nu) + (\partial_\nu\psi_\mu^*)(\partial^\mu\psi^\nu) + m^2\psi_\mu\psi^{\mu*} \\ &= -(\partial_\mu\psi_\nu^*)(\partial^\mu\psi^\nu) + m^2\psi_\mu^*\psi^\mu + \partial_\nu(\psi_\mu^*\partial^\mu\psi^\nu) - \psi_\mu^*\partial^\mu\partial_\nu\psi^\nu. \end{aligned}$$

The last term is zero, by (14.3), and the one preceding it is a total derivative. Omitting this, we obtain the Lagrangian

$$L' = -(\partial_\mu\psi_\nu^*)(\partial^\mu\psi^\nu) + m^2\psi_\mu^*\psi^\mu. \tag{15.1}$$

This has the same form as the Lagrangian (10.9) for a particle with spin zero, the only difference being that the scalar ψ is replaced by the 4-vector ψ_μ and the sign is changed. The change of sign occurs because ψ_μ is a space-like vector, so that $\psi_\mu \psi^{\mu *} < 0$, whereas for a scalar particle $\psi\psi^* > 0$.

On constructing the energy–momentum 4-tensor and the current 4-vector from the Lagrangian (15.1), we obtain expressions of the same form as (10.12) and (10.18) for the scalar field:

$$T_{\mu\nu} = -\partial_\mu \psi^{\lambda *} \cdot \partial_\nu \psi_\lambda - \partial_\nu \psi^{\lambda *} \cdot \partial_\mu \psi_\lambda - L' g_{\mu\nu}, \tag{15.2}$$

$$j_\mu = -i[\psi^*_\lambda \partial_\mu \psi^\lambda - (\partial_\mu \psi^*_\lambda)\psi^\lambda]. \tag{15.3}$$

Thd difference between these and (14.8), (14.10) is again a total derivative. But it has already been stressed that the local values of these quantities have no profound physical significance. Only the volume integrals P_μ (10.15) and Q (10.19) are important, and these will be the same for either choice of $T_{\mu\nu}$ and j_μ.

This method of description can be immediately generalized to particles with any (integral) spin. The wave function of a particle with spin s is an irreducible 4-tensor of rank s, i.e. a tensor symmetrical in all its indices and vanishing on contraction with respect to any pair of indices:

$$\psi_{..\mu.\nu..} = \psi_{..\nu.\mu..}, \qquad \psi_{..\mu.}{}^{\mu}{}_{..} = 0. \tag{15.4}$$

This tensor must satisfy the additional condition of 4-transversality:

$$\hat{p}^\mu \psi_{..\mu..} = 0, \tag{15.5}$$

and each of its components must satisfy the second-order equation

$$(\hat{p}^2 - m^2)\psi_{...} = 0. \tag{15.6}$$

In the rest frame, the condition (15.5) means that every component of the 4-tensor whose indices include a zero must vanish. Thus the wave function in the rest frame (i.e. in the non-relativistic limit) is equivalent, as it should be, to an irreducible 3-tensor of rank s, the number of independent components of which is $2s+1$.

The Lagrangian, the energy–momentum tensor and the current vector for a field of particles with spin s differ from (15.1)–(15.3) only in that ψ_λ is replaced by $\psi_{\lambda\mu...}$.

The normalized plane wave is

$$\psi^{\mu\nu...} = \frac{1}{\sqrt{(2\varepsilon)}} u^{\mu\nu...} e^{-ipx}, \qquad u^*_{\mu\nu...} u^{\mu\nu...} = -1, \tag{15.7}$$

the wave amplitude satisfying the conditions

$$u^{..\mu..} p_\mu = 0. \tag{15.8}$$

There are $2s+1$ independent polarization states.

The quantization of the field is effected by an obvious generalization from the cases of spin zero and one.

The procedure given above is entirely sufficient for the stated purpose: to describe a field of free particles. The situation is different if it is proposed to describe the interaction of the particles with an electromagnetic field. This interaction would have to be included in the Lagrangian in order to yield all the equations without the need to impose additional conditions. In practice, however, this description of the interaction is found to be applicable only for electrons, i.e. particles with spin $\frac{1}{2}$ (see §32). For other spin values, therefore, the problem is only of methodological interest.

For any spin $s > 1$ (integral or half-integral), it proves impossible to formulate a variational principle by means of a single (tensor or spinor) function whose rank corresponds to the given spin. It is necessary to use additional tensor or spinor quantities of lower rank. The Lagrangian is then so chosen that these auxiliary quantities must be zero on account of the free-particle field equations which follow from the variational principle.†

§ 16. Helicity states of a particle‡

In the relativistic theory the orbital angular momentum $\mathbf{l}$ and the spin $\mathbf{s}$ of a moving particle are not separately conserved. Only the total angular momentum $\mathbf{j} = \mathbf{l} + \mathbf{s}$ is conserved. The component of the spin in any fixed direction (taken as the z-axis) is therefore also not conserved, and cannot be used to enumerate the polarization (spin) states of the moving particle.

The component of the spin in the direction of the momentum *is* conserved, however: since $\mathbf{l} = \mathbf{r} \times \mathbf{p}$ the product $\mathbf{s} \cdot \mathbf{n}$ is equal to the conserved product $\mathbf{j} \cdot \mathbf{n}$ ($\mathbf{n} = \mathbf{p}/|\mathbf{p}|$). This quantity is called the *helicity*; it has already been mentioned in §8 in relation to the photon. Its eigenvalues will be denoted by λ ($\lambda = -s, \dots, +s$), and states of a particle having definite values of λ will be called *helicity states*.

Let $\psi_{\mathbf{p}\lambda}$ be the wave function (plane wave) describing the state of a particle with definite values of $\mathbf{p}$ and λ, and $u^{(\lambda)}(\mathbf{p})$ its amplitude; to simplify the notation, we shall omit the indices for the components of this function (4-tensor indices for a particle with integral spin).

It has been shown in earlier sections that a wave function with more than $2s+1$ components is needed in order to give a relativistic description of particles with non-zero (integral) spin. But the number of independent components remains equal to $2s+1$; the "extra" components are eliminated by imposing additional conditions which cause these components to vanish in the rest frame. In Chapter III this will be shown for half-integral s also.

According to the formulae for transformation of the angular momentum (see *Fields*, §14), the helicity is invariant under those Lorentz transformations which do not alter the direction of $\mathbf{p}$ along which the angular momentum component is taken. The number λ therefore remains a good quantum number under such

† See M. Fierz and W. Pauli, *Proceedings of the Royal Society* **A173**, 211, 1939. The procedure indicated above is carried out in this paper for particles with spin 3/2 and 2.

‡ The discussion in this section relates to particles with any spin (integral or half-integral).

transformations, and the symmetry properties of helicity states can be studied by means of a frame of reference in which the momentum $|\mathbf{p}| \ll m$ (in the limit, the rest frame). Then $\psi_{\mathbf{p}\lambda}$ reduces to a non-relativistic wave function with $2s+1$ components. Let its amplitude be denoted by $w^{(\lambda)}(\mathbf{n})$, the argument being the direction $\mathbf{n} = \mathbf{p}/|\mathbf{p}|$ along which the angular momentum is quantized. The amplitude $w^{(\lambda)}$ is an eigenfunction of the operator $\mathbf{n} \cdot \hat{\mathbf{s}}$:

$$(\mathbf{n} \cdot \hat{\mathbf{s}}) w^{(\lambda)}(\mathbf{n}) = \lambda w^{(\lambda)}(\mathbf{n}). \tag{16.1}$$

In the spinor representation, $w^{(\lambda)}$ is a contravariant symmetrical spinor of rank $2s$; according to the correspondence formulae (*QM*, (57.2)), its components can also be enumerated by the corresponding values of the spin component σ along a fixed z-axis.†

In the momentum representation, the wave functions of the states considered are essentially the same as the amplitudes $u^{(\lambda)}(\mathbf{p})$:

$$\psi_{\mathbf{p}\lambda}(\mathbf{k}) = u^{(\lambda)}(\mathbf{k})\delta^{(2)}(\boldsymbol{\nu} - \mathbf{n}) = u^{(\lambda)}(\mathbf{p})\delta^{(2)}(\boldsymbol{\nu} - \mathbf{n}), \tag{16.2}$$

where the momentum as an independent variable is denoted by $\mathbf{k}$, as contrasted with its eigenvalue $\mathbf{p}$, and $\boldsymbol{\nu} = \mathbf{k}/|\mathbf{k}|$, as against $\mathbf{n} = \mathbf{p}/|\mathbf{p}|$.‡ In the non-relativistic limit,

$$\psi_{\mathbf{n}\lambda}(\boldsymbol{\nu}) = w^{(\lambda)}(\boldsymbol{\nu})\delta^{(2)}(\boldsymbol{\nu} - \mathbf{n}) = w^{(\lambda)}(\mathbf{n})\delta^{(2)}(\boldsymbol{\nu} - \mathbf{n}). \tag{16.3}$$

This expression should be written in the more explicit form

$$\psi_{\mathbf{n}\lambda}(\boldsymbol{\nu}, \sigma) = w_\sigma^{(\lambda)}(\boldsymbol{\nu})\delta^{(2)}(\boldsymbol{\nu} - \mathbf{n}),$$

showing the discrete independent variable σ.

The helicity operator $\hat{\mathbf{s}} \cdot \mathbf{n}$ commutes with the operators $\hat{j}_z$ and $\hat{\mathbf{j}}^2$, since the angular momentum operator is related to an infinitesimal rotation of the coordinates, and the scalar product of two vectors is invariant under any rotation. There exist, therefore, stationary states in which the particle simultaneously has definite values of the angular momentum j, its component $j_z = m$, and the helicity λ. Such states will be called *spherical helicity states*.

Let us determine the wave functions of these states in the momentum

† These arguments, like the possible values shown for λ, apply to particles with non-zero mass. For massless particles there is no rest frame, and the helicity can take only the two values $\lambda = \pm s$. This is because of the fact already mentioned in §8, that the states of such a particle are classified by their behaviour with respect to the axial-symmetry group, which allows only twofold degeneracy of levels (as regards the properties of the wave equation, this means that in the limit as $m \to 0$ the set of equations for a particle with spin s separates into independent equations corresponding to massless particles with spins $s, s-1, \ldots$). For example, the photon has $\lambda = \pm 1$, and the corresponding $w^{(\lambda)}$ are the three-dimensional vectors $\mathbf{e}^{(\pm 1)}$ (8.2).

‡ The delta function $\delta^{(2)}$ is defined so that

$$\int \delta^{(2)}(\boldsymbol{\nu} - \mathbf{n})do_\nu = 1.$$

The delta function which imposes a fixed value of the energy is omitted in (16.2), and similarly in (16.4) below.

representation. This may be done by direct analogy with the formulae derived in *QM*, §103 for the wave functions of a symmetrical top. They were obtained there on the basis of the formulae for the transformation of wave functions under finite rotations (*QM*, §58). These in turn were based solely on the symmetry properties with respect to rotation, and are therefore applicable to functions in the momentum representation just as much as to coordinate functions.

In addition to the coordinates x, y, z fixed in space (with respect to which the functions $\psi_{jm\lambda}$ are written), we shall also use "moving" coordinates ξ, η, ζ, with the ζ-axis in the direction of $\boldsymbol{\nu}$. Without repeating the argument (cf. the derivation of *QM*, (103.8)), we can write

$$\psi_{jm\lambda}(\mathbf{k}) = \psi_{j\lambda}^{(0)} D_{\lambda m}^{(j)}(\boldsymbol{\nu}),$$

where $\psi_{j\lambda}^{(0)}$ is the wave function in the moving coordinates, describing the state of a particle with a definite value of the ζ-component of the angular momentum, $j_\zeta = \lambda$; in the momentum representation, of course, this function is the same as the amplitude $u^{(\lambda)}$. The normalized wave function (see below) is

$$\psi_{jm\lambda}(\mathbf{k}) = \sqrt{\frac{2j+1}{4\pi}}\, D_{\lambda m}^{(j)}(\boldsymbol{\nu}) u^{(\lambda)}(\mathbf{k}). \tag{16.4}$$

Here, however, there is a question of the choice of phases, because of the following non-uniqueness: a rotation of the coordinates ξ, η, ζ relative to x, y, z is defined by three Eulerian angles α, β, γ, whereas the direction of $\boldsymbol{\nu}$, on which the particle wave function can alone depend, is defined by the two spherical angles $\alpha \equiv \phi$ and $\beta \equiv \theta$. It is thus necessary to agree on some definite choice of the angle γ. We shall take $\gamma = 0$, defining $D_{\lambda m}^{(j)}(\boldsymbol{\nu})$ as

$$D_{\lambda m}^{(j)}(\boldsymbol{\nu}) = D_{\lambda m}^{(j)}(\phi, \theta, 0) = e^{im\phi}\, d_{\lambda m}^{(j)}(\theta). \tag{16.5}$$

From *QM*, (58.21), the functions (16.5) are seen to satisfy the orthonormality conditions:

$$\int D_{\lambda_1 m_1}^{(j_1)*}(\boldsymbol{\nu}) D_{\lambda_2 m_2}^{(j_2)}(\boldsymbol{\nu}) \frac{do_\nu}{4\pi} = \frac{1}{2j+1} \delta_{j_1 j_2} \delta_{m_1 m_2}, \tag{16.6}$$

where $do_\nu = \sin\theta\, d\theta\, d\phi$. The orthogonality of the functions $\psi_{jm\lambda}$ with respect to the suffix λ is ensured by the factor $u^{(\lambda)}$. Thus the functions $\psi_{jm\lambda}$ are orthogonal in all three suffixes, as they should be, and with the coefficient chosen in (16.4) they are normalized by the condition

$$\int |\psi_{jm\lambda}|^2\, do_\nu = 1. \tag{16.7}$$

Here we assume that the amplitudes $u^{(\lambda)}$ are normalized to unity: $u^{(\lambda)} u^{(\lambda)*} = 1$.

Let us now consider the behaviour of the wave functions of helicity states under inversion of the coordinates. The product of the polar vector $\boldsymbol{\nu}$ and the axial vector $\mathbf{j}$ is a pseudoscalar. It is therefore obvious that inversion will change a state with helicity λ into one with helicity $-\lambda$; all that is necessary is to determine the phase factors in these transformations.

Under inversion, $\boldsymbol{\nu}\to-\boldsymbol{\nu}$. The vector $\boldsymbol{\nu}$ is defined by the two angles ϕ and θ, and the transformation $\boldsymbol{\nu}\to-\boldsymbol{\nu}$ is brought about by the changes $\phi\to\phi+\pi$, $\theta\to\pi-\theta$. This determines the ζ-axis but leaves indefinite the position of the ξ and η axes, which depends also on the third Eulerian angle γ; the transformation of θ and ϕ alone does not distinguish, in this sense, between reflection of the coordinates and rotation of the ζ-axis. Expressed in terms of all three Eulerian angles, inversion is the transformation

$$\alpha\equiv\phi\to\phi+\pi,\quad \beta\equiv\theta\to\pi-\theta,\quad \gamma\to\pi-\gamma. \tag{16.8}$$

Hence, if $D^{(j)}_{\lambda m}(\boldsymbol{\nu})$ is defined as in (16.5) (i.e. with $\gamma=0$), and the transformation $\boldsymbol{\nu}\to-\boldsymbol{\nu}$ is regarded as being the result of inversion, then

$$D^{(j)}_{\lambda m}(-\boldsymbol{\nu})=D^{(j)}_{\lambda m}(\phi+\pi,\pi-\theta,\pi). \tag{16.9}$$

From formulae *QM* (58.9), (58.16) and (58.18) we hence find

$$\begin{aligned}D^{(j)}_{\lambda m}(-\boldsymbol{\nu})&=e^{i\lambda\pi}\,d^{(j)}_{\lambda m}(\pi-\theta)\,e^{im(\phi+\pi)}\\&=(-1)^{j-\lambda}\,e^{im\phi}d^{(j)}_{-\lambda,m}(\theta)\\&=(-1)^{j-\lambda}D^{(j)}_{-\lambda,m}(\phi,\theta,0),\end{aligned}$$

or

$$D^{(j)}_{\lambda m}(-\boldsymbol{\nu})=(-1)^{j-\lambda}D^{(j)}_{-\lambda,m}(\boldsymbol{\nu}), \tag{16.10}$$

where $j-\lambda$ is an integer.

A similar formula for the spinor $w^{(\lambda)}$ can be obtained by noticing that its components $w^{(\lambda)}_\sigma$ are the same, apart from a factor, as the functions

$$w^{(\lambda)}_\sigma(\boldsymbol{\nu})\sim D^{(s)}_{\lambda\sigma}(\boldsymbol{\nu})^*. \tag{16.11}$$

For, by applying the transformation formulae *QM* (58.7) to the spin eigenfunctions and taking the ζ-component of the spin to have a definite value λ (i.e. replacing $\psi_{jm'}$ by $\delta_{m'\lambda}$ on the right-hand side of *QM* (58.7), we find that $D^{(s)}_{\lambda\sigma}(\boldsymbol{\nu})$ are the spin wave functions corresponding to definite values of the z and ζ components (σ and λ) of the spin. The set of these functions with $\sigma=-s,\dots,+s$ forms, according to the correspondence formulae (*QM* (57.6)), a covariant spinor of rank $2s$. The components of the contravariant spinor, which according to the formulae *QM* (57.2) correspond to the components $w^{(\lambda)}_\sigma$, are transformed as the complex conjugates of the components of the covariant spinor of the same rank.

From (16.10) and (16.11), we have

$$w^{(\lambda)}(-\boldsymbol{\nu})=(-1)^{s-\lambda}w^{(-\lambda)}(\boldsymbol{\nu}), \tag{16.12}$$

where $s-\lambda$ is an integer. The inversion operation applied to $w^{(\lambda)}$, however, not only changes $\boldsymbol{\nu}$ into $-\boldsymbol{\nu}$ but also multiplies $w^{(\lambda)}$ by a common phase factor (the "internal parity" of the particle), which we shall denote by η:

$$\hat{P}w^{(\lambda)}(\boldsymbol{\nu}) = \eta w^{(\lambda)}(-\boldsymbol{\nu}) = \eta(-1)^{s-\lambda}w^{(-\lambda)}(\boldsymbol{\nu}). \tag{16.13}$$

For the relativistic amplitude $u^{(\lambda)}(\mathbf{k})$, this transformation becomes

$$\begin{aligned}\hat{P}u^{(\lambda)}(\mathbf{k}) &= \eta\beta u^{(\lambda)}(-\mathbf{k})\\ &= \eta(-1)^{s-\lambda}u^{(-\lambda)}(\mathbf{k}),\end{aligned} \tag{16.14}$$

where β is a certain matrix which is a unit matrix with respect to the components of $u^{(\lambda)}$ which remain in the limit $|\mathbf{p}|\to 0$. It is important to note that this matrix does not depend on the quantum numbers of the state, and in this sense the difference between (16.13) and (16.14) is unimportant.†

On applying (16.14) to (16.2), we obtain the law of transformation of the wave functions of the states $|\mathbf{n}\lambda\rangle$:

$$\hat{P}\psi_{\mathbf{n}\lambda}(\boldsymbol{\nu}) = \eta(-1)^{s-\lambda}\psi_{-\mathbf{n},-\lambda}(\boldsymbol{\nu}). \tag{16.15}$$

For spherical helicity states, using (16.10) and (16.12), we obtain the transformation law

$$\hat{P}\psi_{jm\lambda}(\boldsymbol{\nu}) = \eta(-1)^{j-s}\psi_{jm,-\lambda}(\boldsymbol{\nu}). \tag{16.16}$$

The states ψ_{jm0} are transformed into themselves, according to (16.16), i.e. they have a definite parity. If $\lambda\neq 0$, however, only superpositions of states with opposite helicities have a definite parity:

$$\psi^{(\pm)}_{jm|\lambda|} = \frac{1}{\sqrt{2}}(\psi_{jm\lambda}\pm\psi_{jm,-\lambda}). \tag{16.17}$$

On inversion, these are transformed into themselves:

$$\hat{P}\psi^{(\pm)}_{jm|\lambda|}(\boldsymbol{\nu}) = \pm\,\eta(-1)^{j-s}\psi^{(\pm)}_{jm|\lambda|}(\boldsymbol{\nu}). \tag{16.18}$$

It should be noted that in this section we have arrived at a classification of states of a free particle with a given angular momentum, using only conserved quantities and without invoking the concept of the orbital angular momentum (which was employed, for instance, in §§6 and 7 for classifying photon states).

As an example, let us consider the case of spin one. In the rest frame the amplitudes $u^{(\lambda)}$ (4-vectors) become the three-dimensional vectors $\mathbf{e}^{(\lambda)}$, which here take the place of the amplitudes $w^{(\lambda)}$. The action of the operator of spin one on the vector function $\mathbf{e}$ is given by the formula

$$(\hat{s}_i\mathbf{e})_k = -ie_{ikl}e_l; \tag{16.19}$$

† For example, when $s=1$ the amplitudes $u^{(\lambda)}$ are the 4-vectors (16.22); β is then entirely a unit matrix with respect to the 4-vector indices, $\beta_{\mu\nu}=\delta_{\mu\nu}$. When $s=\frac{1}{2}$, as we shall see in Chapter III, $u^{(\lambda)}$ is a bispinor, the phase factor $\eta = i$, and β is the Dirac matrix γ^0 (see (21.10)).

see *QM*, §57, Problem 2. Thus equation (16.1) becomes

$$i\mathbf{n}\times\mathbf{e}^{(\lambda)} = \lambda\mathbf{e}^{(\lambda)}. \tag{16.20}$$

The solutions of this equation (in $\xi\eta\zeta$ coordinates with the ζ-axis in the direction of $\mathbf{n}$) are the same as the spherical unit vectors (7.14):†

$$\mathbf{e}^{(0)} = i(0, 0, 1), \qquad \mathbf{e}^{(\pm 1)} = \mp\frac{i}{\sqrt{2}}(1, \pm i, 0). \tag{16.21}$$

In a frame of reference in which the particle has momentum $\mathbf{p}$, the helicity state amplitudes are the 4-vectors

$$u^{(0)\mu} = \left(\frac{|\mathbf{p}|}{m}, \frac{\varepsilon}{m}\mathbf{e}^{(0)}\right), \qquad u^{(\pm 1)\mu} = (0, \mathbf{e}^{(\pm 1)}). \tag{16.22}$$

If $\mathbf{e}$ is a polar vector, then $\eta = -1$, and the functions (16.17), which are three-dimensional vectors when $s = 1$, have the following parities:

$$\psi^{(+)}_{jm|\lambda|}: P = (-1)^j,$$
$$\psi^{(-)}_{jm|\lambda|}: P = (-1)^{j+1},$$
$$\psi_{jm0}: P = (-1)^j.$$

On comparing with the definition of the spherical harmonic vectors (7.4), we see that these functions are identical (apart from phase factors) with $\mathbf{Y}^{(e)}_{jm}$, $\mathbf{Y}^{(m)}_{jm}$, $\mathbf{Y}^{(l)}_{jm}$ respectively. After ascertaining the phase factors (by comparing values for $\theta = 0$, say), we obtain the equations

$$\left.\begin{aligned}
\mathbf{Y}^{(e)}_{jm} &= i^{j-1}\sqrt{\frac{2j+1}{8\pi}}\left(\mathbf{e}^{(1)}D^{(j)}_{1m} + \mathbf{e}^{(-1)}D^{(j)}_{-1,m}\right),\\
\mathbf{Y}^{(m)}_{jm} &= i^{j-1}\sqrt{\frac{2j+1}{8\pi}}\left(\mathbf{e}^{(1)'}D^{(j)}_{1m} + \mathbf{e}^{(-1)'}D^{(j)}_{-1,m}\right),\\
\mathbf{Y}^{(l)}_{jm} &= i^{j-1}\sqrt{\frac{2j+1}{4\pi}}\,\mathbf{e}^{(0)}D^{(j)}_{0m},
\end{aligned}\right\} \tag{16.23}$$

where j is an integer; $\mathbf{e}^{(\lambda)'} = \mathbf{n}\times\mathbf{e}^{(\lambda)}$ are spherical unit vectors along axes ξ', η', ζ which are obtained from ξ, η, ζ by a rotation of 90° about the ζ-axis.

The last formula (16.23) is equivalent to the expression *QM* (58.23) for $d^{(j)}_{0m}(\theta)$. The first or second formula (16.23) leads to a simple expression for the functions

† The choice of phase factors is determined by the condition that the spin operator matrix elements calculated with the eigenfunctions (16.21) must be in accordance with the general definitions in *QM*, §§27 and 107.

$d^{(j)}_{\pm 1,m}$. We have

$$i^{j-1}\sqrt{\frac{2j+1}{8\pi}}D^{(j)}_{\pm 1,m} = \mathbf{Y}^{(e)}_{jm}\cdot\mathbf{e}^{(\pm 1)*}$$
$$= \frac{1}{\sqrt{[j(j+1)]}}\mathbf{e}^{(\pm 1)*}\cdot\nabla Y_{jm}.$$

The scalar product on the right can be written explicitly in the coordinates ξ, η, ζ, with

$$\left(\frac{\partial}{\partial\xi}, \frac{\partial}{\partial\eta}\right)\to\left(\frac{\partial}{\partial\theta}, \frac{1}{\sin\theta}\frac{\partial}{\partial\phi}\right).$$

With the definitions (7.2) of Y_{jm} and (16.5), the result is

$$d^{(j)}_{\pm 1,m}(\theta) = (-1)^{m+1}\sqrt{\frac{(j-m)!}{(j+m)!j(j+1)}}\left(\pm\frac{\partial}{\partial\theta}+\frac{m}{\sin\theta}\right)P^m_j(\cos\theta), \quad m\geqslant 0. \tag{16.24}$$

CHAPTER III

FERMIONS

§ 17. Four-dimensional spinors

IN THE non-relativistic theory, a particle with arbitrary spin s is described by a quantity with $2s+1$ components, a symmetrical spinor of rank $2s$. These quantities are, mathematically, realizations of the irreducible representations of the spatial rotation group.

In the relativistic theory, this group is only a subgroup of the wider group of four-dimensional rotations, the Lorentz group. It is therefore necessary to develop the theory of four-dimensional spinors (4-spinors), as quantities which are realizations of the irreducible representations of the Lorentz group. This theory will be given in §§17–19. In §§17 and 18 we shall consider only the proper Lorentz group, which excludes spatial inversion; the latter will be dealt with in §19.

The theory of 4-spinors is analogous in structure of that of three-dimensional spinors (B. L. van der Waerden, 1929; G. E. Uhlenbeck and O. Laporte, 1931).

A spinor ξ^α is a quantity having two components ($\alpha = 1, 2$); as components of the wave function of a particle with spin $\frac{1}{2}$, ξ^1 and ξ^2 correspond to the respective eigenvalues $+\frac{1}{2}$ and $-\frac{1}{2}$ of the z-component of the spin. Under any transformation belonging to the (proper) Lorentz group, the two quantities ξ^1 and ξ^2 are transformed into linear combinations of themselves:

$$\left.\begin{aligned} \xi^{1\prime} &= \alpha\xi^1 + \beta\xi^2, \\ \xi^{2\prime} &= \gamma\xi^1 + \delta\xi^2. \end{aligned}\right\} \tag{17.1}$$

The coefficients α, β, γ, δ are definite functions of the angles of rotation of the 4-coordinate system, and must satisfy the condition

$$\alpha\delta - \beta\gamma = 1; \tag{17.2}$$

that is, the determinant of the binary transformation (17.1) is equal to unity, as are the determinants of the coordinate transformations in the Lorentz group.

Because of the condition (17.2), the bilinear form $\xi^1\Xi^2 - \xi^2\Xi^1$ (where ξ^α and Ξ^α are two spinors) is invariant under the transformation (17.1), and corresponds to a particle with spin zero which "consists" of two particles with spin $\frac{1}{2}$. In order to write such invariant expressions in a natural way, the "covariant" components ξ_α are used as well as the "contravariant" components ξ^α of the spinor. Their relationship is governed by the "metric spinor" $g_{\alpha\beta}$:†

$$\xi_\alpha = g_{\alpha\beta}\xi^\beta, \tag{17.3}$$

† The spinor indices will be denoted by the letters at the beginning of the Greek alphabet: α, β, γ, ...

where

$$g_{\alpha\beta} = \begin{pmatrix} 0 & 1 \\ -1 & 0 \end{pmatrix}, \tag{17.4}$$

so that

$$\xi_1 = \xi^2, \qquad \xi_2 = -\xi^1. \tag{17.5}$$

Then the invariant $\xi^1\Xi^2 - \xi^2\Xi^1$ becomes the scalar product $\xi^\alpha\Xi_\alpha$, and $\xi^\alpha\Xi_\alpha = -\xi_\alpha\Xi^\alpha$.

The properties so far stated are formally the same as those of three-dimensional spinors. A difference arises, however, when complex-conjugate spinors are considered.

In the non-relativistic theory, the sum

$$\psi^1\psi^{1*} + \psi^2\psi^{2*}, \tag{17.6}$$

which determines the probability density for the localization of the particles in space, must be a scalar, and the components $\psi^{\alpha*}$ must therefore be transformed as the covariant components of a spinor; the transformation (17.1) must therefore be unitary ($\alpha = \delta^*$, $\beta = -\gamma^*$). In the relativistic theory, however, the particle density is not a scalar, but is the time component of a 4-vector. The above-mentioned condition therefore no longer applies, and the transformation coefficients need satisfy no condition other than (17.2). The four complex quantities $\alpha, \beta, \gamma, \delta$ under the condition (17.2) alone are equivalent to $8-2=6$ real parameters, in accordance with the number of angles which define a rotation of the 4-coordinate system (rotations in six coordinate planes).

Thus complex-conjugate binary transformations are quite different, and in the relativistic theory there exist two types of spinors. A special notation is customary, in order to distinguish these two types: the indices of spinors which are transformed by the complex conjugate formulae to (17.1) are written with dots over them and are called *dotted indices*. Thus, by definition,

$$\eta^{\dot\alpha} \sim \xi^{\alpha*}, \tag{17.7}$$

where the sign $\sim$ denotes "is transformed as". The transformation formulae for a "dotted" spinor are therefore

$$\eta^{\dot 1\prime} = \alpha^*\eta^{\dot 1} + \beta^*\eta^{\dot 2}, \qquad \eta^{\dot 2\prime} = \gamma^*\eta^{\dot 1} + \delta^*\eta^{\dot 2}. \tag{17.8}$$

The operations of raising and lowering the dotted indices are carried out in the same way as for the undotted indices:

$$\eta_{\dot 1} = \eta^{\dot 2}, \qquad \eta_{\dot 2} = -\eta^{\dot 1}. \tag{17.9}$$

The behaviour of 4-spinors as regards spatial rotation is the same as that of

3-spinors, for which, as we know, $\psi^*_\alpha \sim \psi^\alpha$. According to the definition (17.7), the 4-spinor $\eta_{\dot\alpha}$ therefore behaves under rotations in the same way as the contravariant 3-spinor ψ^α. The covariant components $\eta_{\dot 1}$ and $\eta_{\dot 2}$ therefore correspond, as the components of the wave function of a particle with spin $\frac{1}{2}$, to the eigenvalues $\frac{1}{2}$ and $-\frac{1}{2}$ of the spin component.

Spinors of higher rank are defined as sets of quantities which are transformed as products of the components of a number of spinors of rank one. The indices of these spinors of higher rank may be partly dotted and partly undotted. For example, there exist three types of spinors of rank two:

$$\xi^{\alpha\beta} \sim \xi^\alpha \Xi^\beta, \qquad \zeta^{\alpha\dot\beta} \sim \xi^\alpha \eta^{\dot\beta}, \qquad \eta^{\dot\alpha\dot\beta} \sim \eta^{\dot\alpha} \mathrm{H}^{\dot\beta}.$$

In this respect, the statement of just the total rank of a spinor does not uniquely define it; we shall therefore, where necessary, indicate the rank as a pair of numbers (k, l), the numbers of undotted and dotted indices respectively.

Since the transformations (17.1) and (17.8) are algebraically independent, it is not necessary to specify the sequence of dotted and undotted indices; in this sense the spinors $\zeta^{\alpha\dot\beta}$ and $\zeta^{\dot\beta\alpha}$, for example, are the same.

In order to be invariant, every spinor equation must have on each side the same numbers of undotted and dotted indices, since otherwise the equation could not remain valid when the frame of reference was changed. Here we must remember that taking the complex conjugate implies interchanging dotted and undotted indices. The relationship $\eta^{\dot\alpha\dot\beta} = (\xi^{\alpha\beta})^*$ between two spinors is therefore invariant.

Spinors or their products can be contracted only with respect to pairs of indices of the same kind (dotted or undotted); summation with respect to two indices of different kinds is not an invariant operation. Hence, from the spinor

$$\zeta^{\alpha_1\alpha_2\ldots\alpha_k\dot\beta_1\dot\beta_2\ldots\dot\beta_l}, \tag{17.10}$$

which is symmetrical in all k undotted indices and in all l dotted indices, we can obtain no spinor of lower rank (since contraction with respect to a pair of indices in which the spinor is symmetrical gives zero). Thus we cannot construct from the quantities (17.10) a smaller number of linear combinations of them which in turn are transformed into linear combinations of themselves by every transformation in the group. That is, the symmetrical 4-spinors are realizations of the irreducible representations of the proper Lorentz group. Each irreducible representation is specified by the pair of numbers (k, l).

Each spinor index takes two values, and there are therefore $k+1$ essentially different sets of numbers $\alpha_1, \alpha_2, \ldots, \alpha_k$ in (17.10) (containing $0, 1, 2, \ldots, k$ ones and $k, k-1, \ldots, 0$ twos) and $l+1$ sets of numbers $\dot\beta_1, \dot\beta_2, \ldots \dot\beta_l$. The symmetrical spinor of rank (k, l) thus has a total of $(k+1)(l+1)$ independent components, and this is also the dimension of the corresponding irreducible representation.

§ 18. **The relation between spinors and 4-vectors**

The spinor $\zeta^{\alpha\dot\beta}$, with one dotted and undotted index, has $2\times 2 = 4$ independent components, the same as the number of a 4-vector. It is therefore clear that both are realizations of the same irreducible representation of the proper Lorentz group, and that there must consequently be a certain relation between their components.

In order to ascertain this relation, let us first consider the corresponding relation in the three-dimensional case, using the fact that 3-spinors and 4-spinors must behave in the same manner with respect to purely spatial rotations.

For the three-dimensional spinor $\psi^{\alpha\beta}$, the correspondence formulae are as shown in *QM*, §57; they will here be written as

$$a_x = \tfrac{1}{2}(\psi^{22} - \psi^{11}) = \tfrac{1}{2}(\psi^2{}_1 + \psi^1{}_2),$$
$$a_y = -\tfrac{1}{2}i(\psi^{22} + \psi^{11}) = \tfrac{1}{2}i(\psi^1{}_2 - \psi^2{}_1),$$
$$a_z = \tfrac{1}{2}(\psi^{12} + \psi^{21}) = \tfrac{1}{2}(\psi^1{}_1 - \psi^2{}_2),$$

where a_x, a_y, a_z are the components of a three-dimensional vector $\mathbf{a}$. For the four-dimensional case, the components $\psi^\alpha{}_\beta$ must be replaced by $\zeta^{\alpha\dot\beta}$, and a_x, a_y, a_z must be taken to be the contravariant components a^1, a^2, a^3 of a 4-vector. The form of the expression for the fourth component a^0 is evident from the fact, noted in §17, that the quantity (17.6) must transform as a^0. Hence $a^0 \sim \zeta^{1\dot1} + \zeta^{2\dot2}$, the coefficient of proportionality being determined so that the scalar $\zeta_{\alpha\dot\beta}\zeta^{\alpha\dot\beta}$ is the same as the scalar $2a_\mu a^\mu \equiv 2a^2$.

Thus we obtain the correspondence formulae

$$\left.\begin{aligned} a^1 &= \tfrac{1}{2}(\zeta^{1\dot2} + \zeta^{2\dot1}), & a^2 &= \tfrac{1}{2}i(\zeta^{1\dot2} - \zeta^{2\dot1}), \\ a^3 &= \tfrac{1}{2}(\zeta^{1\dot1} - \zeta^{2\dot2}), & a^0 &= \tfrac{1}{2}(\zeta^{1\dot1} + \zeta^{2\dot2}). \end{aligned}\right\} \tag{18.1}$$

The inverse formulae are

$$\left.\begin{aligned} \zeta^{1\dot1} &= \zeta_{2\dot2} = a^3 + a^0, & \zeta^{2\dot2} &= \zeta_{1\dot1} = a^0 - a^3, \\ \zeta^{1\dot2} &= -\zeta_{2\dot1} = a^1 - ia^2, & \zeta^{2\dot1} &= -\zeta_{1\dot2} = a^1 + ia^2, \end{aligned}\right\} \tag{18.2}$$

with

$$\zeta_{\alpha\dot\beta}\zeta^{\alpha\dot\beta} = 2a^2. \tag{18.3}$$

Moreover

$$\zeta_{\alpha\dot\beta}\zeta^{\gamma\dot\beta} = \delta^\gamma_\alpha a^2, \tag{18.4}$$

as is seen from the fact that the spinor $\zeta_{\alpha\dot\beta}\zeta_\gamma{}^{\dot\beta}$, of rank two, is antisymmetric in the indices α, γ, and is therefore proportional to the metric spinor.

The correspondence between the spinor $\zeta^{\alpha\dot\beta}$ and the 4-vector is a particular case of a general rule: any symmetrical spinor of rank (k, k) is equivalent to a symmetrical 4-tensor of rank k which is irreducible (i.e. which gives zero on contraction with respect to any pair of indices).

The relation between the spinor and the 4-vector may be written in a compact form by means of the two-rowed Pauli matrices†

$$\sigma_x = \begin{pmatrix} 0 & 1 \\ 1 & 0 \end{pmatrix}, \quad \sigma_y = \begin{pmatrix} 0 & -i \\ i & 0 \end{pmatrix}, \quad \sigma_z = \begin{pmatrix} 1 & 0 \\ 0 & -1 \end{pmatrix}. \tag{18.5}$$

If the matrix of the quantities $\zeta^{\alpha\dot{\beta}}$ (with the indices raised and the first undotted) is symbolized by ζ, then formulae (18.2) become

$$\zeta = \mathbf{a}\cdot\boldsymbol{\sigma} + a^0, \tag{18.6}$$

the second term denoting of course the product of a^0 and a unit matrix. The inverse formulae are

$$\mathbf{a} = \tfrac{1}{2}\,\mathrm{tr}\,(\zeta\boldsymbol{\sigma}), \qquad a^0 = \tfrac{1}{2}\,\mathrm{tr}\,\zeta. \tag{18.7}$$

Using formulae (18.6), (18.7), we can determine the relation between the laws of transformation of the 4-vector and the spinor, and thus express the law of transformation of the spinor in terms of the parameters of rotations of the 4-coordinates.

We write the transformation of the spinor ξ^α in the form

$$\xi^{\alpha\prime} = (B\xi)^\alpha, \qquad B = \begin{pmatrix} \alpha & \beta \\ \gamma & \delta \end{pmatrix}, \tag{18.8}$$

where B is a two-rowed matrix formed from the coefficients of the binary transformation. Then the transformation of the dotted spinor is

$$\eta^{\dot{\beta}\prime} = (B^*\eta)^{\dot{\beta}} = (\eta B^+)^{\dot{\beta}}, \tag{18.9}$$

and the transformation of the spinor $\zeta^{\alpha\dot{\beta}} \sim \xi^\alpha\eta^{\dot{\beta}}$, of rank two, may be symbolized as‡ $\zeta' = B\zeta B^+$. For the infinitesimal transformation $B = 1 + \lambda$, where λ is a small matrix, we have as far as first-order quantities

$$\zeta' = \zeta + (\lambda\zeta + \zeta\lambda^+). \tag{18.10}$$

Let us first consider the Lorentz transformation to a frame of reference moving with an infinitesimal velocity $\delta\mathbf{V}$ (without change in direction of the space coordinate axes). Then the 4-vector $a^\mu = (a^0, \mathbf{a})$ is transformed as follows:

$$\mathbf{a}' = \mathbf{a} - a^0\delta\mathbf{V}, \qquad a^{0\prime} = a^0 - \mathbf{a}\cdot\delta\mathbf{V}. \tag{18.11}$$

† To simplify the notation, matrix operators acting on spin variables are written without circumflexes.

‡ For the covariant components we have

$$\xi'_\alpha = (\tilde{B}^{-1}\xi)_\alpha = (\xi B^{-1})_\alpha, \qquad \eta'_{\dot{\alpha}} = (\eta B^{*-1})_{\dot{\alpha}}, \tag{18.8a}$$

so that the product $\xi_\alpha \Xi^\alpha$ of two spinors remains invariant.

We now make use of formulae (18.7). The transformation of a^0 may be represented, firstly, as

$$a^{0\prime} = a^0 - \mathbf{a}\cdot\delta\mathbf{V} = a^0 - \tfrac{1}{2}\,\mathrm{tr}\,(\zeta\boldsymbol{\sigma}\cdot\delta\mathbf{V});$$

secondly, as

$$\begin{aligned} a^{0\prime} &= \tfrac{1}{2}\,\mathrm{tr}\,\zeta' = a^0 + \tfrac{1}{2}\,\mathrm{tr}\,(\lambda\zeta + \zeta\lambda^+) \\ &= a^0 + \tfrac{1}{2}\,\mathrm{tr}\,\zeta(\lambda + \lambda^+). \end{aligned}$$

These two expressions must be identically equal (i.e. equal for all values of ζ). Hence

$$\lambda + \lambda^+ = -\boldsymbol{\sigma}\cdot\delta\mathbf{V}.$$

Treating the transformation of $\mathbf{a}$ in the same way, we find

$$\boldsymbol{\sigma}\lambda + \lambda^+\boldsymbol{\sigma} = -\delta\mathbf{V}.$$

These equations, as equations for λ, have the solution

$$\lambda = \lambda^+ = -\tfrac{1}{2}\boldsymbol{\sigma}\cdot\delta\mathbf{V}.$$

Thus an infinitesimal Lorentz transformation of the spinor ξ^α has the matrix

$$B = 1 - \tfrac{1}{2}\boldsymbol{\sigma}\cdot\mathbf{n}\delta V, \tag{18.12}$$

where $\mathbf{n}$ is a unit vector in the direction of the velocity $\delta\mathbf{V}$. From this we can easily find the transformation for a finite velocity $\mathbf{V}$. To do so, we recall that a Lorentz transformation signifies (geometrically) a rotation of the 4-coordinates in the plane of t and $\mathbf{n}$ through an angle ϕ which is related to the velocity V by† $\tanh\phi = V$. An angle $\delta\phi = \delta V$ corresponds to an infinitesimal transformation, and a rotation through a finite angle ϕ is carried out by a $\phi/\delta\phi$-fold repetition of a rotation through $\delta\phi$. Raising the operator (18.12) to the power $\phi/\delta\phi$ and taking the limit $\delta\phi \to 0$, we obtain

$$B = e^{-\frac{1}{2}\phi\mathbf{n}\cdot\boldsymbol{\sigma}}. \tag{18.13}$$

The mathematical significance of this operator is seen by noticing that, from the properties of the Pauli matrices, all even powers of $\mathbf{n}\cdot\boldsymbol{\sigma}$ are equal to 1, and all odd powers are equal to $\mathbf{n}\cdot\boldsymbol{\sigma}$. Since the expansions of the hyperbolic sine and cosine contain respectively odd and even powers of the argument, we have finally

$$\left.\begin{aligned} B &= \cosh\tfrac{1}{2}\phi - \mathbf{n}\cdot\boldsymbol{\sigma}\sinh\tfrac{1}{2}\phi, \\ \tanh\phi &= V. \end{aligned}\right\} \tag{18.14}$$

† The metric is pseudo-Euclidean in planes containing the time axis.

The matrices B of the Lorentz transformations are Hermitian: $B = B^+$

Let us now consider an infinitesimal rotation of the space coordinates. The three-dimensional vector **a** is transformed as follows:

$$\mathbf{a}' = \mathbf{a} - \delta\boldsymbol{\theta} \times \mathbf{a}, \tag{18.15}$$

where $\delta\boldsymbol{\theta}$ is the vector of the infinitesimal angle of rotation. The corresponding transformation of a spinor may be found similarly. There is no need to do so, however, since the behaviour of 4-spinors under spatial rotations is the same as that of 3-spinors, and the transformation of the latter is known from the general relationship between the spin operator and the operator of an infinitesimal rotation:

$$B = 1 + \tfrac{1}{2} i \boldsymbol{\sigma} \cdot \delta\boldsymbol{\theta}. \tag{18.16}$$

The change to a rotation through a finite angle θ is made in the same way as that from (18.12) to (18.14):

$$B = \exp\left(\tfrac{1}{2} i\theta \mathbf{n} \cdot \boldsymbol{\sigma}\right) = \cos \tfrac{1}{2}\theta + i\mathbf{n} \cdot \boldsymbol{\sigma} \sin \tfrac{1}{2}\theta, \tag{18.17}$$

where **n** is a unit vector along the axis of rotation. This matrix is unitary ($B^+ = B^{-1}$), as it should be for a spatial rotation.

§ 19. **Inversion of spinors**

The discussion (in *QM*) of the three-dimensional theory of spinors did not consider their behaviour under the operation of spatial inversion, since in the non-relativistic theory this would not have led to any new physical results. Here we shall examine the point, however, in order to make clearer the subsequent analysis of the inversion properties of 4-spinors.

The operation of inversion does not alter the sign of the spin vector, or of any axial vector, and the spin component s_z is therefore also unchanged in value. Hence it follows that inversion can change each component of the spinor ψ^α only into a multiple of itself:

$$\psi^\alpha \to P\psi^\alpha, \tag{19.1}$$

where P is a constant factor. On repeating the inversion, we return to the original coordinates. For a spinor, however, a return to the original position can be regarded in two different ways, as a rotation through 0° or 360°. These two definitions are not equivalent with respect to spinors, since ψ^α changes sign on rotation through 360°. Thus two alternative views of inversion are possible: one where

$$P^2 = 1, \qquad P = \pm 1, \tag{19.2}$$

and one where

$$P^2 = -1, \qquad P = \pm i. \tag{19.3}$$

Here it is important to note that the concept of inversion must be defined in the same way for all spinors. It is not permissible for different spinors to behave differently under inversion (i.e. in accordance with both (19.2) and (19.3)), since in that case it would not be possible to construct a scalar (or a pseudoscalar) from every pair of spinors: if the spinor ψ^α were transformed according to (19.2), and ϕ^α according to (19.3), then the quantity $\psi^\alpha\phi_\alpha$ would be multiplied by $\pm i$ under inversion, instead of remaining constant (or simply changing sign).

It should be emphasized that (whatever the definition of inversion) the assignment of a particular parity P to a spinor has no absolute significance, since spinors change sign on rotation through 2π, and this can always be carried out simultaneously with inversion. The "relative parity" of two spinors, defined as the parity of the scalar $\psi^\alpha\phi_\alpha$ formed from them, has absolute significance, however; on rotation through 2π, both spinors change sign, and the indeterminacy therefore does not influence the parity of this scalar.

Let us now go on to discuss four-dimensional spinors, first noting that inversion changes the sign of only three coordinates x, y, z out of four x, y, z, t; it therefore commutes with spatial rotations but not with transformations which rotate the t-axis. If $\hat{L}$ is the Lorentz transformation to a frame of reference moving with velocity $\mathbf{V}$, then $\hat{P}\hat{L} = \hat{L}'\hat{P}$, where $\hat{L}'$ is the transformation to a frame moving with velocity $-\mathbf{V}$.

Hence it follows that the components of the 4-spinor ξ^α cannot be transformed into multiples of themselves under inversion. If the inversion of the spinor ξ^α were given by the transformation (19.1) as before (i.e. if it were represented by a matrix proportional to the unit matrix), it would commute with every Lorentz transformation, and this certainly cannot be true, since the operations $\hat{L}$ and $\hat{L}'$ are not the same when applied to ξ^α.

Thus inversion must transform the components of the spinor ξ^α into expressions involving other quantities. The latter can only be the components of some other spinor $\eta^{\dot{\alpha}}$ whose transformation properties are not the same as those of ξ^α. Since inversion does not affect the z-component of the spin (as mentioned above), the components ξ^1 and ξ^2 can only become $\eta_{\dot{1}}$ and $\eta_{\dot{2}}$ on inversion, these corresponding to the same values $s_z = \frac{1}{2}$ and $s_z = -\frac{1}{2}$. If inversion is taken to be an operation which gives identity when carried out twice, its effect may be expressed by the formulae

$$\xi^\alpha \to \eta_{\dot{\alpha}}, \qquad \eta_{\dot{\alpha}} \to \xi^\alpha. \tag{19.4}$$

For the covariant components ξ_α and contravariant components $\eta^{\dot{\alpha}}$, these transformations change sign:

$$\xi_\alpha \to -\eta^{\dot{\alpha}}, \qquad \eta^{\dot{\alpha}} \to -\xi_\alpha, \tag{19.4a}$$

since the lowering and raising of the same index lead to opposite signs (cf. (17.5) and (17.9)).† If, however, inversion is taken in the sense such that $P^2 = -1$, its effect is given by

$$\xi^\alpha \to i\eta_{\dot\alpha}, \qquad \eta_{\dot\alpha} \to i\xi^\alpha \tag{19.5}$$

or, equivalently,

$$\xi_\alpha \to -i\eta^{\dot\alpha}, \qquad \eta^{\dot\alpha} \to -i\xi_\alpha. \tag{19.5a}$$

There is a certain difference between the two definitions of inversion in that with the second definition complex-conjugate spinors are transformed in the same manner: if $\Xi_\alpha = \eta^*_{\dot\alpha}$, $\mathrm{H}^{\dot\alpha} = \xi^{\alpha *}$, then by (19.5) $\Xi_\alpha \to -i\mathrm{H}^{\dot\alpha}$, $\mathrm{H}^{\dot\alpha} \to -i\Xi_\alpha$, i.e. the rule is the same as for ξ_α, $\eta^{\dot\alpha}$. According to the definition (19.4), however, we should obtain $\Xi_\alpha \to \mathrm{H}^{\dot\alpha}$, $\mathrm{H}^{\dot\alpha} \to \Xi_\alpha$, which is opposite in sign to the transformation of the spinors ξ_α, $\eta^{\dot\alpha}$. We shall return in §27 to some possible physical consequences of this difference.

In the following, the definition (19.5) will be used.

The spinors ξ^α and $\eta_{\dot\alpha}$ are, as we know, transformed in the same way by the rotation subgroup. On taking the combinations

$$\xi^\alpha \pm \eta_{\dot\alpha} \tag{19.6}$$

we obtain quantities which are transformed under inversion according to (19.1) with $P = \pm i$. These combinations, however, do not behave as spinors under all the transformations of the Lorentz group.

Thus the inclusion of inversion in the symmetry group makes necessary the simultaneous treatment of a pair of spinors $(\xi^\alpha, \eta_{\dot\alpha})$; this is called a *bispinor* (of rank one). The four components of a bispinor form a realization of one of the irreducible representations of the extended Lorentz group.

The scalar product of two bispinors $(\xi^\alpha, \eta_{\dot\alpha})$ and $(\Xi^\alpha, \mathrm{H}_{\dot\alpha})$ can be formed in two ways. The quantity

$$\xi^\alpha \Xi_\alpha + \eta_{\dot\alpha}\mathrm{H}^{\dot\alpha} \tag{19.7}$$

is unchanged by inversion, i.e. it is a true scalar. The quantity

$$\xi^\alpha \Xi_\alpha - \eta_{\dot\alpha}\mathrm{H}^{\dot\alpha} \tag{19.8}$$

is also invariant under rotations of the 4-coordinates, but changes sign under inversion, i.e. it is a pseudoscalar.

A spinor of rank two, $\zeta^{\alpha\dot\beta}$, may also be defined in two ways. If it is defined by the transformation rule

$$\zeta^{\alpha\dot\beta} \sim \xi^\alpha \mathrm{H}^{\dot\beta} + \Xi^\alpha \eta^{\dot\beta}, \tag{19.9}$$

† The definition (19.4) is, of course, to some extent arbitrary, since the quantities ξ^α and $\eta_{\dot\alpha}$ are independent. For instance, if $\eta_{\dot\alpha}$ is replaced by a new spinor $\eta'_{\dot\alpha} = e^{i\delta}\eta_{\dot\alpha}$, (19.4) is replaced by the equivalent definition

$$\xi^\alpha \to e^{-i\delta}\,\eta'_{\dot\alpha}, \qquad \eta'_{\dot\alpha} \to e^{i\delta}\,\xi^\alpha.$$

we obtain quantities which are transformed under inversion as follows:

$$\zeta^{\alpha\dot{\beta}} \to \zeta_{\dot{\alpha}\beta}. \tag{19.10}$$

The 4-vector a^{μ} to which such a spinor is equivalent is transformed, according to (18.1), by $(a^0, \mathbf{a}) \to (a^0, -\mathbf{a})$, i.e. it is a true 4-vector, and the three-dimensional vector **a** is a polar vector.

It is also possible, however, to define $\zeta^{\alpha\dot{\beta}}$ thus:

$$\zeta^{\alpha\dot{\beta}} \sim \xi^{\alpha}\mathrm{H}^{\dot{\beta}} - \Xi^{\alpha}\eta^{\dot{\beta}}. \tag{19.11}$$

Then†

$$\zeta^{\alpha\dot{\beta}} \to -\zeta_{\dot{\alpha}\beta}. \tag{19.12}$$

Such a spinor corresponds to a 4-vector such that under inversion $(a^0, \mathbf{a}) \to (-a^0, \mathbf{a})$, i.e. a 4-pseudovector (the three-dimensional vector **a** being an axial vector).

Symmetrical spinors of rank two, with indices of the same type, are defined by

$$\xi^{\alpha\beta} \sim \xi^{\alpha}\Xi^{\beta} + \xi^{\beta}\Xi^{\alpha}, \qquad \eta_{\dot{\alpha}\dot{\beta}} \sim \eta_{\dot{\alpha}}\mathrm{H}_{\dot{\beta}} + \eta_{\dot{\beta}}\mathrm{H}_{\dot{\alpha}}. \tag{19.13}$$

On inversion they are transformed into each other:

$$\xi^{\alpha\beta} \to -\eta_{\dot{\alpha}\dot{\beta}}. \tag{19.14}$$

The pair $(\xi^{\alpha\beta}, \eta_{\dot{\alpha}\dot{\beta}})$ forms a bispinor of rank two. It has $3+3=6$ independent components. The antisymmetric 4-tensor of rank two $a^{\mu\nu}$ also has this number of independent components. There must therefore be a certain correspondence between the bispinor and the tensor; both are realizations of equivalent irreducible representations of the extended Lorentz group.

Since the spinors $\xi^{\alpha\beta}$ and $\eta_{\dot{\alpha}\dot{\beta}}$ are transformed independently by the proper Lorentz group, we can construct from the components of the 4-tensor $a^{\mu\nu}$ two groups of quantities which are transformed only into combinations of one another under any rotation of the 4-coordinates. This division is achieved as follows.

We define a three-dimensional polar vector **p** and a three-dimensional axial vector **a** related to the components of the 4-tensor $a^{\mu\nu}$ by

$$a^{\mu\nu} = \begin{pmatrix} 0 & p_x & p_y & p_z \\ -p_x & 0 & -a_z & a_y \\ -p_y & a_z & 0 & -a_x \\ -p_z & -a_y & a_x & 0 \end{pmatrix} \equiv (\mathbf{p}, \mathbf{a}), \tag{19.15}$$

where $(\mathbf{p}, \mathbf{a})$ is a concise notation which we shall use in order to specify the

† It must be emphasized that the transformation rules (19.10) and (19.12), which differ in the sign on the right, are not equivalent, since components of the same spinor appear on both sides (cf. the last footnote).

components of such a tensor. Then $a_{\mu\nu} = (-\mathbf{p}, \mathbf{a})$, and, of the two quantities

$$\mathbf{a}^2 - \mathbf{p}^2 = \tfrac{1}{2}a_{\mu\nu}a^{\mu\nu}, \qquad \mathbf{a}\cdot\mathbf{p} = \tfrac{1}{8}e_{\mu\nu\rho\sigma}a^{\mu\nu}a^{\rho\sigma},$$

the first is a scalar and the second a pseudoscalar; both are invariant under the proper Lorentz group. The squares of the three-dimensional vectors $\mathbf{f}^{\pm} = \mathbf{p} \pm i\mathbf{a}$ are therefore also invariant. Thus any rotation in 4-space is equivalent, as regards the vectors $\mathbf{f}^{\pm}$, to a "rotation" in 3-space, through angles which are in general complex; the six angles of rotation in 4-space correspond to three complex "angles of rotation" of the three-dimensional coordinates. The operation of spatial inversion changes the sign of $\mathbf{p}$ but not that of $\mathbf{a}$, and converts the vectors $\mathbf{f}^+$ and $-\mathbf{f}^-$ into each other. The components of these vectors are the required two groups of quantities formed from the components of the tensor $a^{\mu\nu}$.

This also makes evident the correspondence between the components of the 4-tensor $a^{\mu\nu}$ and the spinors $\xi^{\alpha\beta}$, $\eta_{\dot\alpha\dot\beta}$. Since the Lorentz group contains as a subgroup the spatial rotations, the relations between the components of the spinor and those of the three-dimensional vector must be the same as for three-dimensional spinors:

$$\left.\begin{aligned} f_x^+ &= \tfrac{1}{2}(\xi^{22} - \xi^{11}), & f_y^+ &= \tfrac{1}{2}i(\xi^{22} + \xi^{11}), & f_z^+ &= \xi^{12}; \\ f_x^- &= \tfrac{1}{2}(\eta_{\dot2\dot2} - \eta_{\dot1\dot1}), & f_y^- &= \tfrac{1}{2}i(\eta_{\dot2\dot2} + \eta_{\dot1\dot1}), & f_z^- &= \eta_{\dot1\dot2}. \end{aligned}\right\} \tag{19.16}$$

PROBLEM

Derive the general correspondence between spinors of even rank and 4-tensors.

SOLUTION. All spinors for which $k+l$ is even are realizations of single-valued irreducible representations of the extended Lorentz group, and are therefore equivalent to the 4-tensors which are realizations of similar representations.†

A spinor of rank (k, k) can be defined so that it is transformed under inversion by

$$\zeta^{\alpha\beta\ldots\dot\gamma\dot\delta\ldots} \to \pm\zeta_{\dot\alpha\dot\beta\ldots\gamma\delta\ldots}. \tag{1}$$

Such a spinor is equivalent to a symmetrical irreducible 4-tensor of rank k, which is a true tensor or a pseudotensor according to the sign in (1).

Spinors of ranks (k, l) and (l, k), forming a bispinor, are transformed under inversion by

$$\zeta^{\overbrace{\alpha\beta\ldots}^{k}\overbrace{\dot\gamma\dot\delta\ldots}^{l}} \to (-1)^{\frac{1}{2}(k-l)}\chi_{\underbrace{\dot\alpha\dot\beta\ldots}_{k}\underbrace{\gamma\delta\ldots}_{l}}. \tag{2}$$

When $l = k+2$, the bispinor is equivalent to an irreducible 4-tensor $a_{[\mu\nu]\rho\sigma\ldots}$ of rank $k+2$, antisymmetric in the indices $[\mu\nu]$ and symmetric in all the other indices. The irreducibility of this tensor signifies that it gives zero on contraction with respect to any pair of indices and on dualization with respect to any three indices (i.e. $e^{\lambda\mu\nu\rho}a_{[\mu\nu]\rho\sigma\ldots} = 0$); the latter condition implies that the result is zero on taking the cyclic sum over three indices, $\mu\nu$ and any one other.

When $l = k+4$, the bispinor is equivalent to an irreducible 4-tensor $a_{[\lambda\mu][\nu\rho]\sigma\tau\ldots}$ of rank $k+4$, having the following properties: it is antisymmetric in the pairs of indices $[\lambda\mu]$ and $[\nu\rho]$, symmetric in all others, symmetric for interchange of $[\lambda\mu]$ with $[\nu\rho]$, and gives zero on contraction with respect to any pair of indices and on dualization with respect to any three indices.

Generally, when $l = k+2n$, the bispinor is equivalent to an irreducible 4-tensor of rank $k+2n$, antisymmetric in n pairs of indices and symmetric in the other k indices. 4-tensors antisymmetric in

† Spinors of odd rank are realizations of *two*-valued representations of the group: a spatial rotation through 360° changes the sign of spinors, so that two matrices of opposite sign correspond to each element of the group.

larger numbers (threes, fours, etc.) of indices do not appear in this classification, for the obvious reason that an antisymmetric tensor of rank 3 is equivalent (dual) to a pseudovector, and an antisymmetric tensor of rank 4 reduces to a scalar (is proportional to the unit pseudotensor $e^{\lambda\mu\nu\rho}$); antisymmetry in a still greater number of indices is not possible in 4-space.

§ 20. Dirac's equation in the spinor representation

A particle with spin $\frac{1}{2}$ is described, in its rest frame, by a two-component wave function, i.e. a three-dimensional spinor. The "four-dimensional origin" of this may be either an undotted or a dotted 4-spinor. Both these 4-spinors appear in the description of the particle in an arbitrary frame of reference; we shall denote them by ξ^α and $\eta_{\dot\alpha}$.†

For a free particle, the only operator which can appear in the wave equation is (as shown in §10) the 4-momentum operator $\hat{p}_\mu = i\partial_\mu$. In the spinor notation, this 4-vector corresponds to the operator spinor $\hat{p}_{\alpha\dot\beta}$, with

$$\left.\begin{aligned} \hat{p}^{1\dot1} &= \hat{p}_{2\dot2} = \hat{p}_z + \hat{p}_0, & \hat{p}^{2\dot2} &= \hat{p}_{1\dot1} = \hat{p}_0 - \hat{p}_z, \\ \hat{p}^{1\dot2} &= -\hat{p}_{2\dot1} = \hat{p}_x - i\hat{p}_y, & \hat{p}^{2\dot1} &= -\hat{p}_{1\dot2} = \hat{p}_x + i\hat{p}_y. \end{aligned}\right\} \tag{20.1}$$

The wave equation is a linear differential relation between the components of spinors, expressed by the operator $\hat{p}_{\alpha\dot\beta}$. The requirement of relativistic invariance leads to the equations

$$\left.\begin{aligned} \hat{p}^{\alpha\dot\beta}\eta_{\dot\beta} &= m\xi^\alpha, \\ \hat{p}_{\dot\beta\alpha}\xi^\alpha &= m\eta_{\dot\beta}, \end{aligned}\right\} \tag{20.2}$$

where m is a dimensional constant. There would be no meaning in using different constants m_1 and m_2 here, or in changing the sign of m, since the equations could still be reduced to the above form by an appropriate transformation of ξ^α or $\eta_{\dot\alpha}$.

By substituting $\eta_{\dot\beta}$ from the second equation (20.2) in the first, we can eliminate one of the two spinors:

$$\hat{p}^{\alpha\dot\beta}\eta_{\dot\beta} = \frac{1}{m}\hat{p}^{\alpha\dot\beta}\hat{p}_{\gamma\dot\beta}\xi^\gamma = m\xi^\alpha.$$

From (18.4), $\hat{p}^{\alpha\dot\beta}\hat{p}_{\gamma\dot\beta} = \hat{p}^2\delta^\alpha_\gamma$, and thus we obtain

$$(\hat{p}^2 - m^2)\xi^\gamma = 0, \tag{20.3}$$

whence it is evident that m is the mass of the particle.

It should be noticed that the need to use the mass in the wave equation implies the simultaneous consideration of two spinors (ξ^α and $\eta_{\dot\alpha}$): with only one of these, it would not be possible to construct a relativistically invariant equation containing

† A three-dimensional spinor of rank one may also "originate" from 4-spinors of higher odd rank which, in the rest frame, become antisymmetric in one or more pairs of indices. These would, however, lead to higher-order equations (cf. the third footnote to §10).

a dimensional parameter. The wave equation is necessarily invariant under spatial inversion if the transformation of the wave function is defined by

$$P:\quad \xi^{\alpha}\to i\eta_{\dot\alpha},\qquad \eta_{\dot\alpha}\to i\xi^{\alpha}. \tag{20.4}$$

It is easily seen that the two equations (20.2) are interchanged by this substitution (together with $\hat{p}^{\dot\alpha\beta}\to\hat{p}_{\alpha\dot\beta}$, which is evident from (20.1)). Two spinors which are interchanged by inversion form a four-component quantity, a bispinor.

The relativistic wave equation given by (20.2) is called *Dirac's equation*, having been first derived by Dirac in 1928. In order to analyse und apply this equation further, let us consider various ways in which it may be written.

Using (18.6), we can rewrite equations (20.2) as

$$\left.\begin{aligned}(\hat{p}_0+\hat{\mathbf{p}}\cdot\boldsymbol{\sigma})\eta &= m\xi,\\ (\hat{p}_0-\hat{\mathbf{p}}\cdot\boldsymbol{\sigma})\xi &= m\eta.\end{aligned}\right\} \tag{20.5}$$

Here the symbols ξ and η denote two-component quantities, the spinors

$$\xi=\begin{pmatrix}\xi^1\\ \xi^2\end{pmatrix},\qquad \eta=\begin{pmatrix}\eta_{\dot1}\\ \eta_{\dot2}\end{pmatrix} \tag{20.6}$$

(the first with upper and the second with lower indices). Here and below, multiplication of the matrices σ by any two-component quantity f means multiplication by the usual matrix rule:

$$(\sigma f)_{\alpha}=\sigma_{\alpha\beta}f_{\beta}. \tag{20.7}$$

The vertical column notation for f is in accordance with the multiplication of each row of σ by the column f.

For subsequent reference, the Pauli matrices may be written once more;

$$\sigma_x=\begin{pmatrix}0&1\\1&0\end{pmatrix},\qquad \sigma_y=\begin{pmatrix}0&-i\\i&0\end{pmatrix},\qquad \sigma_z=\begin{pmatrix}1&0\\0&-1\end{pmatrix}. \tag{20.8}$$

Their fundamental properties are

$$\left.\begin{aligned}\sigma_i\sigma_k+\sigma_k\sigma_i &= 2\delta_{ik},\\ \sigma_i\sigma_k &= ie_{ikl}\sigma_l+\delta_{ik};\end{aligned}\right\} \tag{20.9}$$

see *QM*, §55.

We shall also give the wave equation satisfied by the complex-conjugate wave function formed from the spinors

$$\xi^*=(\xi^{1*},\xi^{2*}),\qquad \eta^*=(\eta^*_{\dot1},\eta^*_{\dot2}). \tag{20.10}$$

Since all the operators $\hat{p}_{\mu}$ contain the factor i, $\hat{p}^*_{\mu}=-\hat{p}_{\mu}$. In taking the complex conjugate of both sides of equation (20.5), we must also use the fact that, since the matrices σ are Hermitian ($\sigma^* = \tilde{\sigma}$),

$$(\sigma f)^*_\alpha = \sigma^*_{\alpha\beta} f^*_\beta = f^*_\beta \sigma_{\beta\alpha} = (f^*\sigma)_\alpha;$$

the resulting equations are

$$\left.\begin{aligned} \eta^*(\hat{p}_0 + \hat{\mathbf{p}}\cdot\boldsymbol{\sigma}) &= -m\xi^*, \\ \xi^*(\hat{p}_0 - \hat{\mathbf{p}}\cdot\boldsymbol{\sigma}) &= -m\eta^*. \end{aligned}\right\} \tag{20.11}$$

Here it is conventionally implied that the operators $\hat{p}^\mu$ act on the function to the left of them. The writing of ξ^* and η^* as horizontal rows is in accordance with the matrix multiplication in these equations: the row f is multiplied by the columns of the matrix σ,

$$(f^*\sigma)_\alpha = f^*_\beta \sigma_{\beta\alpha}. \tag{20.12}$$

The inversion transformation for ξ^*, η^* is defined as the complex conjugate of the transformation (20.4):

$$P: \quad \xi^{\alpha *} \to -i\overset{*}{\eta}_{\dot{\alpha}}, \qquad \overset{*}{\eta}_{\dot{\alpha}} \to -i\xi^{\alpha *}. \tag{20.13}$$

§21. The symmetrical form of Dirac's equation

The spinor form of Dirac's equation is the most natural one, in the sense that its relativistic invariance is immediately apparent. In applications of the equation, however, other forms of the wave equation may be more convenient, which are obtained by a different choice of the four independent components of the wave function.

We shall denote the four-component wave function by the symbol ψ, with components ψ_i ($i = 1, 2, 3, 4$). In the spinor representation, it is a bispinor:

$$\psi = \begin{pmatrix} \xi \\ \eta \end{pmatrix}. \tag{21.1}$$

But the independent components of ψ can equally well be taken as any linearly independent combinations of components of the spinors ξ and η.† We shall arbitrarily limit the acceptable linear transformations by the one condition of unitarity; such transformations leave unchanged the bilinear forms constructed from ψ and ψ^* (§28).

In the general case of an arbitrary choice of the components of ψ, Dirac's equation can be put in the form

$$\hat{p}_\mu \gamma^\mu_{ik} \psi_k = m\psi_i,$$

† For brevity, the four-component quantity ψ will be referred to as a bispinor even in non-spinor representations.

where γ^μ ($\mu = 0, 1, 2, 3$) are certain four-rowed matrices (*Dirac matrices*). We shall usually write this equation in a symbolic form, omitting the matrix indices:

$$(\gamma\hat{p} - m)\psi = 0, \tag{21.2}$$

where

$$\gamma\hat{p} \equiv \gamma^\mu \hat{p}_\mu = \hat{p}_0\gamma^0 - \hat{\mathbf{p}}\cdot\boldsymbol{\gamma}$$
$$= i\gamma^0 \frac{\partial}{\partial t} + i\boldsymbol{\gamma}\cdot\nabla,$$
$$\boldsymbol{\gamma} = (\gamma^1, \gamma^2, \gamma^3).$$

For example, the spinor form of the equation with the components of ψ as in (21.1) corresponds to the matrices†

$$\gamma^0 = \begin{pmatrix} 0 & 1 \\ 1 & 0 \end{pmatrix}, \qquad \boldsymbol{\gamma} = \begin{pmatrix} 0 & -\boldsymbol{\sigma} \\ \boldsymbol{\sigma} & 0 \end{pmatrix}, \tag{21.3}$$

as is easily seen by writing the equations (20.5) as

$$\begin{pmatrix} 0 & \hat{p}_0 + \hat{\mathbf{p}}\cdot\boldsymbol{\sigma} \\ \hat{p}_0 - \hat{\mathbf{p}}\cdot\boldsymbol{\sigma} & 0 \end{pmatrix}\begin{pmatrix} \xi \\ \eta \end{pmatrix} = m\begin{pmatrix} \xi \\ \eta \end{pmatrix}$$

and comparing with (21.2).

In the general case, the matrices γ need satisfy only conditions ensuring that $\hat{p}^2 = m^2$. To find these conditions, we multiply equation (21.2) on the left by $\gamma\hat{p}$:

$$(\gamma^\mu\hat{p}_\mu)(\gamma^\nu\hat{p}_\nu)\psi = m(\hat{p}_\mu\gamma^\mu)\psi = m^2\psi.$$

Since $\hat{p}_\mu\hat{p}_\nu$ is a symmetrical tensor (all the operators $\hat{p}_\mu$ commute), this equation may be rewritten

$$\tfrac{1}{2}\hat{p}_\mu\hat{p}_\nu(\gamma^\mu\gamma^\nu + \gamma^\nu\gamma^\mu)\psi = m^2\psi,$$

and we must therefore have

$$\gamma^\mu\gamma^\nu + \gamma^\nu\gamma^\mu = 2g^{\mu\nu}. \tag{21.4}$$

Thus all the pairs of different matrices γ^μ anticommute, and their squares are

$$(\gamma^1)^2 = (\gamma^2)^2 = (\gamma^3)^2 = -1, \qquad (\gamma^0)^2 = 1. \tag{21.5}$$

Under an arbitrary unitary transformation of the components ψ: $\psi' = U\psi$, where

† Here and below, we use a compact two-rowed notation for four-rowed matrices. Each symbol in (21.3) represents a two-rowed matrix.

U is a unitary four-rowed matrix, the matrices γ are transformed as follows:

$$\gamma' = U\gamma U^{-1} = U\gamma U^{+}, \tag{21.6}$$

so that the equation $(\gamma\hat{p} - m)\psi = 0$ becomes $(\gamma'\hat{p} - m)\psi' = 0$. The commutation relations (21.4) remain unchanged, of course.

The matrix γ^0 (21.3) is Hermitian, and the matrices $\boldsymbol{\gamma}$ are anti-Hermitian. These properties are preserved under any unitary transformation (21.6), and we therefore always have†

$$\boldsymbol{\gamma}^{+} = -\boldsymbol{\gamma}, \qquad \gamma^{0+} = \gamma^0. \tag{21.7}$$

The equation for the complex-conjugate function ψ^* may also be given. Taking the complex conjugate of equation (21.2) and using the properties (21.7), we obtain

$$(-\hat{p}_0\tilde{\gamma}_0 - \hat{\mathbf{p}}\cdot\tilde{\boldsymbol{\gamma}} - m)\psi^* = 0.$$

We commute ψ^* by $\tilde{\gamma}^{\mu}\psi^* = \psi^*\gamma^{\mu}$ and then multiply the whole equation on the right by γ^0; since $\boldsymbol{\gamma}\gamma^0 = -\gamma^0\boldsymbol{\gamma}$, we have in terms of a new bispinor

$$\bar{\psi} = \psi^*\gamma^0, \qquad \psi^* = \bar{\psi}\gamma^0 \tag{21.8}$$

the result

$$\bar{\psi}(\gamma\hat{p} + m) = 0. \tag{21.9}$$

As in (20.11), the operator $\hat{p}$ is here taken to act on the function to its left. The function $\bar{\psi}$ is called the *Dirac conjugate* (or relativistically conjugate) function to ψ. The factor γ^0 in its definition signifies that (in the spinor representation) it interchanges the spinors ξ^* and η^*; thus, in $\bar{\psi} = (\eta^*, \xi^*)$ the first spinor is undotted (as in ψ) and the second is dotted. For this reason $\bar{\psi}$ is a more natural "partner" of ψ than ψ^* is; they appear together, for instance, in various bilinear combinations (see §28).

The inversion transformation for the wave function may be written as

$$P: \quad \psi \to i\gamma^0\psi, \qquad \bar{\psi} = -i\bar{\psi}\gamma^0. \tag{21.10}$$

In the spinor representation of ψ, the matrix γ^0 interchanges the components ξ and η, as should happen on inversion. The invariance of Dirac's equation under the transformation (21.10) in the general case is immediately obvious: changing $\hat{\mathbf{p}}$ into $-\hat{\mathbf{p}}$ and ψ into $i\gamma^0\psi$ in equation (21.2), we have

$$(\hat{p}_0\gamma^0 + \hat{\mathbf{p}}\cdot\boldsymbol{\gamma} - m)\gamma^0\psi = 0.$$

Multiplying this equation on the left by γ^0 and taking into account the fact that γ^0 and $\boldsymbol{\gamma}$ anticommute, we return to the original equation.

† These equations may be written jointly in the form

$$\gamma^{\lambda+} = \gamma^0\gamma^{\lambda}\gamma^0.$$

Multiplying the equation $(\gamma\hat{p} - m)\psi = 0$ on the left by $\bar{\psi}$, and the equation $\bar{\psi}(\gamma\hat{p} + m) = 0$ on the right by ψ, and adding, we obtain

$$\bar{\psi}\gamma^\mu(\hat{p}_\mu\bar{\psi}) + (\hat{p}_\mu\bar{\psi})\gamma^\mu\psi = (\hat{p}_\mu\bar{\psi}\gamma^\mu\psi) = 0,$$

where the parentheses indicate the function on which the operator $\hat{p}$ acts. This equation is in the form of an equation of continuity, $\partial_\mu j^\mu = 0$, so that

$$j^\mu = \bar{\psi}\gamma^\mu\psi$$
$$= (\psi^*\psi, \psi^*\gamma^0\boldsymbol{\gamma}\psi) \tag{21.11}$$

is the particle current density 4-vector. Its time component $j^0 = \psi^*\psi$ is positive-definite.

Dirac's equation may be put in the form of an expression for the time derivative:

$$i\,\partial\psi/\partial t = \hat{H}\psi, \tag{21.12}$$

where $\hat{H}$ is the Hamiltonian of the particle.† To obtain this form, we need only multiply equation (21.2) on the left by γ^0. The resulting expression for the Hamiltonian is

$$\hat{H} = \boldsymbol{\alpha}\cdot\hat{\mathbf{p}} + \beta m, \tag{21.13}$$

where

$$\boldsymbol{\alpha} = \gamma^0\boldsymbol{\gamma}, \qquad \beta = \gamma^0 \tag{21.14}$$

is the customary notation for the matrices concerned.

It may be noted that

$$\alpha_i\alpha_k + \alpha_k\alpha_i = 2\delta_{ik}, \qquad \beta\boldsymbol{\alpha} + \boldsymbol{\alpha}\beta = 0, \qquad \beta^2 = 1, \tag{21.15}$$

i.e. all the matrices $\boldsymbol{\alpha}$, β anticommute and their squares are unity; they are all Hermitian. In the spinor representation,

$$\boldsymbol{\alpha} = \begin{pmatrix} \boldsymbol{\sigma} & 0 \\ 0 & -\boldsymbol{\sigma} \end{pmatrix}, \qquad \beta = \begin{pmatrix} 0 & 1 \\ 1 & 0 \end{pmatrix}. \tag{21.16}$$

In the limit of small velocities the particle must be described, as in the non-relativistic theory, by a single two-component spinor: on taking the limit $\mathbf{p}\to 0$, $\varepsilon \to m$ in equations (20.5), we find $\xi = \eta$, so that the two spinors which form the bispinor are equal. This, however, reveals a defect of the spinor form of Dirac's

† For a particle with spin zero, the wave equation was not capable of being written in this form: the equation (10.5) for the scalar ψ is of the second order in the time, while the first-order equations (10.4) for the five-component quantity (ψ, ψ_μ) contain the time derivatives of only some of the components.

equation: in the limit, all four components of ψ are non-zero, although only two of them are really independent. A more convenient representation of the wave function ψ would be one in which two of its components were zero in the limit.

Accordingly, we replace ξ and η by linear combinations ϕ and χ:

$$\left.\begin{aligned} \psi &= \begin{pmatrix}\phi\\ \chi\end{pmatrix}, \\ \phi = \frac{1}{\sqrt{2}}(\xi+\eta), &\qquad \chi = \frac{1}{\sqrt{2}}(\xi-\eta). \end{aligned}\right\} \tag{21.17}$$

Then $\chi = 0$ for a particle at rest. This will be called the *standard representation* of ψ. On inversion, ϕ and χ are transformed as follows:

$$P:\quad \phi \to i\phi, \qquad \chi \to -i\chi. \tag{21.18}$$

The equations for ϕ and χ are obtained by adding and subtracting equations (20.5):

$$\left.\begin{aligned} \hat{p}_0\phi - \hat{\mathbf{p}}\cdot\boldsymbol{\sigma}\chi &= m\phi, \\ -\hat{p}_0\chi + \hat{\mathbf{p}}\cdot\boldsymbol{\sigma}\phi &= m\chi. \end{aligned}\right\} \tag{21.19}$$

Hence we see that the standard representation corresponds to the matrices

$$\gamma^0 \equiv \beta = \begin{pmatrix}1 & 0\\ 0 & -1\end{pmatrix}, \qquad \boldsymbol{\gamma} = \begin{pmatrix}0 & \boldsymbol{\sigma}\\ -\boldsymbol{\sigma} & 0\end{pmatrix}, \qquad \boldsymbol{\alpha} = \begin{pmatrix}0 & \boldsymbol{\sigma}\\ \boldsymbol{\sigma} & 0\end{pmatrix}. \tag{21.20}$$

Since the first and second components of ξ and η are added separately in (21.17), the components ψ_1 and ψ_3 correspond to the spin component eigenvalue $+\frac{1}{2}$ in both the standard and the spinor representation, and ψ_2 and ψ_4 to $-\frac{1}{2}$. In both representations, therefore, the matrix $\frac{1}{2}\boldsymbol{\Sigma}$, where

$$\boldsymbol{\Sigma} = \begin{pmatrix}\boldsymbol{\sigma} & 0\\ 0 & \boldsymbol{\sigma}\end{pmatrix}, \tag{21.21}$$

in a three-dimensional spin operator: when $\frac{1}{2}\Sigma_z$ acts on a bispinor containing only the components ψ_1, ψ_3, or ψ_2, ψ_4, this bispinor is multiplied by $+\frac{1}{2}$ or $-\frac{1}{2}$. In an arbitrary representation, (21.21) may be written in the form

$$\boldsymbol{\Sigma} = -\boldsymbol{\alpha}\gamma^5 = -\tfrac{1}{2}i\boldsymbol{\alpha}\times\boldsymbol{\alpha}; \tag{21.22}$$

the definition of γ^5 is given in (22.14) below.

PROBLEMS

PROBLEM 1. Find the formulae giving the transformations of the wave function under an infinitesimal Lorentz transformation and an infinitesimal three-dimensional rotation.

SOLUTION. In the spinor representation of ψ, an infinitesimal Lorentz transformation gives

$$\xi' = (1 - \tfrac{1}{2}\boldsymbol{\sigma}\cdot\delta\mathbf{V})\xi, \qquad \eta' = (1 + \tfrac{1}{2}\boldsymbol{\sigma}\cdot\delta\mathbf{V})\eta;$$

see (18.8), (18.8a), (18.10). These formulae may be combined as

$$\psi' = (1 - \tfrac{1}{2}\boldsymbol{\alpha}\cdot\delta\mathbf{V})\psi. \tag{1}$$

Similarly, the transformation under an infinitesimal rotation is

$$\psi' = (1 + \tfrac{1}{2}i\boldsymbol{\Sigma}\cdot\delta\boldsymbol{\theta})\psi. \tag{2}$$

In this form the results are valid for any representation of ψ if $\boldsymbol{\alpha}$ and $\boldsymbol{\Sigma}$ are matrices in that representation.

It is easily verified that the matrices $\boldsymbol{\alpha}$ and $\boldsymbol{\Sigma}$ are the components of an antisymmetric "matrix 4-tensor",

$$\sigma^{\mu\nu} = \tfrac{1}{2}(\gamma^\mu\gamma^\nu - \gamma^\nu\gamma^\mu) = (\boldsymbol{\alpha}, i\boldsymbol{\Sigma});$$

the components are arranged as shown in (19.15). Using also the infinitesimal antisymmetric tensor $\delta\varepsilon^{\mu\nu} = (\delta\mathbf{V}, \delta\boldsymbol{\theta})$, we have

$$\sigma^{\mu\nu}\delta\varepsilon_{\mu\nu} = 2i\boldsymbol{\Sigma}\cdot\delta\boldsymbol{\theta} - 2\boldsymbol{\alpha}\cdot\delta\mathbf{V},$$

and formulae (1) and (2) above may be combined as

$$\psi' = (1 + \tfrac{1}{4}\sigma^{\mu\nu}\delta\varepsilon_{\mu\nu})\psi. \tag{3}$$

PROBLEM 2. Write Dirac's equation in a representation such that it contains no imaginary coefficients (E. Majorana, 1937).

SOLUTION. In the standard representation, the only imaginary quantities in the equation

$$\left(\frac{\partial}{\partial t} + \alpha_x\frac{\partial}{\partial x} + \alpha_y\frac{\partial}{\partial y} + \alpha_z\frac{\partial}{\partial z} + im\beta\right)\psi = 0$$

are the matrices α_y and $i\beta$. These may be eliminated by a transformation $\psi' = U\psi$ such that the imaginary matrix α_y and the real matrix β are interchanged. This is achieved by putting

$$U = \frac{1}{\sqrt{2}}(\alpha_y + \beta) = U^{-1};$$

then $\alpha_x' = U\alpha_x U = -\alpha_x$, $\alpha_y' = \beta$, $\alpha_z' = -\alpha_z$, $\beta' = \alpha_y$, and Dirac's equation becomes

$$\left(\frac{\partial}{\partial t} - \alpha_x\frac{\partial}{\partial x} + \beta\frac{\partial}{\partial y} - \alpha_z\frac{\partial}{\partial z} + im\alpha_y\right)\psi' = 0,$$

in which all the coefficients are real.

§ 22. Algebra of Dirac matrices

In calculations using Dirac's equation, the matrices γ occur repeatedly without reference to their specific form in any particular representation. The rules of operation with these matrices are entirely given by the commutation relations

$$\gamma^\mu\gamma^\nu + \gamma^\nu\gamma^\mu = 2g^{\mu\nu} \quad (\mu, \nu = 0, 1, 2, 3), \tag{22.1}$$

which determine all their general properties.

In this section we shall give various formulae and rules of the algebra of these matrices which are useful in such calculations.

The "scalar product" of the matrices γ with themselves is $g_{\mu\nu}\gamma^\mu\gamma^\nu = 4$. For brevity we use the notation $\gamma_\mu = g_{\mu\nu}\gamma^\nu$ by analogy with the covariant components of 4-vectors. Then

$$\gamma_\mu\gamma^\mu = 4. \tag{22.2}$$

If the matrices γ_μ and γ^μ are separated by one or more factors γ, then they can be brought to adjoining positions by one or more interchanges using the rule (22.1), and the summation over μ is then carried out by means of (22.2). This yields the formulae

$$\left.\begin{aligned}\gamma_\mu\gamma^\nu\gamma^\mu &= -2\gamma^\nu,\\ \gamma_\mu\gamma^\lambda\gamma^\nu\gamma^\mu &= 4g^{\lambda\nu},\\ \gamma_\mu\gamma^\lambda\gamma^\nu\gamma^\rho\gamma^\mu &= -2\gamma^\rho\gamma^\nu\gamma^\lambda,\\ \gamma_\mu\gamma^\lambda\gamma^\nu\gamma^\rho\gamma^\sigma\gamma^\mu &= 2(\gamma^\sigma\gamma^\lambda\gamma^\nu\gamma^\rho + \gamma^\rho\gamma^\nu\gamma^\lambda\gamma^\sigma).\end{aligned}\right\} \tag{22.3}$$

The factors γ^μ, etc., usually appear in combination with various 4-vectors as "scalar products" with the latter,†

$$\gamma a \equiv \gamma^\mu a_\mu. \tag{22.4}$$

For such products, formulae (22.1) become

$$\left.\begin{aligned}(a\gamma)(b\gamma)+(b\gamma)(a\gamma) &= 2(ab),\\ (a\gamma)(a\gamma) &= a^2,\end{aligned}\right\} \tag{22.5}$$

and formulae (22.3) become

$$\left.\begin{aligned}\gamma_\mu(a\gamma)\gamma^\mu &= -2(a\gamma),\\ \gamma_\mu(a\gamma)(b\gamma)\gamma^\mu &= 4(ab),\\ \gamma_\mu(a\gamma)(b\gamma)(c\gamma)\gamma^\mu &= -2(c\gamma)(b\gamma)(a\gamma),\\ \gamma_\mu(a\gamma)(b\gamma)(c\gamma)(d\gamma)\gamma^\mu &= 2[(d\gamma)(a\gamma)(b\gamma)(c\gamma)+(c\gamma)(b\gamma)(a\gamma)(d\gamma)].\end{aligned}\right\} \tag{22.6}$$

A frequent operation is taking the trace of the product of a number of matrices γ. Let us consider the quantities

$$T^{\mu_1\mu_2\ldots\mu_n} \equiv \tfrac{1}{4}\,\mathrm{tr}\,(\gamma^{\mu_1}\gamma^{\mu_2}\ldots\gamma^{\mu_n}). \tag{22.7}$$

On account of a familiar property of the trace of a product of matrices, this tensor is symmetrical with respect to cyclic permutations of the indices $\mu_1, \mu_2, \ldots, \mu_n$.

† In this edition, no special notation is used for such products. Letters with circumflexes or with strokes through them are often found with this meaning in the literature.

Since the matrices γ have the same form in any frame of reference, the quantities T are also independent of this frame, and they therefore form a tensor which can be expressed entirely in terms of the metric tensor $g_{\mu\nu}$, which has this property.

From the tensor $g_{\mu\nu}$ of rank two, however, only tensors of even rank can be constructed. Hence it follows immediately that the trace of the product of any odd number of factors γ is zero. In particular, the trace of each γ is zero:†

$$\text{tr}\,\gamma^{\mu}=0. \tag{22.8}$$

The trace of a unit four-rowed matrix (which is implied on the right-hand side of the commutation rule (22.1)) is 4. Thus, if we take the trace of both sides of (22.1), we find

$$T^{\mu\nu}=g^{\mu\nu}. \tag{22.9}$$

The trace of the four-matrix product is

$$T^{\lambda\mu\nu\rho}=g^{\lambda\mu}g^{\nu\rho}-g^{\lambda\nu}g^{\mu\rho}+g^{\lambda\rho}g^{\mu\nu}. \tag{22.10}$$

This formula may be derived, for instance, by "pulling" the factor γ^{λ} in tr $\gamma^{\lambda}\gamma^{\mu}\gamma^{\nu}\gamma^{\rho}$ to the right by means of the relation (22.1); after each interchange one of the terms in (22.10) appears:

$$\begin{aligned}T^{\lambda\mu\nu\rho}&=2g^{\lambda\mu}T^{\nu\rho}-T^{\mu\lambda\nu\rho}\\&=2g^{\lambda\mu}g^{\nu\rho}-T^{\mu\lambda\nu\rho}\end{aligned}$$

and so on. After all the interchanges there remains on the right $-T^{\mu\nu\rho\lambda}=-T^{\lambda\mu\nu\rho}$, which we take to the left-hand side. The trace of a product of six γ can similarly be reduced to the traces of four-factor products, and so on. For instance,

$$T^{\lambda\mu\nu\rho\sigma\tau}=g^{\lambda\mu}T^{\nu\rho\sigma\tau}-g^{\lambda\nu}T^{\mu\rho\sigma\tau}+g^{\lambda\rho}T^{\mu\nu\sigma\tau}-g^{\lambda\sigma}T^{\mu\nu\rho\tau}+g^{\lambda\tau}T^{\mu\nu\rho\sigma}. \tag{22.11}$$

All the traces $T^{\lambda\mu\ldots}$ are real, and they are non-zero only if each of the matrices $\gamma^0, \gamma^1, \ldots$ appears in the product an even number of times; both these results are obvious from the above formulae. Hence we easily find that the trace is unchanged when the order of the factors is reversed:

$$T^{\lambda\mu\ldots\rho\sigma}=T^{\sigma\rho\ldots\mu\lambda}. \tag{22.12}$$

As already mentioned, the factors γ usually appear as "scalar" products with various 4-vectors. In such cases, formulae (22.9) and (22.10), for example, become

$$\left.\begin{aligned}\tfrac{1}{4}\,\text{tr}(a\gamma)(b\gamma)&=ab,\\ \tfrac{1}{4}\,\text{tr}(a\gamma)(b\gamma)(c\gamma)(d\gamma)&=(ab)(cd)-(ac)(bd)+(ad)(bc).\end{aligned}\right\} \tag{22.13}$$

† The trace of a matrix is invariant under the transformations $\gamma=U\gamma U^{-1}$. Thus the result (22.8) is also evident from the expressions (21.3) for the matrices.

The product $\gamma^0\gamma^1\gamma^2\gamma^3$ is of particular importance. There is a special notation for it which is customarily used:

$$\gamma^5 = -i\gamma^0\gamma^1\gamma^2\gamma^3. \tag{22.14}$$

It is easily seen that

$$\gamma^5\gamma^\mu + \gamma^\mu\gamma^5 = 0, \qquad (\gamma^5)^2 = 1, \tag{22.15}$$

i.e. the matrix γ^5 anticommutes with all the γ^μ. For the matrices $\boldsymbol{\alpha}$ and β, the rules are

$$\boldsymbol{\alpha}\gamma^5 - \gamma^5\boldsymbol{\alpha} = 0, \qquad \beta\gamma^5 + \gamma^5\beta = 0; \tag{22.16}$$

the commutability with $\boldsymbol{\alpha}$ follows because $\boldsymbol{\alpha} = \gamma^0\boldsymbol{\gamma}$ is a product of two matrices γ^μ.

The matrix γ^5 is Hermitian, since

$$\gamma^{5+} = i\gamma^{3+}\gamma^{2+}\gamma^{1+}\gamma^{0+} = -i\gamma^3\gamma^2\gamma^1\gamma^0,$$

and hence

$$\gamma^{5+} = \gamma^5, \tag{22.17}$$

because the sequence 3210 is changed to 0123 by an even number of transpositions.

The form of this matrix in two particular representations is:

$$\left.\begin{aligned} &\text{spinor} \quad \gamma^5 = \begin{pmatrix} -1 & 0 \\ 0 & 1 \end{pmatrix}, \\ &\text{standard} \quad \gamma^5 = \begin{pmatrix} 0 & -1 \\ -1 & 0 \end{pmatrix}. \end{aligned}\right\} \tag{22.18}$$

The trace of the matrix γ^5 is zero:

$$\operatorname{tr} \gamma^5 = 0, \tag{22.19}$$

as can be seen directly from (22.18). The traces of the products $\gamma^5\gamma^\mu\gamma^\nu$ are also zero. For the products of γ^5 with four factors γ^μ we have

$$\tfrac{1}{4} \operatorname{tr} \gamma^5\gamma^\lambda\gamma^\mu\gamma^\nu\gamma^\rho = ie^{\lambda\mu\nu\rho}. \tag{22.20}$$

Another formula is

$$\gamma N = i\gamma^5(\gamma a)(\gamma b)(\gamma c), \qquad N^\lambda = e^{\lambda\mu\nu\rho}\, a_\mu b_\nu c_\rho, \tag{22.21}$$

which is valid for mutually normal 4-vectors a, b, c:

$$ab = ac = bc = 0.$$

In some cases (for problems involving non-relativistic particles), it may be necessary to calculate the traces of products which involve γ^0 and the three-dimensional "vector" $\boldsymbol{\gamma}$ separately. The only non-zero traces are those of products containing even numbers of factors γ^0 and $\boldsymbol{\gamma}$. All the factors γ^0 become unity, and the traces of products with two and four factors $\boldsymbol{\gamma}$ are respectively

$$\left.\begin{aligned} \tfrac{1}{4}\operatorname{tr}(\mathbf{a}\cdot\boldsymbol{\gamma})(\mathbf{b}\cdot\boldsymbol{\gamma}) &= -\mathbf{a}\cdot\mathbf{b}, \\ \tfrac{1}{4}\operatorname{tr}(\mathbf{a}\cdot\boldsymbol{\gamma})(\mathbf{b}\cdot\boldsymbol{\gamma})(\mathbf{c}\cdot\boldsymbol{\gamma})(\mathbf{d}\cdot\boldsymbol{\gamma}) &= (\mathbf{a}\cdot\mathbf{b})(\mathbf{c}\cdot\mathbf{d}) - (\mathbf{a}\cdot\mathbf{c})(\mathbf{b}\cdot\mathbf{d}) + (\mathbf{a}\cdot\mathbf{d})(\mathbf{b}\cdot\mathbf{c}). \end{aligned}\right\} \tag{22.22}$$

§23. **Plane waves**

The state of a free particle having definite values of the momentum and energy is described by a plane wave which may be written in the form

$$\psi_p = \frac{1}{\sqrt{(2\varepsilon)}} u_p i^{-ipx}. \tag{23.1}$$

The suffix p indicates the value of the 4-momentum; the wave amplitude u_p is a suitably normalized bispinor. In proceeding with second quantization we need not only the wave functions (23.1) but also functions with a "negative frequency", which arise in the relativistic theory because of the two-valuedness of the square root $\pm\sqrt{(\mathbf{p}^2+m^2)}$, as shown in §11. As in §11, we shall always take ε to be the positive quantity $\varepsilon = +\sqrt{(\mathbf{p}^2+m^2)}$, so that the "negative frequency" is $-\varepsilon$; on changing also the sign of $\mathbf{p}$, we obtain a function which may naturally be called ψ_{-p}:

$$\psi_{-p} = \frac{1}{\sqrt{(2\varepsilon)}} u_{-p} e^{ipx}. \tag{23.2}$$

The significance of these functions will be explained in §26; here we shall write parallel formulae for ψ_p and ψ_{-p}.

The components of the bispinor amplitudes u_p and u_{-p} satisfy the algebraic equations

$$\left.\begin{aligned} (\gamma p - m)u_p &= 0, \\ (\gamma p + m)u_{-p} &= 0, \end{aligned}\right\} \tag{23.3}$$

which are obtained by substituting (23.1), (23.2) in Dirac's equation (this is equivalent to replacing the operator $\hat{p}$ in that equation by $\pm p$).† The relation $p^2 = m^2$ is then the condition for each such pair of equations to be compatible. We

† There are also similar equations obtained from Dirac's equation (21.9) for the complex-conjugate function:

$$\bar{u}_p(\gamma p - m) = 0, \qquad \bar{u}_{-p}(\gamma p + m) = 0. \tag{23.3a}$$

shall always normalize the bispinor amplitudes by the invariant conditions

$$\left.\begin{aligned} \bar{u}_p u_p &= 2m, \\ \bar{u}_{-p} u_{-p} &= -2m, \end{aligned}\right\} \tag{23.4}$$

where the bar over a letter denotes, as usual, Dirac conjugation: $\bar{u} = u^*\gamma^0$. Multiplying equations (23.3) on the left by $\bar{u}_{\pm p}$, we obtain $(\bar{u}_{\pm p}\gamma u_{\pm p})p = 2m^2 = 2p^2$, whence

$$\bar{u}_p \gamma u_p = \bar{u}_{-p}\gamma u_{-p} = 2p. \tag{23.5}$$

It may be noted that the change from the formulae for u_p to those for u_{-p} is made by changing the sign of m.

The current density 4-vector is

$$j = \bar{\psi}_{\pm p}\gamma\psi_{\pm p} = \frac{1}{2\varepsilon}\bar{u}_{\pm p}\gamma u_{\pm p} = p/\varepsilon, \tag{23.6}$$

i.e. $j^\mu = (1, \mathbf{v})$, where $\mathbf{v} = \mathbf{p}/\varepsilon$ is the velocity of the particle. Hence we see that the functions ψ_p are "normalized to one particle in the volume $V = 1$".

Equations (23.3) show that the components of the wave amplitude are related, but the actual form of the relations depends, of course, on the specific representation of ψ. For the standard representation they are found as follows.

From equations (21.19) we have, for a plane wave,

$$\left.\begin{aligned} (\varepsilon - m)\phi - \mathbf{p}\cdot\boldsymbol{\sigma}\chi &= 0, \\ (\varepsilon + m)\chi - \mathbf{p}\cdot\boldsymbol{\sigma}\phi &= 0. \end{aligned}\right\} \tag{23.7}$$

From these we find the relation between ϕ and χ in two equivalent forms:

$$\phi = \frac{\mathbf{p}\cdot\boldsymbol{\sigma}}{\varepsilon - m}\chi, \qquad \chi = \frac{\mathbf{p}\cdot\boldsymbol{\sigma}}{\varepsilon + m}\phi; \tag{23.8}$$

their equivalence is evident on multiplying the first form on the left by $\mathbf{p}\cdot\boldsymbol{\sigma}/(\varepsilon + m)$ and using the results $(\mathbf{p}\cdot\boldsymbol{\sigma})^2 = \mathbf{p}^2$ and $\varepsilon^2 - m^2 = \mathbf{p}^2$, which gives the second form. The common factor in ϕ and χ is chosen to satisfy the normalization condition (23.4). Thus we obtain for u_p, and correspondingly for u_{-p}, the expressions

$$u_p = \begin{pmatrix} \sqrt{(\varepsilon + m)}w \\ \sqrt{(\varepsilon - m)}(\mathbf{n}\cdot\boldsymbol{\sigma})w \end{pmatrix}, \qquad u_{-p} = \begin{pmatrix} \sqrt{(\varepsilon - m)}(\mathbf{n}\cdot\boldsymbol{\sigma})w' \\ \sqrt{(\varepsilon + m)}w' \end{pmatrix}; \tag{23.9}$$

the second formula is obtained from the first by changing the sign of m and replacing w by $(\mathbf{n}\cdot\boldsymbol{\sigma})w'$. Here $\mathbf{n}$ is a unit vector in the direction of $\mathbf{p}$, and w is an arbitrary two-component quantity subject only to the normalization condition

$$w^* w = 1. \tag{23.10}$$

For $\bar{u} = u^*\gamma^0$ (with γ^0 from (21.20)) we have

$$\left.\begin{aligned}\bar{u}_p &= (\surd(\varepsilon+m)w^*, -\surd(\varepsilon-m)w^*(\mathbf{n}\cdot\boldsymbol{\sigma})),\\ \bar{u}_{-p} &= (\surd(\varepsilon-m)w'^*(\mathbf{n}\cdot\boldsymbol{\sigma}), -\surd(\varepsilon+m)w'^*),\end{aligned}\right\} \tag{23.11}$$

and multiplication shows that in fact $\bar{u}_{\pm p}u_{\pm p} = \pm 2m$.

In the rest frame ($\varepsilon = m$), we have

$$u_p = \surd(2m)\begin{pmatrix} w \\ 0\end{pmatrix}, \qquad u_{-p} = \surd(2m)\begin{pmatrix} 0 \\ w'\end{pmatrix}, \tag{23.12}$$

i.e. w is the three-dimensional spinor to which the amplitude of each wave reduces in the non-relativistic limit. In the bispinor u_{-p}, the first two components, not the second two, vanish in the rest frame. This property of solutions of Dirac's equation having "negative frequencies" is evident, since by putting $\mathbf{p} = 0$ and replacing ε by $-m$ in (23.7), we find $\phi = 0$.†

The amplitude of the plane wave contains one arbitrary two-component quantity. Thus, for a given momentum, there are two different independent states, corresponding to the two possible values of the spin component. But the spin component along an arbitrary z-axis cannot have a definite value. This is evident because the Hamiltonian of a particle with definite $\mathbf{p}$ (i.e. the matrix $H = \boldsymbol{\alpha}\cdot\mathbf{p} + \beta m$) does not commute with the matrix $\Sigma_z = -i\alpha_x\alpha_y$. In accordance with the general conclusions of §16, however, the helicity λ (the component of the spin in the direction of $\mathbf{p}$) is conserved: the Hamiltonian commutes with the matrix $\mathbf{n}\cdot\boldsymbol{\Sigma}$.

Helicity states correspond to plane waves in which the three-dimensional spinor $w = w^{(\lambda)}(\mathbf{n})$ is an eigenfunction of the operator $\mathbf{n}\cdot\boldsymbol{\sigma}$:

$$\tfrac{1}{2}(\mathbf{n}\cdot\boldsymbol{\sigma})w^{(\lambda)} = \lambda w^{(\lambda)}. \tag{23.13}$$

The explicit form of these spinors is

$$\left.\begin{aligned} w^{(\lambda=\frac{1}{2})} &= \begin{pmatrix} e^{-\frac{1}{2}i\phi}\cos\frac{1}{2}\theta \\ e^{\frac{1}{2}i\phi}\sin\frac{1}{2}\theta\end{pmatrix},\\ w^{(\lambda=-\frac{1}{2})} &= \begin{pmatrix} -e^{-\frac{1}{2}i\phi}\sin\frac{1}{2}\theta \\ e^{\frac{1}{2}i\phi}\cos\frac{1}{2}\theta\end{pmatrix},\end{aligned}\right\} \tag{23.14}$$

where θ and ϕ are the polar angle and the azimuth of the direction of $\mathbf{n}$ relative to fixed axes xyz.‡

Another possible choice of the two independent states of a free particle with given $\mathbf{p}$, which is simpler but less clear, corresponds to the two values of the z-component of the spin in the rest frame, which we denote by σ. The spinors are

$$w^{(\sigma=\frac{1}{2})} = \begin{pmatrix} 1 \\ 0\end{pmatrix}, \qquad w^{(\sigma=-\frac{1}{2})} = \begin{pmatrix} 0 \\ 1\end{pmatrix}. \tag{23.15}$$

† In the spinor representation $\xi = -\eta$, instead of $\xi = \eta$ as in the rest frame for solutions having "positive frequencies".

‡ The solution of equation (23.13) can be multiplied by any phase factor, because of the possibility of an arbitrary rotation about the direction of $\mathbf{n}$.

As the two linearly independent solutions with "negative frequency" we take plane waves in which the three-dimensional spinors are

$$w^{(\sigma)\prime} = -\sigma_y w^{(-\sigma)} = 2\sigma i w^{(\sigma)}; \tag{23.16}$$

the significance of this choice will be shown in §26.

We can also find a representation of a plane wave such that in any frame of reference (not only in the rest frame) the wave has only two components corresponding to definite values of the same physical property—the spin component in the rest frame (L. Foldy and S. A. Wouthuysen, 1950).

Starting from the amplitude u_p (23.9) in the standard representation, we seek a unitary transformation to such a representation in the form

$$u'_p = U u_p, \qquad U = e^{W\boldsymbol{\gamma}\cdot\mathbf{n}},$$

where W is real; since $\boldsymbol{\gamma}^+ = -\boldsymbol{\gamma}$, it follows that $U^+ = U^{-1}$. Expanding in series and noting that $(\boldsymbol{\gamma}\cdot\mathbf{n})^2 = -1$, we put

$$U = \cos W + \boldsymbol{\gamma}\cdot\mathbf{n}\sin W;$$

cf. the derivation of (18.14) from (18.13). The condition that the second pair of components in the transformed amplitude u'_p should be zero gives

$$\tan W = |\mathbf{p}|/(m+\varepsilon),$$

so that

$$U = \frac{m+\varepsilon+(\boldsymbol{\gamma}\cdot\mathbf{n})|\mathbf{p}|}{\sqrt{[2\varepsilon(\varepsilon+m)]}}.$$

In the new representation,

$$u'_p = \sqrt{(2\varepsilon)}\begin{pmatrix} w \\ 0 \end{pmatrix}. \tag{23.17}$$

The Hamiltonian of the particle in this representation is

$$\hat{H}' = U(\boldsymbol{\alpha}\cdot\mathbf{p}+\beta m)U^{-1} = \beta\varepsilon, \tag{23.18}$$

where all the matrices $\beta, \boldsymbol{\alpha}$ and $\boldsymbol{\gamma}$ belong to the standard representation. This Hamiltonian commutes with the matrix

$$\boldsymbol{\Sigma} = -\boldsymbol{\alpha}\gamma^5 = \begin{pmatrix} \boldsymbol{\sigma} & 0 \\ 0 & \boldsymbol{\sigma} \end{pmatrix},$$

which is, in the new representation, the operator of a conserved quantity, the spin in the rest frame.

§24. Spherical waves

The wave functions of states of a free particle (with spin $\frac{1}{2}$) having definite values j of the angular momentum are spinor spherical waves. To determine their form, let us first state the corresponding formulae of the non-relativistic theory.

The non-relativistic wave function is a three-dimensional spinor

$$\psi = \begin{pmatrix} \psi^1 \\ \psi^2 \end{pmatrix}.$$

For a state having definite values of the energy ε (and therefore of the momentum† p), the orbital angular momentum l, the total angular momentum j and its component m, the wave function is

$$\psi = R_{pl}(r)\Omega_{jlm}(\theta, \phi). \tag{24.1}$$

The angular functions Ω_{jlm} are three-dimensional spinors whose components (for the two values $j = l \pm \frac{1}{2}$ which are possible for a given l) are

$$\left.\begin{aligned} \Omega_{l+\frac{1}{2},l,m} &= \begin{pmatrix} \sqrt{\dfrac{j+m}{2j}}\, Y_{l,m-\frac{1}{2}} \\[2ex] \sqrt{\dfrac{j-m}{2j}}\, Y_{l,m+\frac{1}{2}} \end{pmatrix}, \\ \Omega_{l-\frac{1}{2},l,m} &= \begin{pmatrix} -\sqrt{\dfrac{j-m+1}{2j+2}}\, Y_{l,m-\frac{1}{2}} \\[2ex] \sqrt{\dfrac{j+m+1}{2j+2}}\, Y_{l,m+\frac{1}{2}} \end{pmatrix} \end{aligned}\right\} \tag{24.2}$$

(see *QM*, §106, Problem). We shall call the Ω_{jlm} *spherical harmonic spinors*. They are normalized by the condition

$$\int \Omega^*_{jlm}\Omega_{j'l'm'}\, do = \delta_{jj'}\delta_{ll'}\delta_{mm'}. \tag{24.3}$$

The radial functions R_{pl} are the common factor in the two components of the spinor ψ, and are given by

$$R_{pl} = \sqrt{\frac{2\pi p}{r}}\, J_{l+\frac{1}{2}}(pr) \tag{24.4}$$

(*QM*, (33.10)). They are normalized by the condition

$$\int_0^\infty r^2 R_{p'l}R_{pl}\, dr = 2\pi\delta(p'-p). \tag{24.5}$$

† In this section, p denotes $|\mathbf{p}|$.

Returning now to the relativistic case, let us note first of all that separate laws of conservation of spin and orbital angular momentum do not exist for a moving particle: the operators $\hat{\mathbf{s}}$ and $\hat{\mathbf{l}}$ do not separately commute with the Hamiltonian. But the parity of the state is still conserved (for a free particle). The quantum number l therefore no longer refers to a definite value of the orbital angular momentum, but it defines the parity of the state (see below).

Let us consider the required wave function (bispinor) in the standard representation:

$$\psi = \begin{pmatrix}\phi\\ \chi\end{pmatrix}.$$

Under rotations, ϕ and χ behave like three-dimensional spinors. Their angular dependence is therefore given by the same spherical harmonic spinors Ω_{jlm}. Let $\phi \propto \Omega_{jlm}$, where l is a certain one of the two values $j+\frac{1}{2}$ and $j-\frac{1}{2}$. Under inversion $\phi(\mathbf{r}) \to i\phi(-\mathbf{r})$ (see (21.18)), and $\Omega_{jlm}(-\mathbf{n}) = (-1)^l\Omega_{jlm}(\mathbf{n})$, so that

$$\phi(\mathbf{r}) \to i(-1)^l\phi(\mathbf{r}).$$

The components $\chi(\mathbf{r})$, under inversion, become $-i\chi(-\mathbf{r})$. In order that the state should have a definite parity (i.e. that all the components should be multiplied by the same factor on inversion), it is therefore necessary that the angular dependence in χ should be given by the spherical harmonic spinor $\Omega_{jl'm}$ with the other of the two possible values of l; since these two values differ by 1, $(-1)^{l'} = -(-1)^l$.

The radial dependence of ϕ and χ will be given by the same functions R_{pl} and $R_{pl'}$ (with the values of l and l' which give the order of the spherical harmonics in Ω_{jlm}). This is clear because each component of ψ satisfies the second-order equation $(\hat{p}^2 - m^2)\psi = 0$, which for a given value of $|\mathbf{p}|$ becomes

$$(\triangle + \mathbf{p}^2)\psi = 0,$$

and this is formally identical with Schrödinger's non-relativistic equation for a free particle.

Thus

$$\phi = AR_{pl}\Omega_{jlm}, \qquad \chi = BR_{pl'}\Omega_{jl'm}, \tag{24.6}$$

and it remains to determine the constant coefficients A and B. To do so, we consider a distant region, where the spherical wave may be regarded as a plane wave. According to the asymptotic formula (*QM*, (33.12)),

$$R_{pl} \approx \frac{1}{ir}\{e^{i(pr-\frac{1}{2}\pi l)} - e^{-i(pr-\frac{1}{2}\pi l)}\}, \tag{24.7}$$

so that ϕ is the difference of two plane waves propagated in the directions $\pm\mathbf{n}$ ($\mathbf{n} = \mathbf{r}/r$). For each plane wave, by (23.8),

$$\chi = \frac{p}{\varepsilon + m}(\pm\mathbf{n}\cdot\boldsymbol{\sigma})\phi.$$

From the previous results (formulae (24.6)) it is obvious that $(\mathbf{n}\cdot\boldsymbol{\sigma})\Omega_{jlm} = a\Omega_{jl'm}$, where a is a constant. This constant is easily found by comparing the values of the two sides of the equation when $m = \frac{1}{2}$ and $\mathbf{n}$ is along the z-axis. Using (7.2a), we find

$$(\mathbf{n}\cdot\boldsymbol{\sigma})\Omega_{jlm} = i^{l'-1}\Omega_{jl'm}. \tag{24.8}$$

These formulae, on comparison with (24.6), show that

$$B = -\frac{p}{\varepsilon + m}A.$$

Finally, the coefficient A is determined by the normalization of ψ. If this is specified by

$$\int \psi^*_{pjlm}\psi_{p'j'l'm'}\,d^3x = 2\pi\delta_{jj'}\delta_{ll'}\delta_{mm'}\delta(p-p'), \tag{24.9}$$

we have

$$\psi_{pjlm} = \frac{1}{\sqrt{(2\varepsilon)}}\begin{pmatrix}\sqrt{(\varepsilon+m)}R_{pl}\Omega_{jlm}\\ -\sqrt{(\varepsilon-m)}R_{pl'}\Omega_{jl'm}\end{pmatrix}, \qquad l' = 2j - l. \tag{24.10}$$

Thus, for given values of j and m (and of the energy ε) there exist two states differing in parity. The parity is uniquely defined by the number l, which takes the values $j \pm \frac{1}{2}$: on inversion, the bispinor (24.10) is multiplied by $i(-1)^l$. The components of this bispinor, however, contain spherical harmonics of both orders l and l', showing that the orbital angular momentum has no definite value.

When $r \to \infty$, the spherical waves (24.7) may be regarded as plane waves in any small region of space, with momentum $\mathbf{p} = \pm p\mathbf{n}$. It is therefore clear that the wave functions in the momentum representation differ from (24.10) essentially only in that the radial factors are absent and $\mathbf{n}$ denotes the direction of the momentum.

In order to make a direct change to the momentum representation, we must carry out a Fourier transformation:

$$\psi(\mathbf{p}') = \int \psi(\mathbf{r})\,e^{-i\mathbf{p}'\cdot\mathbf{r}}\,d^3x. \tag{24.11}$$

The integral is calculated by means of the expansion of a plane wave in spherical waves:

$$e^{i\mathbf{p}\cdot\mathbf{r}} = \frac{2\pi}{p}\sum_{l=0}^{\infty}\sum_{m=-l}^{l} i^l R_{pl}(r) Y^*_{lm}\left(\frac{\mathbf{p}}{p}\right) Y_{lm}\left(\frac{\mathbf{r}}{r}\right). \tag{24.12}$$

With an expansion of this kind for $e^{-i\mathbf{p}'\cdot\mathbf{r}}$ in (24.11) and using (24.5), we find that the Fourier components of the function

$$\psi(\mathbf{r}) = R_{pl}(r)\Omega_{jlm}(\mathbf{r}/r)$$

are

$$\psi(\mathbf{p}') = \frac{(2\pi)^2}{p}\delta(p'-p)i^{-l}Y_{lm'}\left(\frac{\mathbf{p}'}{p'}\right)\int \Omega_{jlm}\left(\frac{\mathbf{r}}{r}\right)Y^*_{lm'}\left(\frac{\mathbf{r}}{r}\right)do.$$

The integral is equal to the coefficient of the spherical harmonic function in the definition (24.2) of the spherical harmonic spinors, and together with the factor $Y_{lm'}(\mathbf{p}'/p')$ it yields the same spherical harmonic spinor, but with argument $\mathbf{p}'/p'$:

$$\psi(\mathbf{p}') = \frac{(2\pi)^2}{p}\delta(p'-p)i^{-l}\Omega_{jlm}\left(\frac{\mathbf{p}'}{p'}\right).$$

Applying this result to the bispinor wave function (24.10), we obtain the momentum representation

$$\psi_{pjlm}(\mathbf{p}') = \delta(p'-p)\frac{(2\pi)^2}{p\sqrt{(2\varepsilon)}}\begin{pmatrix}\sqrt{(\varepsilon+m)}i^{-l}\Omega_{jlm}(\mathbf{p}'/p')\\ \sqrt{(\varepsilon-m)}i^{-l'}\Omega_{jl'm}(\mathbf{p}'/p')\end{pmatrix}. \tag{24.13}$$

The states $|pjlm\rangle$ are the same as the states $|pjm|\lambda|\rangle$ (with $|\lambda| = \frac{1}{2}$) discussed in §16: both have definite values of pjm and the parity. The spherical harmonic spinors Ω_{jlm} are therefore related in a certain way to the functions $D^{(j)}_{\lambda m}$ (both with argument $\mathbf{p}/p$). When $p \to 0$, the wave functions (24.13) reduce to the three-dimensional spinors Ω_{jlm}, the parity of which is $P = \eta(-1)^l$ (where $\eta = i$ is the "internal parity" of the spinor). A comparison with the results of §16 gives the formula

$$\Omega_{jlm} = i^l\sqrt{\frac{2j+1}{8\pi}}\left(w^{(-\frac{1}{2})}D^{(j)}_{-\frac{1}{2},m} \pm w^{(\frac{1}{2})}D^{(j)}_{\frac{1}{2},m}\right) \tag{24.14}$$

where $l = j \mp \frac{1}{2}$, and the $w^{(\lambda)}$ are the three-dimensional spinors (23.14).

§25. The relation between the spin and the statistics

The second quantization of a field of particles with spin $\frac{1}{2}$ (a spinor field) is carried out in a similar way to that of a scalar field in §11.

Without repeating the arguments, we shall immediately write down expressions for the field operators, which are exactly analogous to (11.2):

$$\left.\begin{aligned}\hat{\psi} &= \sum_{\mathbf{p},\sigma}\frac{1}{\sqrt{(2\varepsilon)}}\left(\hat{a}_{\mathbf{p}\sigma}u_{p\sigma}e^{-ipx} + \hat{b}^+_{\mathbf{p}\sigma}u_{-p,-\sigma}e^{ipx}\right),\\ \hat{\bar{\psi}} &\equiv \hat{\psi}^+\gamma^0 = \sum_{\mathbf{p},\sigma}\frac{1}{\sqrt{(2\varepsilon)}}\left(\hat{a}^+_{\mathbf{p}\sigma}\bar{u}_{p\sigma}e^{ipx} + \hat{b}_{\mathbf{p}\sigma}\bar{u}_{-p,-\sigma}e^{-ipx}\right);\end{aligned}\right\} \tag{25.1}$$

the summation is over all values of the momentum $\mathbf{p}$ and over $\sigma = \pm\frac{1}{2}$. The antiparticle annihilation operators $\hat{b}_{\mathbf{p}\sigma}$ (like the particle annihilation operators $\hat{a}_{\mathbf{p}\sigma}$) appear as the coefficients of functions whose coordinate dependence ($e^{i\mathbf{p}\cdot\mathbf{r}}$)

corresponds to a state having momentum **p**.†

To calculate the Hamiltonian of the spinor field, it is not necessary to determine the energy–momentum tensor (as we did for the scalar field), since in this case there exists a particle Hamiltonian which can be used to derive the wave equation (Dirac's equation) (21.12). The mean energy of the particle in a state with wave function ψ is the integral

$$\int \psi^* \hat{H} \psi d^3x = i \int \psi^* \frac{\partial \psi}{\partial t} d^3x$$
$$= i \int \bar{\psi} \gamma^0 \frac{\partial \psi}{\partial t} d^3x. \tag{25.2}$$

It should be noticed that the "energy density" (the integrand) is here not a positive–definite quantity.

Replacing the functions ψ and $\bar{\psi}$ in (25.2) by ψ-operators, using the orthogonality of the wave functions with different **p** or σ, and also using the relation $\bar{u}_{\pm p\sigma}\gamma^0 u_{\pm p\sigma} = 2\varepsilon$ for the wave amplitudes, we obtain the field Hamiltonian in the form

$$\hat{H} = \sum_{\mathbf{p},\sigma} \varepsilon(\hat{a}^+_{\mathbf{p}\sigma}\hat{a}_{\mathbf{p}\sigma} - \hat{b}_{\mathbf{p}\sigma}\hat{b}^+_{\mathbf{p}\sigma}). \tag{25.3}$$

Hence it is seen that in this case Fermi quantization must be used:

$$\{\hat{a}_{\mathbf{p}\sigma}, \hat{a}^+_{\mathbf{p}\sigma}\}_+ = 1, \qquad \{\hat{b}_{\mathbf{p}\sigma}, \hat{b}^+_{\mathbf{p}\sigma}\}_+ = 1, \tag{25.4}$$

and all other pairs of operators $\hat{a}$, $\hat{a}^+$, $\hat{b}$, $\hat{b}^+$ anticommute (see *QM*, §65), since then the Hamiltonian (25.3) may be written

$$\hat{H} = \sum_{\mathbf{p},\sigma} \varepsilon(\hat{a}^+_{\mathbf{p}\sigma}\hat{a}_{\mathbf{p}\sigma} + \hat{b}^+_{\mathbf{p}\sigma}\hat{b}_{\mathbf{p}\sigma} - 1),$$

and the energy eigenvalues are (with the usual omission of an infinite additive constant)

$$E = \sum_{\mathbf{p},\sigma} \varepsilon(N_{\mathbf{p}\sigma} + \overline{N}_{\mathbf{p}\sigma}), \tag{25.5}$$

and are positive–definite, as they should be. With Bose quantization, we should obtain from (25.3) the eigenvalues

$$\sum \varepsilon(N_{\mathbf{p}\sigma} - \overline{N}_{\mathbf{p}\sigma}),$$

which are not positive–definite and have no meaning.

† The two functions also correspond to the same value σ of the spin component in the rest frame; for the functions $\bar{\psi}_{-\mathbf{p},-\sigma}$ this will be proved in §26 (see (26.10)).

An expression analogous to (25.5) is obtained for the momentum of the system, i.e. the eigenvalues of the operator $\int \hat{\psi}^+ \hat{\mathbf{p}} \hat{\psi}\, d^3x$:

$$\mathbf{P} = \sum_{\mathbf{p},\sigma} \mathbf{p}(N_{\mathbf{p}\sigma} + \overline{N}_{\mathbf{p}\sigma}). \tag{25.6}$$

The 4-current operator is

$$\hat{j}^\mu = \hat{\bar{\psi}} \gamma^\mu \hat{\psi}, \tag{25.7}$$

and the "charge" operator of the field is found to be

$$\begin{aligned} \hat{Q} &= \int \hat{\bar{\psi}} \gamma^0 \hat{\psi}\, d^3x \\ &= \sum_{\mathbf{p},\sigma} (\hat{a}^+_{\mathbf{p}\sigma} \hat{a}_{\mathbf{p}\sigma} + \hat{b}_{\mathbf{p}\sigma} \hat{b}^+_{\mathbf{p}\sigma}) \\ &= \sum_{\mathbf{p},\sigma} (\hat{a}^+_{\mathbf{p}\sigma} \hat{a}_{\mathbf{p}\sigma} - \hat{b}^+_{\mathbf{p}\sigma} \hat{b}_{\mathbf{p}\sigma} + 1); \end{aligned} \tag{25.8}$$

its eigenvalues are

$$Q = \sum_{\mathbf{p},\sigma} (N_{\mathbf{p}\sigma} - \overline{N}_{\mathbf{p}\sigma}). \tag{25.9}$$

Thus we again arrive at the concept of particles and antiparticles, and the whole of the discussion of these in §11 is applicable.

But particles with spin $\frac{1}{2}$ are fermions, whereas those with spin zero are bosons. An examination of the formal origin of this difference shows that it is due to the different nature of the expressions for the "energy density" in the scalar and spinor fields. In the scalar field the expression is positive–definite, and the terms $\hat{a}^+\hat{a}$ and $\hat{b}\hat{b}^+$ therefore both have a positive sign in the Hamiltonian (11.3). If the energy eigenvalues are positive, the replacement of $\hat{b}\hat{b}^+$ by $\hat{b}^+\hat{b}$ must occur without change of sign, i.e. in accordance with the Bose commutation rule. For the spinor field, however, the "energy density" is not a positive–definite quantity, and hence the term $\hat{b}\hat{b}^+$ appears with the minus sign in the Hamiltonian (25.3); to obtain positive eigenvalues, the replacement of $\hat{b}\hat{b}^+$ must be accompanied by a change of sign, i.e. must occur in accordance with the Fermi commutation rule.

The form of the energy density is directly related to the transformation properties of the wave function and to the requirements of relativistic invariance. In this sense we may say that the relation between the spin and the statistics obeyed by the particles is likewise a direct consequence of these requirements.

The fact that particles with spin $\frac{1}{2}$ are fermions also leads to a general conclusion: all particles with half-integral spin are fermions, and those with integral spin are bosons (including those with spin zero, as demonstrated in §11).†

† The origin of the relation between the spin of a particle and the statistics which it obeys was first elucidated by W. Pauli (1940).

This is evident because a particle with spin s may be regarded as "composed" of $2s$ particles with spin $\frac{1}{2}$. When s is half-integral, $2s$ is odd; when s is integral, $2s$ is even. A "composite" particle containing an even number of fermions is a boson, and one containing an odd number of fermions is a fermion.†

If a system consists of particles of various kinds, then creation and annihilation operators must be defined separately for each kind of particle. The operators pertaining to different bosons, or to bosons and fermions, commute. Operators pertaining to different fermions may be regarded, in the non-relativistic theory, as either commuting or anticommuting (*QM*, §65). In the relativistic theory, which allows transformations of particles into one another, the creation and annihilation operators of different fermions must be regarded as anticommuting, like those which pertain to different states of the same fermions.

PROBLEM

Find the Lagrangian of the spinor field.

SOLUTION. The Lagrangian corresponding to Dirac's equation is given by the real scalar expression

$$L = \tfrac{1}{2}i(\bar{\psi}\gamma^\mu \partial_\mu \psi - \partial_\mu \bar{\psi} \cdot \gamma^\mu \psi) - m\bar{\psi}\psi. \tag{1}$$

Taking the components of ψ and $\bar{\psi}$ as the "generalized coordinates" q, we easily see that the corresponding Lagrange's equations (10.10) are the same as Dirac's equations for $\bar{\psi}$ and ψ. The overall sign of the Lagrangian (like the common factor in it) is arbitrary. Since L involves the derivatives of ψ and $\bar{\psi}$ linearly, the action $S = \int L\, d^4x$ can in any case have no minimum or maximum. The condition $\delta S = 0$ here gives only a stationary point of the integral, not an extremum.

The Lagrangian of the spinor field is obtained by replacing ψ in (1) by the operator $\hat{\psi}$. Applying formula (12.12) to this Lagrangian, we find the current operator (25.7).

§ 26. Charge conjugation and time reversal of spinors

The coefficients $\psi_{p\sigma} = u_{p\sigma}e^{-ipx}$ which appear with the operators $\hat{a}_{\mathbf{p}\sigma}$ in (25.1) are the wave functions of free particles (electrons, say) having momenta $\mathbf{p}$ and polarizations σ:

$$\psi^{(e)}_{p\sigma} = \psi_{p\sigma}.$$

The coefficients $\bar{\psi}_{-p,-\sigma}$ of the operators $\hat{b}_{\mathbf{p}\sigma}$ are to be regarded as the wave functions of positrons having the same $\mathbf{p}$ and σ. It is found, however, that the electron and positron functions are expressed in different bispinor representations. This is evident from the fact that ψ and $\bar{\psi}$ differ in their transformation properties and their components satisfy different sets of equations. To eliminate this defect, it is necessary to carry out a certain unitary transformation of the components $\bar{\psi}_{-p,-\sigma}$,

† In this argument it is assumed that all particles with the same spin obey the same statistics (whatever the way in which they are "compounded"). The truth of this assumption is seen by analogous arguments. For example, if there existed fermions with spin zero, then a fermion with spin zero and one with spin $\frac{1}{2}$ would yield a particle with spin $\frac{1}{2}$, which would be a boson, in contradiction with the general result demonstrated for spin $\frac{1}{2}$.

such that the new four-component function satisfies the same equation as $\psi_{p\sigma}$.† This will be referred to as the wave function of the positron (with momentum **p** and polarization σ). Denoting the matrix of the required unitary transformation by U_C, we may write

$$\psi^{(p)}_{p\sigma} = U_C \bar{\psi}_{-p,-\sigma}. \tag{26.1}$$

The operation C whereby this function is obtained from $\psi_{-p,-\sigma}$ is called *charge conjugation* of the wave function (H. A. Kramers, 1937). This concept is, of course, not restricted to its application to plane waves: for any function ψ, there exists a "charge-conjugate" function

$$\hat{C}\psi(t, \mathbf{r}) = U_C \bar{\psi}(t, \mathbf{r}), \tag{26.2}$$

which has the same transformation properties as ψ and satisfies the same equation.

The properties of the matrix U_C follow from this definition. If ψ is a solution of Dirac's equation $(\gamma\hat{p} - m)\psi = 0$, then $\bar{\psi}$ satisfies the equation

$$\bar{\psi}(\gamma\hat{p} + m) = 0, \qquad \text{or} \qquad (\tilde{\gamma}\hat{p} + m)\bar{\psi} = 0.$$

Multiplying this equation on the left by U_C:

$$U_C \tilde{\gamma}\hat{p}\bar{\psi} + m U_C \bar{\psi} = 0,$$

we apply the condition that the function $U_C\bar{\psi}$ satisfies the same equation as ψ:

$$(\gamma\hat{p} - m) U_C \bar{\psi} = 0.$$

A comparison of the two equations gives the following "commutation relation" between U_C and the matrices γ^μ:‡

$$U_C \tilde{\gamma}^\mu = -\gamma^\mu U_C. \tag{26.3}$$

We shall further suppose that the wave functions are stated in the spinor or standard representation; the general case of any representation will be considered only at the end of this section. In these representations, we have

$$\left.\begin{aligned} \gamma^{0,2} &= \tilde{\gamma}^{0,2}, & \gamma^{1,3} &= -\tilde{\gamma}^{1,3}, \\ (\gamma^{0,1,3})^* &= \gamma^{0,1,3}, & \gamma^{2*} &= -\gamma^2. \end{aligned}\right\} \tag{26.4}$$

Then the condition (26.3) is satisfied by the matrix $U_C = \eta_C \gamma^2 \gamma^0$, the constant η_C being arbitrary. The condition $\hat{C}^2 = 1$ shows that $|\eta_C|^2 = 1$, and the matrix U_C is

† For particles with spin zero, this problem did not arise, since the scalar functions ψ and ψ^* satisfy the same equation, and ψ^*_{-p} is identical with ψ_p.

‡ From this there follows also the equation

$$U_C \tilde{\gamma}^5 = \gamma^5 U_C. \tag{26.3a}$$

therefore determined apart from a phase factor. We shall take $\eta_C = 1$; thus

$$U_C = \gamma^2\gamma^0 = -\alpha_y. \tag{26.5}$$

Noting also that $\bar{\psi} = \psi^*\gamma^0 = \tilde{\gamma}^0\psi^* = \gamma^0\psi^*$, we may write the effect of the operator $\hat{C}$ as

$$\hat{C}\psi = \gamma^2\gamma^0\bar{\psi} = \gamma^2\psi^*. \tag{26.6}$$

The explicit form of the transformation (26.6) for the spinor representation is

$$C: \quad \xi^\alpha \to -i\eta^{\dot{\alpha}*}, \qquad \eta_{\dot{\alpha}} \to -i\xi^*_{\alpha}, \tag{26.7a}$$

or, equivalently,

$$C: \quad \xi_\alpha \to -i\overset{*}{\eta}_{\dot{\alpha}}, \qquad \eta^{\dot{\alpha}} \to -i\xi^{\alpha*}. \tag{26.7b}$$

The charge-conjugation transformation for the plane waves $\psi_{\pm p\sigma}$ is easily carried out by using the explicit expressions (23.9) for the plane waves and the matrix U_C in the standard representation:

$$U_C = \begin{pmatrix} 0 & -\sigma_y \\ -\sigma_y & 0 \end{pmatrix}. \tag{26.8}$$

Since

$$\sigma_y\boldsymbol{\sigma}^* = -\boldsymbol{\sigma}\sigma_y,$$

we have, if $w^{(\sigma)\prime}$ is defined as in (23.16),

$$U_C\bar{u}_{-p,-\sigma} = u_{p\sigma}, \qquad U_C u_{-p,-\sigma} = \bar{u}_{p\sigma}. \tag{26.9}$$

Thus

$$\hat{C}\psi_{-p,-\sigma} = \psi_{p\sigma}, \tag{26.10}$$

so that the functions $\psi_{-p,-\sigma}$ which appear with the operators $\hat{b}_{\mathbf{p}\sigma}$ in the ψ-operators (25.1) do in fact correspond to states of a particle having momentum $\mathbf{p}$ and polarization σ. We see also that the electron and positron states are described by the same functions:

$$\psi^{(e)}_{p\sigma} = \psi^{(p)}_{p\sigma} = \psi_{p\sigma}.$$

This is to be expected, since the functions $\psi_{p\sigma}$ embody information only as to the momentum and polarization of the particle.

The operation of time reversal may be treated similarly. When the sign of the time is changed, the wave function must change to the complex conjugate. In order to obtain the "time-reversed" wave function ($\hat{T}\psi$) in the same representation as the original ψ, we must also perform some unitary transformation on the components of ψ^* (or $\tilde{\psi}$). Thus the action of the operator $\hat{T}$ on ψ can be written, similarly to (26.2), as

$$\hat{T}\psi(t, \mathbf{r}) = U_T\tilde{\psi}(-t, \mathbf{r}), \tag{26.11}$$

where U_T is a unitary matrix.

Dirac's equation satisfied by ψ is

$$\left(i\gamma^0\frac{\partial}{\partial t} + i\boldsymbol{\gamma}\cdot\nabla - m\right)\psi(t, \mathbf{r}) = 0,$$

and the equation for $\tilde{\psi}$ is

$$\left(i\tilde{\gamma}^0\frac{\partial}{\partial t} + i\tilde{\boldsymbol{\gamma}}\cdot\nabla + m\right)\tilde{\psi}(t, \mathbf{r}) = 0.$$

In the latter equation we change t into $-t$ and multiply on the left by $-U_T$:

$$\left(iU_T\tilde{\gamma}^0\frac{\partial}{\partial t} - iU_T\tilde{\boldsymbol{\gamma}}\cdot\nabla\right)\tilde{\psi}(-t, \mathbf{r}) - mU_T\tilde{\psi}(-t, \mathbf{R}) = 0.$$

We want the function $U_T\tilde{\psi}(-t, \mathbf{r})$ to satisfy the same equation as $\psi(t, \mathbf{r})$:

$$\left(i\gamma^0\frac{\partial}{\partial t} + i\boldsymbol{\gamma}\cdot\nabla\right)U_T\tilde{\psi}(-t, \mathbf{r}) - mU_T\tilde{\psi}(-t, \mathbf{r}) = 0.$$

Comparing the two equations, we find that the matrix U_T must satisfy the conditions

$$U_T\tilde{\gamma}^0 = \gamma^0 U_T, \qquad U_T\tilde{\boldsymbol{\gamma}} = -\boldsymbol{\gamma}U_T. \tag{26.12}$$

In the spinor and standard representations, these conditions are satisfied by the matrix†

$$U_T = i\gamma^3\gamma^1\gamma^0. \tag{26.13}$$

Thus the effect of the operator $\hat{T}$ is given by

$$\hat{T}\psi(t, \mathbf{r}) = i\gamma^3\gamma^1\gamma^0\tilde{\psi}(-t, \mathbf{r}) = i\gamma^3\gamma^1\psi^*(-t, \mathbf{r}). \tag{26.14}$$

The explicit form of this transformation for the spinor representation is

$$T: \quad \xi^\alpha \to -i\xi^*_{\dot{\alpha}}, \qquad \eta_{\dot{\alpha}} \to i\eta^{\dot{\alpha}*}, \tag{26.15a}$$

† The choice of the phase factor in (26.13) depends on that in (26.5); see the second footnote to §27.

or

$$T:\quad \xi_\alpha \to i\xi^{\alpha *}, \qquad \eta^{\dot\alpha} \to -i\overset{*}{\eta}_{\dot\alpha}. \tag{26.15b}$$

In the standard representation,

$$U_T = \begin{pmatrix} \sigma_y & 0 \\ 0 & -\sigma_y \end{pmatrix}. \tag{26.16}$$

To find the effect on ψ of all three operations P, T and C, we write successively

$$\hat{T}\psi(t, \mathbf{r}) = -i\gamma^1\gamma^3\psi^*(-t, \mathbf{r}),$$
$$\hat{P}\hat{T}\psi(t, \mathbf{r}) = i\gamma^0(\hat{T}\psi) = \gamma^0\gamma^1\gamma^3\psi^*(-t, -\mathbf{r}),$$
$$\hat{C}\hat{P}\hat{T}\psi(t, \mathbf{r}) = \gamma^2(\gamma^0\gamma^1\gamma^3\psi^*)^* = \gamma^2\gamma^0\gamma^1\gamma^3\psi(-t, -\mathbf{r}),$$

or

$$\hat{C}\hat{P}\hat{T}\psi(t, \mathbf{r}) = i\gamma^5\psi(-t, -\mathbf{r}). \tag{26.17}$$

In the spinor representation,

$$CPT:\quad \xi^\alpha \to -i\xi^\alpha, \qquad \eta_{\dot\alpha} \to i\eta_{\dot\alpha}, \tag{26.18}$$

as is also easily seen directly from the transformation rules (20.4), (26.7) and (26.15).†

The expressions given above for the matrices U_C and U_T assume the spinor or standard representation of ψ. Let us finally see which properties of these expressions are retained for any representation of ψ.

If ψ is subjected to a unitary transformation:

$$\psi' = U\psi, \qquad \gamma' = U\gamma U^{-1}, \qquad \bar\psi' = \psi'^*\gamma^{0\prime} = \bar\psi U^+ = \bar\psi U^{-1}, \tag{26.19}$$

then in the new representation we have

$$(\hat{C}\psi)' = U(C\psi) = UU_C\bar\psi = UU_C(\bar\psi' U) = UU_C\tilde{U}\bar\psi'.$$

A comparison with the definition of the matrix U'_C in the new representation, $(\hat{C}\psi)' = U'_C\bar\psi'$, shows that

$$U'_C = UU_C\tilde{U}. \tag{26.20}$$

The transformation (26.20) is the same as that of the matrices γ only if U is real. The expression (26.5) also is therefore valid only in representations which are a real transformation of the spinor or standard representation.

† The notation $\hat{C}\hat{P}\hat{T}$ implies that the operators act in the sequence from right to left. The sign of (26.17) and (26.18) depends on this sequence, since $\hat{T}$ does not commute with $\hat{C}$ and $\hat{P}$ (as regards their action on a bispinor).

The matrix (26.5) is unitary, and changes sign when transposed:

$$U_C U_C^+ = 1, \qquad \tilde{U}_C = -U_C. \tag{26.21}$$

These properties are unaffected by the transformation (26.20), and are therefore retained in any representation. The matrix (26.5) is also Hermitian ($U_C = U_C^+$), but this property is in general not preserved by the transformation (26.20).

The above discussion and formulae (26.21) apply likewise to the properties of the matrix U_T.

In the second quantization formalism, the transformations C, P. T for ψ-operators must be formulated as transformation rules for the particle creation and annihilation operators. These rules can be established (as in §13 for particles with spin zero) from the condition that the transformed ψ-operators may be written

$$\left.\begin{aligned} \hat{\psi}^C(t, \mathbf{r}) &= U_C \hat{\bar{\psi}}(t, \mathbf{r}), \\ \hat{\psi}^P(t, \mathbf{r}) &= i\gamma^0 \hat{\psi}(t, -\mathbf{r}), \\ \hat{\psi}^T(t, \mathbf{r}) &= U_T \hat{\bar{\psi}}(-t, \mathbf{r}). \end{aligned}\right\} \tag{26.22}$$

PROBLEM

Find the charge-conjugation operator in the Majorana representation (§21, Problem 2).

SOLUTION. The matrix U_C' in the Majorana representation is obtained from the matrix $U_C = -\alpha_y$ in the standard representation by the transformation (26.20) with $U = (\alpha_y + \beta)/\sqrt{2}$, and is $U_C' = -\alpha_y$ (where α_y and β denote matrices in the standard representation). If primes denote quantities in the Majorana representation, we have $\hat{C}\psi' = U_C'(\psi'^*\beta')$, and since $\beta' = \alpha_y$,

$$\hat{C}\psi' = \alpha_y(\psi'^*\alpha_y) = \alpha_y\tilde{\alpha}_y\psi'^* = \psi'^*,$$

i.e. charge conjugation is the same as complex conjugation.

§ 27. **Internal symmetry of particles and antiparticles**

The wave function of a particle with spin $\frac{1}{2}$, in its rest frame, is a single three-dimensional spinor, which will be denoted by Φ^α. The behaviour of this spinor under inversion is related to the concept of the internal parity of the particle. However, as mentioned in §19, although the two possible laws of transformation of three-dimensional spinors ($\Phi^\alpha \to \pm i\Phi^\alpha$) are not equivalent, there is no absolute significance in assigning a particular parity to a spinor. We therefore cannot speak of the internal parity of any one particle with spin $\frac{1}{2}$, but we can refer to the relative internal parity of two such particles.

From two (three-dimensional) spinors $\Phi^{(1)}$ and $\Phi^{(2)}$, a scalar $\Phi^{(1)}_\alpha\Phi^{(2)\alpha}$ can be formed. If this is a true scalar, the particles described by the spinors are said to have the same internal parity; if it is a pseudoscalar, they are said to have opposite internal parities.

We shall show that particles and antiparticles (with spin $\frac{1}{2}$) have opposite parities (V. B. Berestetskiĭ, 1948).

Firstly, if the operation C (26.7) is applied to both sides of the P transformation

(19.5) (in the spinor representation)

$$P: \quad \xi^\alpha \to i\eta_{\dot\alpha}, \qquad \eta_{\dot\alpha} \to i\xi^\alpha, \tag{27.1}$$

we obtain

$$\eta^{c\dot\alpha *} \to i\xi^{c}_{\alpha}{}^{*}, \qquad \xi^{c}_{\alpha}{}^{*} \to i\eta^{c\dot\alpha *},$$

where the index c marks the components of the bispinor

$$\psi^c = \begin{pmatrix} \xi^c \\ \eta^c \end{pmatrix} \text{ charge-conjugate to } \psi = \begin{pmatrix} \xi \\ \eta \end{pmatrix}.$$

Taking the complex conjugate and interchanging the indices, we find

$$P: \quad \eta^{c}_{\dot\alpha} \to i\xi^{c\alpha}, \qquad \xi^{c\alpha} \to i\eta^{c}_{\dot\alpha}. \tag{27.2}$$

Thus charge-conjugate bispinors are transformed in the same manner by inversion.

Let $\psi^{(e)}$ be the wave function of a particle (say an electron) and $\psi^{(p)}$ that of the antiparticle (a positron). The latter is a bispinor which is the charge conjugate of a "negative-frequency" solution of Dirac's equation. In the rest frame, each function becomes a three-dimensional spinor:

$$\xi^{(e)\alpha} = \eta^{(e)}_{\dot\alpha} = \Phi^{(e)\alpha}, \qquad \xi^{(p)\alpha} = \eta^{(p)}_{\dot\alpha} = \Phi^{(p)\alpha}.$$

According to (27.1), (27.2), these spinors are transformed as follows by inversion:

$$\Phi^\alpha \to i\Phi^\alpha, \tag{27.3}$$

the same for both $\Phi^{(e)}$ and $\Phi^{(p)}$. The product $\Phi^{(e)}\Phi^{(p)}$, however, changes sign, and this proves the result stated above.

A strictly neutral particle is one which coincides with its antiparticle (§12). The ψ-operator of a field of such particles satisfies the condition

$$\hat\psi(t, \mathbf{r}) = \hat\psi^C(t, \mathbf{r}).$$

For particles with spin $\frac{1}{2}$, this implies the conditions (in the spinor representation)†

$$\hat\xi^\alpha = -i\hat\eta^{\dot\alpha +}, \qquad \hat\eta_{\dot\alpha} = -i\hat\xi^{+}_{\alpha}. \tag{27.4}$$

Like any relation which expresses a physical property, these conditions are invariant under the transformation CPT.‡ It is easily verified that they are in fact

† In the Majorana representation, strict neutrality implies simply that the operator $\hat\psi'$ is Hermitian; see §26, Problem.

‡ More precisely, the transformation CPT must here be defined so as to leave relations such as (27.4) unchanged. This is achieved by an appropriate choice of the phase factor in the definition of the matrix U_T; see the third footnote to §26.

invariant not only with respect to *CPT* but also with respect to each of the three transformations separately.

In §19, inversion of spinors was defined to be a transformation for which $\hat{P}^2 = -1$, and this definition has been used so far. The result derived above concerning the relative parity of particles and antiparticles is easily seen to be, as it should be, independent of the way in which inversion is defined.

If inversion is defined by the condition $\hat{P}^2 = 1$, then (27.1) becomes

$$P: \quad \xi^\alpha \to \eta_{\dot\alpha}, \; \eta_{\dot\alpha} \to \xi^\alpha. \tag{27.5}$$

The charge-conjugate function is then transformed according to

$$\xi^{c\alpha} \to -\eta^c_{\dot\alpha}, \qquad \eta^c_{\dot\alpha} \to -\xi^{c\alpha},$$

which differs in sign from (27.5). Accordingly, the three-dimensional spinors Φ will be transformed thus:

$$\Phi^{(e)\alpha} \to \Phi^{(e)\alpha}, \qquad \Phi^{(p)\alpha} \to -\Phi^{(p)\alpha},$$

and the product $\Phi^{(e)}\Phi^{(p)}$ will again be a pseudoscalar.

The only possible difference in the physical consequences of the two views of inversion is that with the definition (27.5) the condition for a strictly neutral field would not be invariant under this transformation (or the transformation *CP*), which would alter the relative sign of the sides of the equations (27.4). Actually, no strictly neutral particle with spin $\frac{1}{2}$ is known, and we cannot yet say whether this difference between the two definitions of inversion has any real physical meaning.†

PROBLEM

Find the charge parity of positronium (a hydrogen-like system consisting of an electron and a positron).

SOLUTION. The wave function of two fermions must be antisymmetric with respect to simultaneous interchange of coordinates, spins and charge variables of the particles (cf. §13, Problem). The interchange of the coordinates multiplies the function by $(-1)^l$, that of the spins by $(-1)^{1+S}$, where S (=0 or 1) is the total spin of the system, and that of the charge variables by the required parity C. The condition

$$(-1)^l(-1)^{1+S}C = -1$$

gives $C = (-1)^{l+S}$. Since the internal parities of the electron and the positron are opposite, the spatial parity of the system is $P = (-1)^{l+1}$. The combined parity $CP = (-1)^{S+1}$.

§ 28. **Bilinear forms**

Let us consider the transformation properties of various bilinear forms which can be constructed from components of the functions ψ and ψ^*. Such forms are of great importance in quantum mechanics. They include the current density 4-vector (21.11).

† The incomplete equivalence of the two definitions of inversion was first noted by G. Racah (1937).

Since ψ and ψ^* have four components each, a total of $4 \times 4 = 16$ independent bilinear combinations can be formed from them. The classification of the transformation properties of these is evident from the ways listed in §19 of multiplying any two bispinors (in this case ψ and ψ^*). We can form a scalar (denoted by S), a pseudoscalar (P), a mixed spinor of rank two equivalent to a true 4-vector V^μ (four independent quantities), a mixed spinor of rank two equivalent to a 4-pseudovector A^μ (four quantities), and a bispinor of rank two equivalent to an antisymmetric 4-tensor $T^{\mu\nu}$ (six quantities).

In a symmetrical form (for any representation of ψ), these combinations may be written

$$\left.\begin{aligned} S = \bar{\psi}\psi, \qquad P = i\bar{\psi}\gamma^5\psi, \\ V^\mu = \bar{\psi}\gamma^\mu\psi, \qquad A^\mu = \bar{\psi}\gamma^\mu\gamma^5\psi, \qquad T^{\mu\nu} = i\bar{\psi}\sigma^{\mu\nu}\psi, \end{aligned}\right\} \tag{28.1}$$

where

$$\sigma^{\mu\nu} = \tfrac{1}{2}(\gamma^\mu\gamma^\nu - \gamma^\nu\gamma^\mu) = (\boldsymbol{\alpha}, i\boldsymbol{\Sigma}); \tag{28.2}$$

the components in (28.2) are stated as in (19.15).† All the expressions given above are real.

The fact that S is a scalar and P a pseudoscalar is evident from their spinor representations:

$$S = \xi^*\eta + \eta^*\xi, \qquad P = i(\xi^*\eta - \eta^*\xi),$$

which agree with (19.7) and (19.8). The fact that the V^μ form a vector is then evident from Dirac's equation: multiplying the equation $\hat{p}_\mu\gamma^\mu\psi = m\psi$ on the left by $\bar{\psi}$, we obtain

$$\bar{\psi}\hat{p}_\mu\gamma^\mu\psi = m\bar{\psi}\psi.$$

Since the right-hand side is a scalar, so is the left-hand side.

The rule whereby the quantities (28.1) are obtained is obvious: they are constructed as if the matrices γ^μ formed a 4-vector, γ^5 were a pseudoscalar, and the $\bar{\psi}$ and ψ on either side together formed a scalar.‡ The non-existence of bilinear forms which are symmetrical 4-tensors is seen from the spinor representation and also from this rule: since the symmetrical combination of matrices is $\gamma^\mu\gamma^\nu + \gamma^\nu\gamma^\mu = 2g^{\mu\nu}$, any such form would reduce to a scalar.

The second-quantized bilinear forms are obtained by replacing the ψ-functions in (28.1) by ψ-operators. For greater generality, we shall assume that the two ψ-operators relate to fields of different particles, denoted by suffixes a and b. Let

† For a unitary transformation of ψ (change of representation), we have $\psi \to U\psi$, $\gamma \to U\gamma U^{-1}$, $\bar{\psi} \to \bar{\psi}U^{-1}$, and the invariance of the bilinear forms under such transformations is obvious.

‡ The "pseudoscalar" nature of γ^5 is itself in accordance with these rules, since

$$\gamma^5 = \frac{i}{24} e_{\lambda\mu\nu\rho}\gamma^\lambda\gamma^\mu\gamma^\nu\gamma^\rho.$$

us see how such operator forms are transformed under charge conjugation. We have†

$$\hat{\psi}^C = U_C \hat{\bar{\psi}}, \qquad \hat{\bar{\psi}}^C = U_C^+ \hat{\psi}, \tag{28.3}$$

and therefore, using (26.3) and (26.21),

$$\begin{aligned}
\hat{\bar{\psi}}_a^C \hat{\psi}_b^C &= \hat{\psi}_a U_C^* U_C \hat{\bar{\psi}}_b \\
&= -\hat{\psi}_a U_C^+ U_C \hat{\bar{\psi}}_b \\
&= -\hat{\psi}_a \hat{\bar{\psi}}_b, \\
\hat{\bar{\psi}}_a^C \gamma^\mu \hat{\psi}_b^C &= \hat{\psi}_a U_C^* \gamma^\mu U_C \hat{\bar{\psi}}_b \\
&= -\hat{\psi}_a U_C^+ \gamma^\mu U_C \hat{\bar{\psi}}_b \\
&= \hat{\psi}_a \tilde{\gamma}^\mu \hat{\bar{\psi}}_b.
\end{aligned}$$

When the operators are restored to their original order ($\hat{\bar{\psi}}$ to the left of $\hat{\psi}$), the Fermi commutation rules (25.4) show that the sign of the product is changed (and moreover terms appear which are independent of the state of the field; we omit these, as in the corresponding treatment in §13). Thus we have

$$\hat{\bar{\psi}}_a^C \hat{\psi}_b^C = \hat{\bar{\psi}}_b \hat{\psi}_a, \qquad \hat{\bar{\psi}}_a^C \gamma^\mu \hat{\psi}_b^C = -\hat{\bar{\psi}}_b \gamma^\mu \hat{\psi}_a.$$

Proceeding similarly with the other forms, we find the results for charge conjugation‡

$$C: \quad \begin{array}{lll} \hat{S}_{ab} \to \hat{S}_{ba}, & \hat{P}_{ab} \to \hat{P}_{ba}, & \hat{V}^\mu_{ab} \to -\hat{V}^\mu_{ba}, \\ \hat{A}^\mu_{ab} \to \hat{A}^\mu_{ba}, & \hat{T}^{\mu\nu}_{ab} \to -\hat{T}^{\mu\nu}_{ba}. & \end{array} \tag{28.4}$$

The behaviour of these forms under time reversal may be ascertained similarly, remembering (see §13) that this operation brings about a change in the order of the operators, so that, for example,

$$(\hat{\bar{\psi}}_a \hat{\psi}_b)^T = \hat{\psi}_b^T \hat{\bar{\psi}}_a^T.$$

† To derive the second of these equations from the first, we write

$$\begin{aligned}
\hat{\bar{\psi}}^C &= [U_C^*(\hat{\psi}\gamma^{0*})]\gamma^0 \\
&= \tilde{\gamma}^0 U_C^* \gamma^0 \hat{\psi} \\
&= -\tilde{\gamma}^0 U_C^+ \gamma^0 \hat{\psi} \\
&= \tilde{\gamma}^0 \gamma^{0*} U_C^+ \hat{\psi} \\
&= U_C^+ \hat{\psi},
\end{aligned}$$

using (26.3), (26.21) and the fact that γ^0 is Hermitian.

‡ It should be noticed that, for bilinear forms constructed from ψ-functions (not ψ-operators), the transformations (28.4) would have the opposite signs, since the return to the original order of the factors $\bar{\psi}$ and ψ would not involve a change of sign.

Substituting here

$$\hat{\psi}^T = U_T\hat{\bar{\psi}}, \qquad \hat{\bar{\psi}}^T = -U_T^+\hat{\psi}, \tag{28.5}$$

we obtain

$$\begin{aligned}(\hat{\bar{\psi}}_a\hat{\psi}_b)^T &= -\hat{\bar{\psi}}_b\tilde{U}_T U_T^+\hat{\psi}_a\\ &= \hat{\bar{\psi}}_b U_T U_T^+\hat{\psi}_a\\ &= \hat{\bar{\psi}}_b\hat{\psi}_a.\end{aligned}$$

Treating the other forms in the same way, we obtian

$$\begin{aligned}T:\quad &\hat{S}_{ab}\to\hat{S}_{ba}, \qquad \hat{P}_{ab}\to-\hat{P}_{ba}, \qquad (\hat{V}^0,\hat{\mathbf{V}})_{ab}\to(\hat{V}^0,-\hat{\mathbf{V}})_{ba},\\ &(\hat{A}^0,\hat{\mathbf{A}})_{ab}\to(-\hat{A}^0\ \hat{\mathbf{A}})_{ba}, \qquad \hat{T}^{\mu\nu}_{ab}=(\hat{\mathbf{p}},\hat{\mathbf{a}})_{ab}\to(\hat{\mathbf{p}},-\hat{\mathbf{a}})_{ba},\end{aligned} \tag{28.6}$$

where $\hat{\mathbf{p}}$ and $\hat{\mathbf{a}}$ are three-dimensional vectors equivalent to the components of $\hat{T}^{\mu\nu}$ as shown in (19.15).

Under spatial inversion we have, in accordance with the tensor properties,†

$$\begin{aligned}P:\quad &\hat{S}_{ab}\to\hat{S}_{ab}, \qquad \hat{P}_{ab}\to-\hat{P}_{ab}, \qquad (\hat{V}^0,\hat{\mathbf{V}})_{ab}\to(\hat{V}^0,-\hat{\mathbf{V}})_{ab},\\ &(\hat{A}^0,\hat{\mathbf{A}})_{ab}\to(-\hat{A}^0,\hat{\mathbf{A}})_{ba}, \qquad \hat{T}^{\mu\nu}_{ab}=(\hat{\mathbf{p}},\hat{\mathbf{a}})_{ab}\to(-\hat{\mathbf{p}},\hat{\mathbf{a}})_{ab}.\end{aligned} \tag{28.7}$$

Finally, the simultaneous application of all three operations leaves all the $\hat{S}_{ab}$, $\hat{P}_{ab}$ and $\hat{T}^{\mu\nu}_{ab}$ unchanged, and changes the sign of all the $\hat{V}^{\mu}_{ab}$ and $\hat{A}^{\mu}_{ab}$, in agreement with the fact that this transformation is a 4-inversion: since 4-inversion is equivalent to a rotation of the 4-coordinates, it creates no difference between true tensors and pseudotensors of any rank.

Let us now consider products of pairs of bilinear forms constructed from four different functions ψ^a, ψ^b, ψ^c, ψ^d. The result depends on which pairs of functions are multiplied together. It is possible, however, to reduce any such product to products of bilinear forms with specified pairs of factors (W. Pauli and M. Fierz, 1936). We shall derive the relationship on which this reduction is based.

If we take the set of four-rowed matrices

$$1,\ \gamma^5,\ \gamma^\mu,\ i\gamma^\mu\gamma^5,\ i\sigma^{\mu\nu}, \tag{28.8}$$

where 1 is the unit matrix, arrange these 16 (=1+1+4+4+6) matrices in any definite order and denote them by γ^A ($A = 1, \dots, 16$), and also denote the same matrices with lowered 4-tensor indices (μ, ν) by γ_A, then they will have the following properties:

$$\left.\begin{aligned}\operatorname{tr}\gamma^A = 0 \quad (\gamma^A\neq 1),\\ \gamma^A\gamma_A = 1, \qquad \tfrac{1}{4}\operatorname{tr}\gamma^A\gamma_B = \delta^A_B.\end{aligned}\right\} \tag{28.9}$$

† To avoid any misunderstanding, it should be mentioned that the transformations T and P also involve a change in the arguments of the functions; the right-hand (transformed) sides in (28.6)–(28.7) are respectively functions of $x^T = (-t, \mathbf{r})$, $x^P = (t, -\mathbf{r})$, when the left-hand sides are functions of $x = (t, \mathbf{r})$.

The last of these shows that the matrices γ^A are linearly independent. Since their number is equal to the number (4×4) of elements of a four-rowed matrix, the matrices γ^A form a complete set in terms of which an arbitrary four-rowed matrix Γ may be expressed:

$$\Gamma = \sum_A c_A\gamma^A, \qquad c_A = \tfrac{1}{4}\operatorname{tr}\gamma_A\Gamma, \tag{28.10}$$

or, in explicit form, with matrix suffixes i, $k = 1, 2, 3, 4$,

$$\Gamma_{ik} = \tfrac{1}{4}\sum_A \Gamma_{lm}\gamma^A_{ml}\gamma_{Aik}.$$

Assuming, in particular, that in the matrix Γ only the element Γ_{lm} is non-zero, we obtain the required relation (the "completeness condition"):

$$\delta_{il}\delta_{km} = \tfrac{1}{4}\sum_A \gamma_{Aik}\gamma^A_{ml}. \tag{28.11}$$

Multiplying both sides of this equation by $\bar\psi^a_i\psi^b_k\bar\psi^c_m\psi^d_l$, we have

$$(\bar\psi^a\psi^d)(\bar\psi^c\psi^b) = \tfrac{1}{4}\sum_A (\bar\psi^a\gamma_A\psi^b)(\bar\psi^c\gamma^A\psi^d). \tag{28.12}$$

This is one equation of the type mentioned above, reducing the product of two scalar bilinear forms to products of forms involving other pairs of factors.†

Other equations of the same type may be obtained from (28.12) by the changes

$$\psi^d \to \gamma^B\psi^d, \qquad \psi^b \to \gamma^C\psi^b,$$

using the expansion

$$\gamma^A\gamma^B = \sum_R c_R\gamma^R, \qquad c_R = \tfrac{1}{4}\operatorname{tr}\gamma^A\gamma^B\gamma_R$$

(see the Problem).

Here we may also give for future reference, the relation for two-rowed matrices which corresponds to (28.11). The complete set of linearly independent two-rowed matrices σ^A $(A = 1, 2, 3, 4)$ is

$$1,\ \sigma_x,\ \sigma_y,\ \sigma_z. \tag{28.13}$$

These have the properties

$$\left.\begin{aligned} \operatorname{tr}\sigma^A &= 0 \quad (\sigma^A \neq 1),\\ \tfrac{1}{2}\operatorname{tr}\sigma^A\sigma^B &= \delta_{AB}. \end{aligned}\right\} \tag{28.14}$$

† It should be mentioned, to avoid any misunderstanding, that we are referring here to forms constructed from ψ-functions. The sign of the transformation would be opposite for forms constructed from the anticommuting ψ-operators.

The completeness condition is

$$\delta_{\alpha\gamma}\delta_{\beta\delta} = \tfrac{1}{2}\sum_{A} \sigma^{A}_{\alpha\beta}\sigma^{A}_{\delta\gamma}$$
$$= \tfrac{1}{2}\boldsymbol{\sigma}_{\alpha\beta}\cdot\boldsymbol{\sigma}_{\delta\gamma} + \tfrac{1}{2}\delta_{\alpha\beta}\delta_{\delta\gamma} \tag{28.15}$$

$(\alpha, \beta, \gamma, \delta = 1, 2)$, or

$$\boldsymbol{\sigma}_{\alpha\beta}\cdot\boldsymbol{\sigma}_{\delta\gamma} = -\tfrac{1}{2}\boldsymbol{\sigma}_{\alpha\gamma}\cdot\boldsymbol{\sigma}_{\delta\beta} + \tfrac{3}{2}\delta_{\alpha\gamma}\delta_{\delta\beta}. \tag{28.16}$$

PROBLEM

Derive formulae analogous to (28.12) for products of pairs of bilinear forms, *P*, *V*, *A*, *T*.

SOLUTION. We use the notation

$$J_S = (\bar\psi^a\psi^b)(\bar\psi^c\psi^d), \qquad J_P = (\bar\psi^a\gamma^5\psi^b)(\bar\psi^c\gamma^5\psi^d),$$
$$J_V = (\bar\psi^a\gamma^\mu\psi^b)(\bar\psi^c\gamma_\mu\psi^d), \qquad J_A = (\bar\psi^a i\gamma^\mu\gamma^5\psi^b)(\bar\psi^c i\gamma_\mu\gamma^5\psi^d),$$
$$J_T = (\bar\psi^a i\sigma^{\mu\nu}\psi^b)(\bar\psi^c i\sigma_{\mu\nu}\psi^d),$$

and the same symbols with primes to denote the products with ψ^b and ψ^d interchanged. The method used above gives

$$\begin{aligned}
4J'_S &= J_S + J_V + J_T + J_A + J_P,\\
4J'_V &= 4J_S - 2J_V + 2J_A - 4J_P,\\
4J'_T &= 6J_S - 2J_T + 6J_P,\\
4J'_A &= 4J_S + 2J_V - 2J_A - 4J_P,\\
4J'_P &= J_S - J_V + J_T - J_A + J_P,
\end{aligned}$$

the first of these equations being the same as (28.12).

§ 29. The polarization density matrix

The dependence on the coordinates of the wave function ψ which describes free motion with momentum $\mathbf{p}$ (a plane wave) reduces to a common factor $e^{i\mathbf{p}\cdot\mathbf{r}}$, and the amplitude u_p acts as a spin wave function. In such a state (a pure state), the particle is completely polarized (see *QM*, §59). In the non-relativistic theory this means that the particle spin has a definite direction in space (more precisely, there exists a direction in which the spin component has the definite value $+\frac{1}{2}$). In the relativistic theory, this description of a state in an arbitrary frame of reference is not possible, because, as already mentioned in §23, the spin vector is not conserved. The term "pure state" signifies only that the spin has a definite direction in the rest frame of the particle.

In a state of partial polarization, there is no definite amplitude, but only a *polarization density matrix* ρ_{ik} ($i, k = 1, 2, 3, 4$ being bispinor indices). We define this matrix in such a way that in a pure state it reduces to products:

$$\rho_{ik} = u_{pi}\bar u_{pk}. \tag{29.1}$$

Accordingly the matrix ρ is normalized by the condition

$$\mathrm{tr}\,\rho = 2m; \tag{29.2}$$

cf. (23.4).

In a pure state, the mean value of the spin is given by the quantity

$$\begin{aligned}\bar{\mathbf{s}} &= \tfrac{1}{2}\int \psi^*\boldsymbol{\Sigma}\psi\, d^3x \\ &= \frac{1}{4\varepsilon}\, u_p^*\boldsymbol{\Sigma} u_p \\ &= \frac{1}{4\varepsilon}\, \bar{u}_p\gamma^0\boldsymbol{\Sigma} u_p.\end{aligned} \tag{29.3}$$

The corresponding expression for a state of partial polarization is

$$\bar{\mathbf{s}} = \frac{1}{4\varepsilon}\,\mathrm{tr}\,(\rho\gamma^0\boldsymbol{\Sigma}) = \frac{1}{4\varepsilon}\,\mathrm{tr}\,(\rho\gamma^5\boldsymbol{\gamma}). \tag{29.4}$$

The amplitudes u_p and $\bar{u}_p$ satisfy the algebraic equations $(\gamma p - m)u_p = 0$, $\bar{u}_p(\gamma p - m) = 0$. The matrix (29.1) therefore satisfies the equations

$$(\gamma p - m)\rho = \rho(\gamma p - m) = 0. \tag{29.5}$$

The density matrix must satisfy similar linear equations in the general case of a state which is mixed (with respect to the spin); cf. the analogous argument in *QM*, §14.

If we consider a free particle in its rest frame, the non-relativistic theory is applicable. In that theory the state of partial polarization is completely defined by three parameters, the components of the mean spin vector $\bar{\mathbf{s}}$ (see *QM*, §59). It is therefore clear that the same parameters will define the polarization state after any Lorentz transformation, i.e. for a moving particle.

Let twice the mean spin vector in the rest frame be denoted by $\boldsymbol{\zeta}$; in a pure state $|\boldsymbol{\zeta}| = 1$, in a mixed state $|\boldsymbol{\zeta}| < 1$. For a four-dimensional description of the polarization state it is convenient to define a 4-vector a^μ which in the rest frame is the same as the three-dimensional vector $\boldsymbol{\zeta}$; since $\boldsymbol{\zeta}$ is an axial vector, a^μ is a 4-pseudovector. This 4-vector is orthogonal to the 4-momentum in the rest frame (in which $a^\mu = (0, \boldsymbol{\zeta})$, $p^\mu = (m, 0)$); in any frame, we therefore have

$$a^\mu p_\mu = 0. \tag{29.6}$$

In any frame, moreover,

$$a_\mu a^\mu = -\boldsymbol{\zeta}^2. \tag{29.7}$$

The components of the 4-vector a^μ in a frame in which the particle is moving with velocity $\mathbf{v} = \mathbf{p}/\varepsilon$ are found by a Lorentz transformation from the rest frame, and are

$$a^0 = \frac{|\mathbf{p}|}{m}\zeta_{\|}, \qquad \mathbf{a}_\perp = \boldsymbol{\zeta}_\perp, \qquad a_{\|} = \frac{\varepsilon}{m}\zeta_{\|}, \tag{29.8}$$

where the suffixes $\|$ and $\perp$ denote the components of the vectors $\boldsymbol{\zeta}$ and $\mathbf{a}$ parallel and perpendicular to the direction of $\mathbf{p}$.† These formulae may be expressed in vector form:

$$\mathbf{a} = \boldsymbol{\zeta} + \frac{\mathbf{p}(\boldsymbol{\zeta}\cdot\mathbf{p})}{m(\varepsilon+m)}, \qquad a^0 = \frac{\mathbf{a}\cdot\mathbf{p}}{\varepsilon} = \frac{\mathbf{p}\cdot\boldsymbol{\zeta}}{m}, \qquad \mathbf{a}^2 = \boldsymbol{\zeta}^2 + \frac{(\mathbf{p}\cdot\boldsymbol{\zeta})^2}{m^2}. \tag{29.9}$$

Let us first consider the unpolarized state ($\boldsymbol{\zeta} = 0$). The density matrix in this case can contain as parameters only the 4-momentum p. The only form for such a matrix which satisfies the equations (29.5) is

$$\rho = \tfrac{1}{2}(\gamma p + m) \tag{29.10}$$

(I. E. Tamm, 1930; H. B. G. Casimir, 1933). The constant coefficient is chosen in accordance with the normalization condition (29.2).

In the general case of partial polarization ($\boldsymbol{\zeta} \neq 0$), we seek the density matrix in the form

$$\rho = \frac{1}{4m}(\gamma p + m)\rho'(\gamma p + m), \tag{29.11}$$

which necessarily satisfies the equations (29.5). When $\boldsymbol{\zeta} = 0$, the auxiliary matrix ρ' must become a unit matrix; since $(\gamma p + m)^2 = 2m(\gamma p + m)$, (29.11) is then the same as (29.10). The matrix ρ' must also contain the 4-vector a linearly as a parameter, i.e. must be of the form

$$\rho' = 1 - A\gamma^5(\gamma a); \tag{29.12}$$

the second term includes the "scalar" product of the pseudovector a and the "matrix 4-pseudovector" $\gamma^5\gamma$. To determine the coefficient A, we write the density

† As regards their transformation properties, the components of the mean spin vector $\bar{\mathbf{s}}$ (like those of any angular momentum) are, in relativistic mechanics, the space components of an antisymmetric tensor $S^{\lambda\mu}$. The 4-vector a^λ is related to this tensor by the equations

$$S^{\lambda\mu} = \frac{1}{2m}e^{\lambda\mu\nu\rho}a_\nu p_\rho, \qquad a^\lambda = -\frac{2}{m}e^{\lambda\mu\nu\rho}S_{\mu\nu}p_\rho.$$

It must be emphasized that, in an arbitrary frame of reference, the spatial part $\mathbf{a}$ of the 4-vector a^λ is not the same as the vector $2\bar{\mathbf{s}}$: we easily see that

$$2\bar{s}_{\|} = \frac{1}{m}(a_{\|}\varepsilon - a^0|\mathbf{p}|) = \zeta_{\|},$$

$$2\bar{\mathbf{s}}_\perp = \frac{\varepsilon}{m}\mathbf{a}_\perp = \frac{\varepsilon}{m}\boldsymbol{\zeta}_\perp.$$

matrix in the rest frame:

$$\rho = \tfrac{1}{4}m(1+\gamma^0)(1+A\gamma^5\boldsymbol{\gamma}\cdot\boldsymbol{\zeta})(1+\gamma^0)$$
$$= \tfrac{1}{2}m(1+\gamma^0)(1+A\gamma^5\boldsymbol{\gamma}\cdot\boldsymbol{\zeta})$$

and calculate the mean spin by (29.4). Using the rules given in §22, we easily find that the only non-zero term in the trace is

$$2\bar{\mathbf{s}} = \frac{1}{2m}\,\mathrm{tr}\,(\rho\gamma^5\boldsymbol{\gamma})$$
$$= -\tfrac{1}{4}A\,\mathrm{tr}\,[(\boldsymbol{\gamma}\cdot\boldsymbol{\zeta})\boldsymbol{\gamma}]$$
$$= A\boldsymbol{\zeta}.$$

Equating this to $\boldsymbol{\zeta}$, we have $A=1$. The final expression for ρ is obtained by substituting (29.12) in (29.11) and interchanging the factors ρ' and $\gamma p+m$; since a and p are orthogonal, the product γp anticommutes with γa:

$$(\gamma a)(\gamma p) = 2ap - (\gamma p)(\gamma a) = -(\gamma p)(\gamma a),$$

and therefore commutes with $\gamma^5(\gamma a)$.

Thus the density matrix of a partially polarized electron is given by the expression

$$\rho = \tfrac{1}{2}(\gamma p + m)(1-\gamma^5(\gamma a)) \tag{29.13}$$

(L. Michel and A. S. Wightman, 1955). If the matrix ρ is known, the 4-vector a which describes the state can be found from

$$a^\mu = \frac{1}{2m}\,\mathrm{tr}\,(\rho\gamma^5\gamma^\mu), \tag{29.14}$$

and the vector $\boldsymbol{\zeta}$ is therefore also known.

The formulae for the density matrix of the positron are entirely analogous to those of the electron. If the positron (with 4-momentum p) were described by a positron amplitude $u_p^{(\mathrm{pos})}$ and by a density matrix $\rho^{(\mathrm{pos})}$ defined on the basis of this amplitude, then there would be no difference from the case of the electron, and the matrix $\rho^{(\mathrm{pos})}$ would be given by the same formula (29.13). But, in actual calculations of cross-sections of scattering processes involving positrons, it is necessary (as we shall see below) to deal not with $u_p^{(\mathrm{pos})}$ but with the "negative frequency" amplitudes u_{-p}. Accordingly, the polarization density matrix (which we denote by $\rho^{(-)}$) must be defined so as to reduce to $u_{-p,i}\bar{u}_{-p,k}$ for a pure state.

According to (26.1), the positron amplitude $u_p^{(\mathrm{pos})} = U_C\bar{u}_{-p}$. Conversely,

$$u_{-p} = U_C\bar{u}_p^{(\mathrm{pos})}, \qquad \bar{u}_{-p} = U_C^+u_p^{(\mathrm{pos})} = u_p^{(\mathrm{pos})}U_C^*;$$

cf. (28.3). If

$$\rho_{ik}^{(-)} = u_{-p,i}\bar{u}_{-p,k}, \qquad \rho_{ik}^{(\mathrm{pos})} = u_{pi}^{(\mathrm{pos})}\bar{u}_{pk}^{(\mathrm{pos})},$$

then these formulae give

$$\rho^{(-)} = U_C\tilde{\rho}^{(\mathrm{pos})}U_C^*. \tag{29.15}$$

Substituting for $\rho^{(\mathrm{pos})}$ the expression (29.13), we obtain, after some simple rearrangements using (26.3) and (26.21),

$$\rho^{(-)} = \tfrac{1}{2}(\gamma p - m)(1 - \gamma^5(\gamma a)) \tag{29.16}$$

In particular, for an unpolarized state

$$\rho^{(-)} = \tfrac{1}{2}(\gamma p - m). \tag{29.17}$$

In referring to positron density matrices below, we shall intend the matrices $\rho^{(-)}$, and the index $(-)$ will be omitted; the matrices $\rho^{(\mathrm{pos})}$ will not be needed in practice.

In various calculations it will often be necessary to average over spin states an expression such as $\bar{u}Fu$ $(\equiv \bar{u}_iF_{ik}u_k)$ where F is a (four-rowed) matrix and u is the bispinor amplitude of a state having a definite 4-momentum p. This averaging is equivalent to replacing the products $u_k\bar{u}_i$ by the density matrix ρ_{ki} of a partially polarized state.

In particular, complete averaging over two independent spin states is equivalent to changing to an unpolarized state, and, by (29.10), we have

$$\tfrac{1}{2}\sum_{\mathrm{polar.}} \bar{u}_pFu_p = \tfrac{1}{2}\,\mathrm{tr}\,(\gamma p + m)F. \tag{29.18}$$

Similarly, for wave functions with negative frequency,

$$\tfrac{1}{2}\sum_{\mathrm{polar.}} \bar{u}_{-p}Fu_{-p} = \tfrac{1}{2}\,\mathrm{tr}\,(\gamma p - m)F. \tag{29.19}$$

If summation over spin states replaces averaging, the result is doubled.

Let us see how the density matrix (29.13) tends to its non-relativistic limit. To do so, we use the rest frame of the electron. In the standard representation of the wave functions, the amplitudes u_p in this frame have two components in the limit, and the density matrix must accordingly have two rows. For, in the rest frame,

$$\rho = \tfrac{1}{2}m(\gamma^0 + 1)(1 - \gamma^5\boldsymbol{\gamma}\cdot\boldsymbol{\zeta}),$$

and we find from the expressions (21.20) and (22.18) for the matrices γ

$$\rho = \begin{pmatrix}\rho_{\text{non-r}} & 0\\ 0 & 0\end{pmatrix}, \qquad \rho_{\text{non-r}} = m(1 + \boldsymbol{\sigma}\cdot\boldsymbol{\zeta}), \tag{29.20}$$

the zeros denoting two-rowed null matrices. If we use the normalization of the density matrix to unity (tr $\rho_{\text{non-r}} = 1$), as is customary in the non-relativistic theory, instead of the normalization to $2m$, then the above expression must be divided by $2m$, giving

$$\tfrac{1}{2}(1 + \boldsymbol{\sigma} \cdot \boldsymbol{\zeta})$$

in agreement with *QM*, (59.4), (59.5).

Similarly, the non-relativistic limit for the positron density matrix is

$$\rho = \begin{pmatrix} 0 & 0 \\ 0 & \rho_{\text{non-r}} \end{pmatrix}, \qquad \rho_{\text{non-r}} = -m(1 + \boldsymbol{\sigma} \cdot \boldsymbol{\zeta}).$$

Finally, there is a simpler expression for the density matrix in the ultra-relativistic case. Putting in (29.8) $|\mathbf{p}| \approx \varepsilon$ (and thereby neglecting small quantities of order $(m/\varepsilon)^2$), substituting these expressions in (29.13) or (29.16), and taking the direction of $\mathbf{p}$ as the x-axis, we can write

$$\rho = \tfrac{1}{2}[\varepsilon(\gamma^0 - \gamma^1) \pm m]\left[1 - \gamma^5\left(\frac{\varepsilon}{m}(\gamma^0 - \gamma^1)\zeta_\parallel + \boldsymbol{\zeta}_\perp \cdot \boldsymbol{\gamma}_\perp\right)\right],$$

where the upper sign refers to the electron and the lower sign to the positron. When the product is expanded, the leading terms cancel, and those of the next order give

$$\rho = \tfrac{1}{2}\varepsilon(\gamma^0 - \gamma^1)[1 - \gamma^5(\pm \zeta_\parallel + \boldsymbol{\zeta}_\perp \cdot \boldsymbol{\gamma}_\perp)]$$

or, again writing $\varepsilon(\gamma^0 - \gamma^1)$ in the form γp,

$$\rho = \tfrac{1}{2}(\gamma p)[1 - \gamma^5(\pm \zeta_\parallel + \boldsymbol{\zeta}_\perp \cdot \boldsymbol{\gamma}_\perp)]. \tag{29.21}$$

This is the required expression for the density matrix in the ultra-relativistic case. It should be noticed that all the components of the polarization vector $\boldsymbol{\zeta}$ appear in this expression equivalently, as terms of the same order of magnitude. It will be recalled that $\zeta_\parallel$ is the component of this vector parallel (if $\zeta_\parallel > 0$) or antiparallel (if $\zeta_\parallel < 0$) to the particle momentum. In particular, for a helicity state of the particle, $\zeta_\parallel = 2\lambda = \pm 1$; the density matrix then has an especially simple form,

$$\rho = \tfrac{1}{2}\gamma p(1 \pm 2\lambda\gamma^5), \tag{29.22}$$

which is, as it should be, the same as the neutrino or antineutrino density matrix, these being particles with zero mass and definite helicity (see §30).

§30. Neutrinos

We have seen in §20 that the necessity of two spinors (ξ and η) to describe a particle with spin $\tfrac{1}{2}$ is due to the mass of the particle. This necessity disappears if the mass is zero. The wave equation which describes such a particle can be derived from a single spinor, say the dotted spinor η:

$$\hat{p}^{\alpha\dot{\beta}}\eta_{\dot{\beta}} = 0, \tag{30.1}$$

or, equivalently,

$$(\hat{p}_0 + \hat{\mathbf{p}}\cdot\boldsymbol{\sigma})\eta = 0. \tag{30.2}$$

It has also been noted in §20 that the wave equation containing the mass m is necessarily symmetrical with respect to inversion (the transformation (20.4)). When the particle is described by a single spinor, this symmetry is lost, but it is not needed, since symmetry with respect to inversion is not a universal property of Nature.

The energy and momentum of a particle with $m = 0$ are related by $\varepsilon = |\mathbf{p}|$. For a plane wave ($\eta_p \sim e^{-ipx}$), equation (30.2) therefore gives

$$(\mathbf{n}\cdot\boldsymbol{\sigma})\eta_p = -\eta_p, \tag{30.3}$$

where $\mathbf{n}$ is a unit vector in the direction of $\mathbf{p}$. A similar equation,

$$(\mathbf{n}\cdot\boldsymbol{\sigma})\eta_{-p} = -\eta_{-p}, \tag{30.4}$$

applies to a wave with "negative frequency" ($\eta_{-p} \sim e^{ipx}$).

The second-quantized ψ-operator is

$$\left.\begin{aligned}\hat{\eta} &= \sum_{\mathbf{p}} (\eta_p \hat{a}_{\mathbf{p}} + \eta_{-p}\hat{b}_{\mathbf{p}}{}^{+}),\\ \hat{\eta}^{+} &= \sum_{\mathbf{p}} (\eta_p^{*}\hat{a}_{\mathbf{p}}^{+} + \eta_{-p}^{*}\hat{b}_{\mathbf{p}}).\end{aligned}\right\} \tag{30.5}$$

Hence it follows, as usual, that η_{-p}^{*} are the wave functions of the antiparticle.

The definition (20.1) of the operators $\hat{p}^{\alpha\dot{\beta}}$ shows that $\hat{p}^{\alpha\dot{\beta}*} = -\hat{p}^{\dot{\alpha}\beta}$. The complex conjugate spinor η^{*} therefore satisfies the equations $\hat{p}^{\dot{\alpha}\beta}\eta_{\dot{\beta}}^{*} = 0$, or, equivalently,

$$\hat{p}_{\dot{\alpha}\beta}\eta^{\dot{\beta}*} = 0.$$

We write $\eta^{\dot{\beta}*} = \xi^{\beta}$, thus expressing the fact that complex conjugation changes the dotted to the undotted spinor. The wave function of the antiparticle then satisfies the equation

$$\hat{p}_{\dot{\alpha}\beta}\xi^{\beta} = 0, \tag{30.6}$$

or

$$(\hat{p}_0 - \hat{\mathbf{p}}\cdot\boldsymbol{\sigma})\xi = 0. \tag{30.7}$$

Hence, for a plane wave,

$$(\mathbf{n}\cdot\boldsymbol{\sigma})\xi_p = \xi_p. \tag{30.8}$$

But $\frac{1}{2}\mathbf{n}\cdot\boldsymbol{\sigma}$ is the operator which projects the spin on the direction of motion. Equations (30.3) and (30.8) therefore signify that states of the particle with a definite momentum are necessarily helicity states, for which the spin component in the direction of motion has a definite value. If the particle spin is opposite to the momentum (helicity $-\frac{1}{2}$), then the antiparticle spin is along the momentum (helicity $+\frac{1}{2}$).

The *neutrinos* which occur in Nature appear to be particles possessing these properties. The particle with helicity $-\frac{1}{2}$ is conventionally called the neutrino, and that with helicity $+\frac{1}{2}$ the antineutrino.†

In connection with the fact that the neutrino states are not degenerate with respect to spin directions, reference should be made to the comment in §8 that a massless particle has only axial symmetry about the direction of the momentum. For a strictly neutral particle (the photon), this symmetry includes both rotations about the axis and reflections in planes passing through the axis. For the neutrino, there is no reflection symmetry, leaving only the group of rotations about the axis, which conserve the angular momentum component along the axis and do not change its sign. The symmetry with respect to reflections exists only if the particle is at the same time replaced by the antiparticle.

It should also be noted that the necessarily longitudinal polarization signifies that the spin of the neutrino cannot be distinguished from its orbital momentum, just as with the necessarily transverse photon fields (§6).

From one spinor η (or ξ), only four bilinear combinations can be constructed, which together form the 4-vector

$$j^\mu = (\eta^*\eta,\ \eta^*\boldsymbol{\sigma}\eta). \tag{30.9}$$

It is easily verified that the equations

$$(\hat{p}_0 + \hat{\mathbf{p}}\cdot\boldsymbol{\sigma})\eta = 0, \qquad \eta^*(\hat{p}_0 - \hat{\mathbf{p}}\cdot\boldsymbol{\sigma}) = 0$$

imply the continuity equation $\partial_\mu j^\mu = 0$, so that j^μ acts as a particle current density 4-vector.

It is convenient to normalize neutrino plane waves in a manner similar to that used in §23 for particles possessing mass:

$$\eta_p = \frac{1}{\sqrt{(2\varepsilon)}}u_p e^{-ipx}, \qquad \eta_{-p} = \frac{1}{\sqrt{(2\varepsilon)}}u_{-p}e^{ipx}, \tag{30.10}$$

the spinor amplitudes being normalized by the invariant condition

$$u^*_{\pm p}(1, \boldsymbol{\sigma})u_{\pm p} = 2(\varepsilon, \mathbf{p}). \tag{30.11}$$

The particle density and particle current density are then $j^0 = 1$, $\mathbf{j} = \mathbf{p}/\varepsilon = \mathbf{n}$.

Since a free neutrino with a given momentum is always completely polarized, there

† The existence of neutrinos was theoretically predicted by W. Pauli (1931) in order to explain the properties of β-decay. Equation (30.1) was first discussed by H. Weyl (1929). The neutrino theory based on these equations was evolved by L. D. Landau, T. D. Lee and C. N. Yang, and A. Salam (1957).

are no states which are mixed (with respect to the spin). It may nevertheless be convienent to define a two-rowed polarization "density matrix" simply as the spinor of rank two

$$\rho_{\dot{\alpha}\beta} = u_{\dot{\alpha}} u^*_{\dot{\beta}}, \tag{30.12}$$

for which tr $\rho = 2\varepsilon$. An expression for this matrix can be obtained by noting that it must satisfy the equations

$$(\varepsilon + \mathbf{p}\cdot\boldsymbol{\sigma})\rho = \rho(\varepsilon + \mathbf{p}\cdot\boldsymbol{\sigma}) = 0.$$

Hence we have

$$\rho = \varepsilon - \mathbf{p}\cdot\boldsymbol{\sigma}. \tag{30.13}$$

In the consideration of various interaction processes, neutrinos may appear together with other particles having spin $\frac{1}{2}$ and possessing mass, which are therefore described by four-component wave functions. In such cases it is convenient to retain uniformity of notation by formally defining for the neutrino also a "bispinor" wave function having two components zero:

$$\psi = \begin{pmatrix} 0 \\ \eta \end{pmatrix}.$$

But this form of ψ is in general not preserved in another (non-spinor) representation. This difficulty can be overcome by noting that in the spinor representation we have identically

$$\frac{1+\gamma^5}{2}\begin{pmatrix} \xi \\ \eta \end{pmatrix} = \begin{pmatrix} 0 \\ \eta \end{pmatrix}, \qquad (\eta^*\ \xi^*)\frac{1-\gamma^5}{2} = (\eta^*\ 0),$$

where ξ is an arbitrary "makeweight" spinor which does not appear in the final result; the matrix γ^5 is given by (22.18). The condition for a true "two-component" neutrino will therefore be maintained when it is described by a four-component ψ in any representation, if ψ is taken to be the solution of Dirac's equation with $m = 0$:

$$(\gamma p)\psi = 0, \tag{30.14}$$

subject to the additional condition $\frac{1}{2}(1+\gamma^5)\psi = \psi$, or

$$\gamma^5\psi = \psi, \tag{30.15}$$

This condition may be taken into account by replacing ψ and $\bar{\psi}$, wherever they would occur, by the following expressions:

$$\psi \to \frac{1+\gamma^5}{2}\psi, \qquad \bar{\psi} \to \bar{\psi}\frac{1-\gamma^5}{2}. \tag{30.16}$$

For example, using these expressions in $\bar{\psi}\gamma^\mu\psi$, the current density 4-vector may be written in the form

$$\begin{aligned} j^\mu &= \tfrac{1}{4}\bar{\psi}(1-\gamma^5)\gamma^\mu(1+\gamma^5)\psi \\ &= \tfrac{1}{2}\bar{\psi}\gamma^\mu(1+\gamma^5)\psi. \end{aligned} \tag{30.17}$$

In the same way, the four-rowed neutrino density matrix becomes

$$\rho = \tfrac{1}{4}(1+\gamma^5)(\gamma p)(1-\gamma^5) = \tfrac{1}{2}(1+\gamma^5)(\gamma p). \tag{30.18}$$

In the spinor representation it reduces, as it should, to the two-rowed matrix (30.13),

$$\rho = \begin{pmatrix} 0 & 0 \\ \varepsilon - \boldsymbol{\sigma}\cdot\mathbf{p} & 0 \end{pmatrix}.$$

The corresponding formulae for the antineutrino differ from those given above by a change of the sign of γ^5.

The neutrino is an electrically neutral particle, but, with the properties described above, it is not a strictly neutral particle. Here it must be noted that the "neutrino field" described by a two-component spinor is equivalent, as regards the number of possible particle states (but not, of course, as regards its other physical properties) to a strictly neutral field described by a four-component bispinor. Instead of states of particles and antiparticles with definite helicity, we should here have the same number of states of one particle with two possible values of the helicity, and the property of symmetry with respect to inversion would automatically occur. However, the zero mass of the "four-component" neutrino would be, so to speak, "accidental", since it would be unrelated to the symmetry properties of the wave equation describing the neutrino (which allows also a non-zero mass). The various interactions of such a particle would therefore necessarily imply the existence of a small but not zero rest mass.

§31. The wave equation for a particle with spin 3/2

A particle with spin 3/2 is described, in its rest frame, by a three-dimensional symmetrical spinor of rank three (having $2s+1=4$ independent components). Correspondingly, in an arbitrary frame of reference, its description may involve the 4-spinors $\xi^{\alpha\dot{\beta}\dot{\gamma}}$, $\eta_{\dot{\alpha}\beta\gamma}$ and $\zeta^{\alpha\beta\gamma}$, $\chi_{\dot{\alpha}\dot{\beta}\dot{\gamma}}$, each of which is symmetrical in all the indices of one kind (dotted or undotted). The spinors in each pair are interchanged by inversion.

In order that the 4-sponors $\xi^{\alpha\dot{\beta}\dot{\gamma}}$ and $\eta_{\dot{\alpha}\beta\gamma}$ should become in the rest frame 3-spinors symmetrical in all three indices, they must satisfy the conditions

$$\hat{p}^{\dot{\alpha}\beta}\eta_{\dot{\alpha}\beta\gamma} = 0, \qquad \hat{p}_{\alpha\dot{\beta}}\xi^{\alpha\dot{\beta}\dot{\gamma}} = 0: \tag{31.1}$$

in the rest frame we have

$$\hat{p}^{\dot{\alpha}\beta} \to \hat{p}_0\delta^\beta_\alpha = m\delta^\beta_\alpha,$$

as is seen from (20.1), and the conditions (31.1) therefore imply the equations

$$\delta^{\beta}_{\alpha}\eta'^{\alpha}{}_{\beta\gamma}=0, \qquad \delta^{\beta}_{\alpha}\xi'^{\alpha}{}_{\beta\gamma}=0,$$

where the primed letters denote the corresponding three-dimensional spinors. Thus these spinors give zero on contraction with respect to the indices α, β, which means that they are symmetrical in these two indices, and hence in all three.

The differential relations between the spinors ξ and η are

$$\left.\begin{aligned} \hat{p}^{\delta\dot{\gamma}}\eta^{\dot{\beta}}_{\alpha\delta} &= m\xi^{\dot{\beta}\dot{\gamma}}_{\alpha}, \\ \hat{p}_{\delta\dot{\gamma}}\xi^{\dot{\beta}\dot{\gamma}}_{\alpha} &= m\eta^{\dot{\beta}}_{\alpha\delta}. \end{aligned}\right\} \tag{31.2}$$

The conditions (31.1) ensure that the left-hand sides of (31.2) vanish on contraction with respect to any other pair of indices, and hence that these quantities are symmetrical in β, $\dot{\gamma}$ or α, δ. In the rest frame, the three-dimensional spinors ξ' and η' are the same according to (31.2). On eliminating η or ξ from (31.2), we find that each component of the spinors ξ and η satisfies the second-order equation

$$(\hat{p}^2-m^2)\xi^{\alpha\dot{\beta}\dot{\gamma}}=0. \tag{31.3}$$

The equations (31.1), (31.2) form a complete set of wave equations for a particle with spin 3/2.† No further results would be obtained by using the spinors ζ and χ. Their structure is as follows:

$$m\zeta^{\alpha\beta\gamma}=\hat{p}^{\alpha\dot{\delta}}\eta^{\beta\gamma}_{\dot{\delta}},$$
$$m\chi_{\dot{\alpha}\dot{\beta}\dot{\gamma}}=\hat{p}_{\dot{\alpha}\delta}\xi^{\delta}_{\dot{\beta}\dot{\gamma}}.$$

The equations of particles with spin 3/2 can also be put in a different form, using the vectorial properties of spinors (W. Rarita and J. Schwinger, 1941; A. S. Davydov and I. E. Tamm, 1942). One four-dimensional vector index μ is assigned to a pair of spinor indices $\alpha\dot{\beta}$. Thus the components $\xi^{\alpha\dot{\beta}\dot{\gamma}}$ of the spinor of rank three can be put into correspondence with the components of the "mixed" quantities $\psi^{\dot{\gamma}}_{\mu}$, which have one vector and one spinor index. Similarly, the spinor $\eta^{\dot{\beta}\alpha\gamma}$ is correlated with ψ^{γ}_{μ}, and the two spinors together correspond to a "vector" bispinor ψ_{μ} (omitting the bispinor index). The wave equation then becomes a "Dirac equation" for each of the vector components ψ_{μ}:

$$(\gamma\hat{p}-m)\psi_{\mu}=0, \tag{31.4}$$

with the added condition

$$\gamma^{\mu}\psi_{\mu}=0. \tag{31.5}$$

Using the expressions for the matrices γ^{μ} in the spinor representation and the

† See the paper by Fierz and Pauli, cited in §15, concerning the Lagrangian formulation of these equations.

relations (18.6), (18.7) between the spinor and vector components, we can easily verify that equation (31.4) implies equations (31.2), and that the condition (31.5) is equivalent to the condition for the spinors $\xi^{\alpha\dot{\beta}\dot{\gamma}}$ and $\eta^{\dot{\alpha}\beta\gamma}$ to be symmetrical in the indices $\dot{\beta}\dot{\gamma}$ or $\beta\gamma$. Multiplying (31.4) by γ^μ and using (31.5), we find

$$\gamma^\mu\gamma^\nu\hat{p}_\nu\psi_\mu = 0,$$

or, with the commutation rules for the matrices γ^μ,

$$2g^{\mu\nu}\hat{p}_\nu\psi_\mu - \gamma^\nu\hat{p}_\nu\gamma^\mu\psi_\mu = 0. \tag{31.6}$$

The second term in turn is zero by (31.5), and the first term gives

$$\hat{p}^\mu\psi_\mu = 0. \tag{31.7}$$

This condition, which is implied by (31.4), (31.5), is easily seen to be equivalent to the conditions (31.1).

Finally, yet another way of expressing the wave equation is to use quantities ψ_{ikl} (i, k, $l = 1, 2, 3, 4$) with three bispinor indices, in which they are symmetrical (V. Bargmann and E. P. Wigner, 1948). The set of these quantities is equivalent to the components of all four spinors ξ, η, ζ, χ. The wave equation becomes a set of "Dirac equations"

$$\hat{p}_\mu\gamma^\mu{}_{im}\psi_{mkl} = m\psi_{ikl}. \tag{31.8}$$

We easily see that these equations yield the necessary number (four) of independent components ψ_{ikl}, and there is therefore no need to impose any further conditions. In the rest frame, (31.8) becomes

$$\gamma^0_{im}\psi_{mkl} = \psi_{ikl},$$

according to which all the components with i, k, $l = 3, 4$ are zero (in the standard representation), i.e. the ψ_{ikl} reduce to the components of a three-dimensional spinor of rank three.

The results given above have an obvious generalization for particles with any half-integral spin s. In the description by equations of the form (31.4), (31.5), the wave function is a symmetrical 4-tensor of rank $\frac{1}{2}(2s-1)$ with one bispinor index. In the description by equations of the form (31.8), the wave function has $2s$ bispinor indices and is symmetrical in these.

CHAPTER IV

PARTICLES IN AN EXTERNAL FIELD

§ 32. **Dirac's equation for an electron in an external field**

THE wave equations of free particles express essentially only those properties which depend on the general requirements of space–time symmetry. Physical processes involving the particles, however, depend on their interaction properties.

The description of the electromagnetic interactions of particles in relativistic quantum theory can be effected by a generalization of the method used in classical non-relativistic quantum theory.

This method, however, is applicable to the description of electromagnetic interactions only of particles that are not capable of strong interactions. These include electrons (and positrons), and the very wide domain of electron quantum electrodynamics is therefore accessible to the existing theory. There are also unstable particles, the muons, which are not capable of strong interactions; they are described by the same quantum electrodynamics as regards phenomena occurring in times short in comparison with their lifetime (with respect to weak interactions).

In this chapter we shall discuss problems of quantum electrodynamics which fall within the scope of single-particle theory. These are problems in which the number of particles is unchanged, and the interaction can be represented in terms of an external electromagnetic field. Besides the conditions which ensure that the external field may be regarded as given, there are conditions arising from "radiative corrections" which also impose limits on the validity of such a theory.

The wave equation of an electron in a given external field can be derived in the same way as in the non-relativistic theory (*QM*, §111). Let $A^{\mu} = (\Phi, \mathbf{A})$ be the 4-potential of the external electromagnetic field ($\mathbf{A}$ being the vector potential and Φ the scalar potential). We obtain the desired equation on replacing the 4-momentum operator $\hat{p}$ in Dirac's equation by $\hat{p} - eA$, where e is the charge on the particle †:

$$[\gamma(\hat{p} - eA) - m]\psi = 0. \tag{32.1}$$

The corresponding Hamiltonian is found by making the same change in (21.13):

$$\hat{H} = \boldsymbol{\alpha} \cdot (\hat{\mathbf{p}} - e\mathbf{A}) + \beta m + e\Phi. \tag{32.2}$$

The invariance of Dirac's equation under gauge transformations of the electromagnetic field potential is shown by the fact that it is unchanged in form if the transformation $A \to A + i\hat{p}\chi$ (where χ is an arbitrary function) is accompanied by

† The charge together with its sign is meant, so that for the electron $e = -|e|$.

the following transformation of the wave function:†

$$\psi \to \psi\, e^{ie\chi}; \tag{32.3}$$

cf. the corresponding transformation for Schrödinger's equation in *QM*, §111.

The current density is expressed in terms of the wave function by the same formula (21.11),

$$j = \bar{\psi}\gamma\psi,$$

as when there is no external field. It is easily seen that, when the calculations used in deriving (21.11) are repeated with the equation (32.1) (and the equation (32.4) below), the external field disappears, and the continuity equation contains the same expression for the current as previously.

Let us apply the operation of charge conjugation to equation (32.1). To do so, we write the equation

$$\bar{\psi}[\gamma(\hat{p} + eA) + m] = 0, \tag{32.4}$$

which is obtained as the complex conjugate of (32.1), in the same way as equation (21.9) was derived in Chapter III, and using the fact that the 4-vector A is real. Putting this equation in the form

$$[\tilde{\gamma}(\hat{p} + eA) + m]\bar{\psi} = 0,$$

multiplying it on the left by the matrix U_C and using the relations (26.3), we find

$$[\gamma(\hat{p} + eA) - m](\hat{C}\psi) = 0. \tag{32.5}$$

Thus the charge-conjugate wave function satisfies an equation which differs from the original equation by a change in the sign of the charge. The operation of charge conjugation, however, corresponds to a change from particles to antiparticles. We see that, if the particles possess an electric charge, the signs of the electron and positron charges are necessarily opposite.

The first-order equations (32.1) can be transformed into second-order equations by applying to them the operator $\gamma(\hat{p} - eA) + m$:

$$[\gamma^\mu\gamma^\nu(\hat{p}_\mu - eA_\mu)(\hat{p}_\nu - eA_\nu) - m^2]\psi = 0.$$

The product $\gamma^\mu\gamma^\nu$ may be written as follows:

$$\begin{aligned}\gamma^\mu\gamma^\nu &= \tfrac{1}{2}(\gamma^\mu\gamma^\nu + \gamma^\nu\gamma^\mu) + \tfrac{1}{2}(\gamma^\mu\gamma^\nu - \gamma^\nu\gamma^\mu)\\ &= g^{\mu\nu} + \sigma^{\mu\nu},\end{aligned}$$

† The transformation (32.3) with a function $\chi(t, \mathbf{r})$ is sometimes called a *local gauge transformation*, in contrast to the *global gauge transformation* (12.10) with a constant phase α.

where $\sigma^{\mu\nu}$ is the antisymmetric "matrix 4-tensor" (28.2). On multiplying by $\sigma^{\mu\nu}$ we can antisymmetrize by the substitution

$$\begin{aligned}(\hat{p}_\mu - eA_\mu)(\hat{p}_\nu - eA_\nu) &\to \tfrac{1}{2}\{(\hat{p}_\mu - eA_\mu)(\hat{p}_\nu - eA_\nu)\}_- \\ &= \tfrac{1}{2}e(-A_\mu\hat{p}_\nu + \hat{p}_\nu A_\mu - \hat{p}_\mu A_\nu + A_\nu\hat{p}_\mu) \\ &= \tfrac{1}{2}ie(\partial_\nu A_\mu - \partial_\mu A_\nu) \\ &= -\tfrac{1}{2}ieF_{\mu\nu}\end{aligned}$$

(with $F_{\mu\nu} = \partial_\mu A_\nu - \partial_\nu A_\mu$ the electromagnetic field tensor). The result is the second-order equation

$$[(\hat{p} - eA)^2 - m^2 - \tfrac{1}{2}ieF_{\mu\nu}\sigma^{\mu\nu}]\psi = 0. \tag{32.6}$$

The product $F_{\mu\nu}\sigma^{\mu\nu}$ may be written in three-dimensional form in terms of the components

$$\sigma^{\mu\nu} = (\boldsymbol{\alpha}, i\boldsymbol{\Sigma}), \qquad F^{\mu\nu} = (-\mathbf{E}, \mathbf{H}).$$

Then

$$[(\hat{p} - eA)^2 - m^2 + e\boldsymbol{\Sigma}\cdot\mathbf{H} - ie\boldsymbol{\alpha}\cdot\mathbf{E}]\psi = 0, \tag{32.7}$$

or, in ordinary units,

$$\left[\left(\frac{i\hbar}{c}\frac{\partial}{\partial t} - \frac{e}{c}\Phi\right)^2 - \left(i\hbar\nabla + \frac{e}{c}\mathbf{A}\right)^2 - m^2c^2 + \frac{e\hbar}{c}\boldsymbol{\Sigma}\cdot\mathbf{H} - i\frac{e\hbar}{c}\boldsymbol{\alpha}\cdot\mathbf{E}\right]\psi = 0. \tag{32.7a}$$

The occurrence in these equations of terms in the fields **E** and **H** is due to the spin of the particle, and will be further discussed in §33.

The solutions of the second-order equation include, of course, "redundant" solutions which do not satisfy the original first-order equation (32.1), being solutions of that equation with the opposite sign of m. The choice of the appropriate solutions in particular cases is usually obvious and causes no difficulty. The customary procedure is that, if ϕ is any solution of the second-order equation, then a solution of the correct first-order equation is

$$\psi = [\gamma(\hat{p} - eA) + m]\phi. \tag{32.8}$$

For, on multiplying this equation by $\gamma(\hat{p} - eA) - m$, we see that the right-hand side vanishes if ϕ satisfies (32.6).

It should be emphasized that the introduction of the external field into the relativistic wave equation by replacing $\hat{p}$ by $\hat{p} - eA$ is not self-evident. In doing so, we have essentially made use of a further principle: this substitution must be applied to first-order equations. For this reason equation (32.6) contains additional terms which would not appear if the substitution were made directly in the second-order equation.

The stationary-state solutions of Dirac's equation in an external field may include states of both the continuous spectrum and the discrete spectrum. As in the non-relativistic theory, states of the continuous spectrum correspond to infinite motion, in which the particle can be at infinity; it may there be regarded as a free particle. Since the eigenvalues of the Hamiltonian of a free particle are $\pm\sqrt{(\mathbf{p}^2+m^2)}$, it is clear that the continuous spectrum of energy eigenvalues is in the ranges $\varepsilon \geqslant m$ and $\varepsilon \leqslant -m$. If $-m < \varepsilon < m$, the particle cannot be at infinity, and the motion is therefore finite and the state belongs to the discrete spectrum.

As in the case of free particles, the wave functions with "positive frequency" ($\varepsilon > 0$) and with "negative frequency" ($\varepsilon < 0$) appear in a definite manner in the second quantization procedure. For particles in an external field, there is a natural generalization of this procedure, the plane waves in formulae (25.1) being replaced by appropriately normalized eigenfunctions $\psi_n^{(+)}$ and $\psi_n^{(-)}$ of Dirac's equation, corresponding to positive and negative frequencies ($\varepsilon_n^{(+)}$ and $-\varepsilon_n^{(-)}$):

$$\left.\begin{aligned} \hat{\psi} &= \sum_n \{\hat{a}_n \psi_n^{(+)} e^{-i\varepsilon_n^{(+)}t} + \hat{b}_n^{+} \psi_n^{(-)} e^{i\varepsilon_n^{(-)}t}\}, \\ \hat{\bar{\psi}} &= \sum_n \{\hat{a}_n^{+} \bar{\psi}_n^{(+)} e^{i\varepsilon_n^{(+)}t} + \hat{b}_n \bar{\psi}_n^{(-)} e^{-i\varepsilon_n^{(-)}t}\}. \end{aligned}\right\} \tag{32.9}$$

However, as the potential well becomes deeper, the energy levels may cross the boundary $\varepsilon = 0$, so that positive levels become negative (or vice versa when the potential has the opposite sign). Nevertheless, for the sake of continuity we must still regard these as electron (not positron) levels. That is, the electron states are to be regarded as all those which approach the positive limit of the continuous spectrum ($\varepsilon = m$) when the field is removed with infinite slowness.

Although Dirac's equation for an electron in an external field does, as already mentioned, yield solutions for a large class of problems in quantum electrodynamics, we must at the same time emphasize that the applicability of the concept of an external field in the one-particle problem of relativistic theory is nevertheless restricted, because of the spontaneous formation of electron–positron pairs in sufficiently strong fields (see §§35 and 36).

We shall not deal in this book with the inclusion of an external field in the wave equations for particles with spin other than $\frac{1}{2}$, since the topic has no immediate physical significance: actual particles having such spins are hadrons, and their electromagnetic interactions cannot be described by wave equations.

PROBLEM

Determine the electron energy levels in a constant magnetic field.

SOLUTION. The vector potential is $A_x = A_z = 0$, $A_y = Hx$ (the field H being along the z-axis). The components p_y, p_z of the generalized momentum (as well as the energy) are conserved.

We use the second-order equation for the auxiliary function ϕ (see (32.8)), and assume that ϕ is an eigenfunction of the operator Σ_z (with eigenvalue $\sigma = \pm 1$) and of the operators $\hat{p}_y$, $\hat{p}_z$. The equation for ϕ is

$$\left\{-\frac{d^2}{dx^2} + (eHx - p_y)^2 - eH\sigma\right\}\phi = (\varepsilon^2 - m^2 - p_z^2)\phi.$$

This equation is the same in form as Schrödinger's equation for a linear oscillator. The eigenvalues of ε are given by

$$\varepsilon^2 - m^2 - p_z^2 = |e|H(2n+1) - eH\sigma, \quad n = 0, 1, 2, \ldots;$$

cf. *QM*, §112. The wave function ψ, which is to be determined from ϕ according to (32.8), is not an eigenfunction of the operator Σ_z, in accordance with the fact that the spin is not conserved for a particle in motion.

§33. Expansion in powers of $1/c$†

We have seen in §21 that, in the non-relativistic limit ($v \to 0$),

two components (χ) of the bispinor $\psi = \begin{pmatrix}\phi\\ \chi\end{pmatrix}$ vanish.

Hence $\chi \ll \phi$ when the electron velocity is small. This leads to an approximate equation involving only the two-component quantity ϕ, obtained by a formal expansion of the wave function in powers of $1/c$.

Dirac's equation for an electron in an external field may be written

$$i\hbar \frac{\partial \psi}{\partial t} = \left\{c\boldsymbol{\alpha}\cdot\left(\hat{\mathbf{p}} - \frac{e}{c}\mathbf{A}\right) + \beta mc^2 + e\Phi\right\}\psi. \tag{33.1}$$

The relativistic energy of the particle includes also its rest energy mc^2. This must be excluded in arriving at the non-relativistic approximation, and we therefore replace ψ by a function ψ' defined as follows:

$$\psi = \psi' e^{-imc^2t/\hbar}.$$

Then

$$\left(i\hbar\frac{\partial}{\partial t} + mc^2\right)\psi' = \left\{c\boldsymbol{\alpha}\cdot\left(\hat{\mathbf{p}} - \frac{e}{c}\mathbf{A}\right) + \beta mc^2 + e\Phi\right\}\psi'.$$

Substituting $\psi' = \begin{pmatrix}\phi'\\ \chi'\end{pmatrix}$, we obtain the equations

$$\left(i\hbar\frac{\partial}{\partial t} - e\Phi\right)\phi' = c\boldsymbol{\sigma}\cdot\left(\hat{\mathbf{p}} - \frac{e}{c}\mathbf{A}\right)\chi', \tag{33.2}$$

$$\left(i\hbar\frac{\partial}{\partial t} - e\Phi + 2mc^2\right)\chi' = c\boldsymbol{\sigma}\cdot\left(\hat{\mathbf{p}} - \frac{e}{c}\mathbf{A}\right)\phi'. \tag{33.3}$$

The primes to ϕ and χ will henceforth be omitted; this will cause no misunderstanding, since only the transformed function ψ' is used in the present section.

† In this section, ordinary units are used.

In the first approximation, only the term $2mc^2\chi$ is retained on the left-hand side of (33.3), which gives

$$\chi = \frac{1}{2mc}\boldsymbol{\sigma}\cdot\left(\hat{\mathbf{p}} - \frac{e}{c}\mathbf{A}\right)\phi \tag{33.4}$$

(thus $\chi \sim \phi/c$). Substitution of this in (33.2) gives

$$\left(i\hbar\frac{\partial}{\partial t} - e\Phi\right)\phi = \frac{1}{2m}\left(\boldsymbol{\sigma}\cdot\left(\hat{\mathbf{p}} - \frac{e}{c}\mathbf{A}\right)\right)^2\phi.$$

For the Pauli matrices we have the relation

$$(\boldsymbol{\sigma}\cdot\mathbf{a})(\boldsymbol{\sigma}\cdot\mathbf{b}) = \mathbf{a}\cdot\mathbf{b} + i\boldsymbol{\sigma}\cdot\mathbf{a}\times\mathbf{b}, \tag{33.5}$$

where $\mathbf{a}$ and $\mathbf{b}$ are arbitrary vectors (see (20.9)). In the present case, $\mathbf{a} = \mathbf{b} = \hat{\mathbf{p}} - e\mathbf{A}/c$, but the vector product $\mathbf{a}\times\mathbf{b}$ is not zero, since $\hat{\mathbf{p}}$ and $\mathbf{A}$ do not commute:

$$\left[\left(\hat{\mathbf{p}} - \frac{e}{c}\mathbf{A}\right)\times\left(\hat{\mathbf{p}} - \frac{e}{c}\mathbf{A}\right)\right]\phi = i\frac{e\hbar}{c}\{\mathbf{A}\times\nabla + \nabla\times\mathbf{A}\}\phi$$

$$= i\frac{e\hbar}{c}\,\mathrm{curl}\,\mathbf{A}\cdot\phi.$$

Thus

$$\left(\boldsymbol{\sigma}\cdot\left(\hat{\mathbf{p}} - \frac{e}{c}\mathbf{A}\right)\right)^2 = \left(\hat{\mathbf{p}} - \frac{e}{c}\mathbf{A}\right)^2 - \frac{e\hbar}{c}\boldsymbol{\sigma}\cdot\mathbf{H}, \tag{33.6}$$

where $\mathbf{H} = \mathrm{curl}\,\mathbf{A}$ is the magnetic field, and for ϕ we obtain the equation

$$i\hbar\frac{\partial\phi}{\partial t} = \hat{H}\phi = \left[\frac{1}{2m}\left(\hat{\mathbf{p}} - \frac{e}{c}\mathbf{A}\right)^2 + e\Phi - \frac{e\hbar}{2mc}\boldsymbol{\sigma}\cdot\mathbf{H}\right]\phi. \tag{33.7}$$

This is *Pauli's equation*. It differs from the non-relativistic Schrödinger's equation by the last term in the Hamiltonian, which has the form of the potential energy of a magnetic dipole in the external field (cf. *QM*, §111). Thus, in the first approximation (with respect to $1/c$), the electron behaves as a particle having both a charge and a magnetic moment

$$\boldsymbol{\mu} = \frac{e}{mc}\hbar\mathbf{s}. \tag{33.8}$$

The gyromagnetic ratio e/mc is twice its value for a magnetic moment due to orbital motion.†

† This remarkable result was first derived by P. A. M. Dirac in 1928. The two-component wave function satisfying equation (33.7) was introduced by W. Pauli (1927), before Dirac's discovery of his equation.

The density $\rho = \psi^*\psi = \phi^*\phi + \chi^*\chi$. The second term must be omitted in the first approximation, so that $\rho = |\phi|^2$, as would be expected for Schrödinger's equation.

The current density is

$$\begin{aligned}\mathbf{j} &= c\psi^*\boldsymbol{\alpha}\psi \\ &= c(\phi^*\boldsymbol{\sigma}\chi + \chi^*\boldsymbol{\sigma}\phi).\end{aligned}$$

We substitute here, from (33.4),

$$\chi = \frac{1}{2mc}\boldsymbol{\sigma}\cdot\left(-i\hbar\nabla - \frac{e}{c}\mathbf{A}\right)\phi,$$

$$\chi^* = \frac{1}{2mc}\left(i\hbar\nabla - \frac{e}{c}\mathbf{A}\right)\phi^*\cdot\boldsymbol{\sigma},$$

and transform the products containing two factors $\boldsymbol{\sigma}$ by means of (33.5) in the form

$$(\boldsymbol{\sigma}\cdot\mathbf{a})\boldsymbol{\sigma} = \mathbf{a} + i\boldsymbol{\sigma}\times\mathbf{a}, \qquad \boldsymbol{\sigma}(\boldsymbol{\sigma}\cdot\mathbf{a}) = \mathbf{a} + i\mathbf{a}\times\boldsymbol{\sigma}. \tag{33.9}$$

The result is

$$\mathbf{j} = \frac{i\hbar}{2m}(\phi\nabla\phi^* - \phi^*\nabla\phi) - \frac{e}{mc}\mathbf{A}\phi^*\phi + \frac{\hbar}{2m}\operatorname{curl}(\phi^*\boldsymbol{\sigma}\phi), \tag{33.10}$$

in agreement with the expression in the non-relativistic theory, *QM* (115.4).

Let us now derive† the second approximation, continuing the expansion as far as terms of order $1/c^2$, and assuming that there is only an electric external field ($\mathbf{A} = 0$).

First, we note that, when terms $\sim 1/c^2$ are included, the density is

$$\rho = |\phi|^2 + |\chi|^2 = |\phi|^2 + \frac{\hbar^2}{4m^2c^2}|\boldsymbol{\sigma}\cdot\nabla\phi|^2.$$

This differs from the Schrödinger expression. In order to find (in the second approximation) the wave equation corresponding to Schrödinger's equation, we must replace ϕ by another (two-component) function ϕ_{Sch}, for which the time-independent integral would be of the form $\int|\phi_{\text{Sch}}|^2\,d^3x$, as it should be for Schrödinger's equation.

To obtain the required transformation, we write the condition

$$\int \phi^*_{\text{Sch}}\phi_{\text{Sch}}\,d^3x = \int\left\{\phi^*\phi + \frac{\hbar^2}{4m^2c^2}(\nabla\phi^*\cdot\boldsymbol{\sigma})(\boldsymbol{\sigma}\cdot\nabla\phi)\right\}d^3x$$

† The method used here is due to V. B. Berestetskiĭ and L. D. Landau (1949).

and integrate by parts:

$$\int (\nabla\phi^* \cdot \boldsymbol{\sigma})(\boldsymbol{\sigma}\cdot\nabla\phi)\, d^3x = -\int \phi^*(\boldsymbol{\sigma}\cdot\nabla)(\boldsymbol{\sigma}\cdot\nabla)\phi\, d^3x$$
$$= -\int \phi^*\triangle\phi\, d^3x$$

(or the same with ϕ and ϕ^* interchanged). Thus

$$\int \phi^*_{\text{Sch}}\phi_{\text{Sch}}\, d^3x = \int \left\{\phi^*\phi - \frac{\hbar^2}{8m^2c^2}(\phi^*\triangle\phi + \phi\triangle\phi^*)\right\} d^3x,$$

whence it is evident that

$$\phi_{\text{Sch}} = \left(1 + \frac{\hat{\mathbf{p}}^2}{8m^2c^2}\right)\phi, \qquad \phi = \left(1 - \frac{\hat{\mathbf{p}}^2}{8m^2c^2}\right)\phi_{\text{Sch}}. \tag{33.11}$$

To simplify the notation we shall assume a stationary state, replacing the operator $i\hbar\,\partial/\partial t$ by the energy ε (with the rest energy subtracted). In the next approximation after (33.4) we have from (33.3)

$$\chi = \frac{1}{2mc}\left(1 - \frac{\varepsilon - e\Phi}{2mc^2}\right)(\boldsymbol{\sigma}\cdot\hat{\mathbf{p}})\phi.$$

This is to be substituted in (33.2) and ϕ then replaced by ϕ_{Sch} according to (33.11), omitting all terms of higher order than $1/c^2$. A simple calculation leads to an equation for ϕ_{Sch} in the form $\varepsilon\phi_{\text{Sch}} = \hat{H}\phi_{\text{Sch}}$, where the Hamiltonian is

$$\hat{H} = \frac{\hat{\mathbf{p}}^2}{2m} + e\Phi - \frac{\hat{\mathbf{p}}^4}{8m^3c^2} + \frac{e}{4m^2c^2}\{(\boldsymbol{\sigma}\cdot\hat{\mathbf{p}})\Phi(\boldsymbol{\sigma}\cdot\hat{\mathbf{p}}) - \tfrac{1}{2}(\hat{\mathbf{p}}^2\Phi + \Phi\hat{\mathbf{p}}^2)\}.$$

The expression in the braces is transformed by means of the formulae

$$(\boldsymbol{\sigma}\cdot\hat{\mathbf{p}})\Phi(\boldsymbol{\sigma}\cdot\hat{\mathbf{p}}) = \Phi\hat{\mathbf{p}}^2 + (\boldsymbol{\sigma}\cdot\hat{\mathbf{p}}\Phi)(\boldsymbol{\sigma}\cdot\hat{\mathbf{p}})$$
$$= \Phi\hat{\mathbf{p}}^2 + i\hbar(\boldsymbol{\sigma}\cdot\mathbf{E})(\boldsymbol{\sigma}\cdot\hat{\mathbf{p}}),$$
$$\hat{\mathbf{p}}^2\Phi - \Phi\hat{\mathbf{p}}^2 = -\hbar^2\triangle\Phi + 2i\hbar\mathbf{E}\cdot\hat{\mathbf{p}},$$

where $\mathbf{E} = -\nabla\Phi$ is the electric field. The final expression for the Hamiltonian is

$$\hat{H} = \frac{\hat{\mathbf{p}}^2}{2m} + e\Phi - \frac{\hat{\mathbf{p}}^4}{8m^3c^2} - \frac{e\hbar}{4m^2c^2}\boldsymbol{\sigma}\cdot\mathbf{E}\times\hat{\mathbf{p}} - \frac{e\hbar^2}{8m^2c^2}\,\text{div}\,\mathbf{E}. \tag{33.12}$$

The last three terms are the required corrections of order $1/c^2$. The first of these three terms is due to the relativistic dependence of the kinetic energy on the momentum (the expansion of the difference $c\sqrt{(\mathbf{p}^2 + m^2c^2)} - mc^2$). The second, which may be called the *spin–orbit interaction* energy, is the energy of the interaction of the moving magnetic moment with the electric field.† The last is zero except at points where there are charges creating the external field (for instance, in the Coulomb field of a point charge Ze, $\triangle\Phi = -4\pi Ze\delta(\mathbf{r})$) (C. G. Darwin, 1928).

If the electric field is centrally symmetric, then

$$\mathbf{E} = -\frac{\mathbf{r}}{r}\frac{d\Phi}{dr},$$

and the spin–orbit interaction operator can be put in the form

$$\frac{e\hbar}{4m^2c^2r}\boldsymbol{\sigma}\cdot\mathbf{r}\times\hat{\mathbf{p}}\frac{d\Phi}{dr} = \frac{\hbar^2}{2m^2c^2r}\frac{dU}{dr}\hat{\mathbf{l}}\cdot\hat{\mathbf{s}}, \tag{33.13}$$

where $\hat{\mathbf{l}}$ is the orbital angular momentum operator, $\hat{\mathbf{s}} = \frac{1}{2}\boldsymbol{\sigma}$ is the electron spin operator, and $U = e\Phi$ is the potential energy of the electron in the field.

§ 34. Fine structure of levels of the hydrogen atom

Let us determine the relativistic corrections to the energy levels of the hydrogen atom—an electron in the Coulomb field of a fixed nucleus.‡ The velocity of the electron in the hydrogen atom is $v/c \sim \alpha \ll 1$. The required corrections can therefore be calculated by the use of perturbation theory, averaging over the unperturbed state (i.e. over the non-relativistic wave function) the relativistic terms in the approximate Hamiltonian (33.12). For somewhat greater generality we shall take the charge of the nucleus as Ze, assuming, however, that $Z\alpha \ll 1$.

The field of the nucleus is $\mathbf{E} = Ze\mathbf{r}/r^3$, and its potential satisfies the equation $\triangle\Phi = -4\pi Ze\delta(\mathbf{r})$. Substituting this in the last three terms in (33.12) and using the fact that the electron charge is negative, we obtain the perturbation operator

$$\hat{V} = -\frac{\hat{\mathbf{p}}^4}{8m^3} + \frac{Z\alpha}{2r^3m^2}\hat{\mathbf{l}}\cdot\hat{\mathbf{s}} + \frac{Z\alpha\pi}{2m^2}\delta(\mathbf{r}). \tag{34.1}$$

Since, according to the non-relativistic Schrödinger's equation,

$$\hat{\mathbf{p}}^2\psi = 2m\left(\varepsilon_0 + \frac{Z\alpha}{r}\right)\psi$$

(where $\varepsilon_0 = -mZ^2\alpha^2/2n^2$ is the unperturbed level and n the principal quantum

† With the magnetic moment (33.8) and the velocity $\mathbf{v} = \mathbf{p}/m$, this energy becomes $-\boldsymbol{\mu}\cdot\mathbf{E}\times\mathbf{v}/2c$. At first sight this result may appear unlikely, because on changing to a frame of reference fixed to the particle there arises a magnetic field $\mathbf{H} = \mathbf{E}\times\mathbf{v}/c$, in which the energy of the magnetic moment should be $-\boldsymbol{\mu}\cdot\mathbf{H}$. The occurrence of the factor $\frac{1}{2}$ (the "Thomas half"; L. H. Thomas, 1926) is due to the general requirements of relativistic invariance together with the particular properties of the electron as a "spinor" particle with the corresponding value of the gyromagnetic ratio (see §41).

‡ The effect of the motion of the nucleus on these corrections is a quantity of a higher order of smallness, which will not be considered here.

number), the mean value is

$$\overline{\mathbf{p}^4} = 4m^2\overline{\left(\varepsilon_0 + \frac{Z\alpha}{r}\right)^2}.$$

This quantity, like the mean value of the second term in (34.1), is calculated by means of the formulae (see *QM*, §36)

$$\left.\begin{aligned}\overline{r^{-1}} &= m\alpha Z/n^2,\\ \overline{r^{-2}} &= (m\alpha Z)^2/n^3(l+\tfrac{1}{2}),\\ \overline{r^{-3}} &= (m\alpha Z)^3/n^3 l(l+\tfrac{1}{2})(l+1),\end{aligned}\right\} \tag{34.2}$$

the last of which is valid if $l \neq 0$; the eigenvalue is

$$\begin{aligned}\mathbf{l}\cdot\mathbf{s} &= \tfrac{1}{2}[j(j+1) - l(l+1) - \tfrac{3}{4}] \quad &&\text{if } l \neq 0,\\ \mathbf{l}\cdot\mathbf{s} &= 0 &&\text{if } l = 0.\end{aligned}$$

Finally, the third term is averaged by means of the formulae

$$\left.\begin{aligned}\psi(0) &= \frac{1}{\sqrt{\pi}}\left(\frac{Z\alpha m}{n}\right)^{3/2}, \quad && l = 0;\\ \psi(0) &= 0, && l \neq 0.\end{aligned}\right\} \tag{34.3}$$

The result of a simple calculation using the above formulae may be written for all cases (for all j and l) as

$$\Delta\varepsilon = -\frac{m(Z\alpha)^4}{2n^3}\left(\frac{1}{j+\frac{1}{2}} - \frac{3}{4n}\right). \tag{34.4}$$

Formula (34.4) gives the required relativistic correction to the energy of the hydrogen levels; that is, it gives the fine-structure energy.† In the non-relativistic theory, there is both degeneracy with respect to spin direction and Coulomb degeneracy with respect to l. The fine structure (spin–orbit interaction) removes this degeneracy, but not completely: there remain levels with mutual double degeneracy, having the same n and j but different $l = j \pm \frac{1}{2}$. The levels with the maximum possible value for j for a given n,

$$j_{\max} = l_{\max} + \tfrac{1}{2} = n - \tfrac{1}{2},$$

are then not degenerate. Thus the sequence of hydrogen levels, with allowance for

† This formula, and the more exact formula (36.10), were first derived by A. Sommerfeld from the old Bohr theory before the development of quantum mechanics.

the fine structure, is

$$
\begin{array}{l}
1s_{1/2};\\
\underbrace{2s_{1/2}, 2p_{1/2}}, 2p_{3/2};\\
\underbrace{3s_{1/2}, 3p_{1/2}}, \underbrace{3p_{3/2}, 3d_{3/2}}, 3d_{5/2};\\
\cdot\ \cdot\ \cdot\ \cdot\ \cdot\ \cdot
\end{array}
$$

The level with principal quantum number n is split into n fine-structure components.

In non-relativistic mechanics, the "accidental" degeneracy of the energy levels in a Coulomb field is due to the existence of a conservation law peculiar to this field, relating to a quantity **A** whose operator is

$$\hat{\mathbf{A}} = \frac{\mathbf{r}}{r} + \frac{1}{2m\alpha}(\hat{\mathbf{l}} \times \hat{\mathbf{p}} - \hat{\mathbf{p}} \times \hat{\mathbf{l}});$$

see *QM*, (36.30). This specific conservation law is also responsible for the twofold degeneracy which remains in the relativistic case: the Hamiltonian $\hat{H} = \boldsymbol{\alpha} \cdot \hat{\mathbf{p}} + \beta m - e^2/r$ of Dirac's equation commutes with the operator

$$\hat{I} = \frac{\mathbf{r}}{r} \cdot \boldsymbol{\Sigma} + \frac{i}{m\alpha}\beta(\boldsymbol{\Sigma} \cdot \hat{\mathbf{l}} + 1)\gamma^5(\hat{H} - m\beta)$$

(M. H. Johnson and B. A. Lippmann, 1950). In the non-relativistic limit, $\hat{I} \to \boldsymbol{\Sigma} \cdot \hat{\mathbf{A}}$.

We shall see later (§123) that this remaining degeneracy is removed by "radiative corrections" (the *Lamb shift*), which are neglected in Dirac's equation for the single-electron problem.

To anticipate, it may be mentioned here that the order of magnitude of these corrections is $mZ^4\alpha^5 \log(1/\alpha)$. The second-order spin–orbit interaction correction would be $\sim m(Z\alpha)^6$, and the ratio of this to the radiative corrections is therefore $\sim Z^2\alpha/\log(1/\alpha)$. For hydrogen ($Z = 1$), this ratio is certainly small, and the exact solution of Dirac's equation is therefore of no significance in that case, but it may be significant as regards the electron energy levels in the field of a nucleus with large Z (§36).

§35. Motion in a centrally symmetric field

Let us now consider the motion of an electron in a centrally symmetric electric field.

Since the angular momentum and the parity (relative to the centre of the field, which is taken as the origin) are conserved in a central field, the discussion in §24 regarding spherical waves of free particles is entirely applicable to the angular dependence of the wave functions of such a motion; only the radial functions vary. Accordingly, we shall seek the wave function of the stationary states (in the standard representation) in the form

$$\psi = \begin{pmatrix} \phi \\ \chi \end{pmatrix} = \begin{pmatrix} f(r)\Omega_{jlm} \\ (-1)^{\frac{1}{2}(1+l-l')} g(r)\Omega_{jl'm} \end{pmatrix}, \tag{35.1}$$

where $l = j \pm \frac{1}{2}$, $l' = 2j - l$, and the power of -1 is chosen for subsequent convenience.

Dirac's equation in the standard representation yields the following equations for ϕ and χ:

$$\left.\begin{aligned} (\varepsilon - m - U)\phi &= \boldsymbol{\sigma} \cdot \hat{\mathbf{p}}\chi, \\ (\varepsilon + m - U)\chi &= \boldsymbol{\sigma} \cdot \hat{\mathbf{p}}\phi, \end{aligned}\right\} \tag{35.2}$$

where $U(r) = e\Phi(r)$ is the potential energy of the electron in the field. The result of substituting the expressions (35.1) is calculated by evaluating the right-hand sides of these equations.

Expressing the spherical harmonic spinor $\Omega_{jl'm}$ in terms of Ω_{jlm} by

$$\Omega_{jl'm} = i^{l-l'}\left(\boldsymbol{\sigma} \cdot \frac{\mathbf{r}}{r}\right)\Omega_{jlm}$$

(see (24.8)), we can write

$$(\boldsymbol{\sigma} \cdot \hat{\mathbf{p}})\chi = -i(\boldsymbol{\sigma} \cdot \hat{\mathbf{p}})(\boldsymbol{\sigma} \cdot \mathbf{r})\frac{g}{r}\Omega_{jlm}.$$

Now transforming the product $(\boldsymbol{\sigma} \cdot \hat{\mathbf{p}})(\boldsymbol{\sigma} \cdot \mathbf{r})$ by means of the formula (33.5), and expanding the vector operators, we have

$$\begin{aligned} (\boldsymbol{\sigma} \cdot \hat{\mathbf{p}})\chi &= -i\{\hat{\mathbf{p}} \cdot \mathbf{r} + i\boldsymbol{\sigma} \cdot \hat{\mathbf{p}} \times \mathbf{r}\}\frac{g}{r}\Omega_{jlm} \\ &= \{-\operatorname{div} \mathbf{r} - (\mathbf{r} \cdot \nabla) - \boldsymbol{\sigma} \cdot \mathbf{r} \times \hat{\mathbf{p}}\}\frac{g}{r}\Omega_{jlm} \\ &= -\left\{g' + \frac{2}{r}g + \frac{g}{r}\boldsymbol{\sigma} \cdot \hat{\mathbf{l}}\right\}\Omega_{jlm}, \end{aligned}$$

where $\hat{\mathbf{l}} = \mathbf{r} \times \hat{\mathbf{p}}$ is the orbital angular momentum operator; the prime denotes differentiation with respect to r. The eigenvalues of the product $\boldsymbol{\sigma} \cdot \hat{\mathbf{l}} = 2\hat{\mathbf{l}} \cdot \hat{\mathbf{s}}$ are

$$\begin{aligned} 2\mathbf{l} \cdot \mathbf{s} &= \mathbf{j}^2 - \mathbf{l}^2 - \mathbf{s}^2 \\ &= j(j+1) - l(l+1) - \tfrac{3}{4} \\ &= \begin{cases} j - \frac{1}{2} & \text{if } l = j - \frac{1}{2}, \\ -j - \frac{3}{2} & \text{if } l = j + \frac{1}{2}. \end{cases} \end{aligned}$$

In order to be able to use the same formulae for both cases ($l = j \pm \frac{1}{2}$), it is convenient to write

$$\left.\begin{aligned}\kappa &= -(j+\tfrac{1}{2}) = -(l+1) \quad \text{for } j = l+\tfrac{1}{2},\\ &= +(j+\tfrac{1}{2}) = l \qquad\qquad \text{for } j = l-\tfrac{1}{2}.\end{aligned}\right\} \tag{35.3}$$

The number κ takes all integral values except zero, the positive numbers corresponding to the case $j = l - \frac{1}{2}$, and the negative numbers to the case $j = l + \frac{1}{2}$. Then $\mathbf{l}\cdot\boldsymbol{\sigma} = -(1+\kappa)$, and

$$(\boldsymbol{\sigma}\cdot\hat{\mathbf{p}})\chi = -\left(g' + \frac{1-\kappa}{r}g\right)\Omega_{jlm}.$$

When this expression is substituted in the first equation (35.2), the spherical harmonic spinor Ω_{jlm} cancels from the two sides. Proceeding similarly with the second equation, we finally obtain the following equations for the radial functions:

$$\left.\begin{aligned}f' + \frac{1+\kappa}{r}f - (\varepsilon + m - U)g = 0,\\ g' + \frac{1-\kappa}{r}g + (\varepsilon - m - U)f = 0,\end{aligned}\right\} \tag{35.4}$$

or

$$\left.\begin{aligned}(fr)' + \frac{\kappa}{r}(fr) - (\varepsilon + m - U)gr = 0,\\ (gr)' - \frac{\kappa}{r}(gr) + (\varepsilon - m - U)fr = 0.\end{aligned}\right\} \tag{35.5}$$

Let us examine the behaviour of f and g at small distances, assuming that the field $U(r)$ increases more rapidly than $1/r$ as $r \to 0$. Then, for small r, equations (35.4) become

$$f' + Ug = 0, \qquad g' - Uf = 0.$$

These have real solutions, of the form

$$\left.\begin{aligned}f = \text{constant} \times \sin\left(\int U\,dr + \delta\right),\\ g = \text{constant} \times \cos\left(\int U\,dr + \delta\right),\end{aligned}\right\} \tag{35.6}$$

where δ is an arbitrary constant. These functions oscillate as $r \to 0$, but tend to no limit. It is easy to see that this situation corresponds, in the non-relativistic theory, to the "fall" of a particle to the centre.

First of all, we note that the smallness of r here places no restrictions on the choice of solution, since there is no condition at $r=0$ for the oscillatory function, and so the choice of δ remains arbitrary (the correct behaviour of the wave function for large r can be ensured by an appropriate choice of δ, for any value of ε). This indeterminacy can be eliminated by regarding the potential with a singularity (at $r=0$) as the limit, when $r_0 \to 0$, of a potential cut off at some r_0, i.e. equal to $U(r)$ for $r > r_0$ and $U(r_0)$ for $r < r_0$. With r_0 finite, we of course obtain a definite set of energy levels, but the energy of the ground state tends to $-\infty$ as $r_0 \to 0$.

In the non-relativistic theory, this signifies a "fall" to the centre, since a particle at a deep level is localized in a small region near $r=0$. In the relativistic theory, such a situation is impossible, since it would imply that the system was unstable with respect to the spontaneous generation of electron–positron pairs. For, whereas an energy exceeding $2m$ is needed to create such a pair in a vacuum, a smaller energy is sufficient in a field. In the presence of an electron bound state with energy $\varepsilon < m$, a pair can be formed by expending an energy $\varepsilon + m < 2m$, the result being a free positron and an electron in a bound state. If, on the other hand, the bound state energy is $\varepsilon < -m$, such a field can create positrons (with energy $-\varepsilon > m$) spontaneously, without taking energy from an external source. In the field under consideration, as $r_0 \to 0$, there is an infinity of such "anomalous" levels with $\varepsilon < -m$. Thus fields whose potential $\Phi(r)$ increases more rapidly than $1/r$ as $r \to 0$ cannot be dealt with by means of Dirac's theory. This applies to potentials of *either* sign. Although a "fall" can occur, of course, only with attraction, the sign of $U = e\Phi$ depends also on that of the charge, so that electron levels behave anomalously in one case and positron levels in the other; in the latter case, the field generates free electrons.

Let us next consider the behaviour of the wave functions at large distances. If the field $U(r)$ decreases sufficiently rapidly as $r \to \infty$, it may be entirely neglected in the equations when determining the asymptotic form of the wave functions at large distances. When $\varepsilon > m$, i.e. in the continuous spectrum, we then return to the equation of free motion, so that the asymptotic form of the wave functions (spherical waves) differs from that for a free particle only by the presence of additional "phase shifts", whose values are determined by the form of the field at short distances.† These shifts depend on the values of j and l; that is, on the number κ defined above (and also, of course, on the energy ε). Denoting the phase shifts by δ_κ and using the expresssion (24.7) for a free spherical wave, we can therefore immediately write down the required asymptotic formula:

$$\psi \approx \frac{2}{r}\frac{1}{\sqrt{(2\varepsilon)}}\begin{pmatrix} \sqrt{(\varepsilon+m)}\Omega_{jlm}\sin(pr-\frac{1}{2}l\pi+\delta_\kappa) \\ -\sqrt{(\varepsilon-m)}\Omega_{jl'm}\sin(pr-\frac{1}{2}l'\pi+\delta_\kappa)\end{pmatrix}, \tag{35.7}$$

or, with the definition (35.1),

$$\left.\begin{matrix} f \\ g \end{matrix}\right\} = \frac{\sqrt{2}}{r}\sqrt{\frac{\varepsilon \pm m}{\varepsilon}}\,\begin{matrix}\sin \\ \cos\end{matrix}\,(pr-\tfrac{1}{2}l\pi+\delta_\kappa), \tag{35.8}$$

† Cf. *QM*, §33. As in the non-relativistic theory, $U(r)$ must decrease more rapidly than $1/r$. The case $U \sim 1/r$ will be discussed separately in §36.

where $p = \sqrt{(\varepsilon^2 - m^2)}$. The common coefficient here corresponds to the normalization of the radial functions by (24.5).

The wave functions of the discrete spectrum ($\varepsilon < m$) decrease exponentially as $r \to \infty$:

$$f = -\sqrt{\frac{m+\varepsilon}{m-\varepsilon}}\, g = \frac{A_0}{r} \exp[-r\sqrt{(m^2 - \varepsilon^2)}], \tag{35.9}$$

where A_0 is a constant.

As in the non-relativistic theory, the phase shifts δ_κ (more precisely, the quantities $e^{2i\delta_\kappa} - 1$) determine the scattering amplitudes in a given field, as will be further discussed in §37. We shall not pause to investigate here the analytical properties of these amplitudes (cf. *QM*, §128), but merely note than $e^{2i\delta_\kappa}$ again has, as a function of energy, poles at the points corresponding to the levels for bound states of the particle. The residue of $e^{2i\delta_\kappa}$ at such a pole is related in a certain manner to the coefficient in the asymptotic expression for the corresponding wave function of the discrete spectrum. This relation may be found by a generalization of the non-relativistic formula, *QM* (128.17). The necessary calculations are entirely similar to those in *QM*, §128.

Differentiation of equations (35.5) with respect to the energy gives

$$\left(\frac{\partial(rf)}{\partial\varepsilon}\right)' + \frac{\kappa}{r}\frac{\partial(rf)}{\partial\varepsilon} - (\varepsilon + m - U)\frac{\partial(rg)}{\partial\varepsilon} = rg,$$

$$\left(\frac{\partial(rg)}{\partial\varepsilon}\right)' - \frac{\kappa}{r}\frac{\partial(rg)}{\partial\varepsilon} + (\varepsilon - m - U)\frac{\partial(rf)}{\partial\varepsilon} = -rf.$$

We multiply these two equations by rg and $-rf$ respectively, and the two equations (35.5) by $-rg$ and rf, and add all four term-by-term. After simplification, we have

$$\frac{\partial}{\partial r}\left[r^2\left(g\frac{\partial f}{\partial\varepsilon} - f\frac{\partial g}{\partial\varepsilon}\right)\right] = r^2(f^2 + g^2).$$

We now integrate with respect to r:

$$r^2\left(g\frac{\partial f}{\partial\varepsilon} - f\frac{\partial g}{\partial\varepsilon}\right) = \int_0^r (f^2 + g^2) r^2\, dr,$$

and take the limit as $r \to \infty$. The integral on the right tends to unity, by the normalization condition. On the left-hand side, we use the fact that in the asymptotic region the functions f and g are related by

$$rg = \frac{(rf)'}{\varepsilon + m},$$

which is derived from (35.5) by neglecting terms in U and in $1/r$. The result is

$$\frac{1}{\varepsilon+m}\left[(rf)'\frac{\partial(rf)}{\partial\varepsilon}-rf\left(\frac{\partial(rf)}{\partial\varepsilon}\right)'\right]=1. \tag{35.10}$$

This formula differs only in the coefficient ($\varepsilon+m$ replacing $2m$) from the corresponding non-relativistic formula (for the function χ). There is therefore no need to repeat the subsequent calculations; the final formula, valid near the point $\varepsilon=\varepsilon_0$ (where ε_0 is the energy level), is

$$e^{2i\delta_\kappa}=(-1)^l\frac{2A_0^2}{\varepsilon-\varepsilon_0}\sqrt{\frac{m-\varepsilon_0}{m+\varepsilon_0}}, \tag{35.11}$$

where A_0 is the coefficient in the asymptotic expression (35.9).

PROBLEM

Find the limiting form of the wave function for small r in a field $U\sim r^{-s}$, $s<1$.

SOLUTION. For a free particle we have, when r is small, $f\sim r^l$, $g\sim r^{l'}$, so that $f\gg g$ if $l<l'$, and $f\ll g$ if $l>l'$. We make the assumption (which is confirmed by the result) that this relationship exists also in the field considered here. If $l<l'$ (i.e. $l=j-\frac{1}{2}$, $\kappa=-l-1$), the term in g may be omitted from the first equation (35.4), so that $f\sim r^l$ as before. The second equation then gives $g\sim rfU$, i.e. $g\sim r^{l+1-s}=r^{l'-s}$. The case $l>l'$ is treated similarly. The result is

$$\text{for } l<l',\quad f\sim r^l,\quad g\sim r^{l'-s};$$
$$\text{for } l>l',\quad f\sim r^{l-s},\quad g\sim r^{l'}.$$

§ 36. **Motion in a Coulomb field**

We shall begin the study of the properties of the motion in the very important case of a Coulomb field by considering the behaviour of the wave functions at short distances, taking the particular case of an attractive field: $U=-Z\alpha/r$.†

For small r, the terms in $\varepsilon\pm m$ may be omitted in equations (35.5), leaving

$$(fr)'+\frac{\kappa}{r}fr-\frac{Z\alpha}{r}gr=0,$$

$$(gr)'-\frac{\kappa}{r}gr+\frac{Z\alpha}{r}fr=0.$$

The two functions fr and gr appear in an equivalent manner in these equations, and we therefore seek them as equal powers of r: $fr=ar^\gamma$, $gr=br^\gamma$. Substitution gives

$$a(\gamma+\kappa)-bZ\alpha=0,\qquad aZ\alpha+b(\gamma-\kappa)=0,$$

whence

$$\gamma^2=\kappa^2-(Z\alpha)^2. \tag{36.1}$$

† In ordinary units, $U=-Ze^2/r$. In relativistic units, e^2 is replaced by the dimensionless quantity α.

Let $(Z\alpha)^2 < \kappa^2$. Then γ is real, and the positive value must be taken: the corresponding solution either does not diverge at $r = 0$ or does so less rapidly than the other. The choice may be justified by considering a potential which is "cut off" (see §35) at a certain small r_0 and then taking the limit $r_0 \to 0$; cf. the analogous discussion in *QM*, §35. Thus

$$\left.\begin{aligned} f &= \frac{Z\alpha}{\gamma+\kappa} g = \text{constant} \times r^{-1+\gamma}, \\ \gamma &= \sqrt{(\kappa^2 - Z^2\alpha^2)} = \sqrt{[(j+\tfrac{1}{2})^2 - Z^2\alpha^2]}. \end{aligned}\right\} \tag{36.2}$$

Although the wave function may become infinite at $r = 0$ (if $\gamma < 1$), the integral of $|\psi|^2$ remains finite, of course.

If $(Z\alpha)^2 > \kappa^2$, both values of γ given by (36.2) are imaginary. The corresponding solutions oscillate as $r^{-1}\cos(|\gamma| \log r)$ when $r \to 0$, and this again corresponds to a situation which is inadmissible in the relativistic theory, as already shown above. Since $\kappa^2 \geqslant 1$, this means that a purely Coulomb field can be discussed in Dirac's theory only if $Z\alpha < 1$, i.e. $Z < 137$.

Let us now give a qualitative description of the situation which arises when $Z > 137$. In order to avoid an indeterminacy in the boundary condition at $r = 0$, we must again consider a potential cut off at a distance r_0 (I. Ya. Pomeranchuk and Ya. A. Smorodinskiĭ, 1945). This has a direct physical significance as well as a formal one. The charge $Z > 137$ can in practice be concentrated only into a "superheavy" nucleus of finite radius. Let us therefore see how the configuration of levels varies as Z increases for a given r_0.

In the Coulomb field when not cut off, the energy ε_1 of the lowest level tends to zero when $Z\alpha = 1$, and the $\varepsilon_1(Z)$ curve terminates, the level ε_1 becoming imaginary for $Z\alpha > 1$; see (36.10) below. In the cut-off field, for a given $r_0 \neq 0$, the level ε_1 passes through zero only for some $Z\alpha > 1$. The value $\varepsilon_1 = 0$ has no physical distinctiveness, and when $r_0 \neq 0$ it also has no formal distinctiveness, the $\varepsilon_1(Z)$ curve not terminating there. When Z increases further, the levels continue to descend, and at a certain "critical" value $Z = Z_c(r_0)$ the energy ε_1 reaches the bottom $(-m)$ of the lowest continuum of levels. As explained in §35, this means that zero energy is required for the creation of a free positron. The critical value Z_c is therefore the maximum charge that the "bare" nucleus can have for a given r_0.

When $Z > Z_c$, the level $\varepsilon_1 < -m$, and the formation of two electron–positron pairs becomes energetically favourable. The positrons go to infinity, carrying kinetic energy $2(|\varepsilon_1| - m)$, and the two electrons occupy the level ε_1. This gives an "ion" with an occupied K shell and a charge $Z_{\text{eff}} = Z - 2$ (S. S. Gershteĭn and Ya. B. Zel'dovich, 1969). The system is stable for $Z > Z_c$, up to values of Z for which the limit $-m$ reaches the next level.†

Lastly, it may be noted that, even for a point charge, the form of the potential at short distances is affected by radiative corrections, but the resulting corrections to $Z_c\alpha$ are only of the order of α.

† For example, if the nuclear charge is uniformly distributed in a sphere with radius $r_0 = 1.2 \times 10^{-12}$ cm, the critical value $Z_c = 170$, and the next level reaches the limit $-m$ when $Z = 185$ (V. S. Popov, 1970). A detailed account of the quantitative theory is in the review article by Ya. B. Zel'dovich and V. S. Popov, *Soviet Physics Uspekhi* **14**, 673, 1972.

Let us now turn to the exact solution of the wave equation (C. G. Darwin, 1928; W. Gordon, 1928).

(a) *Discrete spectrum* ($\varepsilon < m$). We seek the functions f and g in the form

$$\left.\begin{aligned} f &= \sqrt{(m+\varepsilon)}\, e^{-\frac{1}{2}\rho}\rho^{\gamma-1}(Q_1+Q_2), \\ g &= -\sqrt{(m-\varepsilon)}\, e^{-\frac{1}{2}\rho}\rho^{\gamma-1}(Q_1-Q_2), \end{aligned}\right\} \tag{36.3}$$

with the notation

$$\rho = 2\lambda r, \qquad \lambda = \sqrt{(m^2-\varepsilon^2)}, \qquad \gamma = \sqrt{(\kappa^2 - Z^2\alpha^2)}. \tag{36.4}$$

This is reasonable, since we already know the behaviour of the functions as $\rho \to 0$ (36.2) and their exponential decrease ($\sim e^{-\frac{1}{2}\rho}$) as $\rho \to \infty$. Since, as $\rho \to \infty$, the functions f and g must have the same asymptotic behaviour, we must expect that $Q_1 \gg Q_2$ as $\rho \to \infty$.

Substitution of (36.3) in (35.4) yields the equations

$$\rho(Q_1+Q_2)' + (\gamma+\kappa)(Q_1+Q_2) - \rho Q_2 + Z\alpha\sqrt{\frac{m-\varepsilon}{m+\varepsilon}}(Q_1-Q_2) = 0,$$

$$\rho(Q_1-Q_2)' + (\gamma-\kappa)(Q_1-Q_2) + \rho Q_2 - Z\alpha\sqrt{\frac{m+\varepsilon}{m-\varepsilon}}(Q_1+Q_2) = 0,$$

where the prime denotes differentiation with respect to ρ. The sum and difference of these give

$$\left.\begin{aligned} \rho Q_1' + \left(\gamma - \frac{Z\alpha\varepsilon}{\lambda}\right)Q_1 + \left(\kappa - \frac{Z\alpha m}{\lambda}\right)Q_2 &= 0, \\ \rho Q_2' + \left(\gamma + \frac{Z\alpha\varepsilon}{\lambda} - \rho\right)Q_2 + \left(\kappa + \frac{Z\alpha m}{\lambda}\right)Q_1 &= 0, \end{aligned}\right\} \tag{36.5}$$

or, eliminating either Q_1 or Q_2,

$$\rho Q_1'' + (2\gamma+1-\rho)Q_1' - \left(\gamma - \frac{Z\alpha\varepsilon}{\lambda}\right)Q_1 = 0,$$

$$\rho Q_2'' + (2\gamma+1-\rho)Q_2' - \left(\gamma + 1 - \frac{Z\alpha\varepsilon}{\lambda}\right)Q_2 = 0,$$

where we have used the fact that $\gamma^2 - (Z\alpha\varepsilon/\lambda)^2 = \kappa^2 - (Z\alpha m/\lambda)^2$. The solution of these equations which is finite when $\rho = 0$ is

$$\left.\begin{aligned} Q_1 &= AF\left(\gamma - \frac{Z\alpha\varepsilon}{\lambda}, 2\gamma+1, \rho\right), \\ Q_2 &= BF\left(\gamma + 1 - \frac{Z\alpha\varepsilon}{\lambda}, 2\gamma+1, \rho\right), \end{aligned}\right\} \tag{36.6}$$

where $F(\alpha, \beta, z)$ is the confluent hypergeometric function. Putting $\rho = 0$ in either of the equations (36.5), we obtain the relation between the constants A and B:

$$B = -\frac{\gamma - Z\alpha\varepsilon/\lambda}{\kappa - Z\alpha m/\lambda} A. \tag{36.7}$$

The two hypergeometric functions in (36.6) must reduce to polynomials, since otherwise they would increase as e^{ρ} when $\rho \to \infty$, and the wave function itself would increase as $e^{\frac{1}{2}\rho}$. The function $F(\alpha, \beta, z)$ reduces to a polynomial if α is a negative integer or zero. We write

$$\gamma - Z\alpha\varepsilon/\lambda = -n_r. \tag{36.8}$$

If $n_r = 1, 2, \ldots$, the two hypergeometric functions reduce to polynomials. If $n_r = 0$, only one of them does so. In that case, $\gamma = Z\alpha\varepsilon/\lambda$, and $Z\alpha m/\lambda = |\kappa|$, as is easily verified. If $\kappa < 0$, the coefficient B (36.7) is zero, so that $Q_2 = 0$ and the necessary condition is satisfied. If $\kappa > 0$, however, then $B = -A$, and Q_2 remains divergent when $n_r = 0$. Thus the following values are possible for the quantum number n_r:

$$\left.\begin{aligned} n_r &= 0, 1, 2, \ldots \quad \text{for } \kappa < 0, \\ &= 1, 2, 3, \ldots \quad \text{for } \kappa > 0. \end{aligned}\right\} \tag{36.9}$$

The definition (36.8) then yields the following expression for the discrete energy levels:

$$\frac{\varepsilon}{m} = \left[1 + \frac{(Z\alpha)^2}{\{\sqrt{[\kappa^2 - (Z\alpha)^2]} + n_r\}^2}\right]^{-\frac{1}{2}}. \tag{36.10}$$

In particular, the energy of the $1s\frac{1}{2}$ ground level ($|\kappa| = 1$, $n_r = 0$) is

$$\varepsilon_1 = m\sqrt{[1 - (Z\alpha)^2]}.$$

For $Z\alpha \ll 1$, the leading terms in the expansion of (36.10) are

$$\frac{\varepsilon}{m} - 1 = -\frac{(Z\alpha)^2}{2(|\kappa| + n_r)^2}\left\{1 + \frac{(Z\alpha)^2}{|\kappa| + n_r}\left[\frac{1}{|\kappa|} - \frac{3}{4(|\kappa| + n_r)}\right]\right\}.$$

On writing $n_r + |\kappa| = n$ $(= 1, 2, \ldots)$ and noting that $|\kappa| = j + \frac{1}{2}$, we return to formula (34.4), which was previously derived by means of perturbation theory. As already mentioned at the end of §34, the further terms in this expansion have no significance, since they are certainly exceeded by the radiative corrections. Formula (36.10) as it stands, however, is meaningful when $Z\alpha \sim 1$. The double degeneracy of the levels shown by the approximate formula (34.4) exists also in the exact formula, which involves only $|\kappa|$, so that levels with the same j and different l again coincide.

We have still to determine the common normalization factor A in the wave function. The wave function of the discrete spectrum must, as usual, be normalized by the condition $\int |\psi|^2\, d^3x = 1$; the corresponding condition on the functions f and g is

$$\int_0^\infty (f^2+g^2)r^2\, dr = 1.$$

The value of A is most simply determined from the asymptotic form of the functions as $r \to \infty$. Using the asymptotic formula

$$F(-n_r, 2\gamma+1, \rho) \approx \frac{\Gamma(2\gamma+1)}{\Gamma(n_r+2\gamma+1)}(-\rho)^{n_r}$$

(see *QM*, (d. 14)), we find

$$f \approx (-1)^{n_r} A\sqrt{(m+\varepsilon)}\frac{\Gamma(2\gamma+1)}{\Gamma(n_r+2\gamma+1)}\, e^{-\lambda r}(2\lambda r)^{\gamma+n_r-1}.$$

Comparing this with (36.22) derived below, we can find A. Collecting together the formulae, we can write out in full the final expressions for the normalized wave functions:

$$f, g = \frac{\pm(2\lambda)^{3/2}}{\Gamma(2\gamma+1)}\left[\frac{(m\pm\varepsilon)\Gamma(2\gamma+n_r+1)}{4m(Z\alpha m/\lambda)(Z\alpha m/\lambda-\kappa)\cdot n_r!}\right]^{\frac{1}{2}}(2\lambda r)^{\gamma-1}\, e^{-\lambda r}\times$$

$$\times\left\{\left(\frac{Z\alpha m}{\lambda}-\kappa\right)F(-n_r, 2\gamma+1, 2\lambda r)\mp n_r F(1-n_r, 2\gamma+1, 2\lambda r)\right\}, \qquad (36.11)$$

where the upper signs refer to f and the lower signs to g.

(b) *Continuous spectrum* ($\varepsilon > m$). There is no need to solve the wave equation afresh for the states of the continuous spectrum. The wave functions for this case are obtained from those of the discrete spectrum by the substitutions†

$$\sqrt{(m-\varepsilon)} \to -i\sqrt{(\varepsilon-m)}, \qquad \lambda \to -ip, \qquad -n_r \to \gamma - iZ\alpha\varepsilon/p; \qquad (36.12)$$

see *QM*, §128 concerning the choice of sign in the analytical continuation of the square root $\sqrt{(m-\varepsilon)}$. The normalization of the functions must, however, be done again.

Making these substitutions in (36.11), we may write

$$\left.\begin{array}{l} f = \sqrt{(\varepsilon+m)} \\ g = i\sqrt{(\varepsilon-m)}\end{array}\right\}\times A'\, e^{ipr}(2pr)^{\gamma-1}\times$$

$$\times\left[e^{i\xi}F(\gamma-i\nu, 2\gamma+1, -2ipr)\mp e^{-i\xi}F(\gamma+1-i\nu, 2\gamma+1, -2ipr)\right],$$

† In the rest of this section, p denotes $|\mathbf{p}| = \sqrt{(\varepsilon^2 - m^2)}$.

where A' is another normalization constant,

$$\nu = Z\alpha\varepsilon/p, \qquad e^{-2i\xi} = \frac{\gamma - i\nu}{\kappa - i\nu m/\varepsilon}; \tag{36.13}$$

the value of ξ is real, since $\gamma^2 + (Z\alpha\varepsilon/p)^2 = \kappa^2 + (Zam/p)^2$.

According to the formula

$$F(\alpha, \beta, z) = e^z F(\beta - \alpha, \beta, -z)$$

(see *QM*, (d. 10)), we have

$$\begin{aligned} F(\gamma + 1 - i\nu, 2\gamma + 1, -2ipr) &= e^{-2ipr} F(\gamma + i\nu, 2\gamma + 1, 2ipr) \\ &= e^{-2ipr} F^*(\gamma - i\nu, 2\gamma + 1, -2ipr). \end{aligned}$$

Hence

$$f, g = 2iA'\sqrt{(\varepsilon \pm m)}(2pr)^{\gamma-1} \operatorname{im}, \operatorname{re}\{e^{i(pr+\xi)} F(\gamma - i\nu, 2\gamma + 1, -2ipr)\}. \tag{36.14}$$

The normalization coefficient A' is found by comparing the asymptotic expression for this function with the general formula (35.7) for a normalized spherical wave. The resulting expression for the wave functions of the continuous spectrum (which we shall afterwards verify) is†

$$\begin{aligned} f, g = 2^{3/2} \sqrt{\frac{m \pm \varepsilon}{\varepsilon}}\, e^{\frac{1}{2}\pi\nu} \frac{|\Gamma(\gamma + 1 + i\nu)|}{\Gamma(2\gamma + 1)} \frac{(2pr)^{\gamma}}{r} \times \\ \times \operatorname{im}, \operatorname{re}\{e^{i(pr+\xi)} F(\gamma - i\nu, 2\gamma + 1, -2ipr)\}. \end{aligned} \tag{36.15}$$

The asymptotic expression for this function is derived by means of *QM* (d. 14), where only the first term is now significant, the second decreasing as a higher power of $1/r$:

$$f, g = \frac{\sqrt{2}}{r} \sqrt{\frac{\varepsilon \pm m}{\varepsilon}} \sin, \cos(pr + \delta_\kappa + \nu \log 2pr - \tfrac{1}{2} l\pi), \tag{36.16}$$

where

$$\delta_\kappa = \xi - \arg \Gamma(\gamma + 1 + i\nu) - \tfrac{1}{2}\pi\gamma + \tfrac{1}{2} l\pi, \tag{36.17}$$

or

$$e^{2i\delta_\kappa} = \frac{\kappa - i\nu m/\varepsilon}{\gamma - i\nu} \frac{\Gamma(\gamma + 1 - i\nu)}{\Gamma(\gamma + 1 + i\nu)} e^{i\pi(l-\gamma)}. \tag{36.18}$$

For future reference, we shall give an expression for the phase in the ultra-

† The wave functions for a repulsive field are obtained by changing the sign of $Z\alpha$, i.e. that of ν.

relativistic case ($\varepsilon \gg m$, $\nu \approx Z\alpha$):

$$e^{2i\delta_\kappa} = \frac{\kappa}{\gamma - iZ\alpha} \frac{\Gamma(\gamma+1-iZ\alpha)}{\Gamma(\gamma+1+iZ\alpha)} e^{i\pi(l-\gamma)}. \tag{36.19}$$

The expression (36.16) differs from (35.8) only by the logarithmic term in the argument of the trigonometric function. As in Schrödinger's equation, the slowness of the decrease of the Coulomb potential affects the phase of the wave, which becomes a slowly varying function of r.

Analytical continuation into the region $\varepsilon < m$ gives, in place of (36.18),

$$e^{2i\delta_\kappa} = \frac{\kappa - Z\alpha m/\lambda}{\gamma - Z\alpha\varepsilon/\lambda} \frac{\Gamma(\gamma+1-Z\alpha\varepsilon/\lambda)}{\Gamma(\gamma+1+Z\alpha\varepsilon/\lambda)} e^{i\pi(l-\gamma)}. \tag{36.20}$$

This expression has poles at the points where $\gamma+1-Z\alpha\varepsilon/\lambda = 1-n_r$, $n_r = 1, 2, \ldots$ (poles of the gamma function in the numerator), and also at the point $\gamma - Z\alpha\varepsilon = -n_r = 0$ (if also $\kappa < 0$); these points coincide with the discrete energy levels, as they should.

Near any of the poles with $n_r \neq 0$, we have

$$e^{2i\delta_\kappa} \approx \frac{(Z\alpha m/\lambda - \kappa)\, e^{i\pi(l-\gamma)}}{n_r \Gamma(2\gamma+1+n_r)} \Gamma(\gamma+1-Z\alpha\varepsilon/\lambda).$$

The form of the gamma function near its pole is found by means of the familiar formula $\Gamma(z)\Gamma(1-z) = \pi/\sin \pi z$:

$$\Gamma\left(\gamma+1-\frac{Z\alpha\varepsilon}{\lambda}\right) \approx \frac{\pi}{\Gamma(n_r)\sin\pi(\gamma+1-Z\alpha\varepsilon/\lambda)},$$

$$\begin{aligned}\sin\pi(\gamma+1-Z\alpha\varepsilon/\lambda) &\approx \pi\cos\pi n_r \cdot \frac{d}{d\varepsilon}\left(\frac{Z\alpha\varepsilon}{\lambda}\right)\cdot(\varepsilon-\varepsilon_0)\\ &= (-1)^{n_r}(\pi Z\alpha m^2/\lambda^3)(\varepsilon-\varepsilon_0),\end{aligned}$$

where ε_0 is the energy level. Thus we have†

$$e^{2i\delta_\kappa} \approx (-1)^{l+n_r} \frac{e^{-i\pi\gamma}(Z\alpha m/\lambda - \kappa)}{n_r!\Gamma(2\gamma+1+n_r)} \frac{\lambda^3}{Z\alpha m^2} \frac{1}{\varepsilon-\varepsilon_0}. \tag{36.21}$$

At the end of §35 a formula (35.11) was derived which relates the residue of the function $e^{2i\delta_\kappa}$ at its pole to the coefficient in the asymptotic expression for the wave function of the corresponding bound state. For a Coulomb field, however, the formula (35.10) must be slightly modified, because the constant phase shift δ_κ in (35.7) is replaced in (36.16) by the sum $\delta_\kappa + \nu \log 2pr$. We must therefore replace $e^{2i\delta_\kappa}$ on the left-hand side of (35.11) by

$$\exp(2i\delta_\kappa + 2i\nu\log 2pr) \rightarrow e^{2i\delta_\kappa}(2i\lambda r)^{2(n_r+\gamma)}.$$

† This formula is easily seen to be valid even if $n_r = 0$.

Using (36.21) and determining from (35.11) the coefficient A_0 (which will now be a power function of r), we find the asymptotic form of the normalized wave function of the discrete spectrum:

$$f = \left[\frac{(Z\alpha m/\lambda - \kappa)(m+\varepsilon)\lambda^2}{2n_r!Z\alpha m^2\Gamma(2\gamma+1+n_r)}\right]^{\frac{1}{2}}(2\lambda r)^{n_r+\gamma}\frac{e^{-\lambda r}}{r}. \tag{36.22}$$

This has already been used to determine the coefficient in (36.11).

§37. Scattering in a centrally symmetric field

The asymptotic expression for the wave function which describes the scattering of particles in the field of a fixed centre of force may be written†

$$\psi = u_{\varepsilon\mathbf{p}}\, e^{ipz} + u'_{\varepsilon\mathbf{p}'}\frac{e^{ipr}}{r}. \tag{37.1}$$

Here $u_{\varepsilon\mathbf{p}}$ is the bispinor amplitude of the incident plane wave. The bispinor $u'_{\varepsilon\mathbf{p}'}$ is a function of the direction of scattering $\mathbf{n}'$, and for any given value of $\mathbf{n}'$ its form (but not, of course, its normalization) is the same as that of the bispinor amplitude of the plane wave propagated in the direction $\mathbf{n}'$.

We have seen in §24 that the bispinor amplitude of the plane wave is entirely determined by specifying a two-component quantity, the three-dimensional spinor w, which is the non-relativistic wave function in the rest frame of the particle. The flux density is expressed in terms of the same spinor, and is proportional to w^*w, with a proportionality coefficient which depends only on the energy ε and is therefore the same for the incident and scattered particles. The scattering cross-section is $d\sigma = (w'^*w'/w^*w)\, do$ or, if as in §24 the incident wave is normalized by the condition $w^*w = 1$,

$$d\sigma = w'^*w'\, do.$$

We define the scattering operator $\hat{f}$ by

$$w' = \hat{f}w. \tag{37.2}$$

Since the quantities w, w' have two components, the operator thus defined is exactly analogous to the operator scattering amplitude which appears in the non-relativistic scattering theory taking account of spin (*QM*, §140). We can therefore apply immediately the formulae derived there which express the operator in terms of the phase shifts of the wave functions in the scattering field. It is only necessary to transform these phase shifts by expressing δ_l^+ and δ_l^- from *QM*, §140, in terms of the phase shift δ_κ which appears in the relativistic formula (35.7). The phases δ_l^+ and δ_l^- referred to states with orbital angular momentum l and total

† In §§37 and 38 p denotes $|\mathbf{p}|$, and ε and $\mathbf{p}$ will be written separately as suffixes to the amplitude.

angular momentum $j = l + \frac{1}{2}$ and $j = l - \frac{1}{2}$. According to the definition (35.3), $\kappa = -l - 1$ for $j = l + \frac{1}{2}$ and $\kappa = l$ for $j = l - \frac{1}{2}$. We must therefore make the changes

$$\delta_l^+ \to \delta_{-(l+1)}, \qquad \delta_l^- \to \delta_l$$

and remember that the suffix to δ now represents the value of κ. Thus we find

$$\hat{f} = A + B\boldsymbol{\nu}\cdot\boldsymbol{\sigma}, \tag{37.3}$$

$$A = \frac{1}{2ip}\sum_{l=0}^{\infty}[(l+1)(e^{2i\delta_{-l-1}} - 1) + l(e^{2i\delta_l} - 1)]P_l(\cos\theta), \tag{37.4}$$

$$B = \frac{1}{2p}\sum_{l=1}^{\infty}(e^{2i\delta_{-l-1}} - e^{2i\delta_l})P_l^1(\cos\theta), \tag{37.5}$$

where $\boldsymbol{\nu}$ is a unit vector in the direction of $\mathbf{n}\times\mathbf{n}'$.

Since w is the spinor wave function in the rest frame, the polarization properties of the scattering are given in terms of $\hat{f}$ by the same formulae as in *QM*, §140.

For a Coulomb field, it is possible to express both functions $A(\theta)$ and $B(\theta)$ in terms of one function. The calculation is briefly as follows.†

In a Coulomb field, the phases δ_κ are given by (36.18), which we write in the form

$$\left.\begin{aligned} e^{2i\delta_\kappa} &= -\left(\kappa - i\frac{Ze^2m}{p}\right)\frac{\kappa}{|\kappa|}C_\kappa, \\ C_\kappa &= -\frac{\Gamma(\gamma - i\nu)}{\Gamma(\gamma + 1 + i\nu)}e^{i\pi(|\kappa|-\gamma)}; \end{aligned}\right\} \tag{37.6}$$

$e^{i\pi l} = e^{i\pi\kappa}$ when $\kappa > 0$, and $e^{i\pi l} = -e^{i\pi\kappa}$ when $\kappa < 0$. Using the quantities thus defined we can put the series (37.4), (37.5) in the form

$$\left.\begin{aligned} A(\theta) &= \frac{1}{p}G(\theta) - i\frac{Ze^2m}{p^2}F(\theta), \\ B(\theta) &= -\frac{i}{p}\tan\tfrac{1}{2}\theta\cdot G(\theta) + \frac{Ze^2m}{p^2}\cot\tfrac{1}{2}\theta\cdot F(\theta), \end{aligned}\right\} \tag{37.7}$$

where

$$G(\theta) = \tfrac{1}{2}i\sum_{l=1}^{\infty}l^2C_l(P_l + P_{l-1}), \qquad F(\theta) = \tfrac{1}{2}i\sum_{l=1}^{\infty}lC_l(P_l - P_{l-1}). \tag{37.8}$$

In transforming the series $B(\theta)$, we have used the following recurrence relations between Legendre polynomials:

$$P_l^1 + P_{l-1}^1 = -\cot\tfrac{1}{2}\theta\cdot l(P_l - P_{l-1}), \tag{37.9}$$

$$P_l^1 - P_{l-1}^1 = \tan\tfrac{1}{2}\theta\cdot l(P_l + P_{l-1}). \tag{37.10}$$

† R. L. Gluckstern and S. R. Lin, *Journal of Mathematical Physics* **5**, 1594, 1964.

According to the identity

$$(1+\cos\theta)\frac{d}{d\cos\theta}[P_l(\cos\theta)-P_{l-1}(\cos\theta)] = l[P_l(\cos\theta)+P_{l-1}(\cos\theta)], \tag{37.11}$$

the functions $F(\theta)$ and $G(\theta)$ are related by

$$G = (1-\cos\theta)\frac{dF}{d\cos\theta} = -\cot\tfrac{1}{2}\theta\cdot\frac{dF}{d\theta}. \tag{37.12}$$

Thus $A(\theta)$ and $B(\theta)$ are expressed in terms of the single function $F(\theta)$.†

§ 38. Scattering in the ultra-relativistic case

We shall now discuss separately the scattering in the ultra-relativistic case ($\varepsilon \gg m$). In the first approximation, we neglect altogether the mass m in the wave equation. It is convenient to use for ψ the spinor representation

$$\psi = \begin{pmatrix}\xi\\ \eta\end{pmatrix},$$

since the equations for ξ and η are separable when $m = 0$:

$$\left.\begin{aligned} -i\boldsymbol{\sigma}\cdot\nabla\xi &= (\varepsilon - U)\xi,\\ -i\boldsymbol{\sigma}\cdot\nabla\eta &= -(\varepsilon - U)\eta \end{aligned}\right\} \tag{38.1}$$

(the "neutrino" form, §30).

A helicity state of an electron polarized in the direction of **p** corresponds to a wave function $\psi = \begin{pmatrix}\xi\\ 0\end{pmatrix}$, and for polarization opposite to **p** we have $\psi = \begin{pmatrix}0\\ \eta\end{pmatrix}$. Since the equations for ξ and η are separable, it is evident that this property is unaffected by scattering. Thus helicity is conserved in the scattering of ultra-relativistic electrons. From considerations of symmetry (longitudinal polarization) it is obvious that there is no azimuthal asymmetry in the scattering of helical (longitudinally polarized) particles. We can also say that the scattering cross-section of helical electrons is independent of the sign of the helicity; this follows because a central field is invariant under inversion, while the sign of the helicity is reversed.

In the ultra-relativistic case, formulae (37.3)–(37.5) may be considerably simplified (D. R. Yennie, D. G. Ravenhall and R. N. Wilson, 1954).

† The function $F(\theta)$ cannot be expressed in a closed form in terms of the elementary functions, but it can be written as a certain double integral; see the paper cited in the last footnote.

Let the incident electron be polarized, say in the direction of motion $\mathbf{n}$. For a plane wave with a definite value of $\mathbf{n}\cdot\boldsymbol{\sigma}$, the spinor ξ $(=(\phi+\chi)/\sqrt{2})$ is proportional to the same three-dimensional spinor w as appeared in the standard representation of the wave. The relation between the spinor amplitudes of the incident and scattered waves in the new representation is therefore given by the same operator $\hat{f}$.

As a result of the scattering, the polarization is rotated with the momentum to the direction $\mathbf{n}'$. The effect of the operator $\hat{f}$ on the spin wave function of the electron therefore reduces to a rotation of the spin through the angle θ (between $\mathbf{n}$ and $\mathbf{n}'$) about the axis $\boldsymbol{\nu}$. This rotation is itself equivalent to a rotation of the coordinates about that axis but in the opposite direction, i.e. through an angle $-\theta$. Hence it follows that the operator $\hat{f}$ must be the same (apart from a factor) as the operator which transforms the wave function when the coordinates are changed in the way described, i.e. the operator (18.17) with $-\theta$ instead of θ. A comparison of (37.3) with (18.17) shows that

$$B/A = -i\tan\tfrac{1}{2}\theta. \tag{38.2}$$

Thus, in the ultra-relativistic limit,

$$\hat{f} = A(\theta)[1 - i\tan\tfrac{1}{2}\theta\cdot\boldsymbol{\nu}\cdot\boldsymbol{\sigma}]. \tag{38.3}$$

The expression (37.4) for $A(\theta)$ can also be simplified if a relation between δ_κ and $\delta_{-\kappa}$ which exists in the ultra-relativistic limit is used. To derive this relation, we note that, when the terms in m are omitted, the equations (35.4) for the functions f and g become invariant with respect to the changes

$$\kappa\to-\kappa,\qquad f\to g,\qquad g\to-f,$$

which do not affect the parameters of the particle or field itself. We must therefore have $f_\kappa/g_\kappa = -g_{-\kappa}/f_{-\kappa}$, and substitution of the asymptotic expressions gives

$$\tan(pr-\tfrac{1}{2}l\pi+\delta_\kappa) = -\cot(pr-\tfrac{1}{2}l'\pi+\delta_{-\kappa}),$$
$$\delta_\kappa = \delta_{-\kappa}-\tfrac{1}{2}(l'-l)\pi+(n+\tfrac{1}{2})\pi,$$

whence

$$e^{2i\delta_\kappa} = e^{2i\delta_{-\kappa}}. \tag{38.4}$$

From this relation, and replacing the summation variable l by $l-1$ in the first term of the sum in (37.4), we find

$$A(\theta) = \frac{1}{2ip}\sum_{l=1}^{\infty} l(e^{2i\delta_l}-1)][P_l(\cos\theta)+P_{l-1}(\cos\theta)]. \tag{38.5}$$

From (38.2) it follows that $\mathrm{re}(AB^*)=0$. Hence, in the approximation considered, the cross-section is independent of the initial polarization of the particles, and an unpolarized beam remains unpolarized after scattering (see *QM*, (140.8)–(140.10)). We may also note that, when $\theta \to \pi$, the expression (38.5) for $A(\theta)$ tends to zero as $(\pi - \theta)^2$ (since $P_l(-1) = (-1)^l$). The cross-section

$$\frac{d\sigma}{do} = |A|^2 + |B|^2 = |A(\theta)|^2/\cos^2 \tfrac{1}{2}\theta \tag{38.6}$$

therefore tends to zero also. These properties do not occur, of course, in higher approximations with respect to the small quantity m/ε. In particular, analysis shows that as $\theta \to \pi$ the cross-section tends to a limit proportional to $(m/\varepsilon)^2$.

For a Coulomb field in the ultra-relativistic case, the phases δ_κ are independent of the energy, as is seen from (36.19).† Hence, in a purely Coulomb field, the scattering cross-section for $\varepsilon \gg m$ has the form

$$d\sigma = \frac{\tau(\theta)}{\varepsilon^2}\, do, \tag{38.7}$$

where τ is a function of the angle only.

§ 39. **The continuous-spectrum wave functions for scattering in a Coulomb field**

In later sections (§§95, 96) we shall consider various inelastic processes which occur when ultra-relativistic electrons are scattered in the field of a heavy nucleus ($Z\alpha \sim 1$). To calculate the relevant matrix elements, we need wave functions whose asymptotic form (as $r \to \infty$) is the sum of a plane wave and a spherical wave.

We shall see that, in the ultra-relativistic case (electron energy $\varepsilon \gg m$), the most significant values of the momentum transfer from electron to nucleus in scattering are $q = |\mathbf{p}' - \mathbf{p}| \sim m$. These values of q correspond to impact parameters $\rho \sim 1/q \sim 1/m$, the electron being deflected through angles‡

$$\theta \sim q/p \sim m/\varepsilon. \tag{39.1}$$

In terms of the coordinates r (distance from the centre) and $z = r\cos\theta$, this represents the region

$$\rho \equiv r \sin\theta \sim 1/m, \qquad p(r - z) = pr(1 - \cos\theta) \sim 1, \tag{39.2}$$

and $r \sim \varepsilon/m^2$, so that the distances concerned are large.

We write Dirac's equation in the form

$$(\varepsilon - U - m\beta + i\boldsymbol{\alpha}\cdot\nabla)\psi = 0, \qquad U = -Z\alpha/r, \tag{39.3}$$

and transform it into a second-order equation by applying the operator $\varepsilon - U +$

† This is also evident directly from equations (38.1), since for a Coulomb field the energy ε may be eliminated from the equations by the substitution $\mathbf{r} \to \mathbf{r}'/\varepsilon$.

‡ In this section, p denotes $|\mathbf{p}|$.

$m\beta - i\boldsymbol{\alpha}\cdot\nabla$:

$$(\triangle + p^2 - 2\varepsilon U)\psi = (-i\boldsymbol{\alpha}\cdot\nabla U - U^2)\psi. \tag{39.4}$$

Since $r \gg Z\alpha/\varepsilon$ in the region considered, $U \ll \varepsilon$. As a first approximation, the right-hand side of (39.4) may be neglected. The remaining equation,

$$(\triangle + p^2 + 2\varepsilon Z\alpha/r)\psi = 0, \tag{39.5}$$

is of the same form as the non-relativistic Schrödinger's equation in a Coulomb field:

$$\left(\frac{1}{2m}\triangle + \frac{p^2}{2m} + \frac{Z\alpha}{r}\right)\psi = 0, \tag{39.5a}$$

differing only in an obvious change in the notation for the parameters (the "potential energy" containing an extra factor ε/m). We can therefore write down immediately the solution which has the required asymptotic form (see *QM*, §136).

For example, the wave function which asymptotically comprises a plane wave ($\propto e^{i\mathbf{p}\cdot\mathbf{r}}$) and an outgoing spherical wave is

$$\begin{aligned}\psi_{\varepsilon\mathbf{p}}^{(+)} &= C\frac{u_{\varepsilon\mathbf{p}}}{\sqrt{(2\varepsilon)}}e^{i\mathbf{p}\cdot\mathbf{r}}F\left(\frac{iZ\alpha\varepsilon}{p}, 1, i(pr - \mathbf{p}\cdot\mathbf{r})\right)\\ C &= e^{\pi Z\alpha\varepsilon/2p}\Gamma(1 - iZ\alpha\varepsilon/p),\end{aligned} \tag{39.6}$$

where F is the confluent hypergeometric function and $u_{\varepsilon\mathbf{p}}$ the constant bispinor amplitude of the plane wave, normalized by the condition stated earlier (23.4):

$$\bar{u}_{\varepsilon\mathbf{p}}u_{\varepsilon\mathbf{p}} = 2m. \tag{39.7}$$

The wave function (39.6) is normalized in such a way that the plane wave in its asymptotic limit has the usual form,

$$\frac{u_{\varepsilon\mathbf{p}}}{\sqrt{(2\varepsilon)}}e^{i\mathbf{p}\cdot\mathbf{r}},$$

corresponding to "one particle in unit volume". Since $p \approx \varepsilon$ in the ultra-relativistic case, we can write $Z\alpha\varepsilon/p \approx Z\alpha$ in (39.6):

$$\left.\begin{aligned}\psi_{\varepsilon\mathbf{p}}^{(+)} &= C\frac{u_{\varepsilon\mathbf{p}}}{\sqrt{(2\varepsilon)}}e^{i\mathbf{p}\cdot\mathbf{r}}F(iZ\alpha, 1, i(pr - \mathbf{p}\cdot\mathbf{r})),\\ C &= e^{Z\alpha\pi/2}\Gamma(1 - iZ\alpha).\end{aligned}\right\} \tag{39.8}$$

It should be noted that, although we are considering distances so large that $pr \gg 1$, the hypergeometric function in (39.8) cannot be replaced by its asymptotic form: the argument of F is not pr but $pr(1 - \cos\theta)$, which is not assumed large.†

† In *QM*, §135, we were concerned with arbitrarily large r, and this approximation was therefore allowable for all values of θ.

In applications, the next approximation for ψ is also needed, which has a spinor structure different from (39.8) (the latter reducing to the factor $u_{\varepsilon\mathbf{p}}$). To calculate this approximation, we write ψ in the form

$$\psi = \frac{C}{\sqrt{(2\varepsilon)}} e^{i\mathbf{p}\cdot\mathbf{r}}(u_{\varepsilon\mathbf{p}}F + \phi).$$

On the right-hand side of (39.4) we now retain the term linear in U, obtaining for ϕ the equation

$$(\Delta + 2i\mathbf{p}\cdot\nabla - 2\varepsilon U)\phi = -iu_{\varepsilon\mathbf{p}}\boldsymbol{\alpha}\cdot\nabla U. \tag{39.9}$$

The solution of this may be found by noticing that the function F satisfies the equation

$$(\Delta + 2i\mathbf{p}\cdot\nabla - 2\varepsilon U)F = 0,$$

as may be seen by substituting (39.6) in (39.5). Applying the operator ∇ to this equation, we obtain

$$(\Delta + 2i\mathbf{p}\cdot\nabla - 2\varepsilon U)\nabla F = 2\varepsilon F\nabla U.$$

A comparison with (39.9) shows that

$$\phi = -\frac{i}{2\varepsilon}(\boldsymbol{\alpha}\cdot\nabla)u_{\varepsilon\mathbf{p}}F.$$

The final expressions for $\psi^{(+)}$ and for a similar function $\psi^{(-)}$ whose asymptotic form contains an ingoing spherical wave are

$$\left.\begin{aligned}
\psi^{(+)}_{\varepsilon\mathbf{p}} &= \frac{C}{\sqrt{(2\varepsilon)}} e^{i\mathbf{p}\cdot\mathbf{r}}\left(1 - \frac{i}{2\varepsilon}\boldsymbol{\alpha}\cdot\nabla\right)F(iZ\alpha, 1, i(pr - \mathbf{p}\cdot\mathbf{r}))u_{\varepsilon\mathbf{p}},\\
\psi^{(-)}_{\varepsilon\mathbf{p}} &= \frac{C^*}{\sqrt{(2\varepsilon)}} e^{i\mathbf{p}\cdot\mathbf{r}}\left(1 - \frac{i}{2\varepsilon}\boldsymbol{\alpha}\cdot\nabla\right)F(-iZ\alpha, 1, -i(pr + \mathbf{p}\cdot\mathbf{r}))u_{\varepsilon\mathbf{p}},\\
C &= e^{\pi Z\alpha/2}\Gamma(1 - iZ\alpha)
\end{aligned}\right\} \tag{39.10}$$

(W. H. Furry, 1934). We shall also write out the corresponding functions ($\psi_{-\varepsilon,-\mathbf{p}}$) with "negative frequency", which are needed when dealing with processes which involve positrons. These can be derived from the functions $\psi_{\varepsilon\mathbf{p}}$ by the substitutions $\mathbf{p}\to-\mathbf{p}$, $\varepsilon\to-\varepsilon$, with $p = |\mathbf{p}|$ unchanged; the parameter $iZ\alpha$ of the hypergeometric function therefore changes sign, as will be seen from the original expression (39.6), where this parameter occurs in the form $iZ\alpha\varepsilon/p$. Thus we have

$$\left.\begin{aligned}
\psi^{(+)}_{-\varepsilon,-\mathbf{p}} &= \frac{C}{\sqrt{(2\varepsilon)}} e^{-i\mathbf{p}\cdot\mathbf{r}}\left(1 + \frac{i}{2\varepsilon}\boldsymbol{\alpha}\cdot\nabla\right)F(-iZ\alpha, 1, i(pr + \mathbf{p}\cdot\mathbf{r}))u_{-\varepsilon,-\mathbf{p}},\\
\psi^{(-)}_{-\varepsilon,-\mathbf{p}} &= \frac{C^*}{\sqrt{(2\varepsilon)}} e^{-i\mathbf{p}\cdot\mathbf{r}}\left(1 + \frac{i}{2\varepsilon}\boldsymbol{\alpha}\cdot\nabla\right)F(iZ\alpha, 1, -i(pr - \mathbf{p}\cdot\mathbf{r}))u_{-\varepsilon,-\mathbf{p}},\\
C &= e^{-\pi Z\alpha/2}\Gamma(1 + iZ\alpha).
\end{aligned}\right\} \tag{39.11}$$

The following comment is necessary regarding the above calculations. Our asymptotic condition is not in itself sufficient to provide a unique choice of the solution of the wave equation; this is clear, since we can always add to ψ any outgoing Coulomb spherical wave without violating the condition. By writing the solution of equation (39.5) in the form (39.6), we have tacitly presupposed the choice of a solution finite at $r = 0$. This requirement was necessary in *QM*, §§135, 136, where we were considering solutions, valid in all space, of the exact Schrödinger's equation.† In the present case, however, equation (39.5) applies only to large distances, and therefore the choice of solution demands further justification.

This is provided by the fact that large impact parameters $\rho = r \sin\theta$ correspond to large orbital angular momenta l and small scattering angles θ: when $\rho \sim 1/m$, we have

$$l \sim \rho p \sim \rho\varepsilon \sim \varepsilon/m \gg 1,$$

and the angle θ may be estimated by a quasi-classical procedure:

$$\theta \sim \frac{1}{p}\int \frac{dU}{dr}\,dt \sim \frac{U'(\rho)\rho}{p} \sim \frac{m}{\varepsilon} \ll 1.$$

Thus, in the expansion of ψ in terms of spherical waves the main contribution (in this range of r and θ) will come from waves with these large values of l. But a spherical wave with large l will certainly decrease to small values at distances from the origin $r \ll l/\varepsilon$ which are "classically inaccessible" (because of the centrifugal barrier). Hence, if we "join" the solution of equation (39.5) to that of the exact equation (39.4) at short distances $r \sim r_1$, where $l/\varepsilon \gg r_1 \gg Z\alpha/\varepsilon$, then the boundary condition for the solution of equation (39.5) will be that it is small, and this justifies our choice.

PROBLEM

For an attractive Coulomb field with $Z\alpha \ll 1$, find the correction (of relative order $Z\alpha$) to the non-relativistic wave function of the discrete spectrum.

SOLUTION. The electron velocity in a bound state is $v \sim Z\alpha$, and therefore, for $Z\alpha \ll 1$, the wave function is non-relativistic in the zero-order approximation, i.e.

$$\psi = u\psi_{\text{non-r}},$$

where $\psi_{\text{non-r}}$ is the Schrödinger function and u a bispinor of the form

$$u = \begin{pmatrix} w \\ 0 \end{pmatrix},$$

with w a spinor describing the polarization state of the electron. In the next approximation, we write $\psi = u\psi_{\text{non-r}} + \psi^{(1)}$ and, substituting this in (39.4), obtain for $\psi^{(1)}$ the equation

$$\left(\frac{1}{2m}\triangle - |\varepsilon_n| + \frac{Z\alpha}{r}\right)\psi^{(1)} = i\frac{Z\alpha}{2m}\left(\nabla\frac{1}{r}\right)\cdot(\boldsymbol{\alpha}u)\psi_{\text{non-r}},$$

† In the method of solution given in *QM*, §135, this condition was satisfied by taking the particular integral in the form (135.1) instead of as a general sum of integrals with different values of β_1 and β_2.

where ε_n is the non-relativistic discrete energy level. Here we have omitted terms of relative order $\sim(Z\alpha)^2$; in the non-relativistic case, the important distances are of the order of the Bohr radius, $r \sim 1/mZ\alpha$. The solution of this equation is $\psi^{(1)} = -(i/2m)\boldsymbol{\alpha}u \cdot \nabla\psi_{\text{non-r}}$, and therefore

$$\psi = \left(1 - \frac{i}{2m}\boldsymbol{\alpha} \cdot \nabla\right)u\psi_{\text{non-r}}.$$

§ 40. An electron in the field of an electromagnetic plane wave

Dirac's equation can be solved exactly for an electron moving in the field of an electromagnetic plane wave (D. M. Volkov, 1937).

The field of a plane wave with wave 4-vector k ($k^2 = 0$) depends on the 4-coordinates only in the combination $\phi = kx$, so that the 4-potential is

$$A^\mu = A^\mu(\phi), \tag{40.1}$$

and satisfies the Lorentz gauge condition

$$\partial_\mu A^\mu = k_\mu A^{\mu\prime} = 0,$$

the prime denoting differentiation with respect to ϕ. Since the constant term in A is unimportant, we can omit the prime, writing the condition as

$$kA = 0. \tag{40.2}$$

We start from the second-order equation (32.6), in which the field tensor is

$$F_{\mu\nu} = k_\mu A'_\nu - k_\nu A'_\mu. \tag{40.3}$$

When expanding the square $(i\partial - eA)^2$ it must be remembered that, from (40.2), $\partial_\mu(A^\mu\psi) = A^\mu\partial_\mu\psi$. The result is

$$[-\partial^2 - 2ie(A\partial) + e^2A^2 - m^2 - ie(\gamma k)(\gamma A')]\psi = 0, \tag{40.4}$$

where $\partial^2 = \partial_\mu\partial^\mu$.

We seek a solution of this equation in the form

$$\psi = e^{-ipx}F(\phi), \tag{40.5}$$

where p is a constant 4-vector. This form of the function ψ is unaltered by adding to p any constant multiple of the vector k, if the function $F(\phi)$ is appropriately redefined. We can therefore, without loss of generality, impose one further condition on p. Let

$$p^2 = m^2. \tag{40.6}$$

Then, when the field is removed, the quantum numbers p^μ become the components of the free particle 4-momentum. The significance of the components of the 4-vector p, when the field is present, is more clearly seen in a particular frame of reference chosen so that $A_0 = 0$. Let the vector **A** in this frame be along the x^1-axis and **k** along the x^3-axis; the electric field of the wave is then along x^1, the magnetic field along x^2, and the wave itself is propagated along x^3. Then (40.5) will be an eigenfunction of the operators

$$\hat{p}_1 = i\frac{\partial}{\partial x^1}, \qquad \hat{p}_2 = i\frac{\partial}{\partial x^2}, \qquad \hat{p}_0 - \hat{p}_3 = i\left(\frac{\partial}{\partial x^0} - \frac{\partial}{\partial x^3}\right),$$

with eigenvalues p_1, p_2, $p_0 - p_3$; the operators themselves are easily seen to commute with the Hamiltonian of Dirac's equation. Thus, in this frame of reference, p^1 and p^2 are the components of the generalized momentum along the x^1 and x^2 axes; $p^0 - p^3$ is the difference between the total energy and the x^3-component of the generalized momentum.

In substituting (40.5) in (40.4), we note that

$$\partial^\mu F = k^\mu F', \qquad \partial_\mu \partial^\mu F = k^2 F'' = 0,$$

and obtain for $F(\phi)$ the equation

$$2i(kp)F' + [-2e(pA) + e^2 A^2 - ie(\gamma k)(\gamma A')]F = 0.$$

The integral of this equation is

$$F = \exp\left\{-i\int_0^{kx}\left[\frac{e}{(kp)}(pA) - \frac{e^2}{2(kp)}A^2\right]d\phi + \frac{e(\gamma k)(\gamma A)}{2(kp)}\right\}\frac{u}{\sqrt{(2p_0)}},$$

where $u/\sqrt{(2p_0)}$ is an arbitrary constant bispinor; the reason for writing it in this form will be shown below.

All powers of $(\gamma k)(\gamma A)$ above the first are zero, since

$$(\gamma k)(\gamma A)(\gamma k)(\gamma A) = -(\gamma k)(\gamma k)(\gamma A)(\gamma A) + 2(kA)(\gamma k)(\gamma A) = -k^2 A^2 = 0.$$

We can therefore write

$$\exp\frac{e(\gamma k)(\gamma A)}{2(kp)} = 1 + \frac{e}{2(kp)}(\gamma k)(\gamma A),$$

so that ψ becomes

$$\psi_p = \left[1 + \frac{e}{2(kp)}(\gamma k)(\gamma A)\right]\frac{u}{\sqrt{(2p_0)}}e^{iS}, \tag{40.7}$$

where†

$$S = -px - \int_0^{kx} \left[\frac{e}{(kp)}(pA) - \frac{e^2}{2(kp)}A^2\right] d\phi. \tag{40.8}$$

To determine the conditions to be imposed on the constant bispinor u, we must suppose that the wave is "switched on" with infinite slowness, starting from $t = -\infty$. Then $A \to 0$ when $kx \to -\infty$, and ψ must become the solution of the free Dirac's equation. Consequently, $u = u(p)$ must satisfy

$$(\gamma p - m)u = 0. \tag{40.9}$$

This condition rejects the "redundant" solutions of the second-order equation. Since u is independent of time, the condition remains valid for finite kx. Thus $u(p)$ is the same as the bispinor amplitude of the free plane wave; we shall take it to be normalized by the same condition (23.4): $\bar{u}u = 2m$.

The foregoing arguments also show immediately the normalization of the wave functions (40.7). The infinitely slow application of the field does not alter the normalization integral. Hence it follows that the functions (40.7) satisfy the same normalization condition,

$$\int \psi_{p'}^* \psi_p \, d^3x = \int \bar{\psi}_{p'} \gamma^0 \psi_p \, d^3x = (2\pi)^3 \delta(\mathbf{p}' - \mathbf{p}), \tag{40.10}$$

as the free plane waves.

Let us calculate the current density corresponding to the functions (40.7), first noting that

$$\bar{\psi}_p = \frac{\bar{u}}{\sqrt{(2p_0)}}\left[1 + \frac{e}{2(kp)}(\gamma A)(\gamma k)\right]e^{iS},$$

and hence obtaining by direct multiplication

$$j^\mu = \bar{\psi}_p \gamma^\mu \psi_p = \frac{1}{p_0}\left\{p^\mu - eA^\mu + k^\mu\left(\frac{e(pA)}{(kp)} - \frac{e^2A^2}{2(kp)}\right)\right\}. \tag{40.11}$$

If the $A^\mu(\phi)$ are periodic functions, and their time-average value is zero, the mean value of the current density is

$$\overline{j^\mu} = \frac{1}{p_0}\left(p^\mu - \frac{e^2}{2(kp)}\overline{A^2}k^\mu\right). \tag{40.12}$$

We can also find the kinetic momentum density in the state ψ_p. The kinetic

† This S is the same as the classical action for a particle moving in the field of a wave; cf. *Fields*, §47, Problem 2.

momentum operator is the difference $\hat{p} - eA = i\partial - eA$. A direct calculation gives

$$\psi_p^* (\hat{p}^\mu - eA^\mu)\psi_p = \bar{\psi}_p \gamma^0 (\hat{p}^\mu - eA^\mu)\psi_p$$
$$= p^\mu - eA^\mu + k^\mu \left(\frac{e(pA)}{(kp)} - \frac{e^2A^2}{2(kp)}\right) + k^\mu \frac{ie}{8(kp)p_0} F_{\lambda\nu}(u^*\sigma^{\lambda\nu}u). \tag{40.13}$$

The time-average value of this 4-vector, denoted by q^μ, is

$$q^\mu = p^\mu - \frac{e^2\overline{A^2}}{2(kp)} k^\mu. \tag{40.14}$$

Its square is

$$q^2 = m_*^2, \qquad m_* = m\sqrt{\left(1 - \frac{e^2}{m^2}\overline{A^2}\right)}, \tag{40.15}$$

where m_* acts as an "effective mass" of the electron in the field. A comparison of (40.14) with (40.12) shows that

$$\overline{j^\mu} = q^\mu / p_0. \tag{40.16}$$

The normalization condition (40.10), expressed in terms of the vector $\mathbf{q}$, is

$$\int \psi_{p'}^* \psi_p \, d^3x = (2\pi)^3 \frac{q_0}{p_0} \delta(\mathbf{q}' - \mathbf{q}); \tag{40.17}$$

this is most simply proved in the particular frame of reference mentioned above.

§41. Motion of spin in an external field

The quasi-classical approximation in Dirac's equation is reached in the same way as in the non-relativistic theory. In the second-order equation (32.7a) we substitute†

$$\psi = u\, e^{(i/\hbar)S},$$

where S is a scalar and u a slowly varying bispinor. The usual condition of the quasi-classical case is assumed to be satisfied: the momentum of the particle must vary only slightly over distances of the order of the wavelength $\hbar/|\mathbf{p}|$.

In the zero-order approximation with respect to $\hbar$, we obtain the usual classical relativistic Hamilton–Jacobi equation for the action S. All the terms which contain the spin (and are proportional to $\hbar$) are absent from the equations of motion. The spin would appear only in the next approximation with respect to $\hbar$.

† Ordinary units will be used at first.

Thus the influence of the magnetic moment of the electron on its motion is always of the same order of magnitude as the quantum corrections. This is to be expected, since the spin is a purely quantum property and its magnitude is proportional to $\hbar$.

We can therefore reasonably formulate the question of how the electron spin will behave when the electron is executing a given quasi-classical motion in an external field. The answer to this question is contained in the next approximation with respect to $\hbar$ in Dirac's equation. We shall, however, use another method whose significance is more obvious and which does not directly involve Dirac's equation. It has the advantage of allowing a treatment of the motion of any particle, including a particle which has an "anomalous" gyromagnetic ratio not describable by Dirac's equation.

The objective is to derive an "equation of motion" for the spin when the particle moves in any (given) manner. Let us first take the non-relativistic case.

The non-relativistic Hamiltonian of a particle in an external field is

$$\hat{H} = \hat{H}' - \mu \boldsymbol{\sigma} \cdot \mathbf{H}, \tag{41.1}$$

where $\hat{H}'$ includes all terms independent of the spin (see *QM*, §111), and μ is the magnetic moment of the particle. This form of the Hamiltonian relates to any kind of particle. For electrons, $\mu = e\hbar/2mc$ (the electron charge being $e = -|e|$), and for nucleons μ also contains the "anomalous" part†

$$\mu' = \mu - e\hbar/2mc. \tag{41.2}$$

According to the general rules of quantum mechanics, the operator equation of motion for the spin is obtained from the formula

$$\dot{\hat{\mathbf{s}}} = \frac{i}{\hbar}(\hat{H}\hat{\mathbf{s}} - \hat{\mathbf{s}}\hat{H}) = \frac{i}{2\hbar}(\hat{H}\boldsymbol{\sigma} - \boldsymbol{\sigma}\hat{H}). \tag{41.3}$$

Substitution of (41.1) gives

$$\begin{aligned} \dot{\hat{s}}_i &= -\frac{i\mu}{2\hbar} H_k(\sigma_k\sigma_i - \sigma_i\sigma_k) \\ &= -\frac{\mu}{\hbar} e_{ikl} H_k \sigma_l, \end{aligned}$$

or

$$\dot{\hat{\mathbf{s}}} = \frac{2\mu}{\hbar} \hat{\mathbf{s}} \times \mathbf{H}. \tag{41.4}$$

We average this operator equation over the state of the quasi-classical wave packet moving in a given path. This is equivalent to replacing the spin operator by

† When radiative corrections are taken into account the magnetic moment of the electron also contains a very small "anomalous" part.

its mean value $\bar{\mathbf{s}}$, and the vector $\mathbf{H}$ by the function $\mathbf{H}(t)$, which represents the change in the magnetic field at the position of the particle (or wave packet) as the latter moves along its path. In the non-relativistic approximation (i.e. in terms of Pauli's equation), $\hat{\mathbf{s}} = \frac{1}{2}\boldsymbol{\sigma}$ is the spin operator of the particle in its rest frame, whose mean value was denoted in §29 by $\frac{1}{2}\boldsymbol{\zeta}$. Thus we obtain the equation

$$\frac{d\boldsymbol{\zeta}}{dt} = \frac{2\mu}{\hbar}\boldsymbol{\zeta}\times\mathbf{H}(t). \qquad (41.5)$$

This form of the equation is, in essence purely classical. It signifies that the magnetic moment vector precesses about the direction of the field with angular velocity $-2\mu\mathbf{H}/\hbar$, remaining constant in magnitude.†

Again in the non-relativistic case, the velocity $\mathbf{v}$ of the particle varies in accordance with the equation

$$d\mathbf{v}/dt = e\mathbf{v}\times\mathbf{H}/mc,$$

i.e. the vector $\mathbf{v}$ rotates about the direction of $\mathbf{H}$ with angular velocity $-e\mathbf{H}/mc$. If $\mu' = 0$, then $\mu = e\hbar/2mc$, and this angular velocity is the same as the angular velocity $-2\mu\mathbf{H}/\hbar$ with which the vector $\boldsymbol{\zeta}$ rotates; thus the polarization vector is at a constant angle to the direction of motion. We shall see below that this result remains valid in the relativistic case.

Let us now proceed to the relativistic generalization of equation (41.5). For a covariant description of the polarization it is necessary to use the 4-vector a defined in §29, and the equation of motion of the spin will determine its derivative with respect to the proper time τ.‡

The form of this equation is given by considerations of relativistic invariance: its right-hand side must be linear and homogeneous in the electromagnetic field tensor $F^{\mu\nu}$ and in the 4-vector a^μ, and apart from these can include only the 4-velocity $u^\mu = p^\mu/m$. The only form of equation satisfying these conditions is

$$da^\mu/d\tau = \alpha F^{\mu\nu}a_\nu + \beta u^\mu F^{\nu\lambda}u_\nu a_\lambda, \qquad (41.6)$$

where α and β are constant coefficients. It is easily seen, from the condition $a_\mu u^\mu = 0$ and the antisymmetry of the tensor $F^{\mu\nu}$ (whence $F^{\mu\nu}u_\mu u_\nu = 0$), that no other expressions with the required properties can be constructed.

As $v\to 0$, this equation must become the same as (41.5). Putting $a^\mu = (0, \boldsymbol{\zeta})$, $u^\mu = (1, 0)$, $\tau = t$, we have

$$d\boldsymbol{\zeta}/dt = \alpha\boldsymbol{\zeta}\times\mathbf{H}.$$

A comparison with (41.5) shows that $\alpha = 2\mu$.

† The classical equation (41.5) can be derived directly from the equation

$$d\mathbf{M}/dt = \boldsymbol{\mu}\times\mathbf{H},$$

where $\mathbf{M}$ is the angular momentum and $\boldsymbol{\mu}$ the magnetic moment of the system; $\boldsymbol{\mu}\times\mathbf{H}$ is the torque acting on the system. Putting $\mathbf{M} = \frac{1}{2}\hbar\boldsymbol{\zeta}$, $\boldsymbol{\mu} = (\mu/2s)\boldsymbol{\zeta} = \mu\boldsymbol{\zeta}$, we have (41.5).

‡ From here onwards we again take $c = 1$, $\hbar = 1$.

To determine β, we use the fact that $a^\mu u_\mu = 0$. Differentiating this with respect to τ and using the classical equation of motion of a charge in a field,

$$m\, du^\mu/d\tau = eF^{\mu\nu}u_\nu,$$

(see *Fields*, §23), we obtain

$$u_\mu \frac{da^\mu}{d\tau} = -a_\mu \frac{du^\mu}{d\tau} = -a_\mu \frac{e}{m} F^{\mu\nu}u_\nu = \frac{e}{m} F^{\mu\nu}u_\mu a_\nu.$$

Hence, on multiplying both sides of equation (41.6) by u_μ, using the equation $u_\mu u^\mu = 1$ and cancelling the common factor $F^{\mu\nu}u_\mu a_\nu$, we have

$$\beta = -2\left(\mu - \frac{e}{2m}\right) = -2\mu'.$$

Thus the final relativistic equation of motion for the spin is

$$\frac{da^\mu}{d\tau} = 2\mu F^{\mu\nu}a_\nu - 2\mu' u^\mu F^{\nu\lambda}u_\nu a_\lambda \tag{41.7}$$

(V. Bargmann, L. Michel and V. L. Telegdi, 1959).†

We can change from the 4-vector a to the quantity $\boldsymbol{\zeta}$ which directly represents the polarization of the particle in its "instantaneous" rest frame. The relation between a and $\boldsymbol{\zeta}$ is given by formulae (29.7)–(29.9). First of all, from (41.7) we necessarily have $a_\mu da^\mu/d\tau = 0$, and therefore $a_\mu a^\mu =$ constant. Since $a_\mu a^\mu = -\boldsymbol{\zeta}^2$, this is equivalent to the obvious result that the polarization $\boldsymbol{\zeta}$ of the particle remains unchanged in magnitude during its motion.

The equation which shows the change in direction of the polarization is obtained by using three-dimensional notation in (41.7). The space components of this equation are, in explicit form,

$$\frac{d\mathbf{a}}{dt} = \frac{2\mu m}{\varepsilon}\mathbf{a}\times\mathbf{H} + \frac{2\mu m}{\varepsilon}(\mathbf{a}\cdot\mathbf{v})\mathbf{E} - \frac{2\mu'\varepsilon}{m}\mathbf{v}(\mathbf{a}\cdot\mathbf{E}) + \frac{2\mu'\varepsilon}{m}\mathbf{v}(\mathbf{v}\cdot\mathbf{a}\times\mathbf{H}) + \\ + \frac{2\mu'\varepsilon}{\mu}\mathbf{v}(\mathbf{a}\cdot\mathbf{v})(\mathbf{v}\cdot\mathbf{E}).$$

Here we must substitute (29.9), using in the differentiation the equations $\mathbf{p} = \varepsilon\mathbf{v}$, $\varepsilon^2 = \mathbf{p}^2 + m^2$, and the equations of motion

$$d\mathbf{p}/dt = e\mathbf{E} + e\mathbf{v}\times\mathbf{H}, \qquad d\varepsilon/dt = e\mathbf{v}\cdot\mathbf{E}. \tag{41.8}$$

A lengthy but elementary calculation leads to the result‡

† This equation was first derived, in another form, by Ya. I. Frenkel' (1926).

‡ If the gyromagnetic ratio (Landé factor) g is used (as is often done) for charged particles, with $\mu = g(e/2m)\cdot\frac{1}{2}$ $(= g(e/2mc)\cdot\frac{1}{2}\hbar)$, this equation becomes

$$\frac{d\boldsymbol{\zeta}}{dt} = \frac{e}{2m}\left(g - 2 + 2\frac{m}{\varepsilon}\right)\boldsymbol{\zeta}\times\mathbf{H} + \frac{e}{2m}(g-2)\frac{\varepsilon}{\varepsilon+m}(\mathbf{v}\cdot\mathbf{H})\mathbf{v}\times\boldsymbol{\zeta} + \frac{e}{2m}\left(g - \frac{2\varepsilon}{\varepsilon+m}\right)\boldsymbol{\zeta}\times(\mathbf{E}\times\mathbf{v}). \tag{41.9a}$$

$$\frac{d\boldsymbol{\zeta}}{dt}=\frac{2\mu m+2\mu'(\varepsilon-m)}{\varepsilon}\boldsymbol{\zeta}\times\mathbf{H}+\frac{2\mu'\varepsilon}{\varepsilon+m}(\mathbf{v}\cdot\mathbf{H})\mathbf{v}\times\boldsymbol{\zeta}+\frac{2\mu m+2\mu'\varepsilon}{\varepsilon+m}\boldsymbol{\zeta}\times(\mathbf{E}\times\mathbf{v}). \tag{41.9}$$

The variation of the direction of polarization relative to the direction of motion is of more interest than the variation of its absolute position in space. We write

$$\boldsymbol{\zeta}=\mathbf{n}\zeta_{\parallel}+\boldsymbol{\zeta}_{\perp}, \tag{41.10}$$

where $\mathbf{n}=\mathbf{v}/v$, and derive the equation for the component $\zeta_{\parallel}$ of the polarization in the direction of motion. A calculation using (41.8), (41.9) leads to the result†

$$\frac{d\zeta_{\parallel}}{dt}=2\mu'\boldsymbol{\zeta}_{\perp}\cdot\mathbf{H}\times\mathbf{n}+\frac{2}{v}\left(\frac{\mu m^2}{\varepsilon^2}-\mu'\right)\boldsymbol{\zeta}_{\perp}\cdot\mathbf{E}. \tag{41.11}$$

The problems at the end of this section include a number of examples of the application of the above formulae. Here it may be noted that, in motion in a purely magnetic field, the polarization of a particle having no anomalous magnetic moment is at a constant angle to the velocity ($\zeta_{\parallel}=$ constant). Thus this result, already mentioned previously for the non-relativistic case, is in fact a general one.

The conditions for the above formulae to be applicable can be stated more precisely. The requirement specified initially, that the momentum of the particle should vary sufficiently slowly, is equivalent to a certain condition that the fields **E** and **H** should be small; in particular, the Larmor radius in the magnetic field ($\sim p/eH$) must be large compared with the wavelength of the particle. There is also, however, another condition which must, strictly speaking, be fulfilled: the fields must not vary too rapidly in space, and must vary only slightly within the dimensions of the quasi-classical wave packet. That is, the field must vary only slightly over distances of the order of the particle wavelength ($1/p$) and of the Compton wavelength ($1/m$).‡

In practical problems of motion in macroscopic fields, however, the condition of slow variation is certainly satisfied, and only the condition of smallness remains.

In §33 we have derived the first relativistic corrections for the Hamiltonian of an electron moving in an external field. For an electron in an electric field the approximate Hamiltonian is (see (33.12))

$$\hat{H}=\hat{H}'-\frac{e}{4m}\boldsymbol{\sigma}\cdot\mathbf{E}\times\hat{\mathbf{p}}/m, \qquad \hat{\mathbf{p}}=-i\nabla, \tag{41.12}$$

where $\hat{H}'$ includes the terms which do not contain the spin. In our case, since the

† This equation can be obtained a little more directly by writing explicitly the time component of equation (41.7).

‡ The latter requirement arises from the condition that the spread of velocities in the wave packet, in its rest frame, must be small compared with c, since otherwise the non-relativistic formulae could not be applied in this frame.

If the field varies too rapidly, the equations may contain significant additional terms in the derivatives of the field with respect to the coordinates.

field varies slowly, we neglect the term in $\hat{H}'$ which involves derivatives of **E** (i.e. the term in div **E**); the small term in $\hat{\mathbf{p}}^4$ may also be omitted, since it is unrelated to the field effects in question here. Thus $\hat{H}'$, in the absence of a magnetic field, reduces to the non-relativistic Hamiltonian $\hat{H}' = \hat{\mathbf{p}}^2/2m + e\Phi$.

Formula (41.12) can also be derived from (41.9) without making direct use of Dirac's equation. This method will generalize it (in the quasi-classical case) to particles with anomalous magnetic moment.

The equation of motion of the spin in an electric field, as far as first-order terms in the velocity v, is obtained from (41.9) as

$$\frac{d\boldsymbol{\zeta}}{dt} = (\mu + \mu')\boldsymbol{\zeta} \times (\mathbf{E} \times \mathbf{v}) = \left(\frac{e}{2m} + 2\mu'\right)\boldsymbol{\zeta} \times (\mathbf{E} \times \mathbf{v}).$$

If we impose the condition that this equation should be derived quantum-mechanically by commuting the spin operator with the Hamiltonian (as in (41.3)), then it is easily seen that we must put

$$\hat{H} = \hat{H}' - \left(\mu' + \frac{e}{4m}\right)\boldsymbol{\sigma} \cdot \mathbf{E} \times \hat{\mathbf{p}}/m. \tag{41.13}$$

This is the required expression. If $\mu' = 0$, we again obtain (41.12). It should be noted that the "normal" magnetic moment $e/2m$ is multiplied by an extra factor $\frac{1}{2}$ in comparison with the anomalous moment μ'.†

PROBLEMS

PROBLEM 1. Determine the change of the direction of polarization of a particle when it moves in a plane perpendicular to a uniform magnetic field ($\mathbf{v} \perp \mathbf{H}$).

SOLUTION. The right-hand side of equation (41.9) is reduced to its first term, and the vector $\boldsymbol{\zeta}$ therefore precesses about the direction of **H** (the z-axis) with angular velocity

$$-\frac{2\mu m + 2\mu'(\varepsilon - m)}{\varepsilon}\mathbf{H} = -\left(\frac{e}{\varepsilon} + 2\mu'\right)\mathbf{H}.$$

The projection of $\boldsymbol{\zeta}$ on the xy-plane (denoted by $\boldsymbol{\zeta}_\perp$) rotates in that plane with the same angular velocity. The vector **v** rotates in that plane with angular velocity $-e\mathbf{H}/\varepsilon$, as can be seen from the equation of motion $\dot{\mathbf{p}} = \varepsilon\dot{\mathbf{v}} = e\mathbf{v} \times \mathbf{H}$. Hence $\boldsymbol{\zeta}_\perp$ rotates with angular velocity $-2\mu'\mathbf{H}$ relative to the direction of **v**.

PROBLEM 2. The same as Problem 1, but for motion parallel to the magnetic field.

SOLUTION. When **v** and **H** are in the same direction, equation (41.9) reduces to

$$\frac{d\boldsymbol{\zeta}}{dt} = \frac{2\mu m}{\varepsilon}\boldsymbol{\zeta} \times \mathbf{H},$$

so that $\boldsymbol{\zeta}$ precesses about the common direction of **v** and **H** with angular velocity $-2\mu m\mathbf{H}/\varepsilon$.

PROBLEM 3. The same as Problem 1, but for motion in a uniform electric field.

SOLUTION. Let the field **E** be along the x-axis, and let the motion be in the xy-plane (with p_y = constant). According to (41.9), the vector $\boldsymbol{\zeta}$ precesses about the z-axis with instantaneous angular

† This is the "Thomas half" mentioned in the last footnote to §33. Its origin is clearly shown by the derivation given here.

velocity

$$-\left(\frac{e}{\varepsilon+m}+2\mu'\right)E\frac{p_y}{\varepsilon}.$$

We again resolve $\boldsymbol{\zeta}$ into components $\boldsymbol{\zeta}_1$ (in the xy-plane) and ζ_z. Then

$$\zeta_\parallel = \zeta_1 \cos\phi, \qquad \boldsymbol{\zeta}_\perp \cdot \mathbf{E} = -\zeta_1 \sin\phi \cdot v_y/v.$$

From (41.11), $\boldsymbol{\zeta}_1$ rotates relative to the direction of $\mathbf{v}$ with instantaneous angular velocity

$$\dot{\phi} = \frac{2v_y}{v^2}\left(\frac{\mu m^2}{\varepsilon^2}-\mu'\right) = \frac{p_y}{\varepsilon}\left(\frac{em}{p^2}-2\mu'\right).$$

§ 42. Neutron scattering in an electric field

In collisions between neutrons and nuclei, the scattering through large angles is determined by the main interaction, the nuclear forces. In small-angle scattering, however, it can be shown that the interaction of the magnetic moment of the neutron with the electric field of the nucleus becomes important (J. Schwinger, 1948).

We shall assume that the neutron is non-relativistic, so that the interaction in question is described by the approximate Hamiltonian (41.13). The magnetic moment of an electrically neutral particle is wholly "anomalous" and the operator $\hat{H}'$ reduces in this case to the kinetic-energy operator:†

$$\hat{H} = -\frac{\hbar^2}{2m}\triangle + i\frac{\mu\hbar}{mc}\boldsymbol{\sigma}\cdot\mathbf{E}\times\nabla. \tag{42.1}$$

Since the electromagnetic interaction of the neutron is small, the corresponding scattering amplitude f_{em} may be calculated by the Born approximation:

$$f_{\text{em}} = -\frac{m}{2\pi\hbar^2}\int e^{-i\mathbf{p}'\cdot\mathbf{r}/\hbar}\left(i\frac{\mu\hbar}{mc}\boldsymbol{\sigma}\cdot\mathbf{E}\times\nabla\right)e^{i\mathbf{p}\cdot\mathbf{r}/\hbar}\,d^3x$$

(see *QM*, §126), or

$$f_{\text{em}} = \frac{\mu}{2\pi c\hbar^2}\boldsymbol{\sigma}\cdot\mathbf{E}_{\mathbf{q}}\times\mathbf{p}, \qquad \mathbf{E}_{\mathbf{q}} = \int \mathbf{E}(\mathbf{r})\,e^{-i\mathbf{q}\cdot\mathbf{r}}\,d^3x, \tag{42.2}$$

where $\mathbf{p}$ and $\mathbf{p}'$ are the neutron momenta before and after scattering, and $\hbar\mathbf{q} = \mathbf{p}'-\mathbf{p}$. In this form, the amplitude f_{em} is an operator with respect to the spin variable.

Before continuing the calculation, we should note the following point. Formula (42.1) has been derived in §41 for slowly varying fields (which in practice meant neglecting terms in the Hamiltonian containing coordinate derivatives of the field). As applied to the Coulomb field of the nucleus, this means that the wavelength $\hbar/p$

† In this section, ordinary units are used, and m denotes the mass of the neutron.

must be small compared with the distances $r \sim 1/q$ which are important in the integral $\mathbf{E}_q$. Hence $\hbar_q \ll p$, so that the scattering angle $\theta \sim \hbar q/p \ll 1$. Thus the required condition is in fact satisfied for small-angle scattering.

For a Coulomb field with potential $\Phi = Ze/r$, the Fourier component of the field is

$$\mathbf{E}_q = -i\mathbf{q}\Phi_q = -i\mathbf{q}\frac{4\pi Ze}{q^2};$$

see *Fields*, (51.5). Substitution in (42.2) gives

$$f_{em} = i\frac{2Ze\mu}{q^2 c\hbar^3}\boldsymbol{\sigma}\cdot\mathbf{p}\times\mathbf{p}'.$$

For small scattering angles, $\hbar q \approx p\theta$ and $\mathbf{p}\times\mathbf{p}' \approx p^2\theta\boldsymbol{\nu}$, where $\boldsymbol{\nu}$ is a unit vector in the direction of $\mathbf{p}\times\mathbf{p}'$. Thus

$$f_{em} = i\frac{2Ze\mu}{\theta\hbar c}\boldsymbol{\sigma}\cdot\boldsymbol{\nu}.$$

The nuclear scattering amplitude must be added to this expression. Owing to the rapid decrease of the nuclear forces with increasing distance, this amplitude tends for small angles to a finite (energy-dependent) complex limit, which we denote by a. The total scattering amplitude is therefore

$$f = a + i(b/\theta)\boldsymbol{\sigma}\cdot\boldsymbol{\nu}, \qquad b = 2Ze\mu/c\hbar = 2Z\alpha\mu/e. \tag{42.3}$$

We see that the electromagnetic scattering is indeed predominant at sufficiently small angles.

The expression (42.3) is the same in form as that discussed in *QM*, §140. We can therefore make direct use of the formulae derived there. The scattering cross-section summed over all possible final polarization states is

$$\frac{d\sigma}{do} = |a|^2 + \frac{b^2}{\theta^2} + 2b \,\mathrm{im}\, a\cdot\boldsymbol{\nu}\cdot\boldsymbol{\zeta}, \tag{42.4}$$

where $\boldsymbol{\zeta}$ is the initial polarization of the neutron beam (called $\mathbf{P}$ in *QM*, §140). If the initial state is unpolarized ($\boldsymbol{\zeta} = 0$), then the polarization after scattering is

$$\boldsymbol{\zeta}' = \frac{2b\theta \,\mathrm{im}\, a}{|a|^2\theta^2 + b^2}\boldsymbol{\nu}. \tag{42.5}$$

This is a maximum when $\theta = b/|a|$, and $\zeta'_{max} = \mathrm{im}\, a/|a|$.

CHAPTER V

RADIATION

§43. The electromagnetic interaction operator

THE interaction of electrons with an electromagnetic field can, as a rule, be treated by means of perturbation theory. This is because the electromagnetic interaction is comparatively weak, as is shown by the smallness of the corresponding dimensionless "coupling constant", viz. the fine-structure constant $\alpha = e^2/\hbar c = 1/137$. The smallness of this number is of fundamental importance in quantum electrodynamics.

In classical electrodynamics (see *Fields*, §28), the electromagnetic interaction is described by the term

$$-ej^{\mu}A_{\mu} \tag{43.1}$$

in the Lagrangian density of the "field + charge" system (A being the 4-potential of the field and j the particle current density 4-vector). The current density satisfies the equation of continuity,

$$\partial_{\mu}j^{\mu} = 0, \tag{43.2}$$

which expresses the law of conservation of charge. According to *Fields*, §29, the gauge invariance of the theory is closely related to this law: when A_{μ} is replaced by $A_{\mu} + \partial_{\mu}\chi$ (4.1), a term $-ej^{\mu}\partial_{\mu}\chi$ is added to the Lagrangian density (43.1), and this, by (43.2), may be written as the 4-divergence $-e\partial_{\mu}(\chi j^{\mu})$; it therefore disappears on integration over d^4x in the action $S = \int L\, d^4x$.

In quantum electrodynamics, the 4-vectors j and A are replaced by the corresponding second-quantized operators. The current operator is expressed in terms of the ψ-operators by $\hat{j} = \hat{\bar{\psi}}\gamma\hat{\psi}$. The generalized "coordinates" q in the Lagrangian

$$\int \hat{L}_{\text{inter}}\, d^3x = -e\int (\hat{j}\hat{A})\, d^3x$$

are represented by the values of $\hat{\bar{\psi}}$, $\hat{\psi}$ and $\hat{A}$ at each point in space. Since the Lagrangian density is found to depend only on the "coordinates" q themselves (and not on their derivatives with respect to x), the change to the Hamiltonian density by formula (10.11) amounts simply to a change in the sign of the Lagrangian density.† Thus the electromagnetic interaction operator (the space integral of

† Independently of these arguments, it may be noted that, when only the first-order small correction is considered, any small correction in the Lagrangian appears in the Hamiltonian with just a change of sign (see *Mechanics*, §40).

the interaction Hamiltonian density) has the form

$$\hat{V} = e \int (\hat{j}\hat{A})\, d^3x. \tag{43.3}$$

The free electromagnetic field operator is the sum

$$\hat{A} = \sum_n [\hat{c}_n A_n(x) + \hat{c}_n^+ A_n^*(x)], \tag{43.4}$$

which contains the operators of photon creation and annihilation in various states labelled by the suffix n. Each operator has matrix elements only for an increase or decrease of the corresponding occupation number N_n by 1 (the other occupation numbers remaining unchanged). The operator $\hat{A}$ therefore also has matrix elements only for transitions in which the number of photons changes by 1. That is, only processes of the emission or absorption of a single photon occur in the first approximation of perturbation theory.

According to (2.15), the matrix elements are

$$\langle N_n - 1|c_n|N_n\rangle = \langle N_n|c_n^+|N_n - 1\rangle = \sqrt{N_n}. \tag{43.5}$$

If there are no photons (of type n) in the initial state of the field, then $\langle 1|c_n^+|0\rangle = 1$. The matrix element of the operator (43.3) for photon emission is

$$V_{fi}(t) = e \int (\mathbf{j}_{fi} A_n^*)\, d^3x, \tag{43.6}$$

where $A_n(x)$ is the wave function of the emitted photon and j_{fi} the matrix element of the operator $\hat{j}$ for a transition of the emitter from the initial state i to the final state f.† The 4-vector $j_{fi}^{\mu} = (\rho_{fi}, \mathbf{j}_{fi})$ is called the *transition current.*

Similarly, we obtain the matrix element for photon absorption:

$$V_{fi}(t) = e \int (\mathbf{j}_{fi} A_n)\, d^3x. \tag{43.7}$$

This differs from (43.6) only by having $A_n(x)$ in place of $A_n^*(x)$.

The argument t of V_{fi} is shown in order to emphasize that the matrix element is time-dependent. By separating the time factors in the wave functions, we can change in the usual way to time-independent matrix elements:

$$V_{fi}(t) = V_{fi}\, e^{-i(E_i - E_f \mp \omega)t}, \tag{43.8}$$

where E_i, E_f are the initial and final energies of the emitting system, and the sign $\mp$ is for emission and absorption respectively of a photon ω.

† The notation in (43.6) is slightly inconsistent. The suffixes in V_{fi} refer to states of the whole system "emitter + field", those in j_{fi} to states of the emitter only.

The wave function of a photon with a definite momentum **k** and a definite polarization is

$$A^{\mu} = \sqrt{(4\pi)} \frac{e^{\mu}}{\sqrt{(2\omega)}} e^{i\mathbf{k}\cdot\mathbf{r}} \tag{43.9}$$

(see (4.3); the time factor is omitted). Substituting in (43.6), we find the matrix element for the emission of such a photon:

$$V_{fi} = e\sqrt{(4\pi)} \frac{1}{\sqrt{(2\omega)}} e^{*}_{\mu} j^{\mu}_{fi}(\mathbf{k}), \tag{43.10}$$

where $j_{fi}(\mathbf{k})$ is the transition current in the momentum representation, i.e. the Fourier component

$$j_{fi}(\mathbf{k}) = \int j_{fi}(\mathbf{r})\, e^{-i\mathbf{k}\cdot\mathbf{r}}\, d^3x. \tag{43.11}$$

The corresponding formula for photon absorption is

$$V_{fi} = e\sqrt{(4\pi)} \frac{1}{\sqrt{(2\omega)}} e_{\mu} j^{\mu}_{fi}(-\mathbf{k}). \tag{43.12}$$

The equation of conservation of current in the momentum representation is the condition of 4-transversality of the transition currents:

$$k_{\mu} j^{\mu}_{fi} = \omega \rho_{fi}(\mathbf{k}) - \mathbf{k}\cdot\mathbf{j}_{fi}(\mathbf{k}) = 0. \tag{43.13}$$

The formulae given in this section do not assume any particular form of the current operator, and are generally valid for electromagnetic processes involving any charged particles. The existing theory allows the form of the current operator to be determined (and hence, in principle, its matrix elements to be calculated) only for electrons. For applications to systems of strongly interacting particles, including nuclei, a semi-phenomenological theory will be used, in which the transition currents appear as empirically determined quantities subject only to the conditions of space–time symmetry and to the equation of continuity.

§ 44. **Emission and absorption**

The transition probability under the action of a perturbation $\hat{V}$ is given, in the first approximation, by the well-known formulae of perturbation theory (*QM*, §42). Let the initial and final states of the emitting system belong to the discrete spectrum.† Then the probability (per unit time) of the transition $i \to f$ with emission

† This certainly implies that recoil is neglected, the emitter as a whole remaining at rest.

of a photon is

$$dw = 2\pi|V_{fi}|^2\delta(E_i - E_f - \omega)\,d\nu, \tag{44.1}$$

where $d\nu$ arbitrarily denotes the ensemble of quantities describing the state of the photon and taking a continuous sequence of values; the photon wave function is assumed normalized by the delta function "on the ν scale".

If a photon having a definite angular momentum is emitted, the only continuous variable is the frequency ω. Integration of (44.1) with respect to $d\nu \equiv d\omega$ eliminates the delta function, ω being replaced by $E_i - E_f$, and the transition probability is

$$w = 2\pi|V_{fi}|^2. \tag{44.2}$$

If, however, we consider the emission of a photon having a given momentum **k**, then $d\nu \equiv d^3k/(2\pi)^3 = \omega^2\,d\omega\,do/(2\pi)^3$. Here it is presupposed that the photon wave function (plane wave) is "normalized to one photon in the volume $V = 1$", as always in this book; $d\nu$ is the number of states in the phase volume $V\,d^3k$. Thus the probability of emission of a photon with a given momentum is

$$dw = 2\pi|V_{fi}|^2\delta(E_i - E_f - \omega)\,d^3k/(2\pi)^3, \tag{44.3}$$

or, after integration over $d\omega$,

$$dw = \frac{1}{4\pi^2}|V_{fi}|^2\omega^2\,do. \tag{44.4}$$

In this we must substitute the matrix element V_{fi} from (43.10).

In subsequent sections we shall use these formulae to calculate the probability of emission in various specific cases. Here we shall consider certain general relations between radiative processes of various kinds.

If in the initial state of the field there is already a non-zero number N_n of the photons in question, the matrix element for the transition is multiplied by

$$\langle N_n + 1|c_n^+|N_n\rangle = \sqrt{(N_n + 1)}, \tag{44.5}$$

i.e. the transition probability is multiplied by $N_n + 1$. The 1 in this factor corresponds to the spontaneous emission which occurs even if $N_n = 0$. The term N_n represents the *stimulated* or *induced* emission: we see that the presence of photons in the initial state of the field stimulates the further emission of photons of the same kind.

The matrix element V_{if} for the transition with the opposite change of state of the system ($f \to i$) differs from V_{fi} in that (44.5) is replaced by

$$\langle N_n - 1|c_n|N_n\rangle = \sqrt{N_n}$$

(and the other quantities are replaced by their complex conjugates). This opposite transition is a transition of the system from the level E_f to the level E_i with absorption of a photon. Thus the photon emission and absorption probabilities for a given pair of states i, f are related by†

$$w_e/w_a = (N_n + 1)/N_n, \tag{44.6}$$

an expression first derived by A. Einstein (1916).

The number of photons can be related to the intensity of the external radiation incident on the system. Let

$$I_{\mathbf{ke}}\, d\omega\, do \tag{44.7}$$

be the radiation energy incident on unit area per unit time and having polarization $\mathbf{e}$, frequency in the range $d\omega$ and wave-vector direction in the solid-angle element do. These ranges correspond to $k^2\, dk\, do/(2\pi)^3$ field oscillators, each having $N_{\mathbf{ke}}$ photons of the specified polarization. Hence the same energy (44.7) is given by the product

$$c\frac{k^2\, dk\, do}{(2\pi)^3} N_{\mathbf{ke}}\hbar\omega = \frac{\hbar\omega^3}{8\pi^3 c^2} N_{\mathbf{ke}}\, d\omega\, do.$$

From this we find the required relation:

$$N_{\mathbf{ke}} = \frac{8\pi^3 c^2}{\hbar\omega^3} I_{\mathbf{ke}}. \tag{44.8}$$

Let $dw_{\mathbf{ke}}^{(sp)}$ be the probability of spontaneous emission of a photon with polarization $\mathbf{e}$ into the solid angle do, and let the indices (in) and (a) denote the corresponding probabilities for induced emission and for absorption. According to (44.6) and (44.8), these probabilities are related as follows:

$$dw_{\mathbf{ke}}^{(a)} = dw_{\mathbf{ke}}^{(in)} = dw_{\mathbf{ke}}^{(sp)} \cdot \frac{8\pi^3 c^2}{\hbar\omega^3} I_{\mathbf{ke}}. \tag{44.9}$$

If the incident radiation is isotropic and unpolarized ($I_{\mathbf{ke}}$ independent of the directions of $\mathbf{k}$ and $\mathbf{e}$), then the integration of (44.9) with respect to do and summation with respect to $\mathbf{e}$ gives similar relations between the total probabilities of radiative transitions (between given states i and f of the system):

$$w^{(a)} = w^{(in)} = w^{(sp)} \frac{\pi^2 c^2}{\hbar\omega^3} I, \tag{44.10}$$

where $I = 2 \times 4\pi I_{\mathbf{ke}}$ is the total spectral intensity of the incident radiation.

If the states i and f of the emitting (or absorbing) system are degenerate, the total probability of emission (or absorption) of the photons concerned is found by summation over all mutually degenerate final states and averaging over all possible

† In the rest of this section, ordinary units are used.

initial states. Let the degrees of degeneracy (statistical weights) of states i and f be g_i and g_f. For processes of spontaneous or induced emission, the states i are the initial states, and for absorption the states f. Assuming in each case that all g_i or g_f initial states are equally probable, we obviously have instead of (44.10) the relations

$$g_f w^{(a)} = g_i w^{(in)} = g_i w^{(sp)} \frac{\pi^2 c^2}{\hbar\omega^3} I. \tag{44.11}$$

In the literature one frequently meets the *Einstein coefficients*, defined as

$$A_{if} = w^{(sp)}, \quad B_{if} = w^{(in)} c/I, \quad B_{fi} = w^{(a)} c/I, \tag{44.12}$$

where I/c is the spatial spectral density of radiation energy. They are related by the equations

$$g_f B_{fi} = g_i B_{if} = g_i A_{if} \pi^2 c^3/\hbar\omega^3. \tag{44.13}$$

§45. **Dipole radiation**

Let us apply the formulae derived above to the emission of a photon by an electron (in general, a relativistic electron) moving in a given external field. In this case the transition current is the matrix element of the operator

$$\hat{j} = \hat{\bar{\psi}}\gamma\hat{\psi},$$

in which the ψ-operators are assumed expanded in terms of the wave functions of stationary states of the electron in a given field (§32). The matrix element $\langle 0_i 1_f | j | 1_i 0_f \rangle$ corresponds to a transition of the electron from state i to state f. This change in the occupation numbers is brought about by the operator $\hat{a}_f^+ \hat{a}_i$, and the transition current is

$$j_{fi}^{\mu} = \bar{\psi}_f \gamma^{\mu} \psi_i = (\psi_f^* \psi_i, \psi_f^* \boldsymbol{\alpha} \psi_i), \tag{45.1}$$

where ψ_i and ψ_f are the wave functions of the initial and final states of the electron.

Let the wave function of the photon be chosen in the three-dimensionally transverse gauge (the polarization 4-vector $e = (0, \mathbf{e})$). Then the product $j_{fi}e^* = -\mathbf{j}_{fi} \cdot \mathbf{e}^*$ in (43.10). Substituting V_{fi} in (44.4), we obtain the following expression for the probability (per unit time) of emission of a photon with polarization $\mathbf{e}$ into the solid-angle element do:

$$dw_{\mathbf{en}} = e^2(\omega/2\pi)|\mathbf{e}^* \cdot \mathbf{j}_{fi}(\mathbf{k})|^2 \, do, \tag{45.2}$$

where

$$\mathbf{j}_{fi}(\mathbf{k}) = \int \psi_f^* \boldsymbol{\alpha} \psi_i \cdot e^{-i\mathbf{k}\cdot\mathbf{r}} \, d^3x. \tag{45.3}$$

Summation with respect to the polarization of the photon is effected by averaging over the directions of **e** (in a plane perpendicular to the given direction $\mathbf{n} = \mathbf{k}/\omega$), and the result is then doubled because of the two independent possible transverse polarizations of the photon.† Thus the result is

$$dw_{\mathbf{n}} = e^2(\omega/2\pi)|\mathbf{n} \times \mathbf{j}_{fi}(\mathbf{k})|^2 \, do. \tag{45.4}$$

A very important case is that where the photon wavelength λ is large compared with the dimensions a of the radiating system. This usually means that the velocity of the particles is small compared with that of light. In the first approximation in a/λ (corresponding to dipole radiation; cf. *Fields*, §67), the factor $e^{-i\mathbf{k}\cdot\mathbf{r}}$ varies only slightly in the region where ψ_i or ψ_f is appreciably different from zero, and it can be replaced by unity in the transition current (45.3). This implies that the photon momentum is neglected in comparison with the momenta of the particles in the system.

In the same approximation, the integral $\mathbf{j}_{fi}(0)$ may be replaced by its non-relativistic value, which is simply the matrix element $\mathbf{v}_{fi}$ of the electron velocity with respect to the Schrödinger wave functions. In turn, this element $\mathbf{v}_{fi} = -i\omega\mathbf{r}_{fi}$, and $e\mathbf{r}_{fi} = \mathbf{d}_{fi}$, where **d** is the dipole moment of the electron (in its orbital motion). Thus we have the following formula for the probability of dipole radiation:

$$dw_{\mathbf{en}} = (\omega^3/2\pi)|\mathbf{e}^* \cdot \mathbf{d}_{fi}|^2 \, do. \tag{45.5}$$

(Here the direction of **n** occurs implicitly: the vector **e** must be perpendicular to **n**.) Summation with respect to the polarizations gives

$$dw_{\mathbf{n}} = (\omega^3/2\pi)|\mathbf{n} \times \mathbf{d}_{fi}|^2 \, do. \tag{45.6}$$

Since these formulae are non-relativistic (as regards the electron), they can be immediately generalized to any electron system by taking $\mathbf{d}_{fi}$ as the matrix element of the total dipole moment of the system.

Integrating (45.6) over all directions, we have the total probability of radiation:

$$w = (4\omega^3/3)|\mathbf{d}_{fi}|^2, \tag{45.7}$$

or, in ordinary units,

$$w = (4\omega^3/3\hbar c^3)|\mathbf{d}_{fi}|^2. \tag{45.7a}$$

† In the averaging, we use the formula

$$\overline{e_i e_k^*} = \tfrac{1}{2}(\delta_{ik} - n_i n_k) \tag{45.4a}$$

or

$$\begin{aligned}\overline{(\mathbf{a}\cdot\mathbf{e})(\mathbf{b}\cdot\mathbf{e}^*)} &= \tfrac{1}{2}\{\mathbf{a}\cdot\mathbf{b} - (\mathbf{a}\cdot\mathbf{n})(\mathbf{b}\cdot\mathbf{n})\} \\ &= \tfrac{1}{2}[\mathbf{a}\times\mathbf{n}]\cdot[\mathbf{b}\times\mathbf{n}],\end{aligned} \tag{45.4b}$$

where **a** and **b** are constant vectors.

The intensity I is found by multiplying the probability by $\hbar\omega$:

$$I = (4\omega^4/3c^3)|\mathbf{d}_{fi}|^2. \tag{45.8}$$

This is directly analogous to the classical formula (see *Fields*, (67.11)) for the intensity of dipole radiation from a system of periodically moving particles: the intensity of radiation at frequency $\omega_s = s\omega$ (where ω is the frequency of the particle motion and s an integer) is

$$I_s = (4\omega_s^4/3c^3)|\mathbf{d}_s|^2, \tag{45.9}$$

where $\mathbf{d}_s$ are the Fourier components of the dipole moment, i.e. the coefficients in the expansion

$$\mathbf{d}(t) = \sum_{s=-\infty}^{\infty} \mathbf{d}_s\, e^{-is\omega t}. \tag{45.10}$$

The quantum formula (45.8) is got from (45.9) by replacing these Fourier components by the matrix elements of the corresponding transitions. This rule (which is an expression of Bohr's *correspondence principle*) is a particular case of a general relation between the Fourier components of classical quantities and the quantum matrix elements in the quasi-classical case (see *QM*, §48). The radiation is quasi-classical for transitions between states having large quantum numbers; the transition energy $\hbar\omega = E_i - E_f$ is then small in comparison with the energies E_i and E_f of the radiator. This, however, would not lead to any change in the form of (45.8), which is valid for all transitions. This explains the fact (which is something of an accident) that the correspondence principle for the radiation intensity is valid not only in the quasi-classical but in the general quantum case.

§ 46. **Electric multipole radiation**

Instead of considering the emission of a photon in a given direction (i.e. with a given momentum), let us now consider the emission of a photon with definite values of the angular momentum j and its component m in some chosen direction z. We have seen in §6 that such photons can be of two kinds, electric and magnetic. Let us take first the emission of electric photons, and again assume that the dimensions of the radiating system are small in comparison with the wavelength.

The calculations are conveniently carried out by means of the photon wave functions in the momentum representation, i.e. by expressing the 4-vector $A^\mu(\mathbf{r})$ as a Fourier integral. Then the matrix element is

$$\begin{aligned} V_{fi} &= e\int j^{\mu}_{fi}(\mathbf{r})A^*_{\mu}(\mathbf{r})\,d^3x \\ &= e\int d^3x\cdot j^{\mu}_{fi}(\mathbf{r})\int \frac{d^3k}{(2\pi)^3}\,A^*_{\mu}(\mathbf{k})\,e^{-i\mathbf{k}\cdot\mathbf{r}}; \end{aligned} \tag{46.1}$$

for simplicity, we omit the suffixes ωjm to the photon wave functions.

For an Ej photon we take the wave function from (7.10), with the arbitrary constant C having the value

$$C = -\sqrt{\frac{j+1}{j}}.$$

The reason for this choice is to ensure that, in the spatial components of the wave function ($\mathbf{A}$), the terms containing spherical harmonics of order $j-1$ cancel (as is seen from formulae (7.16)). Then $\mathbf{A}$ will include only spherical harmonics of order $j+1$, and therefore the corresponding contribution to V_{fi} is (as will be clear from the subsequent calculation) of a higher order of smallness (in a/λ) than the contribution from the component $A^0 \equiv \Phi$, which includes spherical harmonics of the lower order j.

Thus we put

$$A^\mu = (\Phi, 0), \qquad \Phi = -\sqrt{\frac{j+1}{j}} \frac{4\pi^2}{\omega^{3/2}} \delta(|\mathbf{k}| - \omega) Y_{jm}(\mathbf{n})$$

($\mathbf{n} = \mathbf{k}/\omega$). Substituting this expression in (46.1) and carrying out the integration over $d|\mathbf{k}|$, we obtain

$$V_{fi} = -e\sqrt{\frac{j+1}{j}} \frac{\sqrt{\omega}}{2\pi} \int d^3x \cdot \rho_{fi}(\mathbf{r}) \int do_{\mathbf{n}}\, e^{-i\mathbf{k}\cdot\mathbf{r}} Y^*_{jm}(\mathbf{n}). \tag{46.2}$$

To calculate the inner integral, we use the expansion (24.12), written in the form

$$e^{i\mathbf{k}\cdot\mathbf{r}} = 4\pi \sum_{l=0}^{\infty} \sum_{m=-l}^{l} i^l g_l(kr) Y^*_{lm}(\mathbf{k}/k) Y_{lm}(\mathbf{r}/r), \tag{46.3}$$

where

$$g_l(kr) = \sqrt{\frac{\pi}{2kr}} J_{l+\frac{1}{2}}(kr); \tag{46.4}$$

see *QM*, (34.3).†

Substitution of this expansion in (46.2) gives

$$\int e^{-i\mathbf{k}\cdot\mathbf{r}} Y^*_{jm}(\mathbf{n})\, do_{\mathbf{n}} = 4\pi i^{-j} g_j(kr) Y^*_{jm}(\mathbf{r}/r);$$

the remaining terms are zero because of the orthogonality of the spherical harmonics. On account of the condition $a/\lambda \ll 1$, only distances such that $kr \ll 1$ will be important in the integral with respect to d^3x. We can therefore replace the

† The normalization of the functions g_l is such that their asymptotic form as $kr \to \infty$ is

$$g_l(kr) \approx (1/kr)\sin(kr - \tfrac{1}{2}l\pi). \tag{46.4a}$$

functions $g_j(kr)$ by the first terms of their expansions in powers of kr:†

$$g_j(kr) \approx (kr)^j/(2j+1)!!. \tag{46.5}$$

The result is

$$V_{fi} = (-1)^{m+1} i^j \sqrt{\frac{(2j+1)(j+1)}{\pi j}} \frac{\omega^{j+\frac{1}{2}}}{(2j+1)!!} e(Q^{(e)}_{j,-m})_{fi}, \tag{46.6}$$

with the notation

$$(Q^{(e)}_{jm})_{fi} = \sqrt{\frac{4\pi}{2j+1}} \int \rho_{fi}(\mathbf{r}) r^j Y_{jm}(\mathbf{r}/r)\, d^3x \tag{46.7}$$

($Y_{j,-m} = (-1)^{j-m} Y^*_{jm}$). The quantities (46.7) are called the 2^j-*pole electric transition moments* of the system, by analogy with the corresponding classical quantities (*Fields*, §41).‡

For an electron in an external field, $\rho_{fi} = \psi^*_f \psi_i$, and the quantities (46.7) are then calculated as the matrix elements of the classical quantity

$$Q^{(e)}_{jm} = \sqrt{\frac{4\pi}{2j+1}} r^j Y_{jm}.$$

In the non-relativistic case (as regards the particle velocities), the transition moment can in principle be calculated similarly for any system of N interacting particles. The transition density is expressed in terms of the wave functions of the system by

$$\rho_{fi}(\mathbf{r}) = \int \psi^*_f(\mathbf{r}_1, \ldots \mathbf{r}_N) \psi_i(\mathbf{r}_1, \ldots, \mathbf{r}_N) \sum_{n=1}^{N} \delta(\mathbf{r} - \mathbf{r}_n)\, d^3x_1 \ldots d^3x_N, \tag{46.8}$$

where the integral is taken over the whole of configuration space.§

The photon wave function used here corresponds (in the coordinate representation) to normalization by the delta function on the ω scale, as assumed in formula (44.2). Substituting (46.6), we find the probability of Ej radiation:‖

$$w^{(e)}_{jm} = \frac{2(2j+1)(j+1)}{j[(2j+1)!!]^2} \omega^{2j+1} e^2 |(Q^{(e)}_{j,-m})_{fi}|^2. \tag{46.9}$$

† The power of kr is equal to the order of the function Y_{jm} by which g_j is multiplied. This justifies the neglect of the terms in **A** which contain higher-order spherical harmonics.

‡ The multipole moments are defined without the factor e, since in this book the currents also are defined without the charge factor.

§ A situation can occur where the transition probability vanishes according to the approximate selection rules, valid only when the spin–orbit interaction of the electrons is neglected. Then, to obtain a non-zero result, we must use the wave functions with the relativistic correction which takes account of this interaction.

‖ It might appear at first sight that, owing to the isotropy of space, the total probability of photon emission ought not to depend on the value of m. The incorrectness of this conclusion is easily seen if we notice that different final states of the system (for a given initial state) correspond to the emission of photons with different values of m; cf. the rule (46.16) below.

In particular, for $j = 1$ we have

$$w^{(e)}_{1m} = \frac{4\omega^3}{3} e^2 |(Q^{(e)}_{1,-m})_{fi}|^2. \tag{46.10}$$

The quantities $Q^{(e)}_{1m}$ are related to the components of the electric dipole moment vector by

$$eQ^{(e)}_{10} = id_z, \qquad eQ^{(e)}_{1,\pm1} = \mp\frac{i}{\sqrt{2}}(d_x \pm id_y). \tag{46.11}$$

Summing (46.10) with respect to m, we naturally obtain the earlier formula (45.7) for the total probability of dipole radiation.

The angular distribution of multipole radiation is given by formula (7.11). When this is normalized to the total emission probability w_{jm}, we have

$$\begin{aligned} dw_{jm} &= |\mathbf{Y}^{(e)}_{jm}(\mathbf{n})|^2 w_{jm}\, do \\ &= \frac{w_{jm}}{j(j+1)} |\nabla_{\mathbf{n}} Y_{jm}|^2\, do. \end{aligned} \tag{46.12}$$

In particular, for $j = 1$,

$$Y_{10} = i\sqrt{\frac{3}{4\pi}}\cos\theta, \qquad Y_{1,\pm1} = \mp i\sqrt{\frac{3}{8\pi}}\sin\theta \cdot e^{\pm i\phi},$$

where θ and ϕ are the polar angle and azimuth of the direction $\mathbf{n}$ relative to the z-axis. On calculating the gradient, we find that the angular distribution of dipole radiation with a definite value of m is given by

$$dw_{10} = w_{10}\frac{3}{8\pi}\sin^2\theta\, do, \qquad dw_{1,\pm1} = w_{1,\pm1}\frac{3}{8\pi}\frac{1+\cos^2\theta}{2}\, do. \tag{46.13}$$

These expressions could also, of course, be obtained from formula (45.6) by putting firstly (for $m = 0$) $d_x = d_y = 0$, $d_z = d$, secondly (for $m = \pm 1$) $d_y = \mp id_x = d/\sqrt{2}$, $d_z = 0$.

If the order of magnitude of the dimensions of the system (atom or nucleus) is a, then that of the electric multipole moments is, in general, $Q^{(e)}_{jm} \sim a^j$. The probability of multipole radiation is

$$w^{(e)}_{jm} \sim \alpha k (ka)^{2j}. \tag{46.14}$$

When the multipole order increases by one, the probability decreases by a factor $\sim(ka)^2$.

The laws of conservation of angular momentum and parity imply certain selection rules which restrict the possible changes in the state of the radiating system. If the initial angular momentum of the system is J_i, then, after emission of a photon with angular momentum j, the angular momentum of the system can have only those values J_f which are in accordance with the angular momentum addition rule ($\mathbf{J}_i - \mathbf{J}_f = \mathbf{j}$):

$$|J_i - J_f| \leqslant j \leqslant J_I + J_f. \tag{46.15}$$

For given values of J_i and J_f, the same rule (46.15) specifies the possible values of the photon angular momentum j. But, since the probability of emission decreases rapidly with increasing j, the emission occurs principally with the lowest possible multipole order.

The components M_i and M_f of the angular momenta $\mathbf{J}_i$ and $\mathbf{J}_f$, and m of the photon angular momentum, satisfy the relation

$$M_i - M_f = m, \tag{46.16}$$

which is obvious from the same law of addition of angular momenta.

The parities P_i and P_f of the initial and final states of the radiating system must be such that $P_f P_{\text{ph}} = P_i$, where P_{ph} is the parity of the emitted photon. Since the parities can have only the values ± 1, this condition may also be written

$$P_i P_f = P_{\text{ph}}. \tag{46.17}$$

For an electric photon $P_{\text{ph}} = (-1)^j$, and the parity selection rule for electric multipole radiation is therefore

$$P_i P_f = (-1)^j. \tag{46.18}$$

The selection rules for total angular momentum and for parity are entirely rigorous and must be satisfied in emission by any systems. There may also be other rules which are more restrictive and which arise from certain properties of the structure of particular radiating systems. These latter rules must of necessity be approximate to some extent; they will be discussed in later sections of this chapter.

The dependence of the emission probability on the quantum numbers m, M_i and M_f is entirely determined by the tensor character of the multipole moments. The quantities Q_{jm} with a given j form a spherical tensor of rank j. The dependence of its matrix elements on these quantum numbers is given by the formula

$$|\langle n_f J_f M_f | Q_{j,-m} | n_i J_i M_i \rangle|^2 = \begin{pmatrix} J_f & j & J_i \\ M_f & m & -M_i \end{pmatrix}^2 |\langle n_f J_f \| Q_j \| n_i J_i \rangle|^2 \tag{46.19}$$

(see *QM*, (107.6)), where n conventionally denotes all the quantum numbers specifying the state of the system, other than J and M. The reduced matrix elements on the right of (46.19) do not depend on m, M_i, M_f. On substituting this formula in (46.9), we obtain the required dependence, which is proportional to

$$\begin{pmatrix} J_f & j & J_i \\ M_f & m & -M_i \end{pmatrix}^2;$$

here it is, of course, assumed that the emitter is not in an external field, and that the transition frequency ω is thus independent of M_i and M_f.

Summing the probability over all values of M_f (for a given M_i), we have the total probability of emission of a photon of a given frequency from the initial level n_i, J_i of the system. It is obvious from the isotropy of space that this quantity must also be independent of the initial value M_i. The summation is carried out by means of the formula

$$\sum_{M_f} |\langle n_f J_f M_f | Q_{j,-m} | n_i J_i M_i \rangle|^2 = \frac{1}{2J_i+1} |\langle n_f J_f \| Q_j \| n_i J_i \rangle|^2 \tag{46.20}$$

(see *QM*, (107.11)).

§47. Magnetic multipole radiation

The wave function of a magnetic photon is $A^\mu = (0, \mathbf{A})$, where $\mathbf{A}$ is given by (7.6). Substitution in (46.1) gives for the transition matrix element

$$V_{fi} = -e\frac{\sqrt{\omega}}{2\pi}\int d^3x \cdot \mathbf{j}_{fi}(\mathbf{r}) \int do_{\mathbf{n}} \cdot e^{-i\mathbf{k}\cdot\mathbf{r}}\mathbf{Y}^{(m)*}_{jm}(\mathbf{n}). \tag{47.1}$$

The components of the vector $\mathbf{Y}^{(m)}_{jm}$ can be expressed in terms of the spherical harmonics of order j, as shown in (7.16). Again using the expansion (46.3), we obtain for the inner integral

$$\int e^{-i\mathbf{k}\cdot\mathbf{r}}\mathbf{Y}^{(m)*}_{jm}(\mathbf{n})\, do_{\mathbf{n}} = 4\pi i^{-j} g_j(kr)\mathbf{Y}^{(m)*}_{jm}(\mathbf{r}/r),$$

and, on substituting g_j from (46.5),†

$$V_{fi} = -ei^{-j}\frac{2\omega^{j+\frac{1}{2}}}{(2j+1)!!}\int \mathbf{j}_{fi}(\mathbf{r})r^j \cdot \mathbf{Y}^{(m)*}_{jm}(\mathbf{r}/r)\, d^3x.$$

Here we must substitute, in accordance with the definition (7.4),

$$\mathbf{Y}^{(m)}_{jm}(\mathbf{r}/r) = \frac{1}{\sqrt{[j(j+1)]}}\mathbf{r}\times\nabla Y_{jm};$$

we then transform the integrand by means of the formula

$$r^j\mathbf{j}_{fi}\cdot\mathbf{r}\times\nabla Y^*_{jm} = -\mathbf{r}\times\mathbf{j}_{fi}\cdot\nabla(r^j Y^*_{jm}),$$

obtaining

$$V_{fi} = (-1)^m i^j\sqrt{\frac{(2j+1)(j+1)}{\pi j}}\frac{\omega^{j+\frac{1}{2}}}{(2j+1)!!}e(Q^{(m)}_{j,-m})_{fi}, \tag{47.2}$$

† The current $\mathbf{j}$ must not be confused with the angular momentum j.

with the notation

$$(Q_{jm}^{(m)})_{fi} = \frac{1}{j+1}\sqrt{\frac{4\pi}{2j+1}}\int \mathbf{r}\times\mathbf{j}_{fi}\cdot\nabla(r^j Y_{jm})\,d^3x. \tag{47.3}$$

These are called the 2^j-*pole magnetic transition moments.*

Because of the analogy between the expressions (47.2) and (46.6) for the emission probability, we obtain a formula which differs from (46.10) only in that the electric moments are replaced by magnetic moments. Formula (46.12) for the angular distribution also remains valid (as has already been mentioned in connection with (7.11)).

Let us analyse the form of (47.3) when $j = 1$. In this case, the functions are

$$\sqrt{\frac{4\pi}{3}}\,rY_{10} = iz, \qquad \sqrt{\frac{4\pi}{3}}\,rY_{1,\pm1} = \mp\frac{i}{\sqrt{2}}(x\pm iy),$$

and their gradients are simply the spherical unit vectors $\mathbf{e}^{(0)}$, $\mathbf{e}^{(\pm1)}$ (7.14). The quantities $e(Q_{1m}^{(m)})_{fi}$ are therefore the spherical components of the vector

$$\boldsymbol{\mu}_{fi} = \tfrac{1}{2}e\int \mathbf{r}\times\mathbf{j}_{fi}\,d^3x, \tag{47.4}$$

which is similar in form to the classical magnetic moment (see *Fields*, §44). The total probability of $M1$ radiation is given in terms of this quantity by the formula (in ordinary units)

$$w = (4\omega^3/3\hbar c^3)|\boldsymbol{\mu}_{fi}|^2. \tag{47.5}$$

We shall show how formula (47.4) is related to the usual non-relativistic quantum expression for the magnetic moment operator.

The expression for the transition current is (see *QM*, §115)

$$\mathbf{j}_{fi} = -\frac{i}{2m}(\psi_f^*\nabla\psi_i - \psi_i\nabla\psi_f^*) + \frac{\mu}{es}\,\mathrm{curl}(\psi_f^*\hat{\mathbf{s}}\psi_i), \tag{47.6}$$

where μ is the magnetic moment of the particle and s its spin. Hence

$$\boldsymbol{\mu}_{fi} = -\frac{ie}{4m}\int \psi_f^*(\mathbf{r}\times\nabla)\psi_i\,d^3x + \frac{ie}{4m}\int \psi_i(\mathbf{r}\times\nabla)\psi_f^*\,d^3x + \frac{\mu}{2s}\int \mathbf{r}\times\mathrm{curl}(\psi_f^*\hat{\mathbf{s}}\psi_i)\,d^3x. \tag{47.7}$$

In the second term, we write

$$\int \psi_i(\mathbf{r}\times\nabla)\psi_f^*\,d^3x = -\int \psi_f^*(\mathbf{r}\times\nabla)\psi_i\,d^3x + \int \mathrm{curl}(\mathbf{r}\psi_f^*\psi_i)\,d^3x.$$

The last integral can be transformed into one over an infinitely distant surface, and is zero. Thus the first two terms in (47.7) are equal. In the third term, we transform the integral as follows (temporarily writing $\mathbf{F} = \psi_f^* \hat{\mathbf{s}} \psi_i$):

$$\int \mathbf{r} \times (\nabla \times \mathbf{F})\, d^3x = \oint \mathbf{r} \times (d\mathbf{f} \times \mathbf{F}) - \int (\mathbf{F} \times \nabla) \times \mathbf{r}\, d^3x.$$

The surface integral is zero, and in the last term

$$(\mathbf{F} \times \nabla) \times \mathbf{r} = -\mathbf{F}\, \mathrm{div}\, \mathbf{r} + \mathbf{F} = -2\mathbf{F}.$$

Thus,

$$\int \mathbf{r} \times \mathrm{curl}\, \mathbf{F}\, d^3x = 2 \int \mathbf{F}\, d^3x.$$

The expression for $\boldsymbol{\mu}_{fi}$ therefore becomes

$$\boldsymbol{\mu}_{fi} = \int \psi_f^* \left(\frac{e}{2m} \hat{\mathbf{L}} + \frac{\mu}{s} \hat{\mathbf{s}} \right) \psi_i\, d^3x, \tag{47.8}$$

where $\hat{\mathbf{L}} = -i\mathbf{r} \times \nabla$ is the particle orbital angular momentum operator. This is, as it should be, the matrix element of the operator

$$\hat{\boldsymbol{\mu}} = \frac{e}{2m} \hat{\mathbf{L}} + \frac{\mu}{s} \hat{\mathbf{s}}, \tag{47.9}$$

which contains the operators of the orbital and intrinsic magnetic moments of the particle.

The selection rules for magnetic multipole radiation are analogous to those for the electric case: the rules (46.15), (46.16) again apply to the total angular momentum, and the parity rule is

$$P_i P_f = (-1)^{j+1}, \tag{47.10}$$

which is obtained by substituting in (46.17) the parity of the Mj photon, $P_{\mathrm{ph}} = (-1)^{j+1}$.

§48. Angular distribution and polarization of the radiation

The formulae derived in §§46 and 47 relate to the emission of a photon with definite values of the angular momentum j and component thereof m. It was accordingly assumed that the radiating system (a nucleus, say) has not only definite values of the angular momentum J but also definite polarizations, i.e. values of M, both before and after the emission.

Let us now consider the more general case of emission by a partially polarized nucleus (whose dimensions are again assumed small in comparison with the wavelength). The emitted photon again has a definite angular momentum j, but may be partially polarized. Let us find the emission probability as a function of the direction **n** of the photon. This probability must be expressed in terms of density matrices which describe the polarization states of the nucleus and the photon.

For this purpose, we shall first write down the emission probability as a function of the direction **n** and helicity λ of the photon ($\lambda = \pm 1$), for the case where the initial and final nuclei have definite values J_i, M_i; J_f, M_f.

The matrix element for emission of a photon with definite values j, m is proportional to the matrix element of the (electric or magnetic) 2^j-pole moment of the nucleus:

$$\langle J_f M_f; jm|V|J_i M_i\rangle \propto (-1)^m \langle J_f M_f|Q_{j,-m}|J_i M_i\rangle. \tag{48.1}$$

The wave function of the emitted photon (in the momentum representation) is proportional to $\mathbf{Y}^{(e)}_{jm}(\mathbf{n})$ or $\mathbf{Y}^{(m)}_{jm}(\mathbf{n})$. The wave function of a photon whose momentum is in the direction **n** and whose helicity is λ is proportional to the polarization vector $\mathbf{e}^{(\lambda)}$. The matrix element for emission of a photon **n**, λ is found by multiplying (48.1) by the projection of the wave function of the state $|jm\rangle$ on that of the state $|\mathbf{n}\lambda\rangle$:

$$\langle J_f M_f; \mathbf{n}\lambda|V|J_i M_i\rangle \propto (-1)^m \langle J_f M_f|Q_{j,-m}|J_i M_i\rangle \mathbf{e}^{(\lambda)*}\cdot \mathbf{Y}_{jm}.$$

According to (16.23), for photons of either type

$$\mathbf{e}^{(\lambda)*}\cdot \mathbf{Y}_{jm}(\mathbf{n}) \propto D^{(j)}_{\lambda m}(\mathbf{n}). \tag{48.2}$$

The matrix element of the multipole moment can be expressed in the usual way in terms of the reduced element. Thus we find the transition probability amplitude in the form

$$\langle J_f M_f; \mathbf{n}\lambda|V|J_i M_i\rangle \propto (-1)^{J_f - M_f + m}\begin{pmatrix} J_f & j & J_i \\ -M_f & -m & M_i\end{pmatrix} Q D^{(j)}_{\lambda m}(\mathbf{n}), \tag{48.3}$$

where Q denotes $\langle J_f\|Q\|J_i\rangle$.

We can now proceed to the general case of mixed polarization states. According to the general rules of quantum mechanics, the transition probability is proportional to the expression†

$$\sum_{(m)} \langle J_f M_f; \mathbf{n}\lambda|V|J_i M_i\rangle\langle J_f M'_f; \mathbf{n}\lambda'|V|J_i M'_i\rangle^* \times$$
$$\times \langle M_i|\rho^{(i)}|M'_i\rangle\langle M'_f|\rho^{(f)}|M_f\rangle\langle\lambda'|\rho^{(\gamma)}|\lambda\rangle, \tag{48.4}$$

† If the initial and final states of the system are described by the superpositions

$$\psi^{(i)} = \sum_n a_n \psi^{(i)}_n, \qquad \psi^{(f)} = \sum_m b_m \psi^{(f)}_m,$$

then the matrix element is

$$\langle f|V|i\rangle = \sum_{m,n} b^*_m a_n V_{mn},$$

where $\rho^{(i)}$, $\rho^{(f)}$, $\rho^{(\gamma)}$ are the density matrices of the initial nucleus, the final nucleus and the emitted photon; the symbol (m) beneath the summation sign indicates that the sum is taken over all the m-type quantities which occur twice (M_i, M'_i, M_f, M'_f, λ, λ'). Then (48.3) is to be substituted in (48.4).

Let $w(\mathbf{n})\,do$ denote the probability of emission of a photon into the solid angle do. The total probability of emission, in any direction and with any polarizations of the photon and the final nucleus, is evidently independent of the initial polarization state of the nucleus, is given by formulae already known, and is of no interest here. We shall therefore arbitrarily normalize the probability $w(\mathbf{n})$ to unity. The result is†

$$w(\mathbf{n}) = \frac{(2j+1)(2J_i+1)}{8\pi} \sum_{(m)} (-1)^{2J_i - M_i - M'_i} D^{(j)}_{\lambda m} D^{(j)*}_{\lambda' m'} \times$$
$$\times \begin{pmatrix} J_f & j & J_i \\ -M_f & -m & M_i \end{pmatrix} \begin{pmatrix} J_f & j & J_i \\ -M'_f & -m' & M'_i \end{pmatrix} \times$$
$$\times \langle M_i|\rho^{(i)}|M'_i\rangle \langle M'_f|\rho^{(f)}|M_f\rangle \langle \lambda'|\rho^{(\gamma)}|\lambda\rangle;$$

it will be seen below that the normalization is correct. This formula can be transformed by using the series expansion *QM* (110.2) for the product of the two D functions:

$$D^{(j)}_{\lambda m} D^{(j)*}_{\lambda' m'} = (-1)^{\lambda' + m'} D^{(j)}_{\lambda m} D^{(j)}_{-\lambda', -m'}$$
$$= (-1)^{\lambda + m} \sum_L (2L+1) \begin{pmatrix} j & j & L \\ \lambda & -\lambda' & -\Lambda \end{pmatrix} \begin{pmatrix} j & j & L \\ m & -m' & -\mu \end{pmatrix} D^{(L)}_{\Lambda\mu},$$

where $\Lambda = \lambda - \lambda'$, $\mu = m - m'$ and L takes integral values $\geqslant 2j$. Thus we have finally

$$w(\mathbf{n}) = \frac{(2j+1)(2J_i+1)}{8\pi} \sum_L \sum_{(m)} (-1)^{2J_i - M_i - M'_i + m + 1} (2L+1) \times$$
$$\times \begin{pmatrix} j & j & L \\ \lambda & -\lambda' & -\Lambda \end{pmatrix} \begin{pmatrix} j & j & L \\ m & -m' & -\mu \end{pmatrix} \begin{pmatrix} J_f & j & J_i \\ -M_f & -m & M_i \end{pmatrix} \begin{pmatrix} J_f & j & J_i \\ -M'_f & -m' & M'_i \end{pmatrix} \times$$
$$\times D^{(L)}_{\Lambda\mu}(\mathbf{n}) \langle M_i|\rho^{(i)}|M'_i\rangle \langle M'_f|\rho^{(f)}|M_f\rangle \langle \lambda'|\rho^{(\gamma)}|\lambda\rangle. \tag{48.5}$$

---

and its square is

$$|\langle f|V|i\rangle|^2 = \sum_{n, n', m, m'} V_{mn} V^*_{m'n'} a_n a^*_{n'} b_{m'} b^*_m.$$

The case of mixed states is obtained by making the changes

$$a_n a^*_{n'} \to \rho^{(i)}_{nn'}, \qquad b_{m'} b^*_m \to \rho^{(f)}_{m'm},$$

so that

$$|\langle f|V|i\rangle|^2 \to \sum_{n, n', m, m'} V_{mn} V^*_{m'n'} \rho^{(i)}_{nn'} \rho^{(f)}_{m'm}.$$

† In transforming the sign factor note that the numbers $2J_i$, $2J_f$, $2M_i$, $2M_f$ have the same parity; j and m are integers, and $\lambda = \pm 1$.

As previously, $\Sigma_{(m)}$ denotes summation over all m-type quantities which occur twice. Here it must be noted that λ and λ' differ from the other quantities, since they have only two values, $\lambda, \lambda' = \pm 1$, corresponding to the two polarizations of the photon, and not $2j+1$ values for any given j.

Formula (48.5) embodies all the necessary information about the angular distribution and polarization of the emitted photons, and also about the polarization of the secondary nuclei (i.e. those which have emitted a photon). It is assumed that the initial density matrix is given.

Angular distribution

The angular distribution of the photons is obtained by summation over all polarizations of the photon and the secondary nucleus. The averaging with respect to polarizations is done by substituting the density matrices of the unpolarized states:

$$\langle\lambda|\rho^{(\gamma)}|\lambda'\rangle = \tfrac{1}{2}\delta_{\lambda\lambda'}, \qquad \langle M_f|\rho^{(f)}|M_f'\rangle = \frac{1}{2J_f+1}\delta_{M_fM_f'}, \tag{48.6}$$

after which the summation amounts to a multiplication by 2 for the photon or by $2J_f+1$ for the nucleus. Thus the summation is effected by simply making the changes

$$\langle\lambda|\rho^{(\gamma)}|\lambda'\rangle \to \delta_{\lambda\lambda'}, \qquad \langle M_f|\rho^{(f)}|M_f'\rangle \to \delta_{M_fM_f'}, \tag{48.7}$$

and the angular distribution is

$$\bar{w}(\mathbf{n}) = \frac{(2j+1)(2J_i+1)}{8\pi}\sum_L\sum_{(m)}(-1)^{m'+1}(2L+1)D^{(L)}_{0\mu}(\mathbf{n})\times$$
$$\times\begin{pmatrix} j & j & L\\ \lambda & -\lambda & 0\end{pmatrix}\begin{pmatrix} j & j & L\\ m & -m' & -\mu\end{pmatrix}\begin{pmatrix} J_f & j & J_i\\ -M_f & -m & M_i\end{pmatrix}\begin{pmatrix} J_f & j & J_i\\ -M_f & -m' & M_i'\end{pmatrix}\times$$
$$\times\langle M_i|\rho^{(i)}|M_i'\rangle.$$

This formula can be considerably simplified by carrying out the summation over m-type quantities. First of all, we note that

$$\begin{pmatrix} j & j & L\\ \lambda & -\lambda & 0\end{pmatrix} = (-1)^L\begin{pmatrix} j & j & L\\ -\lambda & \lambda & 0\end{pmatrix}, \tag{48.8}$$

and therefore

$$\sum_{\lambda=\pm1}\begin{pmatrix} j & j & L\\ \lambda & -\lambda & 0\end{pmatrix} = 2\begin{pmatrix} j & j & L\\ 1 & -1 & 0\end{pmatrix} \quad \text{for even } L,$$
$$= 0 \qquad\qquad \text{for odd } L.$$

In the sum over L, therefore, only the terms with even L remain, and it involves only even-order spherical harmonics $D^{(L)}_{0\mu}$. This result is obvious a priori, since, by the conservation of parity, the probability must be unchanged by inversion, i.e. by putting $\mathbf{n}\rightarrow-\mathbf{n}$.

Thus we have

$$\bar{w}(\mathbf{n}) = \frac{(2j+1)(2J_i+1)}{4\pi}\sum_L (2L+1)\begin{pmatrix} j & j & L\\ 1 & -1 & 0\end{pmatrix} D^{(L)}_{0\mu}(\mathbf{n})\times$$

$$\times\sum_{(m)}(-1)^{m'+1}\begin{pmatrix} j & j & L\\ m & -m' & -\mu\end{pmatrix}\begin{pmatrix} J_f & j & J_i\\ -M_f & -m & M_i\end{pmatrix}\begin{pmatrix} J_f & j & J_i\\ -M_f & -m' & M'_i\end{pmatrix}\times$$

$$\times\langle M_i|\rho^{(i)}|M'_i\rangle.$$

The normalization here is easily verified: with the formula

$$\int D^{(L)}_{0\mu}(\mathbf{n})\, do/4\pi = \delta_{L0}\,\delta_{\mu 0},$$

integration over all directions leaves only the term with $L=0$, $\mu=0$, and the formulae

$$\begin{pmatrix} j & j & 0\\ m & -m & 0\end{pmatrix} = (-1)^{j-m}\frac{1}{\sqrt{(2j+1)}},$$

$$\sum_{M_f,m}\begin{pmatrix} J_f & j & J_i\\ -M_f & -m & M_i\end{pmatrix}^2 = \frac{1}{2J_i+1},\quad \operatorname{tr}\rho^{(i)} = 1$$

then show that the integral is equal to unity.

The further summation with respect to m, m', M_f in the inner sum in $\bar{w}(\mathbf{n})$ is effected by means of *QM*, (108.4). The final expression for the photon angular distribution is

$$\bar{w}(\mathbf{n}) = (-1)^{1+J_i+J_f}\frac{(2j+1)\sqrt{(2J_i+1)}}{4\pi}\times$$

$$\times\sum_{L\text{ even}}(-i)^L\sqrt{(2L+1)}\begin{pmatrix} j & j & L\\ 1 & -1 & 0\end{pmatrix}\begin{Bmatrix} J_i & J_i & L\\ j & j & J_f\end{Bmatrix}\sum_\mu \mathcal{P}^{(i)*}_{L\mu}D^{(L)}_{0\mu}(\mathbf{n}), \tag{48.9}$$

where

$$\mathcal{P}^{(i)}_{L\mu} = i^L\sqrt{[(2L+1)(2J_i+1)]}\sum_{M_i,M'_i}(-1)^{J_i-M'_i}\begin{pmatrix} J_i & L & J_i\\ -M'_i & \mu & M_i\end{pmatrix}\langle M_i|\rho^{(i)}|M'_i\rangle, \tag{48.10}$$

$$\mathcal{P}^{(i)*}_{L\mu} = (-1)^{L-\mu}\mathcal{P}^{(i)}_{L,-\mu}.$$

The inner sum in (48.9) is taken over all $|\mu|\leqslant L$, and the outer sum over all even L such that

$$L\leqslant 2j,\qquad L\leqslant 2J_i. \tag{48.11}$$

(These conditions result from the triangle rule which has to be satisfied by the quantities in the $3j$-symbols that appear in (48.9), (48.10).) The number of terms in the sum is therefore usually small. For instance, when $J_i = 0$ or $\frac{1}{2}$, only the term with $L = 0$ remains, and the radiation is isotropic; this term is easily seen to equal $\frac{1}{4}$, as it should by the normalization condition. When $J_i = 1$ or 3/2, or $j = 1$, the two terms with $L = 0$ or 2 remain in the sum over L. If the density matrix $\rho^{(i)}$ is diagonal ($M_i = M_i'$), then $\mu = 0$, and the distribution function (48.9) becomes an expansion in Legendre polynomials; according to (16.5) and *QM*, (58.23), the functions $D^{(L)}_{00}$ are the functions $P_L(\cos\theta)$. Finally, if

$$\langle M_i|\rho^{(i)}|M_i'\rangle = \frac{1}{2J_i+1}\,\delta_{M_iM_i'},$$

i.e. if the initial nucleus is unpolarized, then all the $\mathcal{P}^{(i)}_{L\mu}$ are zero except $\mathcal{P}^{(i)}_{00} = 1$.†

The quantities $\mathcal{P}_{L\mu}$ are convenient characteristics of the polarization state of the nucleus, and will be called *polarization moments*. Formula (48.10) defines them in terms of the density matrix $\rho_{MM'}$. The inverse formula expressing the density matrix in terms of the polarization moments is easily verified:

$$\rho_{MM'} = \sum_{L,\mu}\sqrt{\frac{2L+1}{2J+1}}\,i^{-L}(-1)^{J-M'}\begin{pmatrix} J & L & J \\ -M' & \mu & M\end{pmatrix}\mathcal{P}_{L\mu}. \tag{48.12}$$

Let $f_{L\mu}$ be a spherical tensor depending on the polarization state of the nucleus. According to the general rules (see *QM*, (14.8)), its mean value in a state having the density matrix $\rho_{MM'}$ is

$$\bar{f}_{L\mu} = \sum_{M,M'}\rho_{MM'}\langle JM'|f_{L\mu}|JM\rangle. \tag{48.13}$$

Expressing the matrix elements of the $f_{L\mu}$ in terms of the reduced element $\langle J\|f_L\|J\rangle$ by means of the formula

$$\langle JM'|f_{L\mu}|JM\rangle = i^L(-1)^{J-M'}\begin{pmatrix} J & L & J \\ -M' & \mu & M\end{pmatrix}\langle J\|f_L\|J\rangle,$$

† Using the result that

$$\begin{pmatrix} J & 0 & J \\ -M' & 0 & M\end{pmatrix} = (-1)^{J-M}\frac{1}{\sqrt{(2J+1)}}\,\delta_{MM'},$$

we have

$$\begin{aligned}\sum_{M,M'}(-1)^{J-M'}\begin{pmatrix} J & L & J \\ -M' & \mu & M\end{pmatrix}\delta_{MM'} &= \sqrt{(2J+1)}\sum_{M,M'}\begin{pmatrix} J & L & J \\ -M' & \mu & M\end{pmatrix}\begin{pmatrix} J & 0 & J \\ -M' & 0 & M\end{pmatrix}\\ &= \sqrt{(2J+1)}\,\delta_{L0}\,\delta_{\mu 0},\end{aligned}$$

and the conclusion stated then follows from the definition (48.10).

and using the polarization moments as defined by (48.10), we obtain

$$\bar{f}_{L\mu} = \frac{\langle J\|f_L\|J\rangle}{\sqrt{[(2L+1)(2J+1)]}}\mathscr{P}_{L\mu}. \tag{48.14}$$

Photon polarization

When the matrices $\rho^{(\gamma)}$ and $\rho^{(f)}$, as well as $\rho^{(i)}$, are specified, formula (48.5) determines the probability of a transition in which a photon is emitted, and the nucleus left, in definite polarization states. Such states are essentially characteristic not of the emission process as such, but of the detectors which record the photon and the recoil nucleus and distinguish definite polarizations of these. There is another and more natural formulation of the problem, in which the final state of the "nucleus + photon" system is not specified from the start, and the polarization density matrix of this state is to be determined, with only the direction of the photon emission fixed.

The answer to this problem is given by the same formula (48.5). If this is written as

$$w = \bar{w}(\mathbf{n})\sum_{(m)}\langle M_f;\mathbf{n}\lambda|\rho|M_f';\mathbf{n}\lambda'\rangle\langle\lambda'|\rho^{(\gamma)}|\lambda\rangle\langle M_f'|\rho^{(f)}|M_f\rangle, \tag{48.15}$$

then the expression $\langle M_f;\mathbf{n}\lambda|\rho|M_f';\mathbf{n}\lambda'\rangle$ is the required density matrix, since according to the general rules of quantum mechanics the probability w of a transition to a specified state is given by its "projection" on the given $\rho^{(\gamma)}$, $\rho^{(f)}$. The factor $\bar{w}(\mathbf{n})$ is written in (48.15) so that this matrix shall be normalized by the usual condition,

$$\sum_{\lambda, M_f}\langle M_f;\mathbf{n}\lambda|\rho|M_f;\mathbf{n}\lambda\rangle = 1.$$

If we want the polarization of the photon alone, a summation over $M_f = M_f'$ is necessary:

$$\langle\mathbf{n}\lambda|\rho|\mathbf{n}\lambda'\rangle = \sum_{M_f}\langle M_f;\mathbf{n}\lambda|\rho|M_f;\mathbf{n}\lambda'\rangle.$$

Using a derivation exactly similar to that of (48.9), we obtain

$$\langle\mathbf{n}\lambda|\rho|\mathbf{n}\lambda'\rangle = (-1)^{1+J_i+J_f}\frac{(2j+1)\sqrt{(2J_i+1)}}{8\pi\bar{w}(\mathbf{n})}\times$$
$$\times\sum_L(-i)^L\sqrt{(2L+1)}\begin{pmatrix} j & j & L\\ \lambda & -\lambda' & \Lambda\end{pmatrix}\begin{Bmatrix} J_i & J_i & L\\ j & j & J_f\end{Bmatrix}\sum_\mu\mathscr{P}^{(i)*}_{L\mu}D^L_{\Lambda\mu}(\mathbf{n}), \tag{48.16}$$

where $\Lambda = \lambda-\lambda'$, and the summation is over all integral values of L which satisfy the conditions (48.11).

In particular, circular polarization is determined by the Stokes parameter

$$\xi_2 = \langle\mathbf{n}1|\rho|\mathbf{n}1\rangle - \langle\mathbf{n},-1|\rho|\mathbf{n},-1\rangle;$$

see §8, Problem. Because of the relation (48.8), all terms with even L in this difference are zero, and the resulting formula for ξ_2 differs from (48.9) only in that the summation is over odd instead of even values of L.

Secondary nucleus polarization

Finally, if we are interested only in the final polarization of the nuclei, we must put $\rho^{(\gamma)} \to \delta$. If the integration with respect to directions of the photon is also carried out, the density matrix of the secondary nucleus is

$$\langle M_f|\rho|M_f'\rangle = \int \bar{w}(\mathbf{n})\langle M_f\mathbf{n}|\rho|M_f'\mathbf{n}\rangle\, do$$
$$= (2J_i+1) \sum_{m,\, M_i,\, M_{i'}} (-1)^{2J_i - M_i - M_i'} \times$$
$$\times \begin{pmatrix} J_f & j & J_i \\ -M_f & -m & M_i \end{pmatrix} \begin{pmatrix} J_f & j & J_i \\ -M_f' & -m & M_i' \end{pmatrix} \langle M_i|\rho^{(i)}|M_i'\rangle.$$

The polarization moments calculated by means of this matrix are

$$\mathcal{P}^{(f)}_{L\mu} = (-1)^{J_i+J_f+L+j}\sqrt{[(2J_i+1)(2J_f+1)]} \begin{Bmatrix} J_i & J_i & L \\ J_f & J_f & j \end{Bmatrix} \mathcal{P}^{(i)}_{L\mu}. \tag{48.17}$$

If the initial nucleus is unpolarized, so is the final nucleus, but there exists a correlation polarization, i.e. a polarization of the nucleus after emission in a specified direction. Putting $\rho^{(i)} \to \delta/(2J_i+1)$ (and correspondingly $\bar{w}(\mathbf{n}) = 1/4\pi$) and calculating as in the derivation of (48.9), we obtain for the density matrix describing this polarization

$$\langle M_f; \mathbf{n}|\rho|M_f'; \mathbf{n}\rangle = (2j+1)(-1)^{J_i+M_f'+1} \sum_{L \text{ even}} (2L+1) \begin{pmatrix} j & j & L \\ 1 & -1 & 0 \end{pmatrix} \begin{pmatrix} J_f & L & J_f \\ -M_f' & \mu & M_f \end{pmatrix} \times$$
$$\times \begin{Bmatrix} J_f & J_f & L \\ j & j & J_i \end{Bmatrix} D^{(L)}_{0\mu}(\mathbf{n}). \tag{48.18}$$

The corresponding polarization moments are

$$\mathcal{P}^{(f)}_{L\mu} = i^L(-1)^{1+J_i+J_f}(2j+1)\sqrt{[(2L+1)(2J_f+1)]} \times$$
$$\times \begin{pmatrix} j & j & L \\ 1 & -1 & 0 \end{pmatrix} \begin{Bmatrix} J_f & J_f & L \\ j & j & J_i \end{Bmatrix} D^{(L)}_{0\mu}(\mathbf{n}). \tag{48.19}$$

Only even-order moments occur (which is also a consequence of the conservation of parity already mentioned).

If the secondary nucleus in turn emits a photon, it will generate an anisotropic distribution, being polarized. Since the polarization moments (48.19) depend on the direction $\mathbf{n}$ of the photon emitted in the first decay, there is a certain correlation between the directions of successively emitted photons (with an unpolarized primary nucleus). Other correlation effects (of polarization, etc). in cascade emission can be treated similarly.†

PROBLEM

Find the relation between the polarization moments $\mathcal{P}_{1\mu}$, $\mathcal{P}_{2\mu}$ and the mean values of the angular momentum vector **J** and the quadrupole moment tensor Q_{ik}.

SOLUTION. The reduced elements of the vector **J** and the tensor Q_{ik} are determined from

$$\overline{\mathbf{J}^2} = \langle J\|J\|J\rangle^2/(2J+1),$$
$$\overline{Q_{ik}^2} = \langle J\|Q\|J\rangle^2/(2J+1);$$

cf. *QM*, (107.10), (107.11). The operator $\hat{Q}_{ik}$ is expressed in terms of the angular momentum operators as in *QM* (75.2):

$$\hat{Q}_{ik} = \frac{3Q}{2J(2J-1)}(\hat{J}_i\hat{J}_k + \hat{J}_k\hat{J}_i - \tfrac{2}{3}\hat{\mathbf{J}}^2\delta_{ik}).$$

Hence we find the mean value

$$\overline{Q_{ik}^2} = \frac{3Q^2}{2J^2(2J-1)^2}\mathbf{J}^2(4\mathbf{J}^2-3) = Q^2\left[\frac{3(J+1)(2J+3)}{2J(2J-1)}\right].$$

The reduced matrix elements are

$$\langle J\|J\|J\rangle = \sqrt{[J(J+1)(2J+1)]},$$
$$\langle J\|Q\|J\rangle = Q\left[\frac{3(2J+1)(J+1)(2J+3)}{2J(2J-1)}\right]^{\frac{1}{2}}.$$

From (48.14) we now see that the polarization moments $\mathcal{P}_{1\mu}$ are equal to the spherical components of the vector

$$\sqrt{\frac{3}{J(J+1)}}\,\bar{\mathbf{J}},$$

and the moments $\mathcal{P}_{2\mu}$ are equal to the spherical components of the tensor

$$\left[\frac{10J(2J-1)}{3(J+1)(2J+3)}\right]^{\frac{1}{2}}\frac{\overline{Q_{ik}}}{Q}.$$

§49. Radiation from atoms: the electric type‡

The energies of the outer electrons of an atom (which take part in optical radiative transitions) have, as a rough estimate, the order of magnitude $E \sim me^4/\hbar^2$, so that the radiated wavelengths $\lambda \sim \hbar c/E \sim \hbar^2/\alpha me^2$. The dimension of the atom is $a \sim \hbar^2/me^2$. Thus, in the optical spectra of atoms, we generally have the inequality $a/\lambda \sim \alpha \ll 1$. The ratio $v/c \sim \alpha$, where v is the velocity of the optical electrons, has a similar order of magnitude.

Thus, in the optical spectra of atoms, a condition is satisfied which means that

† A detailed account of these problems is given in the paper by A. Z. Dolginov, in: L. A. Sliv, ed., *Gamma-luchi* (*Gamma Rays*), USSR Academy of Sciences, Moscow 1961, pp. 523–681.

‡ In §§49–51 and 53–55, ordinary units are used.

the probability of electric dipole radiation (if this is allowed by the selection rules) considerably exceeds the probabilities of multipole transitions.† For this reason it is electric dipole transitions which are the most important in atomic spectroscopy.

As has already been mentioned, such transitions are subject to strict selection rules as regards the total angular momentum J of the atom and the parity P:‡

$$|J'-J| \leqslant 1 \leqslant J+J', \tag{49.1}$$

$$PP' = -1. \tag{49.2}$$

The inequality $|J'-J| \leqslant 1$ signifies that the angular momentum J can change only by 0 or ± 1; also, the transition $0 \to 0$ is forbidden by the inequality $J+J' \geqslant 1$. The parities of the initial and final states must be opposite.§

The probability of emission by the transition $nJM \to n'J'M'$ is determined by the corresponding matrix element of the dipole moment of the atom:

$$w(nJM \to n'J'M') = \frac{4\omega^3}{3\hbar c^3} |\langle n'J'M'|d_{-m}|nJM\rangle|^2, \tag{49.3}$$

$$\omega = \omega(nJ \to n'J').$$

On summing (49.3) over all values of $M' = M - m$ (with M given), we obtain the total probability of emission with a given frequency from the atomic level n, J. The summation is carried out by means of (46.20), and the result is

$$w(nJ \to n'J') = \frac{4\omega^3}{3\hbar c^3} \frac{1}{2J+1} |\langle n'J'\|d\|nJ\rangle|^2. \tag{49.4}$$

The squared modulus of the reduced matrix element is sometimes called the *transition line strength*; it is symmetrical as between the initial and final states.

The observed radiation intensity is found by multiplying w by $\hbar\omega$ and by the number N_{nJ} of atoms in the source which are at the excitation level concerned. For example, in a gas at temperature T this number is $N_{nJ} \propto (2J+1)\exp(-E_{nJ}/T)$; the factor $2J+1$ is the statistical weight of the level with angular momentum J.

Further deductions regarding transition probabilities in atomic spectra can be obtained only for specific kinds of atomic states. We shall not here discuss methods of calculating matrix elements where the degree of approximation has no clear theoretical significance, but simply derive some relations valid for a fairly large class of states (especially in light atoms) of the LS coupling type (see *QM*, §72). Such states are described not only by the total angular momentum but also by definite values of the orbital angular momentum L and the spin S, which in this case are conserved.

† Typical values of the dipole transition probability in the optical region of the spectra of atoms are of the order of 10^8 sec^{-1}.

‡ We shall now denote the quantum numbers of the initial and final states by unprimed and primed letters respectively. The letters n, n' will denote all the quantum numbers which define the state of the system, other than those shown explicitly.

§ The parity selection rule was first established by O. Laporte (1924).

Since the dipole moment is a purely orbital quantity, its operator commutes with the spin operator, i.e. its matrix is diagonal with respect to the number S. For the number L, the dipole moment is subject to the same selection rules as any orbital vector (see *QM*, §29). Thus transitions between LS-type states are subject to the following selection rules (in addition to (49.1), (49.2)):

$$S' - S = 0, \tag{49.5}$$

$$|L' - L| \leqslant 1 \leqslant L + L'. \tag{49.6}$$

It should again be stressed that these rules are approximate, and no longer apply when the spin–orbit interaction is taken into account.

The rule (49.5), which forbids transitions between terms of different multiplicity, is valid not only for electric dipole transitions but for all electric transitions: the electric multipole moments of all orders are orbital tensors, and therefore their matrices are diagonal with respect to spin. For instance, for electric quadrupole transitions, in addition to the general rules

$$|J' - J| \leqslant 2 \leqslant J + J', \qquad PP' = 1, \tag{49.7}$$

in the case of LS coupling we have the further rules

$$S' - S = 0, \qquad |L' - L| \leqslant 2 \leqslant L + L'. \tag{49.8}$$

The emission probability can be written in explicit form as a function of the numbers S, J, J'. This is done immediately by means of the matrix elements of spherical tensors in the addition of angular momenta. According to *QM*, (109.3), we have†

$$|\langle n'L'SJ'\|d\|nLSJ\rangle|^2 = (2J+1)(2J'+1)\begin{Bmatrix} L' & J' & S \\ J & L & 1 \end{Bmatrix}^2 |\langle n'L'\|d\|nL\rangle|^2. \tag{49.9}$$

Substitution of this in (49.4) gives

$$w(nLSJ \to n'L'SJ') = \frac{4\omega^3}{3\hbar c^3}(2J'+1)\begin{Bmatrix} L' & J' & S \\ J & L & 1 \end{Bmatrix}^2 |\langle n'L'\|d\|nL\rangle|^2, \tag{49.10}$$

with $\omega = \omega(nLS \to n'L'S)$.‡

A sum rule can be derived for these probabilities. The squares of the $6j$-symbols satisfy the summation formula (see *QM*, (108.7))

$$\sum_{J'} (2J'+1)\begin{Bmatrix} L' & J' & S \\ J & L & 1 \end{Bmatrix}^2 = \frac{1}{2L+1}. \tag{49.11}$$

† The "angular momenta of sub-systems 1 and 2" in the formulae in *QM*, §109, are here to be taken as the orbital angular momentum and spin of the atom, whose interaction is neglected; the quantities $f^{(1)}_{1q}$ are represented by the orbital vector d_q.

‡ In neglecting the spin–orbit interaction in the calculation of the matrix elements, we also neglect the dependence of the frequencies on J and J', i.e. the fine structure of the initial and final levels of the atom.

Using this, we obtain from (49.10)

$$\sum_{J'} w(nLSJ \to n'L'SJ') = \frac{4\omega^3}{3\hbar c^3}\frac{1}{2L+1}|\langle n'L'\|d\|nL\rangle|^2. \tag{49.12}$$

This quantity is thus found to be independent of the initial value of J.

For radiation from a gas whose temperature is much greater than the fine-structure intervals in the atomic term nSL, the states with different J are uniformly occupied, i.e. all values of J are equally probable. The probability that the atom is at a level with some definite value of J is then

$$\frac{2J+1}{(2L+1)(2S+1)}, \tag{49.13}$$

i.e. is equal to the ratio of the statistical weight of the level to the total statistical weight of the term nSL. Averaging the expressions (49.10) or their sums (49.12) with respect to these probabilities is equivalent to multiplying by the factor (49.13). This averaging will be denoted by a bar over the letter w. The total probability of emission of all the lines in a spectral multiplet (formed by all possible transitions between the fine-structure components of the two terms nSL and $n'SL'$) is the sum

$$\bar{w}(nLS \to n'L'S) = \sum_J \sum_{J'} \bar{w}(nLSJ \to n'L'SJ'). \tag{49.14}$$

Since, of course,

$$\sum_J (2J+1) = (2S+1)(2L+1),$$

the result obtained for the total probability agrees with (49.12). Thus the relative probability (which is the same thing as the relative intensity) of a single line is

$$\frac{\bar{w}(nLSJ \to n'L'SJ')}{\bar{w}(nLS \to n'L'S)} = \frac{(2J+1)(2J'+1)}{2S+1}\begin{Bmatrix} L' & J' & S \\ J & L & 1 \end{Bmatrix}^2. \tag{49.15}$$

The analysis of the numerical values given by this formula shows that the strongest lines in the multiplet are those for which $\Delta J = \Delta L$ (called *main lines*, while the remaining components of the multiplet are called *satellites*). The intensity of the main lines increases with the initial value of J.

Summation of the quantities (49.15) with respect to J' and J gives respectively

$$\left.\begin{aligned} \frac{\sum_{J'} \bar{w}(nLSJ \to n'L'SJ')}{\bar{w}(nLS \to n'L'S)} &= \frac{2J+1}{(2L+1)(2S+1)}, \\ \frac{\sum_{J} \bar{w}(nLSJ \to n'L'SJ')}{\bar{w}(nLS \to n'L'S)} &= \frac{2J'+1}{(2L+1)(2S+1)}. \end{aligned}\right\} \tag{49.16}$$

Thus the total intensity of all the lines in a spectral multiplet having a common initial or final level is proportional to the statistical weight of that common level.

We may also consider the hyperfine structure of atomic spectral lines. The hyperfine splitting of atomic levels is due to the interaction of the electrons with the spin of the nucleus if the latter is non-zero (see *QM*, §122). The total angular momentum **F** of the atom (including the nucleus) consists of the total electron angular momentum **J** and the angular momentum **I** of the nucleus. Each component of the hyperfine structure of the level n, J has a different value of the quantum number F.

The rigorous law of conservation of angular momentum now leads to a rigorous selection rule for the total angular momentum F: for electric dipole radiation,

$$|F' - F| \leqslant 1 \leqslant F + F'. \tag{49.17}$$

But, in view of the extreme weakness of the interaction of the electrons with the spin of the nucleus, this interaction may be neglected in calculating the matrix elements of the electric (and magnetic) moments of the electron shell of the atom. Thus the previous selection rules regarding the electron angular momentum J and the electron parity remain valid also. In particular, the latter selection rule prohibits electric dipole transitions between hyperfine structure components of the same term: all these levels have the same parity, whereas such transitions can occur only between states of different parity.

Since the dipole moment operator commutes with the nuclear spin, the dependence of the matrix elements on the numbers I and F can be found explicitly, the calculations differing only by an obvious change of notation from those given above for LS coupling. The probability of emission, summed over the final values of the component of the total angular momentum **F**, is

$$\left.\begin{aligned} w(nJIF \to n'J'IF') &= \frac{4\omega^3}{3\hbar c^3}\frac{1}{2F+1}|\langle n'J'IF'\|d\|nJIF\rangle|^2, \\ \omega &= \omega(nJ \to n'J'), \end{aligned}\right\} \tag{49.18}$$

and the square of the reduced matrix element is

$$|\langle n'J'IF'\|d\|nJIF\rangle|^2 = (2F+1)(2F'+1)\begin{Bmatrix} J' & F' & I \\ F & J & 1 \end{Bmatrix}^2 |\langle n'J'\|d\|nJ\rangle|^2. \tag{49.19}$$

PROBLEM

The majority of the lines in the spectra of the alkali metals can be described as resulting from transitions of a single outer (optical) electron in the self-consistent field of the rest of the atom, which forms a configuration of closed shells; the state of the atom is governed by LS coupling. Under these conditions, determine the relative intensities of the fine structure components of the spectral lines.

SOLUTION. The total angular momenta L and $S = \frac{1}{2}$ of the atom are equal to the orbital angular momentum and spin of the optical electron. The parity of the state is therefore $(-1)^L$ (the parity of the closed configuration of the rest of the atom being positive). The parity selection rules therefore forbid the dipole transition with $L' = L$, and so only transitions with $L' - L = \pm 1$ are possible. The transitions

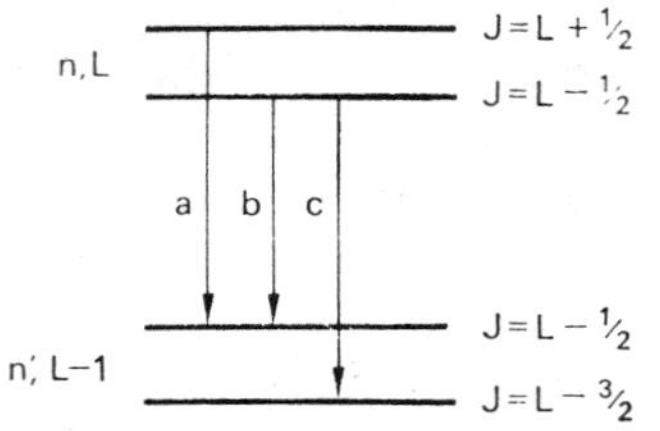

FIG. 1.

between components of the doublet levels n, L and $n', L-1$ give only three lines, because of the selection rule for J (Fig. 1). Their relative intensities (denoted by a, b, c) are most simply determined from the rules (49.16), instead of using (49.15) directly. The ratios of total intensities of lines having each initial (or final) level give two equations

$$\frac{b+c}{a} = \frac{2L}{2L+2}, \qquad \frac{a+b}{c} = \frac{2L}{2L-2},$$

whence

$$a:b:c = (L+1)(2L-1):1:(L-1)(2L+1).$$

If $L = 1$, the lower level is unsplit, line c does not appear and $a/b = 2$.

§ 50. **Radiation from atoms: the magnetic type**

The magnetic moment of an atom is equal, in order of magnitude, to the Bohr magneton: $\mu \sim e\hbar/mc$. This differs by a factor α from the order of magnitude of the electric dipole moment, $d \sim ea \sim \hbar^2/me$ (since $v/c \sim \alpha$, we have $\mu \sim dv/c$, as is to be expected). Hence it follows that the probability of magnetic dipole ($M1$) radiation from the atom is about α^2 times less than that of electric dipole radiation at the same frequency. The magnetic radiation is therefore important in practice only for transitions forbidden by the selection rules for the electric case.

The ratio of the probability of electric quadrupole ($E2$) radiation to that of $M1$ radiation is, in order of magnitude,

$$\frac{E2}{M1} \sim \frac{(ea^2)^2\omega^2/c^2}{\mu^2} \sim \frac{a^4m^2\omega^2}{\hbar^2} \sim \left(\frac{\Delta E}{E}\right)^2; \tag{50.1}$$

the quadrupole moment $\sim ea^2$, $E \sim \hbar^2/ma^2$ is the energy of the atom, and ΔE the change in energy in the transition. We see that, for medium atomic frequencies (i.e. when $\Delta E \sim E$), the probabilities of $E2$ and $M1$ radiation are of the same order of magnitude (assuming, of course, that both are allowed by the selection rules). If, however, $\Delta E \ll E$ (as for transitions between fine structure components of the same term), then $M1$ radiation is more probable than $E2$.

The magnetic dipole transitions are subject to the rigorous selection rules

$$|J'-J| \leq 1 \leq J+J', \tag{50.2}$$

$$PP' = 1. \tag{50.3}$$

For LS coupling, there are additional selection rules, which are even more restrictive than in the electric case. This is because of a particular property of the magnetic moment of the atom, which arises from the fact that all the particles (electrons) in the system are identical: the magnetic moment operator of the atom can be expressed in terms of the total orbital and spin angular momentum operators:

$$\hat{\boldsymbol{\mu}} = \mu_0(\hat{\mathbf{L}} + 2\hat{\mathbf{S}}) = -\mu_0(\hat{\mathbf{J}} + \hat{\mathbf{S}}), \tag{50.4}$$

where $\mu_0 = |e|\hbar/2mc$ is the Bohr magneton (see *QM*, §113). Owing to the conservation of total angular momentum, the operator $\hat{\mathbf{J}}$ has only matrix elements which are diagonal with respect to the energy, and in considering radiative transitions it is therefore sufficient to put $\hat{\boldsymbol{\mu}} = -\mu_0\hat{\mathbf{S}}$.†

The angular momenta **L** and **S** are separately conserved when the spin–orbit interaction is neglected. The spin operator is therefore diagonal in all the quantum numbers n, S, L which belong to the unsplit term. In order for a transition to occur at all, the number J must change. The selection rules are consequently

$$n' = n, \qquad S' = S, \qquad L' = L, \qquad J' - J = \pm 1, \tag{50.5}$$

i.e. transitions are possible only between fine structure components of a single term.

The emission probability can be calculated exactly in this case. By an appropriate change of notation in formula (49.10), we obtain

$$w(nLSJ \to nLSJ') = \frac{4\omega^3\mu_0^2}{3\hbar c^3}(2J'+1)\begin{Bmatrix} S & J' & L \\ J & S & 1 \end{Bmatrix}^2 |\langle S\|S\|S\rangle|^2.$$

The reduced spin matrix element with respect to the spin eigenfunctions is

$$\langle S\|S\|S\rangle = \sqrt{[S(S+1)(2S+1)]}; \tag{50.6}$$

see *QM*, (29.13). The $6j$-symbol is

$$\begin{Bmatrix} S & J-1 & L \\ J & S & 1 \end{Bmatrix}^2 = \frac{(L+S+J+1)(L+S-J+1)(L-S+J)(S-L+J)}{S(2S+1)(2S+2)(2J-1)2J(2J+1)}; \tag{50.7}$$

see *QM*, §108, Table 10. The result is then

$$\begin{aligned} w(nLSJ \to nLS, J-1) &= \frac{2J+1}{2J-1} w(nLS, J-1 \to nLSJ) \\ &= \frac{\omega^3\mu_0^2}{3\hbar c^3(2J+1)J}(L+S+J+1)(L+S-J+1)\times \\ &\quad \times (J+S-L)(J+L-S). \end{aligned} \tag{50.8}$$

† An exception occurs in cases where the electronic angular momentum **J** of the atom is not conserved: when the hyperfine structure is taken into account, when an external field is present, and so on (see the Problems).

Transitions between hyperfine structure components of one level (whose frequencies are in the radio wave range) cannot occur as electric dipole transitions, since all the components have the same parity. $E2$ and $M1$ transitions involve no change of parity. But, owing to the relatively very small intervals in the hyperfine structure, $E2$ radiation has a low probability compared with $M1$ (cf. (50.1)), so that these transitions occur as magnetic dipole transitions.

PROBLEMS

PROBLEM 1. Find the probability of an $M1$ transition between hyperfine structure components of a single level.

SOLUTION. The transition probability is given by formulae (49.18), (49.19), in which the diagonal reduced matrix element $\langle nJ\|\mu\|nJ\rangle$ of the magnetic moment will now appear. Its value can be written down immediately by noting that the total (not reduced) matrix element $\langle nJM|\mu_z|nJM\rangle$ determines the splitting of the relevant level by the Zeeman effect (see *QM*, §113), and is $-\mu_0 gM$, where g is the Landé factor. The reduced matrix element is (see *QM*, (29.7))

$$\begin{aligned}\langle nJ\|\mu\|nJ\rangle &= \frac{1}{M}\sqrt{[J(J+1)(2J+1)]}\langle nJM\|\mu_z|nJM\rangle \\ &= -\mu_0 g\sqrt{[J(J+1)(2J+1)]}.\end{aligned}$$

The required probability is thus found to be†

$$\begin{aligned}w(nJIF\to nJI, F-1) &= \frac{2F+1}{2F-1}w(nJI, F-1\to nJIF) \\ &= \frac{\omega^3\mu_0^2 g^2}{3\hbar c^3(2F+1)F}(J+I+F+1)(J+I-F+1)(F+J-I)(F-J+I).\end{aligned}$$

This expression differs from (50.8) only by an obvious change of notation and the extra factor g^2.

PROBLEM 2. Find the probability of an $M1$ transition between Zeeman components of a single atomic level.

SOLUTION. This is a transition $M\to M-1$ with the values of n and J unchanged; the transition frequency is (see (51.3) below) $\hbar\omega = \mu_0 gH$, where g is the Landé factor. The matrix element of the spherical component μ_{-1} of the vector $\boldsymbol{\mu}$ is

$$\begin{aligned}|\langle nJ, M-1|\mu_{-1}|nJM\rangle| &= \sqrt{\frac{(J-M+1)(J+M)}{2J(J+1)(2J+1)}}|\langle nJ\|\mu\|nJ\rangle| \\ &= -\mu_0 g\sqrt{[\tfrac{1}{2}(J-M+1)(J+M)]}\end{aligned}$$

(see *QM*, (27.12), and Problem 1). The transition probability is

$$\begin{aligned}w &= \frac{4\omega^3}{3\hbar c^3}|\langle nJ, M-1|\mu_{-1}|nJM\rangle|^2 \\ &= \frac{2\mu_0^5 H^3}{3\hbar^4 c^3}(J-M+1)(J+M).\end{aligned}$$

† An interesting example is the transition between the hyperfine structure components of the ground level ($1s\frac{1}{2}$) of the hydrogen atom, where both $E1$ and $E2$ transitions are strictly forbidden, the latter by the rule which prohibits a quadrupole transition with $J+J'=1$. This transition has a frequency $\omega = 2\pi\times 1.42\times 10^9\ \text{sec}^{-1}$ (wavelength $\lambda = 21$ cm). Putting $g=2$, $I=\frac{1}{2}$, $J=\frac{1}{2}$, $F=1$, $F'=0$, we obtain

$$w = 4\omega^3\mu_0^2/3\hbar c^3 = 2.85\times 10^{-15}\ \text{sec}^{-1}.$$

§51. **Radiation from atoms: the Zeeman and Stark effects**

In an external magnetic field H (assumed weak), each atomic level with total angular momentum J is split into $2J+1$ levels,

$$E_M = E^{(0)} + \mu_0 g M H, \tag{51.1}$$

where $E^{(0)}$ is the unperturbed level, μ_0 the Bohr magneton, g the Landé factor, and M the component of J in the direction of the field (see *QM*, §113). Thus the degeneracy with respect to directions of the angular momentum is entirely removed.

Spectral lines resulting from transitions between two split levels are correspondingly split. The number of components of the line is determined by the selection rule for the number M, according to which, for dipole radiation, we must have

$$m = M - M' = 0, \pm 1. \tag{51.2}$$

In addition to this rule, the transitions with $M = M' = 0$ are forbidden if also $J' = J$. This is seen immediately from the general expressions (*QM*, (29.7)) for the matrix elements of an arbitrary vector.

The components resulting from transitions with $m = 0$ and $m = \pm 1$ are called π and σ components respectively. Their frequencies are

$$\left.\begin{aligned} \hbar\omega_\pi &= \hbar\omega^{(0)} + \mu_0 H (g - g') M, \\ \hbar\omega_\sigma &= \hbar\omega^{(0)} + \mu_0 H [gM - g'(M \pm 1)]. \end{aligned}\right\} \tag{51.3}$$

In the particular case where $g = g'$, we have

$$\hbar\omega_\pi = \hbar\omega^{(0)}, \qquad \hbar\omega_\sigma = \hbar\omega^{(0)} \mp \mu_0 g H, \tag{51.4}$$

whatever the value of M; thus, in this case, the line is split into a triplet with an undisplaced π component and two σ components lying symmetrically on either side of it (called the "normal" *Zeeman effect*).

The total probability (for all directions) of emission of radiation is proportional to the squared modulus $|\langle n'J'M'|d_{-m}|nJM\rangle|^2$. Hence, using formula (46.19) with $j = 1$, we see that the relative probability of emission of each of the Zeeman components of the spectral line is

$$\begin{pmatrix} J' & 1 & J \\ M' & m & -M \end{pmatrix}^2. \tag{51.5}$$

In the particular case of the "normal" Zeeman effect there are only three components, each arising from transitions with all initial values of M for given m. Since

$$\sum_{M,M'} \begin{pmatrix} J' & 1 & J \\ M' & m & -M \end{pmatrix}^2 = \frac{1}{3} \tag{51.6}$$

(see *QM*, (106.12)), the emission of all three components is equally probable in this case.

The relative intensity of the Zeeman components when observed in a particular direction (relative to the direction of the magnetic field applied to the source) is of greater interest, however. According to (45.5), the probability of emission (and therefore the line strength) in a given direction **n** is proportional to $\Sigma |\mathbf{e}^* \cdot \mathbf{d}_{fi}|^2$, where the summation is over the two independent polarizations **e** which are possible for a given **n**.

For observation along the field (along the z-axis), this sum is

$$|(d_x)_{fi}|^2 + |(d_y)_{fi}|^2,$$

or, in spherical components,

$$|(d_1)_{fi}|^2 + |(d_{-1})_{fi}|^2.$$

This means that only the two σ components ($m = \pm 1$) are observed in the longitudinal direction (along the field). Their intensities are proportional to

$$\begin{pmatrix} J' & 1 & J \\ M \mp 1 & \pm 1 & -M \end{pmatrix}^2. \tag{51.7}$$

These lines have definite values of the component m of the angular momentum in the direction of propagation, and either right-hand ($m = 1$) or left-hand ($m = -1$) circular polarization (see §8).

For observation in a direction perpendicular to the field (along the x-axis, say), the intensity is proportional to the sum

$$|(d_z)_{fi}|^2 + |(d_y)_{fi}|^2 = |(d_0)_{fi}|^2 + \tfrac{1}{2}\{|(d_1)_{fi}|^2 + |(d_{-1})_{fi}|^2\}.$$

Thus two σ components and a π component are observed in the transverse direction, with respective intensities proportional to

$$\frac{1}{2}\begin{pmatrix} J' & 1 & J \\ M \mp 1 & \pm 1 & -M \end{pmatrix}^2 \quad \text{and} \quad \begin{pmatrix} J' & 1 & J \\ M & 0 & -M \end{pmatrix}^2; \tag{51.8}$$

the intensities of the σ components are half as great as in the case of longitudinal observation. The π component is linearly polarized along the z-axis; the σ components, as observed in this direction, are linearly polarized along the y-axis.

The relative intensities of the Zeeman components are seen to be entirely determined by the initial and final values of J and M, and not by any other properties of the levels.

The selection rules forbid electric dipole transitions between Zeeman components of the same level, since all these have the same parity. Such transitions occur as magnetic dipole transitions, for the same reason as was mentioned at the end of §50 in respect of transitions between hyperfine structure components of a level. Because of the selection rule for the number M, the transitions occur only between adjacent components ($M' - M = \pm 1$).†

The splitting of atomic levels in a weak electric field (the *Stark effect*), unlike that in a magnetic field, does not cause complete removal of the degeneracy with respect to directions of the angular momentum. All the levels except those with $M = 0$ remain doubly degenerate, each corresponding to two states with angular momentum components M and $-M$.

The calculating of the relative intensities of the Stark components of a spectral line is exactly similar to that given above for the Zeeman effect.‡ It must be remembered that the intensity of the π components includes contributions from the transitions $M \to M$ and $-M \to -M$ (when $M \neq 0$), and that of the σ components includes contributions from the transitions $M \to M \pm 1$ and $-M \to -(M \pm 1)$. Hence, for example, in transverse observation the intensities of the π components are proportional to

$$2\begin{pmatrix} J' & 1 & J \\ M & 0 & -M \end{pmatrix}^2,$$

and those of the σ components are proportional to the sums

$$\frac{1}{2}\begin{pmatrix} J' & 1 & J \\ M \pm 1 & \mp 1 & -M \end{pmatrix}^2 + \frac{1}{2}\begin{pmatrix} J' & 1 & J \\ -M \mp 1 & \pm 1 & M \end{pmatrix}^2 = \begin{pmatrix} J' & 1 & J \\ M \pm 1 & \mp 1 & -M \end{pmatrix}^2$$

(when all the numbers in the second row change sign, the $3j$-symbols can at most change sign, so that their squares are unaltered).

In an external field, even if it is weak, the total angular momentum $\mathbf{J}$ is no longer strictly conserved; in a uniform field, only the angular momentum component M is exactly conserved. Thus, in radiative transitions in a weak field, the conservation of angular momentum need not be rigorously maintained, and the atomic spectra may contain lines which are forbidden by the usual selection rules.

The calculation of the intensities of these lines is equivalent to the calculation of the corrections in the dipole moment matrix, which in turn requires the determination of corrections to the wave functions of stationary states. In the first approximation of perturbation theory (with respect to the weak external field), the wave function includes "admixtures" of states which are connected to the initial state by non-zero matrix elements of the perturbation ($-\mathbf{E} \cdot \mathbf{d}$ in the electric field): the admixture of a state ψ_2 in a state ψ_1 is

$$\frac{-\mathbf{E} \cdot \mathbf{d}_{21}}{E_1 - E_2} \psi_2.$$

† The frequencies of these transitions are usually in the range corresponding to centimetre wavelengths, and are observed in absorption and induced emission (electron paramagnetic resonance): the absorbing atoms are in a strong constant magnetic field (which causes the Zeeman splitting) and a weak radio-frequency field of the resonance frequency.

‡ This refers to the quadratic Stark effect, which occurs in all atoms except hydrogen (see *QM*, §76). The field is assumed so weak that the level splitting which it causes is small even in comparison with the fine structure intervals.

Thus the matrix element of the "forbidden" transition contains a term

$$\frac{-(\mathbf{E}\cdot\mathbf{d}_{21})\mathbf{d}_{32}}{E_1-E_2}$$

which is not zero if transitions from the "intermediate" state 2 to the initial state 1 and final state 3 are allowed.

§ 52. **Radiation from atoms: the hydrogen atom**

The hydrogen atom is the only one for which the transition matrix elements can be completely calculated in an analytical form (W. Gordon, 1929).

The parity of a state of the hydrogen atom is $(-1)^l$, i.e. is uniquely determined by the orbital angular momentum of the electron (the number l, which defines the parity of the state, retains its significance for the exact relativistic wave functions, i.e. when the spin–orbit interaction is taken into account). The parity selection rule therefore strictly forbids electric dipole transitions without change of l; only transitions with $l \to l \pm 1$ are possible. There is, however, no restriction on the change in the principal quantum number n.

The dipole moment of the hydrogen atom is equivalent to the position vector of the electron: $\mathbf{d} = e\mathbf{r}$. Since the electron wave function in the hydrogen atom is the product of an angular part and the radial function R_{nl}, the reduced matrix elements of the position vector can also be written as the product

$$\langle n', l-1\|r\|nl\rangle = \langle l-1\|\nu\|l\rangle \int_0^\infty R_{n',l-1} r R_{nl} r^2\, dr,$$

where $\langle l-1\|\nu\|l\rangle$ are the reduced matrix elements of the unit vector $\boldsymbol{\nu}$ in the direction of $\mathbf{r}$. These are

$$\langle l-1\|\nu\|l\rangle = \langle l\|\nu\|l-1\rangle^* = i\sqrt{l};$$

see *QM*, (29.14). Thus

$$\langle n', l-1\|r\|nl\rangle = -\langle nl\|r\|n', l-1\rangle = i\sqrt{l} \int_0^\infty R_{n',l-1} R_{nl} r^3\, dr. \tag{52.1}$$

The non-relativistic radial functions of the discrete spectrum of the hydrogen atom are given in *QM*, (36.13):†

$$R_{nl} = \frac{2}{n^{l+2}(2l+1)!}\sqrt{\frac{(n+l)!}{(n-l-1)!}}(2r)^l e^{-r/n} F(-n+l+1, 2l+2, 2r/n). \tag{52.2}$$

† In the present section, atomic units are used. In ordinary units, the expressions given below for the matrix elements of the coordinate are multiplied by $\hbar^2/me^2$ (or by $\hbar^2/mZe^2$ for a hydrogen-like ion with atomic number Z).

The integral in (52.1), containing the product of two confluent hypergoemetric functions, is calculated by means of the formulae in *QM*, §f.† The result is

$$\langle n', l-1\|r\|nl\rangle = i\sqrt{l}\,\frac{(-1)^{n'-1}}{4(2l-1)!}\sqrt{\frac{(n+l)!(n'+l-1)!}{(n-l-1)!(n'-l)!}}\,\frac{(4nn')^{l+1}(n-n')^{n+n'-2l-2}}{(n+n')^{n+n'}}\times$$
$$\times\Big\{F(-n+l+1, -n'+l, 2l, -4nn'/(n-n')^2)-$$
$$-\left(\frac{n-n'}{n+n'}\right)^2 F(-n+l-1, -n'+l, 2l, -4nn'/(n-n')^2)\Big\}, \qquad (52.3)$$

where the $F(\alpha, \beta, \gamma, z)$ are hypergeometric functions. Since the parameters α and β in this case are negative integers or zero, these functions reduce to polynomials.‡

For reference, the following are the expressions obtained from (52.3) in some particular cases (the values of l being indicated by the spectroscopic symbols $s, p, d, \ldots$):

$$\left.\begin{aligned}
|\langle 1s\|r\|np\rangle|^2 &= \frac{2^8 n^7 (n-1)^{2n-5}}{(n+1)^{2n+5}},\\
|\langle 2s\|r\|np\rangle|^2 &= \frac{2^{17} n^7 (n^2-1)(n-2)^{2n-6}}{(n+2)^{2n+6}},\\
|\langle 2p\|r\|nd\rangle|^2 &= \frac{2^{19} n^9 (n^2-1)(n-2)^{2n-7}}{3(n+2)^{2n+7}},\\
|\langle 2p\|r\|ns\rangle|^2 &= \frac{2^{15} n^9 (n-2)^{2n-6}}{3(n+2)^{2n+6}}.
\end{aligned}\right\} \qquad (52.4)$$

Formula (52.3) is not applicable to transitions with no change in the principal quantum number n (transitions between the fine structure components of a level). In this case ($n = n'$), the integration is carried out by expressing the radial functions in terms of generalized Laguerre polynomials:

$$R_{nl} = -\frac{2}{n^2}\sqrt{\frac{(n-l-1)!}{[(n+l)!]^3}}\, e^{-r/n}\left(\frac{2r}{n}\right)^l L_{n+l}^{2l+1}\left(\frac{2r}{n}\right). \qquad (52.5)$$

In the integral

$$\int_0^\infty R_{n,l-1}R_{nl}r^3\,dr \propto \int_0^\infty e^{-\rho}\rho^{2l+2}L_{n+l}^{2l+1}(\rho)L_{n+l-1}^{2l-1}(\rho)\,d\rho,$$

we replace one of the polynomials by its expression in terms of a generating function (see *QM*, §d):

$$L_{n+l}^{2l+1}(\rho) = -\frac{(n+l)!}{(n-l-1)!}\,e^{\rho}\rho^{-2l-1}\left(\frac{d}{d\rho}\right)^{n-l-1}(e^{-\rho}\rho^{n+l}).$$

† In the notation used there, we have to calculate the integral $J_{2l+2}^{12}(-n+l+1, -n'+l)$. This is done by means of formulae (f.12)–(f.16).

‡ Numerical tabulations of the matrix elements and transition probabilities for hydrogen are given by H. A. Bethe and E. E. Salpeter, *Handbuch der Physik* **35**, 88–436, Springer, Berlin, 1957.

After $n-l-1$ integrations by parts, we obtain an integral of the form

$$\int_0^\infty e^{-\rho}\rho^{n+l}\left(\frac{d}{d\rho}\right)^{n-l-1}(\rho L_{n+l-1}^{2l-1}(\rho))\,d\rho,$$

in which we replace the Laguerre polynomial by its explicit form:

$$L_n^m(\rho) = (-1)^m n! \sum_{k=0}^{n-m} \binom{n}{m+k}\frac{(-\rho)^k}{k!}.$$

After the differentiation in the sum, only three terms remain, and the integration is then elementary. The result is simply

$$\langle n, l-1\|r\|nl\rangle = i\sqrt{l}\cdot\tfrac{3}{2}n\sqrt{(n^2-l^2)}. \tag{52.6}$$

The integral

$$\int_0^\infty R_{n',l-1}R_{nl}r^3\,dr = \int_0^\infty \chi_{n',l-1}(r\chi_{nl})\,dr,$$

where $\chi_{nl} = rR_{nl}$, is the coefficient in the expansion of the function $r\chi_{nl}$ in terms of the orthogonal functions $\chi_{n',l-1}$ ($n' = 1, 2, \ldots$). The sum of the squared moduli of these coefficients is equal to the integral of the square of the expanded function.† Hence

$$\sum_{n'} |\langle n', l-1\|r\|nl\rangle|^2 = l\int_0^\infty r^2\chi_{nl}^2\,dr. \tag{52.7}$$

Using the known expression for the mean square of r in the state nl (see *QM*, (36.16)), we obtain the sum rule

$$\sum_{n'} |\langle n', l-1\|r\|nl\rangle|^2 = \tfrac{1}{2}ln^2[5n^2+1-3l(l+1)]. \tag{52.8}$$

For given n, l and large n', the matrix element of the transition $nl \to n'l'$ decreases according to

$$|\langle n'l'\|r\|nl\rangle|^2 \propto 3/n'^3, \tag{52.9}$$

as can be seen both from the particular expressions (52.4) and from the general formula (52.3). This result is to be expected: the Coulomb levels of energy $E' = -1/2n'^2$ have an almost continuous distribution when n' is large, and the

† The summation is over states of both the discrete spectrum and the continuous spectrum.

probability of a transition to a level in the interval dE' is proportional to the density of these levels, which in turn is proportional to n'^{-3}.

The Stark effect in hydrogen has the peculiarity that the splitting is proportional to the first power of the electric field (*QM*, §77). Here the field is assumed to be both weak enough for perturbation theory to be applicable and such that the level splitting is large compared with the fine structure of the levels. Under these conditions, the magnitude of the angular momentum is not conserved, and the levels have to be classified by the parabolic quantum numbers n_1, n_2, m. The last of these, the magnetic quantum number m, again determines the component of the orbital angular momentum along the z-axis (the direction of the field), which is conserved under the conditions stated (neglecting the spin–orbit interaction). It is therefore governed by the ordinary selection rule

$$m' - m = 0, \pm 1. \qquad (52.10)$$

There is no restriction on the changes in n_1 and n_2.

The matrix elements of the dipole moment in parabolic coordinates can also be calculated analytically. The resulting formulae, however, are very lengthy, and will not be given here.†

PROBLEMS

PROBLEM 1. Find the Stark splitting of hydrogen levels when it is small compared with the fine structure intervals (but large compared with the Lamb shift).

SOLUTION. Under the conditions stated, there remains a twofold degeneracy of the unperturbed levels with $l = j \pm \frac{1}{2}$, and the Stark splitting is therefore again linear in the field. The splitting Δ is determined from the secular equation

$$\begin{vmatrix} -\Delta & -E(d_z)_{12} \\ -E(d_z)_{21} & -\Delta \end{vmatrix} = 0, \qquad \Delta = \pm E|(d_z)_{12}|;$$

the suffixes 1 and 2 correspond to states with $l = j \pm \frac{1}{2}$ and a given magnetic quantum number m; the perturbation $V = -Ed_z$ is diagonal in m and has no elements diagonal in l. The matrix element of the orbital quantity d_z is calculated by means of formulae (29.7) and (109.3) in *QM*:

$$\langle j, l-1, m|d_z|jlm\rangle = \frac{m}{\sqrt{[j(j+1)(2j+1)]}}\langle j, l-1\|d\|jl\rangle,$$

$$\langle j, l-1\|d\|jl\rangle = -(2j+1)\begin{Bmatrix} l-1 & j & \frac{1}{2} \\ j & l & 1 \end{Bmatrix}\langle l-1\|d\|l\rangle,$$

where we must put $l = j + \frac{1}{2}$; the quantity $\langle l-1\|d\|l\rangle$ is taken from (52.6), and the result is

$$\Delta = \pm \tfrac{3}{4}\sqrt{[n^2 - (j+\tfrac{1}{2})^2]}\frac{nm}{j(j+1)}E.$$

PROBLEM 2. Determine the probability of photon emission in the transition $2s_{\frac{1}{2}} \to 1s_{\frac{1}{2}}$ of the hydrogen atom (G. Breit and E. Teller, 1940).

SOLUTION. The process is strictly forbidden by parity for an $E1$ transition, and by the rule (46.15) for an $E2$ transition. We have therefore to calculate the probability of an $M1$ transition, given by (47.5).

† These formulae and the corresponding numerical tabulations are to be found in the work by Bethe and Salpeter cited above.

In the present case ($l = 0$), however, the magnetic moment is a purely spin quantity, and its matrix element is zero if the spin–orbit interaction is neglected, because of the mutual orthogonality of the orbital wave functions with different principal quantum numbers. This means that, to obtain a result other than zero, the Pauli's equation approximation is insufficient, and we must start from the complete Dirac's equation.

In the standard representation of the wave functions, the transition current is†

$$\mathbf{j}_{fi} = \psi_f^* \boldsymbol{\alpha} \psi_i = \phi_f^* \boldsymbol{\sigma} \chi_i + \chi_f^* \boldsymbol{\sigma} \phi_i.$$

According to (35.1), (24.2) and (24.8), the wave functions of the states with $l = 0$, $j = \frac{1}{2}$ are

$$\psi = \begin{pmatrix} \phi \\ \chi \end{pmatrix} = \frac{1}{4\pi} \begin{pmatrix} f(r) w(m) \\ -ig(r)(\boldsymbol{\sigma} \cdot \mathbf{n}) w(m) \end{pmatrix},$$

where $\mathbf{n} = \mathbf{r}/r$, and $w(m)$ is a real three-dimensional unit spinor corresponding to the spin projection value m. Thus

$$\mathbf{j}_{fi} = \frac{1}{4\pi i} \{ f_f g_i w_f \boldsymbol{\sigma} (\boldsymbol{\sigma} \cdot \mathbf{n}) w_i - g_f f_i w_f (\boldsymbol{\sigma} \cdot \mathbf{n}) \boldsymbol{\sigma} w_i \}.$$

Substituting this in (47.4) and carrying out the integration over the directions of n, we find

$$\boldsymbol{\mu}_{fi} = -(e/6i) w_f \boldsymbol{\sigma} \times \boldsymbol{\sigma} w_i I = -\tfrac{1}{3} e w_f \boldsymbol{\sigma} w_i I$$

(from the Pauli matrix commutation rules, $\boldsymbol{\sigma} \times \boldsymbol{\sigma} = 2i\boldsymbol{\sigma}$); here

$$I = \int_0^\infty (f_f g_i + f_i g_f) r^3 \, dr. \tag{1}$$

The photon emission probability (47.5), summed over the values of m_f, is

$$w = (4e^2\omega^3/27) w_i \boldsymbol{\sigma}^2 w_i I^2 = 4e^2\omega^3 I^2/9. \tag{2}$$

From (35.4) we have, with $\kappa = -1$.

$$g = \frac{f'}{\varepsilon + m + \alpha/r} \approx \frac{f'}{2m} - \left(\varepsilon - m + \frac{\alpha}{r}\right) \frac{R'}{4m^2};$$

in the second term, the exact function f is replaced by the non-relativistic radial function R. With the approximation $g = R'/2m$, the integral

$$I = \frac{1}{2m} \int_0^\infty (R_f R_i)' r^3 \, dr = -\frac{3}{2m} \int_0^\infty R_f R_i r^2 \, dr = 0, \tag{3}$$

since R_f and R_i are orthogonal. In the next approximation, using (3), we find

$$I = \frac{1}{2m} \int_0^\infty (f_f f_i)' r^3 \, dr + \frac{1}{4m^2} \int \{ R_f' R_i (\varepsilon_i - \varepsilon_f) - \frac{\alpha}{r} (R_f R_i)' \} r^3 \, dr. \tag{4}$$

Since, from the orthogonality of the exact functions ψ_i and ψ_f, when $\kappa_i = \kappa_f$,

$$\int_0^\infty (f_i f_f + g_i g_f) r^2 \, dr = 0,$$

† In this Problem, relativistic units are used.

the first term in (4) may be written, after integration by parts, as

$$-\frac{3}{2m}\int_0^\infty f_f f_i r^2\,dr = \frac{3}{2m}\int_0^\infty g_f g_i r^2\,dr \approx \frac{3}{8m^3}\int_0^\infty R_f' R_i' r^2\,dr.$$

A calculation of the integral, with the functions

$$R_f = 2(m\alpha)^3 e^{-m\alpha r},\quad R_i = (1/\sqrt{2})(m\alpha)^3(1-\tfrac{1}{2}m\alpha r)e^{-\frac{1}{2}m\alpha r}$$

(see *QM*, §36) and the energy difference

$$\omega = \varepsilon_i - \varepsilon_f = \tfrac{1}{2}m\alpha^2(1-1/2^2) = \tfrac{3}{8}m\alpha^2,$$

gives $I = 2^{3/2}\alpha^2/9m$. Hence the transition probability is (in ordinary units)

$$w = \frac{2^5\alpha^5\hbar^2\omega^3}{3^6 m^2 c^4} = \frac{mc^2\alpha^{11}}{2^4\times 3^3\hbar} = 5.6\times 10^{-6}\ \text{sec}^{-1}.$$

The corresponding lifetime of the $2s\frac{1}{2}$ state is very long, and in practice it is much more likely that de-excitation will occur by the simultaneous emission of two photons; see the next-to-last footnote to §59.

§ 53. Radiation from diatomic molecules: electronic spectra

The specific features of molecular spectra are mainly due to the partition of the energy of the molecule into electronic, vibrational and rotational parts, each of the latter two being small compared with the previous one. The level structure of diatomic molecules has been described in detail in *QM*, Chapter XI. Here we shall consider the resulting pattern of the spectrum and the calculation of line strengths.†

Let us take first the general case, in which the electronic state of the molecule (and therefore also, in general, the vibrational and rotational states) changes in the transition. The frequencies of such transitions lie in the visible and ultra-violet regions of the spectrum. They are spoken of collectively as the *electronic spectrum* of the molecule. We shall always be considering electric dipole transitions; those of other types are of little importance in molecular spectroscopy.

As with dipole transitions in any system, the following selection rule applies to the total angular momentum J of the molecule:

$$|J'-J| \leq 1 \leq J+J'. \tag{53.1}$$

In the present case, the strict selection rule regarding the parity of the system corresponds to a selection rule regarding the *sign* of the level. (In the customary terminology of molecular spectroscopy, states having wave functions which do or do not change sign on inversion, i.e. when the coordinates of the electrons and the nuclei change sign, are called *negative* and *positive* states respectively.) Thus we have the rigorous rule

$$+\rightarrow -,\quad -\rightarrow +. \tag{53.2}$$

† The discussion below is based on *QM*, §§78 and 82–88. For brevity, we shall not make constant reference to those sections.

If the molecule consists of identical atoms (with nuclei of the same isotope), the levels can be classified with respect to interchange of the coordinates of the nuclei: *symmetric* (s) levels, with wave functions which do not change sign under this transformation, and *antisymmetric* (a) levels, with functions which do change sign. Since the electron dipole moment operator is unaffected by this transformation, its matrix elements are non-zero only for transitions without change of this symmetry:†

$$s \to s, \qquad a \to a. \tag{53.3}$$

This rule is not absolutely rigorous, however, since the existence of a given symmetry property of a level depends on a certain definite value of the total spin I of the nuclei in the molecule. Owing to the extreme weakness of the interaction between the nuclear spins and the electrons, the spin I is very nearly conserved, but not exactly. When this interaction is taken into account, I does not have a definite value, the symmetry property (s or a) is not conserved, and the selection rule (53.3) no longer applies.

The electron terms of a molecule consisting of identical atoms are also described by their *parity* (g or u), i.e. the behaviour of the wave functions when the electron coordinates (measured from the centre of the molecule) change sign while the coordinates of the nuclei remain unchanged. This property is closely related to the nuclear symmetry and the sign of the rotational levels belonging to this term. The levels which belong to an even (g) electron term can be $s+$ or $a-$, and those belonging to an odd (u) term can be $s-$ or $a+$. The rules (53.2) and (53.3) therefore give the further rule

$$g \to u, \qquad u \to g. \tag{53.4}$$

The rule (53.4) remains approximately valid for molecules consisting of different isotopes of the same element. Since the nuclear charges are equal, we can consider the electron term with fixed nuclei, and thus have a system of electrons in an electric field which possesses a centre of symmetry at the midpoint of the line joining the nuclei. The symmetry of the electron wave function with respect to inversion in this point determines the parity of the term, and since the electric dipole moment vector changes sign under this transformation, we arrive at (53.4). The rule as derived in this way is only approximate, because the nuclei have been regarded as fixed, and it therefore ceases to be valid when the interaction between the electron state and the rotation of the molecule is taken into account.

Further selection rules depend on specific assumptions concerning the relative magnitude of the different interactions in the molecule (i.e. the type of coupling), and therefore can only be approximate.

The majority of the electron terms in diatomic molecules belong to coupling type a or b. Both these have the property that the coupling of the orbital angular momentum with the axis (the electric interaction between the two atoms in the molecule) is large compared with all other interactions. The quantum numbers Λ

† This rule is clearly valid for transitions of any multipole order.

and S therefore exist, these being respectively the component of the orbital angular momentum of the electrons along the axis of the molecule and the total spin of the electrons. The operator of an orbital quantity, the electron orbital angular momentum, commutes with the spin operator, so that

$$S' - S = 0 \quad \text{(cases } a \text{ and } b\text{)}. \tag{53.5}$$

The change in Λ must satisfy the selection rule

$$\Lambda' - \Lambda = 0, \pm 1 \quad \text{(cases } a \text{ and } b\text{)}, \tag{53.6}$$

and for transitions between states with $\Lambda = 0$ (Σ terms) there is a further rule

$$\Sigma^+ \to \Sigma^+, \quad \Sigma^- \to \Sigma^- \quad \text{(cases } a \text{ and } b\text{)}. \tag{53.7}$$

(The states Σ^+ and Σ^- differ as regards behaviour under reflection in a plane through the axis of the molecule.) The rules (53.6), (53.7) are obtained by considering the molecule in a system of coordinates fixed to the nuclei (see *QM*, §87); the rule (53.6) is analogous to the selection rule for the magnetic quantum number in atoms.

The coupling types a and b differ as regards the relation between the spin–axis interaction energy and the rotation energy (the intervals between rotational levels). In case a the former is greater, in case b it is much smaller. We shall now examine these cases separately.

Case a. Here the quantum number Σ exists, which is the component of the total spin along the axis of the molecule (and therefore so does the number $\Omega = \Sigma + \Lambda$, the component of the total angular momentum). If both the initial state and the final state belong to case a, then we have the rule

$$\Sigma' - \Sigma = 0 \quad \text{(case } a\text{)}, \tag{53.8}$$

which follows from the fact, already mentioned, that the dipole moment commutes with the spin. From (53.6) and (53.8) it follows that†

$$\Omega' - \Omega = 0, \pm 1. \tag{53.9}$$

If $\Omega = \Omega' = 0$, then in addition to the general rule (53.1) the transitions with $J' = J$ are forbidden:‡

$$J' - J = \pm 1 \text{ when } \Omega = \Omega' = 0 \quad \text{(case } a\text{)}. \tag{53.10}$$

† This rule remains valid in case c also (where the coupling of the orbital angular momentum with the axis is small compared with the spin–orbit coupling) although the numbers Λ and Σ do not separately exist.

‡ This rule is analogous to the prohibition of atomic transitions with $J = J'$ when $M = M' = 0$ (see §51), but that rule was of possible interest only in the presence of an external field. Here the rule follows immediately from formula (53.12) below; the 3j-symbol $\begin{pmatrix} J' & 1 & J \\ 0 & 0 & 0 \end{pmatrix}$ is zero if $J' = J$, since the sum $J' + J + 1$ is odd.

Let us consider transitions between any two specified vibrational levels belonging to two different electron terms (of type a). When the fine structure of the electron term is taken into account, each of these levels splits into several components, the number of which, $2S+1$, must be the same for both, according to the rule (53.5). According to the rule (53.8), each component of one level combines with only one component of the other level, having the same value of Σ.

Let us next take a pair of levels with the same Σ; the values of Ω and Ω' can differ (like Λ and Λ') by 0 or ± 1. When rotation is taken into account, each level splits into a series of levels with different values of J and J' in the ranges $J \geqslant |\Omega|$, $J' \geqslant |\Omega'|$. The dependence of the transition probabilities on these numbers can be derived in a general form (H. Hönl and F. London, 1925).

The matrix element of the transition $n\Lambda\Omega JM_J \to n'\Lambda'\Omega'J'M'_J$ (where n denotes the characteristics of the electron term other than Ω and Λ) is

$$|\langle n'\Lambda'\Omega'J'M'_J|d_q|n\Lambda\Omega JM_J\rangle| = \\ = \sqrt{[(2J+1)(2J'+1)]}\begin{pmatrix} J' & 1 & J \\ -\Omega' & q' & \Omega \end{pmatrix}\begin{pmatrix} J' & 1 & J \\ -M'_J & q & M_J \end{pmatrix}|\langle n'\Lambda'|\bar{d}_{q'}|n\Lambda\rangle|, \quad (53.11)$$

where d_q and $\bar{d}_{q'}$ are respectively the spherical components of the dipole moment vector in the fixed coordinate system xyz and in the "moving" system $\xi\eta\zeta$ with the ζ-axis along the axis of the molecule. This formula is derived by means of *QM*, (110.6). The matrix elements $\langle n'\Lambda'|\bar{d}_{q'}|n\Lambda\rangle$ are independent of the rotational quantum numbers J, J', and depend only on the characteristics of the electron terms (and in this case are also independent† of the number Σ; the numbers $\Omega' = \Lambda' + \Sigma$ and $\Omega = \Lambda + \Sigma$ are therefore omitted in the notation for the matrix element.

The probability of the transition $n\Lambda\Omega J \to n'\Lambda'\Omega'J'$ is proportional to the square of the matrix element (53.11) after summation over M'_J. Using the formula *QM* (106.12):

$$\sum_{M'_J}\begin{pmatrix} J' & 1 & J \\ -M'_J & q & M_J \end{pmatrix}^2 = \frac{1}{2J+1},$$

we obtain

$$w(n\Lambda\Omega J \to n'\Lambda'\Omega'J') = (2J'+1)\begin{pmatrix} J' & 1 & J \\ -\Omega' & \Omega'-\Omega & \Omega \end{pmatrix}^2 B(n', n; \Lambda', \Lambda), \quad (53.12)$$

where the coefficients B are independent of J and J' (we are, of course, neglecting the relatively very small difference in the frequencies of transitions with different J and J').‡

† This can be shown in the same way as was done for the scalar f in *QM*, beginning of §29. In the present case the operator of the vector quantity **d** commutes with that of the vector **S**, which is conserved (in the zero-order approximation), and Σ is the component of **S** along the ζ-axis in the rotating coordinate system in which the condition of commutability of **d** and **S** has to be considered.

‡ Each of the rotational levels J considered splits into two when Λ-doubling is taken into account; one of the two is positive and the other negative. Thus, instead of one transition $J \to J'$, we have, using the selection rule (53.2), two transitions: from the positive and negative components of the level J to the negative and positive components, respectively, of the level J'. The probabilities of these transitions are equal.

If we sum (53.12) with respect to J', then (because of the orthogonality of the $3j$-symbols, *QM* (106.13)) the result is simply $B(n', n; \Lambda', \Lambda)$. Thus the total probability of transitions from a rotational level J of the state Ω to all levels J' of the state Ω' is independent of J.

Case b. Here the quantum number K exists, which is the angular momentum of the molecule without regard to its spin, as well as the total angular momentum J. The selection rules for K are the same as the general selection rules for any orbital vector quantity (such as the electric dipole moment):

$$|K' - K| \leq 1 \leq K + K' \quad \text{(case } b\text{)}, \tag{53.13}$$

together with the prohibition of a transition with $K = K'$ when $\Lambda = \Lambda' = 0$ (corresponding to (53.10)):

$$K' - K = \pm 1 \quad \text{when } \Lambda = \Lambda' = 0. \tag{53.14}$$

Let us consider transitions between rotational components of specified vibrational levels of two electron states belonging to type b. The probabilities of such transitions are given by the same formula (53.12), with K and Λ instead of J and Ω. When the fine structure is taken into account (for $S \neq 0$), each rotational level K splits into $2S+1$ components with $J = |K - S|, \ldots, K + S$, and so a multiplet appears in place of the single line $J \to J'$. Since in this case we have addition of the angular momenta $\mathbf{K}$ and $\mathbf{S}$, which are free (i.e. not coupled to the axis of the molecule), the formulae for the relative transition probabilities for the various lines in the multiplet are the same as the corresponding formulae (49.15) for the fine structure components of atomic spectra, where the corresponding angular momenta (in the case of LS coupling) are $\mathbf{L}$ and $\mathbf{S}$.

Thus we have examined the selection rules governing the possible spectral lines in all the fundamental cases that can occur in diatomic molecules.

The group of lines arising from transitions between rotational components of two given electronic-vibrational levels forms what is called in spectroscopy a *band*; the lines in a band are very close together, because the rotational intervals are small. The frequencies of these lines are given by the differences

$$\hbar\omega_{JJ'} = \text{constant} + BJ(J+1) - B'J'(J'+1), \tag{53.15}$$

where B and B' are the rotational constants in the two electronic states; in order to avoid unnecessary complications, the electron terms are assumed to be singlets. For $J' = J, J \pm 1$, formula (53.15) is represented graphically (Fig. 2) by three parabolic branches, whose points for integral J give the values of the frequencies. (The arrangement of the branches in Fig. 2 corresponds to the case $B' < B$. If $B' > B$, their open ends are towards small values of ω, and the branch with $J' = J - 1$ is the highest.)† The existence of a branch which passes through an apex is seen from the diagram to cause the lines to become increasingly dense towards a certain limiting position (the *head* of the band).

† The series of lines corresponding to transitions with $J' = J + 1, J, J - 1$ are called the P, Q and R *branches* respectively.

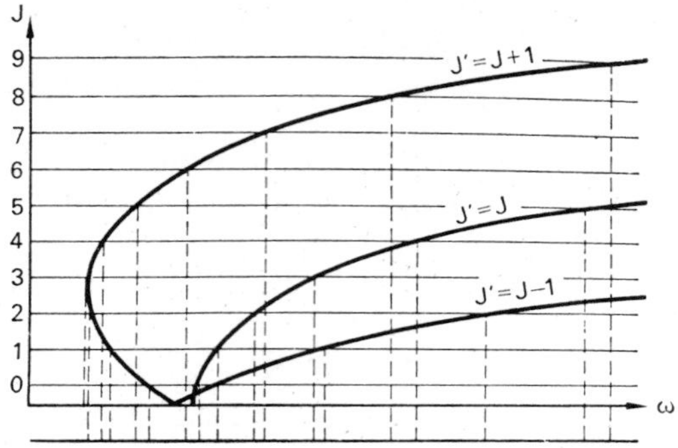

FIG. 2.

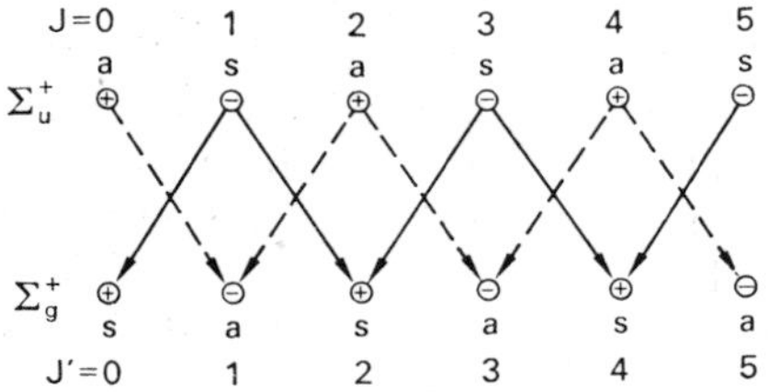

FIG. 3.

In connection with line strengths, mention should be made of the curious effect of *alternating intensities* in certain bands of the electronic spectra of molecules consisting of atoms of the same isotope (W. Heisenberg and F. Hund, 1927). The symmetry conditions pertaining to the nuclear spins have the result that, in the electron Σ terms, the rotational components with even and odd K have opposite symmetry with respect to the nuclei, and therefore different nuclear statistical weights g_s and g_a (see *QM*, §86). According to the rule (53.14), only $J' = J \pm 1$ is allowed in transitions between the two different Σ terms, and according to the rule (53.4), one of the Σ terms must be even and the other odd. The result is that, for a given value of $J' - J$, transitions with successive values of J take place alternately between pairs of symmetric levels and pairs of antisymmetric levels, as shown in Fig. 3 for the example of the states Σ_g^+ and Σ_u^+. The observed line strength is proportional to the number of molecules in the initial state concerned, and therefore to its statistical weight. Thus the intensities of successive lines ($J = 0, 1, 2, \ldots$) will be alternately greater and less, being alternately proportional to g_s and g_a (this behaviour being superimposed on the monotonic variation given by formula (53.12)).†

There are no exact selection rules concerning the change in the vibrational quantum number in transitions between two different electron terms. There is, however, a rule (*Franck and Condon's principle*) whereby the most probable change in the vibrational state may be predicted. It is based on the fact that the

† Here we assume that all the states having different values of the total nuclear spin are uniformly occupied.

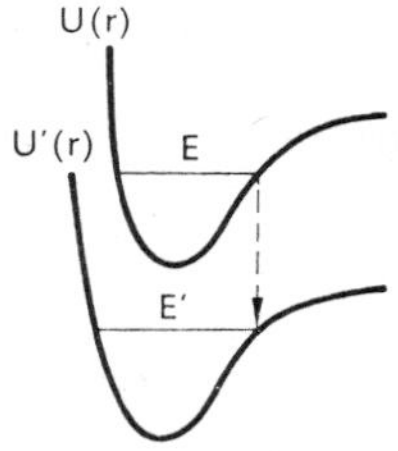

FIG. 4.

motion of the nuclei is quasi-classical, because of their large mass (cf. the discussion of pre-dissociation in *QM*, §90).†

In the integral which determines the matrix element of the transition between vibrational states E and E' of electron terms $U(r)$ and $U'(r)$, the most important range is the neighbourhood of the point $r = r_0$ where

$$U(r_0) - U'(r_0) = E - E', \tag{53.16}$$

i.e. the momenta of the relative motion of the nuclei in the two states are the same, $p = p'$. For a given value of E, the transition probability as a function of the final energy E' increases as each of the differences $E - U$ and $E' - U'$ decreases, and is a maximum when

$$E - U(r_0) = E' - U'(r_0) = 0, \tag{53.17}$$

i.e. when the "transition point" r_0 (the root of equation (53.16)) coincides with the classical turning point of the nuclei. (Fig. 4 illustrates graphically this relationship between E and the most probable E'.) This can be intuitively expressed by saying that the transition is most probable near the turning point of the nuclei, where they spend a relatively large amount of time.

§ 54. Radiation from diatomic molecules: vibrational and rotational spectra

The selection rules and formulae for transition probabilities given in §53 remain valid for transitions in which the electronic state of the molecule is unchanged.‡ Here we shall discuss only some particular features of these transitions.

First of all, the selection rule (53.4) prohibits all (dipole) transitions without change of electronic state in molecules consisting of like atoms, since in such a transition the parity of the electron term would remain unaltered. It follows from the discussion in §53 that such transitions can be allowed only when the interaction between the nuclear spins and the electrons is considered or, for molecules of

† Strictly speaking, it is also necessary that the vibrational quantum number should be sufficiently large.

‡ Transitions in which the vibrational (and therefore the rotational) state changes form what is called the *vibrational spectrum* of the molecule; this lies in the near infra-red (wavelengths $<20\ \mu$m). Transitions in which only the rotational state changes form the *rotational spectrum* in the far infra-red (wavelengths $>20\ \mu$m).

different isotopes of the same element, because of the effect of the rotation on the electronic state.

The calculation of the dipole moment matrix elements is reduced (by the formulae in *QM*, §87) to a calculation in a coordinate system rotating with the molecule. The wave function of the molecule in these coordinates is the product of the wave function of the electrons for a given distance r between the nuclei and the wave function of the vibrational motion of the nuclei in the effective field $U(r)$ of the electrons and the nuclei. When the influence of the motion of the nuclei on the electronic state is entirely neglected, the initial and final electron wave functions for the transitions in question are the same. The integration over the electron coordinates therefore gives, in the matrix element, simply the mean dipole moment $\bar{d}$ of the molecule (which is obviously along its axis) as a function of the distance r. Owing to the smallness of the vibrations, the function $\bar{d}(r)$ can be expanded in powers of the vibrational coordinate $q = r - r_0$. For transitions which involve a change in the vibrational state, the zero-order term in the expansion does not occur in the matrix element, because the wave functions are orthogonal for vibrational motion in the same field $U(q)$, and this leaves the term which is proportional to q. If the vibrations are regarded as harmonic, it follows from the known properties of the linear oscillator (*QM*, §23) that the matrix elements are zero except for transitions between adjacent vibrational states; thus, for the vibrational quantum number v we have the selection rule

$$v' - v = \pm 1. \tag{54.1}$$

This rule is not valid, however, when the vibrations are not harmonic and the subsequent terms in the expansion of the function $\bar{d}(q)$ are taken into account.

For a purely rotational transition (with no change in the vibrational state also), the matrix element of the dipole moment component along the moving ζ-axis can simply be equated to the mean dipole moment of the molecule, $\bar{d} = \bar{d}(0)$.† The probability of the transition $J \to J - 1$ is then

$$w(nJ \to n, J-1) = \frac{4\omega^3}{3\hbar c^3}\bar{d}^2 \frac{J^2 - \Omega^2}{J(2J+1)}, \tag{54.2}$$

and from this formula we can calculate not only the relative probabilities (as in (53.12)) but also the absolute probabilities. Formula (54.2) is for case a; in case b, J and Ω are to be replaced by K and Λ.

The frequencies of the purely rotational transitions are given by the differences of the rotational energies $BJ(J+1)$:

$$\hbar\omega_{J,J-1} = 2BJ. \tag{54.3}$$

Successive lines are at equal distances $2B$.

† It is obvious from symmetry that $\bar{d} = 0$ in a molecule consisting of like atoms.

§55. Radiation from nuclei

For γ radiation from nuclei it is usually true that the dimensions of the system (the radius R of the nucleus) are small compared with the wavelength of the photon. But the intervals between the nuclear levels, and therefore the energy of the γ quantum, are generally small compared with the energy per nucleon in the nucleus. The quantity R/λ is thus not directly related to the velocity v/c of the nucleons in the nucleus, and is in general considerably less than v/c. Accordingly, the probability of Ml radiation is usually greater than that of $E, l+1$ radiation (cf. the beginning of §50).

The general selection rules for the total angular momentum (the "spin") of the nucleus and for the parity are the same as for radiation from any system. The characteristic feature of nuclear radiation is that transitions of high multipole order commonly occur. Unlike atoms, whose radiation is usually of the electric dipole type, nuclei undergo such transitions comparatively rarely at low energies, on account of the selection rules.

If a radiative transition of a nucleus can be regarded as a single-particle transition (a change of state of one nucleon while the state of the rest of the nucleus remains unchanged), then there are additional selection rules regarding the angular momentum of that nucleon, but in practice such "single-particle" selection rules are found to be only approximately obeyed.

The selection rules for the isotopic spin are peculiar to nuclei. The component T_3 of the isotopic spin is determined by the atomic weight and atomic number of the nucleus:

$$T_3 = \tfrac{1}{2}(Z - N) = Z - \tfrac{1}{2}A.$$

When the value of T_3 is specified, the absolute magnitude of the isotopic spin can take any value $T \geqslant |T_3|$. The selection rule for the number T in radiative transitions arises because the electric and magnetic moment operators of the nucleus, expressed in terms of the isotopic spin operators of the nucleons, are the sums of a scalar and the x_3-component of a vector in isotopic space; see *QM*, §116. Their matrix elements are therefore zero unless

$$T' - T = 0, \pm 1. \tag{55.1}$$

This rule itself, however, imposes no special restrictions on transitions in light nuclei (the only ones for which the isotopic spin can be said with reasonable accuracy to be conserved); the low levels of these nuclei in fact include none with $T > 1$.

For $E1$ transitions, however, there is a further rule which arises because there is no isotopic-scalar part for the electric dipole moment, and the operator of this moment is simply the x_3-component of the isotopic vector (see *QM*, §116). Hence, if $T_3 = 0$, transitions with $\Delta T = 0$ are also forbidden, and so, in nuclei having equal numbers of neutrons and protons ($N = Z$, $A = 2Z$), $E1$ transitions are possible only if

$$T' - T = \pm 1 \quad (T_3 = 0). \tag{55.2}$$

The accuracy with which this rule is obeyed depends, of course, on the exactness of conservation of the isotopic spin of the nucleus.

The probability of $E1$ transitions in the nucleus is influenced also by the recoil of the rest of the nucleus when a particular nucleon moves. The result is that the protons contribute to the dipole moment with an effective charge $e(1-Z/A)$ instead of e, and the neutrons with a charge $-eZ/A$ instead of zero (see *QM*, §118). The decreased effective charge of the proton causes some reduction in the probability of $E1$ transitions.

The energy levels of non-spherical nuclei have a rotational structure, and therefore such nuclei show a characteristic rotational structure of the γ-ray spectrum.

The symmetry of the field in which the nucleons move in a "fixed" non-spherical (axial) nucleus is the same as the symmetry of the field in which electrons move in a "fixed" diatomic molecule consisting of like atoms (the point group $C_{\infty h}$). The symmetry properties of the levels of a non-spherical nucleus (and hence the selection rules for the matrix elements) are therefore analogous to those of the diatomic molecule (see *QM*, §119). In particular, electric dipole transitions within a single rotation band (i.e. without change in the internal state of the nucleus) are forbidden, as in a diatomic molecule of like atoms; cf. §54. Such transitions therefore occur as $E2$ or $M1$ transitions. In the former the total angular momentum J of the nucleus can change by 2 or 1, in the latter by 1.

According to (46.9), the probability of a quadrupole transition, summed over values of the component M' of the total angular momentum of the nucleus in the final state, is

$$w_{E2} = \frac{\omega^5}{15\hbar c^5}\sum_{M'} |\langle J'\Omega M'|Q^{(e)}_{2,-m}|J\Omega M\rangle|^2,$$

where J is the total angular momentum of the nucleus, Ω its component along the axis of the nucleus, and $m = M - M'$. By means of *QM*, (110.8) this sum can be expressed in terms of the squares of given quantities, the diagonal (with respect to the internal state of the nucleus) quadrupole transition moments $\bar{Q}_{2\lambda}$, defined relative to coordinates $\xi\eta\zeta$ moving with the nucleus. Here $\lambda = \Omega - \Omega'$, so that in the case considered ($\Omega' = \Omega$) only the component $\bar{Q}_{20}$ appears. The quantity

$$eQ_0 = e\int \rho_{ii}(2\zeta^2 - \xi^2 - \eta^2)\, d\xi\, d\eta\, d\zeta = -2e(\bar{Q}_{20})_{ii}$$

is called simply the quadrupole moment of the nucleus. Hence

$$w_{E2}(\Omega J \to \Omega J') = \frac{\omega^5}{60\hbar c^5} Q_0^2 (2J'+1)\begin{pmatrix} J' & 2 & J \\ -\Omega & 0 & \Omega \end{pmatrix}^2. \qquad (55.3)$$

Explicitly, we have

$$w_{E2}(\Omega J \to \Omega, J-1) = \frac{\omega^5}{20\hbar c^5} Q_0^2 \frac{\Omega^2(J^2-\Omega^2)}{(J-1)J(J+1)(2J+1)},$$

$$w_{E2}(\Omega J \to \Omega, J-2) = \frac{\omega^5}{40\hbar c^5} Q_0^2 \frac{(J^2-\Omega^2)[(J-1)^2-\Omega^2]}{(J-1)J(2J-1)(2J+1)}.$$

The following remark is necessary concerning these formulae. They include matrix elements calculated with wave functions of the form

$$\psi_{J\Omega M} = \text{constant} \times \chi_\Omega D^{(J)}_{\Omega M}(\mathbf{n}),$$

where χ_Ω is the wave function of the internal state of the nucleus. These functions correspond to values of the ζ-component of the angular momentum which are definite in both magnitude and sign. In nuclei, however, states have only a definite parity and a definite magnitude of the angular-momentum component (usually taken as Ω). Hence, if $\Omega \neq 0$, the initial and final wave functions would have to be taken as combinations of the form

$$\frac{1}{\sqrt{2}}(\psi_{J\Omega M} \pm \psi_{J,-\Omega,M}),$$

$$\frac{1}{\sqrt{2}}(\psi_{J'\Omega M'} \pm \psi_{J',-\Omega,M'}).$$

The products of the first terms and of the second terms will give the same value as above for the quadrupole moment matrix elements, but the cross-products will lead to non-vanishing integrals if $2\Omega \leq 2$.† Hence formula (55.3) is not strictly valid if $\Omega = \frac{1}{2}$ or 1. In these cases the transition probability contains an additional term which cannot be expressed in terms of the mean value of the quadrupole moment.‡

In a similar manner to the derivation of (55.3), we obtain for the $M1$ transition probability

$$\begin{aligned} w_{M1}(\Omega J \to \Omega, J-1) &= \frac{4\omega^3}{3\hbar c^3}\mu^2(2J-1)\begin{pmatrix} J-1 & 1 & J \\ -\Omega & 0 & \Omega \end{pmatrix}^2 \\ &= \frac{4\omega^3}{3\hbar c^3}\mu^2\frac{J^2-\Omega^2}{J(2J+1)}, \end{aligned} \qquad (55.4)$$

where μ is the magnetic moment of the nucleus. This formula is not valid if $\Omega = \frac{1}{2}$.

§ 56. The photoelectric effect: non-relativistic case

In §49–52 we have discussed radiative transitions (with emission or absorption of a photon) between atomic levels of the discrete spectrum. The photoelectric effect differs from such a photon absorption process only in that the final state belongs to the continuous spectrum.

The cross-section for the photoelectric effect can be calculated in an exact

† For the matrix elements of 2^l-pole moments, the integrands will include products of the form

$$D^{(J')*}_{-\Omega,M'} D^{(l)}_{q'q} D^{(J)}_{\Omega M}.$$

The angle integral will not be zero if $q' = -2\Omega$, and the range of values of q' is only from $-l$ to $+l$; thus we must have $2\Omega \leq l$.

‡ This term in fact gives a significant correction only in the case $\Omega = \frac{1}{2}$, when the coupling between the rotation and the internal state of the nucleus is especially large (see *QM*, §119).

analytical form for the hydrogen atom and for a hydrogen-like ion (with atomic number $Z \ll 137$).

In the initial state, the electron is at a discrete level $\varepsilon_i \equiv -I$ (where I is the ionization potential of the atom) and the photon has a definite momentum **k**. In the final state, the electron has momentum **p** (and energy $\varepsilon_f \equiv \varepsilon$). Since **p** takes a continuous series of values, cross-section for the photoelectric effect is

$$d\sigma = 2\pi |V_{fi}|^2 \delta(-I + \omega - \varepsilon)\, d^3p/(2\pi)^3 \tag{56.1}$$

(cf. (44.3)); the wave function of the final state of the electron is normalized to "one particle per unit volume". The wave function of the photon is, as before, normalized in the same way; in order to obtain the cross-section $d\sigma$, the probability dw then has to be divided by the photon flux density (which is $c/V = c$ when $V = 1$), but when relativistic units are used this does not affect the form of (56.1).

As in (45.2), we choose the three-dimensionally transverse gauge for the photon. Then

$$V_{fi} = -e\mathbf{A}\cdot\mathbf{j}_{fi} = -e\sqrt{(4\pi)}\frac{1}{\sqrt{(2\omega)}}M_{fi},$$

where

$$M_{fi} = \int \psi'^*(\boldsymbol{\alpha}\cdot\mathbf{e})\, e^{i\mathbf{k}\cdot\mathbf{r}}\psi\, d^3x, \tag{56.2}$$

with $\psi \equiv \psi_i$ and $\psi' \equiv \psi_f$ the initial and final wave functions of the electron. Putting in (56.1) $d^3p \to \mathbf{p}^2\, d|\mathbf{p}|\, do = \varepsilon|\mathbf{p}|\, d\varepsilon\, do$ and integrating to remove the delta function of ε, we can write this formula as

$$d\sigma = e^2 \frac{\varepsilon|\mathbf{p}|}{2\pi\omega}|M_{fi}|^2\, do. \tag{56.3}$$

The calculations will be given in two cases, which differ as regards the magnitude of the photon energy: $\omega \gg I$ and $\omega \ll m$. Since $I \sim me^4Z^2 \ll m$, these two ranges partly overlap (when $I \ll \omega \ll m$), and so an examination of the two cases gives an essentially complete description of the photoelectric effect.

We shall take first the case

$$\omega \ll m. \tag{56.4}$$

The electron velocity is then small in both the initial and the final state, and the problem is therefore entirely non-relativistic as regards the electron. Accordingly, we replace $\boldsymbol{\alpha}$ in (56.2) by the non-relativistic velocity operator $\hat{\mathbf{v}} = -i\nabla/m$ (cf. §45). We can also use the dipole approximation, putting $e^{i\mathbf{k}\cdot\mathbf{r}} \approx 1$, i.e. neglecting the momentum of the photon in comparison with that of the electron. Then

$$\left.\begin{aligned} d\sigma &= e^2 \frac{m|\mathbf{p}|}{2\pi\omega} |\mathbf{e}\cdot\mathbf{v}_{fi}|^2\, do, \\ \mathbf{v}_{fi} &= -\frac{i}{m}\int \psi'^*\nabla\psi \cdot d^3x. \end{aligned}\right\} \tag{56.5}$$

We shall consider the photoelectric effect from the ground level of the hydrogen atom (or of a hydrogen-like ion). Then

$$\psi = \frac{(Ze^2m)^{3/2}}{\sqrt{\pi}} e^{-Ze^2mr}; \tag{56.6}$$

in ordinary units, me^2 becomes $1/a_0$, where $a_0 = \hbar^2/me^2$ is the Bohr radius.

The wave function ψ' must be taken such that its asymptotic form comprises a plane wave ($e^{i\mathbf{p}\cdot\mathbf{r}}$) together with an ingoing spherical wave; cf. *QM*, §136, where this function was denoted by $\psi_{\mathbf{p}}^{(-)}$. On account of the selection rule for l, a transition from an s state can only be to a p state (in the dipole case). Thus, in the expansion†

$$\psi_{\mathbf{p}}^{(-)} = \frac{1}{2p}\sum_{l=0}^{\infty} i^l(2l+1)\, e^{-i\delta_l} R_{pl}(r) P_l(\mathbf{n}\cdot\mathbf{n}_1), \tag{56.7}$$

where $\mathbf{n} = \mathbf{p}/p$, $\mathbf{n}_1 = \mathbf{r}/r$, it is sufficient to retain the term with $l = 1$. Omitting unimportant phase factors, we therefore have

$$\psi' = \frac{3}{2p}(\mathbf{n}\cdot\mathbf{n}_1) R_{p1}(r). \tag{56.8}$$

With the functions ψ and ψ' given by (56.6) and (56.8), we have

$$\begin{aligned} \mathbf{e}\cdot\mathbf{v}_{fi} &= \frac{3(Ze^2m)^{5/2}}{2\sqrt{\pi}mp}\int\!\!\int (\mathbf{n}\cdot\mathbf{n}_1)(\mathbf{n}_1\cdot\mathbf{e})\, e^{-Ze^2mr} R_{p1}(r)\, do_1 \cdot r^2\, dr \\ &= \frac{\sqrt{(2\pi)}(Ze^2m)^{5/2}}{pm}(\mathbf{n}\cdot\mathbf{e})\int_0^{\infty} r^2\, e^{-Ze^2mr} R_{p1}(r)\, dr. \end{aligned}$$

According to *QM* (36.18), (36.24), the radial function is (in the units employed here)

$$R_{p1} = \frac{\sqrt{(8\pi)}Ze^2m}{3}\sqrt{\frac{1+\nu^2}{\nu(1-e^{-2\pi\nu})}}\, pre^{-ipr} F(2+i\nu, 4, 2ipr),$$

with

$$\nu = Ze^2m/p \quad (= Ze^2/\hbar v). \tag{56.9}$$

† In the rest of this section, p denotes $|\mathbf{p}|$.

The integral can be calculated by means of the formula

$$\int_0^\infty e^{-\lambda z} z^{\gamma-1} F(\alpha, \gamma, kz)\, dz = \Gamma(\gamma)\lambda^{\alpha-\gamma}(\lambda - k)^{-\alpha};$$

cf. *QM*, (f.3). Noticing also that

$$\left(\frac{\nu + i}{\nu - i}\right)^{i\nu} = e^{-2\nu \cot^{-1} \nu},$$

we obtain

$$\mathbf{e} \cdot \mathbf{v}_{fi} = \frac{2^{7/2}\pi\nu^3(\mathbf{n}\cdot\mathbf{e})}{\sqrt{p}\cdot m(1+\nu^2)^{3/2}} \frac{e^{-2\nu\cot^{-1}\nu}}{\sqrt{(1-e^{-2\pi\nu})}}.$$

The energy of ionization from the ground level of the hydrogen atom (or a hydrogen-like ion) is $I = Z^2 e^4 m/2$. Hence

$$\omega = \frac{p^2}{2m} + I = \frac{p^2}{2m}(1+\nu^2). \tag{56.10}$$

Using this relation, we obtain as the final expression for the cross-section for the photoelectric effect with emission of an electron into the solid-angle element do

$$d\sigma = 2^7 \pi\alpha a^2 \left(\frac{I}{\hbar\omega}\right)^4 \frac{e^{-4\nu\cot^{-1}\nu}}{1 - e^{-2\pi\nu}} (\mathbf{n}\cdot\mathbf{e})^2\, do, \tag{56.11}$$

where $a = \hbar^2/mZe^2 = a_0/Z$ (ordinary units are used here and below). The angular distribution of the emitted electrons is governed by the factor $(\mathbf{n}\cdot\mathbf{e})^2$. This has maxima in the directions parallel to the direction of polarization of the incident photons, and is zero in directions perpendicular to $\mathbf{e}$, including the direction of incidence. For unpolarized photons, formula (56.11) must be averaged over the directions of $\mathbf{e}$, which is equivalent to substituting

$$(\mathbf{n}\cdot\mathbf{e})^2 \to \tfrac{1}{2}(\mathbf{n}_0 \times \mathbf{n})^2$$

with $\mathbf{n}_0 = \mathbf{k}/k$; see (45.4b).

Integration of formula (56.11) over all angles gives the total cross-section for the photoelectric effect:

$$\sigma = \frac{2^9\pi^2}{3}\alpha a^2 \left(\frac{I}{\hbar\omega}\right)^4 \frac{e^{-4\nu\cot^{-1}\nu}}{1 - e^{-2\pi\nu}} \tag{56.12}$$

(M. Stobbe, 1930).

The limiting value of σ as $\hbar\omega \to I$ (i.e. as $\nu \to \infty$) is

$$\sigma = \frac{2^9\pi^2}{3e^4}\alpha a^2 = \frac{2^9\pi^2}{3e^4}\frac{\alpha a_0^2}{Z^2} = 0.23 a_0^2/Z^2, \tag{56.13}$$

where e in the denominator is the base of natural logarithms. The cross-section for the photoelectric effect tends to a constant limit near the threshold, as it must for a reaction forming charged particles (see *QM*, §147).

The case in which $\hbar\omega \gg I$ (still with $\hbar\omega \ll mc^2$) corresponds to the Born approximation ($\nu = Ze^2/\hbar v \ll 1$). Formula (56.12) becomes

$$\sigma = \frac{2^8\pi}{3}\alpha a_0^2 Z^5\left(\frac{I_0}{\hbar\omega}\right)^{7/2}, \tag{56.14}$$

where $I_0 = e^4m/2\hbar^2$ is the ionization energy of the hydrogen atom.

The process inverse to the photoelectric effect is the radiative recombination of an electron with an ion at rest. The cross-section σ_{rec} for this process can be found from that for the photoelectric effect (σ_{ph}) by means of the principle of detailed balancing (*QM*, §144). This principle states that the cross-sections for the processes $i \to f$ and $f \to i$ (with two particles in each of the states i and f) are related by

$$g_i p_i^2 \sigma_{i\to f} = g_f p_f^2 \sigma_{f\to i},$$

where p_i, p_f are the momenta of the relative motion of the particles, and g_i, g_f the spin statistical weights of the states i and f. Since $g = 2$ for the photon (which has two possible directions of polarization), we find for the hydrogen atom ground state

$$\sigma_{rec} = \sigma_{ph}\cdot 2\mathbf{k}^2/\mathbf{p}^2, \tag{56.15}$$

where $\mathbf{p} = m\mathbf{v}$ is the momentum of the incident electron and $\mathbf{k}$ that of the emitted photon.

PROBLEMS

PROBLEM 1. Derive formula (56.14) by direct use of the Born approximation in the non-relativistic case.

SOLUTION. In the Born approximation, ψ' in (56.5) is simply the plane wave $\psi' = e^{i\mathbf{p}\cdot\mathbf{r}}$, and ψ is again the function (56.6). Then

$$\begin{aligned}\mathbf{v}_{fi} = \mathbf{v}_{if} &= \frac{1}{m}\int \psi\hat{\mathbf{p}}\psi'\, d^3x \\ &= \frac{\mathbf{p}}{m}\frac{(Ze^2m)^{3/2}}{\sqrt{\pi}}(e^{-Ze^2mr})_{\mathbf{p}}.\end{aligned}$$

The Fourier component is given by (57.6b), and so

$$\mathbf{v}_{fi} = 8\sqrt{\pi}p^{-3}m^{3/2}(Ze^2)^{5/2}\mathbf{n}.$$

Substitution in (56.5) and integration over do leads to (56.14) (here, with sufficient accuracy, $p^2/2m \approx \omega$).

PROBLEM 2. Determine the total cross-section for radiative recombination of a fast but non-relativistic electron ($I \ll mv^2 \ll mc^2$) with a nucleus having charge $Z \ll 137$.

SOLUTION. The cross-section for capture to the K shell (principal quantum number $n = 1$) is obtained by substituting (56.14) in (56.15):

$$\sigma_1^{rec} = \frac{2^7\pi}{3}Z^5\alpha^3 a_0^2\left(\frac{I_0}{\varepsilon}\right)^{5/2},$$

where $\varepsilon = \frac{1}{2}mv^2$ is the energy of the incident electron, and $\hbar\omega \approx \varepsilon$. Among the other states of the resulting atom, only s states are important: in the calculation of the matrix element in the Born approximation, the important values are those of the wave function of the bound state when r is small (as will be seen in §57), and when $l > 0$ these values are small compared with those for $l = 0$. It is sufficient to take the first two terms in the expansion of ψ in powers of r. For states with $l = 0$ and any n, these terms are

$$\psi = \frac{1}{\sqrt{(\pi a^3 n^3)}}\left(1 - \frac{r}{a}\right),$$

i.e. they contain n only as a common factor $n^{-3/2}$; this expression is obtained by expansion of *QM*, (36.13). The total recombination cross-section is therefore

$$\sigma^{\text{rec}} = \sum_{n=1}^{\infty} \sigma_n^{\text{rec}} = \sigma_1^{\text{rec}} \sum_{n=1}^{\infty} \frac{1}{n^3} = \zeta(3)\sigma_1^{\text{rec}}.$$

The value of the zeta function is $\zeta(3) = 1.202$.

§ 57. The photoelectric effect: relativistic case

Let us now consider the case

$$\omega \gg I. \tag{57.1}$$

Here we have also $\varepsilon = \omega - I \gg I$, and the influence of the Coulomb field of the nucleus on the wave function ψ' of the emitted electron can be taken into account by means of perturbation theory. We shall write

$$\psi' = \frac{1}{\sqrt{(2\varepsilon)}}(u' e^{i\mathbf{p}\cdot\mathbf{r}} + \psi^{(1)}). \tag{57.2}$$

The electron may be relativistic, and therefore the unperturbed function in (57.2) is written as a relativistic plane wave (23.1).

Although the electron is non-relativistic in the initial state, its wave function ψ must nevertheless, for reasons to be explained below, include the relativistic correction ($\sim Ze^2$). This function is (cf. §39, Problem)

$$\psi = \left(1 - \frac{i}{2m}\gamma^0\boldsymbol{\gamma}\cdot\nabla\right)\frac{u}{\sqrt{(2m)}}\psi_{\text{non-r}}, \tag{57.3}$$

where $\psi_{\text{non-r}}$ is the non-relativistic bound-state function (56.6), and u is the bispinor amplitude of the electron at rest, normalized by the usual condition $\bar{u}u = 2m$.

We substitute the functions (57.2), (57.3) in the matrix element (56.2):†

$$M_{fi} = \frac{1}{2\sqrt{(m\varepsilon)}}\int\left\{\bar{u}'(\boldsymbol{\gamma}\cdot\mathbf{e})\left[\left(1 - \frac{i}{2m}\gamma^0\boldsymbol{\gamma}\cdot\nabla\right)u\psi_{\text{non-r}}\right]e^{-i(\mathbf{p}-\mathbf{k})\cdot\mathbf{r}} + \right.$$
$$\left. + \bar{\psi}^{(1)}(\boldsymbol{\gamma}\cdot\mathbf{e})\,e^{i\mathbf{k}\cdot\mathbf{r}}u\psi_{\text{non-r}}\right\}d^3x. \tag{57.4}$$

† The function (57.3) has been derived for distances $r \sim 1/mZe^2$, at which the relative order of magnitude of the correction term is Ze^2. But for the ground state (and for all s states) formula (57.3) is valid for any r, since the derivative of the purely exponential function (56.6), and therefore the correction term in (57.3), are always proportional to Ze^2. This enables us to use formula (57.3) in the present problem, where (as we shall see below) it is the small values of r that are important.

In order to derive the first term of the expansion of this quantity in powers of Ze^2, we can replace $\psi_{\text{non-r}}$ in the second term in the braces by the constant $(Ze^2m)^{3/2}/\sqrt{\pi}$ simply. The first term would vanish if treated in this way when $\mathbf{p}-\mathbf{k}\neq 0$, and it is for this reason that the first relativistic correction, proportional to Ze^2, has to be included in ψ. When $v\sim 1$, this correction gives a contribution to the cross-section that is of the same order as the contribution of the next term in the expansion of $\psi_{\text{non-r}}$ in powers of Ze^2.

In the first term in (57.4) we integrate by parts, transferring the action of the operator ∇ from $\psi_{\text{non-r}}$ to the exponential factor. The result is

$$M_{fi}=\frac{(Ze^2m)^{3/2}}{2(\pi m\varepsilon)^{1/2}}\left\{\bar{u}'(\gamma\cdot\mathbf{e})\left[1+\frac{1}{2m}\gamma^0\gamma\cdot(\mathbf{p}-\mathbf{k})\right]u(e^{-Ze^2mr})_{\mathbf{p}-\mathbf{k}}+\bar{\psi}^{(1)}_{-\mathbf{k}}(\gamma\cdot\mathbf{e})u\right\},\tag{57.5}$$

where the vector suffix denotes the spatial Fourier component. As far as the Ze^2 term we have†

$$(e^{-Ze^2mr})_{\mathbf{p}-\mathbf{k}}=\frac{8\pi Ze^2m}{(\mathbf{p}-\mathbf{k})^4}.\tag{57.6}$$

To calculate the Fourier components $\psi^{(1)}_{\mathbf{k}}$, we write down the equation satisfied by the function $\psi^{(1)}$:

$$\begin{aligned}(\gamma^0\varepsilon+i\gamma\cdot\nabla-m)\psi^{(1)}&=e(\gamma^\mu A_\mu)u'e^{i\mathbf{p}\cdot\mathbf{r}}\\&=-\frac{Ze^2}{r}\gamma^0u'\,e^{i\mathbf{p}\cdot\mathbf{r}},\end{aligned}$$

obtained by substituting (57.2) in (32.1). Applying the operator $\gamma^0\varepsilon+i\gamma\cdot\nabla+m$ to both sides, we find

$$(\triangle+\mathbf{p}^2)\psi^{(1)}=-Ze^2(\gamma^0\varepsilon+i\gamma\cdot\nabla+m)(\gamma^0u')\frac{1}{r}e^{i\mathbf{p}\cdot\mathbf{r}}.$$

Multiplying this equation by $e^{-i\mathbf{k}\cdot\mathbf{r}}$ and integrating with respect to d^3x, with the

† Taking the Fourier component of each side of the equation

$$(\triangle-\lambda^2)\frac{e^{-\lambda r}}{r}=-4\pi\delta(\mathbf{r}),$$

we obtain

$$\left(\frac{e^{-\lambda r}}{r}\right)_{\mathbf{q}}=\frac{4\pi}{\mathbf{q}^2+\lambda^2},\tag{57.6a}$$

and differentiation with respect to λ gives

$$(e^{-\lambda r})_{\mathbf{q}}=\frac{8\pi\lambda}{(\mathbf{q}^2+\lambda^2)^2}.\tag{57.6b}$$

usual integration by parts in the terms containing $\triangle$ and ∇, gives

$$(\mathbf{p}^2-\mathbf{k}^2)\psi_{\mathbf{k}}^{(1)} = -Ze^2(\gamma^0\varepsilon-\boldsymbol{\gamma}\cdot\mathbf{k}+m)(\gamma^0u')\left(\frac{1}{r}\right)_{\mathbf{k}-\mathbf{p}}$$

$$= -Ze^2(2\varepsilon\gamma^0-\boldsymbol{\gamma}\cdot(\mathbf{k}-\mathbf{p}))(\gamma^0u')\frac{4\pi}{(\mathbf{k}-\mathbf{p})^2}.$$

In the last line we have used the fact that the amplitude u' satisfies the equation

$$(\varepsilon\gamma^0-\mathbf{p}\cdot\boldsymbol{\gamma}-m)u'=0, \quad \text{or} \quad (\varepsilon\gamma^0+\mathbf{p}\cdot\boldsymbol{\gamma}-m)\gamma^0u'=0.$$

Hence

$$\bar{\psi}_{-\mathbf{k}}^{(1)} = \psi_{\mathbf{k}}^{(1)*}\gamma^0 = 4\pi Ze^2\bar{u}'\frac{2\varepsilon\gamma^0+\boldsymbol{\gamma}\cdot(\mathbf{k}-\mathbf{p})}{(\mathbf{k}^2-\mathbf{p}^2)(\mathbf{k}-\mathbf{p})^2}\gamma^0. \tag{57.7}$$

Substituting (57.6) and (57.7) in the matrix element (57.5), we can write it as

$$M_{fi} = \frac{4\pi^{1/2}(Ze^2m)^{5/2}}{(\varepsilon m)^{1/2}(\mathbf{k}-\mathbf{p})^2}\bar{u}'Au,$$

where

$$A = a(\boldsymbol{\gamma}\cdot\mathbf{e})+(\boldsymbol{\gamma}\cdot\mathbf{e})\gamma^0(\boldsymbol{\gamma}\cdot\mathbf{b})+(\boldsymbol{\gamma}\cdot\mathbf{c})\gamma^0(\boldsymbol{\gamma}\cdot\mathbf{e}),$$

$$a = \frac{1}{(\mathbf{k}-\mathbf{p})^2}+\frac{\varepsilon}{m}\frac{1}{\mathbf{k}^2-\mathbf{p}^2}, \quad \mathbf{b} = -\frac{\mathbf{k}-\mathbf{p}}{2m(\mathbf{k}-\mathbf{p})^2}, \quad \mathbf{c} = \frac{\mathbf{k}-\mathbf{p}}{2m(\mathbf{k}^2-\mathbf{p}^2)}.$$

The cross-section is

$$d\sigma = \frac{8e^2(Ze^2m)^5|\mathbf{p}|}{\omega(\mathbf{k}-\mathbf{p})^4m}(\bar{u}'Au)(\bar{u}\bar{A}u')\,do,$$

where $\bar{A}=\gamma^0A^+\gamma^0$; see §65. This expression has to be summed over final directions and averaged over initial directions of the electron spin, using the rules given in §65 below and the polarization density matrices of the initial and final states:

$$\rho = \tfrac{1}{2}m(\gamma^0+1), \quad \rho' = \tfrac{1}{2}(\gamma^0\varepsilon-\boldsymbol{\gamma}\cdot\mathbf{p}+m);$$

in the initial state, $\mathbf{p}=0$ and $\varepsilon=m$. The resulting expression is

$$d\sigma = \frac{16e^2(Ze^2m)^5|\mathbf{p}|}{m\omega(\mathbf{k}-\mathbf{p})^4}\,\mathrm{tr}\,(\rho'A\rho\bar{A})\,do.$$

The calculation of the trace by means of the formulae (22.22) is purely algebraic, and the result is

$$\operatorname{tr}(\rho' A \rho \bar{A}) = \frac{m}{\varepsilon + m}[a\mathbf{p} - (\mathbf{b} - \mathbf{c})(\varepsilon + m)]^2 + 4m(\mathbf{b}\cdot\mathbf{e})[(\varepsilon + m)(\mathbf{c}\cdot\mathbf{e}) + a(\mathbf{p}\cdot\mathbf{e})];$$

the vector $\mathbf{e}$ is assumed real, i.e. the photon is assumed to be linearly polarized.

The formula for the photoelectric effect cross-section will be put in its final form by using the polar angle θ and the azimuth ϕ of the direction of $\mathbf{p}$ when the direction of $\mathbf{k}$ is the z-axis and the plane of $\mathbf{k}$ and $\mathbf{e}$ is the xz plane (so that $\mathbf{p}\cdot\mathbf{e} = |\mathbf{p}|\cos\phi\sin\theta$). When $\omega \gg I$, the conservation of energy may be written in the form $\varepsilon - m = \omega$ instead of $\varepsilon - m = \omega - I$. We then easily see that

$$\mathbf{k}^2 - \mathbf{p}^2 = -2m(\varepsilon - m), \qquad (\mathbf{k} - \mathbf{p})^2 = 2\varepsilon(\varepsilon - m)(1 - v\cos\theta),$$

where $\mathbf{v} = \mathbf{p}/\varepsilon$ is the velocity of the photoelectron. A simple calculation gives finally

$$d\sigma = Z^5\alpha^4 r_e^2 \frac{v^3(1-v^2)^3\sin^2\theta}{[1-\sqrt{(1-v^2)}]^5(1-v\cos\theta)^4}\times$$
$$\times\left\{\frac{[1-\sqrt{(1-v^2)}]^2}{2(1-v^2)^{3/2}}(1-v\cos\theta) + \left[2 - \frac{[1-\sqrt{(1-v^2)}](1-v\cos\theta)}{1-v^2}\right]\cos^2\phi\right\}do, \tag{57.8}$$

where $r_e = e^2/m$.

In the ultra-relativistic case ($\varepsilon \gg m$), the photoelectric effect cross-section has a sharp peak at small angles $\theta \sim \sqrt{(1-v^2)}$, i.e. the electrons are emitted predominantly in the direction of incidence of the photon. Near the maximum

$$1 - v\cos\theta \approx \tfrac{1}{2}[(1-v^2) + \theta^2],$$

and the leading terms in (57.8) give

$$d\sigma \approx 4Z^5\alpha^4 r_e^2 \frac{(1-v^2)^{3/2}\theta^3}{(1-v^2+\theta^2)^3}\,d\theta\,d\phi. \tag{57.9}$$

The integration of (57.8) over angles is elementary but lengthy, and leads to the following expression for the total cross-section (F. Sauter, 1931):

$$\sigma = 2\pi Z^5\alpha^4 r_e^2 \frac{(\gamma^2-1)^{3/2}}{(\gamma-1)^5}\left\{\frac{4}{3} + \frac{\gamma(\gamma-2)}{\gamma+1}\left(1 - \frac{1}{2\gamma\sqrt{(\gamma^2-1)}}\log\frac{\gamma+\sqrt{(\gamma^2-1)}}{\gamma-\sqrt{(\gamma^2-1)}}\right)\right\}, \tag{57.10}$$

where the "Lorentz factor" γ is used for brevity:

$$\gamma = \frac{1}{\sqrt{(1-v^2)}} = \frac{\varepsilon}{m} \approx \frac{m+\omega}{m}. \tag{57.11}$$

In the ultra-relativistic case, this formula reduces to the simple expression

$$\sigma = 2\pi Z^5 \alpha^4 r_e^2/\gamma; \tag{57.12}$$

in the case $I \ll \omega \ll m$, the limit of small $\gamma - 1$ in (57.10) yields the already known result (56.14).

§ 58. **Photodisintegration of the deuteron**

A distinctive property of the deuteron is that its binding energy is small in comparison with the depth of the potential well. This enables reactions involving the deuteron to be described without a detailed knowledge of the behaviour of nuclear forces, using only the binding energy (see *QM*, §133). Here it is assumed that the wavelengths of the colliding particles are large compared with the range a of the action of nuclear forces.

This applies to the disintegration of the deuteron by γ quanta having $ka \ll 1$. It will also be assumed that $pa \ll 1$, where $\mathbf{p}$ is the momentum of relative motion of the neutron and proton released; this is a stronger condition than $ka \ll 1$.†

We start from the non-relativistic formula (56.5) for the photoelectric effect cross-section, integrating over all directions:

$$\sigma = \frac{e^2 p}{2\pi\omega} \frac{M}{2} \frac{4\pi}{3} |(\mathbf{v}_p)_{fi}|^2,$$

where $\mathbf{p}$ is the momentum of relative motion of the proton and neutron,‡ and m in (56.5) has been replaced by their reduced mass $M/2$ (where M is the nucleon mass). The matrix element is that of the proton velocity $\mathbf{v}_p$, since only the proton interacts with the photon. Expressing $\mathbf{v}_p$ in terms of the momentum $\mathbf{p}$ ($\mathbf{v}_p = \frac{1}{2}\mathbf{v} = \mathbf{p}/M$), we have

$$\sigma^{(e)} = \frac{e^2 p}{3M\omega} |\mathbf{p}_{fi}|^2. \tag{58.1}$$

The superscript (e) denotes that this formula corresponds to electric dipole transitions: $e\mathbf{p}/M = e\mathbf{v}_p = \dot{\mathbf{d}}$, so that $e\mathbf{p}_{fi}/M = i\omega \mathbf{d}_{fi}$.

The normalized wave function of the initial (ground) state of the deuteron is

$$\psi = \sqrt{\frac{\kappa}{2\pi}} \frac{e^{-\kappa r}}{r}, \quad \kappa = \sqrt{(MI)}, \tag{58.2}$$

† The photon energy for which $pa \approx 1$ (with $a = 1.5 \times 10^{-13}$ cm) is 15 MeV.
‡ In this section, p denotes $|\mathbf{p}|$.

where $I = 2.23$ MeV is the binding energy (see *QM*, §133).† The wave function of the final state can be taken to be that of free motion, i.e. the plane wave

$$\psi' = e^{i\mathbf{p}\cdot\mathbf{r}}. \tag{58.3}$$

The reason is that, in the theory under consideration, the "size of the deuteron" $1/\kappa$ is assumed large in comparison with the effective interaction radius a. The interaction between the proton and the neutron therefore has to be taken into account only in S states, and can be neglected in states with $l \neq 0$, whose wave functions are small at small distances. According to the selection rules, electric dipole transitions between two S states (the ground state and an S state of the continuous spectrum) are forbidden, and it is therefore possible in this case to neglect the nucleon interaction in the final state.

Integration by parts gives the matrix element

$$\begin{aligned}\mathbf{p}_{fi} &= -i\sqrt{\frac{\kappa}{2\pi}}\int e^{-i\mathbf{p}\cdot\mathbf{r}}\nabla\frac{e^{-\kappa r}}{r}\,d^3x\\ &= \sqrt{\frac{\kappa}{2\pi}}\,\mathbf{p}\left(\frac{e^{-\kappa r}}{r}\right)_{\mathbf{p}}\\ &= \sqrt{\frac{\kappa}{2\pi}}\,\frac{4\pi\mathbf{p}}{p^2+\kappa^2};\end{aligned}$$

see the second footnote to §57.

Using also the equation

$$\frac{1}{M}(\kappa^2+\mathbf{p}^2) = I + \frac{\mathbf{p}^2}{M} = \omega,$$

which expresses the conservation of energy, we finally obtain the photodisintegration cross-section (in ordinary units) as

$$\sigma^{(e)} = \frac{8\pi}{3}\alpha\frac{\hbar^2}{M}\frac{\sqrt{I}(\hbar\omega - I)^{3/2}}{(\hbar\omega)^3} \tag{58.4}$$

(H. A. Bethe and R. Peierls, 1935). It has a maximum at $\hbar\omega = 2I$, and tends to zero as $\hbar\omega \to I$ or $\hbar\omega \to \infty$.

The electric dipole absorption of the photon, described by formula (58.4), does not, however, give the main contribution to the cross-section near the photoelectric

† This function can be made more accurate by including a correction due to the finiteness of a, the normalization coefficient in (58.2) being replaced by

$$\sqrt{\frac{\kappa}{2\pi(1-a\kappa)}};$$

see *QM*, (133.13). A factor $1/(1-a\kappa)$ accordingly appears in the cross-section formulae. This correction is in fact quite large: for the ground state of the deuteron, $a\kappa \approx 0.4$.

The deuteron ground state is $^3S_1$, with a small "admixture" of $^3D_1$ due to the action of the tensor nuclear forces (see *QM*, §117). This admixture will be neglected, and therefore so will the tensor forces.

effect threshold ($\hbar\omega$ close to I). This is because, in this range, the principal effect must come from transitions to an S state, and these do not occur in electric dipole absorption. Nor do they occur in electric quadrupole absorption, since, although they do not then violate the parity selection rule, they are forbidden by the selection rule for orbital angular momentum (the tensor forces are here neglected, and **L** and **S** are therefore separately conserved). To calculate the photodisintegration cross-section near the threshold, we have therefore to consider magnetic dipole absorption, for which the selection rules allow transitions between S states (E. Fermi, 1935).

Replacing the electric moment in (58.1) by the magnetic moment, we have

$$\sigma^{(m)} = \tfrac{1}{3}\omega M p |\boldsymbol{\mu}_{fi}|^2. \tag{58.5}$$

The magnetic moment of the orbital motion makes no contribution to $\boldsymbol{\mu}_{fi}$, since the orbital angular momentum **L** has no matrix elements for transitions between S states. The spin magnetic moment

$$\begin{aligned}\boldsymbol{\mu} &= 2\mu_p \mathbf{s}_p + 2\mu_n \mathbf{s}_n \\ &= 2(\mu_p - \mu_n)\mathbf{s}_p + 2\mu_n \mathbf{S},\end{aligned}$$

where $\mathbf{S} = \mathbf{s}_p + \mathbf{s}_n$, and μ_p, μ_n are the magnetic moments of the proton and the neutron. When the tensor nuclear forces are neglected, the total spin is conserved, and its operator therefore yields no transitions. Hence

$$\boldsymbol{\mu}_{fi} = 2(\mathbf{s}_p)_{fi}(\mu_p - \mu_n).$$

In the same approximation (neglecting the tensor forces), the spin and coordinate variables are separable. The matrix element, like the wave functions, becomes a product of a spin part and a coordinate part:

$$\boldsymbol{\mu}_{fi} = 2(\mu_p - \mu_n)\langle s_p S' M' | \mathbf{s}_p | s_p S M\rangle \int \psi'^*(r)\psi(r)\, d^3x.$$

But the presence of spin–spin nuclear forces has the result that the wave equation for the coordinate functions $\psi(r)$ includes the spin value S as a parameter. If $S' = S$, then $\psi'(r)$ and $\psi(r)$ are eigenfunctions of the same operator, and are therefore orthogonal. Thus a photodisintegration from an initial ^3S state can occur only to a ^1S state of the continuous spectrum.

The square $|\boldsymbol{\mu}_{fi}|^2$ in (58.5) must, of course, be averaged over components M of the spin **S** in the initial state. Thus the problem is to calculate the quantity

$$\frac{1}{2S+1}\sum_M |\langle s_p S' M' | \mathbf{s}_p | s_p S M\rangle|^2,$$

where $s_p = s_n = \tfrac{1}{2}$, $S = 1$, $S' = 0$. The general rules for matrix elements in the addition of angular momenta give

$$\frac{1}{(2S+1)(2S'+1)}|\langle s_pS'\|s_p\|s_pS\rangle|^2 = \begin{Bmatrix} s_p & S' & s_n \\ S & s_p & 1 \end{Bmatrix}^2 |\langle s_p\|s_p\|s_p\rangle|^2$$
$$= \tfrac{1}{6}|\langle s_p\|s_p\|s_p\rangle|^2$$

(see *QM*, (107.11), (109.3)). The reduced matrix element is

$$\langle s_p\|s_p\|s_p\rangle = \sqrt{[s_p(s_p+1)(2s_p+1)]} = \sqrt{(3/2)}.$$

Formula (58.5) then becomes

$$\sigma^{(m)} = \tfrac{1}{3}\omega Mp(\mu_p - \mu_n)^2 \left|\int \psi'^*\psi\, d^3x\right|^2. \tag{58.6}$$

The initial function ψ is given by (58.2); the final function is

$$\psi' = \frac{1}{2p} R_{p0}(r).$$

This is the first term ($l = 0$) in the expansion (56.7) of a function whose asymptotic form comprises a plane wave and an ingoing spherical wave; an unimportant phase factor has been omitted. Since the integration is taken over the region outside the range of action of the nuclear forces, the radial function is

$$R_{p0}(r) = \frac{2\sin(pr+\delta)}{r}.$$

The phase δ is related to the value of the virtual level ($I_1 = 0.067$ MeV) of the proton + neutron system when $S = 0$:

$$\cot\delta = \kappa_1/p, \qquad \kappa_1 = \sqrt{(MI_1)};$$

see *QM*, §133. Then

$$\int \psi'^*\psi\, d^3x = (2\pi)^{3/2}\frac{\sqrt{\kappa}}{p\pi}\,\mathrm{im}\int e^{-\kappa r + ipr}\, e^{i\delta}\, dr$$
$$= (2\pi)^{3/2}\frac{\sqrt{\kappa}}{p\pi}\,\mathrm{im}\,\frac{e^{i\delta}}{\kappa - ip}.$$

After a simple algebraic reduction, we obtain the following expression for the photodisintegration cross-section (in ordinary units):

$$\sigma^{(m)} = \frac{8\pi}{3\hbar c}(\mu_p - \mu_n)^2 \frac{\sqrt{[I(\hbar\omega - I)]}(\sqrt{I} + \sqrt{I_1})^2}{\hbar\omega(\hbar\omega - I + I_1)}. \tag{58.7}$$

When $\hbar\omega \to I$, the cross-section tends to zero as $\sqrt{(\hbar\omega - I)}$, in accordance with the general properties of cross-sections near the reaction threshold (*QM*, §147).

The inverse process to photodisintegration is radiative capture of a proton by a neutron. The capture cross-section (σ_c) is obtained from the photoelectric effect cross-section (σ_{ph}) by means of the principle of detailed balancing; cf. the derivation of (56.15). The spin statistical weights of the neutron and the proton are $2 \times 2 = 4$; those of the deuteron (in a state with $S = 1$) and the photon are $3 \times 2 = 6$. Hence

$$\sigma_c = \frac{3}{2}\frac{(\hbar\omega)^2}{c^2 p^2}\sigma_{ph} = \frac{3(\hbar\omega)^2}{2Mc^2(\hbar\omega - I)}\sigma_{ph}. \tag{58.8}$$

CHAPTER VI

SCATTERING OF RADIATION

§ 59. The scattering tensor

THE scattering of a photon by a system of electrons (which will be referred to below as an atom) consists of the absorption of the initial photon **k** and the simultaneous emission of another photon **k**′. The atom may be left either at its initial energy level or at some other discrete energy level. In the former case the photon frequency is unchanged (*Rayleigh scattering*); in the latter case the frequency changes by

$$\omega' - \omega = E_1 - E_2, \tag{59.1}$$

where E_1 and E_2 are the initial and final energies of the atom (*Raman scattering*).† If the initial state of the atom is the ground state, then $E_2 > E_1$ in Raman scattering, and so $\omega' < \omega$: the frequency is decreased by the scattering (the *Stokes case*). In scattering by an excited atom, either the Stokes case or the *anti-Stokes case* ($\omega' > \omega$) may occur.

Since the electromagnetic perturbation operator has no matrix elements for transitions in which two photon occupation numbers simultaneously change, the scattering effect appears only in the second approximation of perturbation theory. It must be regarded as taking place via certain intermediate states, which may be of one of two types:

(I) The photon **k** is absorbed and the atom enters one of its possible states E_n; in the subsequent transition to the final state, the photon **k**′ is emitted;
(II) The photon **k**′ is emitted and the atom enters the state E_n; in the transition to the final state, the photon **k** is absorbed.

In this process, the matrix element is represented by the sum

$$V_{21} = \sum_n{}' \left(\frac{V'_{2n} V_{n1}}{\mathscr{E}_1 - \mathscr{E}_n^{\mathrm{I}}} + \frac{V_{2n} V'_{n1}}{\mathscr{E}_1 - \mathscr{E}_n^{\mathrm{II}}} \right) \tag{59.2}$$

(see *QM* (43.7)), where the initial energy of the atom + photons system is $\mathscr{E}_1 = E_1 + \omega$, and the energies of the intermediate states are

$$\mathscr{E}_n^{\mathrm{I}} = E_n, \qquad \mathscr{E}_n^{\mathrm{II}} = E_n + \omega + \omega'.$$

† In this chapter, the suffixes 1 and 2 will denote quantities pertaining respectively to the initial and final states of a scattering system.

The $V_{..}$ are the matrix elements for the absorption of the photon **k**, and the $V'_{..}$ are those for the emission of the photon **k**′; the initial state is excluded from the summation over n, this being indicated by the prime to the summation sign. The scattering cross-section is

$$d\sigma = 2\pi|V_{21}|\frac{\omega'^2\, do'}{(2\pi)^3}, \tag{59.3}$$

where do' is a solid-angle element for the directions **k**′. The radiation energy dI' scattered into the solid angle do' per unit time is expressed in terms of the intensity (energy flux density) I of the incident radiation by

$$dI' = I(\omega'/\omega)\, d\sigma.$$

We shall assume that the wavelengths of the initial and final photons are large compared with the dimensions a of the scattering system. All transitions will therefore be considered in the dipole approximation. If the photon states are described by plane waves, this approximation is equivalent to replacing the factors $e^{i\mathbf{k}\cdot\mathbf{r}}$ by unity. Then the wave functions of the photons are (in the three-dimensionally transverse gauge)

$$\mathbf{A}_{\mathbf{e}\omega} = \sqrt{(4\pi)}\frac{\mathbf{e}}{\sqrt{(2\omega)}}e^{-i\omega t}, \qquad \mathbf{A}_{\mathbf{e}'\omega'} = \sqrt{(4\pi)}\frac{\mathbf{e}'}{\sqrt{(2\omega')}}e^{-i\omega' t}.$$

Under the conditions considered, the electromagnetic interaction operator may be written as

$$\hat{V} = -\hat{\mathbf{d}}\cdot\hat{\mathbf{E}}, \tag{59.4}$$

where $\hat{\mathbf{E}} = -\dot{\hat{\mathbf{A}}}$ is the field strength operator and $\hat{\mathbf{d}}$ the dipole moment operator of the atom (similarly to the classical expression for the energy of a small system in an electric field; *Fields*, §42). The matrix elements are

$$V_{n1} = -i\sqrt{(2\pi\omega)}(\mathbf{e}\cdot\mathbf{d}_{n1}), \qquad V'_{2n} = i\sqrt{(2\pi\omega')}(\mathbf{e}'^*\cdot\mathbf{d}_{2n}).$$

Substituting these expressions in (59.2), (59.3), we find as the scattering cross-section (written in ordinary units)†

$$d\sigma = \left|\sum_n\left\{\frac{(\mathbf{d}_{2n}\cdot\mathbf{e}'^*)(\mathbf{d}_{n1}\cdot\mathbf{e})}{\omega_{n1}-\omega-i0}+\frac{(\mathbf{d}_{2n}\cdot\mathbf{e})(\mathbf{d}_{n1}\cdot\mathbf{e}'^*)}{\omega_{n2}+\omega'-i0}\right\}\right|^2\frac{\omega\omega'^3}{\hbar^2c^4}\,do', \tag{59.5}$$

$$\hbar\omega_{n1} = E_n - E_1, \qquad \omega'-\omega = \omega_{12}.$$

The summation is over all possible states of the atom, including those of the continuous spectrum (states 1 and 2 cannot appear in the sum, since the diagonal

† This formula was first derived by H. A. Kramers and W. Heisenberg (1925), before the development of quantum mechanics.

matrix elements $\mathbf{d}_{11}$ and $\mathbf{d}_{22}$ are zero). The infinitesimal imaginary increments in the denominators correspond to the usual rule for pole avoidance in perturbation theory (see *QM*, §43): an infinitesimal negative imaginary part is added to the energies E_n of the intermediate states over which the summation is carried out. The avoidance rule is important when the poles of (59.5) with respect to the variable E_n are in the region of the continuous spectrum; for example, if state 1 is the ground state of the atom, this would occur for $\hbar\omega$ exceeding the ionization threshold of the atom.†

We shall use the notation‡ (in ordinary units)

$$(c_{ik})_{21} = \frac{1}{\hbar}\sum_n \left[\frac{(d_i)_{2n}(d_k)_{n1}}{\omega_{n1}-\omega-i0} + \frac{(d_k)_{2n}(d_i)_{n1}}{\omega_{n2}+\omega'-i0}\right], \tag{59.6}$$

where $i, k = x, y, z$ are three-dimensional vector indices. Then formula (59.5) can be written as

$$d\sigma = \omega(\omega+\omega_{12})^3 |(c_{ik})_{21} e_i'^{*} e_k|^2 \, do'/c^4. \tag{59.7}$$

The notation (59.6) is justifiable in that this sum can in fact be represented as the matrix element of a certain tensor. This is most easily seen by defining a vector quantity $\mathbf{b}$ whose operator satisfies the equation

$$i\dot{\hat{\mathbf{b}}} + \omega\hat{\mathbf{b}} = \mathbf{d}.$$

Its matrix elements are

$$\mathbf{b}_{n1} = \frac{\mathbf{d}_{n1}}{\omega-\omega_{n1}}, \qquad \mathbf{b}_{2n} = \frac{\mathbf{d}_{2n}}{\omega+\omega_{n2}},$$

so that

$$(c_{ik})_{21} = (b_k d_i - d_i b_k)_{21}. \tag{59.8}$$

The matrix elements $(c_{ik})_{21}$ will be called the radiation *scattering tensor*.

It follows from the above that the selection rules for scattering are the same as the selection rules for the matrix elements of an arbitrary tensor of rank two. We can see immediately that, if the system has a centre of symmetry (so that its states can be classified by parity), transitions are possible only between states of the same parity (including transitions without change of state). This rule is the opposite of the parity selection rule for (electric dipole) emission, and so there is an alternate prohibition: transitions allowed in emission are forbidden in scattering, and vice versa.

We can resolve the tensor c_{ik} into irreducible parts:

$$c_{ik} = c^0\delta_{ik} + c^s_{ik} + c^a_{ik}, \tag{59.9}$$

† In a molecule, the threshold for dissociation into atoms here takes the place of the ionization threshold.

‡ Most of the results derived in §§59–61 below are due to G. Placzek (1931–1933).

where

$$\left.\begin{aligned} c^0 &= \tfrac{1}{3}c_{ii}, \\ c^s_{ik} &= \tfrac{1}{2}(c_{ik}+c_{ki}) - c^0\delta_{ik}, \\ c^a_{ik} &= \tfrac{1}{2}(c_{ik}-c_{ki}) \end{aligned}\right\} \qquad (59.10)$$

are respectively a scalar, a symmetric tensor (with zero trace) and an antisymmetric tensor. Their matrix elements are

$$(c^0)_{21} = \tfrac{1}{3}\sum_n \frac{\omega_{n1}+\omega_{n2}}{(\omega_{n1}-\omega)(\omega_{n2}+\omega)}(d_i)_{2n}(d_i)_{n1}, \qquad (59.11)$$

$$(c^s_{ik})_{21} = \tfrac{1}{2}\sum_n \frac{\omega_{n1}+\omega_{n2}}{(\omega_{n1}-\omega)(\omega_{n2}+\omega)}[(d_i)_{2n}(d_k)_{n1}+(d_k)_{2n}(d_i)_{n1}] - (c^0)_{21}\delta_{ik}, \qquad (59.12)$$

$$(c^a_{ik})_{21} = \frac{2\omega+\omega_{12}}{2}\sum_n \frac{(d_i)_{2n}(d_k)_{n1}-(d_k)_{2n}(d_i)_{n1}}{(\omega_{n1}-\omega)(\omega_{n2}+\omega)}; \qquad (59.13)$$

the symbols indicating pole avoidance are omitted, for brevity.

Let us consider some properties of the scattering tensor in the limiting cases of low and high photon frequencies.†

For Rayleigh scattering ($\omega_{12}=0$), the antisymmetric part of the tensor vanishes as $\omega\to 0$, because of the factor ω in front of the sum in (59.13). The scalar and symmetric parts of the scattering tensor, however, tend to finite limits as $\omega\to 0$. The cross-section is therefore proportional to ω^4 when ω is small.

In the opposite case, when the frequency ω is large compared with all the frequencies ω_{n1}, ω_{n2} which are important in (59.6) (but of course the wavelength is still much greater than a), we must arrive at the formulae of the classical theory. The first term in the expansion of the scattering tensor in powers of $1/\omega$ is

$$\frac{1}{\omega}\sum_n [(d_k)_{2n}(d_i)_{n1} - (d_i)_{2n}(d_k)_{n1}] = \frac{1}{\omega}(d_k d_i - d_i d_k)_{21},$$

and is zero, since the operators $\hat{d}_i$ and $\hat{d}_k$ commute. The next term in the expansion is

$$\begin{aligned} (c_{ik})_{21} &= \frac{1}{\omega^2}\sum_n [\omega_{2n}(d_k)_{2n}(d_i)_{n1} - (d_i)_{2n}\omega_{n1}(d_k)_{n1}] \\ &= \frac{1}{i\omega^2}(\dot{d}_k d_i - d_i\dot{d}_k)_{21}. \end{aligned}$$

Using the definition $\mathbf{d} = \Sigma\, e\mathbf{r}$ (with the summation over all the electrons in the atom) and the commutation rules for momenta and coordinates, we obtain

$$(c_{ik})_{11} = -\frac{Ze^2}{m\omega^2}\delta_{ik}, \qquad (c_{ik})_{21} = 0, \qquad (59.14)$$

† The case of resonance (when ω is close to one of the frequencies ω_{n1} and ω_{n2}) will be discussed in §63.

where Z is the total number of electrons in the system, and m the electron mass. Thus, in the limit of high frequencies, there remains in the scattering tensor only the scalar part, and scattering takes place without change in the state of the system, i.e. the scattering is entirely coherent (see below). The scattering cross-section in this case is

$$d\sigma = r_e^2 Z^2 |\mathbf{e}'^* \cdot \mathbf{e}|^2 \, do', \tag{59.15}$$

where $r_e = e^2/m$. After summing over polarizations of the final photon, we have

$$\begin{aligned} d\sigma &= r_e^2 Z^2 \{1 - (\mathbf{e} \cdot \mathbf{n}')^2\} \, do' \\ &= r_e^2 Z^2 \sin^2 \theta \cdot do', \end{aligned} \tag{59.16}$$

which is in fact the same as the classical Thomson's formula (*Fields*, (80.7)); θ is the angle between the direction of scattering and the polarization vector of the incident photon.

Let us consider the scattering of radiation by an assembly of N identical atoms situated in a region small compared with the wavelength. The corresponding scattering tensor is equal to the sum of the tensors for scattering by each atom. It must, however, be remembered that the wave functions (which are used to calculate the dipole moment matrix elements) for several identical atoms taken together are not simply equal functions. The wave functions are essentially defined only to within an arbitrary phase factor, which is different for each atom. The scattering cross-section has to be averaged over the phase factor of each atom separately.

The scattering tensor $(c_{ik})_{21}$ of each atom includes a factor $e^{i(\phi_1 - \phi_2)}$, where ϕ_1 and ϕ_2 are the phases of the wave functions of the initial and final states. For Raman scattering, the states 1 and 2 are different, and this factor is not equal to unity. In the squared modulus

$$|e_i'^* e_k \sum (c_{ik})_{21}|^2,$$

where the sum is over all N atoms, the products of terms pertaining to different atoms will include phase factors which vanish on independent averaging over the phases of the atoms, and only the squared modulus of each term remains. This means that the total cross-section for scattering by N atoms is found by taking N times the cross-section for scattering by one atom; the scattering is incoherent.

If, however, the initial and final states of the atom are the same, then the factors $e^{i(\phi_1 - \phi_2)} = 1$. The amplitude for scattering by the assembly of atoms is N times that for scattering by one atom, and the scattering cross-section consequently differs by a factor N^2; the scattering is coherent.† If the atomic energy level is not degenerate, Rayleigh scattering is therefore entirely coherent. But if the energy level is degenerate, there will also be incoherent Rayleigh scattering arising from

† The factor Z^2 in formulae (59.15) and (59.16) has the same origin: the cross-section for scattering by Z electrons in one atom is Z^2 times that for scattering by one electron.

the transitions of the atom between various mutually degenerate states. This is a purely quantum effect; in the classical theory, any scattering without change of frequency is coherent.

The coherent scattering tensor is given by the diagonal matrix element $(c_{ik})_{11}$, and will be denoted by α_{ik}, omitting for brevity the index which shows the state of the atom. According to (59.6),

$$\alpha_{ik}(\omega) \equiv (c_{ik})_{11} = \sum_n \left[\frac{(d_i)_{1n}(d_k)_{n1}}{\omega_{n1} - \omega - i0} + \frac{(d_k)_{1n}(d_i)_{n1}}{\omega_{n1} + \omega - i0}\right]. \tag{59.17}$$

This expression may also be written as

$$\alpha_{ik}(\omega) = \frac{e^2}{m\omega^2}\left\{-Z\delta_{ik} + \frac{1}{m}\sum_n \left[\frac{(p_i)_{1n}(p_k)_{n1}}{\omega_{n1} - \omega - i0} + \frac{(p_k)_{1n}(p_i)_{n1}}{\omega_{n1} + \omega - i0}\right]\right\}, \tag{59.18}$$

using the limiting form (59.14). Here $\mathbf{p}$ is the total momentum of the electrons in the atom. The equivalence of the two forms is easily seen by noting that the matrix elements of the momentum and the dipole moment are connected by

$$e\mathbf{p}_{1n}/m = i\omega_{1n}\mathbf{d}_{1n}$$

and using the same relationships as in the derivation of (59.14).

If the sum and difference $E_1 \pm \omega$ are not equal to any of the energy levels E_n of the atom (including the continuous spectrum), the terms $i0$ in the denominators may be omitted. Since $p_{1n}^* = p_{n1}$, the tensor α_{ik} is then seen to be Hermitian†:

$$\alpha_{ik} = \alpha_{ki}^*. \tag{59.19}$$

This means that its scalar and symmetric parts are real, and its antisymmetric part is imaginary. The latter is certainly zero if the atom is in a non-degenerate state; the wave function of such a state is real, and therefore the diagonal matrix elements are also real.

The tensor α_{ik} is related to the polarizability of the atom in an external electric field. To show this relation, let us calculate the correction to the mean value of the dipole moment of the system when the latter is placed in an external electric field

$$\tfrac{1}{2}(\mathbf{E}\,e^{-i\omega t} + \mathbf{E}^*\,e^{i\omega t}). \tag{59.20}$$

This can be done by using a well-known formula of perturbation theory (*QM*, §40). If the system is subjected to a perturbation

$$\hat{V} = \hat{F}\,e^{-i\omega t} + \hat{F}^+\,e^{i\omega t},$$

† This result depends on the neglect of the natural line width, and therefore of the possible absorption of the incident radiation; see §62.

then the first-order correction to the diagonal matrix elements of a quantity f is

$$f_{11}^{(1)}(t) = -\sum_n \left\{ \left[\frac{f_{1n}^{(0)} F_{n1}}{\omega_{n1} - \omega - i0} + \frac{f_{n1}^{(0)} F_{1n}}{\omega_{n1} + \omega + i0} \right] e^{-i\omega t} + \left[\frac{f_{1n}^{(0)} F_{1n}^*}{\omega_{n1} + \omega - i0} + \frac{f_{n1}^{(0)} F_{n1}^*}{\omega_{n1} - \omega + i0} \right] e^{i\omega t} \right\}.$$

The perturbation $\hat{V}$ must be regarded as being applied with infinite slowness from $t = -\infty$, so that in the first term ω is to be interpreted as $\omega + i0$, and in the second term as $\omega - i0$; the imaginary increments in the denominators have been written accordingly.

In the present case

$$\hat{F} = -\tfrac{1}{2}\mathbf{d} \cdot \mathbf{E},$$

and the correction to the diagonal matrix element of the dipole moment is found to be

$$\mathbf{d}_{11}^{(1)} = \tfrac{1}{2}(\bar{\mathbf{d}}\, e^{-i\omega t} + \bar{\mathbf{d}}^*\, e^{i\omega t}), \qquad (59.21)$$

where $\bar{\mathbf{d}}$ is a vector whose components are

$$\bar{d}_i = \alpha_{ik}^{(p)} E_k. \qquad (59.22)$$

The expression for the tensor $\alpha_{ik}^{(p)}(\omega)$ differs from (59.17) for α_{ik} by a change in the sign of the imaginary part in the denominator of the second term. By definition, $\alpha_{ik}^{(p)}(\omega)$ is the *polarizability tensor* of the atom in a field of frequency ω. For frequencies such that the imaginary parts in the denominators can be omitted and the tensor α_{ik} is Hermitian, α_{ik} and $\alpha_{ik}^{(p)}$ are identical. In particular, when $\omega = 0$ the formula (59.22) becomes *QM*, (76.4), and the expression *QM*, (76.5) for the static polarizability tensor is the same as $\alpha_{ik}(0)$ from (59.17). Note also that, if state 1 is the ground state,† all $\omega_{n1} > 0$ and the avoidance rule in the first and second terms in (59.17) is important only when $\omega > 0$ and $\omega < 0$ respectively. In that case,

$$\alpha_{ik}(\omega) = \alpha_{ik}^{(p)}(|\omega|). \qquad (59.23)$$

The formulae of scattering theory implicitly have $\omega > 0$; the tensor α_{ik} is then the same as the polarizability tensor.

We shall need not only the cross-section but also the photon scattering amplitude f. As usual in perturbation theory, this is equal, apart from a normalization factor, to minus the matrix element (59.2). Choosing this factor so as to express the cross-section (59.7) in the form $d\sigma = |f|^2\, do'$, we have as the elastic scattering amplitude

$$f = \omega^2 \alpha_{ik} e_i'^* e_k. \qquad (59.24)$$

According to the optical theorem (see (71.10) below), the imaginary part of the

† Only this case (which will be assumed henceforward) allows a completely rigorous treatment, because of the finite lifetime of the excited states; see §62 below.

forward scattering amplitude (without change in momentum and polarization) determines the total cross-section σ_t for all possible elastic and inelastic processes for a given initial state of the photon:

$$\sigma_t = (4\pi/\omega)\,\mathrm{im}(\omega^2\alpha_{ik}e_ie_k^*) = 4\pi\omega(\alpha_{ik} - \alpha_{ki}^*)e_i^*e_k/2i. \tag{59.25}$$

Thus the total cross-section is determined by the anti-Hermitian part of the scattering tensor.

The formula (59.25) has a simple classical significance. The electric field **E** does work $\Sigma e\mathbf{v}\cdot\mathbf{E} = \mathbf{E}\cdot\dot{\mathbf{d}}$ on the system of charges per unit time. Expressing the field in the form (59.20), and the dipole moment in the form (59.21), (59.22), and averaging this work with respect to time, we find

$$\tfrac{1}{2}\omega|E|^2e_i^*e_k(\alpha_{ik} - \alpha_{ki}^*)/2i,$$

with $\mathbf{E} = \mathbf{e}E$. On the other hand, if **E** is the incident radiation field, the mean energy flux density in it is $|E|^2/8\pi$, and the energy absorbed by the atom is $|E|^2\sigma_t/8\pi$. Equating the two expressions, we find (59.25).

If the angular momentum J of the ground state of the atom is zero, then by spherical symmetry $\alpha_{ik} = \alpha\delta_{ik}$, and

$$\sigma_t = 4\pi\omega\ \mathrm{im}\ \alpha. \tag{59.26}$$

For a system having angular momentum, a similar relation holds for quantities averaged over the spatial directions of the angular momentum (see §60).

For photon energies above the ionization threshold of the atom, the principal contribution to the total cross-section σ_t comes from the ionization process (the absorption of a photon in the photoelectric effect). The scattering cross-section is a quantity of higher order in e^2; compare, for instance, (56.13) and (59.16).

If, however, the photon energy is below the ionization threshold (but not too close to resonance, i.e. to any of the discrete excitation frequencies of the atom), then the cross-section (which in this case reduces to the scattering cross-section), and therefore the imaginary part of the amplitude, are of a higher order of smallness than the real part of the amplitude. Neglecting the former, we again obtain (59.19). The situation is different in the neighbourhood of resonance, where the cross-section increases; this case will be discussed in §62.

As well as scattering, the two-photon processes which occur in second-order perturbation theory also include *double emission*, i.e. the simultaneous emission of two quanta by an atom.

The expression for the probability of this process differs from (59.5) only by the changes $\omega \to -\omega$, $\mathbf{e} \to \mathbf{e}^*$ (emission of a photon ω, instead of absorption) and by the extra factor

$$d^3k/(2\pi)^3 = \omega^2\,d\omega\,do/(2\pi)^3,$$

the number of quantum states of the emitted photon in given ranges of the frequency ω and the directions of **k**; the frequency of the second photon is determined from ω by the equation $\omega + \omega' = \omega_{12}$. The emission probability per unit time is therefore†

$$dw = |(b_{ik})_{21} e_i'^* e_k^*|^2 \frac{\omega^3 \omega'^3}{(2\pi)^3 c^6 \hbar^2} \, do \, do' \, d\omega, \tag{59.27}$$

where

$$(b_{ik})_{21} = \sum_n \left[\frac{(d_i)_{2n}(d_k)_{n1}}{\omega_{n1} + \omega - i0} + \frac{(d_k)_{2n}(d_i)_{n1}}{\omega_{n1} + \omega' - i0} \right]$$

differs from $(c_{ik})_{21}$ in (59.6) only by a change in the sign of ω. Summing this expression over the polarizations of the photons and integrating over their directions of emission,‡ we obtain

$$dw = \frac{8\omega^3 \omega'^3}{9\pi\hbar^2 c^6} |(b_{ik})_{21}|^2 \, d\omega. \tag{59.28}$$

The probability of the emission of two photons ω and ω' is usually very smali in comparison with that of the emission of a single photon with frequency $\omega + \omega'$. An exception occurs in cases where the selection rules forbid the latter process but allow the former, such as transitions between two states with $J = 0$, where all processes of single-photon emission are strictly forbidden. Another example is the transition from the first excited state ($2s_{\frac{1}{2}}$) of the hydrogen atom to the ground state ($1s_{\frac{1}{2}}$), which is forbidden for both $E1$ and $M1$ radiation; see §52, Problem 2.§

If the atom is in the field of an incident flux of photons ω, $\mathbf{k}$, there is not only spontaneous double emission, with the probability (59.27), but also induced double emission, in which the field causes emission of a similar photon and a photon ω', $\mathbf{k}'$. The probability of this process differs from that of spontaneous emission by a factor $N_{\mathbf{ke}}$, the number density of incident photons with given $\mathbf{k}$ and $\mathbf{e}$. The incident photon flux density is

$$dI = cN_{\mathbf{ke}} \, d^3k/(2\pi)^3 = N_{\mathbf{ke}}(\omega^2/8\pi^3 c^2) \, d\omega \, do.$$

Expressing $N_{\mathbf{ke}}$ in terms of dI and dividing the probability of the process by dI, we obtain the cross-section

$$d\sigma = \frac{\omega \omega'^3}{\hbar^2 c^4} |(b_{ik})_{12} \, e_i'^* e_k^*|^2 \, do'. \tag{59.29}$$

Similarly, if the atom is in a field of photons ω', $\mathbf{k}'$, the incidence of a photon ω, $\mathbf{k}$ causes *induced Raman scattering*, whose cross-section is proportional to the density of photons ω', $\mathbf{k}'$.

† In the rest of this section, ordinary units are used.

‡ This operation amounts to complete averaging over the directions of $\mathbf{e}$ by $\overline{e_i e_k^*} = \frac{1}{3}\delta_{ik}$, followed by multiplication by $2 \times 2 \times 4\pi \times 4\pi$.

§ The lifetime of the $2s_{\frac{1}{2}}$ level for double emission is 0.15 sec.

The calculation of the tensors $(c_{ik})_{12}$ or $(b_{ik})_{12}$ for specific atoms requires that of sums of the form

$$(M_{ik}^{(2)})_{21} = \sum_n \frac{(d_i)_{2n}(d_k)_{n1}}{E_n - E - i0}, \tag{59.30}$$

with E taking the values $E_1 \pm \hbar\omega$ or $E_1 \pm \hbar\omega'$. To simplify the notation, let us discuss a hydrogen atom. We write the sum (59.30) as the integral

$$(M_{ik}^{(2)})_{21} = \int \psi_2^*(\mathbf{r})\, d_i G(\mathbf{r}, \mathbf{r}')\, d'_k \psi_1(\mathbf{r}')\, d^3x\, d^3x', \tag{59.31}$$

where

$$G(\mathbf{r}, \mathbf{r}'; E) = \sum_n \frac{\psi_n(\mathbf{r})\psi_n^*(\mathbf{r}')}{E_n - E - i0}. \tag{59.32}$$

Let the operator $\hat{H} - E$, where $\hat{H}$ is the Hamiltonian of the atom, act on the function G. Since $\hat{H}\psi_n = E_n\psi_n$, we obtain

$$(\hat{H} - E)G = \sum_n \psi_n(\mathbf{r})\psi_n^*(\mathbf{r}').$$

The sum is the delta function $\delta(\mathbf{r} - \mathbf{r}')$, since the set of functions ψ_n is complete. Thus the function G satisfies the equation

$$(\hat{H} - E)G(\mathbf{r}, \mathbf{r}'; E) = \delta(\mathbf{r} - \mathbf{r}'), \tag{59.33}$$

i.e. it is the Green's function of Schrödinger's equation; the avoidance rule in (59.32) decides which solution of this equation is to be taken. Thus the problem of calculating the sum (59.30) reduces to finding the Green's function of the atom. An exact solution of equation (59.33) is, however, possible only if we know the exact solutions of the homogeneous Schrödinger's equation, i.e. in practice only for the hydrogen atom.†

PROBLEM

Calculate the probability of elastic scattering of a (non-relativistic) electron by an almost monochromatic standing light wave (P. L. Kapitza and P. A. M. Dirac, 1933).

SOLUTION. The standing wave may be regarded as a combination of photons with momenta $\mathbf{k}$ and $-\mathbf{k}$ (and equal polarizations). The scattering of the electron may be regarded as the absorption of a photon $\mathbf{k}$ and induced emission of a photon $-\mathbf{k}$, so that the electron momentum $\mathbf{p}$ is changed by $2\hbar\mathbf{k}$ and rotated (without change of magnitude) through an angle θ such that $|\mathbf{p}| \sin\frac{1}{2}\theta = \hbar\omega/c$. The probability of this process can be obtained from the Thomson scattering cross-section (59.15),

$$\begin{aligned} d\sigma &= r_e^2 |\mathbf{e}'^* \cdot \mathbf{e}|^2\, do' \\ &= r_e^2\, do', \end{aligned}$$

† See L. Hostler, *Journal of Mathematical Physics* **5**, 591, 1964. The application of this Green's function to calculate the scattering amplitude for the hydrogen atom is given by Ya. I. Granovskiĭ, *Soviet Physics JETP* **29**, 333, 1969.

by multiplying by the flux density of photons with momentum **k** and the number of photons with momentum $-\mathbf{k}$.

The flux density of photons having frequencies in the range $d\omega$ is $cU_\omega\, d\omega/2\hbar\omega$, where $U_\omega\, d\omega$ is the energy density in the standing wave in the spectrum interval $d\omega$; the factor $\frac{1}{2}$ appears because the energy of the wave is equally divided between the photons moving in opposite directions. The momenta **k** of all the photons forming the standing wave are parallel to a certain direction **n** (the "direction" of the standing wave). In other words, the energy density as a function of the frequency and direction of the photons $\mathbf{n}'$ is $U_{\omega\mathbf{n}'} = U_\omega \delta^{(2)}(\mathbf{n}' - \mathbf{n})$. Accordingly, the number of $-\mathbf{k}$ photons is

$$\int N_{-\mathbf{k}}\, do' = \frac{8\pi^3 c^3}{\hbar\omega^3} \cdot \frac{U_\omega}{2};$$

cf. (44.8). The electron scattering probability per unit time is then found to be

$$w = \frac{2\pi^3 e^4}{m^2 \hbar^2 \omega^4} \int U_\omega^2\, d\omega.$$

The factor ω^{-4} is taken outside the integral, since the non-monochromaticity $\Delta\omega$ is assumed small. The value of the integral is inversely proportional to $\Delta\omega$ (for a given total intensity).

§ 60. Scattering by freely oriented systems

If an atomic energy level is not degenerate, the polarizability and intensity of coherent scattering are determined by the same tensor $\alpha_{ik} \equiv (c_{ik})_{11}$. If the level is degenerate, however, the observed values of these quantities are averaged over all states belonging to the level in question. The polarizability must be defined as the mean value

$$\alpha_{ik} = \overline{(c_{ik})_{11}}.$$

The observed scattering intensity is determined by the mean values

$$\overline{(c_{ik})_{11}(c_{lm})_{11}}.$$

The relation between the polarizability and the scattering is therefore more indirect.

Although each of the quantities $(c_{ik})_{11}$ may be complex, their mean values (in the absence of absorption, with α_{ik} an Hermitian tensor) are real, since on averaging we can choose arbitrarily the set of independent wave functions (corresponding to a given degenerate level), and we can always ensure that all the functions are real.

For free atoms or molecules (not in an external field), the degeneracy of levels is usually due to an angular momentum which is freely oriented in space. Let the initial state in scattering have angular momentum J_1, and the final state J_2. As usual, the scattering cross-section must be averaged over all values of the component M_1, and summed over the values of M_2. After the averaging, the cross-section is independent of M_2, and the summation is therefore equivalent to multiplying by $2J_2 + 1$. Thus the averaged scattering cross-section is

$$d\bar{\sigma} = \omega\omega'^3 c_{iklm}^{(21)} e_i'^* e_k e_l' e_m^*\, do', \tag{60.1}$$

where

$$c_{iklm}^{(21)} = \frac{1}{2J_1+1}\sum_{M_1, M_2}(c_{ik})_{21}(c_{lm})^*_{21}$$
$$= (2J_2+1)\overline{(c_{ik})_{21}(c_{lm})^*_{21}}^{1}; \qquad (60.2)$$

the bar with index 1 signifies averaging over M_1.

For Rayleigh scattering, states 1 and 2 belong to the same energy level ($\omega_{12} = 0$). If only coherent scattering is considered, then states 1 and 2 must coincide completely, so that $M_1 = M_2$. In that case the summation over M_2, and hence the factor $2J_2+1$ in (60.2), no longer appear:

$$c_{iklm}^{\mathrm{coh}} = \overline{(c_{ik})_{11}(c_{lm})^*_{11}}^{1}. \qquad (60.3)$$

The result of the averaging can be written down without further calculation by using the fact that averaging over M_1 is equivalent to averaging over all orientations of the system, after which the mean value can only be expressed in terms of the unit tensor δ_{ik}, and the only non-zero mean values are those of products of components of either the scalar, the symmetric or the antisymmetric part of the scattering tensor; it is clear that the unit tensor cannot yield expressions with the symmetry properties of cross-products. Thus

$$c_{iklm}^{(21)} = G_{21}^{0}\delta_{ik}\delta_{lm} + c_{iklm}^{(21)s} + c_{iklm}^{(21)a}, \qquad (60.4)$$

where

$$\left.\begin{aligned} G_{21}^{0} &= (2J_2+1)\overline{|(c^0)_{21}|^2}^{1}, \\ c_{iklm}^{(21)s} &= (2J_2+1)\overline{(c^s_{ik})_{21}(c^s_{lm})^*_{21}}^{1}, \\ c_{iklm}^{(21)a} &= (2J_2+1)\overline{(c^a_{ik})_{21}(c^a_{lm})^*_{21}}^{1}. \end{aligned}\right\} \qquad (60.5)$$

The scattering cross-section (and therefore the scattering intensity) for a freely oriented system is therefore a sum of three independent parts, which will be referred to as *scalar, symmetric* and *antisymmetric scattering*.

Each of the three terms in (60.4) can be expressed in terms of one independent quantity: the scalar scattering is expressed in terms of G_{21}^{0}, and for the symmetric and antisymmetric scattering we have

$$\left.\begin{aligned} c_{iklm}^{(21)s} &= \tfrac{1}{10}G_{21}^{s}(\delta_{il}\delta_{km} + \delta_{im}\delta_{kl} - \tfrac{2}{3}\delta_{ik}\delta_{lm}), \\ G_{21}^{s} &= (2J_2+1)\overline{(c^s_{ik})_{21}(c^s_{ik})_{21}}^{1}; \\ c_{iklm}^{(21)a} &= \tfrac{1}{6}G_{21}^{a}(\delta_{il}\delta_{km} - \delta_{im}\delta_{kl}), \\ G_{21}^{a} &= (2J_2+1)\overline{(c^a_{ik})_{21}(c^a_{ik})_{21}}^{1}; \end{aligned}\right\} \qquad (60.6)$$

the combinations of unit tensors are derived from the symmetry properties, and the common factor is then found by contracting with respect to the pairs of indices i, l and k, m.

On substituting (60.4)–(60.6) in (60.1), we obtain for the scattering cross-section

$$d\bar{\sigma} = \omega\omega'^3\{G^0_{21}|\mathbf{e}'^*\cdot\mathbf{e}|^2 + \tfrac{1}{10}G^s_{21}(1+|\mathbf{e}'\cdot\mathbf{e}|^2 - \tfrac{2}{3}|\mathbf{e}'^*\cdot\mathbf{e}|^2) + \tfrac{1}{6}G^a_{21}(1-|\mathbf{e}'\cdot\mathbf{e}|^2)\}\, do'. \tag{60.7}$$

This formula shows explicitly the angular dependences and polarization properties of the scattering.

The total cross-section for scattering in any direction, summed over the polarization of the final photon and averaged over the polarization and direction of incidence of the initial photon, is easily obtained directly from (60.1) by noting that

$$\overline{e_i^* e_k} = \tfrac{1}{3}\delta_{ik}$$

if the averaging is over both the polarization and the direction of propagation of the photon; summation over these would give a corresponding result larger by a factor $2\times 4\pi$. The result is

$$\begin{aligned}\bar{\sigma} &= \frac{8\pi}{9}\omega\omega'^3 c^{(21)}_{ikik} \\ &= \frac{8\pi}{9}\omega\omega'^3(3G^0_{21} + G^s_{21} + G^a_{21}).\end{aligned} \tag{60.8}$$

It has already been mentioned that the selection rules for scattering are the same as those for the matrix elements of an arbitrary tensor of rank two. Because of the separation of the scattering intensity into three independent parts, it is convenient to state the rules for each part separately.

The selection rules for symmetric scattering are the same as those for electric quadrupole radiation, since the latter is likewise determined by an irreducible symmetric tensor (the quadrupole moment tensor). For antisymmetric scattering, the selection rules are the same as those for magnetic dipole radiation, since both are determined by an axial vector (an antisymmetric tensor is equivalent, or dual, to an axial vector).† There is a difference here, however, in that the diagonal matrix elements, which in the case of emission give the mean values of the electric or magnetic moments (and do not correspond to radiative transitions), are important in the case of scattering, since they relate to coherent scattering.

For scalar scattering the selection rules are the same as those for the matrix elements of a scalar. This means that only transitions between states of the same symmetry are possible. In particular, the values of the total angular momentum J and its component M must be the same, and the matrix elements diagonal in M are independent of M; see *QM*, (29.3). For Rayleigh scattering, therefore, states 1 and 2 must coincide completely (as regards M as well as energy), and so scalar Rayleigh scattering is entirely coherent. Conversely, since in scalar scattering all states always combine with themselves, it follows that in coherent scattering there is always a scalar part.

† This refers, of course, to the selection rules based on symmetry, and not due to the specific form of the axial vector in the case of emission; the magnetic moment vector includes a spin part, whereas in scattering we have the matrix elements of orbital (coordinate) quantities.

For a system freely oriented in space, the polarizability tensor also must be averaged over the directions of the angular momentum J_1, in the same way as the scattering cross-section has been averaged above. The averaging is very simply carried out: we evidently have

$$\alpha_{ik} \equiv \overline{(c_{ik})_{11}}^{1} = \overline{(c^0)_{11}}^{1}\,\delta_{ik}.$$

The symmetric and antisymmetric parts of the scattering tensor vanish on averaging, since δ_{ik} is the only isotropic tensor of rank two.

It has been mentioned that the diagonal matrix elements of a scalar are independent of M_1. The mark of averaging of $(c^0)_{11}$ may therefore be omitted, and this quantity calculated for any M_1, so that the polarizability is

$$\alpha_{ik} = (c^0)_{11}\delta_{ik}. \tag{60.9}$$

For the same reason, the averaging sign may be omitted in the quantity G^0_{11}, which determines the scalar part of the coherent scattering:

$$G^0_{11} = \overline{|(c^0)_{11}|^2}^{1} = (c^0)^2_{11}; \tag{60.10}$$

the factor $2J_2+1$ is omitted in accordance with (60.3). Thus there is a simple relation between the mean polarizability and the scalar part of the coherent scattering: both are determined by the quantity

$$(c^0)_{11} = \tfrac{2}{3}\sum_n \frac{\omega_{n1}}{\omega^2_{n1}-\omega^2}|\mathbf{d}_{n1}|^2. \tag{60.11}$$

PROBLEMS

PROBLEM 1. Find the angular distribution and the degree of depolarization in the scattering of linearly polarized radiation.

SOLUTION. Let θ be the angle between the direction of scattering $\mathbf{n}'$ and the direction of polarization $\mathbf{e}$ of the incident radiation. The scattered radiation has two independent components, polarized one in the plane of $\mathbf{n}'$ and $\mathbf{e}$ (intensity I_1) and one perpendicularly to this plane (intensity I_2); the degree of depolarization is I_2/I_1. The intensities I_1 and I_2 are given by (60.7) with the appropriate directions of $\mathbf{e}'$.

In scalar scattering, the radiation remains completely polarized in the same plane ($I_2=0$), and the angular distribution of intensity is

$$I = \tfrac{3}{2}\sin^2\theta.$$

Here and below, the expressions for $I = I_1 + I_2$ are normalized so as to give unity on averaging over directions.

In symmetric scattering

$$I = \tfrac{3}{20}(6+\sin^2\theta), \qquad I_2/I_1 = 3/(3+\sin^2\theta).$$

In antisymmetric scattering

$$I = \tfrac{3}{4}(1+\cos^2\theta), \qquad I_2/I_1 = 1/\cos^2\theta.$$

PROBLEM 2. The same as Problem 1, but for the scattering of natural light.

SOLUTION. Formula (60.7) can be applied to natural (unpolarized) incident light by the substitution

$$e_i e_k^* \to \tfrac{1}{2}(\delta_{ik} - n_i n_k),$$

which corresponds to averaging over the direction of polarization **e** with a given direction of incidence **n**. The scattered light will be partly polarized, and from considerations of symmetry it is evident that its two independent components will be linearly polarized in the scattering plane of **n** and **n**′ (intensity $I_\parallel$) and perpendicularly to this plane (intensity $I_\perp$). The scattering angle between **n** and **n**′ will be denoted by ϑ.

For scalar scattering

$$I = I_\perp + I_\parallel = \tfrac{3}{4}(1+\cos^2\vartheta), \qquad I_\parallel/I_\perp = \cos^2\vartheta.$$

For symmetric scattering

$$I = \tfrac{3}{40}(13+\cos^2\vartheta), \qquad I_\parallel/I_\perp = (6+\cos^2\vartheta)/7.$$

For antisymmetric scattering

$$I = \tfrac{3}{8}(2+\sin^2\vartheta), \qquad I_\parallel/I_\perp = 1+\sin^2\vartheta.$$

PROBLEM 3. For scattering of circularly polarized radiation, determine the reversal factor (the ratio of the intensity of the component circularly polarized in the "reverse" direction to that of the component polarized in the original direction).

SOLUTION. For circularly polarized incident radiation, the angular distribution and the degree of depolarization ($I_\parallel/I_\perp$) are the same as in the scattering of natural light.

Let the vector **e** of the incident radiation have components $(1/\sqrt{2})(1, i, 0)$ in coordinates such that the xz-plane is the scattering plane and the z-axis is along **n**. Then the polarization vectors for the reverse and original circularly polarized components of the scattered radiation are

$$\mathbf{e}' = \frac{1}{\sqrt{2}}(\cos\vartheta, -i, -\sin\vartheta) \quad \text{and} \quad \mathbf{e}' = \frac{1}{\sqrt{2}}(\cos\vartheta, i, -\sin\vartheta).$$

Calculation of the intensity by means of (60.7) gives the reversal factors P for the three types of scattering:

$$P^0 = \tan^4\tfrac{1}{2}\vartheta, \qquad P^s = \frac{13+\cos^2\vartheta+10\cos\vartheta}{13+\cos^2\vartheta-10\cos\vartheta}, \qquad P^a = \frac{1-\cos^4\frac{1}{2}\vartheta}{1-\sin^4\frac{1}{2}\vartheta},$$

where ϑ is the scattering angle.

PROBLEM 4. Calculate the cross-section for scattering of a low-frequency photon by a hydrogen atom in the ground state.

SOLUTION. A low-frequency photon can undergo only elastic scattering. Since the orbital angular momentum l of the hydrogen atom in the ground state is zero, the selection rules (neglecting the spin–orbit interaction) allow only scalar scattering. The static polarizability of the atom is (in ordinary units) $\alpha = (9/2)(\hbar^2/me^2)^3$; see *QM*, §76, Problem 4. Substitution in (60.8) gives the required cross-section:

$$\sigma_t = 54\pi(\omega/c)^4(\hbar^2/me^2)^6.$$

PROBLEM 5. Calculate the cross-section for elastic scattering of γ rays by a deuteron (H. A. Bethe and R. E. Peierls, 1935).

SOLUTION. The wave functions of the deuteron ground state and of its continuous-spectrum states (the dissociated deuteron) are

$$\psi_0 = \sqrt{\frac{\kappa}{2\pi}}\frac{e^{-\kappa r}}{r}, \qquad \psi_{\mathbf{p}} = e^{i\mathbf{p}\cdot\mathbf{r}}, \qquad \kappa = \sqrt{(MI)};$$

see (58.2), (58.3). The matrix element of the dipole moment is $\mathbf{d}_{\mathbf{p}0} = -ie\mathbf{p}_{\mathbf{p}0}/M\omega_{\mathbf{p}0}$ and has been calculated in §58:

$$\mathbf{d}_{\mathbf{p}0} = -\frac{4\pi ie}{M\omega_{\mathbf{p}0}}\sqrt{\frac{\kappa}{2\pi}}\frac{\mathbf{p}}{\kappa^2+\mathbf{p}^2},$$

with the frequencies $\omega_{\mathbf{p}0} = (\mathbf{p}^2+\kappa^2)/M$.

The polarizability tensor is

$$\alpha_{ik} = \left\{\frac{2}{3}\int\frac{\omega_{\mathbf{p}0}}{\omega_{\mathbf{p}0}^2-\omega^2}|\mathbf{d}_{0\mathbf{p}}|^2\frac{d^3p}{(2\pi)^3} - \frac{e^2}{2M\omega^2}\right\}\delta_{ik}.$$

The first term is due to the virtual excitation of the internal degrees of freedom of the deuteron, and is written in the form (60.11). The second term is due to the action of the wave field on the translational motion of the deuteron as a whole. Since this motion is quasi-classical, the corresponding part of the scattering tensor is given by (59.14), with m replaced by the deuteron mass $2M$.

The calculation of α_{ik} depends on that of the integral

$$J = \int_{-\infty}^{\infty}\frac{z^4\,dz}{(z^2+1)^3[(z^2+1)^2-\gamma^2]}, \qquad z = p/\kappa, \qquad \gamma = M\omega/\kappa^2 = \omega/I.$$

We have

$$J = \frac{1}{8}\left[\frac{d}{d\lambda}\left(\frac{1}{\lambda}\frac{dJ_0}{d\lambda}\right)\right]_{\lambda=1},$$

where

$$J_0 = \frac{1}{2}\int_{-\infty}^{\infty}\frac{z^4\,dz}{(z^2+\lambda^2)[(z^2+1)^2-\gamma^2]}.$$

When $\gamma < 1$, the integrand has poles at the points $i\lambda$, $i\sqrt{(1+\gamma)}$, $i\sqrt{(1-\gamma)}$ in the upper half-plane of the complex variable z; the integral J_0 can be calculated from the residues at these poles. The result is

$$J = \frac{1}{2}\pi\left\{\frac{(1+\gamma)^{3/2}}{2\gamma^4}+\frac{(1-\gamma)^{3/2}}{2\gamma^4}-\left(\frac{3}{8\gamma^2}+\frac{1}{\gamma^4}\right)\right\}.$$

The total scattering cross-section is expressed in terms of α_{ik} by (60.8), and is (in ordinary units)

$$\sigma = \frac{8\pi}{3}\left(\frac{e^2}{Mc^2}\right)^2\left|-1-\frac{4}{3\gamma^2}+\frac{2}{3\gamma^2}[(1+\gamma)^{3/2}+(1-\gamma)^{3/2}]\right|^2 \quad \text{for } \gamma = \hbar\omega/I < 1.$$

For $\gamma > 1$ the scattering amplitude (above the deuteron dissociation threshold) is found from that for $\gamma < 1$ by analytical continuation; it has an imaginary part, which must be positive (in accordance with the avoidance rule in (59.17)):

$$\sigma = \frac{8\pi}{3}\left(\frac{e^2}{Mc^2}\right)^2\left|-1-\frac{4}{3\gamma^2}+\frac{2}{3\gamma^2}(\gamma+1)^{3/2}+i\frac{2}{3\gamma^2}(\gamma-1)^{3/2}\right|^2 \quad \text{for } \gamma > 1.$$

When $\gamma \gg 1$ we have $\sigma = (8\pi/3)(e^2/Mc^2)^2$, which agrees, as it should, with (non-relativistic) scattering by a free proton.

The angular distribution of radiation is

$$d\sigma = \sigma\cdot\tfrac{3}{4}(1+\cos^2\theta)\,do/4\pi,$$

where θ is the scattering angle. If the scattering amplitude is defined by (59.24), we have

$$\operatorname{im} f(0) = \frac{2e^2}{3Mc^2}\frac{(\gamma-1)^{3/2}}{\gamma^2} \quad \text{for } \gamma > 1.$$

According to the optical theorem (59.26), this quantity must equal $\omega\sigma_t/4\pi$, where σ_t is the total cross-section for photodissociation (58.4). The elastic scattering cross-section is of a higher order ($\sim e^4$) than the dissociation cross-section ($\sim e^2$; see (58.4)), and therefore σ_t is equal to the dissociation cross-section. For the same reason, in the approximation considered, the scattering amplitude was found to be real for $\gamma < 1$ (i.e. below the dissociation threshold).

§61. Scattering by molecules

The specific properties of molecular scattering are due to the same properties of molecules as form the basis of the theory of molecular spectra, namely the possibility of treating separately the state of the electrons with the nuclei fixed and the motion of the nuclei in a given effective field of the electrons.

Let the frequency ω of the incident radiation be less than the energy ω_e of the first electron excitation. Then the electron terms will not be excited in the scattering process. The scattering will be either Rayleigh scattering, or Raman scattering due to the excitation of rotational or vibrational levels.

Let us further assume that the electron ground term of the molecule is not degenerate (and has no fine structure). That is, we assume that the total spin of the electrons and the component of their total orbital angular momentum along the axis of the molecule (for molecules of the symmetrical-top type) are both zero. For diatomic molecules this means that the electron ground term must be $^1\Sigma$. These conditions are known to be satisfied for the ground states of most molecules.†

Finally, we shall assume the frequency ω large compared with the intervals in the nuclear (rotational and vibrational) structure of the ground term, and the difference $\omega_e - \omega$ to be in a similar relation to the nuclear structure of the excited term. Thus the frequency of the incident radiation must be sufficiently far from resonances. These conditions make it possible, in calculating the scattering tensor, to ignore at first the motion of the nuclei and to discuss the problem with a given configuration of the nuclei.

In such a problem, the scattering tensor is the same as the polarizability tensor, $\alpha_{ik} \equiv (c_{ik})_{11}$, and can in principle be calculated from the general formula (59.17), in which the summation is over all excited electron terms. The quantities α_{ik} thus obtained will be functions of the coordinates q of the nuclear configuration (the energies and wave functions of the electron terms depend on these coordinates as parameters). Since the state is not degenerate, the tensor $\alpha_{ik}(q)$ is real, and therefore symmetrical.

The tensor $\alpha_{ik}(q)$ is the electronic polarizability of a given nuclear configuration in the molecule. To solve an actual problem of scattering, we have also to take into account the motion of the nuclei in the initial and final states. Let $\psi_{s_1}(q)$ and $\psi_{s_2}(q)$ be the nuclear wave functions of these states, s_1 and s_2 being the sets of vibrational and rotational quantum numbers. The required scattering tensor is the matrix

† The results given below are, however, valid (to a certain approximation) also for cases where degeneracy of the electron ground term is due to a non-zero spin, the spin–orbit interaction being small (so that the resulting fine structure may be neglected). In this approximation, states with different spin directions do not combine, and in this sense they behave as if they were not degenerate. The molecule O_2, with ground term $^3\Sigma$, is of this type.

element of the tensor $\alpha_{ik}(q)$ with respect to these functions:

$$\langle s_2|\alpha_{ik}|s_1\rangle = \int \psi^*_{s_2}(q)\alpha_{ik}(q)\psi_{s_1}(q)\, dq. \tag{61.1}$$

Because the tensor $\alpha_{ik}(q)$ is symmetrical, so is the tensor (61.1) (whether s_1 and s_2 are the same or not). Thus we conclude that, under the conditions stated, there will be no antisymmetric part in either Rayleigh or Raman scattering. The scattering will include only scalar and symmetric parts.

The scalar part $\alpha^0(q)$ of the polarizability is independent of the orientation of the molecule, and depends only on the internal configuration of the atoms within it. Let v denote the set of vibrational quantum numbers of the molecule, and r the set of rotational numbers other than the magnetic number m. Then the matrix elements are

$$\langle v_2 r_2 m_2|\alpha^0|v_1 r_1 m_1\rangle = \langle v_2|\alpha^0|v_1\rangle\delta_{r_1 r_2}\delta_{m_1 m_2}. \tag{61.2}$$

The diagonality with respect to the numbers r and m is true of any scalar. The particular property of (61.2) is that here the elements do not depend on these numbers at all. Thus the scalar scattering occurs only for purely vibrational transitions and does not depend on the rotational state.

The symmetric scattering is determined by the matrix elements of the tensor α^s_{ik}. Its components in a fixed coordinate system xyz are expressed in terms of the components $\bar{\alpha}^s_{i'k'}$ in a system $\xi\eta\zeta$ moving with the molecule by

$$\alpha^s_{ik} = \sum_{i',k'} \bar{\alpha}^s_{i'k'} D_{i'i} D_{k'k}, \tag{61.3}$$

where the $D_{i'i}$ are the direction cosines of the new axes relative to the old. The quantities $\bar{\alpha}^s_{i'k'}$ do not depend on the orientation of the molecule, and the $D_{i'i}$ do not depend on the internal coordinates. Hence

$$\langle v_2 r_2 m_2|\alpha^s_{ik}|v_1 r_1 m_1\rangle = \sum_{i',k'} \langle v_2|\bar{\alpha}^s_{i'k'}|v_1\rangle\langle r_2 m_2|D_{i'i}D_{k'k}|r_1 m_1\rangle.$$

The sum of the squared moduli of these quantities over r_2, m_2, i, k is easily seen to be†

$$\sum_{r_2, m_2}\sum_{i,k} |\langle v_2 r_2 m_2|\alpha^s_{ik}|v_1 r_1 m_1\rangle|^2 = \sum_{i',k'} |\langle v_2|\bar{\alpha}^s_{i'k'}|v_1\rangle|^2. \tag{61.4}$$

† In transforming the sum we use the equation

$$\sum_{i,k}\sum_{r_2, m_2} \langle r_1 m_1|D_{il}D_{kg}|r_2 m_2\rangle\langle r_2 m_2|D_{il'}D_{kg'}|r_1 m_1\rangle$$
$$= \left\langle r_1 m_1\left| \sum_{i,k} D_{il}D_{kg}D_{il'}D_{kg'}\right| r_1 m_1\right\rangle = \langle r_1 m_1|\delta_{ll'}\delta_{gg'}|r_1 m_1\rangle = \delta_{ll'}\delta_{gg'},$$

which expresses the unitarity of the matrix D_{ik}.

This means that the total intensity of scattering with transitions from a given vibrational–rotational level v_1, r_1 to all rotational levels of the vibrational state v_2 is independent of r_1.

For molecules of the symmetrical-top type, we can go further and derive a relation between the scattering intensity and the rotational quantum numbers for every transition $v_1r_1 \to v_2r_2$. In this case the numbers r are the angular momentum J and its component k along the axis of the molecule. We replace the Cartesian components of $\overset{s}{\alpha}_{ik}$ by the corresponding spherical tensor of rank two, denoting its components by α_λ ($\lambda = 0, \pm 1, \pm 2$). According to *QM*, (110.7), the squared moduli of its matrix elements are

$$|\langle v_2J_2k_2m_2|\alpha_\lambda|v_1J_1k_1m_1\rangle|^2$$
$$= (2J_1+1)(2J_2+1)\begin{pmatrix} J_2 & 2 & J_1 \\ -k_2 & \lambda' & k_1 \end{pmatrix}^2 \begin{pmatrix} J_2 & 2 & J_1 \\ -m_2 & \lambda & m_1 \end{pmatrix}^2 |\langle v_2|\bar{\alpha}_{\lambda'}|v_1\rangle|^2,$$

where $\bar{\alpha}_{\lambda'}(q)$ is the spherical polarization tensor relative to axes fixed in the molecule, and $\lambda' = k_2 - k_1$. Summing over m_2 and $\lambda = m_2 - m_1$ (with m fixed), we obtain (cf. *QM*, (110.8))

$$\sum_{m_2, \lambda} |\langle v_2J_2k_2m_2|\alpha_\lambda|v_1J_1k_1m_1\rangle|^2 = (2J_2+1)\begin{pmatrix} J_2 & 2 & J_1 \\ -k_2 & \lambda' & k_1 \end{pmatrix}|\langle v_2|\bar{\alpha}_{\lambda'}|v_1\rangle|^2. \quad (61.5)$$

This quantity determines the intensity of scattering with the vibrational–rotational transition $v_1J_1k_1 \to v_2J_2k_2$. Since the matrix elements $\langle v_2|\bar{\alpha}_{\lambda'}|v_1\rangle$ do not depend on the rotation of the molecule, this also defines the dependence of the intensity on J_1, J_2 and on k_1, k_2. The right-hand side of (61.5), it may be noted, involves only one spherical component of the polarizability tensor.

Summation of (61.5) over J_2 and k_2 gives†

$$\sum_{\lambda} \sum_{J_2, k_2, m_2} |\langle v_2J_2k_2m_2|\alpha_\lambda|v_1J_1k_1m_1\rangle|^2 = \sum_{\lambda'} |\langle v_2|\bar{\alpha}_{\lambda'}|v_1\rangle|^2,$$

and we return to the sum rule (61.4).

A special case of the symmetrical top is the rotator, a linear molecule (or, as a particular instance, a diatomic molecule). The angular momentum component along the axis of such a molecule is zero (in a non-degenerate electronic state with zero electronic orbital angular momentum).‡ In this case, therefore, we must put $k_1 = k_2 = 0$ in (61.5).

Finally, let us consider the question of the selection rules in vibrational Raman

† In the summation over J_2 with given k_1, λ' (and $k_2 = k_1 + \lambda'$), we have

$$\sum_{J_2} (2J_2+1)\begin{pmatrix} J_2 & 2 & J_1 \\ -k_2 & \lambda' & k_1 \end{pmatrix}^2 = 1$$

according to *QM*, (106.13). The summation over k_2 (or, equivalently, over $\lambda' = k_2 - k_1$) is then effected for given k_1.

‡ Here we do not include effects due to the interaction between the vibrations and the rotation of the molecule (see *QM*, §104).

scattering, together with the cognate question of vibrational emission (or absorption) spectra of molecules.†

For scattering, the problem is simply to find the conditions under which there are non-zero matrix elements of the tensor $\alpha_{ik}(q)$ with respect to the vibrational wave functions $\psi_v(q)$; the scalar α^0 (for scalar scattering) and the irreducible symmetrical tensor α^s_{ik} (for symmetric scattering) have to be considered separately. A corresponding role in emission (or absorption) is played by the matrix elements of the vector $\mathbf{d}(q)$, the dipole moment of the molecule averaged over the electronic state with a given position of the nuclei. This has already been stated in §54 for diatomic molecules.

The vibrations of a polyatomic molecule are classified according to types of symmetry, the irreducible representations D_a of the corresponding point group, where a numbers the representation (see *QM*, §100). These representations also define the symmetry of wave functions of vibrational states of the molecule (see *QM*, §101). The symmetry of the wave functions of the first vibrational state (quantum number $v_a = 1$) is the same as the symmetry D_a of the vibration type; the symmetry of the higher states ($v_a > 1$) is given by the representations $[D_a^{v_a}]$, which are symmetric products of v_a representations D_a. Finally, the symmetry of states in which different vibrations a and b are simultaneously excited is given by the direct product $[D_a^{v_a}] \times [D_b^{v_b}]$.‡ The selection rules for the various quantities (scalar, vector, tensor) with respect to types of symmetry are found as described in *QM*, §97.

The selection rules resulting from the symmetry properties of the molecule are rigorous. There are also approximate rules based on the assumption that the vibrations are harmonic and that the functions $\alpha_{ik}(q)$ or $\mathbf{d}(q)$ can be expanded in powers of the vibrational coordinates q. These are a consequence of the known selection rule for a harmonic oscillator, according to which the matrix elements of the oscillator coordinate q are zero except for transitions in which the change in the vibrational quantum number $\Delta v = \pm 1$.

§62. Natural width of spectral lines

So far, in the study of emission and scattering of radiation, we have regarded all the levels of the system (an atom, say), as being strictly discrete. But in fact excited levels have a certain probability of decay by emission, and therefore a finite lifetime. According to the general principles of quantum mechanics, this has the result that the levels become quasi-discrete, with a certain small but finite width (see *QM*, §134); they can be written in the form $E - \frac{1}{2}i\Gamma$, where $\Gamma(=\Gamma/\hbar)$ is the total probability (per unit time) of all possible processes of "decay" of the state concerned.

Let us consider how this situation affects the process of emission (V. Weisskopf and E. Wigner, 1930). It is evident that, because of the finite width of the levels, the emitted radiation will not be strictly monochromatic: its frequencies will be spread

† These spectra lie in the infra-red, and are usually observed as absorption spectra.

‡ The symmetry properties of the vibrational wave functions are, of course, independent of the specific form of the vibrational potential energy, and in particular are independent of the assumption made in *QM*, §101, that the vibrations are harmonic.

over a range $\Delta\omega \sim \Gamma$ $(=\Gamma/\hbar)$. But, in order to measure the frequency distribution of the photons with this accuracy, the time needed is $T \gg 1/\Delta\omega \sim 1/\Gamma$. During this time the level will almost certainly decay by emission. We therefore have to deal with the determination of the total probability of emission of a photon of a given frequency, not with the probability per unit time. We shall calculate this total probability, first of all, for a transition of an atom from some excited level $E_1 - \frac{1}{2}i\Gamma_1$ to the ground level E_2, which has an infinite lifetime and is therefore strictly discrete.

Let Ψ be the wave function of the atom and the photon field, and $\hat{H} = \hat{H}^{(0)} + \hat{V}$ the Hamiltonian of the system, where $\hat{V}$ is the atom–field interaction operator. We shall seek a solution of Schrödinger's equation

$$i\frac{\partial \Psi}{\partial t} = (\hat{H}^{(0)} + \hat{V})\Psi \tag{62.1}$$

in the form of an expansion in terms of the wave functions of the unperturbed states of the system:

$$\Psi = \sum_\nu a_\nu(t)\Psi_\nu^{(0)} = \sum_\nu a_\nu(t)\, e^{-i\mathscr{E}_\nu t}\psi_\nu^{(0)}. \tag{62.2}$$

For the coefficients $a_\nu(t)$ we obtain the equations

$$i\frac{\partial a_\nu}{\partial t} = \sum_{\nu'} \langle\nu|V|\nu'\rangle a_{\nu'} \exp\{i(\mathscr{E}_\nu - \mathscr{E}_{\nu'})t\}. \tag{62.3}$$

Let $|\nu\rangle$ be a state with energy $\mathscr{E}_\nu = E_2 + \omega$, in which the atom is at the ground level E_2 and there is one quantum with a definite frequency ω; this state will be symbolized by $|\omega 2\rangle$. At the initial instant, the system is in the state $|1\rangle$, the atom being excited to the level E_1, with no photons present. Thus, for $t = 0$ we must have

$$a_1 = 1, \quad a_{\nu'} = 0 \quad \text{for } |\nu'\rangle \neq |1\rangle. \tag{62.4}$$

The solution of equation (62.3) with this initial condition will give (with the appropriate normalization of the wave functions) the probability that at time t there has been a transition $1 \to 2$ of the atom with emission of a photon in the frequency range $d\omega$: it is $|a_{\omega 2}(t)|^2\, d\omega$. We are interested in the ultimate probability as $t \to \infty$:

$$dw = |a_{\omega 2}(\infty)|^2\, d\omega. \tag{62.5}$$

In order to clarify the problem, it may be recalled that, in finding the ordinary emission probability (per unit time) with a transition $1 \to 2$ (neglecting the level width), equation (62.3) has to be solved with all the $a_{\nu'}(t)$ on the right-hand side replaced, to a first approximation, by the values (62.4). The solution thus obtained is then examined for large t; cf. *QM*, §42. We can now describe this procedure more precisely; it relates to times short in comparison with the lifetime of the excited level, and the large values of t concerned are large compared with $1/(E_1 - E_2)$ but small compared with $1/\Gamma_1$.

In our present case, where times comparable with $1/\Gamma_1$ are considered, the function $a_1(t)$ decreases in time according to

$$a_1(t) = e^{-\frac{1}{2}\Gamma_1 t}. \tag{62.6}$$

The functions $a_{\nu'}(t)$ for states $|\nu'\rangle$ which can result from emission by the atom increase with time, however. If the transition from a given level E_1 can occur to various atomic levels (as well as to E_2), there will be many increasing functions $a_{\nu'}(t)$, each corresponding to a state in which the atom is at a certain level and there is one photon with the appropriate energy. Nevertheless, there still remains on the right of (62.3) only the term with $|\nu'\rangle = |1\rangle$: since the matrix elements are zero except for transitions in which the number of photons with some one energy changes by 1, they are certainly zero for transitions between states containing one photon each, with different energies.

Thus we have for $a_{\omega 2}(t)$ the equation

$$\begin{aligned} i\frac{da_{\omega 2}}{dt} &= \langle \omega 2|V|1\rangle\, e^{i(E_2+\omega-E_1)t}a_1 \\ &= \langle \omega 2|V|1\rangle \exp\{i(\omega-\omega_{12})t - \tfrac{1}{2}\Gamma_1 t\}, \end{aligned} \tag{62.7}$$

where $\omega_{12} = E_1 - E_2$. Integration, with the condition $a_{\omega 2}(0) = 0$, gives

$$a_{\omega 2} = \langle \omega 2|V|1\rangle \frac{1-\exp\{i(\omega-\omega_{12})t - \frac{1}{2}\Gamma_1 t\}}{\omega - \omega_{12} + \frac{1}{2}i\Gamma_1}. \tag{62.8}$$

Hence the probability dw (62.5) is

$$dw = |\langle \omega 2|V|1\rangle^2 \frac{d\omega}{(\omega-\omega_{12})^2 + \frac{1}{4}\Gamma_1^2}.$$

Since the width $\Gamma_1 \ll \omega_{12}$, we can put $\omega = \omega_{12}$ in the factor $|\langle \omega 2|V|1\rangle|^2$. Then the quantity $2\pi|\langle \omega 2|V|1\rangle|^2$ is the ordinary probability (per unit time) for the emission of a photon with frequency ω_{12} and other properties besides the frequency, such as the direction of motion and the polarization, whose existence has so far been ignored in order to simplify the notation. The dependence of the probability on these characteristics is entirely determined by the factor $|\langle \omega 2|V|1\rangle|^2$. Thus the allowance for the level width does not affect the polarization properties or the angular distribution of the radiation.

The sum

$$\Gamma_{1\to 2} = 2\pi \sum |\langle \omega 2|V|1\rangle|^2, \tag{62.9}$$

taken over the polarizations and directions of motion of the photon, is the usual total probability of emission. It is also the part of the width of the level E_1 (the partial width) which is due to the transition $1 \to 2$, as distinct from the total width

Γ_1, which is made up of contributions from all possible modes of "decay" of the quasi-stationary state considered.†

By a similar summation of the probability dw, we obtain the following final formula for the frequency distribution of the emitted radiation:

$$dw = w_t \frac{\Gamma_1}{2\pi} \frac{d\omega}{(\omega_{12} - \omega)^2 + \frac{1}{4}\Gamma_1^2}, \tag{62.10}$$

where $w_t = \Gamma_{1\to 2}/\Gamma_1$ is the total relative probability of the transition $1 \to 2$. This is a dispersion-type distribution. The shape of the spectral line that is given by formula (62.10) is that which occurs for an isolated atom at rest, and is called the *natural* shape.‡

Now let the level E_2 of the atom be also an excited level with a finite width Γ_2. We shall assume for simplicity that this width is due to transitions of the atom to the ground state E_0 with the emission of one photon; the final result (62.12) will not depend on this assumption. The decay of state 1 can then be regarded as an emission of two photons, discussed in §59. The matrix element for this process (not yet taking account of the finite lifetime of state 2) is

$$\langle \omega\omega' 0 | V^{(2)} | 1 \rangle = \frac{\langle \omega\omega' 0 | V | \omega 2 \rangle \langle \omega 2 | V | 1 \rangle}{E_0 - E_2 + \omega' + i0}; \tag{62.11}$$

the state 2 in (59.2) becomes state 0, and in the sum over n the only remaining term corresponding to the atom in state 2 is the one which is large by resonance when ω' is close to $E_2 - E_0$. If we now take account of the finite lifetime of state 2, this simply changes E_2 into $E_2 - \frac{1}{2}i\Gamma_2$ in (62.11), giving

$$\langle \omega\omega' 0 | V^{(2)} | 1 \rangle = \frac{\langle \omega\omega' 0 | V | \omega 2 \rangle \langle \omega 2 | V | 1 \rangle}{E_0 - E_2 + \omega' + \frac{1}{2}i\Gamma_2}.$$

Substituting this value of the matrix element in the equation for $a_{\omega\omega' 2}(t)$ (which differs from (62.7) only as regards notation), we obtain by a derivation exactly similar to that of (62.8)

$$a_{\omega\omega' 0}(\infty) = \frac{\langle \omega\omega' 0 | V | \omega 2 \rangle \langle \omega 2 | V | 1 \rangle}{(\omega' - \omega_{20} + \frac{1}{2}i\Gamma_2)(\omega + \omega' - \omega_{10} + \frac{1}{2}i\Gamma_1)}.$$

The probability of emission of the photons ω and ω' is

$$dw = |a_{\omega\omega' 0}(\infty)|^2 \, d\omega \, d\omega'$$

$$= \frac{\Gamma_{1\to 2}}{2\pi} \frac{\Gamma_{2\to 0}}{2\pi} \frac{d\omega \, d\omega'}{[(\omega' - \omega_{20})^2 + \frac{1}{4}\Gamma_2^2][(\omega + \omega' - \omega_{10})^2 + \frac{1}{4}\Gamma_1^2]}. \tag{62.12}$$

† Formulae (62.6) and (62.9) can, of course, also be obtained by solving the equation for $a_1(t)$ analogous to (62.7).

We may note that transitions to states of the continuous spectrum, causing a finite level width, do not necessarily involve the emission of photons. Highly excited (X-ray) levels can decay with emission of an electron and formation of a positive ion in the ground state (the Auger effect).

‡ As distinct from the broadening caused by the interaction of the atom with other atoms (collision broadening) or by the presence of atoms in the source which move with various velocities (Doppler broadening).

This expression has sharp peaks at $\omega' \approx \omega_{20}$ and at $\omega \approx \omega_{12}$, as it should.

The shape of the spectral line corresponding to the transition $1 \to 2$ is obtained by integrating (62.12) with respect to ω'; the range of integration can extend from $-\infty$ to $+\infty$. The integral is most simply calculated by closing the contour of integration with an infinite semicircle in the upper half of the complex ω'-plane, and is given by the sum of the residues of the integrand at the poles

$$\omega' = \omega_{20} + \tfrac{1}{2}i\Gamma_2, \qquad \omega' = \omega_{10} - \omega + \tfrac{1}{2}i\Gamma_1.$$

The result is

$$dw = w_t \frac{\Gamma_1 + \Gamma_2}{2\pi} \frac{d\omega}{(\omega - \omega_{12})^2 + \frac{1}{4}(\Gamma_1 + \Gamma_2)^2}, \tag{62.13}$$

where $w_t = \Gamma_{1\to2}\Gamma_{2\to0}/\Gamma_1\Gamma_2$ is the total probability of the double transition $1 \to 2 \to 0$.†

The line shape (62.13) differs from (62.10) only in that Γ_1 is replaced by $\Gamma_1 + \Gamma_2$: the line width is equal to the sum of the widths of the initial and final states.

The line width is not, in general, equal to the probability $\Gamma_{1\to2}$ of the transition $1 \to 2$ itself, i.e. is not proportional to the line intensity as in the classical theory. Since $\Gamma_1 + \Gamma_2 > \Gamma_{1\to2}$, the line can have a large width with a relatively small intensity.

§63. **Resonance fluorescence**

The allowance for the finite width of the levels in problems of radiation scattering is important when the frequency ω of the incident radiation is close to one of the "intermediate" frequencies ω_{n1} or ω_{2n}; this is called *resonance fluorescence* (V. Weisskopf, 1931).

Let us consider Rayleigh scattering by a system (an atom, say) in the ground state, so that the initial and final levels are the same and are strictly discrete. Let the frequency of the radiation be close to a certain frequency ω_{n1}, where the level n is an excited level and is therefore quasi-discrete.

This problem could be solved by the method shown in §62, but there is no need to do so, since it is exactly analogous to the problem of non-relativistic resonance scattering at a quasi-discrete level (*QM*, §134). According to the results derived there, the scattering amplitude must contain a pole factor

$$\frac{1}{\omega - (E_n - \frac{1}{2}i\Gamma_n - E_1)}.$$

When $|\omega - \omega_{n1}| \gg \Gamma_n$, on the other hand, the result must tend to the non-resonance formula (59.5). It is therefore clear that the required scattering cross-section is obtained by simply replacing E_n by $E_n - \frac{1}{2}i\Gamma_n$ in (59.5); the sum over n can be

† In more complex cases, w_t is the total probability of all cascades which begin with the transition $1 \to 2$ and finish at the level 0.

restricted to the resonance terms:

$$d\sigma = \frac{\left|\sum_{M_n} (\mathbf{d}_{2n} \cdot \mathbf{e}'^*)(\mathbf{d}_{n1} \cdot \mathbf{e})\right|^2}{(\omega_{n1} - \omega)^2 + \frac{1}{4}\Gamma_n^2} \omega^4 \, do'. \tag{63.1}$$

The summation is over all states (having different angular-momentum components M_n) corresponding to the resonance level E_n; the states 1 and 2 belong to the same level (the ground level), but may differ in the values M_1 and M_2.

The cross-section (63.1) has its maximum value when $\omega = \omega_{n1}$, and this value is, in order of magnitude, $\sigma_{\max} \approx \omega^4 d^4/\Gamma_n^2$. Since the probability of the spontaneous transition $n \to 1$, and hence the width Γ_n, $\sim \omega^3 d^3$, this value is

$$\sigma_{\max} \sim 1/\omega^2 \sim \lambda^2, \tag{63.2}$$

of the order of the square of the wavelength and independent of the fine structure constant, as compared with typical values $\sim r_e^2$ outside the resonance region.

It must be emphasized that, since the atom is at a strictly discrete level (the ground level) before and after the scattering, the frequencies of the primary and secondary photons are exactly the same. If the incident radiation is monochromatic, the scattered line will therefore be monochromatic also. If the incident radiation has a spectral intensity distribution $I(\omega)$ which varies only slightly over the width Γ_n, the intensity of scattered radiation will be proportional to

$$\frac{I(\omega_{n1}) \, d\omega}{(\omega - \omega_{n1})^2 + \frac{1}{4}\Gamma_n^2}. \tag{63.3}$$

Thus the shape of the scattered line will be the same as the natural shape for spontaneous emission from the level E_n.

The cross-section (63.1) corresponds to the scattering tensor

$$(c_{ik})_{21} = \frac{\sum_{M_n} (d_i)_{2n}(d_k)_{n1}}{\omega_{n1} - \omega - \frac{1}{2}i\Gamma_n}. \tag{63.4}$$

In particular, the polarizability tensor is

$$\alpha_{ik} = (c_{ik})_{11} = \frac{\sum_{M_n} (d_i)_{1n}(d_k)_{n1}}{\omega_{n1} - \omega - \frac{1}{2}i\Gamma_n}. \tag{63.5}$$

It can be seen immediately that the addition of an imaginary part to the energy levels of the intermediate excited states makes the polarizability tensor no longer Hermitian, even at frequencies below the ionization threshold. It contains an imaginary part which is directly related to the absorption of radiation.

After absorbing a photon, the atom will sooner or later return to the ground state, emitting one or more photons. The absorption cross-section, viewed in this way, is just the total cross-section σ_t for all possible scattering processes.† On the other hand, according to the optical theorem (59.25), the cross-section can be expressed in terms of the anti-Hermitian part of the polarizability tensor.

Substituting in (59.25) the tensor α_{ik} from (63.5), we find the following formula for the cross-section for absorption of a photon with frequency ω close to ω_{n1}:

$$\sigma_a = 4\pi^2 \sum_{M_n} |\mathbf{d}_{n1} \cdot \mathbf{e}|^2 \omega \frac{\frac{1}{2}\Gamma_n}{\pi[(\omega - \omega_{n1})^2 + \frac{1}{4}\Gamma_n^2]}. \tag{63.6}$$

In the limit as $\Gamma_n \to 0$, the last factor tends to the delta function $\delta(\omega - \omega_{n1})$, in accordance with the fact that in this case only a photon having one particular frequency can be absorbed. Let radiation with a spectral and angular energy flux density $I_{\mathbf{k}e}$ (cf. (44.7)) be incident on the atom. Then the flux density of number of photons is $(I_{\mathbf{k}e}/\omega)\, d\omega\, do$, and the probability of absorption is

$$dw_a = \sigma_a (I_{\mathbf{k}e}/\omega)\, d\omega\, do. \tag{63.7}$$

If the function $I_{\mathbf{k}e}(\omega)$ varies only slightly over the width Γ_n, then we have after the integration over frequencies

$$dw_a = 4\pi^2 \sum_{M_n} |\mathbf{d}_{n1} \cdot \mathbf{e}|^2 I_{\mathbf{k}e}(\omega_{n1})\, do.$$

According to (45.5),

$$dw_{sp} = \frac{\omega^3}{2\pi} \sum_{M_n} |\mathbf{d}_{n1} \cdot \mathbf{e}^*|^2\, do$$
$$= \frac{\omega^3}{2\pi} \sum_{M_n} |\mathbf{d}_{n1} \cdot \mathbf{e}|^2\, do$$

is the probability of spontaneous emission of a photon having the frequency ω_{n1}; thus we return to formula (44.9).

† This discussion, it must be emphasized, refers to absorption by a system in its stable ground state. The problem would have to be stated differently for an excited state, because of the finite duration of the experiment.

CHAPTER VII

THE SCATTERING MATRIX

§ 64. The scattering amplitude

THE general problem concerning collisions is to find, for a given initial state of the system (an assembly of free particles), the probabilities of various possible final states (other assemblies of free particles). If $|i\rangle$ denotes the initial state, the result of the collision can be represented as the superposition

$$\sum_f |f\rangle\langle f|S|i\rangle, \tag{64.1}$$

in which the summation is taken over the various possible final states $|f\rangle$. The coefficients in this expansion $\langle f|S|i\rangle$ (or, more concisely, S_{fi}), form the *scattering matrix* or *S-matrix*. The squares $|S_{fi}|^2$ give the probabilities of transitions to particular states $|f\rangle$.

If there were no interaction between the particles, the state of the system would be unchanged, corresponding to a unit S-matrix (absence of scattering). It is convenient to separate this unit matrix in all cases, writing the scattering matrix in the form

$$S_{fi} = \delta_{fi} + i(2\pi)^4\delta^{(4)}(P_f - P_i)T_{fi}, \tag{64.2}$$

where T_{fi} is another matrix. In the second term we have written separately the four-dimensional delta function which expresses the law of conservation of the 4-momentum (P_i and P_f being the sums of the 4-momenta of all the particles in the initial and final states); the other factors are included for subsequent convenience. In the non-diagonal matrix elements, the first term in (64.2) does not appear, and so, for the transition $i \to f$, the elements of the matrices S and T are related by

$$S_{fi} = i(2\pi)^4\delta^{(4)}(P_f - P_i)T_{fi}. \tag{64.3}$$

The matrix elements T_{fi} which remain after separation of the delta function will be called the *scattering amplitudes*.

When the moduli $|S_{fi}|$ are squared, the square of the delta function appears, and is to be interpreted as follows. The delta function comes from the integral

$$\delta^{(4)}(P_f - P_i) = \frac{1}{(2\pi)^4}\int e^{i(P_f-P_i)x}\,d^4x. \tag{64.4}$$

If another such integral is calculated with $P_f = P_i$ (since one delta function is already present), and if the integration is taken over some large but finite volume V and a time interval t, the result is $Vt/(2\pi)^4$.† Thus we can write

$$|S_{fi}|^2 = (2\pi)^4 \delta^{(4)}(P_f - P_i)|T_{fi}|^2 Vt.$$

Dividing by t, we obtain the transition probability per unit time:

$$w_{f\leftarrow i} = (2\pi)^4 \delta^{(4)}(P_f - P_i)|T_{fi}|^2 V. \tag{64.5}$$

Each of the free particles, initial and final, is described by its own wave function—a plane wave having some amplitude u (a bispinor for an electron, a 4-vector for a photon, and so on). The structure of the scattering amplitude T_{fi} is of the form

$$T_{fi} = u_1^* u_2^* \ldots Q u_1 u_2 \ldots, \tag{64.6}$$

where on the left we have the amplitudes of wave functions of final particles, and on the right those of initial particles; Q is some matrix relating to the indices of the wave amplitude components of all the particles.

The most important cases are those where the initial state comprises only one or two particles. Then we have respectively the decay of one particle or the collision of two particles.

Let us first consider the decay of a particle into any number of other particles having momenta $\mathbf{p}'_a$ in an element $\Pi\, d^3p'_a$ of momentum space; the suffix a labels the particles in the final state, so that $\Sigma\, \mathbf{p}'_a = \mathbf{P}_f$. The number of states in this element and in the normalization volume‡ V is

$$\prod_a V\, d^3p'_a/(2\pi)^3.$$

The expression (64.5) must be multiplied by this quantity:

$$dw = (2\pi)^4 \delta^{(4)}(P_f - P_i)|T_{fi}|^2 V \prod_a \frac{V d^3p'_a}{(2\pi)^3}. \tag{64.7}$$

The wave functions used in calculating the matrix element must be normalized to "one particle in the volume V". For an electron, e.g., the wave function is the plane wave (23.1); for a particle with spin one it is (14.12); for a photon it is (4.3). All these functions include the factor $1/\sqrt{(2\varepsilon V)}$, where ε is the energy of the particle. Henceforth, however, it will be convenient to omit such factors in the wave functions, and include them in the expression for the probability. Thus the

† This can be shown in a different way by first calculating the integral over each coordinate in (64.4) for a finite range and then making the limits tend to infinity by means of *QM*, (42.4):

$$\lim_{\xi\to\infty} \frac{\sin^2 \alpha\xi}{\xi\alpha^2} = \pi\delta(\alpha).$$

‡ For greater clarity, in the calculations in this section, we shall not take V to be unity.

electron plane wave will be

$$\psi = ue^{-ipx}, \qquad \bar{u}u = 2m, \tag{64.8}$$

and the photon wave

$$A = \sqrt{(4\pi)}\, ee^{-ikx}, \qquad ee^* = -1, \qquad ek = 0. \tag{64.9}$$

The scattering amplitude calculated with these functions will be denoted by M_{fi} to distinguish it from T_{fi}. Evidently

$$T_{fi} = \frac{M_{fi}}{(2\varepsilon V \ldots 2\varepsilon_1' V \ldots)^{1/2}}; \tag{64.10}$$

the denominator contains one factor $\sqrt{(2\varepsilon V)}$ for each initial or final particle.

In particular, the decay probability is, instead of (64.7),

$$dw = (2\pi)^4 \delta^{(4)}(P_f - P_i)|M_{fi}|^2 \frac{1}{2\varepsilon} \prod_a \frac{d^3p_a'}{(2\pi)^3 2\varepsilon_a'}, \tag{64.11}$$

where ε is the energy of the decaying particle; as we should expect, the normalization volume does not appear in this formula.†

Formula (64.11) can be given a more definite form by eliminating the delta functions, if the decay produces two particles (with momenta $\mathbf{p}_1'$, $\mathbf{p}_2'$ and energies ε_1', ε_2'). In the rest frame of the decaying particle $\mathbf{p}_1' = -\mathbf{p}_2' \equiv \mathbf{p}'$, $\varepsilon_1' + \varepsilon_2' = m$. We have

$$dw = \frac{1}{(2\pi)^2} |M_{fi}|^2 \frac{1}{2m} \frac{1}{4\varepsilon_1'\varepsilon_2'} \delta(\mathbf{p}_1' + \mathbf{p}_2')\delta(\varepsilon_1' + \varepsilon_2' - m)\, d^3p_1'\, d^3p_2'.$$

The first delta function is eliminated by integration over d^3p_2'; the differential d^3p_1' is written as

$$\begin{aligned} d^3p' &= \mathbf{p}'^2\, d|\mathbf{p}'|\, do \\ &= |\mathbf{p}'|\, do \frac{\varepsilon_1'\varepsilon_2' d(\varepsilon_1' + \varepsilon_2')}{\varepsilon_1' + \varepsilon_2'}. \end{aligned} \tag{64.12}$$

The validity of this is easily seen by noting that $\varepsilon_1'^2 - m_1'^2 = \varepsilon_2'^2 - m_2'^2 = \mathbf{p}'^2$. The integration over $\varepsilon_1' + \varepsilon_2'$ eliminates the second delta function, and the result is

$$dw = \frac{1}{32\pi^2 m^2} |M_{fi}|^2 |\mathbf{p}'|\, do'. \tag{64.13}$$

Let us now consider a collision of two particles (having momenta $\mathbf{p}_1$ and $\mathbf{p}_2$ and

† If the final particles include N which are identical, a factor $1/N!$ must be inserted when integrating over their momenta to obtain the total probability; this factor takes into account the identity of states which differ only by an interchange of the particles.

energies ε_1 and ε_2), in which they are transformed into any number of particles having momenta $\mathbf{p}'_a$. Instead of (64.11) we now have

$$dw = (2\pi)^4 \delta^{(4)}(P_f - P_i)|M_{fi}|^2 \frac{1}{4\varepsilon_1\varepsilon_2 V} \prod_a \frac{d^3p'_a}{(2\pi)^3 2\varepsilon'_a}.$$

The quantity that is of interest in this case is, however, not the probability but the cross-section $d\sigma$. The cross-section invariant under the Lorentz transformations is obtained from dw on dividing by

$$j = I/V\varepsilon_1\varepsilon_2, \tag{64.14}$$

where I denotes the 4-scalar

$$I = \sqrt{[(p_1p_2)^2 - m_1^2 m_2^2]}; \tag{64.15}$$

see *Fields*, §12.† In the centre-of-mass system ($\mathbf{p}_1 = -\mathbf{p}_2 \equiv \mathbf{p}$)

$$I = |\mathbf{p}|(\varepsilon_1 + \varepsilon_2), \tag{64.16}$$

so that

$$j = \frac{|\mathbf{p}|}{V}\left(\frac{1}{\varepsilon_1} + \frac{1}{\varepsilon_2}\right) = \frac{v_1 + v_2}{V}, \tag{64.17}$$

which is the same as the usual definition of the flux density of colliding particles, v_1 and v_2 being their velocities.‡ Thus the cross-section is

$$d\sigma = (2\pi)^4 \delta^{(4)}(P_f - P_i)|M_{fi}|^2 \frac{1}{4I} \prod_a \frac{d^3p'_a}{(2\pi)^3 2\varepsilon'_a}. \tag{64.18}$$

This formula can be put into its final form by eliminating the delta function for the case where in the final state also there are only two particles. Let us consider the process in the centre-of-mass system, and let $\varepsilon = \varepsilon_1 + \varepsilon_2 = \varepsilon'_1 + \varepsilon'_2$ be the total energy; $\mathbf{p}_1 = -\mathbf{p}_2 \equiv \mathbf{p}$ and $\mathbf{p}'_1 = -\mathbf{p}'_2 \equiv \mathbf{p}'$ be the initial and final momenta. The delta function is eliminated in the same way as in the derivation of (64.13), and the result

†For future reference, another form of I is

$$I^2 = \tfrac{1}{4}[s - (m_1 + m_2)^2][s - (m_1 - m_2)^2], \tag{64.15a}$$

where $s = (p_i + p_2)^2$.

‡ In an arbitrary frame of reference,

$$j = \frac{1}{V}\sqrt{[(\mathbf{v}_1 - \mathbf{v}_2)^2 - (\mathbf{v}_1 \times \mathbf{v}_2)^2]}.$$

This expression is the same as the ordinary flux density whenever $\mathbf{v}_1$ is parallel to $\mathbf{v}_2$; then $j = |\mathbf{v}_1 - \mathbf{v}_2|/V$.

is

$$d\sigma = \frac{1}{64\pi^2}|M_{fi}|^2 \frac{|\mathbf{p}'|}{|\mathbf{p}|\varepsilon^2}\,do'; \tag{64.19}$$

in the particular case of elastic scattering, where the nature of the particles is unchanged in the collision, $|\mathbf{p}'| = |\mathbf{p}|$.

This formula can be written in yet another form by using the invariant quantity

$$\begin{aligned} t \equiv (p_1 - p_1')^2 &= m_1^2 + m_1'^2 - 2(p_1 p_1') \\ &= m_1^2 + m_1'^2 - 2\varepsilon_1\varepsilon_1' + 2|\mathbf{p}_1||\mathbf{p}_1'|\cos\theta, \end{aligned} \tag{64.20}$$

where θ is the angle between $\mathbf{p}_1$ and $\mathbf{p}_1'$. In the centre-of-mass system the momenta $|\mathbf{p}_1| \equiv |\mathbf{p}|$ and $|\mathbf{p}_1'| \equiv |\mathbf{p}'|$ are determined only by the total energy ε, and when ε is given we have

$$dt = 2|\mathbf{p}||\mathbf{p}'|\,d\cos\theta. \tag{64.21}$$

Hence, in (64.19),

$$do' = -\,d\phi d\cos\theta = \frac{d\phi\, d(-t)}{2|\mathbf{p}||\mathbf{p}'|},$$

where ϕ is the azimuth of $\mathbf{p}_1'$ relative to $\mathbf{p}_1$.† Thus

$$d\sigma = \frac{1}{64\pi}|M_{fi}|^2 \frac{dt}{I^2}\frac{d\phi}{2\pi}, \tag{64.22}$$

where I is again the invariant (64.16). The azimuth ϕ, and therefore the cross-section in the form (64.22), are invariant under those Lorentz transformations which do not change the direction of relative motion of the particles. If the cross-section is independent of the azimuth, formula (64.22) takes the particularly simple form

$$d\sigma = \frac{1}{64\pi}|M_{fi}|^2 \frac{dt}{I^2}. \tag{64.23}$$

If one of the colliding particles is sufficiently heavy (and its state is unaltered by the collision), it acts only as a fixed source of a constant field in which the other particle is scattered. Since the energy (though not the momentum) of the system is conserved in a constant field, in this treatment of the collision process we can write the S-matrix elements in the form

$$S_{fi} = i \cdot 2\pi\delta(E_f - E_i)T_{fi}. \tag{64.24}$$

† Since the correct sign of the differential in such cases is obvious, we shall henceforward write simply dt for $d(-t)$, and so on.

In the expression $|S_{fi}|^2$, the square of the one-dimensional delta function must be interpreted as

$$[\delta(E_f - E_i)]^2 \to \frac{1}{2\pi}\,\delta(E_f - E_i)t.$$

Now, as in the derivation of (64.11), we change to the amplitude M_{fi} instead of T_{fi}, and obtain the following expression for the probability of a process in which one particle is scattered in a constant field and produces in the final state a certain number of other particles:

$$dw = 2\pi\delta(E_f - \varepsilon)|M_{fi}|^2 \frac{1}{2\varepsilon V} \prod_a \frac{d^3p'_a}{(2\pi)^3 2\varepsilon'_a}.$$

Here ε $(= E_i)$ is again the energy of the initial particle, $\mathbf{p}'_a$ and ε'_a the momenta and energies of the final particles. The scattering cross-section is found by dividing dw by the flux density $j = v/V$, where $v = |\mathbf{p}|/\varepsilon$ is the velocity of the particle that undergoes scattering. The normalization volume again disappears, and the result is

$$d\sigma = 2\pi\delta(E_f - \varepsilon)|M_{fi}|^2 \frac{1}{2|\mathbf{p}|} \prod_a \frac{d^3p'_a}{(2\pi)^3 2\varepsilon'_a}. \tag{64.25}$$

In the particular case of elastic scattering, there is only one particle in the final state, with the same energy and the same momentum (in absolute value). Writing

$$d^3p' \to \mathbf{p}'^2 d|\mathbf{p}'|\, do' = |\mathbf{p}'|\varepsilon'\, d\varepsilon'\, do'$$

and eliminating $\delta(\varepsilon' - \varepsilon)$ by integrating with respect to ε', we find the cross-section in the form

$$d\sigma = \frac{1}{16\pi^2}|M_{fi}|^2\, do'. \tag{64.26}$$

Finally, if the external field is time-dependent, such as the field of a system of particles executing a given motion, the S-matrix also lacks the delta function of energy. Then $S_{fi} = iT_{fi}$ and, after the change from T_{fi} to M_{fi} by (64.10), the probability of (e.g.) a process in which the field creates a given set of particles is

$$dw = |M_{fi}|^2 \prod_a \frac{d^3p'_a}{(2\pi)^3 2\varepsilon'_a}. \tag{64.27}$$

§65. **Reactions involving polarized particles**

In this section we shall show by means of simple examples how the state of polarization of the particles concerned in the reaction is taken into account when calculating the scattering cross-section.

Let there be one electron in the initial state and one in the final state. Then the form of the scattering amplitude is

$$M_{fi} = \bar{u}'Au \ (\equiv \bar{u}'_i A_{ik} u_k), \tag{65.1}$$

where u and u' are the bispinor amplitudes of the initial and final electrons, and A is some matrix, which depends on the momenta and polarizations of the other particles (if any) which take part in the reaction.

The scattering cross-section is proportional to $|M_{fi}|^2$, and

$$\begin{aligned}(\bar{u}'Au)^* &= u'\gamma^{0*}A^*u^* \\ &= u^*A^+\gamma^{0+}u' \\ &= \bar{u}\bar{A}u',\end{aligned} \tag{65.2}$$

where†

$$\bar{A} = \gamma^0 A^+ \gamma^0.$$

Thus

$$\begin{aligned}|M_{fi}|^2 &= (\bar{u}'Au)(\bar{u}\bar{A}u') \\ &\equiv u'_i \bar{u}'_k A_{kl} u_l \bar{u}_m \bar{A}_{mi}.\end{aligned} \tag{65.3}$$

If the initial electron is in a mixed (partially polarized) state with density matrix ρ, and if we wish to find the cross-section for a process in which the final electron is in a specified polarization state ρ', the products of the bispinor amplitude components must be changed as follows: $u'_i\bar{u}'_k \to \rho'_{ik}$, $u_l\bar{u}_m \to \rho_{lm}$. Then

$$|M_{fi}|^2 = \operatorname{tr}(\rho' A \rho \bar{A}). \tag{65.4}$$

The density matrices are given by formula (29.13):

$$\rho = \tfrac{1}{2}(\gamma p + m)(1 - \gamma^5(\gamma a)) \tag{65.5}$$

and similarly for ρ'.

If the initial electron is unpolarized, then

$$\rho = \tfrac{1}{2}(\gamma p + m). \tag{65.6}$$

Substituting this expression is equivalent to averaging over the polarizations of the electron. If it is desired to determine the cross-section for scattering with any

† Since the matrix $\bar{A}$ has to be constructed, we shall note here, for future reference, the following easily verified equations:

$$\left.\begin{aligned}\overline{\gamma^\mu} = \gamma^\mu,\ \overline{\gamma^\mu\gamma^\nu\ldots\gamma^\rho} = \gamma^\rho\ldots\gamma^\nu\gamma^\mu,\\ \overline{\gamma^5} = -\gamma^5,\ \overline{\gamma^5\gamma^\mu} = \gamma^5\gamma^\mu.\end{aligned}\right\} \tag{65.2a}$$

polarization of the final electron, we must also put $\rho' = \frac{1}{2}(\gamma p' + m)$, and double the result; this operation is equivalent to summation over the polarizations of the electron. Thus we have

$$\frac{1}{2} \sum_{\text{polar.}} |M_{fi}|^2 = \frac{1}{2} \operatorname{tr} \{\gamma p' + m)A(\gamma p + m)\bar{A}\}, \tag{65.7}$$

where the sum is taken over initial and final polarizations, and the factor $\frac{1}{2}$ converts one summation into an averaging.

The density matrix ρ' in (65.4) is a secondary quantity which essentially represents the properties of the detector as selecting one or the other polarization of the final electron, not the properties of the scattering process as such. There is the question of the polarization state of the electron resulting from the scattering process itself. If $\rho^{(f)}$ is the density matrix of this state, then the probability of detecting an electron in the state ρ' is obtained by projecting $\rho^{(f)}$ on ρ', i.e. by taking the trace $\operatorname{tr}(\rho^{(f)} \rho')$. This will be proportional to the corresponding cross-section, i.e. to $|M_{fi}|^2$. A comparison with (65.4) shows that

$$\rho^{(f)} \sim A\rho\bar{A}. \tag{65.8}$$

Since we know that $\rho^{(f)}$ must have the form (65.5) with some 4-vector $a^{(f)}$, we need only determine the latter. This could be done by means of formula (29.14), but it is even simpler to proceed as follows.

We have seen in §29 that the components of the 4-vector a can be expressed in terms of those of the 3-vector $\boldsymbol{\zeta}$ which is (twice) the mean value of the electron spin in its rest frame. The polarization states of the electrons are entirely determined by these vectors, and it is convenient to express the scattering cross-section also in terms of them. The square $|M_{fi}|^2$ will clearly be linear in each of the vectors $\boldsymbol{\zeta}$ and $\boldsymbol{\zeta}'$ which relate to the initial and final electrons, and its form as a function of $\boldsymbol{\zeta}'$ will be

$$|M_{fi}|^2 = \alpha + \boldsymbol{\beta} \cdot \boldsymbol{\zeta}', \tag{65.9}$$

where α and $\boldsymbol{\beta}$ are themselves linear functions of $\boldsymbol{\zeta}$.

The vector $\boldsymbol{\zeta}'$ in (65.9) is the particular polarization of the final electron that is selected by the detector. The vector $\boldsymbol{\zeta}^{(f)}$, corresponding to the density matrix $\rho^{(f)}$, is easily found as follows. According to the above argument,

$$|M_{fi}|^2 \sim \operatorname{tr}(\rho'\rho^{(f)}).$$

Since this quantity is relativistically invariant, it may be calculated in any frame of reference. In the rest frame of the final electron we have, by (29.20),

$$\rho'\rho^{(f)} \sim (1 + \boldsymbol{\sigma} \cdot \boldsymbol{\zeta}')(1 + \boldsymbol{\sigma} \cdot \boldsymbol{\zeta}^{(f)}).$$

Hence

$$|M_{fi}|^2 \sim 1 + \boldsymbol{\zeta}' \cdot \boldsymbol{\zeta}^{(f)},$$

and from a comparison with (65.9)

$$\boldsymbol{\zeta}^{(f)} = \boldsymbol{\beta}/\alpha. \tag{65.10}$$

Thus the calculation of the cross-section as a function of the parameter $\boldsymbol{\zeta}'$ also gives the polarization $\boldsymbol{\zeta}^{(f)}$.

In more complex cases, when there is more than one initial or final electron, the calculations are similar to the foregoing.

For instance, if there are two electrons both initially and finally, the form of the scattering amplitude is

$$M_{fi} = (\bar{u}_1' A u_1)(\bar{u}_2' B u_2) + (\bar{u}_2' C u_1)(\bar{u}_1' D u_2),$$

where u_1, u_2 are the bispinor amplitudes of the initial electrons, and u_1', u_2' those of the final electrons. The square $|M_{fi}|^2$ includes terms of the forms

$$|\bar{u}_1' A u_1|^2 |\bar{u}_2' B u_2|^2 \quad \text{and} \quad (\bar{u}_1' A u_1)(\bar{u}_2' B u_2)(\bar{u}_2' C u_1)^* (\bar{u}_1' D u_2)^*.$$

The former reduce to products of two traces like (65.4); the latter reduce to traces having the form

$$\operatorname{tr}(\rho_1' A \rho_1 \bar{C} \rho_2' B \rho_2 \bar{D}).$$

Positrons are described by amplitudes with "negative frequency" $u(-p)$. For reactions involving positrons, the only difference from the preceding analysis is that the expressions to be used for the density matrices differ from (65.5), (65.6) as regards the sign of m; cf. (29.16), (29.17).

Let us now consider the polarization states of photons participating in the reaction.

The polarization of each initial photon appears linearly in the scattering amplitude in the form of a 4-vector e, and that of each final photon as e^*. In each case the 4-tensor $e_\mu e_\nu^*$ occurs in the cross-section (i.e. in the square $|M_{fi}|^2$). To obtain an arbitrary partially polarized state, this tensor must be replaced by the four-dimensional density matrix, the 4-tensor $\rho_{\mu\nu}$:

$$e_\mu e_\nu^* \to \rho_{\mu\nu}. \tag{65.11}$$

In particular, for an unpolarized photon, according to (8.15),

$$\rho_{\mu\nu} = -\tfrac{1}{2} g_{\mu\nu}. \tag{65.12}$$

Thus averaging over polarizations of the photon is equivalent to contracting in $|M_{fi}|^2$ with respect to the corresponding two tensor indices μ, ν.†

If summation over the photon polarizations is desired, not averaging, then we

† The expression (65.12) as it were reduces the averaging over the two actually possible polarizations of the photon to one over the four independent directions of the four-vector e.

must replace $e_\mu e_\nu^*$ by a quantity twice as large:

$$e_\mu e_\nu^* \to -g_{\mu\nu}. \tag{65.13}$$

The density matrix of the polarized photon is given by formula (8.17). The choice of the 4-vectors $e^{(1)}$, $e^{(2)}$ which appear in this expression is usually governed by the particular conditions of the problem. In some cases they may be related to certain spatial directions in a given frame of reference; in other cases, it is more convenient to relate them to the 4-vectors which characterize the problem, namely the 4-momenta of the particles.

In (8.17) the polarization of the photon is described by the Stokes parameters, which form the "vector" $\boldsymbol{\xi} = (\xi_1, \xi_2, \xi_3)$. As with the electron, it is necessary to distinguish the polarization $\boldsymbol{\xi}^{(f)}$ of the final photon as such from the polarization $\boldsymbol{\xi}'$ that is selected by the detector. If the square of the scattering amplitude is known as a function of the parameter $\boldsymbol{\xi}'$:

$$|M_{fi}|^2 = \alpha + \boldsymbol{\beta} \cdot \boldsymbol{\xi}',$$

then the polarization $\boldsymbol{\xi}^{(f)} = \boldsymbol{\beta}/\alpha$, exactly as in (65.10).

§66. Kinematic invariants

Let us consider some kinematic relations for scattering processes in which there are only two particles, both in the initial state and in the final state. The relations in question are deduced from the general conservation laws alone, and are therefore valid for all particles and all laws of interaction.

The law of conservation of 4-momentum, in a general form that does not specify which are the initial and which the final particles, is

$$q_1 + q_2 + q_3 + q_4 = 0. \tag{66.1}$$

Here $\pm q_a$ are the momentum 4-vectors; two of them pertain to the incident particles and two to the scattered particles, the momenta for the latter being $-q_a$. Thus for two of the q_a the time component $q_a^0 > 0$, and for two $q_a^0 < 0$.

The law of charge conservation must be satisfied as well as that of 4-momentum conservation. Here the charge may be interpreted not only as the electric charge but as any other conserved quantity whose sign is opposite for particles and antiparticles.

For given types of particles concerned in the process, the squares of the 4-vectors q_a are the squares of the particle masses, which are fixed ($q_a^2 = m_a^2$). Three different reactions occur, according to the values taken by the time components q_a^0 and the values of the charges. These reactions may be written

$$\left.\begin{aligned} &(\mathrm{I})\ \ 1+2\to 3+4,\\ &(\mathrm{II})\ \ 1+\bar{3}\to \bar{2}+4,\\ &(\mathrm{III})\ \ 1+\bar{4}\to \bar{2}+3. \end{aligned}\right\} \tag{66.2}$$

Here the numbers refer to the particles, and the bar over a number denotes the corresponding antiparticle. The change from one reaction to another, i.e. the transfer of a particle to the opposite side of the formula, corresponds to a change in sign of the corresponding time component q_a^0 and in the sign of the charge (i.e. a replacement of the particle by its antiparticle). The reactions inverse to (66.2) are also possible, of course.

The three processes (66.2) are referred to as three *cross-channels* of a single general reaction.

The following are some examples. If particles 1 and 3 are electrons, and 2 and 4 are photons, then channel I represents the scattering of a photon by an electron; channel III is the same as channel I, since the photon is strictly neutral. Channel II is the conversion of an electron–positron pair into two photons. If all four particles are electrons, then channel I is the scattering of an electron by an electron, and channels II and III the scattering of a positron by an electron. If particles 1 and 3 are electrons, and 2 and 4 are muons, then channel I is the scattering of e by μ, channel III the scattering of e by $\bar{\mu}$, and channel II the conversion of a pair $e\bar{e}$ into a pair $\mu\bar{\mu}$.

In the discussion of scattering processes, the invariant quantities which can be constructed from the 4-momenta are particularly important. The invariant scattering amplitudes are functions of these quantities (§70).

Two independent invariants can be constructed from four 4-momenta, since, according to (66.1), only three of the 4-vectors q_a are independent. Let these be q_1, q_2, q_3. From them, six invariants can be constructed: the three squares q_1^2, q_2^2, q_3^2 and the three products q_1q_2, q_1q_3, q_2q_3. But the first three are the given squares of the masses, and the second three satisfy one relation which follows from the equation†

$$(q_1+q_2+q_3)^2 = q_4^2 = m_4^2.$$

In order to increase the symmetry it is, however, convenient to consider not two but three invariants, which may be taken as

$$\left.\begin{aligned} s &= (q_1+q_2)^2 = (q_3+q_4)^2, \\ t &= (q_1+q_3)^2 = (q_2+q_4)^2, \\ u &= (q_1+q_4)^2 = (q_2+q_3)^2. \end{aligned}\right\} \tag{66.3}$$

These are easily seen to be related by

$$s+t+u = h, \tag{66.4}$$

where

$$h = m_1^2+m_2^2+m_3^2+m_4^2. \tag{66.5}$$

† In the general case of a reaction involving n (≥ 4) particles, the number of functionally independent invariant quantities is $3n-10$. There are altogether $4n$ quantities, the components of the n 4-momenta q_a, between which there are n functional relations $q_a^2 = m_a^2$ and four given by the conservation law $\Sigma q_a = 0$. Arbitrary values can be assigned to six quantities, in accordance with the number of parameters which define the general Lorentz transformation (a general four-dimensional rotation). The number of independent invariants is therefore $4n-n-4-6 = 3n-10$.

In the principal channel (I), the invariant s has a simple physical significance. It is the square of the total energy of the colliding particles (1 and 2) in their centre-of-mass system (for $\mathbf{p}_1+\mathbf{p}_2=0$, $s=(\varepsilon_1+\varepsilon_2)^2$). In channel II, the invariant t has a similar significance, and in channel III the invariant u. The three channels are therefore often called s, t and u channels.

It is easy to express each of the invariants s, t and u in terms of the energies and momenta of the colliding particles in each channel. Let us consider the s channel. In the centre-of-mass system of particles 1 and 2, the time and space components of the 4-vectors q_a are

$$\left.\begin{aligned} q_1 = p_1 = (\varepsilon_1, \mathbf{p}_s), \qquad & q_2 = p_2 = (\varepsilon_2, -\mathbf{p}_s), \\ q_3 = -p_3 = (-\varepsilon_3, -\mathbf{p}'_s), \qquad & q_4 = -p_4 = (-\varepsilon_4, \mathbf{p}'_s); \end{aligned}\right\} \tag{66.6}$$

the suffix s in $\mathbf{p}_s$ and $\mathbf{p}'_s$ indicates that these momenta refer to the reaction in the s channel. Then

$$s = \varepsilon_s^2, \qquad \varepsilon_s = \varepsilon_1+\varepsilon_2 = \varepsilon_3+\varepsilon_4; \tag{66.7}$$

$$\left.\begin{aligned} 4s\mathbf{p}_s^2 &= [s-(m_1+m_2)^2][s-(m_1-m_2)^2], \\ 4s\mathbf{p}_s'^2 &= [s-(m_3+m_4)^2][s-(m_3-m_4)^2]; \end{aligned}\right\} \tag{66.8}$$

$$\left.\begin{aligned} 2t &= h - s + 4\mathbf{p}_s\cdot\mathbf{p}'_s - \frac{1}{s}(m_1^2-m_2^2)(m_3^2-m_4^2), \\ 2u &= h - s - 4\mathbf{p}_s\cdot\mathbf{p}'_s + \frac{1}{s}(m_1^2-m_2^2)(m_3^2-m_4^2). \end{aligned}\right\} \tag{66.9}$$

For elastic scattering ($m_1=m_3$, $m_2=m_4$), we have $|\mathbf{p}_s|=|\mathbf{p}'_s|$, and hence $\varepsilon_1=\varepsilon_3$, $\varepsilon_2=\varepsilon_4$. Instead of (66.9), the simpler formulae

$$\left.\begin{aligned} t &= -(\mathbf{p}_s-\mathbf{p}'_s)^2 = -2\mathbf{p}_s^2(1-\cos\theta_s), \\ u &= -2\mathbf{p}_s^2(1+\cos\theta_s)+(\varepsilon_1-\varepsilon_2)^2 \end{aligned}\right\} \tag{66.10}$$

are then obtained, where θ_s is the angle between $\mathbf{p}_s$ and $\mathbf{p}'_s$. The invariant $-t$ is here the square of the (three-dimensional) momentum transfer in the collision.

Similar formulae for the other channels are found by a straightforward change of notation. For the t channel we must interchange s and t, and 2 and 3, in (66.6)–(66.10); for the u channel, we interchange s and u, and 2 and 4.

§67. **Physical regions**

When considering the scattering amplitudes as functions of the independent variables s, t, u (which are related only by $s+t+u=h$), we encounter the need to distinguish regions in which their values are physically permissible from those in which they are not. Values which can correspond to a physical process of scattering must satisfy certain conditions which follow from the law of conservation of 4-momentum and the fact that the square of each of the 4-vectors q_a is a given quantity m_a^2.

The product of two 4-momenta

$$p_a p_b \geq m_a m_b. \tag{67.1}$$

Hence

$$(q_a + q_b)^2 = (p_a + p_b)^2 \geq (m_a + m_b)^2,$$

if $q_a = p_a$, $q_b = p_b$ (or $q_a = -p_a$, $q_b = -p_b$); or

$$(q_a + q_b)^2 = (p_a - p_b)^2 \leq (m_a - m_b)^2,$$

if $q_a = p_a$, $q_b = -p_b$. Hence, for a reaction in the s channel,

$$\left.\begin{aligned} (m_1 + m_2)^2 &\leq s \geq (m_3 + m_4)^2, \\ (m_1 - m_3) &\geq t \leq (m_2 - m_4)^2, \\ (m_1 - m_4)^2 &\geq u \leq (m_2 - m_3)^2, \end{aligned}\right\} \tag{67.2}$$

and similarly in the t and u channels.

To determine the remaining conditions, we form a 4-vector L which is dual to the product of any three of the 4-vectors q_a, say

$$L_\lambda = e_{\lambda\mu\nu\rho} q_1^\mu q_2^\nu q_3^\rho. \tag{67.3}$$

In the rest frame of particle 1, say, we have $q_1 = (q_1^0, 0)$. Then L has only the spatial components $L_i = e_{i0kl} q_1^0 q_2^k q_3^l$. Thus L is a space-like vector, and $L^2 \leq 0$ in every frame of reference. Expanding L^2, we obtain the condition

$$\begin{vmatrix} q_1^2 & q_1 q_2 & q_1 q_3 \\ q_2 q_1 & q_2^2 & q_2 q_3 \\ q_3 q_1 & q_3 q_2 & q_3^2 \end{vmatrix} \geq 0. \tag{67.4}$$

This can be expressed in terms of the invariants s, t, u in a form which is the same for all channels:

$$stu \geq as + bt + cu, \tag{67.5}$$

where

$$\left.\begin{aligned} ah &= (m_1^2 m_2^2 - m_3^2 m_4^2)(m_1^2 + m_2^2 - m_3^2 - m_4^2), \\ bh &= (m_1^2 m_3^2 - m_2^2 m_4^2)(m_1^2 + m_3^2 - m_2^2 - m_4^2), \\ ch &= (m_1^2 m_4^2 - m_2^2 m_3^2)(m_1^2 + m_4^2 - m_2^2 - m_3^2) \end{aligned}\right\} \tag{67.6}$$

(T. W. B. Kibble, 1960).

For a graphical representation of the regions of variation of s, t and u, it is convenient to use triangular coordinates in a plane, called the *Mandelstam plane* (S. Mandelstam, 1958). The coordinate axes are three straight lines which intersect to form an equilateral triangle. The coordinates s, t, u are measured along directions perpendicular to these three lines; the directions towards the interior of the triangle are reckoned positive, as shown by the arrows in Fig. 5. Thus each point in the plane has corresponding values of s, t and u which are represented (with the appropriate signs) by the lengths of the perpendiculars to the three axes. The condition $s+t+u=h$ is satisfied on account of a known theorem of geometry, h being equal to the altitude of the triangle.†

Let us consider the important case where the principal channel (s) corresponds to elastic scattering. Then the masses of the particles are equal in pairs:

$$m_1 = m_3 \equiv m, \qquad m_2 = m_4 \equiv \mu. \tag{67.7}$$

Let $m > \mu$. The condition (67.5) has

$$h = 2(m^2+\mu^2), \qquad a = c = 0, \qquad b = (m^2-\mu^2)^2,$$

so that

$$sut \geqslant (m^2-\mu^2)^2 t. \tag{67.8}$$

The boundary of the region defined by this inequality comprises the straight line $t=0$ and the hyperbola

$$su = (m^2-\mu^2)^2, \tag{67.9}$$

whose two branches lie in the sectors $u<0$, $s<0$ and $s>0$, $u>0$; the axes $s=0$

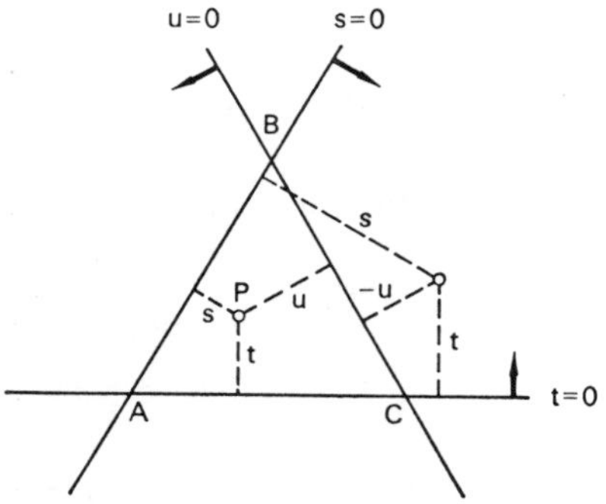

FIG. 5.

† For example, if the point P in Fig. 5 is joined to the three vertices A, B, C of the triangle, the latter is divided into three triangles with altitudes s, t and u; equating the sum of their areas to that of the triangle ABC, we obtain the required relation. The proof is similar when P lies outside the triangle ABC.

and $u = 0$ are the asymptotes of the hyperbola. Instead of (67.8) we can write

$$t > 0, \qquad su > (m^2 - \mu^2)^2$$

or

$$t < 0, \qquad su < (m^2 - \mu^2)^2.$$

Moreover, according to the conditions (67.2) we must apply the inequality $s > (m+\mu)^2$ in the s channel and $u > (m+\mu)^2$ in the u channel; the remaining inequalities are then necessarily satisfied. We thus find that channels I, II, III (s, t, u) correspond to the shaded regions in Fig. 6, which are called *physical regions.*

If $\mu = 0$ (particles 2 and 4 are photons), the lower branch of the hyperbola touches the axis $t = 0$, and the physical regions are as shown in Fig. 7.

If $m = \mu$, the boundaries of the region (67.8) degenerate to the coordinate axes, and the physical regions are the three sectors shown in Fig. 8.

In the general case of four different masses, the equation

$$stu = as + bt + cu \tag{67.10}$$

defines a third-order curve whose branches are the boundaries of the physical regions

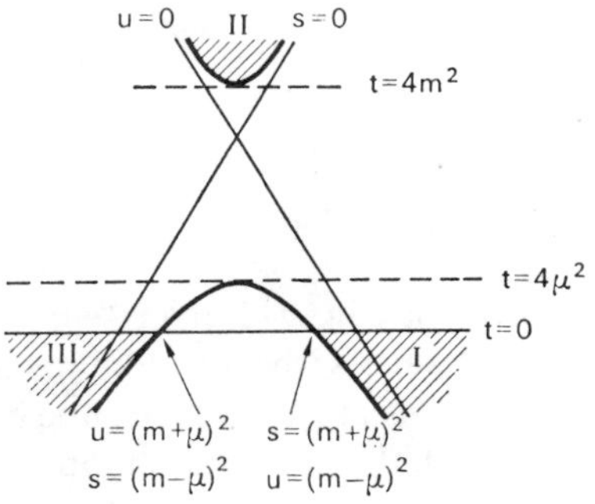

FIG. 6.

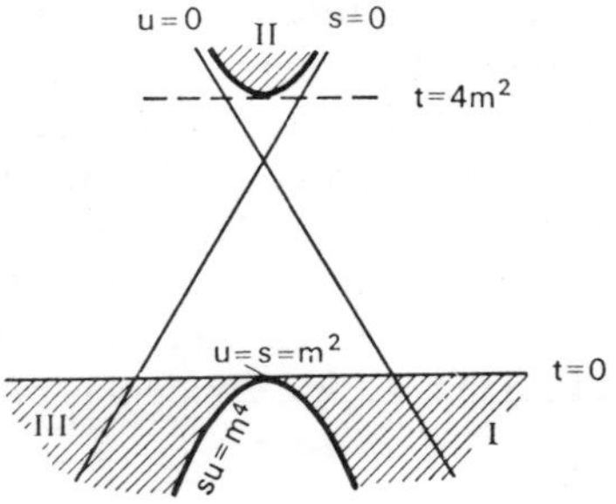

FIG. 7.

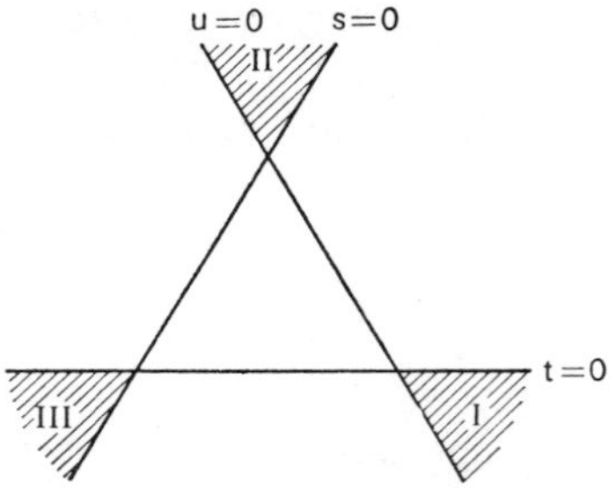

FIG. 8.

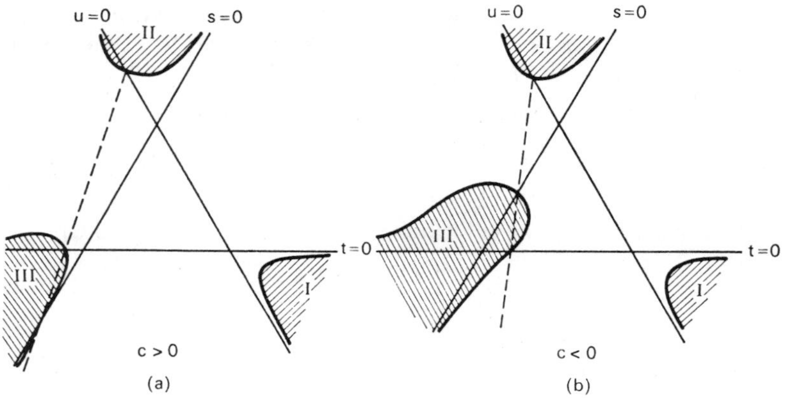

FIG. 9.

of the three channels, as shown in Fig. 9. Let $m_1 \geqslant m_2 \geqslant m_3 \geqslant m_4$. Then $a \geqslant b \geqslant c$, $a > 0$, $b > 0$. The curve (67.10) meets the coordinate axes at points on the line $as + bt + cu = 0$ (see the broken lines in Fig. 9). This line is as shown in Fig. 9a and 9b, depending on the sign of c. If $c < 0$, the physical region of the u channel includes part of the area of the coordinate triangle. In this case, therefore, the quantities s, t and u may all be positive at the same time. All three branches of the boundary curve have the appropriate coordinate axes as asymptotes; this may be seen by eliminating one of the variables from (67.10) by means of the relation $s + t + u = h$, and then making one of the other variables tend to infinity. In general, the conditions (67.2) yield nothing in addition to the limits defined by equation (67.10). The straight lines which correspond to the equality signs in (67.2) do not intersect the physical regions shown by the shaded areas in Fig. 9; some of them touch the boundaries of these regions, corresponding to extreme values of the variable s, t or u in the corresponding channel.

When the mass of one of the particles exceeds the sum of the masses of the other three ($m_1 > m_2 + m_3 + m_4$), a fourth reaction channel is possible, corresponding to the disintegration

$$\text{(IV)}\ 1 \to \bar{2} + 3 + 4. \tag{67.11}$$

For this channel, in the rest frame of the disintegrating particle,

$$q_1 = (m_1, 0), \qquad q_2 = (-\varepsilon_2, -\mathbf{p}_2), \qquad q_3 = (-\varepsilon_3, -\mathbf{p}_3),$$
$$q_4 = (-\varepsilon_4, -\mathbf{p}_4), \qquad \varepsilon_2 + \varepsilon_3 + \varepsilon_4 = m_1, \qquad \mathbf{p}_2 + \mathbf{p}_3 + \mathbf{p}_4 = 0.$$

The invariants are

$$\left.\begin{aligned} s &= m_1^2 + m_2^2 - 2m_1\varepsilon_2, \\ t &= m_1^2 + m_3^2 - 2m_1\varepsilon_3, \\ u &= m_1^2 + m_4^2 - 2m_1\varepsilon_4. \end{aligned}\right\} \tag{67.12}$$

We then have from (67.1)

$$\left.\begin{aligned} (m_3 + m_4)^2 &\leqslant s \leqslant (m_1 - m_2)^2, \\ (m_2 + m_4)^2 &\leqslant t \leqslant (m_1 - m_3)^2, \\ (m_2 + m_3)^2 &\leqslant u \leqslant (m_1 - m_4)^2. \end{aligned}\right\} \tag{67.13}$$

Thus all three invariants are positive, and the physical region of the disintegration channel is within the coordinate triangle.

PROBLEMS

PROBLEM 1. Find the physical regions for the case of three equal masses: $m_1 \equiv m$, $m_2 = m_3 = m_4 \equiv \mu$ (for example, the reaction $K + \pi \to \pi + \pi$).

SOLUTION. Equation (67.10) becomes

$$stu = \mu^2(m^2 - \mu^2)^2, \tag{1}$$

with

$$s + t + u = 3\mu^2 + m^2.$$

Regions I, II and III are bounded by curves of the same shape, with $s > 0$, $t < 0$, $u < 0$ for region I, and so on. If $m > 3\mu$, equation (1) also has a branch in the form of a closed curve with $s > 0$, $t > 0$, $u > 0$, which bounds the region of channel IV (Fig. 10).

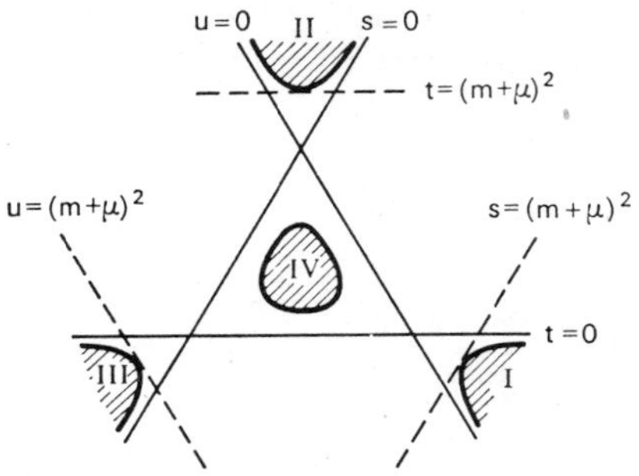

FIG. 10.

PROBLEM 2. The same as Problem 1, but for the case $m_1 \equiv m, m_2 \equiv \mu, m_3 = m_4 = 0, m > \mu$ (for example, the reaction $\mu + \nu \to e + \nu$).

SOLUTION. The condition (67.5) becomes

$$stu \geqslant m^2\mu^2 s,$$

with $s + t + u = m^2 + \mu^2$. The physical regions are bounded by the axis $s = 0$ and the two branches of the hyperbola $tu = m^2\mu^2$ (Fig. 11).

PROBLEM 3. The same as Problem 1, but for the case $m_1 = m_3 \equiv m, m_2 = 0, m_4 \equiv \mu$, with $m > 2\mu$ (for example, the reaction $p + \gamma \to p + \pi^0$).

SOLUTION. The boundary equation (67.10) becomes

$$stu = a(s+u) + bt,$$

$$ah = m^2\mu^4, \qquad bh = m^4(2m^2 - \mu^2), \qquad h = 2m^2 + \mu^2.$$

Elimination of u gives

$$t^2 + \left(\frac{b-a}{s} + s - h\right)t + \frac{ah}{s} = 0.$$

For a given value of s, this is a quadratic in t. If $s > (m+\mu)^2$ (the region of the s channel), there are two negative values of t for each value of s. If $s = (m+\mu)^2$, these two roots of the quadratic coincide at $t = -m\mu^2/(m+\mu)$. The boundary of the s channel region is then as shown in Fig. 12. The lower branch of the boundary tends asymptotically to the axis $u = 0$, and the upper branch crosses this axis at the point $t = \mu^4/(\mu^2 - m^2)$.

The u channel region is symmetrical with the s channel region; the t channel region is situated as shown in Fig. 12.

§68. Expansion in partial amplitudes

An important step in the analysis of a reaction of the form

$$a + b \to c + d \tag{68.1}$$

is the expansion of the scattering amplitude in partial amplitudes, each of which

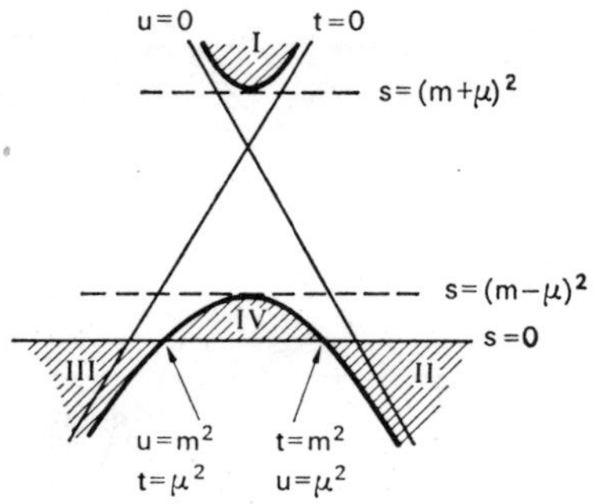

FIG. 11.

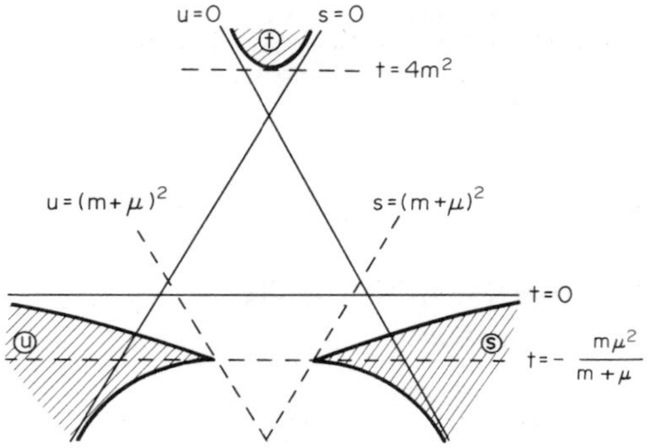

FIG. 12.

corresponds (for a given total energy ε) to a definite value of the total angular momentum of the particles J in their centre-of-mass system.†

These partial amplitudes are, therefore, elements of the S-matrix in the angular momentum representation:

$$\langle \varepsilon J'M'|S|\varepsilon JM\rangle.$$

Since the angular momentum J and its component M along a specified z-axis are conserved, the S-matrix is diagonal with respect to these numbers (and also with respect to the energy ε). Because of the isotropy of space, the diagonal elements are independent of the value of M. For given J, M and ε, the scattering matrix is still a matrix with respect to the spin quantum numbers; the elements of this matrix will be written in a more concise form:

$$\langle \varepsilon JM\lambda'|S|\varepsilon JM\lambda\rangle \equiv \langle \lambda'|S^J(\varepsilon)|\lambda\rangle, \tag{68.2}$$

where λ and λ' are the sets of spin quantum numbers. These can most naturally be taken to be the helicities of the particles. The helicity, unlike the spin component along an arbitrary axis in space, is conserved for a free particle, and it commutes with both the momentum and the angular momentum of the particle (§16). The helicities may therefore be used in both the momentum and the angular momentum representation of the scattering matrix.

The elements of the S-matrix with respect to the helicity indices will be called the *helicity scattering amplitudes*, and λ, λ' will be taken to include the helicities of the initial and final particles respectively:

$$\lambda = (\lambda_a, \lambda_b), \qquad \lambda' = (\lambda_c, \lambda_d).$$

In the momentum representation, the scattering matrix elements are defined with respect to the states $|\varepsilon\mathbf{n}\lambda\rangle$ (where $\mathbf{n} = \mathbf{p}/|\mathbf{p}|$ is the direction of the momentum of relative motion in the centre-of-mass system); in the angular momentum representation, they

† Most of the results in §§68 and 69 are due to M. Jacob and G. C. Wick (1959).

are defined with respect to the states $|\varepsilon JM\lambda\rangle$. They can be related by means of the expansions

$$|JM\lambda\rangle = \int |\mathbf{n}\lambda\rangle\langle\mathbf{n}\lambda|JM\lambda\rangle\, do_{\mathbf{n}}, \tag{68.3}$$

where the integration is over the directions $\mathbf{n}$; the energy ε is, for brevity, omitted from the state symbols. Since this transformation is unitary (see *QM*, § 12), the coefficients of the inverse transformation are

$$\langle JM\lambda|\mathbf{n}\lambda\rangle = \langle\mathbf{n}\lambda|JM\lambda\rangle^*. \tag{68.4}$$

By the general rule of matrix transformation, the same coefficients give the relation between the S-matrix elements in the two representations:

$$\langle\mathbf{n}'\lambda'|S|\mathbf{n}\lambda\rangle = \sum_{J,M} \langle\mathbf{n}'\lambda'|JM\lambda'\rangle\langle JM\lambda'|S|JM\lambda\rangle\langle JM\lambda|\mathbf{n}\lambda\rangle. \tag{68.5}$$

The coefficients in the expansion (68.3) are easily found by means of the results of §16. Let the wave functions of all states be expressed in the momentum representation, i.e. as functions of the direction of the momentum (for a given energy); this direction, as an independent variable, will be denoted by $\boldsymbol{\nu}$ to distinguish it from the direction $\mathbf{n}$ as a quantum number of the state. In this representation, the wave function has the form (16.2):

$$\psi_{\mathbf{n}\lambda}(\boldsymbol{\nu}) = u^{(\lambda)}\delta^{(2)}(\boldsymbol{\nu}-\mathbf{n}). \tag{68.6}$$

When (68.6) is substituted in the expansion (68.3), the latter reduces to a single term:

$$\psi_{JM\lambda} = \langle\boldsymbol{\nu}\lambda|JM\lambda\rangle u^{(\lambda)}. \tag{68.7}$$

The helicities λ_a and λ_b of the two particles are defined as the components of their spins in the directions of their respective momenta. If the momenta are $\mathbf{p}_a \equiv \mathbf{p}$, $\mathbf{p}_b = -\mathbf{p}$, then these directions are $\mathbf{n}$ for the first particle and $-\mathbf{n}$ for the second particle. If now the system is regarded as a single particle with helicity Λ in the direction $\mathbf{n}$, then $\Lambda = \lambda_a - \lambda_b$. Its wave function (in the momentum representation) can be written, according to (16.4), in the form

$$\psi_{JM\lambda}(\boldsymbol{\nu}) = u^{(\lambda)}D^{(J)}_{\Lambda M}(\boldsymbol{\nu})\sqrt{\frac{2J+1}{4\pi}}. \tag{68.8}$$

Comparison of (68.7) and (68.8), with the variable $\boldsymbol{\nu}$ replaced by $\mathbf{n}$, gives the required coefficients:

$$\langle\mathbf{n}\lambda|JM\lambda\rangle = \sqrt{\frac{2J+1}{4\pi}}\, D^{(J)}_{\Lambda M}(\mathbf{n}). \tag{68.9}$$

Substituting these coefficients in (68.5), we have

$$\left.\begin{aligned} \langle \mathbf{n}'\lambda'|S|\mathbf{n}\lambda\rangle &= \sum_{J,M} \frac{2J+1}{4\pi} D^{(J)}_{\Lambda'M}(\mathbf{n}') D^{(J)*}_{\Lambda M}(\mathbf{n}) \langle\lambda'|S^J|\lambda\rangle, \\ \Lambda = \lambda_a - \lambda_b, &\qquad \Lambda' = \lambda_c - \lambda_d, \end{aligned}\right\} \tag{68.10}$$

with the abbreviated notation (68.2). If the direction **n** is taken as that of the z-axis, then

$$D^{(J)}_{\Lambda M}(\mathbf{n}) = \delta_{\Lambda M},$$

and (68.10) becomes

$$\langle \mathbf{n}'\lambda'|S|\mathbf{n}\lambda\rangle = \sum_{J} \frac{2J+1}{4\pi} D^{(J)}_{\Lambda'\Lambda}(\mathbf{n}') \langle\lambda'|S^J|\lambda\rangle. \tag{68.11}$$

We see that the expansion in partial amplitudes has the functions $D^{(J)}_{\Lambda'\Lambda}$ as coefficients. For a reaction of the form (68.1), it is convenient to define the scattering amplitude f in such a way that the cross-section (in the centre-of-mass system) is

$$d\sigma = |\langle \mathbf{n}'\lambda'|f|\mathbf{n}\lambda\rangle|^2 \, do'; \tag{68.12}$$

by comparison with (64.19), we can relate this amplitude to the matrix element M_{fi}. The expansion of the amplitude in partial amplitudes may be written

$$\langle \mathbf{n}'\lambda'|f|\mathbf{n}\lambda\rangle = \sum_{J,M} (2J+1) D^{(J)}_{\Lambda'M}(\mathbf{n}') D^{(J)}_{\Lambda M}{}^*(\mathbf{n}) \langle\lambda'|f^J|\lambda\rangle, \tag{68.13}$$

or, taking the z-axis in the direction **n**,

$$\langle \mathbf{n}'\lambda'|f|\mathbf{n}\lambda\rangle = \sum_{J} (2J+1) D^{(J)}_{\Lambda'\Lambda}(\mathbf{n}') \langle\lambda'|f^J|\lambda\rangle. \tag{68.14}$$

This is a generalization of the usual expansion in partial amplitudes for the scattering of spinless particles; see *QM*, (123.14). Since $D^{(L)}_{00} = P_L(\cos\theta)$, (68.14) reduces in the case of zero spins to an expansion in Legendre polynomials

$$f(\theta) = \sum_{L} (2L+1) f_L P_L(\cos\theta).$$

The cross-section (68.12) is valid when all the particles have definite helicities. If they are in mixed polarization states, the cross-section is found by averaging the product

$$\langle\lambda_c\lambda_d|f|\lambda_a\lambda_b\rangle\langle\lambda_c'\lambda_d'|f|\lambda_a'\lambda_b'\rangle^*$$

over the polarization density matrices of the particles,

$$\langle\lambda_a|\rho^{(a)}|\lambda_a'\rangle\langle\lambda_b|\rho^{(b)}|\lambda_b'\rangle\langle\lambda_c'|\rho^{(c)}|\lambda_c\rangle\langle\lambda_d'|\rho^{(d)}|\lambda_d\rangle;$$

see the first footnote to §48. For example, in a reaction between unpolarized particles a, b to form unpolarized particles c, d, we have

$$d\sigma = \frac{do}{(2s_a+1)(2s_b+1)} \sum_{(\lambda)} \sum_{J,J'} (2J+1)(2J'+1) \times$$
$$\times \langle\lambda_c\lambda_d|f^J|\lambda_a\lambda_b\rangle\langle\lambda_c\lambda_d|f^{J'}|\lambda_a\lambda_b\rangle^* D^{(J)}_{\Lambda'\Lambda}(\mathbf{n}')D^{(J')*}_{\Lambda'\Lambda}(\boldsymbol{n}'); \tag{68.15}$$

the z-axis is along $\mathbf{n}$, and the first summation is over $\lambda_a, \lambda_b, \lambda_c, \lambda_d$. Using *QM*, (58.19) for the function $D^{(J')*}_{\Lambda'\Lambda}$ and then the expansion *QM*, (110.2), we have finally

$$d\sigma = \frac{do}{(2s_a+1)(2s_b+1)} \sum_{(\lambda)J,J'} (-1)^{\Lambda-\Lambda'}(2J+1)(2J'+1) \times$$
$$\times \langle\lambda_c\lambda_d|f^J|\lambda_a\lambda_b\rangle \langle\lambda_c\lambda_d|f^{J'}|\lambda_a\lambda_b\rangle^* \times$$
$$\times \sum_L (2L+1) \begin{pmatrix} J & J' & L \\ \Lambda & -\Lambda & 0 \end{pmatrix} \begin{pmatrix} J & J' & L \\ \Lambda' & -\Lambda' & 0 \end{pmatrix} P_L(\cos\theta), \tag{68.16}$$

where θ is the angle between $\mathbf{n}'$ and the z-axis; the summation with respect to L is over all integers which can occur when $\mathbf{J}$ and $\mathbf{J}'$ are added vectorially. The expansion of the scattering amplitude in partial amplitudes gives a full expression of all properties of the angular distribution of scattering that are due to the symmetry with respect to spatial rotations. But it does not explicitly reveal the properties that are due to the symmetry with respect to spatial inversion. The P invariance (if possessed by the interaction) leads to certain relations between the various helicity amplitudes (see §69).

§69. Symmetry of helicity scattering amplitudes

The conditions imposed by the symmetry with respect to the transformations P, T, C (if, of course, the particle interaction process in question in fact possesses such symmetry) lead to certain relations between the helicity scattering amplitudes, and therefore reduce the number of independent amplitudes.†

To establish these relations, we shall first determine the symmetry properties of the helicity states of a system of two particles.

Let us consider the particles in their centre-of-mass system. One particle has momentum $\mathbf{p}_1 \equiv \mathbf{p}$ and helicity λ_1 with respect to the direction of $\mathbf{p}$; the other has momentum $\mathbf{p}_2 = -\mathbf{p}$ and helicity λ_2 with respect to the direction of $-\mathbf{p}$. If the helicity is defined with respect to the same direction, that of $\mathbf{p}$, its values are λ_1 and $-\lambda_2$, and the particles will thus be described by plane waves with amplitudes $u_{\mathbf{p}}^{(\lambda_1)}$ and $u_{\mathbf{p}}^{(-\lambda_2)}$. The two-particle system is described by a (multi-component) function $u_{\mathbf{p}}^{(\lambda_1\lambda_2)}$ formed from the products of the amplitudes $u_{\mathbf{p}}^{(\lambda_1)}$ and $u_{\mathbf{p}}^{(-\lambda_2)}$.

Let us next regard the system as a single particle with helicity $\Lambda = \lambda_1 - \lambda_2$ in the direction $\mathbf{n} = \mathbf{p}/|\mathbf{p}|$; we can then write the wave function (in the momentum

† This number does not, of course, depend on the specific representation of the matrix S^J, and is the same for any choice of the spin variables.

representation, i.e. as a function of **n**) for a state with definite values of J, M, λ_1, λ_2 (and of the total energy ε):

$$\psi_{JM\lambda_1\lambda_2} = u_{\mathbf{p}}^{(\lambda_1\lambda_2)} D_{\Lambda M}^{(J)}(\mathbf{n}) \sqrt{\frac{2J+1}{4\pi}}, \qquad \Lambda = \lambda_1 - \lambda_2; \tag{69.1}$$

cf. (68.8). Since Λ is the component of the total angular momentum in the direction of **p**, we must have

$$|\Lambda| \leq J. \tag{69.2}$$

According to (16.14), under inversion

$$\begin{aligned} \hat{P} u^{(\lambda_1\lambda_2)}(\mathbf{n}) &= \eta_1 \eta_2 u^{(\lambda_1\lambda_2)}(-\mathbf{n}) \\ &= \eta_1 \eta_2 (-1)^{s_1+s_2-\lambda_1+\lambda_2} u^{(-\lambda_1-\lambda_2)}(\mathbf{n}), \end{aligned} \tag{69.3}$$

where η_1 and η_2 are the internal parities of the particles. Using also (16.10), we find the transformation law for the functions (69.1):

$$\hat{P}\psi_{JM\lambda_1\lambda_2} = \eta_1\eta_2(-1)^{s_1+s_2-J}\psi_{JM,-\lambda_1,-\lambda_2}. \tag{69.4}$$

If the two particles are identical, the question arises of the symmetry with respect to their interchange. This interchange implies interchanging their momenta and their spins. To show the significance of this operation as applied to the function (69.1), we note that its definition contains an asymmetry, in that the angular momenta of the two particles are projected on the direction of the same vector $\mathbf{p}_1 \equiv \mathbf{p}$, the momentum of the first particle. After the interchange, this vector is replaced by $\mathbf{p}_2 \equiv -\mathbf{p}$, and the components of the angular momenta $\mathbf{j}_1$ and $\mathbf{j}_2$ along this vector are $-\lambda_1$ and λ_2 (instead of λ_1 and $-\lambda_2$ along **p**). The result of applying the particle interchange operator $\hat{P}_{12}$ to the function (69.1) may therefore be written

$$\hat{P}_{12}\psi_{JM\lambda_1\lambda_2} = u^{(-\lambda_2-\lambda_1)}(-\mathbf{n}) D_{\Lambda M}^{(J)}(-\mathbf{n}) \sqrt{\frac{2J+1}{4\pi}},$$

where again $\Lambda = \lambda_1 - \lambda_2$. Then, using (69.3) and (16.10), we find that

$$\hat{P}_{12}\psi_{JM\lambda_1\lambda_2} = (-1)^{2s-J}\psi_{JM\lambda_2\lambda_1}, \tag{69.5}$$

where $s_1 = s_2 \equiv s$.

For identical particles, the permissible states must be either symmetric (for bosons) or antisymmetric (for fermions) with respect to interchange. Since the former case occurs when the particle spin s is integral and the latter case when it is half-integral, in either case the permissible helicity states of the two-particle system can be written as linear combinations

$$[1 + (-1)^{2s}\hat{P}_{12}]\psi_{JM\lambda_1\lambda_2},$$

or, according to (69.5),

$$\psi_{JM\lambda_1\lambda_2} + (-1)^J \psi_{JM\lambda_2\lambda_1}. \tag{69.6}$$

It is noteworthy that this combination is the same for both bosons and fermions.

For a particle–antiparticle system, the result of the interchange is expressed by the same formula (69.5), but, unlike the case of identical particles, states of either symmetry under interchange are here permissible, i.e. both combinations

$$\psi^{\pm} = \psi_{JM\lambda_1\lambda_2} \pm (-1)^J \psi_{JM\lambda_2\lambda_1} \tag{69.7}$$

can occur. These states have certain charge parities C. The operation of charge conjugation may be regarded as the result of a total interchange of all variables (spin and charge) of the two particles, followed by reverse interchange of the spin variables (helicities). The result of the first operation must be the same as that of interchange in a system of two identical particles. Hence it is clear that, with the upper sign in (69.7) (which is the same as the sign in the state (69.6) permissible for identical particles), the system will be charge-even, and with the lower sign charge-odd:

$$\hat{C}\psi^{\pm} = \pm \psi^{\pm}.$$

Finally, let us consider the operation of time reversal. The wave function of a particle at rest with spin s and component thereof σ is transformed according to

$$\hat{T}\psi_{s\sigma} = (-1)^{s-\sigma}\psi_{s,-\sigma};$$

see *QM*, (60.2). The wave function of two particles in their centre-of-mass system may also be regarded (in respect ot its transformation properties) as that of a "particle" at rest, with angular momentum J and component thereof M. The helicities λ_1, λ_2 are unchanged: time reversal changes the sign of the momentum and angular momentum vectors, and the products $\mathbf{j}\cdot\mathbf{p}$ are therefore unaffected. Hence

$$\hat{T}\psi_{JM\lambda_1\lambda_2} = (-1)^{J-M}\psi_{JM\lambda_1\lambda_2}. \tag{69.8}$$

We can now write down immediately the symmetry relations for the helicity amplitudes.

If the interaction is P-invariant, then for the reaction $a + b \to c + d$ the amplitudes of the transitions

$$|\lambda_a\lambda_b\rangle \to |\lambda_c\lambda_d\rangle \quad \text{and} \quad \hat{P}|\lambda_a\lambda_b\rangle \to \hat{P}|\lambda_c\lambda_d\rangle$$

must be the same (for given J and ε). Hence, using (69.4), we find

$$\langle\lambda_c\lambda_d|S^J|\lambda_a\lambda_b\rangle = \frac{\eta_c\eta_d}{\eta_a\eta_b}(-1)^{s_c+s_d-s_a-s_b}\langle -\lambda_c, -\lambda_d|S^J|-\lambda_a, -\lambda_b\rangle. \tag{69.9}$$

If states with definite parities, i.e. the combinations

$$\frac{1}{\sqrt{2}}(\psi_{JM\lambda_1\lambda_2} \pm \hat{P}\psi_{JM\lambda_1\lambda_2}),$$

where $\lambda_1, \lambda_2 = \lambda_a, \lambda_b$ or λ_c, λ_d, are chosen instead of those with definite helicities, then the amplitudes of transitions in which parity is not conserved are zero.

Time reversal transforms each state in accordance with (69.8), and also interchanges initial and final states. Thus T invariance leads to the relations

$$\langle\lambda_c\lambda_d|S^J(\varepsilon)|\lambda_a\lambda_b\rangle = \langle\lambda_a\lambda_b|S^J(\varepsilon)|\lambda_c\lambda_d\rangle. \tag{69.10}$$

These two amplitudes, however, pertain to different processes, the direct and reverse reactions. These two processes are essentially equivalent only in the case of elastic scattering, and (69.10) is then a relation between helicity amplitudes for the same reaction.

In elastic scattering of two identical particles, the number of different amplitudes is further reduced because of the symmetry with respect to interchange. We have seen that, for a given J, the states which occur are either all symmetric or all antisymmetric in λ_1, λ_2. The conservation of angular momentum therefore implies that of the symmetry with respect to interchange of helicities.

A similar situation occurs in the elastic scattering of a particle by its antiparticle, or the conversion of one particle–antiparticle pair into another, i.e. a reaction $a + \bar{a} \to b + \bar{b}$. For given J, there are both symmetric and antisymmetric states with regard to λ_1, λ_2, but they correspond to different values of the charge parity of the system. Hence it follows that, if the interaction of the particles is C-invariant, so that the charge parity is conserved, transitions between states of different symmetry with regard to λ_1, λ_2 are forbidden.† It must be emphasized, however, that there is a difference from the case of identical particles, in which states of one symmetry are entirely absent for any given J. In the "particle–antiparticle" case, only transitions between states of different symmetry are forbidden; the states themselves exist for every J.

Because of the universal CPT invariance, the existence of T invariance implies that of CP invariance. The latter brings about the equality of amplitudes for two reactions, one obtained from the other by replacing all particles by antiparticles (and changing the sign of the helicities):

$$\langle\lambda_c\lambda_d|S^J|\lambda_a\lambda_b\rangle = \langle\lambda_{\bar{c}}\lambda_{\bar{d}}|S^J|\lambda_{\bar{a}}\lambda_{\bar{b}}\rangle, \tag{69.11}$$

where $\lambda_{\bar{a}} = -\lambda_a$ and so on.‡

The number of independent amplitudes is the same for all the cross-channels of

† A similar prohibition can also arise from isotopic invariance of the interaction of non-identical particles. For instance, transitions between states of different symmetry with regard to λ_1, λ_2 are forbidden, to the extent that this invariance holds, in the scattering of a neutron by a proton.

‡ Since these two amplitudes relate to different reactions, interference between which is not possible, the phase factor in (69.11) would have no significance, and can be taken as unity. Only the equality of cross-sections which follows from (69.11) is actually meaningful.

one generalized reaction, and therefore this number can be determined from any channel. For example, the elastic scattering $a+b\to a+b$ and the annihilation $a+\bar{a}\to b+\bar{b}$ are described by the same number of independent amplitudes. The restrictions imposed by T invariance in the first case are equivalent to those imposed by C invariance in the second case.

Let us also consider a reaction in which one particle disintegrates into two: $a\to b+c$. In the centre-of-mass system (the rest frame for particle a), we have $\mathbf{p}_b=-\mathbf{p}_c$. Scalar multiplication of the equation $\mathbf{j}_a=\mathbf{j}_b+\mathbf{j}_c$ by $\mathbf{p}_b$ gives

$$\lambda_a=\lambda_b-\lambda_c \tag{69.12}$$

(the helicity λ_a of particle a is defined as the component of its spin in the direction of the momentum of one of the secondary particles). This relation can be regarded as a consequence of the additional symmetry present in the process considered, namely the axial symmetry about the directions of $\mathbf{p}_b$ and $\mathbf{p}_c$. If the spin s_a of particle a is less than s_b+s_c, the relation (69.12) reduces the number of possible sets of values of λ_a, λ_b, λ_c and therefore the number of independent helicity amplitudes of the disintegration. The total angular momentum J is then equal to the spin s_a of the primary particle, and is consequently fixed.

The P invariance in the disintegration is expressed by the relation

$$\langle\lambda_b\lambda_c|S^J|\lambda_a\rangle=\frac{\eta_b\eta_c}{\eta_a}(-1)^{s_a-s_b-s_c}\langle-\lambda_b,-\lambda_c|S^J|-\lambda_a\rangle, \tag{69.13}$$

where we have used (69.4) and also the transformation (16.16) for the wave function of a single particle.

If the primary particle is strictly neutral, further limitations arise if C parity is conserved. Three cases are to be distinguished here. If the disintegration products are also strictly neutral, we must have $C_a=C_bC_c$; this condition either prohibits the disintegration altogether, or is satisfied and causes no further restriction. If the particles b and c are different, then C invariance implies a relation between the amplitudes of the different processes $a\to\bar{b}+\bar{c}$ and $a\to b+c$. Finally, for the disintegration $a\to b+\bar{b}$, there is a restriction because, for a given charge parity C and a given total angular momentum $J=s_a$, the system may be in states either symmetric or antisymmetric with respect to the helicities, depending on the parity of the number J and on the sign of C.

CP invariance implies the equality of amplitudes for the disintegrations $a\to b+c$ and $\bar{a}\to\bar{b}+\bar{c}$:

$$\langle\lambda_b\lambda_c|S^J|\lambda_a\rangle=\langle\lambda_{\bar{b}}\lambda_{\bar{c}}|S^J|\lambda_{\bar{a}}\rangle, \tag{69.14}$$

where $\lambda_{\bar{a}}=-\lambda_a$ and so on; i.e. it implies equal probabilities of disintegration for the particle and the antiparticle. If the particle can disintegrate in more than one way (through various channels), this equality applies to each channel. This conclusion, it must be emphasized, is based on the existence of CP invariance, which is not a universal property of Nature. Only CPT invariance is universal, and this by itself would lead only to the equation

$$\langle \lambda_b \lambda_c | S^J | \lambda_a \rangle = \langle \lambda_{\bar{a}} | S^J | \lambda_{\bar{b}} \lambda_{\bar{c}} \rangle,$$

in which the right-hand side refers to the process inverse to disintegration. We shall see later (§71) that the condition of *CPT* invariance, together with unitarity requirements, does lead to a relation, although a more restricted one, between the disintegration probabilities for the particle and the antiparticle.

PROBLEMS

PROBLEM 1. Using (69.6), obtain a classification of the possible states of a two-photon system.

SOLUTION. In this case $\lambda_1, \lambda_2 = \pm 1$. For even J (>0), according to (69.6), three states symmetric in λ_1, λ_2 are allowed:

$$\text{(a) } \psi_{JM11}, \qquad \text{(b) } \psi_{JM,-1,-1}, \qquad \text{(c) } \psi_{JM1,-1} + \psi_{JM,-1,1}.$$

For odd J (>1), one antisymmetric state is allowed:

$$\text{(d) } \psi_{JM1,-1} - \psi_{JM,-1,1}.$$

States (c) and (d) also have a definite parity (+1): according to (69.4),

$$\hat{P}(\psi_{JM1,-1} \pm \psi_{JM,-1,1}) = \pm(-1)^J (\psi_{JM1,-1} \pm \psi_{JM,-1,1});$$

the factor $\pm(-1)^J = 1$, since the upper sign refers to even values of J and the lower sign to odd values. States (a) and (b) themselves have no definite parity, but even and odd states are obtained by taking the combinations

$$\text{(a}') \ \psi_{JM11} + \psi_{JM,-1,-1}, \qquad \text{(b}') \ \psi_{JM11} - \psi_{JM,-1,-1}.$$

When $J = 0$, only $\lambda_1 = \lambda_2$ is allowed by the condition $|\lambda_1 - \lambda_2| \leq J$, so that state (c) does not occur, leaving one even and one odd state, (a′) and (b′). Finally, if $J = 1$, state (d), which is the only possible state for odd J, is forbidden because it has $\lambda = 2 > J$. Thus we arrive at the table (9.5) for the permissible states.

PROBLEM 2. In the non-relativistic approximation, the total angular momentum J of the system is found by adding the spin S and the orbital angular momentum L. For a system of two particles, find the relation between the states $|JLSM\rangle$ and $|JM\lambda_1\lambda_2\rangle$.

SOLUTION. According to the rule for constructing wave functions when adding angular momenta, we have

$$\psi_{JLSM} = \Sigma\{\psi_{s_1\sigma_1}\psi_{s_2\sigma_2}\langle\sigma_1\sigma_2|SM_S\rangle\}\psi_{LM_L}\langle M_L M_S|JM\rangle, \tag{1}$$

where $\psi_{s\sigma}$ are the eigenfunctions of the spin s with component σ along a fixed z-axis, ψ_{LM_L} those of the orbital angular momentum L with component M_L; the expression in the braces corresponds to the addition of s_1 and s_2 to give S, after which S is added to L to give J; the summation is over all m-type indices. Let all functions be expressed in the momentum representation, as functions of the direction $\mathbf{n}$ of the momentum $\mathbf{p} \equiv \mathbf{p}_1$, and let the functions $\psi_{s\sigma}$ be expressed in terms of the functions $\psi_{\mathbf{n}\lambda}$ of the helicity states by means of *QM*, (58.7):

$$\psi_{s_1\sigma_1} = \sum_{\lambda_1} D^{(s_1)}_{\lambda_1\sigma_1}(\mathbf{n})\psi_{\mathbf{n}\lambda_1},$$

$$\psi_{s_2\sigma_2} = \sum_{\lambda_2} D^{(s_2)}_{-\lambda_2\sigma_2}(\mathbf{n})\psi_{\mathbf{n},-\lambda_2}.$$

For the function ψ_{LM_L}, we have

$$\psi_{LM_L} = Y_{LM_L}(\mathbf{n}) = i^L \sqrt{\frac{2L+1}{4\pi}}\, D^{(L)}_{0M_L}(\mathbf{n}),$$

using *QM* (58.25) and the definition (16.5). Substituting these functions in equation (1), and twice using the expansion *QM*, (110.1), together with the orthogonality of the Clebsch–Gordan coefficients (*QM*, (106.13)), we obtain the expansion

$$\psi_{JLSM} = \sum_{\lambda_1,\lambda_2} \psi_{JM\lambda_1\lambda_2}\langle JM\lambda_1\lambda_2|JLSM\rangle, \tag{2}$$

where

$$\psi_{JM\lambda_1\lambda_2} = \psi_{\mathbf{n}\lambda_1}\psi_{\mathbf{n},-\lambda_2}D^{(J)}_{\Lambda M}(\mathbf{n})\sqrt{\frac{2J+1}{4\pi}}, \qquad \Lambda = \lambda_1 - \lambda_2,$$

and the coefficients are

$$\langle JM\lambda_1\lambda_2|JLSM\rangle = (-i)^L(-1)^{s_1-s_2+S}\sqrt{[(2L+1)(2S+1)]}\begin{pmatrix} s_1 & s_2 & S \\ \lambda_1 & -\lambda_2 & -\Lambda\end{pmatrix}\begin{pmatrix} L & S & J \\ 0 & \Lambda & -\Lambda\end{pmatrix}. \tag{3}$$

Since the transformation (2) is unitary, we have

$$\langle JLSM|JM\lambda_1\lambda_2\rangle = \langle JM\lambda_1\lambda_2|JLSM\rangle^*.$$

§ 70. **Invariant amplitudes**

In the helicity amplitudes, a particular frame of reference is used, namely the centre-of-mass system. But, in order to calculate the scattering amplitudes by means of invariant perturbation theory (and also to examine their general analytical properties), it is convenient to write them in an explicitly invariant form.

If the particles concerned in the reaction have no spin, the scattering amplitude depends only on the invariant products of the 4-momenta of the particles. For a reaction of the form

$$a + b \to c + d, \tag{70.1}$$

these invariants may be taken as any two of the quantities s, t, u defined in §66. Then the scattering amplitude reduces to a single function $M_{fi} = f(s, t)$.

If the particles have spins, then, besides the kinematic invariants s, t, u, there are also invariants which can be constructed from the wave amplitudes of the particles (bispinors, 4-tensors, etc.). The scattering amplitudes must then have the form

$$M_{fi} = \sum_n f_n(s, t)F_n, \tag{70.2}$$

where the F_n are invariants which depend linearly on the wave amplitudes of all the particles concerned (and also on their 4-momenta). The coefficients $f_n(s, t)$ are called *invariant amplitudes*.

By choosing the wave amplitudes in such a way as to correspond to particles with definite helicities, we obtain definite values of the invariants $F_n = F_n(\lambda_i, \lambda_f)$. Then the helicity scattering amplitudes are linear homogeneous combinations of the invariant amplitudes f_n. Hence we see that the number of independent functions $f_n(s, t)$ is equal to the number of independent helicity amplitudes. Since the latter number is easily determined, as shown in §69, this makes easier the construction of the invariants F_n, their number being known in advance.

Let us consider some examples, assuming in every case that the interaction is T-invariant and P-invariant. The latter property implies that the invariants F_n must be true scalars, not pseudoscalars.

Scattering of a particle with spin 0 by one with spin $\frac{1}{2}$

To find the number of invariants (that is, the number of independent helicity amplitudes), we note that the total number of elements of the matrix S^J (i.e. of different sets $\lambda_1, \lambda_2, \lambda_1', \lambda_2'$) is in this case four: $\lambda_1 = \lambda_1' = 0$, $\lambda_2, \lambda_2' = \pm\frac{1}{2}$. When the P invariance is taken into account, the number of independent elements is reduced to two, and this is unchanged by the inclusion of T invariance.

The two independent invariants may be taken as

$$F_1 = \bar{u}'u, \qquad F_2 = \bar{u}'(\gamma K)u, \tag{70.3}$$

where $u = u(p)$, $u' = u(p')$ are the bispinor amplitudes of the initial and final fermions; $K = k + k'$, where k and k' are the 4-momenta of the initial and final bosons.†

The T invariance of the quantities (70.3) is evident if we note that under time reversal the products $\bar{u}'u$ and $\bar{u}'\gamma^\mu u$ are transformed according to the same rule (28.6) as the operators $\hat{\bar{\psi}}\hat{\psi}$ and $\hat{\bar{\psi}}\gamma^\mu\hat{\psi}$, whose matrix elements they are: $\bar{u}'u$ is invariant, and the 4-vector $\bar{u}'\gamma u$ is transformed according to

$$\bar{u}'\gamma^0 u \to \bar{u}'\gamma^0 u, \qquad \bar{u}'\boldsymbol{\gamma} u \to -\bar{u}'\boldsymbol{\gamma} u.$$

The 4-momenta are transformed similarly: $(K^0, \mathbf{K}) \to (K^0, -\mathbf{K})$, and the scalar product $F_2 = K_\mu(\bar{u}'\gamma^\mu u)$ is therefore invariant.

Elastic scattering of two identical particles with spin $\frac{1}{2}$

To find the number of independent helicity amplitudes, it is convenient to start from linear combinations of the helicity states:

$$\psi_{1g} = \psi_{++} + \psi_{--}, \qquad \psi_{2g} = \psi_{++} - \psi_{--}.$$
$$\psi_{3g} = \psi_{+-} + \psi_{-+}, \qquad \psi_u = \psi_{+-} - \psi_{-+},$$

where the suffixes $\pm$ denote the values $\pm\frac{1}{2}$ of the helicities of the two particles. The states $1g$, $2g$, $3g$ are even, and u is odd, with respect to interchange of the particles. The transitions $g \leftrightarrow u$ are forbidden, so that there remain $16 - 6 = 10$ matrix elements, when the interchange symmetry is taken into account. The functions ψ_{1g} and ψ_{3g}, and ψ_{2g} have opposite parities with respect to the inversion P; the prohibition of transitions between them reduces the number of independent amplitudes to six. Lastly, the T invariance equalizes the amplitudes of the transitions $1g \to 3g$ and $3g \to 1g$, leaving only five independent amplitudes. The five independent invariants may be taken as

$$\left.\begin{aligned} F_1 &= (\bar{u}_1'u_1)(\bar{u}_2'u_2), & F_2 &= (\bar{u}_1'\gamma^5 u_1)(\bar{u}_2'\gamma^5 u_2),\\ F_3 &= (\bar{u}_1'\gamma^\mu u_1)(\bar{u}_2'\gamma_\mu u_2), & F_4 &= (\bar{u}_1'\gamma^\mu\gamma^5 u_1)(\bar{u}_2'\gamma_\mu\gamma^5 u_2),\\ & & F_5 &= (\bar{u}_1'\sigma^{\mu\nu}u_1)(\bar{u}_2'\sigma_{\mu\nu}u_2), \end{aligned}\right\} \tag{70.4}$$

† At first sight, there might appear to be another invariant of the form $\bar{u}'\sigma_{\mu\nu}k^\mu k'^\nu u$ (with the matrices $\sigma_{\mu\nu}$ defined by (28.2)), but this is easily seen to reduce to the invariants (70.3) by means of the conservation law $k' = p + k - p'$ and the equations $(\gamma p)u = mu$, $\bar{u}'(\gamma p') = m\bar{u}'$ satisfied by the bispinor amplitudes.

where u_1, u_2 are the bispinor amplitudes of the initial particles and u_1', u_2' those of the final particles. Interchange of the initial (or of the final) particles gives no new invariants: the invariants obtained can be expressed in terms of the previous ones (§28, Problem). But the expression (70.2), with the F_n given by (70.4), does not explicitly take account of the requirement that interchange of two identical fermions must change the sign of the scattering amplitude. An expression which satisfies this condition may be written

$$M_{fi} = [(\bar{u}_1' u_1)(\bar{u}_2' u_2) f_1(t, u) - (\bar{u}_2' u_1)(\bar{u}_1' u_2) f_1(u, t)] + \cdots. \tag{70.5}$$

When p_1' and p_2' (or p_1 and p_2) are interchanged, the kinematic invariants $s \to s$, $t \to u$, $u \to t$, so that the condition is necessarily satisfied.

Elastic scattering of a photon by particles with spin 0 or $\frac{1}{2}$

The amplitude of this process is conveniently expressed by means of the space-like unit 4-vectors $e^{(1)}$, $e^{(2)}$ which satisfy the conditions

$$\left.\begin{aligned} &e^{(1)2} = e^{(2)2} = -1, \qquad && e^{(1)}e^{(2)} = 0, \\ &e^{(1)}k = e^{(2)}k = 0, && e^{(1)}k' = e^{(2)}k' = 0; \end{aligned}\right\} \tag{70.6}$$

for each of the two photons, these 4-vectors can be the unit 4-vectors by means of which an invariant description of their polarization properties is obtained (§8).

Let k and k' be the initial and final 4-momenta of the photon; p and p' those of the scattering particle. The 4-vectors

$$\left.\begin{aligned} P^\lambda &= p^\lambda + p'^\lambda - K^\lambda \frac{pK + p'K}{K^2}, \\ N^\lambda &= e^{\lambda\mu\nu\rho} P_\mu q_\nu K_\rho, \end{aligned}\right\} \tag{70.7}$$

where

$$K = k + k', \qquad q = p - p' = k' - k,$$

are evidently orthogonal to one another and also to the 4-vectors K and q, and therefore to k and k'. Being orthogonal to the time-like 4-vector K ($K^2 = 2kk' > 0$), they must themselves be space-like: in a frame of reference for which $\mathbf{K} = 0$, it follows from $KP = 0$ that $P_0 = 0$ and hence $P^2 < 0$. Normalizing P and N by putting

$$e^{(1)\lambda} = \frac{N^\lambda}{\sqrt{(-N^2)}}, \qquad e^{(2)\lambda} = \frac{P^\lambda}{\sqrt{(-P^2)}}, \tag{70.8}$$

we obtain a pair of 4-vectors which have all the required properties. It may be noted that $e^{(2)}$ is a true vector and $e^{(1)}$ a pseudovector.

The photon scattering amplitude may be written

$$M_{fi} = F^{\lambda\mu} e'^*_\lambda e_\mu, \tag{70.9}$$

in terms of the polarization 4-vectors e and e' of the initial and final photons.

The photon helicity has only two values, ± 1. Hence, for the scattering of a photon by a particle with spin zero, the number of independent helicity amplitudes is the same as for the mutual scattering of particles with spin 0 and $\frac{1}{2}$, namely two. The tensor $F^{\lambda\mu}$ in (70.9) has to be constructed from the particle 4-momenta only. It can be written

$$F^{\lambda\mu} = f_1 e^{(1)\lambda} e^{(1)\mu} + f_2 e^{(2)\lambda} e^{(2)\mu}, \tag{70.10}$$

where f_1 and f_2 are invariant amplitudes. It should be noted that no term containing a product $e^{(1)\lambda} e^{(2)\mu}$ can appear in $F^{\lambda\mu}$, since this product is a pseudotensor and would give a pseudoscalar on substitution in (70.9).

Lastly, let us consider the scattering of a photon by a particle with spin $\frac{1}{2}$. To find the number of independent helicity amplitudes, we note that the total number of elements of the matrix S^J in this case is sixteen; the helicity of each of two initial and two final particles has two values. The condition of P invariance reduces this number to eight, and that of T invariance brings it down to six.

Here, we write the tensor $F_{\lambda\mu}$ in the form

$$F_{\lambda\mu} = G_0(e^{(1)}_\lambda e^{(1)}_\mu + e^{(2)}_\lambda e^{(2)}_\mu) + G_1(e^{(1)}_\lambda e^{(2)}_\mu + e^{(2)}_\lambda e^{(1)}_\mu) + \\ + G_2(e^{(1)}_\lambda e^{(2)}_\mu - e^{(2)}_\lambda e^{(1)}_\mu) + G_3(e^{(1)}_\lambda e^{(1)}_\mu - e^{(2)}_\lambda e^{(2)}_\mu), \tag{70.11}$$

where G_0 and G_3 are true scalars, G_1 and G_2 are pseudoscalars, and all four are bilinear in the bispinor fermion amplitudes $\bar{u}(p')$ and $u(p)$, being of the form

$$G_n = \bar{u}(p') Q_n u(p). \tag{70.12}$$

The general form of the matrices (with respect to the bispinor indices) Q_n is

$$\left.\begin{aligned} Q_0 &= f_1 + f_2(\gamma K), & Q_1 &= \gamma^5(f_3 + f_4(\gamma K)), \\ Q_2 &= \gamma^5(f_5 + f_6(\gamma K)), & Q_3 &= f_7 + f_8(\gamma K), \end{aligned}\right\} \tag{70.13}$$

where $K = k + k'$. The coefficients $f_1, \ldots, f_8$ are invariant amplitudes, in this case eight in number (instead of the correct value of six), because the condition of T invariance has not yet been imposed.

Time reversal interchanges the initial and final 4-momenta of the particles, and also changes the sign of their space components:

$$(k_0, \mathbf{k}) \leftrightarrow (k'_0, -\mathbf{k}'), \qquad (p_0, \mathbf{p}) \leftrightarrow (p'_0, -\mathbf{p}'). \tag{70.14}$$

The photon polarization 4-vectors are transformed according to

$$(e_0, \mathbf{e}) \leftrightarrow (e'^*_0, -\mathbf{e}'^*) \tag{70.15}$$

(cf. (8.11a)); hence

$$(e_0'^* e_0,\ e_i'^* e_0,\ e_i'^* e_k) \to (e_0'^* e_0,\ -e_0'^* e_i,\ e_k'^* e_i).$$

By virtue of the last transformation, the condition of invariance of the scattering amplitude (70.9) is equivalent to

$$(F_{00},\ F_{i0},\ F_{ik}) \to (F_{00},\ -F_{0i},\ F_{ki}).$$

On the other hand, the changes (70.14) imply

$$(K_0, \mathbf{K}) \to (K_0, -\mathbf{K}), \qquad (q_0, \mathbf{q}) \to (-q_0, \mathbf{q}),$$
$$(P_0, \mathbf{P}) \to (P_0, -\mathbf{P}), \qquad (N_0, \mathbf{N}) \to (N_0, -\mathbf{N}),$$

so that

$$(e_0^{(1,2)}, \mathbf{e}^{(1,2)}) \to (e_0^{(1,2)}, -\mathbf{e}^{(1,2)}). \tag{70.16}$$

Hence, from (70.11), we must have

$$G_{0,1,3} \to G_{0,1,3}, \qquad G_2 \to -G_2.$$

Under time reversal,

$$\bar{u}'\gamma^5 u \to -\bar{u}'\gamma^5 u, \qquad \bar{u}'\gamma^5(\gamma K)u \to \bar{u}'\gamma^5(\gamma K)u,$$

as is evident from the transformation laws for pseudoscalar and pseudovector bilinear forms (28.6). From (70.12), (70.13) it is now evident that, because of the T invariance of the scattering amplitude,

$$f_3 = f_6 = 0. \tag{70.17}$$

§71. **The unitarity condition**

The scattering matrix must be unitary: $\hat{S}\hat{S}^+ = 1$, or in terms of matrix elements,

$$(SS^+)_{fi} = \sum_n S_{fn} S_{in}^* = \delta_{fi}, \tag{71.1}$$

where the suffix n labels the possible intermediate states.† This is the most general property of the S-matrix, which ensures that the orthonormality of the states is preserved in the reaction; cf. *QM*, §§125 and 144. In particular, the diagonal elements of equation (71.1) simply express the fact that the sum of the transition

† The actual meaning of δ_{fi} in (71.1) depends, of course, on the specific choice of quantum numbers and on the normalization of the wave functions of the system. It must be defined so that $\Sigma_f\, \delta_{if} = 1$.

probabilities from a given initial state to all final states is unity:

$$\sum_n |S_{ni}|^2 = 1.$$

Substituting in (71.1) the matrix elements in the form (64.2), we obtain

$$\begin{aligned} T_{fi} - T^*_{if} &= i(2\pi)^4 \sum_n \delta^{(4)}(P_f - P_n) T_{fn} T^*_{in} \\ &= i(2\pi)^4 \sum_n \delta^{(4)}(P_f - P_n) T^*_{nf} T_{ni}. \end{aligned} \tag{71.2}$$

The two equivalent forms on the right are obtained by writing the unitarity condition respectively as $\hat{S}\hat{S}^+ = 1$ and $\hat{S}^+\hat{S} = 1$, with opposite orders of the factors $\hat{S}$ and $\hat{S}^+$.

It should be noticed that the left-hand side of the equation is linear in the matrix elements of T, but the right-hand side is quadratic. If the interaction contains a small parameter (e.g. the electromagnetic interaction), the left-hand side is therefore of the first order of smallness and the right-hand side is of the second order. The latter may consequently be neglected in a first approximation; then

$$T_{fi} = T^*_{if}, \tag{71.3}$$

i.e. the matrix T is Hermitian.

In order to make the unitary condition (71.2) more specific, we must understand precisely what is meant by the summation over n. Let us do this for a two-particle collision, assuming that the conservation laws allow only elastic scattering. Then all the intermediate states in (71.2) are likewise "two-particle" states. Summation over these signifies integration over the intermediate momenta $\mathbf{p}''_1$, $\mathbf{p}''_2$, and summation over the spin quantum numbers (for example, the helicities) of the two particles, which we denote by λ'':

$$\sum_n = \int \frac{V^2 d^3p''_1 d^3p''_2}{(2\pi)^6} \sum_{\lambda''}.$$

Eliminating the delta functions in the same way as in §64, we obtain the "two-particle" unitarity condition in the form

$$T_{fi} - T^*_{if} = \frac{iV^2}{(2\pi)^2} \sum_{\lambda''} \frac{|\mathbf{p}|}{\varepsilon} \int T_{fn} T^*_{in} \varepsilon''_1 \varepsilon''_2 \, do'',$$

where $\mathbf{p}$ is the momentum and ε the total energy in the centre-of-mass system. The normalization volume does not appear after changing from the amplitudes T_{fi} to M_{fi} in accordance with (64.10):

$$M_{fi} - M^*_{if} = \frac{i}{(4\pi)^2} \sum_{\lambda''} \frac{|\mathbf{p}|}{\varepsilon} \int M_{fn} M^*_{in} \, do''. \tag{71.4}$$

Let the elastic scattering amplitude be defined so that

$$d\sigma = |\langle \mathbf{n}'\lambda'|f|\mathbf{n}\lambda\rangle|^2\, do', \tag{71.5}$$

where $\mathbf{n}$ and $\mathbf{n}'$ are the directions of the initial and final momenta, λ and λ' the initial and final spin quantum numbers. Comparison with (64.19) shows that

$$\langle \mathbf{n}'\lambda'|f|\mathbf{n}\lambda\rangle = \frac{1}{8\pi\varepsilon} M_{fi}, \tag{71.6}$$

and the unitarity condition (72.4) becomes

$$\langle \mathbf{n}'\lambda'|f|\mathbf{n}\lambda\rangle - \langle \mathbf{n}\lambda|f|\mathbf{n}'\lambda'\rangle^* = \frac{i|\mathbf{p}|}{2\pi}\sum_{\lambda''}\int \langle \mathbf{n}'\lambda'|f|\mathbf{n}''\lambda''\rangle\langle \mathbf{n}\lambda|f|\mathbf{n}''\lambda''\rangle^*\, do'', \tag{71.7}$$

which generalizes the familiar formula of the non-relativistic theory, *QM* (125.8).

The "amplitude of zero-angle elastic scattering" is the diagonal matrix element T_{ii}, in which the final states of the particles are the same as their initial states.† For this amplitude the unitarity condition (71.2) becomes

$$2\,\mathrm{im}\, T_{ii} = (2\pi)^4 \sum_n |T_{in}|^2 \delta^{(4)}(P_i - P_n). \tag{71.8}$$

The right-hand side of this equation differs only by a factor from the total cross-section for all possible processes of scattering from the given initial state i. For let this cross-section be denoted by σ_t; then summation of the probability (64.5) over states f and division by the flux density j gives

$$\sigma_t = \frac{(2\pi)^4 V}{j}\sum_n |T_{in}|^2 \delta^{(4)}(P_i - P_n),$$

whence

$$(2V/j)\,\mathrm{im}\, T_{ii} = \sigma_t.$$

The normalization volume is eliminated by putting $T_{ii} = M_{ii}/(2\varepsilon_1 V \cdot 2\varepsilon_2 V)$ (where ε_1 and ε_2 are the energies of the particles in the centre-of-mass system) and substituting j from (64.17):

$$\mathrm{im}\, M_{ii} = 2|\mathbf{p}|\varepsilon\sigma_t. \tag{71.9}$$

This formula expresses the *optical theorem*. If the elastic scattering amplitude (71.6) is used, the theorem takes the customary form

$$\mathrm{im}\langle \mathbf{n}\lambda|f|\mathbf{n}\lambda\rangle = |\mathbf{p}|\sigma_t/4\pi; \tag{71.10}$$

cf. *QM*, (142.10).

† It must be stressed that the matrix elements of T are concerned, not those of S; that is, the diagonal element is taken after subtracting the unit matrix from S.

If the S-matrix is given in the angular-momentum representation (partial amplitudes), it is diagonal with respect to J, and the unitarity condition can therefore be written separately for each value of J.

For example, if only elastic scattering is possible, the unitarity condition is

$$\sum_{\lambda''} \langle\lambda'|S^J|\lambda''\rangle\langle\lambda|S^J|\lambda''\rangle^* = \delta_{\lambda\lambda'}. \tag{71.11}$$

Because of the T invariance, the elastic scattering matrix is symmetric (cf. (69.10)), and hence can be reduced to diagonal form. The unitarity condition then requires that the diagonal elements should be of unit modulus, and they are customarily written in the form

$$S_n^J = \exp(2i\delta_{Jn}), \tag{71.12}$$

where the δ_{Jn} are real constants, depending on the energy (the suffix n labelling the diagonal elements for a given J). In the general case, when the number N of independent amplitudes exceeds the order of the (square) matrix S^J, the coefficients of the transformation which diagonalizes S^J depend on J and E (these coefficients then comprise not only the principal values of the matrix but also independent quantities equivalent to the original N quantities). If, however, the number N is equal to the order of the matrix S^J (and therefore to the number of its principal values), the diagonalization coefficients are universal constants, and the diagonalizing states have definite parities (but not, of course, definite helicities).

The condition (71.11), expressed in terms of the partial amplitudes $\langle\lambda'|f^J|\lambda\rangle$, is

$$\langle\lambda'|f^J|\lambda\rangle - \langle\lambda|f^J|\lambda'\rangle^* = 2i|\mathbf{p}| \sum_{\lambda''} \langle\lambda'|f^J|\lambda''\rangle\langle\lambda|f^J|\lambda''\rangle^*, \tag{71.13}$$

as is easily seen by substituting the expansion (68.13) in (71.7) and using the orthonormality of the D functions. If there is T invariance, the matrix $\langle\lambda'|f^J|\lambda\rangle$ is symmetric, and (71.13) becomes

$$\operatorname{im} \langle\lambda'|f^J|\lambda\rangle = |\mathbf{p}|\langle\lambda'|f^J f^{J+}|\lambda\rangle. \tag{71.14}$$

If the matrix is diagonalized, the diagonal elements are

$$f_n^J = \frac{1}{2i|\mathbf{p}|}(e^{2i\delta_{Jn}} - 1) = \frac{1}{|\mathbf{p}|} e^{i\delta_{Jn}} \sin \delta_{Jn}. \tag{71.15}$$

Finally, we may mention some consequences which follow from the unitary condition together with the requirement of CPT invariance. The latter shows that

$$T_{fi} = T_{\bar{i}\bar{f}}, \tag{71.16}$$

where $\bar{i}$ and $\bar{f}$ are states which differ from i and f in that all the particles are replaced by antiparticles (and helicities are reversed, and also angular momentum components if spherical waves are used). In particular, for the diagonal elements,

$$T_{ii} = T_{\bar{i}\bar{i}}.$$

It therefore follows from (71.8) and (71.9) that the total cross-section for all possible processes (with a given initial state) is the same for reactions of particles and of antiparticles.

In particular, the total disintegration probabilities (i.e. the lifetimes) of the particle and the antiparticle are equal. These results, together with the equality of particle and anti-particle masses (§11), are most important consequences of the *CPT* invariance of the interactions. A similar statement for each possible disintegration channel separately would require *CP* invariance also (see the end of §69).

PROBLEM

From the unitarity condition, find the relation between the phases of the partial amplitudes for photoproduction of pions from nucleons ($\gamma + N \to \pi + N$) and elastic scattering of pions by nucleons ($\pi + N \to \pi + N$), using the fact that πN scattering depends on strong interactions but photoproduction and γN scattering depend on an electromagnetic interaction.

SOLUTION. Let the partial amplitudes be denoted by

$$\langle \pi N|S|\gamma N\rangle = S_{\pi\gamma}, \qquad \langle \gamma N|S|\gamma N\rangle = S_{\gamma\gamma}, \qquad \langle \pi N|S|\pi N\rangle = S_{\pi\pi},$$

the suffix J and the helicity suffixes being omitted. Photoproduction is a first-order process with respect to the charge e, and γN scattering a second-order process; hence $S_{\pi\gamma} \sim e$, $S_{\gamma\gamma} - 1 \sim e^2$. The amplitude $S_{\pi\pi}$ is not small. The conditions (71.1) give, as far as terms in e,

$$S_{\pi\gamma}S^*_{\gamma\gamma} + S_{\pi\pi}S^*_{\gamma\pi} \approx S_{\pi\gamma} + S_{\pi\pi}S^*_{\gamma\pi} = 0, \tag{1}$$

$$S_{\pi\gamma}S^*_{\pi\gamma} + S_{\pi\pi}S^*_{\pi\pi} \approx S_{\pi\pi}S^*_{\pi\pi} = 1; \tag{2}$$

on the right-hand side of (2), 1 denotes a unit matrix in the spin variables. Because of T invariance the matrix $S_{\pi\pi}$ is symmetric, and $S_{\gamma\pi} = S_{\pi\gamma}$. Let us take the matrix $S_{\pi\pi}$ in diagonal form, i.e. with respect to pion states having definite parities; then it follows from (2) that the diagonal elements have the form $e^{2i\delta_\pi}$ with various constants δ_π. Then (1) gives for each element of the matrix $S_{\pi\gamma}$

$$S_{\pi\gamma}/S^*_{\pi\gamma} = -e^{2i\delta_\pi},$$

whence

$$S_{\pi\gamma} = \pm|S_{\pi\gamma}|ie^{i\delta_\pi}.$$

Thus the phase of the partial amplitude for photoproduction (in a state having a definite parity) is determined by the phase of elastic πN scattering.

CHAPTER VIII

INVARIANT PERTURBATION THEORY

§ 72. The chronological product

THE probabilities of various processes in collisions between particles whose interaction may be regarded as small are calculated by means of perturbation theory. In its ordinary form (in non-relativistic quantum mechanics), however, the formalism of this theory has the defect of not exhibiting explicitly the conditions of relativistic invariance. Although, when this formalism is applied to relativistic problems, the final result will satisfy these conditions, the calculations are considerably complicated by the non-invariant form of the intermediate expressions. The present chapter will deal with the development of a consistent relativistic perturbation theory free from this defect, first established by R. P. Feynman (1948–1949).

With a view to a second-quantization description of the system, let Φ denote its wave function in the "space" of occupation numbers for the various states of free particles. The Hamiltonian of the system is $\hat{H} = \hat{H}_0 + \hat{V}$, where $\hat{V}$ is the interaction operator. Let Φ_n be the eigenfunctions of the unperturbed Hamiltonian, each corresponding to certain definite values of all the occupation numbers. Any function Φ can be expanded as $\Phi = \Sigma\, C_n \Phi_n$. Then the exact wave equation

$$i\,\partial\Phi/\partial t = (\hat{H}_0 + \hat{V})\Phi \tag{72.1}$$

becomes a set of equations for the coefficients C_n:

$$i\dot{C}_n = \sum_m V_{nm}\, e^{i(E_n - E_m)t} C_m, \tag{72.2}$$

where V_{nm} are the time-independent matrix elements of the operator $\hat{V}$, and E_n the energy levels of the unperturbed system (cf. *QM*, §40).

By definition, the operator $\hat{V}$ does not depend explicitly on the time. The quantities

$$V_{nm}(t) = V_{nm}\, e^{i(E_n - E_m)t}, \tag{72.3}$$

on the other hand, may be regarded as matrix elements of the time-dependent operator

$$\hat{V}(t) = e^{i\hat{H}_0 t}\, \hat{V}\, e^{-i\hat{H}_0 t}. \tag{72.4}$$

This is said to be an operator in the *interaction representation*, as opposed to the original time-independent Schrödinger operator $\hat{V}$.† Now denoting the wave function in this new representation by the same letter Φ, we can write equations (72.2) symbolically as

$$i\dot{\Phi} = \hat{V}(t)\Phi. \tag{72.5}$$

The change in the wave function in this representation is due entirely to the action of the perturbation, i.e. it corresponds to processes which result from the interaction of the particles.

If $\Phi(t)$ and $\Phi(t+\delta t)$ are the values of Φ at two successive instants, (72.5) shows that

$$\begin{aligned}\Phi(t+\delta t) &= [1 - i\delta t \cdot \hat{V}(t)]\Phi(t)\\ &= e^{-i\delta t \cdot \hat{V}(t)}\Phi(t).\end{aligned}$$

Accordingly the value of Φ at any instant t_f can be expressed in terms of its value at some initial instant t_i ($<t_f$) by

$$\Phi(t_f) = \left(\prod_i^f e^{-i\delta t_\alpha \hat{V}(t_\alpha)}\right)\Phi(t_i), \tag{72.6}$$

where the product $\prod$ is the limit of the product over all the infinitesimal intervals δt_α between t_i and t_f. If $V(t)$ were an ordinary function, this limit would reduce simply to

$$\exp\left(-i\int_{t_i}^{t_f} V(t)\,dt\right),$$

but this result depends on the commutativity of the factors pertaining to different instants, which is assumed in changing from the product in (72.6) to the summation in the exponent. For the operator $\hat{V}(t)$ there is no such commutativity, and the reduction to an ordinary integral is not possible.

We can write (72.6) in the symbolic form

$$\Phi(t_f) = \mathrm{T}\exp\left\{-i\int_{t_i}^{t_f} \hat{V}(t)\,dt\right\}\Phi(t_i), \tag{72.7}$$

where T is the *chronological operator*, implying a certain "chronological" sequence

† It must be emphasized that the definition (72.4) makes use of the unperturbed Hamiltonian $\hat{H}_0$. In this it differs from the *Heisenberg representation* of operators, where

$$\hat{V}^H(t) = e^{i\hat{H}t}\hat{V}e^{-i\hat{H}t};$$

see *QM*, §13.

of time instants in the successive factors of the product (72.6). In particular, putting $t_i \to -\infty$, $t_f \to +\infty$, we have

$$\Phi(+\infty) = \hat{S}\Phi(-\infty), \tag{72.8}$$

where

$$\hat{S} = \mathrm{T}\exp\left\{-i\int_{-\infty}^{\infty} \hat{V}(t)\,dt\right\}. \tag{72.9}$$

The significance of writing the formally exact solution of the wave equation in the form (72.7)–(72.9) is that it easily leads to the series in powers of the perturbation

$$\hat{S} = \sum_{k=0}^{\infty} \frac{(-i)^k}{k!} \int_{-\infty}^{\infty} dt_1 \int_{-\infty}^{\infty} dt_2 \ldots \int_{-\infty}^{\infty} dt_k \cdot \mathrm{T}\{\hat{V}(t_1)\hat{V}(t_2)\ldots\hat{V}(t_k)\}. \tag{72.10}$$

Here, in each term, the kth power of the integral is written as a k-fold integral, and the symbol T signifies that in each range of values of the variables $t_1, t_2, \ldots, t_k$ the corresponding operators must be put in chronological order, with the value of t increasing from right to left.†

It is evident from the definition (72.8) that, if the system was in a state Φ_i (an assembly of free particles) before the collision, the probability amplitude for a transition to a state Φ_f (another assembly of free particles) is the matrix element S_{fi}. Thus these form the S-matrix.

The electromagnetic interaction operator has already been given in §43:

$$\hat{V} = e\int (\hat{j}\hat{A})\,d^3x. \tag{72.11}$$

Substitution of this in (72.9) gives

$$\hat{S} = \mathrm{T}\exp\left\{-ie\int (\hat{j}\hat{A})\,d^4x\right\}. \tag{72.12}$$

It is important to note that the operator (72.12) is relativistically invariant. This is seen from the facts that the integrand is a scalar, the integration over d^4x is invariant, and the time-ordering operation is invariant. The last point, however, needs further explanation.

The order of two time instants t_1 and t_2, i.e. the sign of $t_2 - t_1$, is independent of the frame of reference chosen if these instants relate to world points x_1 and x_2 separated by a time-like interval: $(x_2 - x_1)^2 > 0$. In such a case the invariance of

† The derivation of the rules of relativistic perturbation theory by means of the expansion (72.10) is due to F. J. Dyson (1949).

time-ordering necessarily follows. But if $(x_2 - x_1)^2 < 0$ (a space-like interval), we may have both $t_2 > t_1$ and $t_2 < t_1$ in different frames of reference.† Now two such points correspond to events between which there can be no causal connection. It is therefore evident that the operators of two physical quantities relating to such points must commute, since the non-commutativity of operators signifies, physically, that the corresponding quantities cannot be measured simultaneously, and this presupposes a physical connection between the two measurements. Thus the time-ordering of the product remains invariant in this case also: though a Lorentz transformation may reverse the sequence of time instants, the factors commute and can therefore be restored to their chronological order.‡

It is easy to see that the definition of the S-matrix given in this section necessarily satisfies the unitarity condition. Writing $\hat{S}$ as the chronological product in (72.6) and using the fact that $\hat{V}$ is Hermitian, we find that $\hat{S}^+$ is given by the product of similar factors, $\exp[i\delta t_\alpha \cdot \hat{V}(t_\alpha)]$ (with the opposite sign of the exponent), in the reverse of the chronological order. Thus all the factors cancel in pairs when $\hat{S}$ is multiplied by $\hat{S}^+$.

It should be noted that the unitarity of the operator $\hat{S}$ is ensured in this case because the Hamiltonian is Hermitian. The unitarity condition is actually more general than the assumptions on which the theory given here is based. It must be satisfied even in a quantum-mechanical description which makes no use of the concepts of the Hamiltonian and the wave functions.

§73. **Feynman diagrams for electron scattering**

We shall show by means of specific examples how the scattering matrix elements are calculated. These examples will facilitate the subsequent formulation of the general rules of invariant perturbation theory.

The current operator $\hat{j}$ contains the product of two electron ψ-operators. Hence processes might occur in the first order of perturbation theory which involve (in the initial and final states) only three particles: two electrons (the operator $\hat{j}$) and one photon (the operator $\hat{A}$). It is easily seen, however, that such processes cannot occur between free particles, being forbidden by the laws of conservation of energy and momentum. If p_1 and p_2 are the 4-momenta of the electrons, and k that of the photon, the conservation of 4-momentum would be represented by $k = p_2 - p_1$ or $k = p_2 + p_1$. But such equations are impossible, since for a photon $k^2 = 0$, whereas the square $(p_2 \pm p_1)^2$ is certainly not zero: if we calculate this invariant in

† Instead of using the terms "time-like" and "space-like", we often refer briefly to regions respectively inside and outside the light cone: all points x separated from a point x' by an interval such that $(x - x')^2 > 0$ lie within a double cone having its vertex at x'; points for which $(x - x')^2 < 0$ lie outside this cone.

‡ This statement needs refinement to avoid misunderstanding in its application to the product $\hat{V}(t_1)\hat{V}(t_2)\ldots$. Since the operator $\hat{V}$ itself is not gauge-invariant (it varies with $\hat{A}$), the factors $\hat{V}(t_1)$, $\hat{V}(t_2), \ldots$, though commuting in one gauge of the potential, may be non-commutative in some other gauge. The statements made above must therefore be formulated as asserting the possibility of choosing a gauge for the potential in which $\hat{V}(t_1)$ and $\hat{V}(t_2)$ commute outside the light cone. This reservation clearly has no effect on the invariance of the S-matrix: the scattering amplitudes, which are actual physical quantities, cannot depend on the gauge of the potential, a result which formally follows from the gauge invariance of the action integral (§43).

the rest frame of one of the electrons, we have

$$\begin{aligned}(p_2 \pm p_1)^2 &= 2(m^2 \pm p_1 p_2)\\ &= 2(m^2 \pm \varepsilon_1\varepsilon_2 \mp \mathbf{p}_1 \cdot \mathbf{p}_2)\\ &= 2m(m \pm \varepsilon_2),\end{aligned}$$

and, since $\varepsilon_2 > m$, it follows that

$$(p_2 + p_1)^2 > 0, \qquad (p_2 - p_1)^2 < 0. \tag{73.1}$$

Thus the first non-vanishing (non-diagonal) elements of the S-matrix can appear only in the second order of perturbation theory. All the relevant processes are comprised in the second-order operator obtained by expanding the expression (72.12):

$$\hat{S}^{(2)} = -\frac{e^2}{2!}\int\int d^4x\, d^4x' \cdot \mathrm{T}(\hat{j}^\mu(x)\hat{A}_\mu(x)\hat{j}^\nu(x')\hat{A}_\nu(x')).$$

Since the electron and photon operators commute, the T product can be resolved into two:

$$\hat{S}^{(2)} = -\frac{e^2}{2!}\int\int d^4x\, d^4x' \cdot \mathrm{T}(\hat{j}^\mu(x)\hat{j}^\nu(x'))\mathrm{T}(\hat{A}_\mu(x)\hat{A}_\nu(x')). \tag{73.2}$$

As a first example, let us consider elastic scattering of two electrons. In the initial state there are two electrons with 4-momenta p_1 and p_2, in the final state two electrons with other 4-momenta p_3 and p_4. It is also assumed that all the electrons are in definite spin states; the spin variable indices will be everywhere omitted, for brevity.

Since there are no photons in either state, the required matrix element of the T product of the photon operators is the diagonal element $\langle 0|\ldots|0\rangle$, where $|0\rangle$ denotes the photon vacuum state. This value of the T product averaged over the vacuum is (for each pair of indices μ, ν) a definite function of the coordinates of the two points x and x'. Since 4-space is homogeneous, the coordinates can appear only as the difference $x - x'$. The tensor

$$D_{\mu\nu}(x - x') = i\langle 0|\mathrm{T}A_\mu(x)A_\nu(x')|0\rangle \tag{73.3}$$

is called the *photon propagation function* or *photon propagator.* It will be calculated in §76.

For the T product of the electron operators, we have to calculate the matrix element

$$\langle 34|\mathrm{T}j^\mu(x)j^\nu(x')|12\rangle, \tag{73.4}$$

where the symbols $|12\rangle$, $|34\rangle$ denote states in which pairs of electrons have the corresponding momenta. This element also can be represented as a vacuum expectation value, by using the obvious relation

$$\langle 2|F|1\rangle = \langle 0|a_2 F a_1^+|0\rangle,$$

where $\hat{F}$ is any operator, $\hat{a}_1^+$ the creation operator for the first electron and $\hat{a}_2$ the annihilation operator for the second electron. Hence, instead of (73.4), we can calculate the quantity

$$\langle 0|a_3 a_4 \mathrm{T}(j^\mu(x) j^\nu(x')) a_2^+ a_1^+|0\rangle, \tag{73.5}$$

the indices 1, 2, ... being abbreviations for $p_1, p_2, \ldots$.

Each of the two current operators is a product, $\hat{j} = \hat{\bar{\psi}}\gamma\hat{\psi}$, and each of the ψ-operators is a sum:

$$\hat{\psi} = \sum_{\mathbf{p}} (\hat{a}_{\mathbf{p}}\psi_p + \hat{b}_{\mathbf{p}}^+\psi_{-p}), \qquad \hat{\bar{\psi}} = \sum_{\mathbf{p}} (\hat{a}_{\mathbf{p}}^+\bar{\psi}_p + \hat{b}_{\mathbf{p}}\bar{\psi}_{-p}); \tag{73.6}$$

the second term in each expression contains the positron operators, which in the present case "do not act". Hence the product $\hat{j}^\mu(x)\hat{j}^\nu(x')$ is a sum of terms, each containing the product of two operators $\hat{a}_{\mathbf{p}}$ and two $\hat{a}_{\mathbf{p}}^+$. These operators must annihilate electrons 1 and 2, and create electrons 3 and 4. They must therefore be the operators $\hat{a}_1$, $\hat{a}_2$, $\hat{a}_3^+$, $\hat{a}_4^+$, which are said to *contract* with the "external" operators $\hat{a}_1^+$, $\hat{a}_2^+$, $\hat{a}_3$, $\hat{a}_4$ in (73.5) and cancel according to the equations

$$\langle 0|a_{\mathbf{p}} a_{\mathbf{p}}^+|0\rangle = 1. \tag{73.7}$$

Four terms result, according to the ψ-operators from which $\hat{a}_1$, $\hat{a}_2$, $\hat{a}_3^+$, $\hat{a}_4^+$ in (73.5) are taken:

$$(73.5) = a_3 a_4(\bar{\psi}\gamma^\mu\psi)(\bar{\psi}'\gamma^\nu\psi')a_2^+a_1^+ + a_3 a_4(\bar{\psi}\gamma^\mu\psi)(\bar{\psi}'\gamma^\nu\psi')a_2^+a_1^+ +$$
$$+ a_3 a_4(\bar{\psi}\gamma^\mu\psi)(\bar{\psi}'\gamma^\nu\psi')a_2^+a_1^+ + a_3 a_4(\bar{\psi}\gamma^\mu\psi)(\bar{\psi}'\gamma^\nu\psi')a_2^+a_1^+, \tag{73.8}$$

where $\psi = \psi(x)$, $\psi' = \psi(x')$, and the brackets join operators which contract, i.e. those from which a pair of operators $\hat{a}$, $\hat{a}^+$ is taken for the cancellation according to (73.7). In each term we can bring the conjugate operators together in pairs ($\hat{a}_1\hat{a}_1^+$, etc.) by successive interchanges of $\hat{a}_1, \hat{a}_2, \ldots$, and the mean value of their product is then equal to the product of the mean values (73.7). Since all these operators anticommute (1, 2, 3, 4 being different states),† we find that the matrix element (73.4) is

$$\langle 34|\mathrm{T}j^\mu(x)j^\nu(x')|12\rangle = (\bar{\psi}_4\gamma^\mu\psi_2)(\bar{\psi}_3'\gamma^\nu\psi_1') + (\bar{\psi}_3\gamma^\mu\psi_1)(\bar{\psi}_4'\gamma^\nu\psi_2') -$$
$$- (\bar{\psi}_3\gamma^\mu\psi_2)(\bar{\psi}_4'\gamma^\nu\psi_1') - (\bar{\psi}_4\gamma^\mu\psi_1)(\bar{\psi}_3'\gamma^\nu\psi_2'). \tag{73.9}$$

† Because of this anticommutativity, the operators $\hat{j}(x)$ and $\hat{j}(x')$ may here be considered to commute (in the calculation of the matrix element), and the T product symbol may therefore be omitted.

The sign of the entire sum is arbitrary, and depends on the order of the "external" electron operators in (73.5). This is in accordance with the fact that the sign of the matrix element for scattering of identical fermions is itself arbitrary. The relative sign of the various terms in (73.9), of course, does not depend on the order of the external operators.

The two terms in each line of (73.9) differ only by a simultaneous interchange of the indices μ, ν and the arguments x, x'. This interchange clearly does not affect the matrix element (73.3), in which the order of factors is still established by the symbol T. Hence when (73.3) is multiplied by (73.9) and integrated over $d^4x\,d^4x'$, the four terms in (73.9) give two pairs of equal results, and the matrix element is therefore

$$S_{fi} = ie^2 \int\int d^4x\,d^4x'\,D_{\mu\nu}(x-x')\{(\bar{\psi}_4\gamma^\mu\psi_2)(\bar{\psi}_3'\gamma^\nu\psi_1') - (\bar{\psi}_4\gamma^\mu\psi_1)(\bar{\psi}_3'\gamma^\nu\psi_2')\}; \tag{73.10}$$

the factor $\frac{1}{2}$ has now disappeared.

The electron wave functions are the plane waves (64.8). The expression in the braces is therefore

$$\begin{aligned}\{\ldots\} &= (\bar{u}_4\gamma^\mu u_2)(\bar{u}_3\gamma^\nu u_1)\,e^{-i(p_2-p_4)x-i(p_1-p_3)x'} - \\ &\quad -(\bar{u}_4\gamma^\mu u_1)(\bar{u}_3\gamma^\nu u_2)\,e^{-i(p_1-p_4)x-i(p_2-p_3)x'} \\ &= \{(\bar{u}_4\gamma^\mu u_2)(\bar{u}_3\gamma^\nu u_1)\,e^{-i[(p_2-p_4)+(p_3-p_1)]\xi/2} - \\ &\quad -(\bar{u}_4\gamma^\mu u_1)(\bar{u}_3\gamma^\nu u_2)\,e^{-i[(p_1-p_4)+(p_3-p_2)]\xi/2}\}\,e^{-i(p_1+p_2-p_3-p_4)X},\end{aligned}$$

where $X = \frac{1}{2}(x+x')$, $\xi = x - x'$. The integration over $d^4x\,d^4x'$ is replaced by one over $d^4\xi\,d^4X$. The integral over d^4X gives a delta function, so that $p_1+p_2 = p_3+p_4$. Then, changing from the matrix S to the matrix M (§64), we have finally for the scattering amplitude

$$M_{fi} = e^2\{(\bar{u}_4\gamma^\mu u_2)D_{\mu\nu}(p_4-p_2)(\bar{u}_3\gamma^\nu u_1) - (\bar{u}_4\gamma^\mu u_1)D_{\mu\nu}(p_4-p_1)(\bar{u}_3\gamma^\nu u_2)\}. \tag{73.11}$$

Here we have used the photon propagation function in the momentum representation:

$$D_{\mu\nu}(k) = \int D_{\mu\nu}(\xi)\,e^{ik\xi}\,d^4\xi. \tag{73.12}$$

Each of the two terms in the amplitude (73.11) can be symbolically represented by means of a *Feynman diagram*: the first term by

$$e^2(\bar{u}_4\gamma^\mu u_2)\,D_{\mu\nu}(k)\,(\bar{u}_3\gamma^\nu u_1) =$$

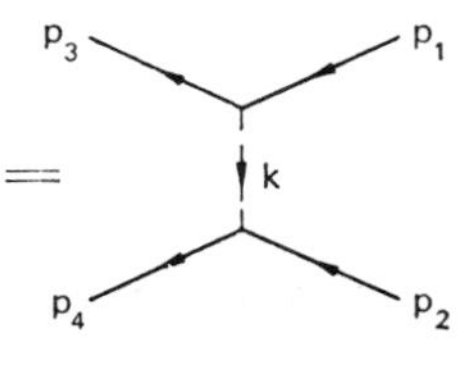

(73.13)

Each point of intersection of lines (a *vertex* of the diagram) has a corresponding factor γ. The "incoming" continuous lines towards a vertex represent the initial electrons, which are associated with the factors u, the bispinor amplitudes of the corresponding electron states. The "outgoing" continuous lines leaving a vertex are the final electrons, and correspond to the factors $\bar{u}$. When the diagram is "read", these factors are written from left to right in the order of movement along the continuous lines against the direction of the arrows. The two vertices are joined by a broken line which represents a *virtual* (intermediate) photon "emitted" at one vertex and "absorbed" at the other, and corresponds to the factor $-iD_{\mu\nu}(k)$. The 4-momentum of the virtual photon k is determined by the "conservation of 4-momentum" at the vertex: the total momenta of the incoming and outgoing lines are equal. In this case $k = p_1 - p_3 = p_4 - p_2$. As well as the factors mentioned, the whole diagram is also assigned a factor $(-ie)^2$ (the exponent being the number of vertices in the diagram), and then represents a term in iM_{fi}. Similarly, the second term in (73.11) is represented by the diagram

$e^2(\bar{u}_4\gamma^\mu u_1)D_{\mu\nu}(k')(\bar{u}_3\gamma^\nu u_2) =$ 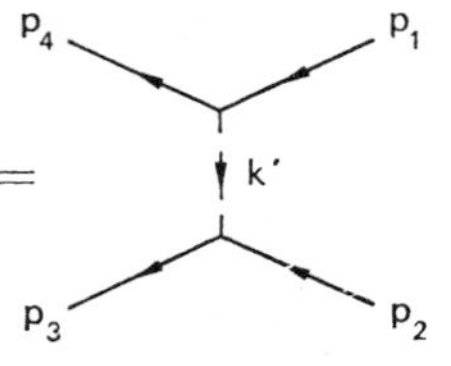 (73.14)

with $k' = p_1 - p_4 = p_3 - p_2$. It does not matter whether the diagram is read from the end of p_3 or from that of p_4. The resulting expressions are equal, because the tensor $D_{\mu\nu}$ is symmetrical. The choice of direction for the virtual photon line is also immaterial: a change in its direction simply reverses the sign of k, which does not matter, since the functions $D_{\mu\nu}(k)$ are even (see §76).

The lines corresponding to the initial and final particles are called the *external lines* or *free ends* of the diagram. The diagrams (73.13) and (73.14) differ by the interchange of two electron free ends (p_3 and p_4). This interchange of two fermions reverses the sign of the diagram, in accordance with the fact that the two terms appear with opposite signs in the amplitude (73.11).

We shall everywhere use Feynman diagrams in this momentum representation, but they can also be associated with the terms in the scattering amplitude in the original coordinate representation (the integrals (73.10)). Here the electron amplitudes are replaced by the corresponding coordinate wave functions, and the propagators are in the coordinate representation. Each vertex corresponds to one of the variables of integration (x or x' in (73.10)); the factors assigned to the lines that meet at a vertex are taken as functions of the corresponding variable.

Let us now consider the mutual scattering of an electron and a positron; their initial momenta will be denoted by p_- and p_+ respectively, and their final momenta by p'_- and p'_+.

The positron creation and annihilation operators appear in the ψ-operators (73.6) together with the electron annihilation and creation operators respectively. Whereas in the previous case the operator $\hat{\psi}$ annihilated the two initial particles and $\hat{\bar{\psi}}$ created the two final particles, here these operators act oppositely with regard to electrons and positrons. The conjugate function $\bar{\psi}(-p_+)$ will therefore now describe the initial positron, and $\psi(-p'_+)$ the final positron, both being functions of the 4-momentum with reversed sign. Taking account of this difference, we obtain the scattering amplitude†

$$M_{fi} = -e^2(\bar{u}(p'_-)\gamma^\mu u(p_-))D_{\mu\nu}(p_- - p'_-)(\bar{u}(-p_+)\gamma^\nu u(-p'_+)) + \\ + e^2(\bar{u}(-p_+)\gamma^\mu u(p_-))D_{\mu\nu}(p_- + p_+)(\bar{u}(p'_-)\gamma^\nu u(-p'_+)). \quad (73.15)$$

The two terms in this expression are represented by the following diagrams:

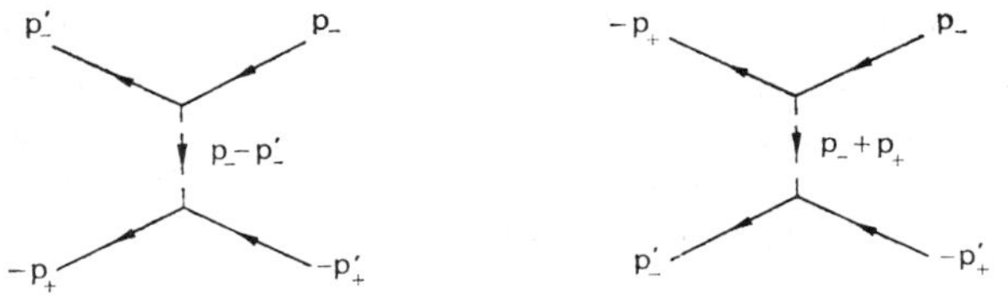

(73.16)

The rules for constructing the diagrams are altered only as regards the positrons. The incoming and outgoing continuous lines are again associated with factors u and $\bar{u}$ respectively. Now, however, the incoming lines correspond to final positrons and the outgoing lines to initial positrons, the momenta of all the positrons being taken with reversed sign.

The difference between the two diagrams (73.16) should be noted. In the first diagram, lines of the initial and final electrons meet at one vertex, and those of the two positrons at the other. In the second diagram, initial electron and positron lines meet at one vertex, and final lines at the other. The upper vertex represents annihilation of a pair with emission of a virtual photon; the lower vertex represents the creation of a pair from this photon.

This difference affects the properties of the virtual photons in the two diagrams. In the first diagram ("scattering" type), the 4-momentum of the virtual photon is the difference between those of the two electrons (or positrons); hence $k^2 < 0$ (cf. (73.1)). In the second diagram ("annihilation" type), $k' = p_- + p_+$, and hence $k'^2 > 0$. Here it should be noted that for a virtual photon we always have $k^2 \neq 0$, unlike a real photon, for which $k^2 = 0$.

If the colliding particles are not identical and also not a particle and its antiparticle (for instance, an electron and a muon), then the scattering amplitude is represented by a single diagram:

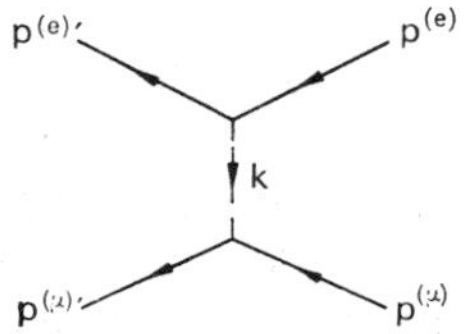

(73.17)

† The sign of the whole amplitude is definite in the scattering of non-identical particles, being determined by the fact that in (73.5) the "external" operators must be arranged so that the two electron operators are both at the ends: $\langle 0|a'b' \ldots b^+a^+|0\rangle$ (or both in the middle); this condition ensures the "same sign" of the initial and final vacuum states. The sign of the amplitude can also be verified from the non-relativistic limit: we shall see later (§81) that, in this limit, the second term in (73.15) tends to zero, and the first term tends to the Born amplitude of Rutherford scattering.

There can be no annihilation or exchange type diagram in this case. The same result can be obtained analytically by writing the current operator as the sum of electron and muon currents:

$$\hat{j} = \hat{j}^{(e)} + \hat{j}^{(\mu)}$$
$$= (\hat{\bar{\psi}}^{(e)}\gamma\hat{\psi}^{(e)}) + (\hat{\bar{\psi}}^{(\mu)}\gamma\hat{\psi}^{(\mu)})$$

and taking, in the product $\hat{j}^{(\mu)}(x)\hat{j}^{(\nu)}(x')$, the matrix elements of terms which give the required annihilations and creations of particles.

Let us now consider first-order processes, which, as mentioned at the beginning of this section, are forbidden by the conservation of 4-momentum. The matrix elements of the operator

$$\hat{S}^{(1)} = -ie\int \hat{j}(x)\hat{A}(x)\, d^4x \tag{73.18}$$

for such transitions correspond to the creation or annihilation of three real particles (two electrons and one photon) at "the same point x". They occur by the contraction of the operators $\hat{\psi}(x)$ and $\hat{\bar{\psi}}(x)$ at the same point, and are expressed, for example in the case of photon emission, by integrals of the form

$$S_{fi} = -ie\int \bar{\psi}_2(x)\psi_1(x)(\gamma A^*(x))\, d^4x,$$

which vanish because the integrand includes the factor $\exp[-i(p_1-p_2-k)x]$ with a non-zero exponent. In the language of Feynman diagrams, this means that diagrams with three free ends such as

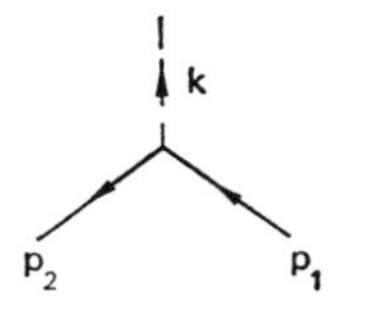

(73.19)

are zero.

For the same reason, second-order processes involving six particles in the initial state (or in the final state) are impossible. In the matrix element S_{fi} for such a transition, the integral over $d^4x\, d^4x'$ would separate into a product of two vanishing integrals over d^4x and d^4x' of products of three wave functions taken at the same point. In other words, the corresponding diagram would separate into two independent diagrams of the type (73.19).

§ 74. Feynman diagrams for photon scattering

Let us now consider another second-order effect: the scattering of a photon by an electron (the *Compton effect*). In the initial state let the photon and the electron have 4-momenta k_1 and p_1, and in the final state k_2 and p_2 (and also definite polarizations, which will be omitted for brevity).

The photon matrix element is

$$\langle 2|\mathrm{T}A_\mu(x)A_\nu(x')|1\rangle = \langle 0|c_2 \mathrm{T}A_\mu(x)A_\nu(x')c_1^+|0\rangle, \tag{74.1}$$

where

$$\hat{A} = \sum_{\mathbf{k}} (\hat{c}_{\mathbf{k}} A_{\mathbf{k}} + \hat{c}_{\mathbf{k}}^+ A_{\mathbf{k}}^*).$$

Contraction of the external and internal operators gives

$$\begin{aligned}(74.1) &= \underbrace{c_2 A_\mu}\underbrace{A'_\nu c_1^+} + \underbrace{c_2 \overbrace{A_\mu A'_\nu} c_1^+} \\ &= A^*_{2\mu} A'_{1\nu} + A_{1\mu} A^{*\prime}_{2\nu},\end{aligned} \tag{74.2}$$

where we have used the commutativity of the operators $\hat{c}_1$ and $\hat{c}_2^+$; for the same reason, the symbol T can here be omitted.

The electron matrix element is

$$\langle 2|\mathrm{T}j^\mu(x)j^\nu(x')|1\rangle = \langle 0|a_2 \mathrm{T}(\bar{\psi}\gamma^\mu\psi)(\bar{\psi}'\gamma^\nu\psi')a_1^+|0\rangle. \tag{74.3}$$

This involves four ψ-operators. Only two are concerned with the annihilation of electron 1 and the creation of electron 2, and will be contracted with the operators $\hat{a}_1^+$ and $\hat{a}_2$. These may be $\hat{\bar{\psi}}'$, $\hat{\psi}$ or $\hat{\psi}'$, $\hat{\bar{\psi}}$ (but not $\hat{\psi}$, $\hat{\bar{\psi}}$ or $\hat{\psi}'$, $\hat{\bar{\psi}}'$: the creation and annihilation of two real electrons and one real photon at the same point x or x' would give an expression equal to zero). By carrying out the two possible ways of contraction, we obtain two terms in the matrix element (74.3). These will first be written on the assumption that $t > t'$:

$$(74.3) = \underbrace{a_2(\bar{\psi}}\gamma^\mu\psi)(\bar{\psi}'\gamma^\nu\underbrace{\psi')a_1^+} + \underbrace{a_2(\bar{\psi}\gamma^\mu\overbrace{\psi)(\bar{\psi}'}}\gamma^\nu\psi')a_1^+. \tag{74.4}$$

In the first term the contracted operators are $\hat{a}_2\hat{\bar{\psi}} \to \hat{a}_2\hat{a}_2^+\bar{\psi}_2$, $\hat{\psi}'\hat{a}_1^+ \to \hat{a}_1\hat{a}_1^+\psi'_1$. Since the operators $\hat{a}_2\hat{a}_2^+$ and $\hat{a}_1\hat{a}_1^+$ are diagonal and appear at the end of the products, they can be replaced by the vacuum expectation value, i.e. unity. To make a similar transformation in the second term of (74.4), the operator $\hat{a}_2^+$ must first be "pulled" to the left, and $\hat{a}_1$ to the right. This is done by means of the commutation rules for the operators $\hat{a}_{\mathbf{p}}$, $\hat{a}_{\mathbf{p}}^+$:

$$\left.\begin{aligned}&\{\hat{a}_{\mathbf{p}}, \hat{\psi}\}_+ = \{\hat{a}_{\mathbf{p}}^+, \hat{\bar{\psi}}\}_+ = 0, \\ &\{\hat{a}_{\mathbf{p}}, \hat{\bar{\psi}}\}_+ = \bar{\psi}_p, \quad \{\hat{a}_{\mathbf{p}}^+, \hat{\psi}\}_+ = \psi_p.\end{aligned}\right\} \tag{74.5}$$

Then (74.4) becomes

$$\langle 0|(\bar{\psi}_2\gamma^\mu\hat{\psi})(\hat{\bar{\psi}}'\gamma^\nu\psi'_1) - (\hat{\bar{\psi}}\gamma^\mu\psi_1)(\bar{\psi}'_2\gamma^\nu\hat{\psi}')|0\rangle, \qquad t > t'; \tag{74.6}$$

only the operator factors are averaged, of course. Similarly, for $t < t'$, we obtain an expression differing by the interchange of μ and ν and of the primed and unprimed symbols:

$$\langle 0| - (\hat{\bar{\psi}}'\gamma^\nu\psi_1')(\bar{\psi}_2\gamma^\mu\hat{\psi}) + (\bar{\psi}_2'\gamma^\nu\hat{\psi}')(\hat{\bar{\psi}}\gamma^\mu\psi_1)|0\rangle, \quad t < t'. \tag{74.7}$$

The two expressions (74.6) and (74.7) can be written as one by using the chronological product of the ψ-operators:

$$\left.\begin{aligned} \mathrm{T}\hat{\psi}_i(x)\hat{\bar{\psi}}_k(x') &= \hat{\psi}_i(x)\hat{\bar{\psi}}_k(x'), \qquad & t' < t; \\ &= -\hat{\bar{\psi}}_k(x')\hat{\psi}_i(x), \qquad & t' > t, \end{aligned}\right\} \tag{74.8}$$

where i and k are bispinor indices. Then the first and second terms in (74.6), (74.7) can be combined in the form

$$\bar{\psi}_2\gamma^\mu\langle 0|\mathrm{T}\psi\cdot\bar{\psi}'|0\rangle\gamma^\nu\psi_1' + \bar{\psi}_2'\gamma^\nu\langle 0|\mathrm{T}\psi'\cdot\bar{\psi}|0\rangle\gamma^\mu\psi_1, \tag{74.9}$$

where $\psi\cdot\bar{\psi}$ denotes the matrix $\psi_i\bar{\psi}_k$.

It should be noted that, in the natural definition (74.8), the operator products are taken with opposite signs for $t < t'$ and $t > t'$. In this respect it differs from the definition of the T product which has been used for the operators $\hat{A}$ and $\hat{j}$. This difference arises because the fermion operators $\hat{\psi}$ and $\hat{\bar{\psi}}$ anticommute outside the light cone, unlike the commuting boson operators $\hat{A}$ and the bilinear operators $\hat{j} = \hat{\bar{\psi}}\gamma\hat{\psi}$.† This procedure ensures the relativistic invariance of the definition (74.8). A formal proof of the commutation rules for the ψ-operators will be given in §75.‡

We shall define the *electron propagation function* or *electron propagator*, a bispinor of rank two, as

$$G_{ik}(x - x') = -i\langle 0|\mathrm{T}\psi_i(x)\bar{\psi}_k(x')|0\rangle. \tag{74.10}$$

Then the electron matrix element becomes

$$\langle 2|\mathrm{T}j^\mu(x)j^\nu(x')|1\rangle = i\bar{\psi}_2\gamma^\mu G\gamma^\nu\psi_1' + i\bar{\psi}_2'\gamma^\nu G\gamma^\mu\psi_1. \tag{74.11}$$

On multiplication by the photon matrix element (74.1) and integration over $d^4x\, d^4x'$, the two terms in (74.11) give the same result, and so we have

$$\begin{aligned} S_{fi} = -ie^2 \int\int d^4x\, d^4x'\, \bar{\psi}_2(x)\gamma^\mu G(x - x')\gamma^\nu\psi_1(x') \times \\ \times \{A_{2\mu}^*(x)A_{1\nu}(x') + A_{2\nu}^*(x')A_{1\mu}(x)\}. \end{aligned} \tag{74.12}$$

† The ψ-operators themselves, it will be recalled, correspond to no measurable physical quantities, and therefore need not commute outside the light cone.

‡ The T product of any number of ψ-operators may be defined similarly. It is equal to the product of all the operators arranged in order of increasing time from right to left, the sign being determined by the parity of the interchange needed to obtain this order from the order shown under the T product symbol. Accordingly, this sign changes when any two ψ-operators are interchanged; for example,

$$T\hat{\psi}_i(x)\hat{\bar{\psi}}_k(x') = -T\hat{\bar{\psi}}_k(x')\hat{\psi}_i(x).$$

Substituting the plane waves (64.8), (64.9) for the electron and photon wave functions and separating the delta function as in (73.10), we obtain finally the scattering amplitude

$$M_{fi} = -4\pi e^2 \bar{u}_2\{(\gamma e_2^*)G(p_1+k_1)(\gamma e_1) + (\gamma e_1)G(p_1-k_2)(\gamma e_2^*)\}u_1, \tag{74.13}$$

where e_1, e_2 are the photon polarization 4-vectors and $G(p)$ the electron propagator in the momentum representation.

The two terms in this expression are represented by the following Feynman diagrams:

$$4\pi e^2 \bar{u}_2 (\gamma e_2^*) G(f) (\gamma e_1) u_1 =$$

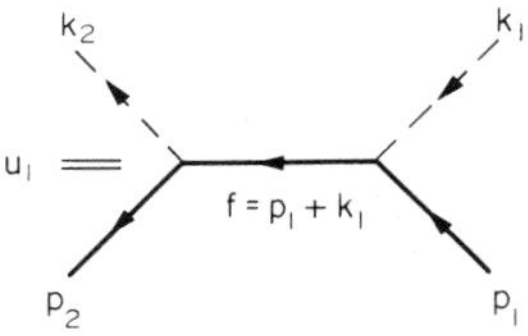

(74.14)

$$4\pi e^2 \bar{u}_2 (\gamma e_1) G(f') (\gamma e_2^*) u_1 =$$

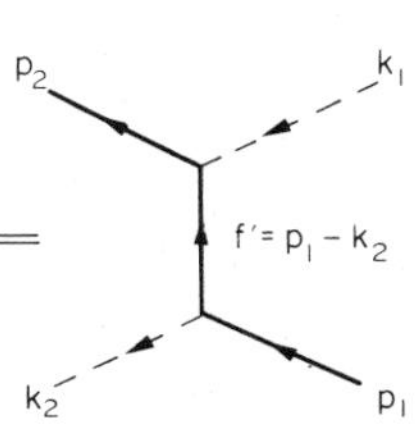

The broken-line free ends of the diagrams correspond to real photons; the incoming lines (initial photon) are associated with a factor $\sqrt{(4\pi)}e$, and the outgoing lines (final photon) with a factor $\sqrt{(4\pi)}e^*$, where e is the polarization 4-vector. In the first diagram, the initial photon is absorbed together with the initial electron, and the final photon is emitted together with the final electron. In the second diagram, the final photon is emitted together with the annihilation of the initial electron, and the initial photon is absorbed together with the creation of the final electron.

The continuous internal line joining the two vertices represents a virtual electron whose 4-momentum is determined by the conservation of the 4-momentum at the vertices. This line is associated with a factor $iG(f)$. Unlike the 4-momentum of a real particle, that of the virtual electron has a square which is not equal to m^2. If the invariant f^2 is considered, for example, in the rest frame of the electron, we easily find that

$$f^2 = (p_1+k_1)^2 > m^2, \qquad f'^2 = (p_1-k_2)^2 < m^2. \tag{74.15}$$

§75. **The electron propagator**

The propagation functions or propagators defined in §§73 and 74 are of fundamental importance in the formalism of quantum electrodynamics. The photon propagator $D_{\mu\nu}$ is a basic characteristic of the interaction of two electrons, as is shown by its position in the electron scattering amplitude, in which it is multiplied by the transition currents of the two particles. The electron propagator plays a similar part in the electron–photon interaction.

Let us now calculate the actual values of the propagators, taking first the electron propagator. Let the operator $\gamma\hat{p} - m$, where $\hat{p}_\mu = i\partial_\mu$, act on the function

$$G_{ik}(x - x') = -i\langle 0|\mathrm{T}\psi_i(x)\bar{\psi}_k(x')|0\rangle, \tag{75.1}$$

i and k being bispinor indices. Since $\hat{\psi}(x)$ satisfies Dirac's equation $(\gamma\hat{p} - m)\hat{\psi}(x) = 0$, we find that the result is zero at all points x, except those for which $t = t'$. The reason is that $G(x - x')$ tends to different limits as $t \to t' + 0$ and $t \to t' - 0$: according to the definition (74.8) these limits are respectively

$$-i\langle 0|\psi_i(\mathbf{r}, t)\bar{\psi}_k(\mathbf{r}', t)|0\rangle \quad \text{and} \quad +i\langle 0|\bar{\psi}_k(\mathbf{r}', t)\psi_i(\mathbf{r}, t)|0\rangle,$$

and, as we shall see, they are not the same on the light cone. This causes an additional delta-function term to appear in the derivative $\partial G/\partial t$:

$$\frac{\partial G}{\partial t} = -i\left\langle 0\left|\mathrm{T}\,\frac{\partial\psi_i(x)}{\partial t}\,\bar{\psi}_k(x')\right|0\right\rangle + \delta(t - t')(G_{t\to t'+0} - G_{t\to t'-0}). \tag{75.2}$$

Since the derivative with respect to t appears in the operator $\gamma\hat{p} - m$ in the form $i\gamma^0\,\partial/\partial t$, we therefore have

$$(\gamma\hat{p} - m)_{ik}G_{kl}(x - x') = \delta(t - t')\gamma^0_{ik}\langle 0|\{\psi_k(\mathbf{r}, t), \bar{\psi}_l(\mathbf{r}', t)\}_+|0\rangle. \tag{75.3}$$

The anticommutator is calculated as follows. On multiplying the operators $\hat{\psi}(\mathbf{r}, t)$ and $\hat{\bar{\psi}}(\mathbf{r}', t)$ (see (73.6)) and using the rules for interchange of the fermion operators $\hat{a}_\mathbf{p}$, $\hat{b}_\mathbf{p}$, we find

$$\{\hat{\psi}_i(\mathbf{r}, t), \hat{\bar{\psi}}_k(\mathbf{r}', t)\}_+ = \sum_\mathbf{p} [\psi_{pi}(\mathbf{r})\bar{\psi}_{pk}(\mathbf{r}') + \psi_{-p,i}(\mathbf{r})\bar{\psi}_{-p,k}(\mathbf{r}')], \tag{75.4}$$

where $\psi_{\pm p}(\mathbf{r})$ are wave functions without the time factor; as in §§73 and 74, the polarization indices are omitted for brevity. The set of all functions $\psi_{\pm p}(\mathbf{r})$, which are eigenfunctions of the electron Hamiltonian, forms a complete set of normalized functions, and according to the general properties of these (cf. *QM*, (5.12)) we have

$$\sum_\mathbf{p} [\psi_{pi}(\mathbf{r})\psi^*_{pk}(\mathbf{r}') + \psi_{-p,i}(\mathbf{r})\psi^*_{-p,k}(\mathbf{r}')] = \delta_{ik}\delta(\mathbf{r} - \mathbf{r}'). \tag{75.5}$$

The sum on the right-hand side of (75.4) differs from that in (75.5) in that ψ^*_k is replaced by $(\psi^*\gamma^0)_k$, and its value is $\gamma^0_{ik}\delta(\mathbf{r} - \mathbf{r}')$. Thus

$$\{\hat{\psi}_i(\mathbf{r}, t), \hat{\bar{\psi}}_k(\mathbf{r}', t)\}_+ = \delta(\mathbf{r} - \mathbf{r}')\gamma^0_{ik}. \tag{75.6}$$

From this formula it follows, in particular, that the operators $\hat{\psi}$ and $\hat{\bar{\psi}}$ anticommute outside the light cone, as stated in §74. When $(x-x')^2<0$ there is always a frame of reference in which $t=t'$; if then $\mathbf{r}\neq\mathbf{r}'$, the anticommutator (75.6) is in fact zero.

Substituting (75.6) in (75.3) (and omitting the bispinor indices), we have finally†

$$(\gamma\hat{p}-m)G(x-x')=\delta^{(4)}(x-x'). \tag{75.7}$$

Thus the electron propagator satisfies Dirac's equation with a delta function on the right-hand side. Mathematically speaking, therefore, it is the Green's function for Dirac's equation.

We shall later be concerned not with the function $G(\xi)$ itself ($\xi=x-x'$), but with its Fourier components:

$$G(p)=\int G(\xi)\,e^{ip\xi}\,d^4\xi \tag{75.8}$$

(the propagator in the momentum representation). Taking the Fourier component of each side of (75.7), we find that $G(p)$ satisfies the algebraic equations

$$(\gamma p-m)G(p)=1, \tag{75.9}$$

the solution of which is

$$G(p)=\frac{\gamma p+m}{p^2-m^2}. \tag{75.10}$$

The four components of the 4-vector p in $G(p)$ are independent variables, not related by $p^2\equiv p_0^2-\mathbf{p}^2=m^2$. Writing the denominator in (75.10) as $p_0^2-(\mathbf{p}^2+m^2)$, we see that $G(p)$, as a function of p_0 for given $\mathbf{p}^2$, has two poles at $p_0=\pm\varepsilon$, where $\varepsilon=\sqrt{(\mathbf{p}^2+m^2)}$. Thus, in the integration with respect to p_0 in the integral

$$\begin{aligned}G(\xi)&=\frac{1}{(2\pi)^4}\int e^{-ip\xi}\,G(p)\,d^4p\\&=\frac{1}{(2\pi)^4}\int d^3p\cdot e^{i\mathbf{p}\cdot\mathbf{r}}\int dp_0\cdot e^{-ip_0\tau}\,G(p)\end{aligned} \tag{75.11}$$

(where $\tau=t-t'$), the question of avoiding the poles arises; until this is decided, the expression (75.10) remains essentially indeterminate.

To settle this question, we go back to the original definition (75.1), and substitute in it the ψ-operators as the sums (73.6), noting that the only non-zero vacuum expectation values are those of the following products of creation and annihilation operators:

$$\langle 0|a_{\mathbf{p}}a_{\mathbf{p}}^{+}|0\rangle=1,\qquad \langle 0|b_{\mathbf{p}}b_{\mathbf{p}}^{+}|0\rangle=1.$$

† The explicit form, including the bispinor indices, is

$$(\gamma\hat{p}-m)_{il}G_{lk}(x-x')=\delta^{(4)}(x-x')\delta_{ik}. \tag{75.7a}$$

(Since in the vacuum state there are no particles, a particle has to be "created" by the operator $\hat{a}_{\mathbf{p}}^{+}$ or $\hat{b}_{\mathbf{p}}^{+}$ before it can be "annihilated" by $\hat{a}_{\mathbf{p}}$ or $\hat{b}_{\mathbf{p}}$.) The result is

$$\left.\begin{aligned} G_{ik}(x-x') &= -i\sum_{\mathbf{p}} \psi_{pi}(\mathbf{r},t)\bar{\psi}_{pk}(\mathbf{r}',t') \\ &= -i\sum_{\mathbf{p}} e^{-i\varepsilon(t-t')}\psi_{pi}(\mathbf{r})\bar{\psi}_{pk}(\mathbf{r}') \qquad \text{for } t-t'>0; \\ G_{ik}(x-x') &= i\sum_{\mathbf{p}} \bar{\psi}_{-p,k}(\mathbf{r}',t')\psi_{-p,i}(\mathbf{r},t) \\ &= i\sum_{\mathbf{p}} e^{i\varepsilon(t-t')}\psi_{-p,i}(\mathbf{r})\bar{\psi}_{-p,k}(\mathbf{r}') \qquad \text{for } t-t'<0. \end{aligned}\right\} \tag{75.12}$$

For $t>t'$ only the electron terms contribute to G, and for $t<t'$ only the positron terms.

If the summation over $\mathbf{p}$ is replaced by an integration over d^3p, a comparison of (75.12) and (75.11) shows that the integral

$$\int e^{-ip_0\tau}\, G(p)\, dp_0 \tag{75.13}$$

must have a phase factor $e^{-i\varepsilon\tau}$ for $\tau>0$ and $e^{i\varepsilon\tau}$ for $\tau<0$. This can be achieved by passing above the pole $p_0=\varepsilon$ and below $p_0=-\varepsilon$ in the plane of the complex variable p_0:

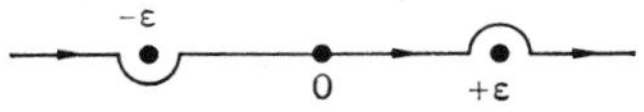

(75.14)

For, when $\tau>0$, the path of integration is closed by an infinite semicircle in the lower half-plane, so that the value of the integral (75.13) is given by the residue at the pole $p_0=+\varepsilon$; when $\tau<0$, the path is closed in the upper half-plane, and the integral is given by the residue at the pole $p_0=-\varepsilon$. The desired result is thus obtained in each case.

This rule for avoiding the poles (*Feynman's rule*) can be differently stated as follows: the integration is everywhere along the real axis, but the mass m of the particle is given an infinitesimal negative imaginary part:

$$m \to m - i0. \tag{75.15}$$

We then have

$$\begin{aligned} \varepsilon &\to \sqrt{[\mathbf{p}^2+(m-i0)^2]} \\ &= \sqrt{[\mathbf{p}^2+m^2-i0]} \\ &= \varepsilon - i0. \end{aligned}$$

The poles $p_0 = \pm\varepsilon$ are therefore moved off the real axis:

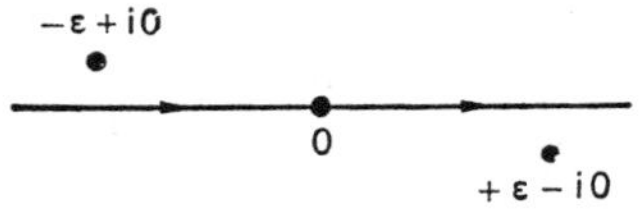

(75.16)

and the integration along this axis is equivalent to integration along the path (75.14).† Using the rule (75.15), we can write the propagator (75.10) in the form

$$G(p) = \frac{\gamma p + m}{p^2 - m^2 + i0}. \tag{75.17}$$

The rule of integration with displaced poles can be proved by means of the relation

$$\frac{1}{x + i0} = P\frac{1}{x} - i\pi\delta(x), \tag{75.18}$$

which is to be taken in the sense that multiplication by any function $f(x)$ and integration gives

$$\int_{-\infty}^{\infty} \frac{f(x)}{x + i0}\,dx = P\int_{-\infty}^{\infty} \frac{f(x)}{x}\,dx - i\pi f(0) \tag{75.19}$$

(the symbol P denoting the principal value).

The Green's function (75.10) is the product of the bispinor factor $\gamma p + m$ and a scalar,

$$G^{(0)}(p) = 1/(p^2 - m^2). \tag{75.20}$$

The corresponding coordinate function $G^{(0)}(\xi)$ is evidently a solution of the equation

$$(\hat{p}^2 - m^2)G^{(0)}(x - x') = \delta^{(4)}(x - x'), \tag{75.21}$$

i.e. it is the Green's function of the equation $(\hat{p}^2 - m^2)\psi = 0$. In this sense we can say that $G^{(0)}(x - x')$ is the scalar-particle propagator. It is easily seen by calculation, in the same manner as above, that the scalar field propagation function can be expressed in terms of the ψ-operators (11.2) by

$$G^{(0)}(x - x') = -i\langle 0|\mathrm{T}\psi(x)\psi^{+}(x')|0\rangle, \tag{75.22}$$

which is analogous to the definition (75.1). The chronological product is defined (as

† It is useful to note that the rule for moving the poles corresponds to an infinitesimal damping of $G(x - x')$ with respect to $|\tau| \equiv |t - t'|$: if the value of p_0 at the displaced poles is written as $-(\varepsilon - i\delta)$ and $+(\varepsilon - i\delta)$, with $\delta \to +0$, then the time factor in the integral (75.13) becomes $\exp(-i\varepsilon|\tau| - \delta|\tau|)$.

for all boson operators) by

$$T\hat{\psi}(x)\hat{\psi}^+(x') = \hat{\psi}(x)\hat{\psi}^+(x'), \quad t > t'; \\ = \hat{\psi}^+(x')\hat{\psi}(x), \quad t < t', \tag{75.23}$$

with the same sign for both $t > t'$ and $t < t'$.

§76. The photon propagator

Hitherto we have been concerned (in §§43 and 74) with the explicit form of the electromagnetic field operator $\hat{A}$ only in finding the matrix elements with respect to a change in the number of real photons. For this purpose it was sufficient to use the representation (§2) of the free field potentials in terms of transverse plane waves.

This representation, however, does not give a complete description of every field, as is clear from the fact that the scattering diagrams (73.13), (73.14) must also take account of the Coulomb interaction of the electrons. The latter is described by the scalar potential Φ and certainly cannot be reduced to an exchange between transverse virtual photons (describable by a vector potential such that $\operatorname{div}\mathbf{A} = 0$).†

Thus we have as yet essentially no complete definition of the operators $\hat{A}$, and without this it is impossible to carry out a direct calculation of the photon propagator by means of the formula

$$D_{\mu\nu}(x - x') = i\langle 0|\mathrm{T}A_\mu(x)A_\nu(x')|0\rangle. \tag{76.1}$$

On the other hand, the fact that the potentials are not gauge-invariant deprives of much of their physical meaning the operators which would be needed for a complete quantization of the electromagnetic field.

These difficulties, however, are purely formal, not physical, and can be avoided by using certain general properties of the propagator, which are evident from the requirements of relativistic invariance and gauge invariance.

The most general 4-tensor of rank two which depends only on the 4-vector $\xi = x - x'$ is

$$D_{\mu\nu}(\xi) = g_{\mu\nu}D(\xi^2) - \partial_\mu\partial_\nu D^{(l)}(\xi^2), \tag{76.2}$$

where D and $D^{(l)}$ are scalar functions of the invariant ξ^2.‡ This tensor is necessarily symmetrical.

† With the condition $\operatorname{div}\mathbf{A} = 0$, Maxwell's equations lead to the following equations for $\mathbf{A}$ and Φ:

$$\Box\mathbf{A} = -4\pi\mathbf{j} + \nabla\frac{\partial\Phi}{\partial t}, \qquad \Delta\Phi = -4\pi\rho.$$

In this gauge, the potential Φ satisfies the static Poisson's equation; cf. (76.13) below with D_{00} in the same gauge.

‡ These functions are different in the three ranges of values of the argument which are not mutually interchanged by Lorentz transformations: the regions outside the light cone ($\xi^2 < 0$), and within its two parts ($\xi^2 > 0$; $\xi_0 > 0$, $\xi_0 < 0$).

In the momentum representation, we correspondingly have

$$D_{\mu\nu}(k) = D(k^2)g_{\mu\nu} + k_\mu k_\nu D^{(l)}(k^2). \tag{76.3}$$

where $D(k^2)$, $D^{(l)}(k^2)$ are the Fourier components of the functions $D(\xi^2)$, $D^{(l)}(\xi^2)$.

The photon propagation function, in physical quantities (scattering amplitudes), is multiplied by the transition currents of two electrons, i.e. it appears in combinations of the form $(j^\mu)_{21}D_{\mu\nu}(j^\nu)_{43}$; see, for instance, (73.13). But, because of the conservation of current ($\partial_\mu j^\mu = 0$), the matrix elements $j_{21} = \bar{\psi}_2\gamma\psi_1$ satisfy the condition of 4-transversality,

$$k_\mu (j^\mu)_{21} = 0, \tag{76.4}$$

where $k = p_2 - p_1$; cf. (43.13). It is therefore clear that all physical results are unchanged by the substitution

$$D_{\mu\nu} \to D_{\mu\nu} + \chi_\mu k_\nu + \chi_\nu k_\mu, \tag{76.5}$$

where the χ_μ are any functions of $\mathbf{k}$ and k_0. This arbitrariness in the choice of $D_{\mu\nu}$ corresponds to that in the field potential gauge.

The arbitrary gauge transformation (76.5) can violate the relativistically invariant form $D_{\mu\nu}$ assumed in (76.3) if the quantities χ_μ do not make up a 4-vector. But, even considering only relativistically invariant forms of the propagator, we see that the choice of the function $D^{(l)}(k^2)$ in (76.3) is entirely arbitrary; it does not affect any physical results, and can be made in any convenient manner (L. D. Landau, A. A. Abrikosov and I. M. Khalatnikov, 1954).

Thus the determination of the propagation function amounts to that of a single gauge-invariant function $D(k^2)$. If we take a given value of k^2, and the z-axis in the direction of $\mathbf{k}$, the transformations (76.5) will not affect the components $D_{xx} = D_{yy} = -D(k^2)$. It is therefore sufficient to calculate the component D_{xx}, using any gauge for the potentials.

We shall use a gauge in which $\operatorname{div}\hat{\mathbf{A}} = 0$ and the operator $\hat{\mathbf{A}}$ is given by the expansion (2.17), (2.18):

$$\hat{\mathbf{A}} = \sum_{\mathbf{k},\alpha} \sqrt{\frac{2\pi}{\omega}} \left(\hat{c}_{\mathbf{k}\alpha}\mathbf{e}^{(\alpha)} e^{-ikx} + \hat{c}^+_{\mathbf{k}\alpha}\mathbf{e}^{(\alpha)*} e^{ikx}\right), \qquad \omega = |\mathbf{k}|; \tag{76.6}$$

the index $\alpha = 1, 2$ labels the polarizations. The only non-zero vacuum expectation values of products of the operators $\hat{c}$, $\hat{c}^+$ are $\langle 0|c_{\mathbf{k}\alpha}c^+_{\mathbf{k}\alpha}|0\rangle = 1$. Then, by the definition (76.1), we have

$$D_{ik}(\xi) = \frac{1}{(2\pi)^3}\int \frac{2\pi i\, d^3k}{\omega}\left(\sum_\alpha e_i^{(\alpha)}e_k^{(\alpha)*}\right) e^{-i\omega|\tau| + i\mathbf{k}\cdot\boldsymbol{\xi}}, \tag{76.7}$$

where i, k are three-dimensional vector indices; the summation over $\mathbf{k}$ has been replaced by an integration over $d^3k/(2\pi)^3$. The absolute value of $\tau = t - t'$ appears in the exponent, because the operator product in (76.1) is chronological.

It is evident from (76.7) that the integrand without the factor $e^{i\mathbf{k}\cdot\boldsymbol{\xi}}$ is the component of the three-dimensional Fourier expansion of the function $D_{ik}(\mathbf{r}, t)$. For $D_{xx} = -D$, it is

$$(2\pi i/\omega)\, e^{-i\omega|\tau|} \sum_\alpha |e_x^{(\alpha)}|^2 = (2\pi i/\omega)\, e^{-i\omega|\tau|}.$$

To find $D_{xx}(k^2)$ we now have to represent this function as a Fourier integral over time. The appropriate formula is

$$\frac{2\pi i}{\omega} e^{-i\omega|\tau|} = -\frac{1}{2\pi} \int_{-\infty}^{\infty} \frac{4\pi}{k_0^2 - \mathbf{k}^2 + i0} e^{-ik_0\tau}\, dk_0.$$

As explained in §75, this integration is understood to be taken along a contour passing below the pole $k_0 = |\mathbf{k}| = \omega$ and above the pole $k_0 = -|\mathbf{k}| = -\omega$; for $\tau > 0$ the value of the integral is determined by the residue at the pole $k_0 = +\omega$, and for $\tau < 0$ by that at $k_0 = -\omega$.

Thus we have finally

$$D(k^2) = 4\pi/(k^2 + i0). \tag{76.8}$$

The term $+i0$ in the denominator which results from this proof is in accordance with the rule (75.15), $i0$ being subtracted from the (zero) mass of the photon. It is evident from (76.8) that the corresponding coordinate function $D(\xi^2)$ satisfies the equation

$$-\partial_\mu \partial^\mu D(x - x') = 4\pi\delta^{(4)}(x - x'), \tag{76.9}$$

i.e. it is the Green's function of the wave equation.

We shall generally take $D^{(l)} = 0$, i.e. use the propagation function

$$D_{\mu\nu} = g_{\mu\nu} D(k^2) = \frac{4\pi}{k^2 + i0} g_{\mu\nu} \tag{76.10}$$

(the *Feynman gauge*).

There are also other gauges which may be advantageous in certain applications. Putting $D^{(l)} = -D/k^2$, we obtain the propagator in the form

$$D_{\mu\nu} = \frac{4\pi}{k^2}\left(g_{\mu\nu} - \frac{k_\mu k_\nu}{k^2}\right) \tag{76.11}$$

(the *Landau gauge*), with $D_{\mu\nu}k^\nu = 0$. This choice is similar to the Lorentz gauge for potentials ($A_\mu k^\mu = 0$).

The propagator gauge conditions $D_{il}k^l = 0$, $D_{0l}k^l = 0$ are analogous to the three-dimensional gauge condition div $\mathbf{A} = 0$ for the potentials. Together with

$$D_{xx} = -D = -4\pi/k^2,$$

these conditions give

$$D_{il} = -\frac{4\pi}{\omega^2 - \mathbf{k}^2}\left(\delta_{il} - \frac{k_i k_l}{\mathbf{k}^2}\right). \tag{76.12}$$

In order to obtain this D_{il}, we must apply to the propagator (76.10) the transformation (76.5), putting

$$\chi_0 = -\frac{4\pi\omega}{2(\omega^2 - \mathbf{k}^2)\mathbf{k}^2}, \qquad \chi_i = \frac{4\pi k_i}{2(\omega^2 - \mathbf{k}^2)\mathbf{k}^2}.$$

The remaining components $D_{\mu\nu}$ are then found to be

$$D_{00} = -4\pi/\mathbf{k}^2, \qquad D_{0i} = 0. \tag{76.13}$$

This is called the *Coulomb gauge* (E. E. Salpeter, 1952). D_{00} is here the Fourier component of the Coulomb potential.

Finally, the propagator gauge in which

$$D_{il} = -\frac{4\pi}{\omega^2 - \mathbf{k}^2}\left(\delta_{il} - \frac{k_i k_l}{\omega^2}\right), \qquad D_{0i} = D_{00} = 0, \tag{76.14}$$

is analogous to the potential gauge condition $\Phi = 0$. This is a convenient form for use in non-relativistic problems (I. E. Dzyaloshinskiĭ and L. P. Pitaevskiĭ, 1959).

All the above expressions relate to the momentum representation of the propagator. In some cases, it is convenient to use the mixed frequency–coordinate representation, i.e. the function

$$D_{\mu\nu}(\omega, \mathbf{r}) = \int D_{\mu\nu}(\omega, \mathbf{k})\, e^{i\mathbf{k}\cdot\mathbf{r}}\, d^3k/(2\pi)^3. \tag{76.15}$$

In the Feynman gauge (76.10)

$$D_{\mu\nu}(\omega, \mathbf{r}) = g_{\mu\nu} D(\omega, \mathbf{r}),$$

where

$$D(\omega, \mathbf{r}) = 4\pi \int \frac{e^{i\mathbf{k}\cdot\mathbf{r}}}{\omega^2 - \mathbf{k}^2 + i0} \frac{d^3k}{(2\pi)^3}$$

$$= -\frac{i}{\pi r} \int_0^\infty \frac{e^{ikr} - e^{-ikr}}{\omega^2 - k^2 + i0} k\, dk$$

or, changing k to $-k$ in the second term in the integrand,

$$D(\omega, \mathbf{r}) = -\frac{i}{\pi r} \int_{-\infty}^{\infty} \frac{e^{ikr} k\, dk}{\omega^2 - k^2 + i0}.$$

The integration here is carried out by closing the contour of integration with an infinite semicircle in the upper half-plane of the complex variable k, and amounts to taking the residue at the pole $k = |\omega| + i0$. The final result is

$$D(\omega, \mathbf{r}) = -\frac{1}{r} e^{i|\omega| r}. \tag{76.16}$$

The following comment may be made regarding this expression. The process described by the diagrams (73.13) and (73.14) may be intuitively regarded as the scattering of electron 2 in the field due to electron 1 (or vice versa). The function (76.16) corresponds to the usual "retarded" potential $\propto e^{i\omega r}$ (see *Fields*, (64.1), (64.2)) only when $\omega > 0$. The sign of ω, however, depends on the arbitrary choice of the direction of the arrow of k in the diagram. The above-mentioned property of $D(\omega, \mathbf{r})$ signifies that in quantum electrodynamics the source of the field is to be regarded as the particle which loses energy, i.e. emits a virtual photon.

To conclude this section, let us also consider the problem of the propagator for particles with spin 1 and non-zero mass. There is then no arbitrariness of the gauge, and the choice of the propagator is unambiguous.

Substituting the ψ-operators (14.16) in the definition

$$G_{\mu\nu} = -i\langle 0|\mathrm{T}\psi_\mu(x)\psi_\nu^+(x')|0\rangle, \tag{76.17}$$

we obtain an expression that differs from (76.7) only in that the sum over polarizations in the integrand is replaced by

$$\sum_\alpha u_\mu^{(\alpha)} u_\nu^{(\alpha)*}.$$

Summation over polarizations is equivalent to averaging and multiplication by 3, the number of independent polarizations. Averaging gives the density matrix of unpolarized particles (14.15). Thus we find for the propagator of vector particles

$$G_{\mu\nu}(p) = -\frac{1}{p^2 - m^2 + i0}\left(g_{\mu\nu} - \frac{p_\mu p_\nu}{m^2}\right). \tag{76.18}$$

The propagators (75.17) and (76.18) have similar structures: the denominator contains the difference $p^2 - m^2$, and the numerator is (apart from a factor) the density matrix of unpolarized particles with a given spin.

§77. General rules of the diagram technique

The calculation of the scattering matrix elements that has been given for some simple cases in §§73 and 74 contains all the fundamental features of the general method. There is no particular difficulty in deriving the corresponding general rules for calculating the matrix elements in any order of perturbation theory.

As has already been mentioned, the matrix element of the scattering operator $\hat{S}$ for the transition between any initial and final states is equal to the vacuum expectation value of the operator obtained by multiplying $\hat{S}$ on the right by the creation operators of all the initial particles and on the left by the annihilation operators of all the final particles.

This treatment puts the S-matrix element in the following form in the nth order of perturbation theory:

$$\langle f|S^{(n)}|i\rangle = \frac{1}{n!}\langle 0|\ldots b_{2f}b_{1f}\ldots a_{1f}\ldots c_{1f}\times$$

$$\times\int d^4x_1\ldots d^4x_n \mathrm{T}\{\bar{\psi}_1(-ie\gamma A_1)\psi_1)\ldots(\bar{\psi}_n(-ie\gamma A_n)\psi_n)\}c_{1i}^{+}\ldots a_{1i}^{+}\ldots b_{1i}^{+}\ldots|0\rangle; \tag{77.1}$$

the suffixes $1i$, $2i,\ldots$ label the initial particles (positrons, electrons and photons separately), and the suffixes $1f$, $2f,\ldots$ label the final particles. The suffixes $1, 2,\ldots$ to the operators $\hat{\psi}$ and $\hat{A}$ signify that $\hat{\psi}_1 = \hat{\psi}(x_1)$ and so on. The operators $\hat{\psi}$ and $\hat{A}$ which appear here are linear combinations of the creation and annihilation operators of the corresponding particles in various states. Thus we obtain expressions for the matrix elements which are the vacuum expectation values of the products of the particle creation and annihilation operators and of their linear combinations. The calculation of such expectation values is effected by means of the following results, which constitute *Wick's theorem* (G. C. Wick, 1950).

(1) The vacuum expectation value of the product of any number of boson operators $\hat{c}^+$ and $\hat{c}$ is equal to the sum of the products of all possible expectation values of these operators taken in pairs (contraction). In each pair, the factors must be placed in the same order as in the original product.

(2) For the fermion operators $\hat{a}^+$, $\hat{a}$, $\hat{b}^+$, $\hat{b}$ (of the same or different particles), the rule is the same except that each term appears in the sum with positive or negative sign according to the parity of the number of interchanges of fermion operators needed to bring together all the operators that are averaged in pairs.

The expectation value must obviously be zero unless the product contains a factor $\hat{a}^+$, $\hat{b}^+$, $\hat{c}^+$ for each factor $\hat{a}$, $\hat{b}$, $\hat{c}$. Then only pairs of operators $(\hat{a}, \hat{a}^+),\ldots,$ pertaining to the same states are to be contracted, and moreover only those pairs in which $\hat{a}^+$, etc., is to the right of $\hat{a}$, etc.: the particle is first created and then annihilated (whereas $\langle 0|a^+a|0\rangle = 0$, etc.).

If each pair $(\hat{a}, \hat{a}^+)$, etc. appears only once in the product, Wick's theorem is obviously true, the expectation value then reducing to a single product of pairwise expectation values. Its validity is also evident when all the annihilation operators in the product are to the right of the creation operators; this is called a *normal product*. The expectation value is then zero. Wick's theorem is now easily proved by induction for the general case where one pair of operators appears k times in the product, as follows.

Let us consider the expectation value $\langle 0|\ldots cc^+\ldots|0\rangle$, in which the pair of boson operators appears k times; the argument is entirely similar for fermion operators. If we interchange the factors $\hat{c}$ and $\hat{c}^+$ in one pair, the commutation rules give

$$\langle 0|\ldots cc^+\ldots|0\rangle = \langle 0|\ldots c^+c\ldots|0\rangle + \langle 0|\ldots 1\ldots|0\rangle. \tag{77.2}$$

The expectation value $\langle 0|\ldots 1\ldots|0\rangle$ contains $k-1$ pairs, and Wick's theorem is assumed to be valid for it. If the expectation value $\langle 0|\ldots cc^+\ldots|0\rangle$ is expanded by Wick's theorem, it differs from $\langle 0|\ldots c^+c\ldots|0\rangle$ by just the term

$$\langle 0|\ldots 1\ldots|0\rangle\langle 0|cc^+|0\rangle = \langle 0|\ldots 1\ldots|0\rangle;$$

in the expansion of $\langle 0|\ldots c^+c\ldots|0\rangle$, the corresponding term $\langle 0|\ldots 1\ldots|0\rangle\langle 0|c^+c|0\rangle$ is zero. Hence it follows from (77.2) that, if Wick's theorem is valid for a matrix element $\langle 0|\ldots c^+c\ldots|0\rangle$, it is still valid when $\hat{c}$ and $\hat{c}^+$ are interchanged. Since the theorem is known to be valid for one particular order of factors (the normal order), it is therefore true in every case.

Since Wick's theorem is valid for products of operators $\hat{a}$, $\hat{b}, \ldots$, it is also true for all products which contain the linear combinations $\hat{\psi}$, $\hat{\bar{\psi}}$, $\hat{A}$ of $\hat{a}$, $\hat{b}, \ldots$, as well as the latter operators themselves. On applying this theorem to the matrix element (77.1), we bring it to the form of a sum of terms, each term being the product of a number of pairwise expectation values. The latter will include contractions of the operators $\hat{\psi}$, $\hat{\bar{\psi}}$, $\hat{A}$ with "external" operators—those which create the initial particles or annihilate the final particles. These contractions are expressed in terms of the wave functions of the initial and final particles by the formulae

$$\left.\begin{aligned} &\langle 0|Ac_{\mathbf{p}}^+|0\rangle = A_p, && \langle 0|c_{\mathbf{p}}A|0\rangle = A_p^*,\\ &\langle 0|\psi a_{\mathbf{p}}^+|0\rangle = \psi_p, && \langle 0|a_{\mathbf{p}}\bar{\psi}|0\rangle = \psi_p^*,\\ &\langle 0|b_{\mathbf{p}}\psi|0\rangle = \psi_{-p}, && \langle 0|\bar{\psi}b_{\mathbf{p}}^+|0\rangle = \bar{\psi}_{-p}, \end{aligned}\right\} \tag{77.3}$$

where A_p and ψ_p are the photon and electron wave functions with momentum $\mathbf{p}$ (the polarization indices are omitted for brevity, as in §§73 and 74). Contractions of the "internal" operators in the T product will also occur. Since the sequence of factors in each contracted pair is preserved when applying Wick's theorem, the chronological sequence of operators is preserved in these contractions, and they are therefore replaced by the corresponding propagators.†

Each term of the sum obtained from the matrix element by applying Wick's theorem is represented by a particular Feynman diagram. In the nth-order diagram there are n vertices, each corresponding to one of the variables of integration (the 4-vectors x_1, $x_2, \ldots$). Three lines meet at each vertex, two being continuous (electron lines) and one broken (photon line); these correspond to the electron operators $\hat{\psi}$ and $\hat{\bar{\psi}}$ and the photon operator $\hat{A}$ as functions of the same variable x. The operator $\hat{\psi}$ corresponds to the incoming line and $\hat{\bar{\psi}}$ to the outgoing line.

By way of illustration, we shall give some examples of the correlation between the terms of the matrix element in the third approximation and the diagrams.

† The following comment must be made regarding this last statement. In proving Wick's theorem, we have made use of the commutation rules for the operators $\hat{c}$ and $\hat{c}^+$, which are meaningful only for real ("transverse") photons. The "external" operators $\hat{c}_i^+$, $\hat{c}_f$ do in fact, of course, correspond to such (initial and final) photons, but the operators $\hat{A}$ (which appear within the T product) describe, as shown in §76, not only transverse photons. The situation here is similar to that in the calculation of $D_{\mu\nu}$ (§76). Owing to the relativistic and gauge invariance, it is sufficient to prove the theorem for those products (i.e. components of the tensors $\langle 0|\mathrm{T}A_\mu A_\nu\ldots|0\rangle$) which are determined by the transverse parts of the potentials. The theorem is then valid for all products.

Omitting the integral sign and the symbol T, and also the operator symbols, the factors $-ie\gamma$, and the arguments of the operators, we can symbolically write these terms as

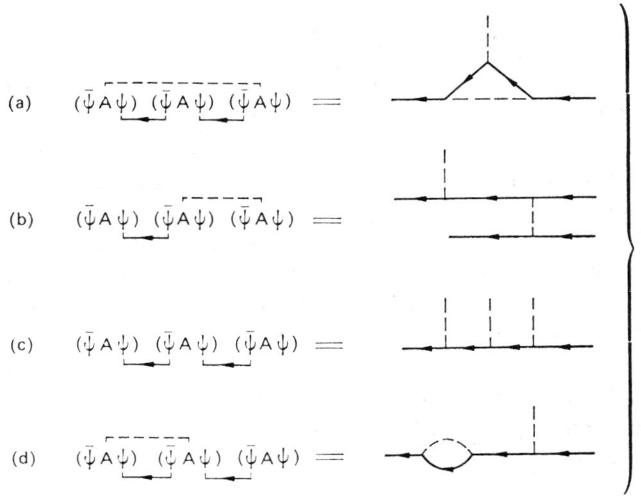

(77.4)

For clarity, the electron and photon contractions are shown by continuous and broken lines as in the diagrams. The direction of the arrows from $\hat{\bar{\psi}}$ to $\hat{\psi}$ for the electron contractions is the same as in the diagrams. For the internal photon contractions the direction is immaterial (the photon propagator is an even function of $x - x'$).

The terms thus obtained include equivalent terms which differ only in that the vertices are renumbered, i.e. that the correlation between the vertices and the variables is changed, or simply that the variables of integration are renamed. The number of such interchanges is $n!$; this cancels the factor $1/n!$ in (77.1), and there is then no need to consider diagrams differing only by interchange of vertices. This has already been noted in §§73 and 74. For example, there are two equivalent diagrams in the second approximation:

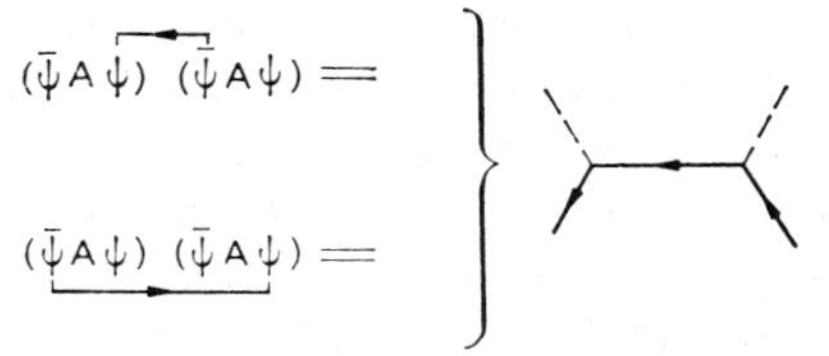

(77.5)

In (77.4) and (77.5) only internal contractions which correspond to internal diagram lines are shown (virtual electrons and photons). The operators still free are contracted with external operators, and this establishes a correlation between the free ends of the diagrams and certain initial and final particles. Then $\hat{\bar{\psi}}$ (contracting with operators $\hat{a}_f$ or $\hat{b}_i^+$) gives the final electron line or the initial positron line, and $\hat{\psi}$ (contracting with $\hat{a}_i^+$ or $\hat{b}_f$) gives the initial electron line or the final positron line.

The free operator $\hat{A}$ (contracting with $\hat{c}_i^+$ or $\hat{c}_f$) can correspond to either an initial or a final photon. Thus we obtain sets of several topologically identical diagrams (i.e. diagrams having the same number of lines arranged in the same way), differing only by interchanges of initial and final particles between incoming and outgoing free ends.

Each such interchange is clearly equivalent to a certain interchange of the external operators $\hat{a}$, $\hat{b}$, ... in (77.1). It is therefore evident that, if the initial particles or the final particles include identical fermions, diagrams which differ by an odd number of interchanges of free ends must have opposite signs.

An uninterrupted sequence of continuous lines in the diagrams constitutes an electron line along which the arrows maintain a constant direction. Such a line may have two free ends or form a closed loop. For example, the diagram

$$(\bar{\psi} A \psi)\ (\bar{\psi} A \psi) \quad =$$ 

has a loop with two vertices. The maintenance of direction along the electron line is the graphical expression of the conservation of charge: the "incoming" charge at each vertex is equal to the "outgoing" charge.

The arrangement of the bispinor indices along the continuous electron line corresponds to writing the matrices from left to right in motion contrary to the arrows. The bispinor indices of different electron lines can never become confused. Along an open line, the sequence of indices terminates at the free ends with electron (or positron) wave functions; in a closed loop, the sequence of indices is itself closed, and the loop corresponds to the trace of the product of the matrices found on it. This trace must be taken with negative sign, as is easily seen. A loop with k vertices corresponds to a set of k contractions:

$$(\bar{\psi} A \psi)\ \ (\bar{\psi} A \psi)\ \cdots\ (\bar{\psi} A \psi)$$

(or to another which is equivalent, differing only in an interchange of the vertices). In the $(k-1)$th contraction the operators $\hat{\psi}$ and $\hat{\bar{\psi}}$ are together in the order ($\hat{\bar{\psi}}$ to the right of $\hat{\psi}$) in which they must appear in the electron propagator. The operators at the ends are brought together by an even number of interchanges with other ψ-operators, and are then in the order $\hat{\bar{\psi}}\hat{\psi}$.

Since

$$\langle 0|\mathrm{T}\bar{\psi}'\psi|0\rangle = -\langle 0|\mathrm{T}\psi\bar{\psi}'|0\rangle$$

(see the second footnote to §74), the replacement of this contraction by the corresponding propagator means a change in the sign of the whole expression.

In general, the change to the momentum representation is made in an exactly similar manner to that in §§73 and 74. As well as the general law of conservation of 4-momentum, "conservation laws" must also be satisfied at each vertex. But all these laws may not suffice to determine uniquely the momenta of all the internal lines in the diagram. In such cases, there remain integrations over $d^4p/(2\pi)^4$ for all the undetermined internal momenta; these integrations extend throughout p-space, including p_0 from $-\infty$ to $+\infty$.

In the above discussion it has been assumed that the perturbation is represented by the interaction between those particles which are "actively" concerned in the reaction (i.e. between particles whose state is altered as a result of the process). A similar treatment can be given for the case where there is an external electromagnetic field, i.e. a field generated by "passive" particles, whose state is not altered in the process.

Let $A^{(e)}(x)$ be the 4-potential of the external field. It appears in the Lagrangian of the interaction together with the photon operator $\hat{A}$, as the sum $\hat{A}+A^{(e)}$ (which is multiplied by the current operator $\hat{j}$). Since $A^{(e)}$ does not involve any operators, it cannot contract with other operators. Thus only external lines in Feynman diagrams can correspond to an external field.

If $A^{(e)}$ is expressed as a Fourier integral:

$$\left.\begin{aligned} A^{(e)}(x) &= \int A^{(e)}(q)\, e^{-iqx}\, d^4q/(2\pi)^4, \\ A^{(e)}(q) &= \int A^{(e)}(x)\, e^{iqx}\, d^4x, \end{aligned}\right\} \tag{77.6}$$

the expressions for the matrix elements in the momentum representation will contain the 4-vector q together with the 4-momenta of other external lines corresponding to real particles. Each such external-field line can be correlated with a factor $A^{(e)}(q)$, and the line is to be regarded as "incoming" (in accordance with the sign of the exponent in the factor e^{-iqx} which accompanies $A^{(e)}(q)$ in the Fourier integral; an "outgoing line" would be correlated with a factor $A^{(e)*}(q)$). If the 4-momenta of all the external-field lines are not uniquely defined (for given 4-momenta of all the real particles) by the law of conservation of 4-momentum, then there remain integrations over $d^4q/(2\pi)^4$ for all the "free" q and over all the other undetermined 4-momenta of the diagram lines.

If the external field is independent of time, then

$$A^{(e)}(q) = 2\pi\delta(q^0)A^{(e)}(\mathbf{q}), \tag{77.7}$$

where $A^{(e)}(\mathbf{q})$ is the three-dimensional Fourier component:

$$A^{(e)}(\mathbf{q}) = \int A^{(e)}(\mathbf{r})\, e^{-i\mathbf{q}\cdot\mathbf{r}}\, d^3x. \tag{77.8}$$

In this case the external line is correlated with $A^{(e)}(\mathbf{q})$ and assigned a 4-momentum $q^\mu = (0, \mathbf{q})$; the energies of the electron lines which (together with the field line) meet at a vertex will be equal by virtue of the conservation law. Integration over $d^3p/(2\pi)^3$ is necessary for the other "free" three-dimensional momenta $\mathbf{p}$ of the internal lines. The amplitude M_{fi} thus calculated determines, for example, the scattering cross-section by (64.25).

We may give a list of final rules for the *diagram technique* whereby an expression may be obtained for the scattering amplitude (or rather for iM_{fi}) in the momentum representation.

(1) The nth approximation of perturbation theory corresponds to diagrams with n vertices, each of which is the meeting point of one incoming and one outgoing electron line (continuous) and one photon line (broken). The amplitude of the scattering process involves all the diagrams having free ends (external lines) equal in number to the initial and final particles.

(2) Each incoming continuous external line is associated with the amplitude $u(p)$ of an initial electron or $u(-p)$ of a final positron (where p is the 4-momentum of the particle). Each outgoing continuous line is associated with the amplitude $\bar{u}(p)$ of a final electron or $\bar{u}(-p)$ of an initial positron.

(3) Each vertex is associated with a 4-vector $-ie\gamma^\mu$.

(4) Each incoming broken external line is associated with the amplitude $\sqrt{(4\pi)}e_\mu$ of an initial photon, and each such outgoing line with the amplitude $\sqrt{(4\pi)}e_\mu^*$ of a final photon, where e is the polarization 4-vector. The vector index μ is the same as the index of the matrix γ^μ at the corresponding vertex, so that the scalar product γe or γe^* is obtained.

(5) Each continuous internal line is associated with a factor $iG(p)$, and each broken internal line with a factor $-iD_{\mu\nu}(p)$. The tensor indices μ, ν are the same as the indices of the matrices γ^μ, γ^ν at the vertices joined by the broken line.

(6) The arrows have a constant direction along any continuous sequence of electron lines, and the arrangement of the bispinor indices along them corresponds to writing the matrices from left to right in motion contrary to the arrows. A closed electron loop corresponds to the trace of the product of the matrices found on it.

(7) At each vertex, the 4-momenta of the lines which meet there satisfy a conservation law, i.e. the sum of the momenta of the incoming lines is equal to the sum of the momenta of the outgoing lines. The momenta of the free ends are given quantities (subject to the general conservation law), with momentum $-p$ assigned to the positron line. Integration over $d^4p/(2\pi)^4$ is carried out for the momenta of internal lines which remain undetermined after application of the conservation laws at every vertex.

(8) An incoming free end corresponding to an external field is associated with a factor $A^{(e)}(q)$; the 4-vector q is related to the 4-momenta of the other lines by the conservation law at the vertex. If the field is constant, the line is associated with a factor $A^{(e)}(\mathbf{q})$, and integration over $d^3p/(2\pi)^3$ is carried out for the three-dimensional momenta of internal lines which remain undetermined.

(9) An additional factor -1 is included in iM_{fi} for each closed electron loop in the diagram and for each pair of positron external lines if these are the beginning and end of a single sequence of continuous lines. If the initial particles or the final particles include more than one electron or positron, the diagrams differing by an odd number of interchanges of identical particles (i.e. of the corresponding external lines) must have opposite signs.

To clarify the last rule, it may be added that diagrams having the same continuous lines, i.e. diagrams which would be identical after removal of all photon lines, must always have the same sign. When identical fermions are present, the sign of the amplitude as a whole is arbitrary.

§ 78. Crossing invariance

The representation of the scattering amplitudes M_{fi} by Feynman integrals reveals the following noteworthy symmetry property of these amplitudes.

Any of the incoming external lines in a Feynman diagram may be regarded (without changing the direction of its arrow) as either an initial particle or a final antiparticle, and any outgoing line as either a final particle or an initial antiparticle. When the change is made from particle to antiparticle, the significance of the 4-momentum p assigned to the line also changes: $p = p_e$ for the electron (say), and $p = -p_p$ for the positron. The polarization assigned to the particle is also changed. Since an incoming external line must correspond to an amplitude u and an outgoing one to u^*, we have $u = u_e$ for the electron and $u = u_p^*$ for the positron; and the change from u to u^* signifies a change in the sign of the spin component (or the helicity) of the particle.

For the photon, a strictly neutral particle, this is simply a change from emission to absorption or vice versa: an external photon line with momentum k corresponds either to the absorption of a photon with momentum $k_a = k$, or to the emission of a photon with momentum $k_e = -k$ and the opposite helicity.

This change in the significance of the external lines is equivalent to a change from one cross-channel of the reaction to others. Hence it follows that the same amplitude, as a function of the momenta of the free ends of the diagrams, describes every channel of the reaction.† Only the meaning of the arguments of the function varies with the channel: the change from particle to antiparticle implies $p_i \to -p_f$, where p_i is the 4-momentum of the initial particle (in one channel) and p_f the 4-momentum of the final particle (in the other channel). This property of the scattering amplitude is called *crossing symmetry* or *crossing invariance.*

In terms of the invariant amplitudes defined in §70 as functions of the kinematic invariants, we can say that these functions will be the same for all channels, but for each channel their arguments will take values in the corresponding physical region. Thus the Feynman integrals determine the invariant amplitudes as analytic functions; their values in the various physical regions are the analytical continuation of a function specified in one region. Since the integrands in the Feynman integrals have singularities, so do the invariant amplitudes, and their singularities can be determined from the expressions for the integrals, using the rule of pole avoidance. If the invariant amplitudes are calculated for any one channel from the Feynman integrals, their analytical continuation to the other channels will necessarily take account of these singularities.

It should be emphasized that crossing invariance goes beyond the properties of the scattering matrix which follow from the general requirements of space–time symmetry. The latter imply the equality of amplitudes for processes which differ by the interchange of initial and final states and the replacement of all particles by antiparticles (with the momenta **p** of all particles unchanged and the signs of their angular momentum components reversed). This is the condition of *CPT* in-

† If a particular channel is forbidden by the conservation of 4-momentum, the transition probability is necessarily zero because of the delta function which appears as a factor in (64.5).

variance.† Crossing invariance, however, allows this transformation not only for all the particles at once but also for any one particle.

§79. **Virtual particles**

The internal lines in the Feynman diagrams play a role in invariant perturbation theory analogous to that of the intermediate states in the "ordinary" theory, but the nature of these states is different in the two theories. In the ordinary theory the (three-dimensional) momentum is conserved in the intermediate states, but the energy is not, and for this reason they are said to be *virtual states*. In the invariant theory, the momentum and the energy appear on an equal footing: in the intermediate states, the whole 4-momentum is conserved (this results from the fact that the integration in the S-matrix elements is over both coordinates and time, thus ensuring the invariance of the theory). But the relation between energy and momentum which holds for real particles and is expressed by the equation $p^2 = m^2$ is no longer satisfied in the intermediate states, which are therefore spoken of as intermediate *virtual particles*. The relation between the momentum and energy of a virtual particle may be anything required by the conservation of 4-momentum at the vertices.

Let us consider a diagram consisting of two parts I and II, joined by a single line. Ignoring the internal structure of these parts, we can represent the diagram in the schematic form

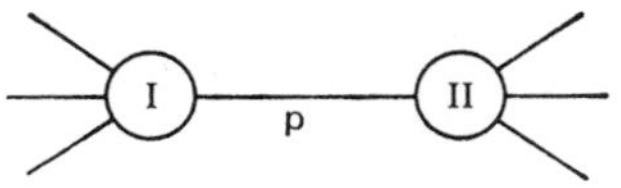

(79.1)

(the lines shown may be either continuous or broken lines). By the general conservation law, the sums of the 4-momenta of the external lines for parts I and II are equal. Because of the conservation at each vertex, they are also equal to the 4-momentum p of the internal line joining parts I and II. Thus this momentum is uniquely defined, and there is therefore no integration with respect to it in the matrix element.

The quantity p^2 may be either positive or negative, depending on the reaction channel. There is always a channel in which $p^2 > 0$.‡ Then the virtual particle is entirely analogous, as regards its formal properties, to a real particle with real mass $M = \sqrt{p^2}$. Its rest frame can be defined, its spin determined, and so on.

The tensor structure of the photon propagator (76.11) is the same as that of the density matrix of an unpolarized particle with spin 1 and non-zero mass:

$$\rho_{\mu\nu} = -\tfrac{1}{3}(g_{\mu\nu} - p_\mu p_\nu / m^2)$$

† The formal description of the change from one of these reactions to the other by reversing the signs of all the 4-momenta in the Feynman diagrams corresponds to the significance of the operation CPT as 4-inversion.

‡ For example, the channel (if it is allowable on energy grounds) in which all the free ends of part I correspond to initial particles and those of part II to final particles. Then $p = P_i$ (the sum of the 4-momenta of all the initial particles), and in the centre-of-mass system $p = (P_i^0, 0)$, so that $p^2 > 0$.

(see (14.15)). For a virtual particle the propagator (a quantity obtained from a quadratic combination of the field operators) plays a role analogous to that of the density matrix for a real particle. Thus a virtual photon, like a real photon, must be assigned spin 1. But, unlike the two independent polarizations of the real photon, all three polarizations are possible for the virtual photon, which is a "particle" with finite mass.

The electron propagation function is

$$G \propto \gamma p + m,$$

where m is the mass of the real electron, the "mass" of the virtual particle being $M = \sqrt{p^2}$. Putting

$$\gamma p + m = \frac{M+m}{2M}(\gamma p + M) + \frac{M-m}{2M}(\gamma p - M), \tag{79.2}$$

we see that the first term corresponds to the density matrix of a particle with mass M and spin $\frac{1}{2}$, and the second term to that of a similar "antiparticle"; cf. (29.10) and (29.17). Since the particle and the antiparticle have different internal parities (§27), we conclude that the same spin $\frac{1}{2}$ must be assigned to the virtual electron, but that no definite parity can be assigned to it.

A characteristic feature of the diagram (79.1) is that it can be cut into two unconnected parts by dividing only one internal line.† This line corresponds, in such cases, to a *one-particle* intermediate state, i.e. a state having only one virtual particle. The scattering amplitude corresponding to such a diagram contains the characteristic factor (which does not undergo integration)

$$\frac{1}{p^2 - m^2 + i0},$$

arising from the internal line p; m is the electron mass for an electron line and zero for a photon line. Thus the scattering amplitude has poles at the values of p for which the virtual particle would become a physical one ($p^2 = m^2$). This situation is similar to the one in non-relativistic quantum mechanics, where the scattering amplitude has poles for energy values corresponding to bound states of the system of colliding particles (*QM*, §128).

Let us consider the diagram (79.1) for the reaction channel in which all the free ends on the right correspond to initial particles, and all those on the left to final particles; then $p^2 > 0$. Then we can say that, in the intermediate state, all the initial particles are converted into one virtual particle. This is possible only if such a conversion would not contradict the necessary conservation laws (not including the conservation of 4-momentum), namely the conservation of angular momentum, charge, charge parity, etc. This is the necessary condition for the occurrence of what are called *pole diagrams*. If these exist for one reaction channel, they exist also for the remaining channels, because of crossing invariance.

† This property occurs for the diagrams of almost all processes in the first non-vanishing approximation.

For example, the conservation laws mentioned do not preclude the formation of a virtual electron by $e+\gamma\to e$. This corresponds to a pole of the Compton effect amplitude (and therefore a pole of the other channel of this reaction, namely two-photon annihilation of an electron–positron pair). The formation of a virtual photon by $e^-+e^+\to\gamma$ corresponds to a pole of the amplitude for the scattering of an electron by a positron, and therefore that of an electron by an electron. Two photons can give neither a virtual electron nor a virtual photon: the conversion $\gamma+\gamma\to e$ is forbidden by the conservation of charge or angular momentum, and $\gamma+\gamma\to\gamma$ by that of charge parity. Accordingly, the photon–photon scattering amplitude cannot involve pole diagrams.

The origin of the pole singularities of the scattering amplitudes, which has been discussed above on the basis of Feynman integrals, is really more general and is not dependent on perturbation theory. We shall show that such singularities arise simply as a consequence of the unitarity condition (71.2).

Let us assume that the intermediate states n which appear in (71.2) include a one-particle state. The contribution of this state is

$$(T_{fi}-T^*_{if})^{(\text{one-p})}=i(2\pi)^4\sum_\lambda\int\delta^{(4)}(P_f-p)T_{fn}T^*_{in}\frac{V\,d^3p}{(2\pi)^3},$$

where p and λ are the 4-momentum and helicity of the intermediate particle. The integration over d^3p is replaced by one over d^4p (in the range $p_0\equiv\varepsilon>0$):

$$d^3p\to 2\varepsilon\delta(p^2-M^2)\,d^4p,$$

where M is the mass of the intermediate particle. The integration eliminates the delta function $\delta^{(4)}(P_f-p)$; we then change from the amplitudes T_{fi} to M_{fi} by (64.10), obtaining

$$(M_{fi}-M^*_{if})^{(\text{one-p})}=2\pi i\delta(p^2-M^2)\sum_\lambda M_{fn}M^*_{in}. \tag{79.3}$$

Assuming T and P invariance, we have (apart from a phase factor) $M_{if}=M_{f'i'}$, where the states i', f' differ from i, f only in the sign of the particle helicities (with the same momenta). Taking the sum of equation (79.3) and the corresponding equation for $M_{f'i'}-M^*_{i'f'}$, we have

$$\operatorname{im}\overline{M}{}^{(\text{one-p})}_{fi}=-\pi\delta(p^2-M^2)R, \tag{79.4}$$

where

$$\overline{M}_{fi}=M_{fi}+M_{f'i'},$$

$$R=-\sum_\lambda(M_{fn}M^*_{in}+M_{f'n}M^*_{i'n}).$$

Hence it follows that $\overline{M}_{fi}$, as an analytic function of $p^2=P_i^2=P_f^2$, has a pole at

$p^2 = M^2$. According to (75.18) the pole part is

$$\overline{M}_{fi}^{\text{(one-p)}} = \frac{R}{p^2 - M^2 + i0}. \tag{79.5}$$

Real transitions to a one-particle state are possible only for one value of $P_i^2 = P_f^2$, namely M^2. Thus we in fact obtain the scattering amplitude structure corresponding to a diagram of the form (79.1).

Finally, let us consider an important property of diagrams containing closed electron loops. This property is easily derived by applying the concept of charge parity to a virtual photon: a virtual photon, like a real photon, must be assigned a definite (negative) charge parity.†

If a diagram contains a closed loop (with number of vertices $N > 2$), the amplitude for the process concerned must include not only that diagram but also another which differs only in the direction of traversal of the loop (if $N = 2$, there is evidently no distinguishable "direction of traversal"). If these loops are "cut out" along the broken lines which come to them, we obtain two loops, Π_{I} and Π_{II}:

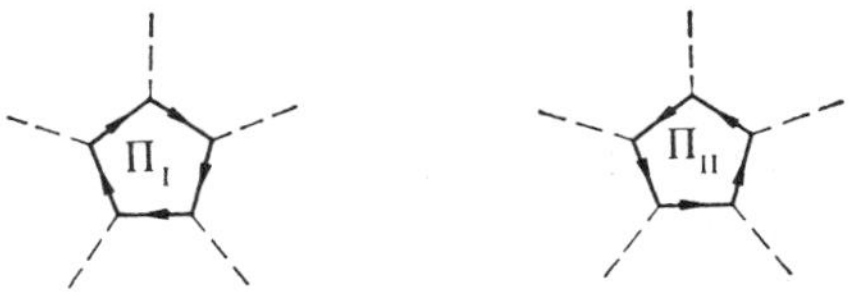

(79.6)

which may be regarded as diagrams determining the amplitude for the process of conversion of one set of photons (real or virtual) into another; the number N is the sum of the numbers of initial and final photons. But the conservation of charge parity forbids the conversion of an even number of photons into an odd number. When N is odd, therefore, the sum of the expressions corresponding to the loops (79.6) must be zero. The total contribution to the scattering amplitude from two diagrams containing these loops as constituent parts is consequently zero also, a result known as *Furry's theorem* (W. H. Furry, 1937).

Thus, in constructing the amplitude for a given process, we can ignore diagrams containing loops with an odd number of vertices.

This cancellation of diagrams occurs for the following reason. A closed electron loop corresponds to an expression (with given momenta $k_1, k_2, \ldots, k_N$ of the photon lines)

$$\int d^4p \cdot \text{tr}[(\gamma e_1)G(p)(\gamma e_2)G(p + k_1)\ldots], \tag{79.7}$$

where $p, p + k_1, \ldots$ are the momenta of the electron lines (which are not completely determined after the conservation laws have been applied at the vertices). Let the operation of charge conjugation be applied to all matrices γ^μ and G, replacing them

† This follows from the same arguments as were given at the end of §13 for a real photon, concerning the electromagnetic interaction operator acting at each vertex.

by $U_C^{-1}\gamma^\mu U_C$ and $U_C^{-1}GU_C$. The expression (79.7) is then unchanged, since the trace of a product of matrices is unaffected by such a transformation. According to (26.3),

$$U_C^{-1}\gamma^\mu U_C = -\tilde{\gamma}^\mu, \tag{79.8}$$

and hence

$$U_C^{-1}G(p)U_C = \frac{-p\tilde{\gamma}+m}{p^2-m^2} = \tilde{G}(-p). \tag{79.9}$$

But the replacement of $G(p)$ by the transposed matrix with the sign of p changed is clearly equivalent to a change in the direction of traversal of the loop, all the arrows being reversed. Thus this transformation changes one loop into the other, and there is a factor $(-1)^N$ from the change (79.8) at each vertex. Hence

$$\Pi_{\mathrm{I}} = (-1)^N \Pi_{\mathrm{II}}, \tag{79.10}$$

i.e. the contributions from the two loops are the same when the number of vertices is even, but equal and opposite when this number is odd.

CHAPTER IX

INTERACTION OF ELECTRONS

§ 80. Scattering of an electron in an external field

ELASTIC scattering of an electron in a constant external field is a simple process which occurs even in the first approximation of perturbation theory (the first Born approximation). It corresponds to a diagram with one vertex:

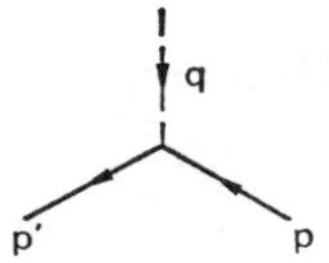

(80.1)

where p and p' are the initial and final 4-momenta of the electron, and $q = p' - p$. Since the electron energy is conserved in scattering in a constant field ($\varepsilon = \varepsilon'$), we have $q = (0, \mathbf{q})$.†

The corresponding scattering amplitude is

$$M_{fi} = -e\bar{u}(p')[\gamma A^{(e)}(\mathbf{q})]u(p), \tag{80.2}$$

where $A^{(e)}(\mathbf{q})$ is the component of the spatial Fourier resolution of the external field. The scattering cross-section is, according to (64.26),

$$d\sigma = \frac{1}{16\pi^2}|M_{fi}|^2\, do'. \tag{80.3}$$

For an electrostatic field, $A^{(e)} = (A_0^{(e)}, 0)$, and hence

$$\begin{aligned} M_{fi} &= -e\bar{u}(p')\gamma^0 u(p) A_0^{(e)}(\mathbf{q}) \\ &= -eu^*(p')u(p)A_0^{(e)}(\mathbf{q}). \end{aligned} \tag{80.4}$$

In the non-relativistic case, the bispinor amplitudes $u(p)$ of the plane waves reduce to the non-relativistic (two-component) amplitudes. For scattering without change of polarization, $u' = u$, and $u^*u = 2m$ by the normalization condition chosen. Thus

$$d\sigma = \left|-\frac{m}{2\pi}U(\mathbf{q})\right|^2 do',$$

† When there is an external field, such a diagram is, of course, not forbidden by the law of conservation of 4-momentum, as the diagram (73.19) with a real photon was: q^2, unlike the square of the 4-momentum of a real photon, need not be zero, and the component with the necessary q is automatically taken from the Fourier integral which represents the external field.

where $U(\mathbf{q}) = eA_0^{(e)}(\mathbf{q})$ is the Fourier component of the potential energy of the electron in the field; this expression is the same as the familiar Born's formula (*QM*, (126.4)).

In the general relativistic case, the cross-section for scattering of unpolarized electrons is obtained by averaging $|M_{fi}|^2$ over initial polarizations and summing over final polarizations, i.e. by taking the quantity

$$\tfrac{1}{2}\sum_{\text{polar.}} |M_{fi}|^2,$$

where the summation is over the spin directions of the initial and final electrons; the factor $\frac{1}{2}$ changes one of these summations into an averaging. According to the rules given in §65, we obtain

$$\begin{aligned}\tfrac{1}{2}\sum_{\text{polar.}} |M_{fi}|^2 &= 2\,\mathrm{tr}\,\rho'(\gamma A^{(e)})\rho(\gamma A^{(e)*})\\ &= \tfrac{1}{2}|A_0^{(e)}(\mathbf{q})|^2\,\mathrm{tr}\,(m+\gamma p')\gamma^0(m+\gamma p)\gamma^0.\end{aligned}$$

To calculate the trace, we note that $\gamma^0(\gamma p)\gamma^0 = (\gamma\tilde{p})$, where $\tilde{p} = (\varepsilon, -\mathbf{p})$, and therefore

$$\begin{aligned}\tfrac{1}{4}\mathrm{tr}\,(m+\gamma p')\gamma^0(m+\gamma p)\gamma^0 &= \tfrac{1}{4}\,\mathrm{tr}\,(m+\gamma p')(m+\gamma\tilde{p})\\ &= m^2 + p'\tilde{p}\\ &= \varepsilon^2 + m^2 + \mathbf{p}\cdot\mathbf{p}'\\ &= 2\varepsilon^2 - \tfrac{1}{2}\mathbf{q}^2.\end{aligned}$$

Hence the cross-section is

$$d\sigma = \frac{e^2|A_0^{(e)}(\mathbf{q})|^2}{4\pi^2}\,\varepsilon^2\left(1 - \frac{\mathbf{q}^2}{4\varepsilon^2}\right)do'. \tag{80.5}$$

For a field due to a static distribution of charge with density $\rho(\mathbf{r})$, we have

$$A_0^{(e)}(\mathbf{q}) = 4\pi\rho(\mathbf{q})/\mathbf{q}^2, \tag{80.6}$$

where $\rho(\mathbf{q})$ is the Fourier transform of the distribution $\rho(\mathbf{r})$ (the form factor). In particular, for the Coulomb field of a point charge Ze we have $\rho(\mathbf{q}) = Ze$. The cross-section is then

$$d\sigma = do'\,\frac{4(Ze^2)^2\varepsilon^2}{\mathbf{q}^4}\left(1 - \frac{\mathbf{q}^2}{4\varepsilon^2}\right) \tag{80.7}$$

(N. F. Mott, 1929). The quantity $\mathbf{q}^2 = 4\mathbf{p}^2\sin^2\frac{1}{2}\theta$, where θ is the scattering angle. The angular dependence of the quantity preceding the parenthesis is therefore that of a

Rutherford cross-section:

$$d\sigma_{\mathrm{Ru}} = do\,\frac{4(Ze^2)^2\varepsilon^2}{\mathbf{q}^4}$$

$$= do\,\frac{(Ze^2)^2\varepsilon^2}{4\mathbf{p}^4}\sin^{-4}\tfrac{1}{2}\theta; \tag{80.8}$$

in the non-relativistic limit, $\varepsilon^2/\mathbf{p}^4 \to 1/m^2v^4$. Thus†

$$d\sigma = d\sigma_{\mathrm{Ru}}(1 - v^2\sin^2\tfrac{1}{2}\theta). \tag{80.9}$$

In the ultra-relativistic case, the angular distribution differs from the non-relativistic case in that there is much less backward scattering: as $\theta \to \pi$, $d\sigma/d\sigma_{\mathrm{Ru}} \to m^2/\varepsilon^2$.

In the ultra-relativistic case, formula (80.7) gives for small-angle scattering

$$d\sigma = \frac{4(Ze^2)^2}{\varepsilon^2\theta^4}\,do'. \tag{80.10}$$

Although this formula has been derived in the Born approximation (i.e. on the assumption that $Ze^2 \ll 1$), it remains valid (for angles $\theta \lesssim m/\varepsilon$) even if $Ze^2 \sim 1$. This can be seen by using the ultra-relativistic wave function $\psi^{(+)}_{\varepsilon\mathbf{p}}$ (39.10), which is exact as regards Ze^2. This solution, which is valid in the range (39.2), of course remains valid in the asymptotic range $r \to \infty$. Here

$$F \sim 1 + \text{constant} \times e^{i(pr - \mathbf{p}\cdot\mathbf{r})}, \qquad \frac{\boldsymbol{\alpha}\cdot\nabla F}{\varepsilon} \sim 1 - \cos\theta \sim \theta^2 \ll 1,$$

so that the correction term remains small, as it should. The wave function of the form $e^{i\mathbf{p}\cdot\mathbf{r}}F$, which has the same form as the non-relativistic function (with an obvious change of parameters), has the same asymptotic expression, and therefore the cross-section is given by the Rutherford formula.

To calculate the scattering cross-section for electrons with any polarization, we could use the density matrix (29.13), following the general procedure. In this case, however, the result can be more readily obtained by expressing the bispinor amplitudes $u(p')$ and $u(p)$ in the form (23.9). Multiplication gives

$$u^*(p')u(p) = w'^*\{\varepsilon + m + (\varepsilon - m)(\mathbf{n}'\cdot\boldsymbol{\sigma})(\mathbf{n}\cdot\boldsymbol{\sigma})\}w,$$

or, using (33.5),

$$u^*(p')u(p) = w'^*\hat{f}w, \tag{80.11}$$

† The difference between $d\sigma$ and $d\sigma_{\mathrm{Ru}}$ shown by this formula is specific to particles with spin $\frac{1}{2}$. In the scattering of particles with spin 0, if their motion in the electromagnetic field is described by the wave equation, the result is $d\sigma = d\sigma_{\mathrm{Ru}}$. At first sight it might appear puzzling that the factor expressing this purely quantum effect does not contain $\hbar$. However, it must be remembered that the condition for the Born approximation to be valid ($e^2/\hbar v \ll 1$) is contrary to the condition for quasi-classical motion in a Coulomb field, and therefore formula (80.9) cannot be taken to the classical limit.

where†

$$\left.\begin{aligned}\hat{f} &= A + B\boldsymbol{\nu}\cdot\boldsymbol{\sigma},\\ A &= (\varepsilon+m)+(\varepsilon-m)\cos\theta,\\ B &= -i(\varepsilon-m)\sin\theta,\\ \boldsymbol{\nu} &= \mathbf{n}\times\mathbf{n}'/\sin\theta.\end{aligned}\right\} \tag{80.12}$$

The two-component quantity (three-dimensional spinor) w is the non-relativistic spin wave function of the electron. The change to the partially polarized states is therefore made by replacing the products $w_\alpha w_\beta^*$ (where α, β are spinor indices) by the non-relativistic two-rowed density matrix $\rho_{\alpha\beta}$. Thus we must put

$$|M_{fi}|^2 \to e^2|A_0^{(e)}(\mathbf{q})|^2 \operatorname{tr} \rho(A - B\boldsymbol{\nu}\cdot\boldsymbol{\sigma})\rho'(A + B\boldsymbol{\nu}\cdot\boldsymbol{\sigma}),$$

where

$$\rho = \tfrac{1}{2}(1+\boldsymbol{\sigma}\cdot\boldsymbol{\zeta}), \qquad \rho' = \tfrac{1}{2}(1+\boldsymbol{\sigma}\cdot\boldsymbol{\zeta}'),$$

and $\boldsymbol{\zeta}$, $\boldsymbol{\zeta}'$ are the vectors of the initial polarization and the final polarization selected by the detector. The result of calculating the trace is

$$d\sigma = d\sigma_0\left\{1 + \frac{(A^2-|B|^2)\boldsymbol{\zeta}\cdot\boldsymbol{\zeta}' + 2|B|^2(\boldsymbol{\nu}\cdot\boldsymbol{\zeta})(\boldsymbol{\nu}\cdot\boldsymbol{\zeta}') + 2A|B|\boldsymbol{\nu}\cdot\boldsymbol{\zeta}\times\boldsymbol{\zeta}'}{A^2+|B|^2}\right\}, \tag{80.13}$$

where $d\sigma_0$ is the scattering cross-section for unpolarized electrons.

Expressing the quantity in the braces in (80.13) in the form $\{1+\boldsymbol{\zeta}^{(f)}\cdot\boldsymbol{\zeta}'\}$, we find the polarization $\boldsymbol{\zeta}^{(f)}$ of the final electron itself, as opposed to the detected polarization ζ' (see §65):‡

$$\boldsymbol{\zeta}^{(f)} = \frac{(A^2-|B|^2)\boldsymbol{\zeta} + 2|B|^2(\boldsymbol{\nu}\cdot\boldsymbol{\zeta})\boldsymbol{\nu} + 2A|B|\boldsymbol{\nu}\times\boldsymbol{\zeta}}{A^2+|B|^2}. \tag{80.14}$$

We see that the scattered electrons are polarized only if the incident electrons are polarized. This is a general property of the first Born approximation; cf. *QM*, §140.

In the non-relativistic case ($\varepsilon \to m$), (80.14) gives $\boldsymbol{\zeta}^{(f)} = \boldsymbol{\zeta}$, i.e. the electron retains its polarization on scattering, a natural consequence of the neglect of spin–orbit interaction.

In the opposite (ultra-relativistic) case, we have

$$A = \varepsilon(1+\cos\theta), \qquad B = -i\varepsilon\sin\theta,$$

in accordance with the general formula (38.2).

† The definition of $\hat{f}$ used here differs by a factor from that in §§37 and 38.

‡ Formula (80.14) corresponds to that derived in *QM*, §140, Problem 1, and is obtained from it by taking A real and B imaginary.

If the incident electron has a definite helicity ($\boldsymbol{\zeta} = 2\lambda\mathbf{n}$, $\lambda = \pm\frac{1}{2}$), (80.14) gives after a simple calculation

$$\boldsymbol{\zeta}^{(f)} = 2\lambda\mathbf{n}'.$$

Thus the electron remains helical after scattering, with the same value (λ) of the helicity.

This property occurs because, as already mentioned in §38, when the mass is neglected Dirac's equation in the spinor representation separates into two independent equations for the functions ξ and η. The result has also a more general significance, since the current

$$\hat{j} = (\xi^*\xi + \eta^*\eta, \quad \xi^*\boldsymbol{\sigma}\xi - \eta^*\boldsymbol{\sigma}\eta),$$

and therefore the electromagnetic perturbation operator $\hat{V} = e\hat{j}\hat{A}$, do not contain mixed terms in ξ and η, and thus have no matrix elements for transitions between ξ states and η states. Hence it follows that, if an ultra-relativistic electron has a definite helicity (i.e. if either η or ξ is non-zero), this helicity is conserved in interaction processes in an approximation corresponding to completely neglecting the electron mass.

§ 81. Scattering of electrons and positrons by an electron

Let us consider the scattering of an electron by an electron, in which two electrons with 4-momenta p_1, p_2 collide and emerge with 4-momenta p_1', p_2'. The conservation of 4-momentum is expressed by

$$p_1 + p_2 = p_1' + p_2'. \tag{81.1}$$

We shall use the kinematic invariants of §66, defined by

$$\left.\begin{aligned} s &= (p_1 + p_2)^2 = 2(m^2 + p_1p_2), \\ t &= (p_1 - p_1')^2 = 2(m^2 - p_1p_1'), \\ u &= (p_1 - p_2')^2 = 2(m^2 - p_1p_2'), \\ s &+ t + u = 4m^2. \end{aligned}\right\} \tag{81.2}$$

The process in question is represented by the two Feynman diagrams (73.13), (73.14), and its amplitude is†

$$M_{fi} = 4\pi e^2\left\{\frac{1}{t}(\bar{u}_2'\gamma^\mu u_2)(\bar{u}_1'\gamma_\mu u_1) - \frac{1}{u}(\bar{u}_1'\gamma^\nu u_2)(\bar{u}_2'\gamma_\nu u_1)\right\}. \tag{81.3}$$

† This form of M_{fi} is in accordance with the general expression (70.5). In the first non-vanishing approximation of perturbation theory, only one of the five invariant amplitudes is non-zero: $f_3(t, u) = 4\pi e^2/t$.

According to the rules given in §65 for the states of initial and final particles described by polarization density matrices $\rho_1, \rho_1', \ldots$, we make the change

$$|M_{fi}|^2 \to 16\pi^2 e^4 \left\{ \frac{1}{t^2} \operatorname{tr}(\rho_2'\gamma^\mu \rho_2 \gamma^\nu) \operatorname{tr}(\rho_1'\gamma_\mu \rho_1 \gamma_\nu) + \frac{1}{u^2} \operatorname{tr}(\rho_1'\gamma^\mu \rho_2 \gamma^\nu) \operatorname{tr}(\rho_2'\gamma_\mu \rho_1 \gamma_\nu) - \right.$$

$$\left. - \frac{1}{tu} \operatorname{tr}(\rho_2'\gamma^\mu \rho_2 \gamma^\nu \rho_1' \gamma_\mu \rho_1 \gamma_\nu) - \frac{1}{tu} \operatorname{tr}(\rho_1'\gamma^\mu \rho_2 \gamma^\nu \rho_2' \gamma_\mu \rho_1 \gamma_\nu) \right\}. \quad (81.4)$$

For the scattering of unpolarized electrons (without regard to their polarization after scattering), we must put for all the density matrices $\rho = \frac{1}{2}(\gamma p + m)$, and multiply the result by $2 \times 2 = 4$ (averaging over the polarizations of the two initial electrons, and summation over the polarizations of the two final electrons). The scattering cross-section is given by formula (64.23), in which, by (64.15a), $I^2 = \frac{1}{4}s(s - 4m^2)$. It may be written

$$\left. \begin{aligned} d\sigma &= dt \frac{4\pi e^4}{s(s-4m^2)} \{f(t, u) + g(t, u) + f(u, t) + g(u, t)\}, \\ f(t, u) &= \frac{1}{16t^2} \operatorname{tr}[(\gamma p_2' + m)\gamma^\mu (\gamma p_2 + m)\gamma^\nu] \operatorname{tr}[(\gamma p_1' + m)\gamma_\mu (\gamma p_1 + m)\gamma_\nu], \\ g(t, u) &= -\frac{1}{16tu} \operatorname{tr}[(\gamma p_2' + m)\gamma^\mu (\gamma p_2 + m)\gamma^\nu (\gamma p_1' + m)\gamma_\mu (\gamma p_1 + m)\gamma_\nu]. \end{aligned} \right\} \quad (81.5)$$

In $f(t, u)$ the traces are first calculated (using (22.9), (22.10)), followed by summation over μ and ν;† in $g(t, u)$ the summation over μ and ν is taken first, using formulae (22.6). The result is

$$f(t, u) = \frac{2}{t^2} [(p_1 p_2)^2 + (p_1 p_2')^2 + 2m^2(m^2 - p_1 p_1')],$$

$$g(t, u) = \frac{2}{tu} (p_1 p_2 - 2m^2)(p_1 p_2),$$

or, in terms of the invariants (81.2),

$$\left. \begin{aligned} f(t, u) &= \frac{1}{t^2} [\tfrac{1}{2}(s^2 + u^2) + 4m^2(t - m^2)], \\ g(t, u) &= g(u, t) = \frac{2}{tu} (\tfrac{1}{2}s - m^2)(\tfrac{1}{2}s - 3m^2). \end{aligned} \right\} \quad (81.6)$$

† The following formula is given for future reference:

$$\tfrac{1}{4} \operatorname{tr}(\gamma p_1 + m)\gamma^\mu (\gamma p_2 + m)\gamma^\nu = g^{\mu\nu}(m^2 - p_1 p_2) + p_1^\mu p_2^\nu + p_1^\nu p_2^\mu.$$

Thus the cross-section is

$$d\sigma = r_e^2 \frac{4\pi m^2\, dt}{s(s-4m^2)} \left\{ \frac{1}{t^2} [\tfrac{1}{2}(s^2+u^2) + 4m^2(t-m^2)] + \right.$$

$$\left. + \frac{1}{u^2} [\tfrac{1}{2}(s^2+t^2) + 4m^2(u-m^2)] + \frac{4}{tu} (\tfrac{1}{2}s - m^2)(\tfrac{1}{2}s - 3m^2) \right\}, \tag{81.7}$$

where $r_e = e^2/m$.

In the centre-of-mass system, we have

$$\left. \begin{aligned} s = 4\varepsilon^2, \quad t = -4\mathbf{p}^2 \sin^2 \tfrac{1}{2}\theta, \quad u = -4\mathbf{p}^2 \cos^2 \tfrac{1}{2}\theta, \\ -dt = -2\mathbf{p}^2\, d\cos\theta = (\mathbf{p}^2/\pi)\, do, \end{aligned} \right\} \tag{81.8}$$

where $|\mathbf{p}|$ and ε are the magnitude of the momentum and energy of the electrons, which are unchanged in the scattering, and θ is the scattering angle. In the non-relativistic case ($\varepsilon \approx m$),† we obtain for the cross-section

$$\begin{aligned} d\sigma &= r_e^2 \frac{\pi m^4\, dt}{\mathbf{p}^2} \left(\frac{1}{t^2} + \frac{1}{u^2} - \frac{1}{tu} \right) \\ &= \left(\frac{e^2}{mv^2} \right)^2 \left(\frac{1}{\sin^4 \frac{1}{2}\theta} + \frac{1}{\cos^4 \frac{1}{2}\theta} - \frac{1}{\sin^2 \frac{1}{2}\theta \cos^2 \frac{1}{2}\theta} \right) do \\ &= \left(\frac{e^2}{mv^2} \right)^2 \frac{4(1+3\cos^2\theta)}{\sin^4\theta}\, do \quad \text{(non-relativistic)}, \end{aligned} \tag{81.9}$$

where $\mathbf{v} = 2\mathbf{p}/m$ is the relative velocity of the electrons, in accordance with the non-relativistic theory (see *QM*, §137). In the general case of arbitrary velocities, formula (81.7) with the substitution (81.8) can easily be brought to the form

$$d\sigma = r_e^2 \frac{m^2(\varepsilon^2+\mathbf{p}^2)^2}{4\mathbf{p}^4\varepsilon^2} \left[\frac{4}{\sin^4\theta} - \frac{3}{\sin^2\theta} + \left(\frac{\mathbf{p}^2}{\varepsilon^2+\mathbf{p}^2} \right)^2 \left(1 + \frac{4}{\sin^2\theta} \right) \right] do \tag{81.10}$$

(C. Møller, 1932). In the ultra-relativistic case ($\mathbf{p}^2 \approx \varepsilon^2$),

$$d\sigma = r_e^2 \frac{m^2}{\varepsilon^2} \frac{(3+\cos^2\theta)^2}{4\sin^4\theta}\, do \quad \text{(ultra-relativistic)}. \tag{81.11}$$

In the laboratory system, where one of the electrons (say electron 2) is at rest before the collision, the cross-section can be expressed in terms of the quantity

$$\Delta = \frac{\varepsilon_1 - \varepsilon_1'}{m} = \frac{\varepsilon_2' - m}{m}, \tag{81.12}$$

† The velocity v is assumed small ($v \ll 1$) but such that the condition for perturbation theory to be applicable is still satisfied: e^2/v $(=e^2/\hbar v) \ll 1$.

the energy (in units of m) transferred by the incident electron (electron 1) to electron 2.† The invariants are

$$s = 2m(m + \varepsilon_1), \qquad t = -2m^2\Delta, \qquad u = -2m(\varepsilon_1 - m - m\Delta). \tag{81.13}$$

Substitution of these expressions in (81.7) gives the following formula for the energy distribution of the secondary electrons (called *δ electrons*) formed in the scattering of fast primary electrons:

$$d\sigma = 2\pi r_e^2 \frac{d\Delta}{\gamma^2 - 1}\left\{\frac{(\gamma-1)^2\gamma^2}{\Delta^2(\gamma-1-\Delta)^2} - \frac{2\gamma^2 + 2\gamma - 1}{\Delta(\gamma-1-\Delta)} + 1\right\}, \tag{81.14}$$

where $\gamma = \varepsilon_1/m$. The quantities $m\Delta$ and $m(\gamma - 1 - \Delta)$ are the kinetic energies of the two electrons after the collision; the identity of the two particles is shown here by the symmetry of the formula with respect to these quantities. If the term "recoil electron" is arbitrarily applied to the electron with the smaller energy, Δ takes values from 0 to $\frac{1}{2}(\gamma - 1)$. When Δ is small, formula (81.14) becomes

$$d\sigma = 2\pi r_e^2 \frac{\gamma^2}{\gamma^2 - 1}\frac{d\Delta}{\Delta^2} = \frac{2\pi r_e^2}{v_1^2}\frac{d\Delta}{\Delta^2}, \quad \Delta \ll \gamma - 1. \tag{81.15}$$

This formula, if expressed in terms of the velocity of the incident electron ($v_1 = |\mathbf{p}_1|/\varepsilon_1$), retains the same form in the non-relativistic case. Its form is naturally, therefore, the same as that of the result given by the non-relativistic theory (cf. *QM*, (148.17)).

Let us now consider the scattering of a positron by an electron (H. J. Bhabha, 1936). This is another cross-channel of the same general reaction as the electron–electron scattering. If p_-, p_+ are the initial momenta of the electron and positron, and p_-', p_+' their final momenta, the change from one case to the other is made by the substitutions

$$p_1 \to -p_+', \qquad p_2 \to p_-, \qquad p_1' \to -p_+, \qquad p_2' \to p_-'.$$

The kinematic invariants (81.2) become

$$s = (p_- - p_+')^2, \qquad t = (p_+ - p_+')^2, \qquad u = (p_- + p_+)^2. \tag{81.16}$$

If *ee* scattering is the s channel, $\bar{e}e$ scattering is the u channel of the reaction. The square of the scattering amplitude, expressed in terms of s, t and u, remains as before; in the denominator of (81.5), s must be replaced by u. Thus the cross-section for scattering of a positron by an electron is, instead of (81.7),

$$d\sigma = r_e^2 \frac{4\pi m^2\, dt}{u(u - 4m^2)}\left\{\frac{1}{t^2}[\tfrac{1}{2}(s^2 + u^2) + 4m^2(t - m^2)] + \right.$$
$$\left. + \frac{1}{u^2}[\tfrac{1}{2}(s^2 + t^2) + 4m^2(u - m^2)] + \frac{4}{tu}(\tfrac{1}{2}s - m^2)(\tfrac{1}{2}s - 3m^2)\right\}. \tag{81.17}$$

† The kinematic relations for elastic collisions in various frames of reference are given in *Fields*, §13.

In the centre-of-mass system, the values of the invariants s, t, u differ from (81.8) by the interchange of s and u:

$$s = -4\mathbf{p}^2\cos^2\tfrac{1}{2}\theta, \qquad t = -4\mathbf{p}^2\sin^2\tfrac{1}{2}\theta, \qquad u = 4\varepsilon^2. \tag{81.18}$$

In the non-relativistic limit, formula (81.17) reduces to Rutherford's formula:

$$d\sigma = \left(\frac{e^2}{mv^2}\right)^2 \frac{do}{\sin^4\frac{1}{2}\theta} \quad \text{(non-relativistic)}, \tag{81.19}$$

where $\mathbf{v} = 2\mathbf{p}/m$. This comes from the first term in the braces in (81.17), which originates from the "scattering"-type diagram (see §73). The contributions from the "annihilation" diagram (the second term in (81.17)) and from its interference with the scattering diagram (the third term) vanish in the non-relativistic limit.†

In the general case of arbitrary velocities, the contributions of all three terms in (81.17) are of the same order of magnitude; the first term predominates only at small angles, because of the factor $t^{-2} \propto \sin^{-4}\frac{1}{2}\theta$. Combining like terms, we can write the cross-section for scattering of a positron by an electron (in the centre-of-mass system) in the form

$$d\sigma = do\,\frac{r_e^2}{16}\frac{m^2}{\varepsilon^2}\left\{\frac{(\varepsilon^2+\mathbf{p}^2)^2}{\mathbf{p}^4}\frac{1}{\sin^4\frac{1}{2}\theta} - \frac{8\varepsilon^4 - m^4}{\mathbf{p}^2\varepsilon^2}\frac{1}{\sin^2\frac{1}{2}\theta} + \right.$$
$$\left. + \frac{12\varepsilon^4 + m^4}{\varepsilon^4} - \frac{4\mathbf{p}^2(\varepsilon^2+\mathbf{p}^2)}{\varepsilon^4}\sin^2\tfrac{1}{2}\theta + \frac{4\mathbf{p}^4}{\varepsilon^4}\sin^4\tfrac{1}{2}\theta\right\}. \tag{81.20}$$

The symmetry with respect to θ and $\pi - \theta$ which is typical of scattering involving identical particles does not, of course, occur when a positron is scattered by an electron. In the ultra-relativistic limit, the expression (81.20) differs from the electron–electron cross-section only by the factor $\cos^4\frac{1}{2}\theta$:

$$d\sigma_{e\bar{e}} = \cos^4\tfrac{1}{2}\theta\, d\sigma_{ee} \quad \text{(ultra-relativistic)}. \tag{81.21}$$

In the laboratory system, where one of the particles (say the electron) is at rest before the collision, we again define

$$\Delta = \frac{\varepsilon_+ - \varepsilon_+'}{m} = \frac{\varepsilon_-' - m}{m}, \tag{81.22}$$

i.e. the energy transferred by the positron to the electron. As in (81.13), we now have

$$s = -2m(\varepsilon_+ - m - m\Delta),$$
$$t = -2m^2\Delta, \qquad u = 2m(m + \varepsilon_+).$$

Substitution of these expressions in (81.17) easily gives the following formula for

† See (83.4) and (83.20) for the passage to the non-relativistic limit in the scattering and annihilation terms in the scattering amplitude. The latter term contains a factor $1/c^2$, and therefore tends to zero.

the energy distribution of the secondary electrons:

$$d\sigma = 2\pi r_e^2 \frac{d\Delta}{\gamma^2-1}\left\{\frac{\gamma^2}{\Delta^2} - \frac{2\gamma^2+4\gamma+1}{\gamma+1}\frac{1}{\Delta} + \frac{3\gamma^2+6\gamma+4}{(\gamma+1)^2} - \frac{2\gamma}{(\gamma+1)^2}\Delta + \frac{1}{(\gamma+1)^2}\Delta^2\right\}, \tag{81.23}$$

where $\gamma = \varepsilon_+/m$; Δ varies from 0 to $\gamma - 1$. When $\Delta \ll \gamma - 1$, (81.23) leads to the same formula (81.15) as for electron scattering.

The polarization effects in the scattering of electrons or positrons are calculated by the general rules given in §65. In all but special cases, the resulting formulae are lengthy. Here we shall give only some comments.†

In the approximation considered (the first non-vanishing approximation of perturbation theory), the cross-section contains no terms linear in the polarization vectors of the initial or final particles. As in the non-relativistic theory (*QM*, §140), such terms are forbidden in consequence of the requirement for the scattering matrix to be Hermitian. The scattering cross-section is therefore unchanged if only one of the colliding particles is polarized; and unpolarized particles do not become polarized as a result of scattering.

The same conditions prohibit correlation terms in the cross-section which contain the products of the polarizations of three of the particles (initial and final) concerned in the process. The cross-section does, however, contain double and quadruple correlation terms. In the scattering of unlike particles (electron and positron, electron and muon), these terms vanish in the non-relativistic limit, since there is no spin–orbit interaction. In collisions of like particles, however, there are correlation terms even in the non-relativistic case, because of exchange effects.

PROBLEMS

PROBLEM 1. Determine the scattering cross-section for polarized electrons in the non-relativistic case.

SOLUTION. In the non-relativistic case, the bispinor amplitudes in the standard representation have two components, and the density matrices are the two-rowed matrices (29.20). In the scattering amplitude (81.3), the only non-zero terms are those with $\mu = \nu = 0$, which contain matrices γ^0 that are diagonal (in the standard representation). Instead of (81.4) we have

$$\sum_{\text{polar.}} |M_{fi}|^2 = 16\pi^2 e^4 \cdot 4m^4 \left\{\left(\frac{1}{t^2}+\frac{1}{u^2}\right)\operatorname{tr}(1+\boldsymbol{\sigma}\cdot\boldsymbol{\zeta}_1)\operatorname{tr}(1+\boldsymbol{\sigma}\cdot\boldsymbol{\zeta}_2) - \frac{2}{tu}\operatorname{tr}(1+\boldsymbol{\sigma}\cdot\boldsymbol{\zeta}_1)(1+\boldsymbol{\sigma}\cdot\boldsymbol{\zeta}_2)\right\}$$

$$= 16\pi^2 e^4 \cdot 4m^4 \cdot 4\left[\frac{1}{t^2}+\frac{1}{u^2}-\frac{1}{tu}(1+\boldsymbol{\zeta}_1\cdot\boldsymbol{\zeta}_2)\right],$$

the summation being over the polarizations of the final electrons. Hence the scattering cross-section is

$$d\sigma = d\sigma_0\left(1 - \frac{\sin^2\theta}{1+3\cos^2\theta}\boldsymbol{\zeta}_1\cdot\boldsymbol{\zeta}_2\right),$$

where θ is the scattering angle in the centre-of-mass system and $d\sigma_0$ the scattering cross-section (81.9) for unpolarized particles. For completely polarized electrons, this formula is the same as the result in

† For further details see the paper by W. H. McMaster, *Reviews of Modern Physics* **33**, 8, 1961.

QM, §137, Problem, with $|\boldsymbol{\zeta}_1| = |\boldsymbol{\zeta}_2| = 1$, $\boldsymbol{\zeta}_1 \cdot \boldsymbol{\zeta}_2 = \cos\alpha$, where α is the angle between the directions of polarization of the electrons.

For the scattering of positrons by electrons, there is no dependence on the polarization in this approximation ($d\sigma = d\sigma_0$); this is easily seen by noticing that, in the non-relativistic limit, different pairs of components are non-zero in the electron and positron amplitudes u_p and u_{-p}.

PROBLEM 2. In the non-relativistic case, determine the polarization of scattered electrons in the scattering of an unpolarized beam by a polarized target.

SOLUTION. We can calculate the scattering cross-section for given initial polarization $\boldsymbol{\zeta}_2$ and detected final polarization $\boldsymbol{\zeta}_1'$; only the polarization of one final electron is detected. By the same method as in Problem 1, we find

$$d\sigma = \tfrac{1}{2}d\sigma_0\left[1 - \boldsymbol{\zeta}_1' \cdot \boldsymbol{\zeta}_2 \frac{2\cos\theta(1-\cos\theta)}{1+3\cos^2\theta}\right].$$

The polarization vector of the scattered electron is therefore

$$\boldsymbol{\zeta}_1^{(f)} = -\boldsymbol{\zeta}_2 \frac{2\cos\theta(1-\cos\theta)}{1+3\cos^2\theta}.$$

PROBLEM 3. In the non-relativistic case, determine the probability of spin reversal of a completely polarized electron scattered by an unpolarized electron.

SOLUTION. We similarly find the cross-section for given polarizations $\boldsymbol{\zeta}_1$ and $\boldsymbol{\zeta}_1'$:

$$d\sigma = \tfrac{1}{2}d\sigma_0\left[1 + \boldsymbol{\zeta}_1 \cdot \boldsymbol{\zeta}_1' \frac{2\cos\theta(1+\cos\theta)}{1+3\cos^2\theta}\right].$$

Putting $\boldsymbol{\zeta}_1 \cdot \boldsymbol{\zeta}_1' = -1$, we then find the probability of reversal of the spin direction:

$$\frac{d\sigma}{d\sigma_0} = \frac{(1-\cos\theta)^2}{2(1+3\cos^2\theta)}.$$

PROBLEM 4. Determine the ratio of the scattering cross-sections for helical electrons with parallel and antiparallel spins, in the ultra-relativistic case.

SOLUTION. In (81.4) we must put, according to (29.22),

$$\rho_1 = \tfrac{1}{2}(\gamma p_1)(1 - 2\lambda_1\gamma^5), \qquad \rho_2 = \tfrac{1}{2}(\gamma p_2)(1 - 2\lambda_2\gamma^5),$$
$$\rho_1' = \tfrac{1}{2}\gamma p_1', \qquad \rho_2' = \tfrac{1}{2}\gamma p_2',$$

where $\lambda_1, \lambda_2 = \pm\frac{1}{2}$. The traces are calculated by the formulae given in §22; in particular,

$$\begin{aligned}\mathrm{tr}\,(\gamma^5(\gamma a)\gamma^\mu(\gamma b)\gamma^\nu)\,\mathrm{tr}\,(\gamma^5(\gamma c)\gamma_\mu(\gamma d)\gamma_\nu) &= i^2(e^{\rho\mu\lambda\nu}a_\rho b_\lambda)(e_{\sigma\mu\tau\nu}c^\sigma d^\tau)\\ &= 2(\delta^\rho_\sigma\delta^\lambda_\tau - \delta^\rho_\tau\delta^\lambda_\sigma)a_\rho b_\lambda c^\sigma d^\tau\\ &= 2(ac)(bd) - 2(ad)(bc).\end{aligned}$$

The result is

$$\frac{d\sigma}{dt} \propto \left(\frac{s^2+u^2}{t^2} + \frac{s^2+t^2}{u^2} + \frac{2s^2}{tu}\right) + 4\lambda_1\lambda_2\left(\frac{s^2-u^2}{t^2} + \frac{s^2-t^2}{u^2} + \frac{2s^2}{tu}\right).$$

Since the momenta of the colliding electrons (in the centre-of-mass system) are opposite, antiparallel spins correspond to like helicities ($\lambda_1 = \lambda_2$), and parallel spins to unlike helicities ($\lambda_1 = -\lambda_2$). Substituting s, t, u from (81.8) (with $\mathbf{p}^2 \approx \varepsilon^2$), we find the required ratio:

$$d\sigma_{\uparrow\uparrow}/d\sigma_{\uparrow\downarrow} = \tfrac{1}{8}(1 + 6\cos^2\theta + \cos^4\theta). \tag{1}$$

This has its least value, $\frac{1}{8}$, when $\theta = \frac{1}{2}\pi$.

PROBLEM 5. The same as Problem 4, but for the scattering of positrons by electrons.

SOLUTION. In this case we have to calculate, instead of (81.4),

$$|M_{fi}|^2 \to 16\pi^2 e^4 \left\{ \frac{1}{t^2} \operatorname{tr}(\rho'_- \gamma^\mu \rho_- \gamma^\nu) \operatorname{tr}(\rho_+ \gamma_\mu \rho'_+ \gamma_\nu) - \frac{1}{tu} \operatorname{tr}(\rho'_- \gamma^\mu \rho_- \gamma^\nu \rho_+ \gamma_\mu \rho'_+ \gamma_\nu) + \dots \right\};$$

the remaining terms are obtained by interchanging ρ_+ and ρ'_-. The density matrices are

$$\rho_- = \tfrac{1}{2}(\gamma p_-)(1 - 2\lambda_- \gamma^5), \qquad \rho_+ = \tfrac{1}{2}(\gamma p_+)(1 + 2\lambda_+ \gamma^5),$$
$$\rho'_- = \tfrac{1}{2}\gamma p'_-, \qquad \rho'_+ = \tfrac{1}{2}\gamma p'_+,$$

where $\lambda_+, \lambda_- = \pm\frac{1}{2}$ (and for the positron, as for the electron, $\lambda_+ = \frac{1}{2}$ denotes that the spin is parallel to the momentum). The result of the calculation is

$$\frac{d\sigma}{dt} \propto \left(\frac{s^2+u^2}{t^2} + \frac{s^2+t^2}{u^2} + \frac{2s^2}{tu} \right) - 4\lambda_+\lambda_- \left(\frac{s^2-u^2}{t^2} + \frac{s^2-t^2}{u^2} + \frac{2s^2}{tu} \right).$$

Hence we find for the ratio of cross-sections the same value as formula (1), Problem 4.

PROBLEM 6. Determine the cross-section for scattering of muons by electrons.

SOLUTION. The process is described by the one diagram (73.17). Instead of (81.5) we have

$$\left.\begin{aligned} d\sigma &= \frac{\pi e^4\, dt}{(p_e p_\mu)^2 - m^2\mu^2} f(t,u) \\ &= \frac{4\pi e^4\, dt}{[s-(m+\mu)^2][s-(m-\mu)^2]} f(t,u), \\ f(t,u) &= \frac{1}{16t^2} \operatorname{tr}[(\gamma p'_\mu + \mu)\gamma^\lambda(\gamma p_\mu + \mu)\gamma^\nu] \operatorname{tr}[(\gamma p'_e + m)\gamma_\lambda(\gamma p_e + m)\gamma_\nu]; \end{aligned}\right\} \tag{1}$$

p_e, p_μ and p'_e, p'_μ are the initial and final 4-momenta of the electron and the muon, and m, μ their masses. The invariants are

$$\begin{aligned} s &= (p_e + p_\mu)^2 = m^2 + \mu^2 + 2p_e p_\mu; \\ t &= (p_e - p'_e)^2 = 2(m^2 - p_e p'_e) \\ &= 2(\mu^2 - p_\mu p'_\mu), \\ u &= (p_e - p'_\mu)^2 = m^2 + \mu^2 - 2p_e p'_\mu, \\ s &+ t + u = 2(m^2 + \mu^2). \end{aligned}$$

The result of the calculation is

$$\begin{aligned} f &= \frac{2}{t^2}\{(p_e p_\mu)^2 + (p_e p'_\mu)^2 + \tfrac{1}{2}(m^2+\mu^2)t\} \\ &= \frac{1}{t^2}\{\tfrac{1}{2}(s^2+u^2) + (m^2+\mu^2)(2t - m^2 - \mu^2)\}. \end{aligned} \tag{2}$$

Formulae (1) and (2) give the solution of the problem. In the centre-of-mass system,

$$d\sigma = \frac{e^4\, do}{8(\varepsilon_e + \varepsilon_\mu)^2 \mathbf{p}^4 \sin^4 \frac{1}{2}\theta} [(\varepsilon_e\varepsilon_\mu + \mathbf{p}^2)^2 + (\varepsilon_e\varepsilon_\mu + \mathbf{p}^2\cos\theta)^2 - 2(m^2+\mu^2)\mathbf{p}^2 \sin^2\tfrac{1}{2}\theta],$$

where $do = 2\pi \sin\theta\, d\theta$; ε_e, ε_μ are the energies of the electron and the muon; $\mathbf{p}^2 = \varepsilon_e^2 - m^2 = \varepsilon_\mu^2 - \mu^2$. If $\mathbf{p}^2 \ll \mu^2$, we return to formula (80.9) for scattering by a fixed centre of Coulomb force. In the ultra-relativistic case ($\mathbf{p}^2 \gg \mu^2$),

$$d\sigma = \frac{e^4}{8\mathbf{p}^4} \frac{1 + \cos^4 \frac{1}{2}\theta}{\sin^4 \frac{1}{2}\theta}\, do.$$

In the laboratory system (where the electron is at rest before the collision),

$$d\sigma = 2\pi\left(\frac{e^2}{m}\right)^2 \frac{d\Delta}{\mathbf{v}_\mu^2\Delta^2}\left(1 - \mathbf{v}_\mu^2\frac{\Delta}{\Delta_{\max}} + \frac{m^2}{2\varepsilon_\mu^2}\Delta^2\right),$$

where ε_μ is the energy of the incident muon, and $\mathbf{v}_\mu = \mathbf{p}_\mu/\varepsilon_\mu$ is its velocity; $m\Delta = \varepsilon_e' - m = \varepsilon_\mu - \varepsilon_\mu'$ is the energy of the recoil electron; and

$$\Delta_{\max} = \frac{2\mathbf{p}_\mu^2}{m^2 + \mu^2 + 2m\varepsilon_\mu}$$

is the maximum value of Δ.

PROBLEM 7. Determine the ratio of cross-sections for the mutual scattering of helical electrons and muons with parallel and with antiparallel spins, in the ultra-relativistic case ($\varepsilon_\mu \gg \mu$, $\varepsilon_e \gg m$).

SOLUTION.† As in Problem 4, we find

$$d\sigma\uparrow\uparrow/d\sigma\uparrow\downarrow = \cos^4\tfrac{1}{2}\theta.$$

where θ is the scattering angle in the centre-of-mass system.

PROBLEM 8. Determine the cross-section for the conversion of an electron pair into a muon pair (V. B. Berestetskiĭ and I. Ya. Pomeranchuk, 1955).

SOLUTION. This is another cross-channel of the reaction to which μe scattering belongs. In this channel,

$$s = (p_e - \bar{p}_\mu)^2, \qquad t = (p_e + \bar{p}_e)^2, \qquad u = (p_e - p_\mu)^2,$$

where p_e, $\bar{p}_e$ are the 4-momenta of the electron and the positron, and p_μ, $\bar{p}_\mu$ those of the muon and the antimuon. The reaction threshold corresponds to an energy 2μ of the electron pair (in the centre-of-mass system), so that we must have $t > 4\mu^2$. In the laboratory system, where the electron is at rest before the collision and the positron has energy ε_+,

$$t = 2m(\varepsilon_+ + m) \approx 2m\varepsilon_+,$$

so that we must have $\varepsilon_+ > \varepsilon_t$, where the threshold energy $\varepsilon_t = 2\mu^2/m$; here and below, all approximations allowed by the inequality $\mu \gg m$ are made.

The differential cross-section is (instead of formulae (1) and (2), Problem 6)

$$d\sigma = \frac{4\pi e^4\,ds}{(t - 4m^2)t}f(t, u)$$
$$\approx 4\pi e^4\frac{ds}{t^4}[\tfrac{1}{2}(s^2 + u^2) + 2\mu^2 t - \mu^4].$$

For given t, the quantity s takes values between the limits determined by the equations $su \approx \mu^4$, $s + t + u \approx 2\mu^2$, i.e.

$$\mu^2 - \tfrac{1}{2}t - \tfrac{1}{2}\sqrt{[t(t - 4\mu^2)]} \le s \le \mu^2 - \tfrac{1}{2}t + \tfrac{1}{2}\sqrt{[t(t - 4\mu^2)]}.$$

An elementary integration gives

$$\sigma = \frac{4\pi}{3}r_e^2\frac{m^2}{t}\left(1 - \frac{4\mu^2}{t}\right)^{\frac{1}{2}}\left(1 + \frac{2\mu^2}{t}\right), \qquad r_e = e^2/m; \tag{1}$$

in the laboratory system, $t = 2m\varepsilon_+$. This formula is not valid in the immediate neighbourhood of the threshold: when $\varepsilon_+ - \varepsilon_t \sim \mu e^4$, the muons formed cannot be regarded as free particles; when the Coulomb interaction between them is taken into account, the cross-section tends not to zero but to a constant value as $\varepsilon_+ \to \varepsilon_t$ (see *QM*, §147).

The cross-section (1) has its maximum value when $\varepsilon_+ = 1.7\varepsilon_t$. Its maximum value is about 20 times less than the cross-section for two-photon annihilation at the same energy.

† Another method of solving this problem is given at the end of §144.

§82. Ionization losses of fast particles

Let us consider a collision between a fast relativistic particle and an atom, accompanied by excitation or ionization of the atom. Such inelastic collisions for the non-relativistic case have been discussed in *QM*, §§148–150; here we shall derive the relativistic generalization of the formulae obtained in *QM* (H. A. Bethe, 1933).

The velocity of the particle incident on the atom is assumed large in comparison with those of the atomic electrons; thus we always assume that $Z\alpha \ll 1$, i.e. that the atomic number is fairly small. This condition ensures the applicability of the Born approximation to the process under consideration. The solution of the problem depends to some extent on whether the fast particle is light (electron or positron) or heavy (meson, proton, α-particle, etc.). The second case is the simpler, and will be taken here.

Let $p = (\varepsilon, \mathbf{p})$ and $p' = (\varepsilon', \mathbf{p}')$ be the initial and final momenta of the fast particle in the laboratory system, where the atom is at rest before the collision; the difference $q = p' - p$ gives the energy and momentum transferred to the atom by the particle. The range of possible momentum transfers can be divided into two parts:

$$\text{(I) } \mathbf{q}^2/m \ll m, \qquad \text{(II) } \mathbf{q}^2/m \gg I, \tag{82.1}$$

where m is the electron mass and I is a mean energy of the atom, its ionization potential. The two parts overlap with $I \ll \mathbf{q}^2/m \ll m$; this allows an exact joining of the results for each part separately. The momentum transfer will be said to be respectively small and large in the two parts of the range.

Small momentum transfer

In this range, the atomic electrons may be regarded as non-relativistic in both the initial and the final state of the atom.

The amplitude of the process is

$$M_{fi}^{(n)} = e^2 J_{n0}^{\mu}(-q) J_{p'p}^{\nu}(q) D_{\mu\nu}(q), \tag{82.2}$$

where J_{n0} is the transition 4-current of the atom from the initial state (0) to the final state (n), and $J_{p'p}$ is the transition 4-current of the fast particle; these currents here replace $\bar{u}'\gamma u$, which would occur, for example, in the scattering amplitude of two "elementary" particles such as the electron and muon in (73.17); cf. also (139.3). The transition currents are taken in the momentum representation (see (43.11)). The cross-section of the process in the laboratory system is

$$d\sigma_n = 2\pi\delta(\varepsilon - \varepsilon' - \omega_{n0})\, |M_{fi}^{(n)}|^2 \frac{d^3p'}{2\,|\mathbf{p}| \cdot 2\varepsilon'(2\pi)^3}, \tag{82.3}$$

where $\omega_{n0} = E_n - E_0$ is the transition frequency between the states of the atom. The final state may belong to either the discrete or the continuous spectrum, corresponding to excitation and ionization of the atom respectively. In the law of conservation of energy (represented by the delta function in (82.3)), the recoil energy of the atom is neglected; this is certainly permissible for a small momentum transfer.

The photon propagator is here conveniently taken in the gauge (76.14), in which only its space components are non-zero:

$$D_{ik}(q) = -\frac{4\pi}{\omega^2 - \mathbf{q}^2}\left(\delta_{ik} - \frac{q_i q_k}{\omega^2}\right). \tag{82.4}$$

Then only the space components of the transition currents in (82.2) are likewise needed.

The atomic transition current $\mathbf{J}_{n0}(\mathbf{q})$ is here the Fourier component of the usual non-relativistic expression:

$$\mathbf{J}_{n0}(\mathbf{q}) = \frac{i}{2m}\int e^{-i\mathbf{q}\cdot\mathbf{r}}(\psi_0 \nabla \psi_n^* - \psi_n^* \nabla \psi_0)\, d^3x, \tag{82.5}$$

where ψ_0 and ψ_n are the atomic wave functions; for simplicity, the sign of summation over the electrons in the atom will be omitted henceforward, i.e. the formulae will be written as if there were only one electron in the atom. Integrating by parts in the first term, we can rewrite this expression as a matrix element:

$$\mathbf{J}_{n0}(\mathbf{q}) = \tfrac{1}{2}(\mathbf{v}e^{-i\mathbf{q}\cdot\mathbf{r}} + e^{-i\mathbf{q}\cdot\mathbf{r}}\mathbf{v})_{n0}, \tag{82.6}$$

where $\hat{\mathbf{v}} = -(i/m)\nabla$ is the electron velocity operator.

Since the momentum lost by the scattered particle is relatively small ($|\mathbf{q}| \ll |\mathbf{p}|$), the transition current for this particle can be replaced simply by the diagonal element

$$\mathbf{J}_{pp}(0) = 2\mathbf{p}z, \tag{82.7}$$

corresponding to classical motion in a straight line (cf. (99.5)); a factor z is included to take account of a possible difference between the particle charge ze and the electron charge e.

Since $\mathbf{q}$ is small, so is the angle of deviation ϑ of the particle. The longitudinal and transverse components of $\mathbf{q}$ (relative to $\mathbf{p}$) are

$$-q_{\parallel} \approx (dp/d\varepsilon)\omega_{n0} = \omega_{n0}/v, \qquad q_{\perp} \approx |\mathbf{p}|\vartheta, \tag{82.8}$$

and hence $\mathbf{q}\cdot\mathbf{p} \approx -\varepsilon\omega_{n0}$.

Substitution of (82.4)–(82.8) in (82.2) gives

$$M_{fi}^{(n)} = -\frac{4\pi z e^2}{q^2}\left\langle n \left| \frac{\varepsilon}{\omega_{n0}}(\mathbf{q}\cdot\mathbf{v}\,e^{-i\mathbf{q}\cdot\mathbf{r}} + e^{-i\mathbf{q}\cdot\mathbf{r}}\mathbf{q}\cdot\mathbf{v}) + (\mathbf{p}\cdot\mathbf{v}e^{-i\mathbf{q}\cdot\mathbf{r}} + e^{-i\mathbf{q}\cdot\mathbf{r}}\mathbf{p}\cdot\mathbf{v}) \right| 0 \right\rangle.$$

In the first term, since

$$\mathbf{q}\cdot\hat{\mathbf{v}}f + f\mathbf{q}\cdot\hat{\mathbf{v}} = 2i\hat{\dot{f}},$$

where $f \equiv e^{-i\mathbf{q}\cdot\mathbf{r}}$ (see *QM*, §149), the matrix element of this operator is $2i(\dot{f})_{n0} = 2\omega_{n0}f_{n0}$. In the second term, $e^{-i\mathbf{q}\cdot\mathbf{r}}$ can be taken as unity, since $\mathbf{q}$ is small. Then

$$M_{fi}^{(n)} = -\frac{8\pi ze^2}{q}\{\varepsilon(e^{-i\mathbf{q}\cdot\mathbf{r}})_{n0} - i\mathbf{p}\cdot\mathbf{r}_{n0}\omega_{n0}\}.$$

The squared modulus is

$$|M_{fi}^{(n)}|^2 = \frac{64\pi^2(ze^2)^2}{(q^2)^2}\{\varepsilon^2|(e^{-i\mathbf{q}\cdot\mathbf{r}})_{n0}|^2 + 2(\mathbf{q}\cdot\mathbf{r}_{n0})(\mathbf{p}\cdot\mathbf{r}_{n0})\varepsilon\omega_{n0} + (\mathbf{p}\cdot\mathbf{r}_{n0})^2\omega_{n0}^2\}; \quad (82.9)$$

here, in the second term, we have put $e^{-i\mathbf{q}\cdot\mathbf{r}} \approx 1 - i\mathbf{q}\cdot\mathbf{r}$, but this cannot be done in the first term, for a reason explained in the next footnote but two.

The energy lost by the fast particle in its inelastic collisions with atoms† is given by

$$\kappa = \sum_n \int \omega_{n0}\, d\sigma_n = \frac{1}{16\pi^2}\sum_n \int \omega_{n0}|M_{fi}^{(n)}|^2\, do', \quad (82.10)$$

where the summation is over all possible final states of the atom, and the integration is over the directions of the scattered particle; this quantity will be called the *effective retardation* (κ/ε is known as the *energy loss cross-section*).

The integration in (82.10) can be carried out in two stages, by averaging over the azimuth of the direction of $\mathbf{p}'$ relative to $\mathbf{p}$ and then integrating over $do' \approx 2\pi\vartheta\, d\vartheta$, where ϑ is the small scattering angle. The first stage makes the change

$$\mathbf{q}\cdot\mathbf{r}_{n0} \to q_{\|}x_{n0} = -(\omega_{n0}/v)x_{n0},$$

where x_{n0} is the matrix element of one of the Cartesian coordinates of the atomic electrons.‡ The integration over $d\vartheta$ can be replaced by integration with respect to q^2, since

$$-q^2 = -\omega_{n0}^2 + \mathbf{q}^2 \approx -\omega_{n0}^2 + \frac{\omega_{n0}^2}{v^2} + \mathbf{p}^2\vartheta^2 = \frac{\omega_{n0}^2 M^2}{\mathbf{p}^2} + \mathbf{p}^2\vartheta^2 \quad (82.11)$$

and therefore $2\vartheta\, d\vartheta = d|q^2|/\mathbf{p}^2$, M being the mass of the fast particle. The result is

$$\kappa = 4\pi(ze^2)^2\sum_n \int \left\{|(e^{-i\mathbf{q}\cdot\mathbf{r}})_{n0}|^2\frac{\omega_{n0}}{v^2} - \omega_{n0}^3|x_{n0}|^2\left(\frac{M^2}{p^2} + \frac{1}{v^2}\right)\right\}\frac{d|q^2|}{|q^2|^2}. \quad (82.12)$$

The lower limit of the integration with respect to q^2 is

$$|q^2|_{\min} = (M^2/\mathbf{p}^2)\omega_{n0}^2. \quad (82.13)$$

† These are often called *ionization losses*, although they are due to excitation as well as ionization of atoms.

‡ It does not matter which coordinate: after the summation over the directions of the angular momentum of the atom in the final state, which is implied below, the matrix element does not depend on the direction of the x-axis.

As the upper limit, we take a value $|q^2|_1$ such that

$$I \ll |q^2|_1/m \ll m, \tag{82.14}$$

which thus lies in the overlap of ranges I and II (82.1).

The integration and summation in (82.12) are carried out in the same way as was done for the non-relativistic case in *QM*, §149. The entire range of integration is divided into two parts: (a) from $|q^2|_{\min}$ to $|q^2|_0$ and (b) from $|q^2|_0$ to $|q^2|_1$, where $|q^2|_0$ is such that

$$IM/|\mathbf{p}| \ll \sqrt{|q^2|_0} \ll m\alpha; \tag{82.15}$$

the quantity $m\alpha$ on the right is of the order of the momenta of the atomic electrons. In part (a) we can put $e^{-i\mathbf{q}\cdot\mathbf{r}} \approx 1 - i\mathbf{q}\cdot\mathbf{r}$, and the contribution of this part to κ is

$$4\pi(ze^2)^2 \sum_n \int\limits_{|q^2|_{\min}}^{|q^2|_0} \left\{\frac{1}{v^2}\,\omega_{n0}|x_{n0}|^2\frac{1}{|q^2|} - \frac{M^2}{\mathbf{p}^2}\,\omega_{n0}^3|x_{n0}|^2\frac{1}{|q^2|^2}\right\} d|q^2|$$

$$\approx \frac{4\pi(ze^2)^2}{v^2}\sum_n \omega_{n0}|x_{n0}|^2\left[\log\frac{|q^2|_0\mathbf{p}^2}{M^2\omega_{n0}^2} - v^2\right].$$

In the second term, the integration can be extended to infinity.

The summation is carried out by means of the formula

$$\sum_n \omega_{n0}|x_{n0}|^2 = Z/2m, \tag{82.16}$$

where Z is the number of electrons in the atom; see *QM*, (149.10). The result can be written

$$\frac{2\pi(ze^2)^2 Z}{mv^2}\left[\log\frac{|q^2|_0\mathbf{p}^2}{M^2 I^2} - v^2\right], \tag{82.17}$$

where I is an average energy of the atom, defined by

$$\log I = \frac{\sum_n \omega_{n0}|x_{n0}|^2 \log\omega_{n0}}{\sum_n \omega_{n0}|x_{n0}|^2}$$

$$= \frac{2m}{Z}\sum_n \omega_{n0}|x_{n0}|^2\log\omega_{n0}. \tag{82.18}$$

In part (b), (82.11) shows that $|q^2| \approx \mathbf{p}^2\vartheta^2$, i.e. $|q^2|$ is independent of the particular final state n of the atom, and the limits of integration are also independent of n. The summation over n can therefore be taken inside the integral in (82.12). In the first term, the summation is carried out by means of the formula

$$\sum_n |(e^{-i\mathbf{q}\cdot\mathbf{r}})_{n0}|^2 \omega_{n0} = (Z/2m)\mathbf{q}^2 \tag{82.19}$$

(see *QM*, (149.5)), and the integral is†

$$\frac{2\pi Z(ze^2)^2}{mv^2}\log\frac{|q^2|_1}{|q^2|_0}.$$

The integral of the second term in (82.12) over this part of the range makes a negligible contribution to κ.

Adding the last formula to (82.17), we find as the contribution to κ from the whole range of small momentum transfers

$$\frac{2\pi Z(ze^2)^2}{mv^2}\left[\log\frac{|q^2|_1\mathbf{p}^2}{M^2I^2} - v^2\right]. \tag{82.20}$$

Large momentum transfer

Let us now consider collisions with a momentum transfer which is large compared with the momentum of the atomic electrons ($\mathbf{q}^2 \gg mI$). Here we can evidently neglect the binding of the electrons in the atom, regarding them as free. Accordingly, the collision between the fast particle and the atom may be taken as an elastic collision with each of the Z atomic electrons. Because of the high speed of the particle, the atomic electrons may be assumed to be originally at rest.

Let $m\Delta$ denote the energy transferred from the fast particle to an atomic electron, and let $d\sigma_\Delta$ be the cross-section for elastic scattering with this energy transfer. The differential effective retardation by the whole atom is then

$$d\kappa = Zm\Delta d\sigma_\Delta. \tag{82.21}$$

The maximum energy which can be transferred to an electron at rest by the impact of a particle with mass $M \gg m$ is

$$m\Delta_{\max} = \frac{2m\mathbf{p}^2}{m^2+M^2+2m\varepsilon} \approx \frac{2m\mathbf{p}^2}{M^2+2m\varepsilon},$$

where ε and $\mathbf{p}$ are the energy and momentum of the incident particle; see *Fields*, (13.13). We shall also suppose that the energy ε, though ultra-relativistic ($\varepsilon \gg M$), is nevertheless such that

$$\varepsilon \ll M^2/m. \tag{82.22}$$

† The logarithmic divergence of the integral at the upper limit is the reason why $e^{-i\mathbf{q}\cdot\mathbf{r}}$ cannot be expanded in powers of $\mathbf{q}$ in the first term in (82.12).

Then even the maximum energy transfer

$$m\Delta_{\max} \approx 2m\mathbf{p}^2/M^2 = 2mv^2\gamma^2 \quad (\gamma = \varepsilon/M = 1/\sqrt{(1-v^2)}) \tag{82.23}$$

is small in comparison with the initial kinetic energy of the incident particle ($m\Delta_{\max} \ll \varepsilon - M$). Correspondingly, the momentum transfer $\mathbf{q}$ is always small in comparison with the initial momentum $\mathbf{p}$ of the particle. This enables us to regard the motion of the particle as being unaltered by the collision, i.e. the particle itself as having infinite mass. Then the scattering cross-section is found by simply transforming the cross-section (80.7) for electron scattering by a fixed centre to the laboratory system, in which the electron is initially at rest. This is easily done by noting that, in the approximation used,

$$-q^2 \approx \mathbf{q}^2 = 4\mathbf{p}^2 \sin^2 \tfrac{1}{2}\vartheta, \qquad do' = \pi d|q^2|/\mathbf{p}^2,$$

and the relative velocity is v in both systems. Formula (80.7) becomes

$$d\sigma = \frac{4\pi(ze^2)^2}{v^2}\left(1 - \frac{|q^2|}{4m^2\gamma^2}\right)\frac{d|q^2|}{|q^2|^2}.$$

The energy transfer Δ is expressed in terms of the same invariant q^2: $-q^2 = 2m^2\Delta$, and therefore†

$$d\sigma_\Delta = \frac{2\pi(ze^2)^2}{m^2v^2}\left(1 - v^2\frac{\Delta}{\Delta_{\max}}\right)\frac{d\Delta}{\Delta^2}. \tag{82.24}$$

The contribution to the effective retardation from this range of momentum transfers is found by integrating (82.21) from the limit $|q^2|_1$ defined above to $|q^2|_{\max} = 2m^2\Delta_{\max}$. The result is

$$\frac{2\pi(ze^2)^2Z}{mv^2}\left(\log\frac{2\Delta_{\max}m^2}{|q^2|_1} - v^2\right). \tag{82.25}$$

Finally, adding the contributions (82.20) and (82.25), we have for the total ionization losses by the fast heavy particle (in ordinary units)

$$\kappa = \frac{4\pi Z(ze^2)^2}{mv^2}\left(\log\frac{2mv^2}{I(1-v^2/c^2)} - \frac{v^2}{c^2}\right). \tag{82.26}$$

In the non-relativistic case, this reduces to *QM*, (150.10):

$$\kappa = \frac{4\pi Z(ze^2)^2}{mv^2}\log\frac{2mv^2}{I}, \tag{82.27}$$

† In this expression, of course, no account is taken of the specific effects of strong interactions when the heavy particle is a hadron. Such effects (corresponding to the hadron form factor) are, however, important only when $|q^2| \propto 1/M^2$, and such momentum transfers are excluded by the condition (82.22).

and in the ultra-relativistic case

$$\kappa = \frac{4\pi Z(ze^2)^2}{mc^2}\left(\log\frac{2mc^2}{I(1-v^2/c^2)} - 1\right). \tag{82.28}$$

The retardation depends only on the velocity of the fast particle, and not on its mass. The decrease of the retardation with increasing velocity (82.27) changes to a slow (logarithmic) increase in the ultra-relativistic range.

PROBLEMS

PROBLEM 1. Determine the effective retardation of a relativistic electron.

SOLUTION. The contribution of the range of small momentum transfers is again given by (82.20). For large momentum transfers, (82.24) must be replaced by (81.14), which includes exchange effects. Integrating $\Delta \cdot d\sigma_\Delta$ over $d\Delta$ from $|q^2|_1/2m^2$ to $\frac{1}{2}(\gamma - 1)$ and adding to (82.20), we get

$$\kappa = \frac{2\pi Ze^4}{mv^2}\left[\log\frac{m^2(\gamma^2-1)(\gamma-1)c^4}{2I^2} - \left(\frac{2}{\gamma} - \frac{1}{\gamma^2}\right)\log 2 + \frac{1}{\gamma^2} + \frac{(\gamma-1)^2}{8\gamma^2}\right], \tag{1}$$

where

$$\gamma = (1 - v^2/c^2)^{-\frac{1}{2}}.$$

In the non-relativistic case, we get the formula given in *QM*, §149, Problem, and in the ultra-relativistic case ($\gamma \gg 1$)

$$\kappa = \frac{2\pi Ze^4}{mc^2}\left(\log\frac{m^2c^4\gamma^3}{2I^2} + \frac{1}{8}\right). \tag{2}$$

PROBLEM 2. The same as Problem 1, but for a positron.

SOLUTION. For $d\sigma_\Delta$ in the range of large momentum transfers, the expression (81.23) must be used, the upper limit for Δ being $\gamma - 1$. In the ultra-relativistic case, the result is

$$\kappa = \frac{2\pi Ze^4}{mc^2}\left(\log\frac{2m^2c^4\gamma^3}{I^2} - \frac{23}{12}\right).$$

§83. Breit's equation

In classical electrodynamics, a system of interacting particles can be described by means of a Lagrangian function depending only on the coordinates and velocities of the particles, and correct as far as terms $\sim 1/c^2$ (*Fields*, §65). This is because radiation appears only as an effect of order $1/c^3$.

In the quantum theory, this corresponds to the possibility of describing the system by Schrödinger's equation including second-order terms. For an electron moving in an external electromagnetic field such an equation has been derived in §33. We shall now derive a similar equation describing a system of interacting particles.

We start from the relativistic expression for the scattering amplitude for two particles. In the non-relativistic approximation, this becomes the usual Born amplitude, proportional to the Fourier component of the potential of electrostatic interaction of two charges. By calculating the amplitude as far as second-order terms, we can establish the form of the corresponding potential, taking account of terms $\sim 1/c^2$.

Let us first assume that the two particles are different, with masses m_1 and m_2 (say an electron and muon). Then the scattering process is represented by a single diagram,

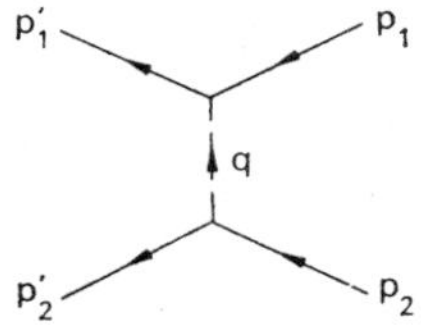

The corresponding amplitude is

$$\left.\begin{aligned} M_{fi} &= e^2(\bar{u}_1'\gamma^\mu u_1)D_{\mu\nu}(q)(\bar{u}_2'\gamma^\nu u_2), \\ q &= p_1' - p_1 = p_2 - p_2'; \end{aligned}\right\} \tag{83.1}$$

here it is assumed that the charges have the same sign. If the signs are different, e^2 becomes $-e^2$.

The subsequent calculations are considerably simplified if the photon propagator $D_{\mu\nu}$ is chosen not in the ordinary gauge but in the Coulomb gauge (76.12), (76.13):†

$$D_{00} = -\frac{4\pi}{\mathbf{q}^2}, \qquad D_{0i} = 0, \qquad D_{ik} = \frac{4\pi}{\mathbf{q}^2 - \omega^2/c^2 - i0}\left(\delta_{ik} - \frac{q_i q_k}{\mathbf{q}^2}\right). \tag{83.2}$$

Then the scattering amplitude is

$$M_{fi} = e^2\{(\bar{u}_1'\gamma^0 u_1)(\bar{u}_2'\gamma^0 u_2)D_{00} + (\bar{u}_1'\gamma^i u_1)(\bar{u}_2'\gamma^k u_2)D_{ik}\}. \tag{83.3}$$

If all terms in $1/c$ are neglected, the second term in the braces vanishes, and the first term gives

$$M_{fi} = -2m_1 \cdot 2m_2(w_1^{(0)\prime *} w_1^{(0)})(w_2^{(0)\prime *} w_2^{(0)})U(\mathbf{q}), \tag{83.4}$$

where

$$U(\mathbf{q}) = 4\pi e^2/\mathbf{q}^2, \tag{83.5}$$

and $w_1^{(0)}, w_2^{(0)}, \ldots$ denote the spinor (two-component) amplitudes of the non-relativistic plane waves, as defined in §23. The function $U(\mathbf{q})$ is the Fourier component of the Coulomb interaction potential energy, $U(r) = e^2/r$.

In the next approximation (with respect to $1/c$), the "Schrödinger" wave function of the free particle ϕ_{Sch} (normalized by the integral $\int |\phi_{\text{Sch}}|^2\, d^3x$) satisfies

† In this section, factors of c will be written in all formulae, and factors of $\hbar$ in the final formulae.

the equation

$$\left.\begin{aligned}&\hat{H}^{(0)}\phi_{\text{Sch}} = (\varepsilon - mc^2)\phi_{\text{Sch}},\\ &\hat{H}^{(0)} = \frac{\hat{\mathbf{p}}^2}{2m} - \frac{\hat{\mathbf{p}}^4}{8m^3c^2}, \qquad \hat{\mathbf{p}} = -i\nabla,\end{aligned}\right\} \tag{83.6}$$

which includes the next term in the expansion of the relativistic expression for the kinetic energy. The (spinor) amplitude of this plane wave will be denoted by w, which tends to $w^{(0)}$ as $1/c \to 0$. The required scattering amplitude must be expressed in terms of these amplitudes, in order to determine from its form the "Schrödinger" interaction potential of the particles in the approximation considered.

In accordance with formula (33.11), the bispinor amplitude u of the free particle can be expressed in terms of the "Schrödinger" amplitude w, with sufficient accuracy, by

$$u = \sqrt{(2m)}\begin{pmatrix}(1 - \mathbf{p}^2/8m^2c^2)w\\ (\boldsymbol{\sigma}\cdot\mathbf{p}/2mc)w\end{pmatrix}. \tag{83.7}$$

This formula gives

$$\begin{aligned}\bar{u}_1'\gamma^0 u_1 &= u_1'^* u_1\\ &= 2m_1\left(1 - \frac{\mathbf{p}_1'^2 + \mathbf{p}_1^2}{8m_1^2c^2}\right) w_1'^* w_1 + \frac{1}{2m_1c^2} w_1'^*(\boldsymbol{\sigma}\cdot\mathbf{p}_1')(\boldsymbol{\sigma}\cdot\mathbf{p}_1)w_1\\ &= 2m_1 w_1'^*\left\{1 - \frac{\mathbf{q}^2}{8m_1^2c^2} + \frac{i\boldsymbol{\sigma}\cdot\mathbf{q}\times\mathbf{p}_1}{4m_1^2c^2}\right\} w_1,\\ \bar{u}_1'\boldsymbol{\gamma} u_1 &= u_1'^*\boldsymbol{\alpha} u_1\\ &= (1/c)w_1'^*\{\boldsymbol{\sigma}(\boldsymbol{\sigma}\cdot\mathbf{p}_1) + (\boldsymbol{\sigma}\cdot\mathbf{p}_1')\boldsymbol{\sigma}\}w_1\\ &= (1/c)w_1'^*\{i\boldsymbol{\sigma}\times\mathbf{q} + 2\mathbf{p}_1 + \mathbf{q}\}w_1,\end{aligned}$$

where $\mathbf{q} = \mathbf{p}_1' - \mathbf{p}_1 = \mathbf{p}_2 - \mathbf{p}_2'$. The corresponding expressions for $(\bar{u}_2'\gamma^0 u_2)$ and $(\bar{u}_2'\boldsymbol{\gamma} u_2)$ differ in that the suffix 1 is replaced by 2 and $\mathbf{q}$ by $-\mathbf{q}$.

We now substitute these expressions in (83.3). Since the product $(\bar{u}_1'\boldsymbol{\gamma} u_1)(\bar{u}_2'\boldsymbol{\gamma} u_2)$ already contains the factor $1/c^2$, the term ω^2/c^2 in the denominator of D_{ik} may be neglected. The scattering amplitude is then

$$M_{fi} = -2m_1\cdot 2m_2(w_1'^* w_2'^* U(\mathbf{p}_1, \mathbf{p}_2, \mathbf{q})w_1 w_2), \tag{83.8}$$

where

$$\begin{aligned}U(\mathbf{p}_1, \mathbf{p}_2, \mathbf{q}) = 4\pi e^2\bigg\{&\frac{1}{\mathbf{q}^2} - \frac{1}{8m_1^2c^2} - \frac{1}{8m_2^2c^2} + \frac{(\mathbf{q}\cdot\mathbf{p}_1)(\mathbf{q}\cdot\mathbf{p}_2)}{m_1m_2\mathbf{q}^4} -\\ &- \frac{\mathbf{p}_1\cdot\mathbf{p}_2}{m_1m_2\mathbf{q}^2} + \frac{i\boldsymbol{\sigma}_1\cdot\mathbf{q}\times\mathbf{p}_1}{4m_1^2c^2\mathbf{q}^2} - \frac{i\boldsymbol{\sigma}_1\cdot\mathbf{q}\times\mathbf{p}_2}{2m_1m_2c^2\mathbf{q}^2} - \frac{i\boldsymbol{\sigma}_2\cdot\mathbf{q}\times\mathbf{p}_2}{4m_2^2c^2\mathbf{q}^2} +\\ &+ \frac{i\boldsymbol{\sigma}_2\cdot\mathbf{q}\times\mathbf{p}_1}{2m_1m_2c^2\mathbf{q}^2} + \frac{(\boldsymbol{\sigma}_1\cdot\mathbf{q})(\boldsymbol{\sigma}_2\cdot\mathbf{q})}{4m_1m_2c^2\mathbf{q}^2} - \frac{\boldsymbol{\sigma}_1\cdot\boldsymbol{\sigma}_2}{4m_1m_2c^2}\bigg\};\end{aligned} \tag{83.9}$$

the suffixes 1 and 2 to the Pauli matrices indicate the spinor indices on which they act, $\boldsymbol{\sigma}_1$ acting on w_1 and $\boldsymbol{\sigma}_2$ on w_2.

The function $U(\mathbf{p}_1, \mathbf{p}_2, \mathbf{q})$ is the particle interaction operator in the momentum representation. It is then related to the operator $\hat{U}(\hat{\mathbf{p}}_1, \hat{\mathbf{p}}_2, \mathbf{r})$ in the coordinate representation by

$$\int e^{-i(\mathbf{p}_1'\cdot\mathbf{r}_1+\mathbf{p}_2'\cdot\mathbf{r}_2)}\hat{U}(\hat{\mathbf{p}}_1, \hat{\mathbf{p}}_2, \mathbf{r})\, e^{i(\mathbf{p}_1\cdot\mathbf{r}_1+\mathbf{p}_2\cdot\mathbf{r}_2)}\, d^3x_1\, d^3x_2$$

$$= (2\pi)^3\delta(\mathbf{p}_1+\mathbf{p}_2-\mathbf{p}_1'-\mathbf{p}_2')U(\mathbf{p}_1, \mathbf{p}_2, \mathbf{q}). \tag{83.10}$$

If the operator $\hat{U}$ is simply a function $U(\mathbf{r})$ $(\mathbf{r} = \mathbf{r}_1 - \mathbf{r}_2)$, then $U(\mathbf{p}_1, \mathbf{p}_2, \mathbf{q})$ is independent of $\mathbf{p}_1$ and $\mathbf{p}_2$, and formula (83.10) reduces to the usual definition of the Fourier component:

$$\int e^{-i\mathbf{q}\cdot\mathbf{r}}U(\mathbf{r})\, d^3x = U(\mathbf{q}).$$

Hence it is clear that, to find $\hat{U}(\hat{\mathbf{p}}_1, \hat{\mathbf{p}}_2, \mathbf{r})$, we must calculate the integral

$$\int e^{i\mathbf{q}\cdot\mathbf{r}}\, U(\mathbf{p}_1, \mathbf{p}_2, \mathbf{q})\, d^3q/(2\pi)^3,$$

and then replace $\mathbf{p}_1$ and $\mathbf{p}_2$ by the operators $\hat{\mathbf{p}}_1 = -i\nabla_1$, $\hat{\mathbf{p}}_2 = -i\nabla_2$, writing these to the right of all the other factors.

The required integrals are found by differentiation of the formula

$$\int e^{i\mathbf{q}\cdot\mathbf{r}}\frac{4\pi}{\mathbf{q}^2}\frac{d^3q}{(2\pi)^3} = \frac{1}{r}. \tag{83.11}$$

For example, taking the gradient gives

$$\int e^{i\mathbf{q}\cdot\mathbf{r}}\frac{4\pi\mathbf{q}}{\mathbf{q}^2}\frac{d^3q}{(2\pi)^3} = -i\nabla\frac{1}{r} = \frac{i\mathbf{r}}{r^3}. \tag{83.12}$$

Next, with $\mathbf{a}$ and $\mathbf{b}$ constant vectors, we have

$$\int \frac{4\pi(\mathbf{a}\cdot\mathbf{q})(\mathbf{b}\cdot\mathbf{q})}{\mathbf{q}^4}e^{i\mathbf{q}\cdot\mathbf{r}}\frac{d^3q}{(2\pi)^3} = \tfrac{1}{2}i\left(\mathbf{a}\cdot\frac{\partial}{\partial\mathbf{r}}\right)\int e^{i\mathbf{q}\cdot\mathbf{r}}\left(\mathbf{b}\cdot\frac{\partial}{\partial\mathbf{q}}\right)\frac{1}{\mathbf{q}^2}\frac{d^3q}{(2\pi)^3};$$

the resulting integral, after integration by parts, reduces to (83.12), so that

$$\int \frac{4\pi(\mathbf{a}\cdot\mathbf{q})(\mathbf{b}\cdot\mathbf{q})}{\mathbf{q}^4}e^{i\mathbf{q}\cdot\mathbf{r}}\frac{d^3q}{(2\pi)^3} = \tfrac{1}{2}(\mathbf{a}\cdot\nabla)\frac{\mathbf{b}\cdot\mathbf{r}}{r}$$

$$= \frac{1}{2r}\left[\mathbf{a}\cdot\mathbf{b} - \frac{(\mathbf{a}\cdot\mathbf{r})(\mathbf{b}\cdot\mathbf{r})}{r^2}\right]. \tag{83.13}$$

Finally,

$$\int \frac{4\pi(\mathbf{a}\cdot\mathbf{q})(\mathbf{b}\cdot\mathbf{q})}{\mathbf{q}^2} e^{i\mathbf{q}\cdot\mathbf{r}} \frac{d^3q}{(2\pi)^3} = -(\mathbf{a}\cdot\nabla)(\mathbf{b}\cdot\nabla)\frac{1}{r}.$$

In expanding the derivatives, it must be remembered that these expressions include the delta function $\delta(\mathbf{r})$. To separate this, we note that, after averaging over the directions of $\mathbf{r}$,

$$\overline{-(\mathbf{a}\cdot\nabla)(\mathbf{b}\cdot\nabla)\frac{1}{r}} = -\tfrac{1}{3}(\mathbf{a}\cdot\mathbf{b})\triangle\frac{1}{r} = \frac{4\pi}{3}(\mathbf{a}\cdot\mathbf{b})\delta(\mathbf{r}).$$

Now expanding the derivatives in the usual manner, we find

$$\int \frac{4\pi(\mathbf{a}\cdot\mathbf{q})(\mathbf{b}\cdot\mathbf{q})}{\mathbf{q}^2} e^{i\mathbf{q}\cdot\mathbf{r}} \frac{d^3q}{(2\pi)^3} = \frac{1}{r^3}\left\{\mathbf{a}\cdot\mathbf{b} - 3\frac{(\mathbf{a}\cdot\mathbf{r})(\mathbf{b}\cdot\mathbf{r})}{r^2}\right\} + \frac{4\pi}{3}\mathbf{a}\cdot\mathbf{b}\delta(\mathbf{r}); \quad (83.14)$$

on averaging over the directions of $\mathbf{r}$, the first term vanishes and only the delta-function term remains.

Using these formulae, we obtain the following final expression for the particle interaction operator:

$$\begin{aligned}\hat{U}(\hat{\mathbf{p}}_1, \hat{\mathbf{p}}_2, \mathbf{r}) = \frac{e^2}{r} - \frac{\pi e^2\hbar^2}{2c^2}\left(\frac{1}{m_1^2}+\frac{1}{m_2^2}\right)\delta(\mathbf{r}) - \\ - \frac{e^2}{2m_1m_2c^2r}\left[\hat{\mathbf{p}}_1\cdot\hat{\mathbf{p}}_2 + \frac{\mathbf{r}\cdot(\mathbf{r}\cdot\hat{\mathbf{p}}_1)\hat{\mathbf{p}}_2}{r^2}\right] - \frac{e^2\hbar}{4m_1^2c^2r^3}\mathbf{r}\times\hat{\mathbf{p}}_1\cdot\boldsymbol{\sigma}_1 + \\ + \frac{e^2\hbar}{4m_2^2c^2r^3}\mathbf{r}\times\hat{\mathbf{p}}_2\cdot\boldsymbol{\sigma}_2 - \frac{e^2\hbar}{2m_1m_2c^2r^3}\{\mathbf{r}\times\hat{\mathbf{p}}_1\cdot\boldsymbol{\sigma}_2 - \mathbf{r}\times\hat{\mathbf{p}}_2\cdot\boldsymbol{\sigma}_1\} + \\ + \frac{e^2\hbar^2}{4m_1m_2c^2}\left\{\frac{\boldsymbol{\sigma}_1\cdot\boldsymbol{\sigma}_2}{r^3} - 3\frac{(\boldsymbol{\sigma}_1\cdot\mathbf{r})(\boldsymbol{\sigma}_2\cdot\mathbf{r})}{r^5} - \frac{8\pi}{3}\boldsymbol{\sigma}_1\cdot\boldsymbol{\sigma}_2\,\delta(\mathbf{r})\right\}. \quad (83.15)\end{aligned}$$

The total Hamiltonian of the two-particle system in this approximation is

$$\hat{H} = \hat{H}_1^{(0)} + \hat{H}_2^{(0)} + \hat{U}, \quad (83.16)$$

where $\hat{H}^{(0)}$ is the free-particle Hamiltonian (83.6).

Two electrons

If the two particles are identical (two electrons), then the scattering amplitude includes a second term which is represented by the "exchange" diagram

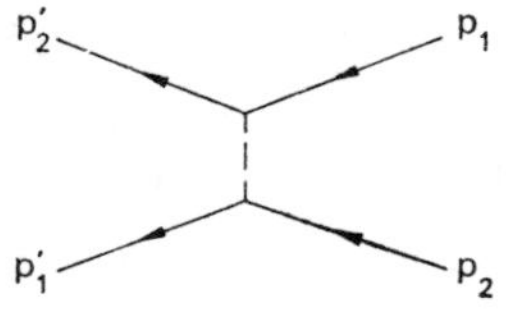

There is, however, no need to calculate the contribution of this term to the interaction operator. The reason is that the description of a system of identical particles by means of Schrödinger's equation can be achieved with an interaction operator similar to that for non-identical particles, if the solutions of the equation are appropriately symmetrized. In particular, for particle scattering this symmetrization will automatically take account of the contributions to the amplitude which correspond to the two Feynman diagrams.

Thus the Hamiltonian of the two-electron system is obtained from formulae (83.15), (83.16) by simply putting $m_1 = m_2$:†

$$\hat{H} = \frac{1}{2m}(\hat{\mathbf{p}}_1^2 + \hat{\mathbf{p}}_2^2) - \frac{1}{8m^3c^2}(\hat{\mathbf{p}}_1^4 + \hat{\mathbf{p}}_2^4) + \hat{U}(\hat{\mathbf{p}}_1, \hat{\mathbf{p}}_2, \mathbf{r}),$$

$$\hat{U}(\hat{\mathbf{p}}_1, \hat{\mathbf{p}}_2, \mathbf{r}) = \frac{e^2}{r} - \pi\left(\frac{e\hbar}{mc}\right)^2 \delta(\mathbf{r}) - \frac{e^2}{2m^2c^2r}\left(\hat{\mathbf{p}}_1 \cdot \hat{\mathbf{p}}_2 + \frac{\mathbf{r}\cdot(\mathbf{r}\cdot\hat{\mathbf{p}}_1)\hat{\mathbf{p}}_2}{r^2}\right) +$$
$$+ \frac{e^2\hbar}{4m^2c^2r^3}\{-(\boldsymbol{\sigma}_1 + 2\boldsymbol{\sigma}_2)\cdot\mathbf{r}\times\hat{\mathbf{p}}_1 + (\boldsymbol{\sigma}_2 + 2\boldsymbol{\sigma}_1)\cdot\mathbf{r}\times\hat{\mathbf{p}}_2\} +$$
$$+ \frac{1}{4}\left(\frac{e\hbar}{mc}\right)^2\left\{\frac{\boldsymbol{\sigma}_1\cdot\boldsymbol{\sigma}_2}{r^3} - \frac{3(\boldsymbol{\sigma}_1\cdot\mathbf{r})(\boldsymbol{\sigma}_2\cdot\mathbf{r})}{r^5} - \frac{8\pi}{3}\boldsymbol{\sigma}_1\cdot\boldsymbol{\sigma}_2\,\delta(\mathbf{r})\right\}. \tag{83.17}$$

The presence of terms in $\delta(\mathbf{r})$ does not, of course, imply that there is a particularly strong interaction. The value of all the correction terms after integration is of the same order, and according to the sense of the expansion used they are all to be regarded as small compared with the first term (the Coulomb interaction).

The different groups of terms in the interaction operator (83.17) are of different types. The first three terms have a purely orbital origin. The next term is linear in the spin operators of the particles, and corresponds to the spin–orbit interaction. The last term, which is quadratic in the spin operators, describes the spin–spin interaction.‡

Electron and positron

The electron–positron system needs special consideration. The scattering amplitude in this case consists of two terms:

$$M_{fi} = -e^2[\bar{u}(p_-')\gamma^\mu u(p_-)]D_{\mu\nu}(p_- - p_-')[\bar{u}(-p_+)\gamma^\nu u(-p_+')] +$$
$$+ e^2[\bar{u}(-p_+)\gamma^\mu u(p_-)]D_{\mu\nu}(p_- + p_+)[\bar{u}(p_-')\gamma^\nu u(-p_+')]; \tag{83.18}$$

the first term corresponds to the scattering diagram and the second to the annihilation diagram. Since the wave function of the "electron + positron" system need

† The wave equation with the Hamiltonian (83.17) was first derived by G. Breit (1929); a consistent quantum-mechanical derivation was given by L. D. Landau (1932).

‡ This interaction has been mentioned in *QM*, §72, in connection with the fine structure of the atomic levels, and the spin–spin interaction between the electrons and the nucleus is considered in *QM*, §121, in connection with the hyperfine structure of levels. In particular, the formula *QM* (121.9) corresponds to the delta-function term in the spin–spin interaction operator.

not be antisymmetric, the two terms make independent contributions to the interaction operator.

The first term (which has the same structure as the amplitude (83.1)) leads, of course, to an operator differing only in sign from (83.17). Let us now consider the transformation of the second term.

Here we use the photon propagator in the ordinary gauge:

$$D_{\mu\nu} = \frac{4\pi}{k^2} g_{\mu\nu} = \frac{4\pi}{\omega^2/c^2 - \mathbf{k}^2} g_{\mu\nu}.$$

In the present case $k = p_+ + p_-$, and since the particles are "almost non-relativistic", we have

$$\omega^2/c^2 \equiv (\varepsilon_+ + \varepsilon_-)^2/c^2 \approx 4m^2c^2 \gg (\mathbf{p}_+ + \mathbf{p}_-)^2 \equiv \mathbf{k}^2. \tag{83.19}$$

For the photon propagator it is therefore sufficient to write

$$D_{\mu\nu} \approx (\pi/m^2c^2) g_{\mu\nu}.$$

This already contains a factor $1/c^2$. It is therefore sufficient to take the amplitudes $u(p)$ in the zero-order approximation:

$$u(p_-) = \sqrt{(2m)} \begin{pmatrix} w_-^{(0)} \\ 0 \end{pmatrix}, \qquad u(-p_+) = \sqrt{(2m)} \begin{pmatrix} 0 \\ w^{(0)} \end{pmatrix},$$

where $w_-^{(0)}$, $w^{(0)}$ are the three-dimensional spinors which appear in (23.12); the index (0) will henceforward be omitted. With these amplitudes we have

$$\bar{u}(-p_+)\gamma^0 u(p_-) = u^*(-p_+)u(p_-) = 0,$$

$$\bar{u}(-p_+)\boldsymbol{\gamma} u(p_-) = u^*(-p_+)\boldsymbol{\alpha} u(p_-) = 2m(w^*\boldsymbol{\sigma} w_-).$$

On substitution of these expressions, the "annihilation" term in the scattering amplitude becomes

$$M_{fi}^{(\text{ann})} = -e^2 \frac{\pi}{m^2c^2} (2m)^2 (w^*\boldsymbol{\sigma} w_-)(w_-'^*\boldsymbol{\sigma} w'). \tag{83.20}$$

It is not yet possible, however, to draw from this any immediate conclusions as to the form of the interaction operator. Firstly, the spinors w in terms of which the amplitudes $u(-p_+)$ are expressed are not yet literally positron spinors. The positron amplitudes are got from $u(-p_+)$ by charge conjugation, and according to (26.6) the corresponding spinors (which we denote by w_+) are related to w by $w_+ = \sigma_y w^*$, whence

$$w^* = \sigma_y w_+ = -w_+\sigma_y, \qquad w = -\sigma_y w_+^*. \tag{83.21}$$

Secondly, the scattering amplitude must be brought to a form in which the electron spinors (w_- and w'_-) are contracted, and likewise the positron spinors (w_+ and w'_+). This is achieved by means of the formula

$$(w^*\boldsymbol{\sigma} w_-)(w'^*_-\boldsymbol{\sigma} w') = \tfrac{3}{2}(w'^*_- w_-)(w^* w') - \tfrac{1}{2}(w'^*_-\boldsymbol{\sigma} w_-)(w^*\boldsymbol{\sigma} w'), \qquad (83.22)$$

which follows from (28.17)

Finally, expressing w and w' in terms of w_+ and w'_+ by (83.21), we easily find

$$\left.\begin{aligned}(w^* w') &= (w'^*_+ w_+),\\ (w^*\boldsymbol{\sigma} w') &= -(w'^*_+\boldsymbol{\sigma} w_+).\end{aligned}\right\} \qquad (83.23)$$

Substituting (83.23) in (83.22) and then in (83.20), we obtain the final expression for the annihilation part of the scattering amplitude:

$$M_{fi}^{(\mathrm{ann})} = -4m^2\left(w'^*_- w'^*_+\left[\frac{\pi e^2}{2m^2c^2}(3+\boldsymbol{\sigma}_+\cdot\boldsymbol{\sigma}_-)\right]w_- w_+\right),$$

the matrices $\boldsymbol{\sigma}_-$ and $\boldsymbol{\sigma}_+$ acting on w_- and w_+ respectively. The expression in the square brackets is the interaction operator in the momentum representation. The corresponding coordinate operator is

$$\hat{U}^{(\mathrm{ann})}(\mathbf{r}) = \frac{\pi\hbar^2e^2}{2m^2c^2}(3+\boldsymbol{\sigma}_+\cdot\boldsymbol{\sigma}_-)\,\delta(\mathbf{r}), \qquad \mathbf{r}=\mathbf{r}_- - \mathbf{r}_+ \qquad (83.24)$$

(J. Pirenne, 1947; V. B. Berestetskiĭ and L. D. Landau, 1949). The total electron–positron interaction operator is $-\hat{U}+\hat{U}^{(\mathrm{ann})}$, with $\hat{U}$ given by (83.17).

§84. Positronium

The results obtained in §83 can be applied to *positronium*, a hydrogen-like system consisting of an electron and a positron.

In the centre-of-mass system, the electron and positron momentum operators in positronium are $\hat{\mathbf{p}}_- = -\hat{\mathbf{p}}_+ \equiv \hat{\mathbf{p}}$, where $\hat{\mathbf{p}} = -i\hbar\nabla$ is the operator of the momentum of relative motion corresponding to the relative position vector $\mathbf{r}=\mathbf{r}_- - \mathbf{r}_+$. The total Hamiltonian for positronium is†

$$\left.\begin{aligned}\hat{H} &= \frac{\hat{\mathbf{p}}^2}{m} - \frac{e^2}{r} + \hat{V}_1 + \hat{V}_2 + \hat{V}_3,\\ \hat{V}_1 &= -\frac{\hat{\mathbf{p}}^4}{4m^3c^2} + 4\pi\mu_0^2\delta(\mathbf{r}) - \frac{e^2}{2m^2c^2r}\left\{\hat{\mathbf{p}}^2 + \frac{\mathbf{r}\cdot(\mathbf{r}\cdot\hat{\mathbf{p}})\hat{\mathbf{p}}}{r^2}\right\},\\ \hat{V}_2 &= 6\mu_0^2\frac{1}{r^3}\hat{\mathbf{l}}\cdot\hat{\mathbf{S}},\\ \hat{V}_3 &= 6\mu_0^2\frac{1}{r^3}\left\{\frac{(\hat{\mathbf{S}}\cdot\mathbf{r})(\hat{\mathbf{S}}\cdot\mathbf{r})}{r^2} - \tfrac{1}{3}\hat{\mathbf{S}}^2\right\} + 4\pi\mu_0^2(\tfrac{7}{3}\hat{\mathbf{S}}^2 - 2)\,\delta(\mathbf{r}).\end{aligned}\right\} \qquad (84.1)$$

† In ordinary units.

Here $\mu_0 = e\hbar/2mc$ is the Bohr magneton, $\hbar\hat{\mathbf{l}} = \mathbf{r} \times \hat{\mathbf{p}}$ is the orbital angular momentum operator, $\hat{\mathbf{S}} = \frac{1}{2}(\boldsymbol{\sigma}_+ + \boldsymbol{\sigma}_-)$ the total spin operator of the system, whose square $\hat{\mathbf{S}}^2 = \frac{1}{2}(3 + \boldsymbol{\sigma}_+ \cdot \boldsymbol{\sigma}_-)$. $\hat{V}_1$ includes all the purely orbital correction terms, $\hat{V}_2$ the spin–orbit interaction, and $\hat{V}_3$ the spin–spin and "annihilation" interactions.

The "unperturbed" Hamiltonian

$$\hat{H} = \hat{\mathbf{p}}^2/m - e^2/r$$

naturally differs from the Hamiltonian of the hydrogen atom only in that the electron mass is replaced by the reduced mass $\frac{1}{2}m$. The energy levels of positronium therefore have absolute values which are half those of hydrogen:

$$E = -me^4/4\hbar^2 n^2, \tag{84.2}$$

where n is the principal quantum number.

The remaining terms in (84.1) cause a splitting of the levels (84.2), i.e. the appearance of a fine structure. The resulting levels are classified primarily by the values of the total angular momentum j. We also see that the particle spin operators appear in the Hamiltonian (84.1) only through the sum $\hat{\mathbf{S}}$. This means that the Hamiltonian commutes with the squared total spin operator $\hat{\mathbf{S}}^2$, i.e. the value of the total spin continues to be conserved in the approximation considered (the second approximation with respect to $1/c$). The energy levels of positronium can therefore be classified by the total spin, which takes values $S = 0$ and $S = 1$. The levels with spin 0 are called *parapositronium* levels, and those with spin 1 *orthopositronium* levels.

It must be emphasized that the conservation of the total spin in positronium is actually exact, and does not depend on any particular approximation with respect to $1/c$; it follows from the *CP* invariance of electromagnetic interactions. Positronium is a strictly neutral system, and its states therefore have definite charge parity and combined parity. The latter is equal to $(-1)^{S+1}$ (see §27, Problem); since S can take only two values, 0 and 1, the conservation of combined parity is equivalent to that of total spin.

When $S = 0$ the total angular momentum j is equal to the orbital angular momentum, but when $S = 1$ and j is given, the number l can take the values j, $j \pm 1$, so that in general each level (n, j) of orthopositronium is split into three. Since the values $l = j$ and $l = j \pm 1$ correspond to opposite parities, the Hamiltonian has no matrix elements between these states. But the perturbation operator (the first term in $\hat{V}_3$) in general has non-diagonal elements between states with $l = j + 1$ and $l = j - 1$; the number l then, of course, no longer has the strict significance of an orbital angular momentum.

The Zeeman effect in positronium has some unusual features (V. B. Berestetskiĭ and I. Ya. Pomerachuk, 1949).

The orbital magnetic moment of positronium is always zero: since in positronium $\mathbf{r}_+ \times \mathbf{p}_+ = \mathbf{r}_- \times \mathbf{p}_-$, we have the operator

$$\hat{\boldsymbol{\mu}}_l = \mu_0(\mathbf{r}_+ \times \hat{\mathbf{p}}_+ - \mathbf{r}_- \times \hat{\mathbf{p}}_-) = 0.$$

The spin magnetic moment operator is

$$\hat{\boldsymbol{\mu}}_s = \mu_0(\boldsymbol{\sigma}_+ - \boldsymbol{\sigma}_-); \tag{84.3}$$

it is not proportional to the total spin operator $\hat{\mathbf{S}} = \frac{1}{2}(\boldsymbol{\sigma}_+ + \boldsymbol{\sigma}_-)$, and the operators $\hat{\mathbf{S}}^2$ and $\hat{\boldsymbol{\mu}}^2$ do not commute. The states with definite values of the total spin S and its component S_z are therefore not, in general, eigenstates for the magnetic moment.

States with given S kand S_z are described by spin functions χ_{SS_z} having the form

$$\left.\begin{aligned} &\chi_{11} = \alpha_+\alpha_-, \qquad \chi_{1,-1} = \beta_+\beta_-, \\ &\chi_{10} = \frac{1}{\sqrt{2}}(\alpha_+\beta_- + \alpha_-\beta_+), \\ &\chi_{00} = \frac{1}{\sqrt{2}}(\alpha_+\beta_- - \alpha_-\beta_+), \end{aligned}\right\} \tag{84.4}$$

where α and β are the spin functions of one particle corresponding to spin projections $+\frac{1}{2}$ and $-\frac{1}{2}$; the suffixes + and − indicate that the function belongs to the positron and the electron respectively. The first two spin functions, χ_{11} and $\chi_{1,-1}$, are also eigenfunctions of the operator μ_z, corresponding to the eigenvalue zero. The functions χ_{10} and χ_{00} are not eigenfunctions of μ_z, but the following combinations are eigenfunctions:

$$\frac{1}{\sqrt{2}}(\chi_{10} + \chi_{00}) = \alpha_+\beta_-, \qquad \frac{1}{\sqrt{2}}(\chi_{10} - \chi_{00}) = \alpha_-\beta_+. \tag{84.5}$$

It is easy to see that the only non-zero matrix elements $\langle S'S_z'|\mu_z|SS_z\rangle$ calculated from the functions (84.4) are

$$\langle 00|\mu_z|10\rangle = \langle 10|\mu_z|00\rangle = 2\mu_0. \tag{84.6}$$

In weak magnetic fields (when $\mu_0 H \ll \Delta$, where Δ is the difference between the level energies with $S = 0$ and $S = 1$) the initial approximation for the calculation of the Zeeman splitting is formed by states with definite values of the total spin. In the first approximation, this splitting is given by the mean value of the perturbation energy operator

$$\hat{V}_H = -\hat{\mu}_z H.$$

But all the diagonal matrix elements of the operator $\hat{\mu}_z$, and therefore $\hat{V}_H$, as calculated from the functions (84.4), are zero. Thus, in weak fields, there is no linear Zeeman effect in positronium.

In the opposite limiting case of strong fields ($\mu_0 H \gg \Delta$), we can neglect the spin interaction which brings about definite values of S. The components of the split level will then correspond to states with definite values of $\mu_z = \pm 2\mu_0$ (described by the functions (84.5)), and the displacement of these components will be $\pm 2\mu_0 H$.

PROBLEMS

PROBLEM 1. Determine the fine structure of the levels of parapositronium (V. B. Berestetskiĭ, 1949).†

SOLUTION. The required level splitting energy is given by the mean values of the correction terms in the Hamiltonian (84.1), calculated by means of the wave functions of the unperturbed states with different values of $j = l$ $(=0, 1, \ldots, n-1)$. When $S = 0$, the only non-zero contributions come from $\hat{V}_1$ and the second term in $\hat{V}_3$.

The unperturbed wave functions, which we denote by ψ, satisfy Schrödinger's equation‡

$$\hat{\mathbf{p}}^2\psi = -\triangle\psi = \left(E + \frac{1}{r}\right)\psi, \qquad E = -1/4n^2.$$

Hence

$$\begin{aligned}\hat{\mathbf{p}}^4\psi = \hat{\mathbf{p}}^2\left(E + \frac{1}{r}\right)\psi &= \left(E + \frac{1}{r}\right)^2\psi - \psi\triangle\frac{1}{r} + 2\left(\nabla\frac{1}{r}\right)\cdot(\nabla\psi)\\ &= \left(E + \frac{1}{r}\right)^2\psi + 4\pi\delta(\mathbf{r})\psi + \frac{2}{r^2}\frac{\partial\psi}{\partial r}.\end{aligned}$$

The mean value is

$$\overline{\mathbf{p}^4} = \overline{\left(E + \frac{1}{r}\right)^2} + 4\pi|\psi(0)|^2 + \int\int_0^\infty \frac{\partial|\psi|^2}{\partial r}\,dr\,do.$$

The integral is equal to $-\int|\psi(0)|^2\,do$; since $\psi(0) = 0$ except when $l = 0$, and the wave functions of S states are spherically symmetric, the integral is $-4\pi|\psi(0)|^2$ and cancels with the second term.

Using the orbital angular momentum operator $\hat{\mathbf{l}} = \mathbf{r}\times\hat{\mathbf{p}}$, we can write

$$-\hat{\mathbf{p}}^2\psi = \frac{\partial^2\psi}{\partial r^2} + \frac{2}{r}\frac{\partial\psi}{\partial r} - \frac{\hat{\mathbf{l}}^2\psi}{r^2} = -\left(E + \frac{1}{r}\right)\psi.$$

The other required mean value is therefore

$$\begin{aligned}\int\psi^*\frac{\mathbf{r}}{r^3}\cdot(\mathbf{r}\cdot\hat{\mathbf{p}})\hat{\mathbf{p}}\psi\,d^3x &= -\int\psi^*\frac{1}{r}\frac{\partial^2\psi}{\partial r^2}\,d^3x\\ &= \overline{\frac{1}{r}\left(E + \frac{1}{r}\right)} - 4\pi|\psi(0)|^2 - (+1)\overline{r^{-3}};\end{aligned}$$

if $l = 0$, the last term does not appear.

According to the familiar formulae in the theory of the hydrogen atom (*QM* (36.14), (36.16)), with the electron mass m replaced by $\frac{1}{2}m$, we have

$$|\psi(0)|^2 = \frac{1}{8\pi n^3}\delta_{l0}, \qquad \overline{r^{-1}} = \frac{1}{2n^2}, \qquad \overline{r^{-2}} = \frac{1}{2n^3(2l+1)},$$

$$\overline{r^{-3}} = \frac{1}{4n^3 l(l+1)(2l+1)} \qquad (l \neq 0).$$

From these formulae, we find the required energy levels of parapositronium:

$$E_{nl} = -\frac{1}{4n^2} - \alpha^2\frac{me^4}{\hbar^2}\frac{1}{2n^3}\left(\frac{1}{2l+1} - \frac{11}{32n}\right).$$

† The fine structure of orthopositronium has been discussed by A. A. Sokolov and V. N. Tsytovich, *Zhurnal éksperimental'noĭ i teoreticheskoĭ fiziki* **24**, 253, 1953.

‡ In the calculation it is convenient to use atomic units.

PROBLEM 2. Determine the difference between the energies of the ground states ($n = 1, l = 0$) of orthopositronium and parapositronium.

SOLUTION. The dependence of the energy on the total spin S when $l = 0$ arises only from the mean value of the second term in $\hat{V}_3$; the first term gives zero on averaging over angles in the spherically symmetric S state.† The ground level of orthopositronium ($^3S_1$) lies above that of parapositronium ($^1S_0$) by an amount

$$E(^3S_1) - E(^1S_0) = \frac{7}{12}\alpha^2 \frac{me^4}{\hbar^2} = 8.2 \times 10^{-4}\ \text{eV}.$$

§ 85. The interaction of atoms at large distances

Attractive forces act between two neutral atoms at a distance r apart which is large compared with the dimensions of the atoms themselves. The usual quantum-mechanical calculation of these forces (see *QM*, §89) is, however, inapplicable at very large distances, because this calculation considered only the electrostatic interaction, i.e. retardation effects are ignored. Such a treatment is valid only if the distance r is small in comparison with the characteristic wavelengths λ_0 of the interacting atoms. In this section we shall give a calculation not subject to that limitation.

The procedure is much the same as in §83: the amplitude of elastic scattering (i.e. scattering without change of internal state) for two different atoms is calculated in the first non-vanishing approximation. The resulting expression is compared with the amplitude which would result if the interaction between the atoms were described by the potential energy $U(r)$.

In the latter case, the first non-vanishing S-matrix element describing the process in question would be the first-approximation element

$$S_{fi} = -i \int \psi_1'^*(\mathbf{r}_1)\psi_2'^*(\mathbf{r}_2)U(r)\psi_1(\mathbf{r}_1)\psi_2(\mathbf{r}_2)\, d^3x_1\, d^3x_2 \times$$
$$\times \int \exp\{-i(\varepsilon_1 + \varepsilon_2 - \varepsilon_1' - \varepsilon_2')t\}\, dt. \tag{85.1}$$

Here ψ_1, ψ_2 and ψ_1', ψ_2' are the time-independent parts of the wave functions (plane waves), describing the translational motion of the two atoms with initial and final momenta; ε_1, ε_2 and ε_1', ε_2' are the kinetic energies of this motion; the coordinates $\mathbf{r}_1$ and $\mathbf{r}_2$ of the atoms as a whole can be regarded as the coordinates of their nuclei, and the distance $r = |\mathbf{r}_1 - \mathbf{r}_2|$. The time integral in (85.1) gives, as usual, the delta function which expresses the law of conservation of energy. For convenience in the subsequent comparison, however, it is better to consider formally the limiting case of atoms of infinite mass; for given momenta, this limit corresponds to zero energies ε. Or we can say that the times considered are small in comparison with the periods $1/\varepsilon$.

† The averaging over angles must precede the integration over r, as is evident from the manner of calculation of the integral (83.14) which leads to the first term in $\hat{V}_3$.

Then (85.1) becomes

$$S_{fi} = -it \iint \psi_1'^* \psi_2'^* U(r) \psi_1 \psi_2 \, d^3x_1 \, d^3x_2, \tag{85.2}$$

where t is the time integration range.

The actual calculation of the elastic-scattering amplitude, under these assumptions, can be divided into two stages. We first average the S-operator over the wave functions of the unchanged (ground) states of the two atoms (for given coordinates $\mathbf{r}_1$ and $\mathbf{r}_2$ of their nuclei) and over the photon vacuum: no photons are present at the beginning and end of the process. We then obtain a quantity which is a function of the distance between the nuclei, and which we denote by $\langle S(r)\rangle$.† In order to find the required transition matrix element, we have then to calculate the integral

$$S_{fi} = \iint \psi_1'^* \psi_2'^* \langle S(r)\rangle \psi_1 \psi_2 \, d^3x_1 d^3x_2. \tag{85.3}$$

Comparison with (85.2) shows that, if $\langle S(r)\rangle$ is obtained in the form $\langle S(r)\rangle = -itU(r)$, then the function $U(r)$ is the required energy of interaction of the atoms.

Since we are here concerned with a collision not of elementary particles but of more complicated systems, namely atoms, which may be excited in the intermediate states, the usual formal rules of the diagram technique are not directly applicable, and we shall begin from the expression of the S-operator as the expansion (72.10).

In the interaction of atoms, the important field components are those whose frequencies are of the order of atomic frequencies or less. The corresponding wavelengths are large compared with atomic dimensions. The electromagnetic interaction operator can therefore be taken in the form

$$\hat{V} = -\hat{\mathbf{E}}(\mathbf{r}_1)\cdot\hat{\mathbf{d}}_1 - \hat{\mathbf{E}}(\mathbf{r}_2)\cdot\hat{\mathbf{d}}_2, \tag{85.4}$$

where $\hat{\mathbf{d}}_1$, $\hat{\mathbf{d}}_2$ are the dipole moment operators of the atoms (i.e. the time-dependent or Heisenberg operators) and $\hat{\mathbf{E}}(\mathbf{r})$ is the electric field operator at the positions of the corresponding atoms.

The mean values of the dipole moment of the atom in its stationary states are zero (*QM*, §75). Hence it follows that a non-zero amplitude occurs only in the fourth approximation of perturbation theory, i.e. as the matrix element of the operator

$$\hat{S}^{(4)} = \frac{(-i)^4}{4!} \int dt_1 \ldots \int dt_4 .\, \mathrm{T}\{\hat{V}(t_1)\hat{V}(t_2)\hat{V}(t_3)\hat{V}(t_4)\}: \tag{85.5}$$

in lower orders, every term in the product of operators $\hat{V}$ will contain at least one of the operators $\hat{\mathbf{d}}_1$ and $\hat{\mathbf{d}}_2$ in the first degree, and on averaging over the state of the corresponding atom the result is zero.

† In place of the more lengthy notation for a diagonal matrix element, indicating the states of the atom and of the photon field.

Let us now average the operator (85.5) over the photon vacuum. According to Wick's theorem, the expectation value of the product of four field operators $\hat{\mathbf{E}}$ is the sum of products of pairwise expectation values (contractions). The division into pairs can be made in three ways, which may be represented by the diagrams

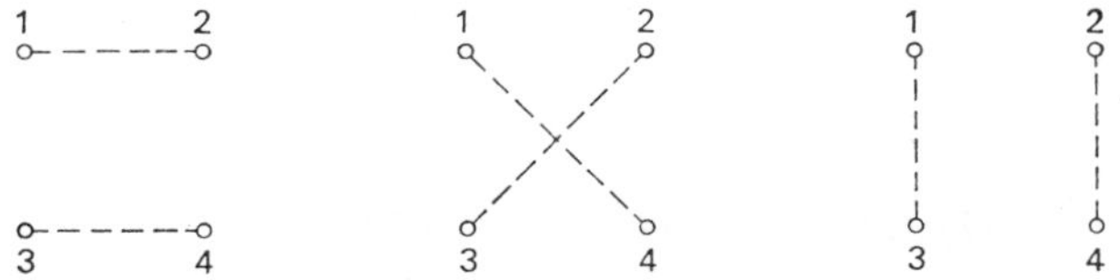

(85.6)

where the broken lines represent contractions and the numbers correspond to the arguments t_1, t_2, t_3, t_4. Moreover, spatial coordinates $\mathbf{r}_1$ or $\mathbf{r}_2$ may correspond to each point, with two points having $\mathbf{r}_1$ and two $\mathbf{r}_2$, since otherwise, in the relevant term of the sum, one of the operators $\hat{\mathbf{d}}_1$ and $\hat{\mathbf{d}}_2$ will appear in the first degree, giving zero on averaging with respect to the state of the atom. It is clear that there must be one $\mathbf{r}_1$ and one $\mathbf{r}_2$ at the ends of each line, since otherwise the diagram (i.e. the corresponding term in the matrix element) will reduce to a product of independent functions of $\mathbf{r}_1$ and of $\mathbf{r}_2$ instead of being a function of the difference $\mathbf{r}_1 - \mathbf{r}_2$; such terms do not pertain to scattering.† In accordance with these conditions, the arguments $\mathbf{r}_1$ and $\mathbf{r}_2$ can be assigned to the four points in the diagram in four ways. Using also the commutativity of the operators $\hat{\mathbf{d}}_1$ and $\hat{\mathbf{d}}_2$ and averaging over the states of each atom, we find that all the $3 \times 4 = 12$ terms thus obtained are equal, differing only in the naming of the variables of integration. The result is

$$\langle S(r)\rangle = \tfrac{1}{2}\int dt_1 \ldots \int dt_4.\ \langle \mathrm{T}(E_i(\mathbf{r}_1, t_1)E_k(\mathbf{r}_2, t_2))\rangle \times$$
$$\times \langle \mathrm{T}(E_l(\mathbf{r}_2, t_3)E_m(\mathbf{r}_1, t_4))\rangle\langle \mathrm{T}(d_{1i}(t_1)d_{1m}(t_4))\rangle\langle \mathrm{T}(d_{2k}(t_2)d_{2l}(t_3))\rangle, \qquad (85.7)$$

where i, k, l, m are three-dimensional vector indices.

To calculate the quantities

$$D^E_{ik}(x_1 - x_2) = \langle \mathrm{T}(E_i(x_1)E_k(x_2))\rangle \qquad (85.8)$$

we use the gauge in which the scalar potential $\Phi = 0$. Then $\hat{\mathbf{E}} = -\partial\hat{\mathbf{A}}/\partial t$, and we have

$$D^E_{ik}(x_1 - x_2) = \frac{\partial^2}{\partial t_1 \partial t_2}\langle \mathrm{T}(A_i(x_1)A_k(x_2))\rangle$$
$$= i\frac{\partial^2}{\partial t^2} D_{ik}(x),$$

where $x = x_1 - x_2$ and $D_{ik}(x)$ is the photon propagator in this gauge.‡

† They give corrections, of no interest here, to the energy eigenvalues of each atom.

‡ The first derivative $\partial D_{ik}(t)/\partial t$ has a finite discontinuity at $t = 0$. The second derivative, i.e. the function $D^E_{ik}(t)$, therefore includes a delta-function term $\sim \delta^{(4)}(x_2 - x_1)$. This term, however, is zero for all $\mathbf{r}_1 \neq \mathbf{r}_2$ and is of no interest here.

We shall find it more convenient to use the propagator $D_{ik}(\omega, \mathbf{r})$ in the mixed ω–$\mathbf{r}$ representation, related to $D_{ik}(t, \mathbf{r})$ by

$$D_{ik}(t, \mathbf{r}) = \int D_{ik}(\omega, \mathbf{r})\, e^{-i\omega t}\, d\omega/2\pi,$$

with

$$D^{E}_{ik}(t, \mathbf{r}) = -i \int \omega^2 D_{ik}(\omega, \mathbf{r})\, e^{-i\omega t}\, d\omega/2\pi. \tag{85.9}$$

The quantities

$$\alpha_{ik}(t_1 - t_2) = i\langle \mathrm{T}(d_i(t_1) d_k(t_2))\rangle \tag{85.10}$$

can be expressed as a Fourier integral

$$\alpha_{ik}(t) = \int_{-\infty}^{\infty} e^{-\omega t}\, \alpha_{ik}(\omega)\, d\omega/2\pi.$$

Putting for convenience $t_2 = 0$, $t_1 = t$, and using the definition of the T product, we can write

$$\begin{aligned} \alpha_{ik}(\omega) &= \int_{-\infty}^{\infty} e^{i\omega t}\, \alpha_{ik}(t)\, dt \\ &= i \int_{-\infty}^{0} e^{i\omega t}\, \langle d_k(0) d_i(t)\rangle\, dt + i \int_{0}^{\infty} e^{i\omega t} \langle d_i(t) d_k(0)\rangle\, dt. \end{aligned} \tag{85.11}$$

The mean values (with respect to the ground state of the atom) which appear here can be expressed in terms of the matrix elements of the dipole moment:

$$\langle d_k(0) d_i(t)\rangle = \sum_n (d_k)_{0n} (d_i)_{n0}\, e^{i\omega_{n0} t},$$

$$\langle d_i(t) d_k(0)\rangle = \sum_n (d_i)_{0n} (d_k)_{n0}\, e^{-i\omega_{n0} t}.$$

For convergence of the integrals in (85.11) it is necessary to take ω in the first integral as $\omega - i0$, and in the second as $\omega + i0$. Carrying out the integrations, we obtain

$$\alpha_{ik}(\omega) = \sum_n \left(\frac{(d_i)_{0n} (d_k)_{n0}}{\omega_{n0} - \omega - i0} + \frac{(d_k)_{0n} (d_i)_{n0}}{\omega_{n0} + \omega - i0} \right). \tag{85.12}$$

If the ground state is an S state, this tensor is simply a scalar, $\alpha_{ik}(\omega) = \alpha\delta_{ik}$, where

$$\alpha(\omega) = \frac{1}{3}\sum_n |\mathbf{d}_{0n}|^2 \left(\frac{1}{\omega_{n0} - \omega - i0} + \frac{1}{\omega_{n0} + \omega - i0}\right). \tag{85.13}$$

If, however, the atom has an angular momentum, the same result is obtained on averaging over the directions of this angular momentum, and it will be assumed that this has been done; we are, of course, interested in the interaction of atoms averaged over their mutual orientations.

Comparison of (85.12) with (59.17) shows that $\alpha_{ik}(\omega)$ is the same as the tensor for coherent scattering of a photon of frequency ω by an atom. According to (59.23), $\alpha(\omega)$ for $\omega > 0$ is the polarizability of the atom. Its values for $\omega < 0$ are expressed in terms of those for $\omega > 0$ by means of the relation $\alpha(-\omega) = \alpha(\omega)$, which is obvious from (85.13).

Substitution of these expressions in (85.7) gives

$$\langle S(r)\rangle = \frac{1}{2}\int dt_1 \ldots dt_4 \frac{d\Omega_1}{2\pi}\frac{d\Omega_2}{2\pi}\frac{d\omega_1}{2\pi}\frac{d\omega_2}{2\pi} \times$$
$$\times\, \alpha_1(\Omega_1)\alpha_2(\Omega_2)\omega_1^2 D_{ik}(\omega_1, \mathbf{r})\omega_2^2 D_{ik}(\omega_2, \mathbf{r}) \times$$
$$\times \exp\{-i\omega_1(t_1 - t_2) - i\omega_2(t_3 - t_4) - i\Omega_1(t_1 - t_4) - i\Omega_2(t_2 - t_3)\},$$

where $\mathbf{r} = \mathbf{r}_1 - \mathbf{r}_2$, and we have used the fact that $D_{ik}(\omega, \mathbf{r})$ is an even function of $\mathbf{r}$. The integration over three times gives delta functions (whereby $-\Omega_1 = \Omega_2 = \omega_2 = \omega_1$), and that over the fourth time gives a factor t:

$$\langle S(r)\rangle = -itU(r),$$

where

$$U(r) = \tfrac{1}{2}i \int_{-\infty}^{\infty} \omega^4 \alpha_1(\omega)\alpha_2(\omega)[D_{ik}(\omega, \mathbf{r})]^2\, d\omega/2\pi. \tag{85.14}$$

This formula gives the energy of interaction of two atoms at any distance large compared with the atomic dimensions a. We have now to find and insert an explicit expression for $D_{ik}(\omega, \mathbf{r})$.

Comparison of the expressions (76.14) and (76.8) shows that

$$D_{ik}(\omega, \mathbf{k}) = -\left(\delta_{ik} - \frac{k_i k_k}{\omega^2}\right)D(\omega, \mathbf{k}),$$

where $D(\omega, \mathbf{k})$ is given by (76.8). In the $\omega - \mathbf{r}$ representation, the relationship is correspondingly

$$D_{ik}(\omega, \mathbf{r}) = -\left(\delta_{ik} + \frac{1}{\omega^2}\frac{\partial^2}{\partial x_i \partial x_k}\right) D(\omega, \mathbf{r}). \tag{85.15}$$

Substitution of $D(\omega, \mathbf{r})$ from (76.16), and carrying out the differentiations, gives

$$D_{ik}(\omega, \mathbf{r}) = \left[\delta_{ik}\left(1 + \frac{i}{|\omega| r} - \frac{1}{\omega^2 r^2}\right) + \right.$$

$$\left. + \frac{x_i x_k}{r^2}\left(\frac{3}{\omega^2 r^2} - \frac{3i}{|\omega| r} - 1\right)\right]\frac{e^{i|\omega| r}}{r}. \tag{85.16}$$

Then, substituting this expression in (85.14), we find by a simple calculation, using the fact that $\alpha(\omega)$ is even, the final expression for the interaction energy of the atoms:

$$U(r) = \frac{i}{\pi r^2}\int_0^\infty \omega^4 \alpha_1(\omega)\alpha_2(\omega) e^{2i\omega r}\left[1 + \frac{2i}{\omega r} - \frac{5}{(\omega r)^2} - \frac{6i}{(\omega r)^3} + \frac{3}{(\omega r)^4}\right] d\omega. \tag{85.17}$$

This general result can be simplified in the limiting cases of "small" distances ($a \ll r \ll \lambda_0$) and "large" distances ($r \gg \lambda_0$).

When $r \ll \lambda_0$, the important values in the integral are (see below) $\omega \sim \omega_0$, where $\omega_0 \sim c/\lambda_0$ are the atomic frequencies, and therefore $\omega r \ll 1$. Then only the last term in the bracket need be retained, and the exponential may be replaced by unity. Writing the integral as one from $-\infty$ to ∞ (with a view to the subsequent calculations), we find

$$U(r) = \frac{3i}{2\pi r^6}\int_{-\infty}^{\infty} \alpha_1(\omega)\alpha_2(\omega)\, d\omega. \tag{85.18}$$

The interaction law at these distances proves to be $1/r^6$, as it should. The integral in (85.18) is easily calculated, after substitution of $\alpha(\omega)$ from (85.13), by closing the contour of integration with an infinite semicircle in the lower half of the complex ω-plane; the integral is determined from the residues of the integrand at the poles $\omega = \omega_{n0} \sim \omega_0$. Assuming (to simplify the result) that the two atoms are identical, we find (in ordinary units)

$$U(r) = -\frac{2}{3r^6}\sum_{n, n'} \frac{|\mathbf{d}_{0n}|^2 |\mathbf{d}_{0n'}|^2}{\hbar(\omega_{n0} + \omega_{n'0})}, \tag{85.19}$$

the same as the familiar London's formula (see *QM*, §89, Problem).

In the limit of large distances ($r \gg \lambda_0$), the important values in the integral are $\omega \lesssim c/r \ll \omega_0$; when $\omega \gtrsim \omega_0$, the integral is made small by the rapidly oscillating factor $\exp 2i\omega r$. We can therefore replace the polarizabilities $\alpha_1(\omega)$ and $\alpha_2(\omega)$ by their static values $\alpha_1(0)$ and $\alpha_2(0)$. The integration is then elementary. (To ensure convergence, r in the exponential is to be replaced by $r + i0$.) The final result is (in ordinary units)

$$U(r) = -\frac{23}{4\pi}\frac{\hbar c\alpha_1(0)\alpha_2(0)}{r^7} \tag{85.20}$$

(H. B. G. Casimir and D. Polder, 1948).†

† The derivation given here is due to I. E. Dzyaloshinskiĭ.

CHAPTER X

INTERACTION OF ELECTRONS WITH PHOTONS

§ 86. Scattering of a photon by an electron

THE conservation of 4-momentum in the scattering of a photon by a free electron (the *Compton effect*) is expressed by the equation

$$p + k = p' + k', \tag{86.1}$$

where p and k are the 4-momenta of the electron and the photon before the collision, and p' and k' their 4-momenta after the collision. The kinematic invariants defined in §66 are

$$\left.\begin{aligned} s &= (p+k)^2 = (p'+k')^2 = m^2 + 2pk = m^2 + 2p'k', \\ t &= (p-p')^2 = (k'-k)^2 = 2(m^2 - pp') = -2kk', \\ u &= (p-k')^2 = (p'-k)^2 = m^2 - 2pk' = m^2 - 2p'k, \\ & s + t + u = 2m^2. \end{aligned}\right\} \tag{86.2}$$

The process in question is represented by the two Feynman diagrams (74.14), and its amplitude is

$$M_{fi} = -4\pi e^2 e'^*_\mu e_\nu (\bar{u}' Q^{\mu\nu} u), \tag{86.3}$$

where

$$Q^{\mu\nu} = \frac{1}{s-m^2}\gamma^\mu(\gamma p + \gamma k + m)\gamma^\nu + \frac{1}{u-m^2}\gamma^\nu(\gamma p - \gamma k' + m)\gamma^\mu. \tag{86.4}$$

Here e, e' are the polarization 4-vectors of the initial and final photons; u, u' the bispinor amplitudes of the initial and final electrons.

According to the rules given in §65, for arbitrary polarization states of the particles $|M_{fi}|^2$ is replaced by

$$|M_{fi}|^2 \to 16\pi^2 e^4 \operatorname{tr}\{\rho^{(e)\prime}\rho^{(\gamma)\prime}_{\lambda\mu} Q^{\mu\nu}\rho^{(e)}\rho^{(\gamma)}_{\nu\sigma}\bar{Q}^{\lambda\sigma}\}, \tag{86.5}$$

where $\rho^{(e)}$, $\rho^{(e)\prime}$ are the density matrices of the initial and final electrons, $\rho^{(\gamma)}$, $\rho^{(\gamma)\prime}$ those of the photons. The photon (tensor) indices are written explicitly, but the electron (bispinor) indices are not. The trace symbol refers to the latter indices, as does the superscript plus in the definiton $\bar{Q}_{\mu\nu} = \gamma^0 Q^+_{\mu\nu}\gamma^0$.

Let us consider the scattering of an unpolarized photon by an unpolarized electron, without regard to their polarizations after the scattering. The averaging with respect to the polarizations of all particles is given by the density matrices:

$$\rho^{(\gamma)}_{\lambda\mu} = \rho^{(\gamma)\prime}_{\lambda\mu} = -\tfrac{1}{2}g_{\lambda\mu}, \qquad \rho^{(e)} = \tfrac{1}{2}(\gamma p + m), \qquad \rho^{(e)\prime} = \tfrac{1}{2}(\gamma p' + m);$$

the change to summation over the polarizations of the final particles involves a further multiplication by $2\times 2 = 4$.

From formula (64.23), in which we must now put $I^2 = \tfrac{1}{4}(s-m^2)^2$ (see (64.15a)), we find the cross-section

$$d\sigma = \frac{\pi e^4}{4}\frac{dt}{(s-m^2)^2}\,\mathrm{tr}\{(\gamma p' + m)Q^{\lambda\mu}(\gamma p + m)\bar{Q}_{\lambda\mu}\}.$$

From (65.2a), $\bar{Q}_{\mu\lambda} = Q_{\lambda\mu}$. Separating the terms which differ only by the changes $k \leftrightarrow -k'$ (and accordingly $s \leftrightarrow u$), we can put the cross-section in the form

$$d\sigma = dt\,\frac{\pi e^4}{(s-m^2)^2}[f(s,u) + g(s,u) + f(u,s) + g(u,s)],$$

with the notation

$$f(s,u) = \frac{1}{4(s-m^2)^2}\,\mathrm{tr}\{(\gamma p' + m)\gamma^\mu(\gamma p + \gamma k + m)\gamma^\nu(\gamma p + m)\gamma_\nu(\gamma p + \gamma k + m)\gamma_\mu\},$$

$$g(s,u) = \frac{1}{4(s-m^2)(u-m^2)}\mathrm{tr}\{(\gamma p' + m)\gamma^\mu(\gamma p + \gamma k + m)\gamma^\nu(\gamma p + m)\gamma_\mu \times$$
$$\times(\gamma p - \gamma k' + m)\gamma_\nu\};$$

this notation takes account of the fact that the result will depend only on the invariant quantities.

The summation over μ and ν is effected by means of formulae (22.6); then, omitting terms which contain an odd number of factors γ, we obtain

$$f(s,u) = \frac{1}{(s-m^2)^2}\mathrm{tr}\{(\gamma p')(\gamma p + \gamma k)(\gamma p)(\gamma p + \gamma k) + 4m^2(\gamma p + \gamma k)(\gamma k - \gamma p') +$$
$$+ m^2(\gamma p)(\gamma p') + 4m^4\}.$$

The trace is calculated by means of formulae (22.13); expressing all quantities in terms of the invariants s and u, we easily obtain

$$f(s,u) = \frac{2}{(s-m^2)^2}\{4m^4 - (s-m^2)(u-m^2) + 2m^2(s-m^2)\}.$$

Similarly,

$$g(s,u) = \frac{2m^2}{(s-m^2)(u-m^2)}\{4m^2 + (s-m^2) + (u-m^2)\}.$$

The cross-section is thus

$$d\sigma = 8\pi r_e^2 \frac{m^2\,dt}{(s-m^2)^2}\left\{\left(\frac{m^2}{s-m^2}+\frac{m^2}{u-m^2}\right)^2 + \right.$$
$$\left. + \left(\frac{m^2}{s-m^2}+\frac{m^2}{u-m^2}\right) - \frac{1}{4}\left(\frac{s-m^2}{u-m^2}+\frac{u-m^2}{s-m^2}\right)\right\}, \tag{86.6}$$

where $r_e = e^2/m$. This formula expresses the cross-section in terms of invariant quantities, and can easily be used to express it in terms of the collision parameters in any specified frame of reference.

Let us do this for the laboratory system, in which the electron is at rest before the collision: $p = (m, 0)$. Here

$$s - m^2 = 2m\omega, \qquad u - m^2 = -2m\omega'. \tag{86.7}$$

Squaring the equation of conservation of 4-momentum in the form $p + k - k' = p'$, we have

$$pk - pk' - kk' = 0,$$

whence (in the laboratory system)

$$m(\omega - \omega') - \omega\omega'(1 - \cos\vartheta) = 0,$$

where ϑ is the angle of scattering of the photon. This equation gives the relation between the photon energy change and the scattering angle:

$$\frac{1}{\omega'} - \frac{1}{\omega} = \frac{1}{m}(1 - \cos\vartheta). \tag{86.8}$$

The invariant t is

$$t = -2kk' = -2\omega\omega'(1 - \cos\vartheta).$$

For a given energy ω we find, using (86.8),

$$dt = 2\omega'^2\, d\cos\vartheta = (1/\pi)\omega'^2\, do' \quad (do' = 2\pi \sin\vartheta\, d\vartheta).$$

Substitution of these expressions in (86.6) gives the following formula for the scattering cross-section in the laboratory system:

$$d\sigma = \tfrac{1}{2} r_e^2 \left(\frac{\omega'}{\omega}\right)^2 \left(\frac{\omega}{\omega'} + \frac{\omega'}{\omega} - \sin^2\vartheta\right) do' \tag{86.9}$$

(O. Klein and Y. Nishina, 1929; I. E. Tamm, 1930).

Since the angle ϑ is unambiguously related to ω' by (86.8), the cross-section can be expressed in terms of the energy ω' of the scattered photon:

$$d\sigma = \pi r_e^2 \frac{m\,d\omega'}{\omega^2}\left[\frac{\omega}{\omega'}+\frac{\omega'}{\omega}+\left(\frac{m}{\omega'}-\frac{m}{\omega}\right)^2-2m\left(\frac{1}{\omega'}-\frac{1}{\omega}\right)\right], \tag{86.10}$$

with ω' varying in the range

$$\frac{\omega}{1+2\omega/m} \leqslant \omega' \leqslant \omega. \tag{86.11}$$

When $\omega \ll m$, we can put $\omega' \approx \omega$ in (86.9), and the result is, as it should be, the classical non-relativistic Thomson's formula

$$d\sigma = \tfrac{1}{2} r_e^2 (1+\cos^2\vartheta)\, do'; \tag{86.12}$$

see *Fields*, (78.7).

To calculate the total cross-section, we return to formula (86.6). The invariants s, t, u there take values satisfying the inequalities

$$s \geqslant m^2, \qquad t \leqslant 0, \qquad us \leqslant m^4. \tag{86.13}$$

These have already been derived in §67; the corresponding physical region is I in Fig. 7 (§67). They are also easily obtained directly from the expressions for the invariants in the centre-of-mass system. Here $\mathbf{p}+\mathbf{k}=0$, and the energies ε of the electron and ω of the photon are related by $\varepsilon = \sqrt{(\omega^2+m^2)}$. The invariants are

$$\left.\begin{aligned} s &= (\varepsilon+\omega)^2 = m^2 + 2\omega(\omega+\varepsilon), \\ u &= m^2 - 2\omega(\varepsilon + \omega\cos\theta), \\ t &= -2\omega^2(1-\cos\theta), \end{aligned}\right\} \tag{86.14}$$

where θ is the scattering angle (the angle between $\mathbf{p}$ and $\mathbf{p}'$ or between $\mathbf{k}$ and $\mathbf{k}'$). The three inequalities (86.13) then result from the conditions $\omega \geqslant 0$ and $-1 \leqslant \cos\theta \leqslant 1$.

For a given s (i.e. a given energy of the particles), the integration with respect to t can be replaced by one with respect to $u = 2m^2 - s - t$ over the range

$$m^4/s \leqslant u \leqslant 2m^2 - s.$$

Using instead of s and u the quantities

$$x = (s-m^2)/m^2, \qquad y = (m^2-u)/m^2, \tag{86.15}$$

we obtain

$$\sigma = \frac{8\pi r_e^2}{x^2} \int\limits_{x/(x+1)}^{x} \left[\left(\frac{1}{x}-\frac{1}{y}\right)^2 + \frac{1}{x} - \frac{1}{y} + \frac{1}{4}\left(\frac{x}{y}+\frac{y}{x}\right)\right] dy,$$

and after the elementary integration

$$\sigma = 2\pi r_e^2 \frac{1}{x}\left\{\left(1-\frac{4}{x}-\frac{8}{x^2}\right)\log(1+x)+\frac{1}{2}+\frac{8}{x}-\frac{1}{2(1+x)^2}\right\}. \quad (86.16)$$

The leading terms in the expansion for $x \ll 1$ (the non-relativistic case) are

$$\sigma = \frac{8\pi r_e^2}{3}(1-x). \quad (86.17)$$

The first term is the classical Thomson cross-section. In the opposite, ultra-relativistic, case ($x \gg 1$), the expansion of (86.16) gives

$$\sigma = 2\pi r_e^2 \frac{1}{x}(\log x + \tfrac{1}{2}). \quad (86.18)$$

In the laboratory system,

$$x = 2\omega/m, \quad (86.19)$$

so that formulae (86.16)–(86.18) give immediately the photon energy dependence of the cross-section for scattering by an electron at rest. Figure 13 shows σ as a function of ω/m.

In the ultra-relativistic case, the cross-section decreases with increasing energy both in the laboratory system ($\sigma \propto \omega^{-1}\log\omega$) and in the centre-of-mass system ($x \approx 4\omega^2/m^2$, $\sigma \propto \omega^{-2}\log\omega$). But the angular distribution in the ultra-relativistic case has quite different forms in these two frames of reference.

In the laboratory system, the differential cross-section has a sharp peak in the forward direction. In a narrow cone $\vartheta \lesssim \sqrt{(m/\omega)}$ we have $\omega' \sim \omega$ and the cross-

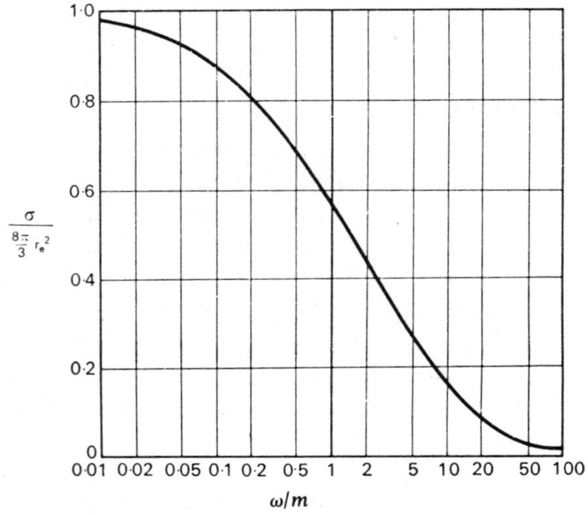

FIG. 13.

section $d\sigma/do' \sim r_e^2$, reaching the value r_e^2 as $\vartheta \to 0$. Outside this cone, the cross-section decreases, and in the range $\vartheta^2 \gg m/\omega$ (where $\omega' \approx m/(1-\cos\vartheta)$) we have

$$\frac{d\sigma}{do'} = \tfrac{1}{2} r_e^2 \frac{m}{\omega(1-\cos\vartheta)},$$

i.e. the cross-section is reduced by a factor $\sim\omega/m$.

In the centre-of-mass system, on the other hand, the differential cross-section has a peak in the backward direction. For $\pi - \theta \ll 1$ we have from (86.14)

$$\frac{s-m^2}{m^2} \approx \frac{4\omega^2}{m^2}, \qquad \frac{m^2-u}{m^2} \approx 1 + \frac{\omega^2}{m^2}(\pi-\theta)^2.$$

The largest term in the cross-section (86.6) is

$$d\sigma \approx 8\pi r_e^2 \frac{m^2\, dt}{4(s-m^2)(m^2-u)},$$

whence

$$d\sigma = \tfrac{1}{2} r_e^2 \frac{do'}{1+(\pi-\theta)^2\omega^2/m^2}. \tag{86.20}$$

The cross-section $d\sigma/do' \sim r_e^2$ in a narrow cone $\pi - \theta \lesssim m/\omega$; outside this cone it is reduced by a factor of the order of $\sim\omega^2/m^2$.

§87. Scattering of a photon by an electron. Polarization effects

We shall now go back to the original formulae of §86 and show how the calculations must be made in order to take account of the polarization of the initial and final photons and electrons.

The density matrix of the photon can be expressed, according to (8.17), by means of a pair of unit 4-vectors $e^{(1)}$, $e^{(2)}$ which satisfy the conditions (8.16). In the present case, these vectors can be taken to be, for both photons, the 4-vectors defined in §70†

$$e^{(1)} = N/\sqrt{(-N^2)}, \qquad e^{(2)} = P/\sqrt{(-P^2)}, \tag{87.1}$$

where

$$\left.\begin{aligned} P^\lambda &= (p^\lambda + p'^\lambda) - K^\lambda (pK + p'K)/K^2, \\ N^\lambda &= e^{\lambda\mu\nu\rho} P_\mu q_\nu K_\rho, \\ K^\lambda &= k^\lambda + k'^\lambda, \\ q^\lambda &= k'^\lambda - k^\lambda = p^\lambda - p'^\lambda. \end{aligned}\right\} \tag{87.2}$$

† An alternative procedure is to consider from the start a specified frame of reference (say the laboratory system) and take for each photon as $e^{(1)}$, $e^{(2)}$ purely spatial unit vectors $e = (0, \mathbf{e})$ which are orthogonal to the photon momenta and to each other. In that case, however, the calculations will be entirely in three-dimensional form, and the result will not be invariant.

The quantities $Q^{\mu\nu}$ in (86.5) are given by (86.4). They may be regarded as components of a 4-tensor (in the sense that they form a 4-tensor after being contracted with spinors as the quantities $\bar{u}'Q^{\mu\nu}u$). All the components of a 4-tensor can be obtained by projecting it on four mutually orthogonal 4-vectors, for instance on P, N, q and K defined above. Since the tensors $\rho^{(\gamma)'}_{\mu\nu}$, $\rho^{(\gamma)}_{\mu\nu}$ contain only components along P and N, we need in fact only the components of $Q_{\mu\nu}$ along these 4-vectors. In other words, it is sufficient to find in $Q_{\mu\nu}$ the terms of the form

$$Q_{\mu\nu} = Q_0(e^{(1)}_\mu e^{(1)}_\nu + e^{(2)}_\mu e^{(2)}_\nu) + Q_1(e^{(1)}_\mu e^{(2)}_\nu + e^{(2)}_\mu e^{(1)}_\nu) - \\ - iQ_2(e^{(1)}_\mu e^{(2)}_\nu - e^{(2)}_\mu e^{(1)}_\nu) + Q_3(e^{(1)}_\mu e^{(1)}_\nu - e^{(2)}_\mu e^{(2)}_\nu); \qquad (87.3)$$

the remaining terms would disappear on substitution in (86.5). The quantities Q_0 and Q_3 are scalars in the same sense that $Q_{\mu\nu}$ is a 4-tensor; they therefore contain the matrices γ only in the "invariant" combinations γK, etc. In the same sense, Q_1 and Q_2 are pseudoscalars (N is a pseudovector), and hence must contain the matrix γ^5.

By direct projection of the tensor $Q_{\mu\nu}$ we find

$$Q_0 = \tfrac{1}{2}Q^{\mu\nu}(e^{(1)}_\mu e^{(1)}_\nu + e^{(2)}_\mu e^{(2)}_\nu),$$

etc. In the calculation it is convenient first to express $Q_{\mu\nu}$ in terms of the mutually orthogonal 4-vectors P, N, q, K:

$$Q^{\mu\nu} = \gamma^\mu \frac{\frac{1}{2}\gamma P + m}{s - m^2}\gamma^\nu + \gamma^\nu \frac{\frac{1}{2}\gamma P + m}{u - m^2}\gamma^\mu - \frac{1}{t}(\gamma^\mu(\gamma K)\gamma^\nu - \gamma^\nu(\gamma K)\gamma^\mu).$$

There then remain some purely algebraic calculations using the formulae given in §22. It is also possible to make changes in $Q^{\mu\nu}$ which do not affect the result after the subsequent construction of the product $\bar{u}'Q^{\mu\nu}u$. For instance, since

$$\bar{u}'(\gamma p + \gamma p')u = 2m\bar{u}'u,$$
$$\bar{u}'\gamma^5(\gamma q)u = \bar{u}'(\gamma^5(\gamma p) + (\gamma p')\gamma^5)u = 2m\bar{u}'\gamma^5 u,$$

we can make in $Q^{\mu\nu}$ the changes

$$\gamma p + \gamma p' \to 2m, \qquad \gamma^5(\gamma q) \to 2m\gamma^5. \qquad (87.4)$$

The detailed calculations are omitted here; the final result is†

$$\left.\begin{aligned} Q_0 &= -ma_+, & Q_1 &= \tfrac{1}{2}ia_+\gamma^5(\gamma K), \\ Q_2 &= -ma_+\gamma^5, & Q_3 &= ma_+ + \tfrac{1}{2}a_-(\gamma K), \end{aligned}\right\} \qquad (87.5)$$

† The expression (87.3) with the values (87.5) corresponds to the formulae (70.11)–(70.13) derived in §70 from general considerations. Besides the equations $f_3 = f_6 = 0$ which follow from T invariance, another invariant amplitude (f_2) is here zero also. This is a property of the approximation of perturbation theory used here, and would not occur in higher approximations.

where

$$a_{\pm} = \frac{1}{s - m^2} \pm \frac{1}{u - m^2}.$$

In the subsequent calculations, it is convenient to apply to $Q_{\mu\nu}$ the same formal treatment as has been described in §8 for the photon density matrix: the four components of the tensor (87.3) in the directions $e^{(1)}$, $e^{(2)}$ are combined to form a two-rowed matrix Q which is then expanded in terms of Pauli matrices. Similarly to (8.18), we obtain

$$Q = Q_0 + \mathbf{Q} \cdot \boldsymbol{\sigma}, \qquad \mathbf{Q} = (Q_1, Q_2, Q_3). \tag{87.6}$$

The components of the tensor $\bar{Q}_{\mu\nu} = \gamma^0 Q^+_{\mu\nu} \gamma^0$ in (86.5) are easily seen from (87.3), (87.5) and the rules (65.2a) to be obtained from those of $Q_{\mu\nu}$ on replacing $Q_0, Q_1, \ldots$ by $\bar{Q}_0, \bar{Q}_1, \ldots$, where

$$\bar{Q}_0 = Q_0, \quad \bar{Q}_1 = -Q_1, \quad \bar{Q}_2 = -Q_2, \quad \bar{Q}_3 = Q_3, \tag{87.7}$$

and simultaneously interchanging the indices μ, ν.† In matrix form,

$$\bar{Q} = \bar{Q}_0 + \bar{\mathbf{Q}} \cdot \tilde{\boldsymbol{\sigma}}. \tag{87.8}$$

Let us now define more precisely the sense of the 4-vectors $e^{(1)}$, $e^{(2)}$ in relation to the polarization of the photons. For each photon, the independent directions of polarization will be determined by the components of the 3-vectors $\mathbf{e}^{(1)}$, $\mathbf{e}^{(2)}$ transverse to the photon momentum $\mathbf{k}$.‡ It is easily seen that, in both the centre-of-mass system and the laboratory system (in which the initial electron is at rest), the vector $\mathbf{P}$ is in the plane of $\mathbf{k}$ and $\mathbf{k}'$, and $\mathbf{N}$ perpendicular to that plane. The direction $\mathbf{e}^{(1)}$ is therefore that of the polarization perpendicular to the plane of scattering, and $\mathbf{e}^{(2)}$ is that of the polarization in the plane of scattering. It must also be noted that the Stokes parameters ξ_1, ξ_2, ξ_3 are defined with respect to the axes xyz, which form a right-handed set with the z-axis in the direction of $\mathbf{k}$. It is easily seen that for the initial photon the vectors $\mathbf{N}$, $\mathbf{P}_\perp$, $\mathbf{k}$ form such a set, and for the final photon the vectors $\mathbf{N}$, $-\mathbf{P}'_\perp$, $\mathbf{k}'$ (where $\mathbf{P}_\perp$ and $\mathbf{P}'_\perp$ are the components of $\mathbf{P}$ perpendicular to $\mathbf{k}$ and $\mathbf{k}'$ respectively). A change of sign $e^{(2)}$ in the photon density matrix (8.17) is equivalent to a change of sign of ξ_1 and ξ_2. The density matrices of the initial and final photons, referred to the unit 4-vectors $e^{(1)}$ and $e^{(2)}$, are therefore

$$\left.\begin{aligned} \rho^{(\gamma)} &= \tfrac{1}{2}(1 + \boldsymbol{\xi} \cdot \boldsymbol{\sigma}), \qquad & \boldsymbol{\xi} &= (\xi_1, \xi_2, \xi_3); \\ \rho^{(\gamma)\prime} &= \tfrac{1}{2}(1 + \boldsymbol{\xi}' \cdot \boldsymbol{\sigma}), \qquad & \boldsymbol{\xi}' &= (-\xi'_1, -\xi'_2, \xi'_3). \end{aligned}\right\} \tag{87.9}$$

† For the matrix $Q_{\mu\nu}$ in the original form (86.4) we should have simply $\bar{Q}_{\mu\nu} = Q_{\nu\mu}$. This property, however, is lost as a result of transformations such as (87.4).

‡ The longitudinal components of $\mathbf{e}$, like the time components of the 4-vectors e, can here be simply ignored; this is permissible, owing to gauge invariance.

The tensor trace

$$\rho^{(\gamma)\prime}_{\lambda\mu} Q^{\mu\nu} \rho^{(\gamma)}_{\nu\rho} \bar{\bar{Q}}^{\rho\lambda}$$

is now calculated as the trace of the matrix product of the matrices (87.6)–(87.9), using (33.5). The result is

$$\begin{aligned}|M_{fi}|^2 = 8\pi^2 e^4 \operatorname{tr}\{&(\rho^{(e)\prime} Q_0 \rho^{(e)} \bar{Q}_0 + \rho^{(e)\prime} \mathbf{Q} \cdot \rho^{(e)} \bar{\mathbf{Q}}) + \\ &+ (\boldsymbol{\xi} + \boldsymbol{\xi}') \cdot (\rho^{(e)\prime} Q_0 \rho^{(e)} \bar{\mathbf{Q}} + \rho^{(e)\prime} \mathbf{Q} \rho^{(e)} \bar{Q}_0) - i(\boldsymbol{\xi} - \boldsymbol{\xi}') \cdot \rho^{(e)\prime} \mathbf{Q} \times \rho^{(e)} \bar{\mathbf{Q}} + \\ &+ (\boldsymbol{\xi} \cdot \boldsymbol{\xi}')(\rho^{(e)\prime} Q_0 \rho^{(e)} \bar{Q}_0 - \rho^{(e)\prime} \mathbf{Q} \cdot \rho^{(e)} \bar{\mathbf{Q}}) + \\ &+ \rho^{(e)\prime}(\boldsymbol{\xi}' \cdot \mathbf{Q}) \rho^{(e)} (\boldsymbol{\xi} \cdot \bar{\mathbf{Q}}) + \rho^{(e)\prime}(\boldsymbol{\xi} \cdot \mathbf{Q}) \rho^{(e)} (\boldsymbol{\xi}' \cdot \bar{\mathbf{Q}}) - \\ &- i\boldsymbol{\xi} \times \boldsymbol{\xi}' \cdot (\rho^{(e)\prime} Q_0 \rho^{(e)} \bar{\mathbf{Q}} - \rho^{(e)\prime} \mathbf{Q} \rho^{(e)} \bar{Q}_0)\}.\end{aligned} \tag{87.10}$$

SCATTERING BY UNPOLARIZED ELECTRONS

We shall complete the calculation of the cross-section for the scattering of polarized photons by an unpolarized electron, summed over polarizations of the final electron. To do so, we must put in (87.10)

$$\rho^{(e)} = \tfrac{1}{2}(\gamma p + m), \qquad \rho^{(e)\prime} = \tfrac{1}{2}(\gamma p' + m),$$

double the result, and substitute it in place of $|M_{fi}|^2$ in the formula (64.22) for the cross-section:

$$d\sigma = \frac{1}{32\pi^2} \frac{dt\, d\phi}{(s - m^2)^2} |M_{fi}|^2,$$

where ϕ is the azimuth in the centre-of-mass or laboratory system. Some of the terms in (87.10) are identically zero; the calculation of the other terms gives the final result (with the notation (86.15))

$$\begin{aligned}d\sigma = \tfrac{1}{2} d\bar{\sigma} + 2r_e^2 \frac{dy\, d\phi}{x^2} \Big\{ &(\xi_3 + \xi_3') \Big[-\Big(\frac{1}{x} - \frac{1}{y}\Big)^2 - \Big(\frac{1}{x} - \frac{1}{y}\Big) \Big] + \\ &+ \xi_1 \xi_1' \Big(\frac{1}{x} - \frac{1}{y} + \frac{1}{2}\Big) + \xi_2 \xi_2' \cdot \frac{1}{4}\Big(\frac{x}{y} + \frac{y}{x}\Big)\Big(1 + \frac{2}{x} - \frac{2}{y}\Big) + \\ &+ \xi_3 \xi_3' \Big[\Big(\frac{1}{x} - \frac{1}{y}\Big)^2 + \Big(\frac{1}{x} - \frac{1}{y}\Big) + \frac{1}{2} \Big] \Big\}.\end{aligned} \tag{87.11}$$

Here $d\bar{\sigma}$ is the scattering cross-section for unpolarized photons given by (86.9); the factor $\frac{1}{2}$ appears because there is no summation over the polarizations of the final photon in (87.11).

In the laboratory system, formula (87.11) becomes

$$d\sigma = \tfrac{1}{4}r_e^2\left(\frac{\omega'}{\omega}\right)^2 do'\{F_0 + F_3(\xi_3 + \xi_3') + F_{11}\xi_1\xi_1' + F_{22}\xi_2\xi_2' + F_{33}\xi_3\xi_3'\},$$

$$do' = \sin\vartheta\, d\vartheta\, d\phi, \tag{87.12}$$

where

$$\left.\begin{aligned} F_0 &= \frac{\omega}{\omega'} + \frac{\omega'}{\omega} - \sin^2\vartheta, \qquad F_3 = \sin^2\vartheta, \\ F_{11} &= 2\cos\vartheta, \qquad F_{22} = \left(\frac{\omega}{\omega'} + \frac{\omega'}{\omega}\right)\cos\vartheta, \qquad F_{33} = 1 + \cos^2\vartheta \end{aligned}\right\} \tag{87.13}$$

(U. Fano, 1949). Although (87.12) shows no explicit dependence on the azimuth ϕ of the scattering plane, there is an implicit dependence, since the parameters ξ_1, ξ_2, ξ_3 are defined with respect to the axes xyz, which are fixed to the scattering plane. The x-axis is the same for both photons and perpendicular to the scattering plane:

$$x \parallel \mathbf{k}\times\mathbf{k}',$$

and the y-axes are in that plane:

$$y \parallel \mathbf{k}\times(\mathbf{k}\times\mathbf{k}'), \qquad y' \parallel \mathbf{k}'\times(\mathbf{k}\times\mathbf{k}').$$

Taking the sum of cross-sections differing in the sign of $\boldsymbol{\xi}'$ (i.e. putting $\boldsymbol{\xi}' = 0$ and doubling the result), we obtain the total cross-section (summed over polarizations of the final photon) for scattering of a polarized photon by an unpolarized electron. Denoting this cross-section by $d\sigma(\boldsymbol{\xi})$, we have

$$d\sigma(\boldsymbol{\xi}) = \tfrac{1}{2}r_e^2(\omega'/\omega)^2 F\, do', \tag{87.14}$$

where

$$F = F_0 + \xi_3 F_3 = \frac{\omega}{\omega'} + \frac{\omega'}{\omega} - (1 - \xi_3)\sin^2\vartheta. \tag{87.15}$$

We see that the scattering cross-section for photons polarized perpendicular to the scattering plane ($\xi_3 = 1$) is greater than that for photons polarized in the scattering plane ($\xi_3 = -1$). The cross-section is independent of circular polarization and of the parameter ξ_1. The scattering cross-section is therefore equal to that for unpolarized photons if there is no linear polarization relative to the x and y axes ($\xi_3 = 0$) or even if there is polarization relative to directions at 45° to these axes.

The cross-section for scattering of unpolarized photons with detection of a polarized photon has similar properties. This cross-section, which we denote by $d\sigma(\boldsymbol{\xi}')$, is obtained from (87.12) by putting $\boldsymbol{\xi} = 0$:

$$d\sigma(\boldsymbol{\xi}') = \tfrac{1}{4}r_e^2(\omega'/\omega)^2 F'\, do', \qquad F' = F_0 + \xi_3' F_3. \tag{87.16}$$

From formula (87.12) it is also possible to deduce the polarization of the secondary photon itself; we shall denote the parameters of this polarization by $\boldsymbol{\xi}^{(f)}$ to distinguish them from the detected polarization $\boldsymbol{\xi}'$. According to the rules given in §65, the quantities $\xi_i^{(f)}$ are equal to the ratios of the coefficients of the ξ_i' to the term independent of $\boldsymbol{\xi}'$:

$$\xi_1^{(f)} = (F_{11}/F)\xi_1, \qquad \xi_2^{(f)} = (F_{22}/F)\xi_2, \qquad \xi_3^{(f)} = (F_3 + F_{33}\xi_3)/F. \tag{87.17}$$

In particular, for the scattering of an unpolarized photon

$$\xi_1^{(f)} = \xi_2^{(f)} = 0, \qquad \xi_3^{(f)} = \frac{\sin^2\vartheta}{\omega/\omega' + \omega'/\omega - \sin^2\vartheta}. \tag{87.18}$$

Here $\xi_3^{(f)} > 0$, i.e. the secondary photon is polarized perpendicular to the scattering plane. Circular polarization of the secondary photon occurs only if the primary photon is circularly polarized: $\xi_2^{(f)} \neq 0$ only if $\xi_2 \neq 0$.

Let us consider the case of complete linear polarization of the incident photon ($\xi_2 = 0$, $\xi_1^2 + \xi_3^2 = 1$), and find the cross-section for scattering with detection of a linearly polarized secondary photon. Expressing the parameters ξ_i and ξ_i' in terms of the components of the photon polarization vectors $\mathbf{e}$ and $\mathbf{e}'$, we obtain the following expression for the scattering cross-section:

$$d\sigma = \tfrac{1}{4} r_e^2 \left(\frac{\omega'}{\omega}\right)^2 \left(\frac{\omega}{\omega'} + \frac{\omega'}{\omega} - 2 + 4\cos^2\Theta\right) do', \tag{87.19}$$

where Θ is the angle between the directions of polarization of the incident and scattered photons.†

According to this formula, the cross-section behaves quite differently when the polarizations $\mathbf{e}$ and $\mathbf{e}'$ are perpendicular and when they are parallel. Distinguishing these two cases by the suffixes $\perp$ and $\|$, we have in the non-relativistic limit ($\omega \ll m$, $\omega' \approx \omega$)

$$d\sigma_\perp = 0, \qquad d\sigma_\| = r_e^2 \cos^2\Theta \, do', \tag{87.20}$$

in agreement with the classical formulae. In the opposite, ultra-relativistic, case we have $\omega \gg m$, $\omega' \approx m/(1 - \cos\vartheta)$. Here the two ranges of large and small angles (large and small ω/ω') must be distinguished:

$$\left.\begin{aligned} d\sigma_\perp = d\sigma_\| = \tfrac{1}{4} r_e^2 \frac{\omega'}{\omega} do' = \tfrac{1}{4} r_e^2 \frac{m\, do'}{\omega(1-\cos\vartheta)} \quad &\text{for } \vartheta^2 \gg m/\omega; \\ d\sigma_\perp = 0, \qquad d\sigma_\| = r_e^2 \cos^2\Theta \, do' \quad &\text{for } \vartheta^2 \ll m/\omega. \end{aligned}\right\} \tag{87.21}$$

† Formula (87.19) itself could be more simply derived by writing from the start $e = (0, \mathbf{e})$, $e' = (0, \mathbf{e}')$ in the scattering amplitude (86.3) and continuing the calculation of the squared amplitude in three-dimensional form (i.e. separating the time and space components of the 4-vectors).

On averaging $\cos^2\Theta = (\mathbf{e}\cdot\mathbf{e}')^2$ over the directions of $\mathbf{e}$ and $\mathbf{e}'$ (using (45.4a)), and doubling the cross-section (to sum over $\mathbf{e}'$), we of course return to (86.9).

We see that the scattering cross-section has its classical value at very small angles. The approximate equality of $d\sigma_\perp$ and $d\sigma_\parallel$ at angles which are not very small signifies that in this range, in the ultra-relativistic case, the scattered radiation is unpolarized; but it must be emphasized that this conclusion applies specifically to a linearly polarized incident photon. From (87.17) it is evident that, for a circularly polarized photon in the ultra-relativistic case, $\xi_2^{(f)} \approx \xi_2 \cdot \cos\vartheta$.

Scattering by polarized electrons

For polarized electrons, the calculation of the traces in formula (87.10) becomes very laborious, though not difficult in principle. Here we shall give only some of the final results of the calculation.†

In general, the cross-section depends both on the polarization parameters $\boldsymbol{\xi}$ and $\boldsymbol{\xi}'$ of the initial and final photons, and on the polarizations of the initial and final electrons, described by vectors $\boldsymbol{\zeta}$ and $\boldsymbol{\zeta}'$. The dependence on each of these parameters is linear. The cross-section has the form

$$d\sigma = \tfrac{1}{2}d\sigma(\boldsymbol{\xi}, \boldsymbol{\xi}') + \tfrac{1}{8}r_e^2(\omega'/\omega)^2\, do'\{\mathbf{f}\cdot\boldsymbol{\zeta}\xi_2 + \mathbf{f}'\cdot\boldsymbol{\zeta}\xi_2' + \mathbf{g}\cdot\boldsymbol{\zeta}'\xi_2 + \mathbf{g}'\cdot\boldsymbol{\zeta}'\xi_2' + G_{ik}\zeta_i\zeta_k' + \cdots\}, \tag{87.22}$$

where $d\sigma(\boldsymbol{\xi}, \boldsymbol{\xi}')$ is the cross-section (87.12). All the terms which contain products of two polarization parameters have been written out in (87.22). Terms containing products of three or four parameters have been omitted; they are unimportant as regards correlations between the polarizations of only two particles, and disappear when the polarization parameters of the other two particles are equated to zero. The following are the values of some of the coefficients in the laboratory system:

$$\left.\begin{aligned}
\mathbf{f} &= -\frac{1}{m}(1-\cos\vartheta)(\mathbf{k}\cos\vartheta + \mathbf{k}'),\\
\mathbf{f}' &= -\frac{1}{m}(1-\cos\vartheta)(\mathbf{k} + \mathbf{k}'\cos\vartheta),\\
\mathbf{g} &= -\frac{1}{m}(1-\cos\vartheta)\left[(\mathbf{k}\cos\vartheta + \mathbf{k}') - (1+\cos\vartheta)\frac{\omega+\omega'}{\omega-\omega'+2m}(\mathbf{k}-\mathbf{k}')\right],\\
\mathbf{g}' &= -\frac{1}{m}(1-\cos\vartheta)\left[(\mathbf{k} + \mathbf{k}'\cos\vartheta) - (1+\cos\vartheta)\frac{\omega+\omega'}{\omega-\omega'+2m}(\mathbf{k}-\mathbf{k}').\right].
\end{aligned}\right\} \tag{87.23}$$

The cross-section (87.22) contains no term of the form $\mathbf{G}\cdot\boldsymbol{\zeta}$. This signifies that the polarization of the electron does not affect the total cross-section (summed over $\boldsymbol{\xi}'$ and $\boldsymbol{\zeta}'$) for the scattering of unpolarized photons. There is also no term of the form $\mathbf{G}'\cdot\boldsymbol{\zeta}'$. This signifies that, in the scattering of unpolarized photons, the recoil electron is unpolarized.

We see also that the terms bilinear in the polarizations of the electron and photon contain only the parameters ξ_2 and ξ_2' which correspond to circular

† Further details may be found in the review articles by H. A. Tolhoek, *Reviews of Modern Physics* **28**, 277, 1956; W. H. McMaster, *ibid.* **33**, 8, 1961.

polarization of the photon. The polarization vectors $\boldsymbol{\zeta}$ and $\boldsymbol{\zeta}'$ of the electrons appear in the form of scalar products $\mathbf{f}\cdot\boldsymbol{\zeta}$, etc., which contain only the projections of these vectors on the scattering plane. Hence, for example, the cross-section for scattering of a polarized photon by a polarized electron,

$$d\sigma(\boldsymbol{\xi}, \boldsymbol{\zeta}) = d\sigma(\boldsymbol{\xi}) + \tfrac{1}{2}r_e^2(\omega'/\omega)^2\xi_2\mathbf{f}\cdot\boldsymbol{\zeta}\, do', \tag{87.24}$$

differs from $d\sigma(\boldsymbol{\xi})$ only in that the photon is circularly polarized and the electrons have a non-zero projection of the mean spin on the scattering plane. For the same reason, the recoil electron is polarized only if the photon is circularly polarized; the resulting electron polarization vector is then in the scattering plane:

$$\boldsymbol{\zeta}^{(f)} = \xi_2\mathbf{g}/F. \tag{87.25}$$

Symmetry relations

Finally, we shall show that the qualitative properties of the polarization effects in the scattering of photons by electrons follow from the general requirements of symmetry.

The parameter ξ_2 of circular polarization is a pseudoscalar (see §8). Hence, from the requirement of P invariance, terms $\propto\xi_2$ (or $\propto\xi_2'$) in the scattering cross-section could occur only as the product of ξ_2 with some pseudoscalar formed from the available vectors $\mathbf{k}$ and $\mathbf{k}'$.† But a pseudoscalar cannot be formed from two polar vectors. It therefore follows that no such terms can appear in the cross-section.

The parameters ξ_1 and ξ_3 of linear polarization are related to the components of the two-dimensional (in a plane perpendicular to $\mathbf{k}$) symmetric tensor

$$\begin{aligned} S_{\alpha\beta} &= \tfrac{1}{2}(\rho_{\alpha\beta}^{(\gamma)} + \rho_{\beta\alpha}^{(\gamma)}) \\ &= \frac{1}{2}\begin{pmatrix} 1+\xi_3 & \xi_1 \\ \xi_1 & 1-\xi_3 \end{pmatrix}. \end{aligned}$$

In the present case, one of the polarization axes is taken to lie along the vector $\boldsymbol{\nu} = \mathbf{k}\times\mathbf{k}'$, and the other lies in the plane of $\mathbf{k}$ and $\mathbf{k}'$ (along $\mathbf{k}\times\boldsymbol{\nu}$ for one photon and along $\mathbf{k}'\times\boldsymbol{\nu}$ for the other). Terms $\propto\xi_1$ could occur in the cross-section only as products $S_{\alpha\beta}\nu_\alpha(\mathbf{k}'\times\boldsymbol{\nu})_\beta$ (or, equivalently, $S_{\alpha\beta}\nu_\alpha k_\beta'$), etc. But, since $\boldsymbol{\nu}$ is an axial vector, $\mathbf{k}$ a polar vector, and $S_{\alpha\beta}$ a true tensor, such products are not invariant with respect to inversion. There are therefore also no terms $\propto\xi_1$ (or $\propto\xi_1'$) in the cross-section. Terms $\propto\xi_3$ (or $\propto\xi_3'$), however, occur as products $S_{\alpha\beta}\nu_\alpha\nu_\beta$, etc., and are not forbidden by considerations of symmetry.

Terms in the cross-section that are proportional to the electron polarization $\boldsymbol{\zeta}$ are not forbidden by parity: such terms could arise from the products $\boldsymbol{\zeta}\cdot\boldsymbol{\nu}$ of two axial vectors. They must, however, be absent in the first non-vanishing approximation of perturbation theory, here considered, because the scattering matrix

† We are considering the process in the laboratory system, where $\mathbf{p} = 0$, $\mathbf{p}' = \mathbf{k} - \mathbf{k}'$. It is evident that the relevant consequences of the symmetry requirements (the presence or absence of particular terms in the cross-section) will not depend on the choice of the frame of reference.

is Hermitian in that approximation (§71). Owing to this property, the square of the scattering amplitude (and therefore the cross-section) is unchanged when the initial and final states are interchanged. At the same time the cross-section must be invariant under time reversal, i.e. interchange of the initial and final states together with a change of sign of the momentum and angular momentum vectors of all the particles; the Stokes parameters ξ_1, ξ_2, ξ_3 are then unaltered (see §8). On combining these two requirements, we find that in the approximation considered the cross-section must be unchanged by a change of sign of all the momenta and angular momenta without interchange of the initial and final states, i.e. by the transformation

$$\mathbf{k} \to -\mathbf{k}, \qquad \mathbf{k}' \to -\mathbf{k}', \qquad \boldsymbol{\zeta} \to -\boldsymbol{\zeta}, \qquad \boldsymbol{\zeta}' \to -\boldsymbol{\zeta}' \tag{87.26}$$

with $\boldsymbol{\xi}$ and $\boldsymbol{\xi}'$ unaltered.

The transformation (87.26) changes the sign of the product $\boldsymbol{\zeta} \cdot \boldsymbol{\nu}$, and such terms therefore cannot appear in the cross-section. It must be emphasized, however, that this prohibition is not a consequence of strict requirements of symmetry, and may therefore no longer apply in higher approximations of perturbation theory.

Among the terms of the binary correlation between the polarizations of the photons, only those of the form $\xi_1\xi_3$ and $\xi_2\xi_3$ are forbidden by parity, and none of those of the photon–electron correlation are forbidden. But all terms of the form $\xi_1\xi_2$, $\xi_1\boldsymbol{\zeta}$, $\xi_3\boldsymbol{\zeta}$ are forbidden in the first approximation by the requirement of invariance under the transformation (87.26). For instance, terms of the form $\xi_1\xi'_2$ and $\xi_1\boldsymbol{\zeta}$ could be formed (so far as parity is concerned) as scalars such as $\xi'_2 S_{\alpha\beta}k'_\alpha\nu_\beta$ and $S_{\alpha\beta}k'_\alpha\nu_\beta\boldsymbol{\zeta}\cdot\mathbf{k}$, but such combinations change sign under the transformation (87.26).

The allowed correlation terms of the form $\xi_2\boldsymbol{\zeta}$ can be formed as products of the type $\xi_2\boldsymbol{\zeta}\cdot\mathbf{k}$. The electron polarization vectors appear in them only as projections on the scattering plane.

Finally, a number of relations between the coefficients in the allowed terms result from the requirements of crossing symmetry. Reaction channels which differ by an interchange of initial and final photons correspond to the same process—scattering of a photon by an electron. The squared modulus of the amplitude, and therefore the scattering cross-section, must consequently be invariant under a transformation which expresses the change from one of these channels to the other:

$$k \leftrightarrow -k', \qquad e \leftrightarrow e'^*$$

with the electron momenta and polarizations unchanged. In three-dimensional form, this transformation is

$$\left.\begin{aligned} \omega \leftrightarrow -\omega', \qquad \mathbf{k} \leftrightarrow -\mathbf{k}', \\ \xi_1 \leftrightarrow \xi'_1, \qquad \xi_2 \leftrightarrow -\xi'_2, \qquad \xi_3 \leftrightarrow \xi'_3. \end{aligned}\right\} \tag{87.27}$$

The change in the sign of ξ_2 is evident from the expression $\xi_2 = i\mathbf{e}\times\mathbf{e}^*\cdot\mathbf{n}$, in which the vector $\mathbf{e}\times\mathbf{e}^*$ changes sign when $\mathbf{e}$ and $\mathbf{e}^*$ are interchanged, while the vector

$\mathbf{n} = \mathbf{k}/\omega$ is unchanged when $\mathbf{k} \leftrightarrow -\mathbf{k}$, $\omega \leftrightarrow -\omega$. The transformation (87.27) does not affect the electron momenta and therefore leaves the laboratory system unaltered. Hence the cross-section (87.22) cannot change its form under this transformation, and in fact the formulae (87.12), (87.22), (87.23) comply with this requirement.

§88. Two-photon annihilation of an electron pair

The annihilation of an electron and a positron (with 4-momenta p_- and p_+) to form two photons (k_1 and k_2) corresponds to two diagrams

(88.1)

These differ from the diagrams for scattering of a photon by an electron as follows:

$$p \to p_-, \qquad p' \to -p_+, \qquad k \to -k_1, \qquad k' \to k_2. \tag{88.2}$$

The two processes are two cross-channels of the same (generalized) reaction. After the changes (88.2), the kinematic invariants (86.2) become

$$\left.\begin{aligned} s &= (p_- - k_1)^2, \\ t &= (p_- + p_+)^2 = (k_1 + k_2)^2, \\ u &= (p_- - k_2)^2. \end{aligned}\right\} \tag{88.3}$$

If the photon scattering is the s channel, then the annihilation is the t channel.

The quantity $|M_{fi}|^2$ for annihilation (averaged over polarizations of the electrons and summed over those of the photons), when expressed in terms of the invariants s and u, is the same as corresponding quantity for scattering, only the meaning of the invariants being changed.† In the formula (64.23) for the cross-section, the change $s \leftrightarrow t$ is needed in the coefficients of $|M_{fi}|^2$, and I^2 is now, according to (64.15a), equal to $\frac{1}{4}t(t-4m^2)$. Making the appropriate alterations in formula (86.6), we find the annihilation cross-section

$$d\sigma = 8\pi r_e^2 \frac{m^2\, ds}{t(t-4m^2)} \left\{ \left(\frac{m^2}{s-m^2} + \frac{m^2}{u-m^2} \right)^2 + \right.$$
$$\left. + \left(\frac{m^2}{s-m^2} + \frac{m^2}{u-m^2} \right) - \frac{1}{4} \left(\frac{s-m^2}{u-m^2} + \frac{u-m^2}{s-m^2} \right) \right\}. \tag{88.4}$$

The physical region of the annihilation channel is region II in Fig. 7 (§67). For given t (given energy in the centre-of-mass system), the range of variation of s is

† This takes account of the fact that the photons and the electrons have the same number of independent polarizations (two), and it is therefore immaterial which correspond to the averaging of $|M_{fi}|^2$ and which to the summation.

determined by the equation of the boundary $su = m^4$. Together with the relation $s+t+u = 2m^2$, this gives

$$-\tfrac{1}{2}t - \tfrac{1}{2}\sqrt{[t(t-4m^2)]} \leq s - m^2 \leq -\tfrac{1}{2}t + \tfrac{1}{2}\sqrt{[t(t-4m^2)]}. \tag{88.5}$$

The integration of (88.4) is elementary; the result must be divided by two to take account of the identity of the two final particles (the photons). Thus we have

$$\sigma = \frac{\pi r_e^2}{2\tau^2(\tau-1)}\left\{(\tau^2+\tau-\tfrac{1}{2})\log\frac{\sqrt{\tau}+\sqrt{(\tau-1)}}{\sqrt{\tau}-\sqrt{(\tau-1)}} - (\tau+1)\sqrt{[\tau(\tau-1)]}\right\}, \tag{88.6}$$

where $\tau = \frac{1}{4}t/m^2$ (P. A. M. Dirac, 1930).

In the non-relativistic limit ($\tau \to 1$), this gives

$$\sigma = \tfrac{1}{2}\pi r_e^2/\sqrt{(\tau-1)}. \tag{88.7}$$

In the ultra-relativistic case ($\tau \to \infty$),

$$\sigma \doteq (\tfrac{1}{2}\pi r_e^2/\tau)(\log 4\tau - 1). \tag{88.8}$$

In the laboratory system, in which one particle (say the electron) is at rest before the collision, the invariant τ is

$$\tau = \tfrac{1}{2}(1+\gamma), \qquad \gamma = \varepsilon_+/m. \tag{88.9}$$

Formulae (88.6)–(88.8) give as the dependence of the total cross-section on the energy of the incident positron

$$\sigma = \frac{\pi r_e^2}{\gamma+1}\left\{\frac{\gamma^2+4\gamma+1}{\gamma^2-1}\log[\gamma+\sqrt{(\gamma^2-1)}] - \frac{\gamma+3}{\sqrt{(\gamma^2-1)}}\right\}. \tag{88.10}$$

In particular, in the non-relativistic limit†

$$\sigma = \pi r_e^2/v_+ \quad \text{(non-relativistic)}, \tag{88.11}$$

where v_+ is the velocity of the positron.

In the centre-of-mass system the electron, the positron and the two photons have equal energies, $\varepsilon = \omega$. The invariants are

$$\left.\begin{aligned} m^2 - s &= 2\varepsilon(\varepsilon - |\mathbf{p}|\cos\theta), \\ m^2 - u &= 2\varepsilon(\varepsilon + |\mathbf{p}|\cos\theta), \\ t &= 4\varepsilon^2, \end{aligned}\right\} \tag{88.12}$$

where θ is the angle between the momentum of the electron and that of one of the

† This formula becomes inapplicable, however, when $v_+ \lesssim \alpha$ and the Coulomb interaction of the components of the pair cannot be neglected; cf. the end of §94.

photons. Substituting in (88.4), we find the angular distribution of the annihilation photons:

$$d\sigma = \frac{r_e^2 m^2}{4\varepsilon|\mathbf{p}|}\left[\frac{\varepsilon^2+\mathbf{p}^2(1+\sin^2\theta)}{\varepsilon^2-\mathbf{p}^2\cos^2\theta}-\frac{2\mathbf{p}^4\sin^4\theta}{(\varepsilon^2-\mathbf{p}^2\cos^2\theta)^2}\right]do. \tag{88.13}$$

In the ultra-relativistic case this has symmetrical maxima in the directions $\theta = 0$ and $\theta = \pi$. Near $\theta = 0$, we have

$$d\sigma \approx \frac{r_e^2 m^2\, do}{2\varepsilon^2(\theta^2+m^2/\varepsilon^2)} \quad \text{(ultra-relativistic)}. \tag{88.14}$$

The total cross-section is obtained from (88.6):

$$\sigma = \pi r_e^2 \frac{1-v^2}{4v}\left[\frac{3-v^4}{v}\log\frac{1+v}{1-v}-2(2-v^2)\right], \tag{88.15}$$

where $v = |\mathbf{p}|/\varepsilon = \sqrt{(\varepsilon^2-m^2)}/\varepsilon$ is the velocity of the colliding particles.

We shall not discuss here the details of the polarization effects in annihilation,† but merely consider certain qualitative features of these effects in the limiting cases where the velocity v of the colliding particles is large or small. The process will be considered in the centre-of-mass system.

In the limit $v \to 0$, only the state with orbital angular momentum of relative motion $l = 0$ gives a non-zero contribution to the cross-section. But the S state of the electron + positron system has negative parity (§27, Problem). In odd states of a two-photon system, their polarizations are orthogonal (§9). The same must therefore be true of the annihilation photons in the non-relativistic case.

If the electron and positron are polarized, their annihilation is possible (again in the non-relativistic case) only if their spins are antiparallel: since the annihilation occurs in the S state, the total angular momentum of the system is equal to the total spin of the particles, which is 1 when the spins are parallel. The two-photon system, however, has no state with total angular momentum 1 (see §9).

In the ultra-relativistic limit ($v \to 1$), the annihilation of a longitudinally polarized (helical) electron and positron is possible only when their helicities have opposite signs.‡ In this limit, helical particles behave as neutrinos (see the end of §80), and the electron and positron undergoing annihilation must be analogous to a neutrino and an antineutrino, whence the result stated follows.

The annihilation of an electron and a positron with the same helicity occurs, in the ultra-relativistic case, only when terms containing m are taken into account. The amplitude of this process differs, in order of magnitude, by a factor m/ε from that of the annihilation of a pair with parallel spins; the cross-section accordingly differs by a factor $(m/\varepsilon)^2$.

† See W. H. McMaster, *Reviews of Modern Physics* **33**, 8, 1961.

‡ Since the directions of the particle momenta are also opposite (in the centre-of-mass system), helicities of opposite sign correspond to parallel spins.

PROBLEM

Find the cross-section for the formation of an electron pair in the collision of two photons (G. Breit and J. A. Wheeler, 1934).

SOLUTION. This is the process inverse to the two-photon annihilation of an electron pair. The squared amplitudes are the same for the two processes, and their relationship to the cross-section differs only in that here $I^2 = (k_1k_2)^2 = \frac{1}{4}t^2$. Hence

$$d\sigma_{\text{form}} = d\sigma_{\text{ann}}\frac{t-4m^2}{t}.$$

In the centre-of-mass system ($t = 4\varepsilon^2 = 4\omega^2$),

$$d\sigma_{\text{form}} = v^2\,d\sigma_{\text{ann}},$$

where v is the velocity of the components of the pair. In integrating to obtain the total cross-section, the result is not to be divided by 2 (as in the case of annihilation), because the two final particles (electron and positron) are not identical. Hence, in the centre-of-mass system,

$$\begin{aligned}\sigma_{\text{form}} &= 2v^2\sigma_{\text{ann}}\\ &= \tfrac{1}{2}\pi r_e^2(1-v^2)\left\{(3-v^4)\log\frac{1+v}{1-v}-2v(2-v^2)\right\}.\end{aligned}\qquad(1)$$

In an arbitrary frame of reference K, in which the two photons k_1 and k_2 are moving in opposite directions, we have (from the invariance of k_1k_2)

$$\omega_1\omega_2 = \omega^2,$$

where ω is the energy of the photons in the centre-of-mass system. Since this energy is equal to that of the pair components, we have $\omega = \varepsilon = m/\sqrt{(1-v^2)}$. To change to the frame K, we must therefore put in (1)

$$v = \sqrt{(1-m^2/\omega_1\omega_2)}.$$

§ 89. Annihilation of positronium

Owing to the conservation of momentum, the annihilation of the electron and positron in positronium must be accompanied by the emission of at least two photons. Such a decay is possible (in the ground state), however, only for parapositronium. In §9 we have shown that the total angular momentum of a two-photon system cannot be 1. Hence orthopositronium in the $^3S_1$ state cannot decay into two photons. Moreover, since positronium in the $^3S_1$ state is a charge-odd system (see §27, Problem), Furry's theorem (§79) shows that it cannot decay into any even number of photons. In the $^1S_0$ state, on the other hand, positronium is charge-even, and the decay of parapositronium into any odd number of photons is therefore forbidden.

The main process which determines the lifetime of positronium is therefore two-photon annihilation for parapositronium and three-photon annihilation for orthopositronium (I. Ya. Pomeranchuk, 1948). The decay probability can be related to the cross-section for annihilation of a free pair.

The electron and positron momenta in positronium are $\sim me^2/\hbar$, i.e. small compared with mc. Hence, in calculating the probability of annihilation, we can take the limit of two particles at rest at the origin. Let $\bar{\sigma}_{2\gamma}$ be the cross-section for two-photon annihilation of a free pair, averaged over the spin directions of both particles. In the non-relativistic limit, according to (88.11),†

$$\bar{\sigma}_{2\gamma} = \pi(e^2/mc^2)^2 c/v, \tag{89.1}$$

where v is the relative velocity of the particles. The annihilation probability $\bar{w}_{2\gamma}$ is obtained on multiplying $\bar{\sigma}_{2\gamma}$ by the flux density $v|\psi(0)|^2$. Here $\psi(r)$ is the wave function, normalized to unity, of the positronium ground state:

$$\psi(r) = \frac{1}{\sqrt{(\pi a^3)}} e^{-r/a}, \qquad a = 2\hbar^2/me^2; \tag{89.2}$$

the Bohr radius a for positronium is twice that for the hydrogen atom, because its reduced mass is half as great. This probability, however, corresponds to the initial state averaged over spins, whereas in positronium, of the four possible spin states of a two-particle system, only one (with total spin 0) can undergo two-photon annihilation. Hence the mean decay probability $\bar{w}_{2\gamma}$ is related to the parapositronium decay probability w_0 by $\bar{w}_{2\gamma} = \frac{1}{4}w_0$, and so

$$w_0 = 4|\psi(0)|^2(v\bar{\sigma}_{2\gamma})_{v\to 0}. \tag{89.3}$$

Substituting the values from (89.1), (89.2), we obtain for the lifetime of parapositronium

$$\tau_0 = 2\hbar/mc^2\alpha^5 = 1.23 \times 10^{-10} \text{ sec}. \tag{89.4}$$

It should be noticed that the level width $\Gamma_0 = \hbar/\tau_0$ is small compared with the level energy

$$|E_{\text{gr}}| = me^4/4\hbar^2 = mc^2\alpha^2/4.$$

For this reason positronium may be regarded as a system in a quasi-stationary state.

Similarly we find that the decay probability for orthopositronium is related to the spin-averaged cross-section for three-photon annihilation of a free pair by

$$w_1 = \tfrac{4}{3}\bar{w}_{3\gamma} = \tfrac{4}{3}|\psi(0)|^2(v\bar{\sigma}_{3\gamma})_{v\to 0}, \tag{89.5}$$

the statistical weight of a state with spin 1 being $\frac{3}{4}$. Anticipating, we may mention that

$$\bar{\sigma}_{3\gamma} = \frac{4(\pi^2-9)c}{3v}\alpha\left(\frac{e^2}{mc^2}\right)^2. \tag{89.6}$$

The lifetime of orthopositronium is therefore

$$\tau_1 = \frac{9\pi}{2(\pi^2-9)}\frac{\hbar}{mc^2\alpha^6} = 1.4 \times 10^{-7} \text{ sec}. \tag{89.7}$$

† Formulae (89.1)–(89.7) are written in ordinary units.

The inequality $\Gamma_1 \ll |E_{gr}|$ is here, of course, satisfied even more markedly than for parapositronium.

Let us now calculate the cross-section for three-photon annihilation of a free pair (A. Ore and J. L. Powell, 1949). According to (64.18), the cross-section in the centre-of-mass system is expressed in terms of the squared amplitude by

$$d\sigma_{3\gamma} = \frac{(2\pi)^4|M_{fi}|^2}{4I}\,\delta(\mathbf{k}_1+\mathbf{k}_2+\mathbf{k}_3)\delta(\omega_1+\omega_2+\omega_3-2m)\frac{d^3k_1\,d^3k_2\,d^3k_3}{(2\pi)^9 2\omega_1\cdot 2\omega_2\cdot 2\omega_3}, \tag{89.8}$$

where, according to (64.16), $I = 2m\cdot\frac{1}{2}mv = m^2v$, v being the relative velocity (assumed small) of the positron and the electron; $\mathbf{k}_1$, $\mathbf{k}_2$, $\mathbf{k}_3$ and ω_1, ω_2, ω_3 are the wave vectors and frequencies of the photons formed; the delta functions express the laws of conservation of energy and momentum. Because of these laws, the three frequencies ω_1, ω_2, ω_3 must be represented by the lengths of the sides of a triangle with perimeter $2m$. Thus the magnitudes of the momenta $\mathbf{k}_1$, $\mathbf{k}_2$, $\mathbf{k}_3$ and the angles between them are entirely determined by specifying two frequencies.

The three-photon annihilation corresponds to the diagram

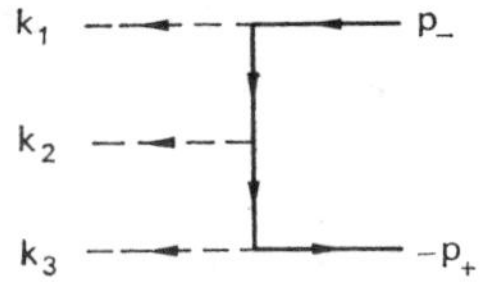

and a further five diagrams obtained from it by interchanging the photons k_1, k_2, k_3. The amplitude may be written

$$M_{fi} = (4\pi)^{3/2} e_\lambda^{(3)*} e_\mu^{(2)*} e_\nu^{(1)*} \bar{u}(-p_+) Q^{\lambda\mu\nu} u(p_-), \tag{89.9}$$

where

$$Q^{\lambda\mu\nu} = \sum_{\text{int.}} \gamma^\lambda G(k_3-p_+)\gamma^\mu G(p_- - k_1)\gamma^\nu, \tag{89.10}$$

the sum being taken over all interchanges of the photon numbers 1, 2, 3 together with corresponding simultaneous interchanges of the tensor indices λ, μ, ν. The squared modulus of the amplitude, averaged over the polarizations of the electron and the positron and summed over those of the photons, is

$$\frac{1}{4}\sum_{\text{polar.}} |M_{fi}|^2 = (4\pi)^3\,\text{tr}\{\rho_+ Q^{\lambda\mu\nu}\rho_- \bar{Q}_{\lambda\mu\nu}\}, \tag{89.11}$$

where

$$\rho_- = \tfrac{1}{2}(\gamma p_- + m), \qquad \rho_+ = \tfrac{1}{2}(\gamma p_+ - m).$$

The matrices $\bar{Q}^{\lambda\mu\nu}$ differ from the matrices $Q^{\lambda\mu\nu}$ in that the order of the factors is reversed in each term of the sum. In the limiting case considered, where the electron and positron velocities are small, their 3-momenta $\mathbf{p}_-$ and $\mathbf{p}_+$ may be taken as zero, putting $p_- = p_+ = (m, 0)$. Then the electron Green's functions are

$$G(p_- - k_1) = \frac{\gamma p - \gamma k_1 + m}{(p_- - k_1)^2 - m^2} \approx \frac{-\gamma k_1 + m(\gamma^0 + 1)}{-2m\omega_1},$$

etc., and the density matrices reduce to

$$\rho_\mp = \tfrac{1}{2}m(\gamma^0 \pm 1).$$

A large number of terms arise on carrying out the multiplication in (89.11), but the number that need to be calculated can be greatly reduced by making full use of the symmetry with respect to interchanges of photons. For example, it is sufficient to multiply out the six terms in $Q^{\lambda\mu\nu}$ (89.10) each with only one term in $\bar{Q}_{\lambda\mu\nu}$. In the six traces then remaining, we can again select certain parts which are transformed into one another by various interchanges of photons. The products of the 4-vectors p, k_1, k_2, k_3 which occur when the traces are expanded can all be expressed in terms of the frequencies ω_1, ω_2, ω_3. Since $p = (m, 0)$, we have $pk_1 = m\omega_1, \ldots$. The products $k_1k_2, \ldots$ are determined from the equation of conservation of 4-momentum: $2p = k_1 + k_2 + k_3$; for example, writing this equation in the form $2p - k_3 = k_1 + k_2$ and squaring, we have

$$k_1k_2 = 2m(m - \omega_3), \ldots. \tag{89.12}$$

The result of the calculation, which is still fairly lengthy, is

$$\tfrac{1}{4}\sum_{\text{polar.}} |M_{fi}|^2 = (4\pi)^3 e^6 \cdot 16\left[\left(\frac{m-\omega_1}{\omega_2\omega_3}\right)^2 + \left(\frac{m-\omega_2}{\omega_1\omega_3}\right)^2 + \left(\frac{m-\omega_3}{\omega_1\omega_2}\right)^2\right].$$

Substituting this expression in (89.8), we obtain the differential cross-section for three-photon annihilation:

$$d\bar{\sigma}_{3\gamma} = \frac{e^6}{\pi^2 m^2 v}\left[\left(\frac{m-\omega_1}{\omega_2\omega_3}\right)^2 + \left(\frac{m-\omega_2}{\omega_1\omega_3}\right)^2 + \left(\frac{m-\omega_3}{\omega_1\omega_2}\right)^2\right] \times$$
$$\times\, \delta(\mathbf{k}_1 + \mathbf{k}_2 + \mathbf{k}_3)\delta(\omega_1 + \omega_2 + \omega_3 - 2m)\frac{d^3k_1\, d^3k_2\, d^3k_3}{\omega_1\omega_2\omega_3}. \tag{89.13}$$

The delta functions have still to be eliminated. The first is removed by integrating over d^3k_3, and we then write

$$d^3k_1\, d^3k_2 \to 4\pi\omega_1^2\, d\omega_1 \cdot 2\pi\omega_2^2\, d(\cos\theta_{12})\, d\omega_2,$$

where θ_{12} is the angle between $\mathbf{k}_1$ and $\mathbf{k}_2$; it is assumed that the integration has already been performed over the directions of $\mathbf{k}_1$ and the azimuth of $\mathbf{k}_2$ relative to

$\mathbf{k}_1$. Differentiating the equation

$$\omega_3 = \sqrt{(\omega_1^2 + \omega_2^2 + 2\omega_1\omega_2 \cos \theta_{12})},$$

we find

$$d \cos \theta_{12} = (\omega_3/\omega_1\omega_2)\, d\omega_3.$$

The second delta function is removed by integrating over $d\omega_3$. The resulting cross-section for annihilation with formation of photons having specified energies is

$$d\bar{\sigma}_{3\gamma} = \frac{1}{6}\frac{8e^6}{vm^2}\left\{\left(\frac{m-\omega_3}{\omega_1\omega_2}\right)^2 + \left(\frac{m-\omega_2}{\omega_1\omega_3}\right)^2 + \left(\frac{m-\omega_1}{\omega_2\omega_3}\right)^2\right\} d\omega_1\, d\omega_2; \tag{89.14}$$

the factor 1/6 has been included in order to take account of the identity of the photons in the subsequent integration over frequencies (cf. the third footnote to §64).

Each of the frequencies ω_1, ω_2, ω_3 can take values between 0 and m; the latter can be reached by two frequencies when the third is zero. For given ω_1, the frequency ω_2 varies between $m - \omega_1$ and m. Integrating (89.14) over $d\omega_2$ between these limits, we obtain the spectral distribution of decay photons:

$$d\bar{\sigma}_{3\gamma} = (8e^6/3vm^3)F(\omega_1)\, d\omega_1,$$

$$F(\omega_1) = \frac{\omega_1(m-\omega_1)}{(2m-\omega_1)^2} + \frac{2m-\omega_1}{\omega_1} + \left[\frac{2m(m-\omega_1)}{\omega_1^2} - \frac{2m(m-\omega_1)^2}{(2m-\omega_1)^3}\right] \log \frac{m-\omega_1}{m}.$$

The function $F(\omega_1)$ increases monotonically from zero when $\omega_1 = 0$ to unity when $\omega_1 = m$, and is shown graphically in Fig. 14.

The total annihilation cross-section is obtained by integrating (89.14) over both

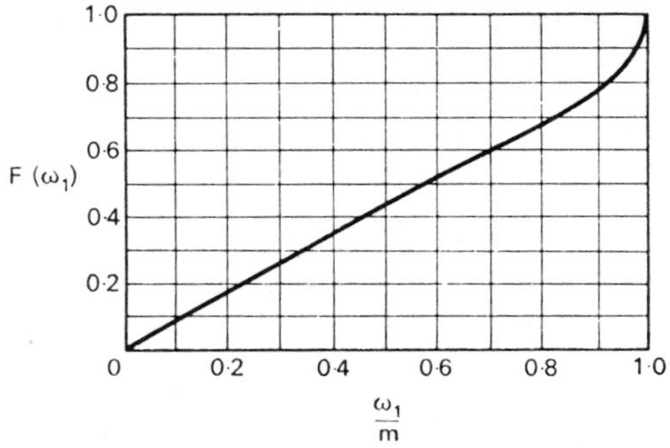

FIG. 14.

frequencies:

$$\bar{\sigma}_{3\gamma} = \frac{4e^6}{3vm^2} \cdot 3 \int_0^m \int_{m-\omega_1}^m \frac{(\omega_1 + \omega_2 - m)^2}{\omega_1^2 \omega_2^2} \, d\omega_1 \, d\omega_2.$$

The value of the integral is $(\pi^2 - 9)/3$, and we thus return to formula (89.6).

§90. Synchrotron radiation

According to the classical theory (*Fields*, §74), an ultra-relativistic electron moving in a constant magnetic field $\mathbf{H}$ emits a quasi-continuous spectrum with a maximum at the frequency

$$\omega \sim \omega_0 (\varepsilon/m)^3, \tag{90.1}$$

where

$$\begin{aligned} \omega_0 &= v|e|H/|\mathbf{p}| \\ &\approx |e|H/\varepsilon \end{aligned} \tag{90.2}$$

is the frequency of revolution of an electron having energy ε in a circular orbit (in a plane perpendicular to the field).† We shall assume that the longitudinal velocity of the electron (parallel to $\mathbf{H}$) is zero, as can always be achieved by a suitable choice of the frame of reference.

Quantum effects in synchrotron radiation originate in two ways: from the quantization of the motion of the electron, and from the quantum recoil when a photon is emitted. The latter is determined by the ratio $\hbar\omega/\varepsilon$, and this must be small if the classical theory is applicable. It is therefore convenient to use the parameter

$$\chi = \frac{H}{H_0} \frac{|\mathbf{p}|}{m} \approx \frac{H\varepsilon}{H_0 m} \approx \frac{\hbar\omega_0}{\varepsilon} \left(\frac{\varepsilon}{m}\right)^3, \tag{90.3}$$

where $H_0 = m^2/|e|\hbar$ $(= m^2c^3/|e|\hbar) = 4.4 \times 10^{13}$ G. In the classical case, $\chi \sim \hbar\omega/\varepsilon \ll 1$. In the opposite limit $(\chi \gg 1)$, the energy of the emitted photon $\hbar\omega \sim \varepsilon$, and (as we shall see below) the significant region of the spectrum extends to frequencies at which the electron energy after the emission is

$$\varepsilon' \sim mH_0/H. \tag{90.4}$$

If the electron remains ultra-relativistic, the field must satisfy the condition

$$H/H_0 \ll 1. \tag{90.5}$$

The quantization of the electron motion itself is expressed by the ratio $\hbar\omega_0/\varepsilon$; $\hbar\omega_0$ is the interval between adjacent energy levels for motion in a magnetic field.

† In this section we shall put $c = 1$ but retain factors of $\hbar$.

Since

$$\hbar\omega_0/\varepsilon = (H/H_0)(m/\varepsilon)^2,$$

it follows from (90.5) that $\hbar\omega_0 \ll \varepsilon$, i.e. the motion of the electron is quasi-classical for all values of χ. That is, the non-commutativity between the operators of dynamical variables of the electron (quantities of order $\hbar\omega_0/\varepsilon$) may be neglected, while the non-commutativity of these operators with those of the photon field (quantities of order $\hbar\omega/\varepsilon$) is not neglected.†

The quasi-classical wave functions of stationary states of an electron in an external field can be put in the symbolic form

$$\psi = \frac{1}{\sqrt{(2\hat{H})}} u(\hat{p})\, e^{-(i/\hbar)\hat{H}t} \phi(\mathbf{r}), \tag{90.6}$$

where $\phi(\mathbf{r}) \sim \exp(iS/\hbar)$ are the quasi-classical wave functions of a spinless particle ($S(\mathbf{r})$ being its classical action); $u(\mathbf{p})$ is the operator bispinor

$$u(\hat{p}) = \begin{pmatrix} \sqrt{(\hat{H}+m)}w \\ \dfrac{1}{\sqrt{(\hat{H}+m)}}(\boldsymbol{\sigma}\cdot\hat{\mathbf{p}})w \end{pmatrix},$$

obtained from the bispinor plane-wave amplitude $u(p)$ (23.9) on replacing $\mathbf{p}$ and ε by the operators‡

$$\hat{\mathbf{p}} = \hat{\mathbf{P}} - e\mathbf{A} = -i\hbar\nabla - e\mathbf{A}, \qquad \hat{H} = \sqrt{(\hat{\mathbf{p}}^2 + m^2)},$$

where $\mathbf{P}$ is the generalized momentum of the particle in a field with vector potential $\mathbf{A}(\mathbf{r})$. The order of the operator factors in ψ is immaterial, since their non-commutativity is neglected, and the spin state of the electron is determined by the three-dimensional spinor w.

In order to calculate the probability of photon emission in the quasi-classical case, it is more convenient to start not from the final formula (44.3) of perturbation theory but from a formula in which the integration with respect to time has not yet been carried out. For the total (over all time) differential probability we have§

$$dw = \sum_f |a_{fi}|^2 \frac{d^3k}{(2\pi)^3}, \qquad a_{fi} = \int_{-\infty}^{\infty} V_{fi}(t)\, dt \tag{90.7}$$

† The full solution of the quantum problem of synchrotron radiation was first given by N. P. Klepikov (1954), and the first quantum correction to the classical formula by A. A. Sokolov, N. P. Klepikov and I. M. Ternov (1952). The derivation given here, which explicitly makes use of the fact that the motion is quasi-classical, is due to V. N. Baĭer and V. M. Katkov (1967). A similar method had been used earlier by J. Schwinger (1954) to derive the first quantum correction in the radiation intensity.

‡ In this section, unlike Chapter IV, the generalized momentum is denoted by the capital letter $\mathbf{P}$, while $\mathbf{p}$ denotes the ordinary (kinetic) momentum.

§ Putting $V_{fi}(t) = V_{fi}\exp(i\omega_{fi}t)$, we find $a_{fi} = 2\pi V_{fi}\delta(\omega_{fi})$, and, since the squared delta function is to be taken as $[\delta(\omega)]^2 \to (t/2\pi)\delta(\omega)$, where t is the total observation time (cf. the derivation of (64.5)), we obtain from (90.7) the formula (44.3) for the probability per unit time.

(cf. *QM*, (41.2)); the summation is over final states of the electron.

Using (90.6), we can write the matrix element $V_{fi}(t)$ for emission of a photon ω, $\mathbf{k}$ in the operator form

$$V_{fi}(t) = -\frac{e\sqrt{(4\pi)}}{\sqrt{(2\hbar\omega)}}\int\left[\phi_f^* e^{(i/\hbar)\hat{H}t}\frac{u^+(\hat{p})}{\sqrt{(2\hat{H})}}\right]e^{i\omega t - i\mathbf{k}\cdot\mathbf{r}}(\mathbf{e}^*\cdot\boldsymbol{\alpha})\frac{u(\hat{p})}{\sqrt{(2\hat{H})}}e^{-(i/\hbar)\hat{H}t}\phi_i\, d^3x,$$

where the operators in the square brackets act to the left; the photon field is taken in the three-dimensionally transverse gauge. The factors $\exp(\pm i\hat{H}t/\hbar)$ convert the Schrödinger operators between them into explicitly time-dependent operators of the Heisenberg representation. We can write $V_{fi}(t)$ in the form

$$V_{fi}(t) = \frac{e\sqrt{(2\pi)}}{\sqrt{(\hbar\omega)}}\langle f|Q(t)|i\rangle\, e^{i\omega t},$$

where $\hat{Q}(t)$ denotes the Heisenberg operator

$$\hat{Q}(t) = \frac{u_f^+(\hat{p})}{\sqrt{(2\hat{H})}}(\boldsymbol{\alpha}\cdot\mathbf{e}^*)e^{-i\mathbf{k}\cdot\hat{\mathbf{r}}(t)}\frac{u_i(\hat{p})}{\sqrt{(2\hat{H})}}, \tag{90.8}$$

and the matrix element is taken with respect to the functions ϕ_f, ϕ_i.

The summation in (90.7) is taken over all final wave functions ϕ_f, and is effected by means of the equation

$$\sum_f \phi_f^*(\mathbf{r}')\phi_f(\mathbf{r}) = \delta(\mathbf{r}'-\mathbf{r}),$$

which expresses the completeness of the set of functions ϕ_f. The result is

$$dw = \frac{e^2}{\hbar\omega}\frac{d^3k}{4\pi^2}\int dt_1\int dt_2\cdot e^{i\omega(t_1-t_2)}\langle i|Q^+(t_2)Q(t_1)|i\rangle. \tag{90.9}$$

If the integration is over a sufficiently long time interval, t_1 and t_2 can be replaced by new variables

$$\tau = t_2 - t_1, \qquad t = \tfrac{1}{2}(t_1 + t_2),$$

and in the integral over t the integrand may be regarded as the probability of emission per unit time. Multiplying by $\hbar\omega$, we obtain the intensity

$$dI = \frac{e^2}{4\pi^2}d^3k\int e^{-i\omega\tau}\langle i|Q^+(t+\tfrac{1}{2}\tau)Q(t-\tfrac{1}{2}\tau)|i\rangle\, d\tau. \tag{90.10}$$

An ultra-relativistic electron radiates into a narrow cone at angles $\theta \sim m/\varepsilon$ relative to its velocity $\mathbf{v}$. The emission in a given direction $\mathbf{n} = \mathbf{k}/\omega$ therefore occurs over a section of the path in which $\mathbf{v}$ turns through an angle $\sim m/\varepsilon$. This section is traversed in a time τ such that $\tau|\dot{\mathbf{v}}| \approx \tau\omega_0 \sim m/\varepsilon \ll 1$. This region gives the principal contribution to the integral over τ. In the subsequent calculations, we shall therefore expand all quantities in powers of $\omega_0\tau$. It may, however, be necessary to retain more than just the leading term in the expansion, because of cancellations which occur since $1-\mathbf{n}\cdot\mathbf{v}\sim\theta^2\sim(m/\varepsilon)^2$.

If the operator $\hat{Q}^+\hat{Q}$ is reduced to a product of operators which commute (to the necessary degree of accuracy), the taking of the diagonal matrix element $\langle i|\dots|i\rangle$ is equivalent to replacing these operators by the classical (time-dependent) values of the corresponding quantities. This is achieved in the following way.

According to the foregoing discussion, in the expression for $\hat{Q}(t)$ only the non-commutativity of the electron operators with the photon field operator $\exp(-i\mathbf{k}\cdot\hat{\mathbf{r}}(t))$ need be taken into account. We have

$$\left.\begin{aligned}\hat{\mathbf{p}}\,e^{-i\mathbf{k}\cdot\hat{\mathbf{r}}} &= e^{-i\mathbf{k}\cdot\hat{\mathbf{r}}}(\hat{\mathbf{p}}-\hbar\mathbf{k}),\\ H(\hat{\mathbf{p}})\,e^{-i\mathbf{k}\cdot\hat{\mathbf{r}}} &= e^{-i\mathbf{k}\cdot\hat{\mathbf{r}}}H(\hat{\mathbf{p}}-\hbar\mathbf{k}).\end{aligned}\right\}\tag{90.11}$$

These formulae follow because $e^{-i\mathbf{k}\cdot\hat{\mathbf{r}}}$ is the displacement operator in momentum space. Using (90.11), we can take the operator $e^{-i\mathbf{k}\cdot\hat{\mathbf{r}}(t)}$ out on the left in (90.8), and write $\hat{Q}(t)$ in the form

$$\hat{Q}(t)=e^{-i\mathbf{k}\cdot\hat{\mathbf{r}}(t)}\hat{R}(t),\qquad \hat{R}(t)=\frac{u_f^+(\hat{p}')}{\sqrt{(2\hat{H}')}}(\boldsymbol{\alpha}\cdot\mathbf{e}^*)\frac{u_i(\hat{p})}{\sqrt{(2\hat{H})}},\tag{90.12}$$

where $\hat{H}'=\hat{H}-\hbar\omega$, $\hat{\mathbf{p}}'=\hat{\mathbf{p}}-\hbar\mathbf{k}$.

Then

$$\hat{Q}_2^+\hat{Q}_1=\hat{R}_2\,e^{i\mathbf{k}\cdot\hat{\mathbf{r}}_2}\,e^{-i\mathbf{k}\cdot\hat{\mathbf{r}}_1}\hat{R}_1;\tag{90.13}$$

here and henceforward, the suffixes 1 and 2 denote the values of quantities at the times $t_1=t-\frac{1}{2}\tau$ and $t_2=t+\frac{1}{2}\tau$. It remains to calculate the product of the two non-commuting operators $e^{i\mathbf{k}\cdot\hat{\mathbf{r}}_2}$ and $e^{-i\mathbf{k}\cdot\hat{\mathbf{r}}_1}$. This product itself may be regarded as commuting with the remaining factors.

We write

$$\hat{L}(\tau)=e^{-i\omega\tau}\,e^{i\mathbf{k}\cdot\hat{\mathbf{r}}_2}\,e^{-i\mathbf{k}\cdot\hat{\mathbf{r}}_1},\tag{90.14}$$

this being the combination of operators which appears in (90.10). The operator $e^{i\hat{H}\tau/\hbar}$ is a time-shift operator, and so

$$e^{i\mathbf{k}\cdot\hat{\mathbf{r}}_2}=e^{i\hat{H}\tau/\hbar}\,e^{i\mathbf{k}\cdot\hat{\mathbf{r}}_1}\,e^{-i\hat{H}\tau/\hbar}.$$

Substituting this in (90.14) and noting that $e^{i\mathbf{k}\cdot\hat{\mathbf{r}}_1}$ is a displacement operator in momentum space, we find

$$\hat{L}(\tau)=\exp\{i[\hat{H}-\hbar\omega]\tau/\hbar\}\exp\{-i\hat{H}(\hat{\mathbf{p}}_1-\hbar\mathbf{k})\tau/\hbar\}.\tag{90.15}$$

Differentiating (90.15) with respect to τ and again using the properties of the time-shift operator, we have†

$$d\hat{L}/d\tau = (i/\hbar)\exp\{i(\hat{H}-\hbar\omega)\tau/\hbar\}[\hat{H}-\hbar\omega-\hat{H}(\hat{\mathbf{p}}_1-\hbar\mathbf{k})]\times$$
$$\times\exp\{-i\hat{H}(\hat{\mathbf{p}}_1-\hbar\mathbf{k})\tau/\hbar\} = (i/\hbar)[\hat{H}-\hbar\omega-\hat{H}(\hat{\mathbf{p}}_2-\hbar\mathbf{k})]\hat{L}(\tau). \quad (90.16)$$

Having thus made use of the non-commutativity of the operators, we can replace all the operators by the corresponding classical quantities (the Hamiltonian $\hat{H}$ by the electron energy ε). We have identically

$$\varepsilon(\mathbf{p}_2-\hbar\mathbf{k}) = [(\mathbf{p}_2-\hbar\mathbf{k})^2+m^2]^{1/2}$$
$$= [(\varepsilon-\hbar\omega)^2+2\hbar(\omega\varepsilon-\mathbf{k}\cdot\mathbf{p}_2)]^{1/2}.$$

The difference

$$\omega\varepsilon-\mathbf{k}\cdot\mathbf{p}_2 = \omega\varepsilon(1-\mathbf{n}\cdot\mathbf{v}_2)$$

is small, since from the above analysis $1-\mathbf{v}\cdot\mathbf{n}\sim(m/\varepsilon)^2$. As far as the first order in this difference,

$$\varepsilon(\mathbf{p}_2-\hbar\mathbf{k})\approx\varepsilon'+(\varepsilon/\varepsilon')\hbar(\omega-\mathbf{k}\cdot\mathbf{v}_2),$$

where $\varepsilon'=\varepsilon-\hbar\omega$. From (90.16), we now find the differential equation for $L(\tau)$:

$$i\hbar dL/d\tau = (\varepsilon/\varepsilon')\hbar(\omega-\mathbf{k}\cdot\mathbf{v}_2)L. \quad (90.17)$$

This equation is to be solved with the obvious initial condition $L(0)=1$. Since

$$\int_0^\tau \mathbf{v}_2\,d\tau = \mathbf{r}_2-\mathbf{r}_1,$$

we have

$$L(\tau) = \exp\{i(\varepsilon/\varepsilon')(\mathbf{k}\cdot\mathbf{r}_2-\mathbf{k}\cdot\mathbf{r}_1-\omega\tau)\}. \quad (90.18)$$

So far, no use has been made of the specific form of the electron trajectory. Now expressing $\mathbf{r}_2-\mathbf{r}_1$ in (90.18) in terms of $\mathbf{p}_1$ by means of the equation of motion of the electron in the plane perpendicular to the field $\mathbf{H}$ (see *Fields*, §21):

$$\mathbf{r}_2-\mathbf{r}_1 = \frac{\mathbf{p}_1}{eH}\sin\frac{eH\tau}{\varepsilon}+\frac{\mathbf{p}_1\times\mathbf{H}}{eH^2}\left(1-\cos\frac{eH\tau}{\varepsilon}\right),$$

and expanding in powers of τ gives

$$\mathbf{k}\cdot(\mathbf{r}_2-\mathbf{r}_1)-\omega\tau\approx\omega\tau\left\{(\mathbf{n}\cdot\mathbf{v}_1-1)+\tau\frac{e\mathbf{n}\cdot\mathbf{p}_1\times\mathbf{H}}{2\varepsilon^2}-\tau^2\frac{e^2H^2}{6e^2}\right\}, \quad (90.19)$$

where in the last term we have put $\mathbf{n}\cdot\mathbf{v}_1\approx 1$.

† Because of the conservation of energy, the Heisenberg operators $\hat{H}(\hat{\mathbf{p}}_1)$ and $\hat{H}(\hat{\mathbf{p}}_2)$ are the same; we therefore omit the arguments of $\hat{H}$ in such cases. However, $\hat{H}(\hat{\mathbf{p}}_1-\hbar\mathbf{k})$ is of course not the same as $\hat{H}(\hat{\mathbf{p}}_2-\hbar\mathbf{k})$.

We next transform the remaining factors in (90.13). A direct expansion of the product in $R(t)$, using the matrix $\boldsymbol{\alpha}$ from (21.20), leads to

$$\left.\begin{aligned} R(t) &= w_f^*\mathbf{e}^*\cdot(\mathbf{A}+i\mathbf{B}\times\boldsymbol{\sigma})w_i,\\ \mathbf{A} &= \tfrac{1}{2}\mathbf{p}\left(\frac{1}{\varepsilon}+\frac{1}{\varepsilon'}\right)=\frac{\varepsilon+\varepsilon'}{2\varepsilon'}\mathbf{v},\\ \mathbf{B} &= \tfrac{1}{2}\left(\frac{\mathbf{p}}{\varepsilon+m}-\frac{\mathbf{p}'}{\varepsilon'+m}\right)\approx\frac{\hbar\omega}{2\varepsilon'}(\mathbf{n}-\mathbf{v}+\mathbf{v}m/\varepsilon), \end{aligned}\right\} \tag{90.20}$$

where $\mathbf{p}'(t)=\mathbf{p}(t)-\hbar\mathbf{k}$; terms of higher order in m/ε are omitted. Thus we have finally

$$\left.\begin{aligned} e^{-i\omega\tau}\langle i|Q_2^+Q_1|i\rangle &= R_2^*R_1L(\tau),\\ R_2^*R_1 &= \mathrm{tr}\,\tfrac{1}{2}(1+\boldsymbol{\zeta}_i\cdot\boldsymbol{\sigma})(\mathbf{A}_2-i\mathbf{B}_2\times\boldsymbol{\sigma})\cdot\mathbf{e}\cdot\tfrac{1}{2}(1+\boldsymbol{\zeta}_f\cdot\boldsymbol{\sigma})(\mathbf{A}_1+i\mathbf{B}_1\times\boldsymbol{\sigma})\cdot\mathbf{e}^*. \end{aligned}\right\} \tag{90.21}$$

The factors $\frac{1}{2}(1+\boldsymbol{\zeta}\cdot\boldsymbol{\sigma})$ are two-rowed polarization density matrices of the initial and final electron.

Let us consider the radiation intensity summed over the polarizations of the photon and of the final electron, and averaged over the polarizations of the initial electron. These operations give, after a simple calculation,†

$$\frac{1}{2}\sum_{\text{polar.}} R_2^*R_1=\frac{\varepsilon^2+\varepsilon'^2}{2\varepsilon'^2}(\mathbf{v}_1\cdot\mathbf{v}_2-1)+\frac{1}{2}\left(\frac{\hbar\omega}{\varepsilon'}\right)^2\left(\frac{m}{\varepsilon}\right)^2.$$

With sufficient accuracy we can put

$$\begin{aligned} \mathbf{v}_1\cdot\mathbf{v}_2 &= \mathbf{v}^2-\tfrac{1}{4}\tau^2\dot{\mathbf{v}}^2+\tfrac{1}{4}\tau^2\mathbf{v}\cdot\ddot{\mathbf{v}}\\ &= 1-\frac{m^2}{\varepsilon^2}-\tfrac{1}{2}\omega_0^2\tau^2. \end{aligned}$$

Substitution of these expressions in (90.21) and thence in (90.10) gives

$$dI=-\frac{e^2}{4\pi^2}\omega^2\,d\omega\,do_{\mathbf{n}}\times$$
$$\times\int_{-\infty}^{\infty}\left(\frac{m^2}{\varepsilon\varepsilon'}+\frac{\varepsilon^2+\varepsilon'^2}{4\varepsilon'^2}\omega_0^2\tau^2\right)\exp\left\{-\frac{i\omega\tau\varepsilon}{\varepsilon'}\left(1-\mathbf{n}\cdot\mathbf{v}+\frac{\tau^2}{24}\omega_0^2\right)\right\}d\tau. \tag{90.22}$$

† This calculation makes use also of the following result. In the summation over $\mathbf{e}$,

$$\sum_{\mathbf{e}}(\mathbf{v}_1\cdot\mathbf{e})(\mathbf{v}_2\cdot\mathbf{e}^*)=\mathbf{v}_1\cdot\mathbf{v}_2-(\mathbf{v}_1\cdot\mathbf{n})(\mathbf{v}_2\cdot\mathbf{n}).$$

On substituting (90.21) in (90.10), we can integrate by parts, noting that

$$(\mathbf{v}_1\cdot\mathbf{n})\exp\left(-i\frac{\varepsilon}{\varepsilon'}\mathbf{k}\cdot\mathbf{r}_1\right)=\frac{i\varepsilon'}{\varepsilon\omega}\frac{d}{dt_1}\exp\left(-i\frac{\varepsilon}{\varepsilon'}\mathbf{k}\cdot\mathbf{r}_1\right),$$

and similarly for $\mathbf{v}_2\cdot\mathbf{n}$. Consequently, in the remaining integration $\mathbf{v}_1\cdot\mathbf{n}$ and $\mathbf{v}_2\cdot\mathbf{n}$ can be replaced by unity.

This formula shows the frequency and angular distribution of the radiation intensity.

To find the frequency distribution, we integrate over $do_{\mathbf{n}}$. If the direction of $\mathbf{v}$ is taken as the polar axis, with an angle ϑ between $\mathbf{n}$ and $\mathbf{v}$, then

$$\mathbf{n}\cdot\mathbf{v} = v\cos\vartheta, \qquad do_{\mathbf{n}} = \sin\vartheta\, d\vartheta\, d\phi,$$

and

$$\int \exp\left\{\frac{i\omega\tau\varepsilon}{\varepsilon'}\mathbf{n}\cdot\mathbf{v}\right\} do_{\mathbf{n}} = \frac{2\pi\varepsilon'}{i\omega\tau\varepsilon v}\left\{\exp\left(\frac{i\omega\tau\varepsilon}{\varepsilon'}\right) - \exp\left(-\frac{i\omega\tau\varepsilon}{\varepsilon'}\right)\right\}.$$

When this is substituted in (90.22), only the first term need be retained, since the second term yields a faster varying exponential (with a factor $1+v\approx 2$ instead of the small $1-v\approx m^2/2\varepsilon^2$). Hence

$$\frac{dI}{d\omega} = \frac{ie^2\omega}{2\pi}\int_{-\infty}^{\infty}\left(\frac{m^2}{\varepsilon^2\tau} + \frac{\varepsilon^2+\varepsilon'^2}{2\varepsilon\varepsilon'}\tau\right)\exp\left\{-\frac{i\omega\tau\varepsilon}{\varepsilon'}\left(1 - v + \frac{\tau^2}{24}\omega_0^2\right)\right\} d\tau.$$

According to the integral representation of the Airy function Φ (see *QM*, §b), the first term reduces to the integral of the Airy function, and the second term to its derivative. The final result is

$$\frac{dI}{d\omega} = -\frac{e^2m^2\omega}{\sqrt{\pi}\varepsilon^2}\left\{\int_x^{\infty}\Phi(\xi)\,d\xi + \left(\frac{2}{x} + \frac{\hbar\omega}{\varepsilon}\chi\sqrt{x}\right)\Phi'(x)\right\}, \tag{90.23}$$

$$x = (\hbar\omega/\varepsilon'\chi)^{2/3} = (m^2/\varepsilon^2)(\varepsilon\omega/\varepsilon'\omega_0)^{2/3} \tag{90.24}$$

(A. I. Nikishov and V. I. Ritus, 1967). The frequency distribution has a maximum when $x\sim 1$; for $\chi\ll 1$ we find (90.1), and for $\chi\gg 1$ (90.4). In the classical limit, $\hbar\omega\ll\varepsilon'\approx\varepsilon$, $x\approx(\omega/\omega_0)^{\frac{2}{3}}(m/\varepsilon)^2$; the second term in the round brackets is small, and (90.23) becomes the classical formula (*Fields*, (74.13)).

Figure 15 shows diagrams of the frequency distribution for various values of χ. The quantity

$$\frac{1}{3I_{\text{cl}}/2}\frac{dI}{d(\omega/\omega_c)}$$

is plotted against ω/ω_c, where

$$\hbar\omega_c = \frac{\varepsilon\chi}{\frac{2}{3}+\chi}, \qquad I_{\text{cl}} = \frac{2e^2m^2\chi^2}{3\hbar^2} = \frac{2e^4H^2\varepsilon^2}{3m^4}.$$

The quantity I_{cl} is the classical value of the total radiation intensity; cf. *Fields*, (74.2).

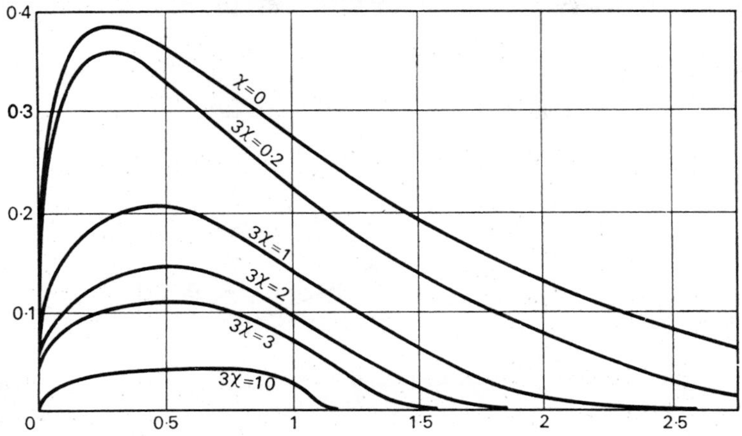

FIG. 15.

To calculate the total radiation intensity, (90.23) must be integrated with respect to ω from 0 to ε. We change to integration with respect to x, noting that

$$\hbar\omega = \varepsilon\left(1 - \frac{1}{1+\chi x^{3/2}}\right),$$

and x therefore varies from 0 to ∞. With two integrations by parts in the first term in (90.23), we find

$$I = -\frac{e^2m^2\chi^2}{2\sqrt{\pi}\hbar^2}\int_0^\infty \frac{4+5\chi x^{3/2}+4\chi^2x^3}{(1+\chi x^{3/2})^4}\Phi'(x)x\,dx. \tag{90.25}$$

Figure 16 shows a graph of the function $I(\chi)/I_{cl}$.

When $\chi \ll 1$, the important region in the integral is $x \sim 1$. Expanding the integrand in powers of χ and integrating by means of the formula

$$\int_0^\infty x^\nu \Phi'(x)\,dx = -\frac{1}{2\sqrt{\pi}}3^{(4\nu-1)/6}\Gamma(\tfrac{1}{3}\nu+1)\Gamma(\tfrac{1}{3}\nu+\tfrac{1}{3}),$$

we obtain

$$I = I_{cl}\left(1 - \frac{55\sqrt{3}}{16}\chi + 48\chi^2 - \cdots\right). \tag{90.26}$$

When $\chi \gg 1$, the important region is that in which $\chi x^{3/2} \sim 1$, i.e., $x \ll 1$. In the first approximation, we can therefore replace $\Phi'(x)$ by $\Phi'(0) = -3^{1/6}\Gamma(\tfrac{2}{3})/2\sqrt{\pi}$, and the

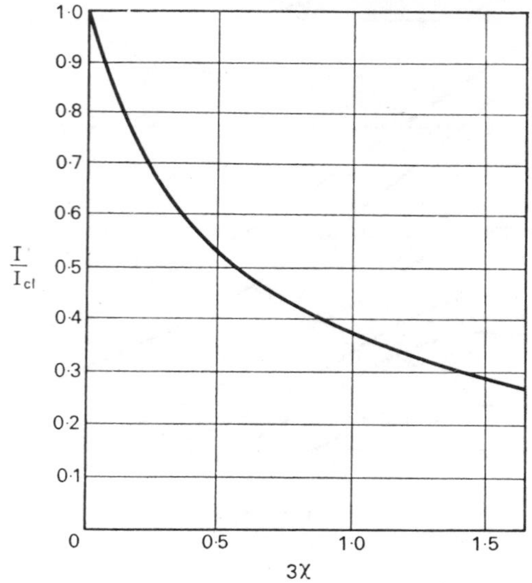

FIG. 16.

integration then leads to the result

$$I \approx \frac{32\Gamma(\frac{2}{3})e^2m^2}{243\hbar^2}(3\chi)^{2/3}$$

$$= 0.37\frac{e^2m^2}{\hbar^2}\left(\frac{H\varepsilon}{H_0 m}\right)^{2/3}. \tag{90.27}$$

Synchrotron emission causes the occurrence of a polarization of electrons moving in the field (A. A. Sokolov and I. M. Ternov, 1963). To discuss this, we have to find the probability of a radiative transition with spin reversal.

Putting in (90.21) $\boldsymbol{\zeta}_i = -\boldsymbol{\zeta}_f \equiv \boldsymbol{\zeta}$, $|\boldsymbol{\zeta}| = 1$, we have

$$R_2^*R_1 = \mathbf{B}_1\cdot\mathbf{B}_2 - (\mathbf{e}^*\cdot\mathbf{B}_1)(\mathbf{e}\cdot\mathbf{B}_2) - (\mathbf{e}^*\cdot\mathbf{B}_1\times\boldsymbol{\zeta})(\mathbf{e}\cdot\mathbf{B}_2\times\boldsymbol{\zeta}) - i(\boldsymbol{\zeta}\cdot\mathbf{e}^*)(\mathbf{e}\cdot\mathbf{B}_1\times\mathbf{B}_2).$$

Summation over polarizations of the photon gives, after a simple calculation,

$$\sum_{\mathbf{e}} R_2^*R_1 = (\mathbf{B}_1\cdot\mathbf{B}_2)(1-(\boldsymbol{\zeta}\cdot\mathbf{n})^2) + (\boldsymbol{\zeta}\cdot\mathbf{n})(\mathbf{n}\cdot\mathbf{B}_1)(\boldsymbol{\zeta}\cdot\mathbf{B}_2) +$$

$$+ (\boldsymbol{\zeta}\cdot\mathbf{n})(\mathbf{n}\cdot\mathbf{B}_2)(\boldsymbol{\zeta}\cdot\mathbf{B}_1) - i(\boldsymbol{\zeta}-\mathbf{n}(\mathbf{n}\cdot\boldsymbol{\zeta}))\cdot\mathbf{B}_1\times\mathbf{B}_2. \tag{90.28}$$

We shall assume that $\chi \ll 1$ and seek only the principal term in the expansion of the probability in powers of $\hbar$. Since the expression (90.28) (with **B** given by (90.20)) already contains $\hbar^2$, all the remaining quantities ε', including those in the exponent in (90.18), can be replaced by ε.

With the expansions

$$\mathbf{B}_1 = \frac{\omega}{2\varepsilon}\left(\mathbf{n} - \mathbf{v} + \tfrac{1}{2}\tau\dot{\mathbf{v}} + \mathbf{v}\frac{m}{\varepsilon}\right),$$

$$\mathbf{B}_2 = \frac{\omega}{2\varepsilon}\left(\mathbf{n} - \mathbf{v} - \tfrac{1}{2}\tau\dot{\mathbf{v}} + \mathbf{v}\frac{m}{\varepsilon}\right),$$

$$\mathbf{r}_2 - \mathbf{r}_1 = \tau\mathbf{v} + \frac{\tau^3}{24}\ddot{\mathbf{v}},$$

and substituting (90.28) in (90.21) and thence in (90.10), we find the differential transition probability per unit time ($dw = dI/\hbar\omega$). The integration over d^3k is carried out by means of the formula

$$\int f(k_\mu)\, e^{-ikx}\frac{d^3k}{\omega} = -f(i\partial_\mu)\frac{4\pi}{(x_0 - i0)^2 - \mathbf{x}^2}, \tag{90.29}$$

where in this case

$$x_0 = \tau, \qquad \mathbf{x} = \mathbf{r}_2 - \mathbf{r}_1, \qquad x^2 = x_0^2 - \mathbf{x}^2 = \tau^2\left(\frac{m^2}{\varepsilon^2} + \frac{\tau^2\omega_0^2}{12}\right).$$

The result of the calculation is

$$w = \frac{\alpha}{\pi}\frac{\hbar^2}{m^2}\left(\frac{\varepsilon}{m}\right)^5\omega_0^3 \oint \frac{dz}{(1 + z^2/12)^3}\left[\frac{3}{z^4} - \frac{5}{12z^2} + \left(\frac{1}{z^4} + \frac{5}{12z^2}\right)(\boldsymbol{\zeta}\cdot\mathbf{v})^2 - \frac{2i}{z^3\omega_0}\boldsymbol{\zeta}\cdot\dot{\mathbf{v}}\times\mathbf{v}\right],$$

where $z = \tau\omega_0\varepsilon/m$ and the contour of integration passes below the real axis and is closed in the lower half-plane. After this integration we finally obtain for the total probability of a radiative transition with spin reversal

$$w = \frac{5\sqrt{3}\alpha}{16}\frac{\hbar^2}{m^2}\left(\frac{\varepsilon}{m}\right)^5\omega_0^3\left(1 - \tfrac{2}{9}\zeta_\parallel^2 - \frac{8\sqrt{3}}{15}\frac{e}{|e|}\zeta_\perp\right), \tag{90.30}$$

where $\zeta_\parallel = \boldsymbol{\zeta}\cdot\mathbf{v}$, $\zeta_\perp = \boldsymbol{\zeta}\cdot\mathbf{H}/H$. This formula is valid for both electrons ($e < 0$) and positrons ($e > 0$).

The probability (90.30) is independent of the sign of the longitudinal polarization $\zeta_\parallel$ but depends on that of $\zeta_\perp$. The polarization resulting from the emission is therefore transverse.† For electrons, the probability of a transition from a state with the spin parallel to the field ($\zeta_\perp = 1$) to a state with the spin antiparallel to the field is greater than that of the opposite transition. The radiative polarization of the electrons is therefore antiparallel to the field, and the degree of polarization in a

† This is also evident from the fact that the axial vector of the resultant polarization must be along **H**, which is the only axial vector occurring in the problem.

stationary state is (when $\zeta_\parallel = 0$)

$$\frac{w(\zeta_\perp = -1) - w(\zeta_\perp = 1)}{w(\zeta_\perp = -1) + w(\zeta_\perp = 1)} = \frac{8\sqrt{3}}{15} = 0.92.$$

Positrons are polarized, to the same degree, parallel to the field.

§91. Pair production by a photon in a magnetic field

The production of an electron–positron pair by a photon in a magnetic field, and synchrotron radiation, are two cross-channels of the same reaction. The amplitude M_{fi} of the pair production process is therefore found from the synchrotron radiation amplitude by simply making the changes

$$\varepsilon, \mathbf{p} \to -\varepsilon_+, -\mathbf{p}_+; \qquad \varepsilon', \mathbf{p}' \to \varepsilon_-, \mathbf{p}_-; \qquad \omega, \mathbf{k} \to -\omega, -\mathbf{k}. \tag{91.1}$$

Here ε_-, $\mathbf{p}_-$ and ε_+, $\mathbf{p}_+$ are the electron and positron energies and momenta; ε, $\mathbf{p}$ and ε', $\mathbf{p}'$ the initial and final energies and momenta of the electron in synchrotron radiation. In terms of angles and magnitudes, the momenta are transformed according to

$$|\mathbf{p}| \to |\mathbf{p}_+|, \quad |\mathbf{p}'| \to |\mathbf{p}_-|, \quad \theta \to \pi - \theta_+, \quad \theta' \to \theta_-, \quad \phi \to \phi - \pi, \tag{91.2}$$

where $\theta_\pm$ are the angles between $\mathbf{p}_\pm$ and $\mathbf{k}$, ϕ the angle between the $\mathbf{kp}_+$ and $\mathbf{kp}_-$ planes.

For synchrotron radiation, the cross-section is given in terms of the amplitude by†

$$d\sigma = |M_{fi}|^2 \frac{1}{8|\mathbf{p}|\varepsilon'\omega} \delta(\varepsilon - \varepsilon' - \omega) \frac{d^3p'\, d^3k}{(2\pi)^5}; \tag{91.3}$$

see (64.25). The delta function is eliminated by integrating with respect to ε'. Since, in the present case, $\mathbf{p}'$ and $\mathbf{k}$ are independent variables, and

$$d^3p' = |\mathbf{p}'|\varepsilon'\, d\varepsilon'\, do', \qquad d^3k = \omega^2\, d\omega do_{\mathbf{k}},$$

we have simply to substitute

$$\delta(\varepsilon - \varepsilon' - \omega)\, d^3p'\, d^3k \to \omega^2|\mathbf{p}'|\varepsilon'\, do_{\mathbf{k}}\, do'\, d\omega.$$

Then

$$d\sigma = |M_{fi}|^2 \frac{\omega|\mathbf{p}'|}{8(2\pi)^5|\mathbf{p}|}\, do_{\mathbf{k}}\, do\, d\omega. \tag{91.4}$$

† In this section we again put $\hbar = 1$ as well as $c = 1$.

For pair production by a photon, the cross-section is given in terms of the amplitude by

$$d\sigma = |M_{fi}|^2 \frac{1}{8\omega\varepsilon_-\varepsilon_+}\,\delta(\omega - \varepsilon_+ - \varepsilon_-)\,\frac{d^3p_+\,d^3p_-}{(2\pi)^5}$$

or, after elimination of the delta function,

$$d\sigma = |M_{fi}|^2 \frac{|\mathbf{p}_+||\mathbf{p}_-|}{8(2\pi)^5\omega}\,do_+\,do_-\,d\varepsilon_+. \tag{91.5}$$

Comparison with (91.4) shows that, to obtain the pair production cross-section from the synchrotron radiation cross-section, we have to make in (91.4) the changes (91.1), multiply by

$$(\mathbf{p}_+^2/\omega^2)\,d\varepsilon_+/d\omega, \tag{91.6}$$

and replace $do'\,do_{\mathbf{k}}$ by $do_+\,do_-$.

In the ultra-relativistic case $\omega \gg m$,† this can be done in the formulae derived in §90. Here it is assumed that both particles in the pair are ultra-relativistic; it is easily verified that all the approximations used in §90 then remain valid.

In particular, the probability of pair production by an unpolarized photon, summed over the electron and positron spin projections and integrated over the directions of emergence of the electron, is found on making the changes (91.1) in (90.22) (or rather in the expression for $dI/d\omega$), with $d^3k = \omega^2\,d\omega\,do_{\mathbf{n}}$ replaced by d^3p_+:

$$dw = \frac{e^2}{4\pi^2}\frac{d^3p_+}{\omega}\int\limits_{-\infty}^{\infty}\left(\frac{m^2}{\varepsilon_+\varepsilon_-} - \frac{\varepsilon_+^2+\varepsilon_-^2}{4\varepsilon_-^2}\,\omega_{0+}^2\tau^2\right)\times$$
$$\times\exp\left\{\frac{i\omega\tau\varepsilon_+}{\varepsilon_-}\left(1 - \mathbf{n}\cdot\mathbf{v}_+ + \frac{\tau^2}{24}\,\omega_{0+}^2\right)\right\}d\tau. \tag{91.7}$$

where $\omega_{0+} = |e|H/\varepsilon_+$, and $\mathbf{n}$ is a unit vector parallel to the photon momentum, which lies in the plane perpendicular to the magnetic field. The integration is carried out in the same way as in §90, and (since (91.7) depends only on the angle between $\mathbf{n}$ and $\mathbf{v}_+$) it does not matter whether we integrate over do_+ or over $do_{\mathbf{n}}$. The result can therefore be obtained directly by analogy with (90.23):

$$dw = \frac{m^2e^2}{\sqrt{\pi}}\frac{d\varepsilon_+}{\omega^2}\left\{\int\limits_x^{\infty}\Phi(\xi)\,d\xi + \left(\frac{2}{x} - \kappa\sqrt{x}\right)\Phi'(x)\right\}, \tag{91.8}$$

† More precisely, we must have $\omega \sin\vartheta \gg m$, where ϑ is the angle between $\mathbf{k}$ and $\mathbf{H}$; when $\vartheta = 0$, no pairs are formed. In what follows, we shall take $\vartheta = \frac{1}{2}\pi$.

where now

$$\left.\begin{aligned} x &= (m^3\omega/|e|H\varepsilon_+\varepsilon_-)^{\frac{2}{3}}, \\ \kappa &= |e|H\omega/m^3 \ (= \hbar^2|e|H\omega/m^3c^5). \end{aligned}\right\} \tag{91.9}$$

The total pair production probability per unit time is found by integrating (91.8) with respect to ε_+; in view of the obvious symmetry in ε_+ and $\varepsilon_- = \omega - \varepsilon_+$, it is sufficient to take twice the integral from 0 to $\frac{1}{2}\omega$. Changing variables from ε_+ to x and integrating by parts in the first term in (91.8), we obtain

$$w = \frac{|e|^3 H}{m\kappa\sqrt{\pi}} \int\limits_{(4/\kappa)^{2/3}}^{\infty} \left\{ \frac{(x^{3/2} - 4/\kappa)^{1/2}}{x^{3/4}} \Phi(x) - \frac{3(x^{3/2} - 2/\kappa)\Phi'(x)}{x^{11/4}(x^{3/2} - 4/\kappa)^{1/2}} \right\} dx \tag{91.10}$$

(A. I. Nikishov and V. I. Ritus, 1967).

In the limit of weak fields ($\kappa \ll 1$), values of x near the lower limit are important in the integral (91.10). Since these values are large, we can use the asymptotic expression for the Airy function,

$$\Phi(x) \approx \frac{1}{2x^{1/4}} \exp(-\tfrac{2}{3}x^{3/2});$$

see *QM*, §b. With the variable of integration $y = x^{3/2} - 4/\kappa$, and putting $y = 0$ wherever possible, we find by calculation

$$w = \frac{3^{3/2}|e|^3 H}{2^{9/2} m} e^{-8/3\kappa}, \qquad \kappa \ll 1. \tag{91.11}$$

The exponential decrease of the probability as $\kappa \to 0$ corresponds to the impossibility of pair production in the classical limit.

In the opposite limit of strong fields ($\kappa \gg 1$), only the second term in (91.10) is important, and it is governed by the range of x for which $x^{3/2} \sim 1/\kappa \ll 1$. In that range, $\Phi'(x)$ may be replaced by $\Phi'(0) = -3^{1/6}\Gamma(\frac{2}{3})/2\pi^{\frac{1}{2}}$. With the value of the integral

$$\int\limits_1^{\infty} y^{-\nu}(y-1)^{\mu-1}\, dy = \Gamma(\nu - \mu)\Gamma(\mu)/\Gamma(\nu),$$

we find

$$w = \frac{3^{1/6} \times 5\Gamma^2(\frac{2}{3})}{2^{4/3} \times 7\pi^{1/2}\Gamma(7/6)} \frac{|e|^3 H}{m\kappa^{1/3}} = 0.38|e|^3 H/m\kappa^{1/3}, \quad \kappa \gg 1. \tag{91.12}$$

The function $mw(\kappa)/|e|^3 H$ has a maximum value of 0.11 at $\kappa \approx 11$.

§92. Electron–nucleus bremsstrahlung. The non-relativistic case

This section and those following are concerned with the important phenomenon of *bremsstrahlung*, the radiation emitted in a collision between particles. We shall first consider a non-relativistic collision between an electron and a nucleus, assuming that the nucleus remains at rest; that is, we consider radiation from the scattering of an electron in the Coulomb field of a fixed centre (A. Sommerfeld, 1931).

We begin from formula (45.5) for the probability of dipole radiation:

$$dw = (\omega^3/2\pi)|\mathbf{e}^* \cdot \mathbf{d}_{fi}|^2 \, do_{\mathbf{k}}. \tag{92.1}$$

In the present case, the initial and final states of the electron belong to the continuous spectrum, and the photon frequency

$$\omega = (1/2m)(\mathbf{p}^2 - \mathbf{p}'^2), \tag{92.2}$$

where $\mathbf{p} = m\mathbf{v}$ and $\mathbf{p}' = m\mathbf{v}'$ are the initial and final momenta of the electron. If the initial and final wave functions of the electron are normalized to "one particle per unit volume" ($V = 1$) the expression (92.1), on multiplication by $d^3p'/(2\pi)^3$ and division by the incident flux density $v/V = v$, will give the cross-section $d\sigma_{\mathbf{kp}'}$ for emission of a photon $\mathbf{k}$ into the solid angle $do_{\mathbf{k}}$ with scattering of the electron into the range of states d^3p'. Replacing the matrix element of the dipole moment $\mathbf{d} = e\mathbf{r}$ by that of the momentum:

$$\mathbf{d}_{fi} = -\frac{1}{i\omega}\frac{e}{m}\mathbf{p}_{fi},$$

we can write the expression for the cross-section in the form†

$$d\sigma_{\mathbf{kp}'} = \frac{\omega e^2}{(2\pi)^4 mp}|\mathbf{e}^* \cdot \mathbf{p}_{fi}|^2 \, do_{\mathbf{k}} \, d^3p', \tag{92.3}$$

where

$$\mathbf{p}_{fi} = \int \psi_f^* \hat{\mathbf{p}} \psi_i \, d^3x = -i \int \psi_f^* \nabla \psi_i \, d^3x.$$

For ψ_i and ψ_f we must use the exact wave functions in an attractive Coulomb field, whose asymptotic form consists of a plane wave and a spherical wave. The spherical wave must be ingoing in ψ_f and outgoing in ψ_i (see *QM*, §136). These functions are

$$\left.\begin{aligned} \psi_i &= A_i \, e^{i\mathbf{p}\cdot\mathbf{r}} F(i\nu, 1, i(pr - \mathbf{p}\cdot\mathbf{r})), & \nu &= Ze^2m/p; \\ \psi_f &= A_f \, e^{i\mathbf{p}'\cdot\mathbf{r}} F(-i\nu', 1, -i(p'r + \mathbf{p}'\cdot\mathbf{r})), & \nu' &= Ze^2m/p', \end{aligned}\right\} \tag{92.4}$$

† In this section, p and p' denote $|\mathbf{p}|$ and $|\mathbf{p}'|$ respectively.

with the normalization factors

$$A_i = e^{\pi\nu/2}\Gamma(1-i\nu), \qquad A_f = e^{\pi\nu'/2}\Gamma(1+i\nu'). \tag{92.5}$$

Since

$$\begin{aligned}\nabla F(i\nu, 1, i(pr-\mathbf{p}\cdot\mathbf{r})) &= i(p\mathbf{r}/r-\mathbf{p})F' \\ &= -\frac{p}{r}\left(\frac{\partial F}{\partial \mathbf{p}}\right)_\nu,\end{aligned}$$

we can write the gradient $\nabla\psi_i$ as

$$\nabla\psi_i = i\mathbf{p}\psi_i - A_i e^{i\mathbf{p}\cdot\mathbf{r}}\frac{p}{r}\left(\frac{\partial F}{\partial \mathbf{p}}\right)_\nu.$$

On multiplication by ψ_f^* and integration, the first term vanishes, because ψ_i and ψ_f are orthogonal. The matrix element $\mathbf{p}_{fi}$ is therefore

$$\mathbf{p}_{fi} = iA_iA_f p\,\partial J/\partial\mathbf{p}, \tag{92.6}$$

where J denotes the integral

$$\left.\begin{aligned}J &= \int \frac{e^{-i\mathbf{q}\cdot\mathbf{r}}}{r}F(i\nu', 1, i(p'r+\mathbf{p}'\cdot\mathbf{r}))F(i\nu, 1, i(pr-\mathbf{p}\cdot\mathbf{r}))\,d^3x, \\ \mathbf{q} &= \mathbf{p}'-\mathbf{p}.\end{aligned}\right\} \tag{92.7}$$

The symbol $\partial/\partial\mathbf{p}$ has been taken outside the integral, with the understanding that, in the differentiation of J, the quantities ν, ν', $\mathbf{q}$ are to be regarded as independent parameters, ν and $\mathbf{q}$ being expressed in terms of $\mathbf{p}$ only after the differentiation.

The integral is calculated by replacing the confluent hypergeometric functions by their expressions as contour integrals. Here we shall give only the result:†

$$\left.\begin{aligned}J &= BF(i\nu', i\nu, 1, z) \\ B &= 4\pi e^{-\pi\nu}(-\mathbf{q}^2-2\mathbf{q}\cdot\mathbf{p})^{-i\nu}(\mathbf{q}^2-2\mathbf{q}\cdot\mathbf{p}')^{-i\nu'}(\mathbf{q}^2)^{i\nu+i\nu'-1}, \\ z &= 2\frac{\mathbf{q}^2(pp'+\mathbf{p}\cdot\mathbf{p}')-2(\mathbf{q}\cdot\mathbf{p})(\mathbf{q}\cdot\mathbf{p}')}{(\mathbf{q}^2-2\mathbf{q}\cdot\mathbf{p}')(\mathbf{q}^2+2\mathbf{q}\cdot\mathbf{p})}.\end{aligned}\right\} \tag{92.8}$$

Here $F(i\nu', i\nu, 1, z)$ is the complete hypergeometric function.

After differentiating in (92.6), we can put $\mathbf{q} = \mathbf{p}'-\mathbf{p}$; then

$$z = -2\frac{pp'-\mathbf{p}\cdot\mathbf{p}'}{(p-p')^2}, \qquad \mathbf{q}^2 = (p-p')^2(1-z) \tag{92.9}$$

†The calculations are given by A. Nordsieck, *Physical Review* **93**, 785, 1954.

($z<0$). Also,

$$-\mathbf{q}^2-2\mathbf{q}\cdot\mathbf{p}=\mathbf{q}^2-2\mathbf{q}\cdot\mathbf{p}'=\mathbf{p}^2-\mathbf{p}'^2>0.$$

The matrix element is thus finally found to be

$$\mathbf{p}_{fi}=A_iA_f\frac{8\pi i\,e^{-\pi\nu}}{(p-p')^3(p+p')}\left(\frac{p+p'}{p-p'}\right)^{-i(\nu+\nu')}\times$$
$$\times(1-z)^{i(\nu+\nu')-1}[i\nu p\,\mathbf{q}F(z)+(1-z)F'(z)(p'\mathbf{p}-p\mathbf{p}')], \tag{92.10}$$

where we have put for brevity

$$F(z)=F(i\nu', i\nu, 1, z). \tag{92.11}$$

The cross-section is obtained by substituting (92.10) in (92.3), but the general formula is very lengthy and obscure. We shall therefore go on immediately to calculate the spectral distribution of the radiation, i.e. to integrate the cross-section over the directions of the photon and the final electron.

The integration over $do_{\mathbf{k}}$ and the summation over the polarizations of the photon are equivalent to averaging over all directions $\mathbf{e}$ and multiplication by $2\times4\pi$, i.e. to the substitution

$$e_ie_k^*\,do_{\mathbf{k}}\to(8\pi/3)\delta_{ik}.$$

The cross-section is then

$$d\sigma_{\mathbf{p}'}=\frac{4\omega e^2}{3pm}|\mathbf{p}_{fi}|^2\frac{d^3\mathbf{p}'}{(2\pi)^3}$$
$$=\frac{\omega e^2p'}{6\pi^3p}|\mathbf{p}_{fi}|^2\,d\omega\,do_{\mathbf{p}'}. \tag{92.12}$$

The value of $|\mathbf{p}_{fi}|^2$ is calculated by using (92.9)–(92.11) and the formula

$$|\Gamma(1-i\nu)|^2=\pi\nu/\sinh\pi\nu.$$

The result is

$$|\mathbf{p}_{fi}|^2=\frac{32\pi(Ze^2)^2m^3}{p(p+p')^2(p-p')^4(1-e^{-2\pi\nu'})(e^{2\pi\nu}-1)}\times$$
$$\times\left\{\frac{\nu\nu'}{1-z}|F|^2-z|F'|^2+\tfrac{1}{2}i(\nu+\nu')\frac{z}{1-z}(FF'^*-F^*F')\right\}. \tag{92.13}$$

To integrate the cross-section (92.12) over $do_{\mathbf{p}'}=2\pi\sin\vartheta\,d\vartheta$, we change from the variable ϑ (the scattering angle) to

$$z=-\frac{2pp'}{(p-p')^2}(1-\cos\vartheta),\qquad do_{\mathbf{p}'}\to\frac{\pi(p-p')^2}{pp'}dz.$$

In order to integrate with respect to z, we transform the expression in the braces in (92.13) as follows. According to the differential equation of the hypergeometric function (see *QM*, (e.2)), we have

$$z(1-z)F''+[1-(1+i\nu+i\nu')z]F'+\nu\nu'F=0,$$

$$z(1-z)F''^{*}+[1-(1-i\nu-i\nu')z]F'^{*}+\nu\nu'F^{*}=0.$$

Multiplying these equations by F^* and F respectively and adding, we obtain

$$(1-z)\left[\frac{d}{dz}z(F'F^*+F'^*F)-2z|F'|^2+\frac{i(\nu+\nu')z}{1-z}(F'^*F-F'F^*)+\frac{2\nu\nu'}{1-z}|F|^2\right]=0.$$

Hence the expression in the braces in (92.13) is seen to be

$$\{\cdots\}=-\frac{1}{2}\frac{d}{dz}z(F'F^*+FF'^*), \tag{92.14}$$

and the integration is immediate.

Collecting the above formulae, we find as the final expression for the bremsstrahlung emission cross-section in the frequency range $d\omega$†

$$d\sigma_\omega=\frac{64\pi^2}{3}Z^2\alpha r_e^2\frac{m^2c^2}{(p-p')^2}\frac{p'}{p}\frac{1}{(1-e^{-2\pi\nu'})(e^{2\pi\nu}-1)}\left(-\frac{d}{d\xi}|F(\xi)|^2\right)\frac{d\omega}{\omega}, \tag{92.15}$$

where

$$\nu=Z\alpha mc/p=Ze^2/\hbar v,\qquad \nu'=Ze^2/\hbar v',\qquad p'=\sqrt{(p^2-2m\hbar\omega)},$$

$$F(\xi)=F(i\nu',i\nu,1,\xi),\qquad \xi=-4pp'/(p-p')^2.$$

Let us consider the limiting case where both velocities v and v' are so large that $\nu\ll1$, $\nu'\ll1$ (but, of course, still with $v\ll1$, so that $Z\alpha\ll\nu\ll1$; this is possible only if Z is small). To calculate the derivative $F'(\xi)$ in this case, we use the formula

$$\frac{d}{dz}F(\alpha,\beta,\gamma,z)=\frac{\alpha\beta}{\gamma}F(\alpha+1,\beta+1,\gamma+1,z),$$

which is easily obtained by simple differentiation of the hypergeometric series. Then

$$F'(\xi)\approx i\nu\cdot i\nu'F(1,1,2,\xi)$$
$$=(\nu\nu'/\xi)\log(1-\xi);$$

the last equation is evident from a direct comparison of the corresponding series.

† Formulae (92.15)–(92.25) are given in ordinary units.

For the function $F(\xi)$ itself, we have simply

$$F(\xi) \approx F(0, 0, 1, \xi) = 1.$$

Then, from (92.15),

$$\left.\begin{aligned} d\sigma_\omega &= \tfrac{16}{3} Z^2 \alpha r_e^2 \frac{c^2}{v^2} \log \frac{v + v'}{v - v'} \frac{d\omega}{\omega}, \\ Ze^2/\hbar v &\ll 1, \qquad Ze^2/\hbar v' \ll 1. \end{aligned}\right\} \tag{92.16}$$

The smallness of ν and ν' is just the condition for the Born approximation to be valid in the case of Coulomb interaction. Formula (92.16) itself can therefore be more simply obtained directly by means of perturbation theory (see Problem 1).

Now let a fast electron ($\nu \ll 1$) lose a considerable fraction of its energy by radiation, so that $v' \ll v$ and ν' may not be small. Then

$$\begin{gathered} -\xi \approx 4p'/p = 4\nu/\nu' \ll 1, \\ F(\xi) \approx F(i\nu', 0, 1, \xi) = 1, \\ F'(\xi) \approx -\nu\nu' F(1 + i\nu', 1, 2, \xi) \approx -\nu\nu', \end{gathered}$$

and the cross-section is

$$\left.\begin{aligned} d\sigma_\omega &= \frac{64\pi}{3} Z^3 \alpha^2 r_e^2 \left(\frac{c}{v}\right)^3 \frac{1}{1 - \exp[-2\pi Z e^2/\hbar v']} \frac{d\omega}{\omega}, \\ Ze^2/\hbar v &\ll 1, \qquad Ze^2/\hbar v' \gtrsim 1. \end{aligned}\right\} \tag{92.17}$$

When $\nu' \ll 1$, this formula yields the same limiting expression,

$$d\sigma_\omega = \tfrac{32}{3} Z^2 \alpha r_e^2 \frac{c^2 v'}{v^3} \frac{d\omega}{\omega},$$

as (92.16) does when $v' \ll v$. Hence formulae (92.16) and (92.17) jointly cover the whole range of ν' (when $\nu \ll 1$).

When $\omega \to \omega_0$, where $\hbar\omega_0 = \frac{1}{2} m v^2$, the velocity $v' \to 0$ and $\nu' \to \infty$. In this limiting case, (92.17) gives

$$d\sigma_\omega = \frac{128\pi}{3} Z^3 \alpha^2 r_e^2 \left(\frac{c}{v}\right)^3 \frac{\hbar\, d\omega}{m v^2}. \tag{92.18}$$

Thus $d\sigma_\omega/d\omega$ tends to a finite limit as $\omega \to \omega_0$. This can be explained in a general manner by arguments similar to those given in *QM*, §147. The physical reason is that the frequency ω_0 is the limit only of the continuous bremsstrahlung spectrum. The electron can also go to a bound state with emission of a frequency $\omega > \omega_0$. But highly excited bound states in a Coulomb field have properties almost the same as those of the free states near their limit. Hence the boundary between the continuous and the discrete spectrum is not essentially a physically distinctive point.

Let us now consider the case where both parameters ν, $\nu' \gg 1$. The motion of both the initial and the final electrons is then quasi-classical. If the condition $\hbar\omega \ll p^2/2m$ is also satisfied, the matrix element too is quasi-classical. Then the formula of quantum mechanics must become the result given by the classical theory (see *Fields*, §70). We shall, however, suppose that $p^2/2m \sim \hbar\omega$, so that we need an asymptotic expression for the function $F(\xi)$ when $\nu, \nu' \to \infty$ and $\xi \sim 1$; a more exact condition will be stated below, (92.24).

To derive this expression, we start from the integral representation of the hypergeometric function, *QM*, (e.3), writing it as

$$F(i\rho\nu', i\nu', 1, \xi) = \frac{e^{-\pi\rho\nu'}}{2\pi i} \oint_{C'} t^{i\rho\nu'-1}(1-t)^{-i\rho\nu'}(1-t\xi)^{-i\nu'}\, dt', \tag{92.19}$$

where

$$\rho = \nu/\nu', \qquad 0 < \rho < 1,$$

so that

$$\xi = -4\rho/(1-\rho)^2. \tag{92.20}$$

The contour of integration is taken as shown in Fig. 17, passing along part of the real axis and avoiding the points $t = 0$ and $t = 1$.†

When ν, $\nu' \gg 1$, the value of the integrand is small on the lower part of this contour, and may be neglected. On passing downwards round the point $t = 0$, the integrand is multiplied by the small factor $\exp(-2\pi\rho\nu')$, and on passing upwards round $t = 1$ it is multiplied by $\exp(2\pi\rho\nu')$. The integral

$$F = \frac{e^{-\pi\rho\nu'}}{2\pi i} \int e^{\nu' f(t)}\, dt/t, \quad f(t) = i \log \frac{t^\rho}{(1-t)^\rho(1-\xi t)}, \tag{92.21}$$

may be calculated by the saddle-point method. The saddle point t_0 is given by the condition $f'(t_0) = 0$, whence $t_0 = \frac{1}{2}(1-\rho)$. At this point, however, the derivative $f''(t_0)$ is also zero, so that we must write

$$f(t) \approx f(t_0) + \tfrac{1}{3} i a \tau^3, \qquad \tau = t - t_0,$$

FIG. 17.

† For the hypergeometric function $F(\alpha, \beta, \gamma, \xi)$, the contour is to be chosen so that the function

$$V(t) = e^t t^{\alpha-\gamma+1}(t-1)^{1-\alpha}$$

returns to its initial value on passing round the contour. When γ is integral (here, $\gamma = 1$), the contour chosen satisfies this condition.

where

$$f(t_0) = 2\pi\rho + i(1+\rho)\log\frac{1-\rho}{1+\rho}, \quad a = \frac{1}{2i}f'''(t_0) = \frac{16\rho}{(1-\rho^2)^2}.$$

The coefficient $1/t$ of the exponential in the integrand may be written

$$1/t \approx 1/t_0 - \tau/t_0^2;$$

we cannot here take simply the term $1/t_0$, as this would reduce to zero the derivative $d|F(\xi)|^2/d\xi$ in (92.15). Thus we have, after an obvious substitution in the integrals,

$$F \approx \frac{1}{2\pi i t_0(\alpha\nu')^{1/3}}\exp\{-\pi\rho\nu' + \nu' f(t_0)\}\times$$
$$\times\left\{-\int_{-\infty}^{\infty} e^{\frac{1}{3}ix^3}\,dx + \frac{i}{t_0(\alpha\nu')^{1/3}}\int_{-\infty}^{\infty} xe^{\frac{1}{3}ix^3}\,dx\right\}. \qquad (92.22)$$

The two integrals here are, respectively,

$$2\int_0^\infty \cos\tfrac{1}{3}x^3\,dx = \frac{2\pi}{3^{2/3}\Gamma(\frac{2}{3})},$$

$$2\int_0^\infty x\sin\tfrac{1}{3}x^3\,dx = 3^{1/6}(\tfrac{2}{3}).$$

The derivative $F'(\xi)$ is calculated similarly; according to (92.19), it is given by an integral which differs from (92.21) only in that the coefficient $1/it$ of the exponential is replaced by $\nu'/(1-\xi t)$. A simple calculation then leads to the result

$$-\frac{d}{d\xi}|F(\xi)|^2 = \frac{(1-\rho)^2 e^{2\pi\nu}}{4\sqrt{3}\pi\rho}.$$

Finally, substitution of this in (92.15) gives, with the necessary accuracy, the simple expression

$$d\sigma_\omega = \frac{16\pi}{3^{3/2}}Z^2\alpha r_e^2\frac{m^2c^2}{p^2}\frac{d\omega}{\omega}. \qquad (92.23)$$

The condition for this to be valid, i.e. for the asymptotic formula (92.22) to be valid, is that the second term in (92.22) should be much less than the first: $(1-\rho)\nu \gg 1$, or, expressing the parameters of the hypergeometric function in terms of physical quantities,

$$\hbar\omega \gg (\hbar v/Ze^2)\cdot\tfrac{1}{2}mv^2. \qquad (92.24)$$

This inequality is compatible with the quasi-classicality condition $\hbar\omega \ll \frac{1}{2}mv^2$, i.e. $1-\rho \ll 1$. When the latter condition is satisfied, the result must be the same as the corresponding formula in the classical theory, since on multiplication by $\hbar\omega$ the expression (92.23) becomes the classical formula (*Fields*, (70.22)) for the "effective retardation" in the high-frequency limit.† In order to go to the classical formulae throughout the range $(1-\rho)\nu \sim 1$, $\nu \gg 1$, we should have to find the asymptotic form of the hypergeometric function when the saddle point is close to the singularity $t=0$; we shall not deal with this here, since the final result is obvious.

All the formulae given above refer to an attractive Coulomb field. The cross-section for emission in a repulsive field is obtained from (92.15) by changing the signs of ν and ν'. Then, in particular, the limiting Born formula (92.16) remains unaltered, but in the limit $\nu \ll 1$, $\nu' \to \infty$ we have instead of (92.18)

$$d\sigma_\omega = \frac{128\pi}{3} Z^3\alpha^2 r_e^2 \left(\frac{c}{v}\right)^3 \exp\left(-\frac{\sqrt{(2mc^2)}\pi Z\alpha}{\sqrt{(\hbar\omega_0-\hbar\omega)}}\right)\frac{\hbar\, d\omega}{mv^2}, \tag{92.25}$$

i.e. the differential cross-section tends exponentially to zero as $\omega \to \omega_0$. This result also is reasonable: in a repulsive field, there are no bound states, and the frequency ω_0 is the true boundary of the radiation spectrum.

PROBLEMS

PROBLEM 1. In the Born approximation, find the bremsstrahlung cross-section for a non-relativistic collision of two particles having different values of the ratio e/m.

SOLUTION. The dipole moment of two particles with charges e_1, e_2 and masses m_1, m_2, in their centre-of-mass system, is

$$\mathbf{d} = \mu\left(\frac{e_1}{m_1}-\frac{e_2}{m_2}\right)\mathbf{r},$$

where $\mu = m_1m_2/(m_1+m_2)$, $\mathbf{r} = \mathbf{r}_1 - \mathbf{r}_2$. Hence

$$\ddot{\mathbf{d}} = \left(\frac{e_1}{m_1}-\frac{e_2}{m_2}\right)\mu\ddot{\mathbf{r}} = -\left(\frac{e_1}{m_1}-\frac{e_2}{m_2}\right)\nabla\frac{e_1e_2}{r}.$$

The matrix element is

$$\mathbf{d}_{\mathbf{p}'\mathbf{p}} = -\frac{1}{\omega^2}(\ddot{\mathbf{d}})_{\mathbf{p}'\mathbf{p}}, \qquad \omega = (p^2-p'^2)/2\mu,$$

where $\mathbf{p} = \mu\mathbf{v}$, $\mathbf{p}' = \mu\mathbf{v}'$ are the momenta of relative motion, and it is calculated from the plane waves‡

$$\psi_{\mathbf{p}} = e^{i\mathbf{p}\cdot\mathbf{r}}, \qquad \psi_{\mathbf{p}'} = e^{i\mathbf{p}'\cdot\mathbf{r}}$$

by means of the formula

$$\left(\nabla\frac{1}{r}\right)_{\mathbf{p}'\mathbf{p}} = \frac{4\pi i\mathbf{q}}{q^2}, \qquad \mathbf{q} = \mathbf{p}'-\mathbf{p}.$$

† The agreement of the formula (92.23) for $\hbar\omega\, d\sigma_\omega$ with the classical formula, when the one condition (92.24) is satisfied, is to some extent accidental. In the classical formula, the difference between v and v' is beyond the accuracy postulated, and there is no reason to identify v_0 in *Fields*, (70.22) with v specifically; if v_0 is identified with v', there is no longer agreement with (92.23).

‡ The replacement of two particles by a single particle having the reduced mass is, of course, permissible only in the non-relativistic case.

The result is

$$d\sigma_{\mathbf{k}\mathbf{p}'} = \frac{e_1^2 e_2^2}{\pi^2}\left(\frac{e_1}{m_1} - \frac{e_2}{m_2}\right)^2 \frac{v'}{v}\frac{\mu^2}{q^4}(\mathbf{e}\cdot\mathbf{q})(\mathbf{e}^*\cdot\mathbf{q})\frac{d\omega}{\omega}\, do_{\mathbf{p}'}\, do_{\mathbf{k}}.$$

After summation over polarizations, the angular distribution of the radiation is given by a factor $\sin^2\Theta$, where Θ is the angle between the direction of the photon $\mathbf{k}$ and the vector $\mathbf{q}$, which lies in the scattering plane (see (45.4a)).

After integration over the directions of the photon we have

$$d\sigma_{\omega\theta} = \tfrac{16}{3} e_1^2 e_2^2 \left(\frac{e_1}{m_1} - \frac{e_2}{m_2}\right)^2 \frac{v'}{v}\frac{d\omega}{\omega}\frac{\sin\theta\, d\theta}{v^2 + v'^2 - 2vv'\cos\theta},$$

where θ is the scattering angle. Finally, integration with respect to θ gives

$$d\sigma_\omega = \tfrac{16}{3} e_1^2 e_2^2 \left(\frac{e_1}{m_1} - \frac{e_2}{m_2}\right)^2 \frac{1}{v^2}\log\frac{v+v'}{v-v'}\frac{d\omega}{\omega}.$$

For radiation in the field of a fixed centre of Coulomb force, this formula is equivalent to (92.16).

PROBLEM 2. In the Born approximation, find the bremsstrahlung cross-section for a non-relativistic collision of two electrons.†

SOLUTION. In this case there is no dipole radiation, and we must therefore consider quadrupole radiation. In the classical theory, the spectral distribution of the total intensity of quadrupole radiation is given by

$$I_\omega = \tfrac{1}{90}|(\dddot{D}_{ik})_\omega|^2,$$

where $D_{ik} = \Sigma\, e(3x_i x_k - r^2\delta_{ik})$ is the quadrupole moment tensor of a system of charges.‡ For two electrons we have, in their centre-of-mass system,

$$D_{ik} = \tfrac{1}{2}e(3x_i x_k - r^2\delta_{ik}), \qquad \mathbf{r} = \mathbf{r}_1 - \mathbf{r}_2.$$

In the quantum theory, the Fourier components must be replaced by the matrix elements (cf. the discussion of dipole radiation in §45), and, with appropriate normalization of the wave functions (plane waves) and division by the photon energy ω, we obtain the cross-section for emission of radiation with scattering of the electrons into the range of states d^3p':

$$d\sigma_{\mathbf{p}'} = \frac{1}{90\omega}|(\dddot{D}_{ik})_{\mathbf{p}'\mathbf{p}}|^2 \frac{d^3p'}{v(2\pi)^3},$$

where $v = 2p/m$ is the initial velocity of relative motion; the emitted frequency $\omega = (p^2 - p'^2)/m$.

The operator $\hat{\dddot{D}}_{ik}$ is calculated by threefold commutation of the operator $\hat{D}_{ik}$ with the Hamiltonian

$$\hat{H} = \frac{\hat{\mathbf{p}}^2}{m} + \frac{e^2}{r},$$

and is§

$$\hat{\dddot{D}}_{ik} = \frac{2e^3}{m}\left[6\left(\frac{x_i}{r^3}\hat{p}_k + \hat{p}_k\frac{x_i}{r^3}\right) + 6\left(\frac{x_k}{r^3}\hat{p}_i + \hat{p}_i\frac{x_k}{r^3}\right) - 9\left(\frac{x_i x_k x_l}{r^5}\hat{p}_l + \hat{p}_l\frac{x_i x_k x_l}{r^5}\right) - \delta_{ik}\left(\frac{x_l}{r^3}\hat{p}_l + \hat{p}_l\frac{x_l}{r^3}\right)\right].$$

† The collision velocity v satisfies the conditions $\alpha \ll e^2/\hbar v \ll 1$. The classical case ($e^2/\hbar v \gg 1$) is discussed in *Fields*, §71, Problem.

‡ This formula is obtained from *Fields* (71.5) in the same way as (67.11) in that book is derived from (67.8).

§ This expression is analogous to the classical form

$$\dddot{D}_{ik} = \frac{4e^3}{m^2}\left[6\frac{x_i}{r^3}p_k + 6\frac{x_k}{r^3}p_i - 9\frac{x_i x_k}{r^5}\mathbf{p}\cdot\mathbf{r} - \frac{1}{r^3}\delta_{ik}\mathbf{p}\cdot\mathbf{r}\right],$$

which would be obtained on differentiating D_{ik} and using the classical equation of motion:

$$\tfrac{1}{2}m\ddot{\mathbf{r}} = e^2\mathbf{r}/r^3.$$

Since the two particles (electrons) are identical, the matrix elements are calculated from the wave functions

$$\psi_{\mathbf{p}} = \frac{1}{\sqrt{2}}(e^{i\mathbf{p}\cdot\mathbf{r}} \pm e^{-i\mathbf{p}\cdot\mathbf{r}}), \qquad \psi_{\mathbf{p}'} = \frac{1}{\sqrt{2}}(e^{i\mathbf{p}'\cdot\mathbf{r}} \pm e^{-i\mathbf{p}'\cdot\mathbf{r}}),$$

where the signs + and − correspond to total spins 0 and 1 of the electrons (interchange of the electrons corresponds to changing the sign of **r**).

The lengthy calculations lead to the following formula for the spectral distribution of the radiation:

$$d\sigma_\omega = \tfrac{4}{15}\alpha r_e^2\left\{17 - \frac{3x^2}{(2-x)^2} + \frac{12(2-x)^4 - 7(2-x)^2x^2 - 3x^4}{(2-x)^3\sqrt{(1-x)}}\cosh^{-1}\frac{1}{\sqrt{x}}\right\}\frac{\sqrt{(1-x)}}{x}\,dx,$$

where $x = \omega/\varepsilon$ and $\varepsilon = p^2/m$ is the initial energy of relative motion of the electrons; the cross-section is averaged over values of the total spin of the electrons. The cross-section for energy loss by radiation is

$$\frac{1}{\varepsilon}\int_0^\varepsilon \omega\, d\sigma_\omega = 8.1\,\alpha r_e^2$$

(B. K. Fedyushin, 1952).

PROBLEM 3. Determine the energy of the radiation resulting from the emission by a nucleus of a non-relativistic electron in the *s* state.

SOLUTION. The wave function of the emitted electron is an outgoing spherical *s* wave normalized to unit total flux:

$$\psi_i = \frac{1}{\sqrt{(4\pi v)}}\frac{e^{ipr}}{r};$$

see *QM*, (33.14). As the wave function of the final state of the electron (after emission of the photon), we choose the plane wave

$$\psi_f = e^{i\mathbf{p}'\cdot\mathbf{r}}.$$

The transition matrix element is

$$\begin{aligned}\mathbf{p}_{fi} = (\mathbf{p}_{if})^* &= \left(\int \psi_i^*\hat{\mathbf{p}}\psi_f\, d^3x\right)^* \\ &= \frac{\mathbf{p}'}{\sqrt{(4\pi v)}}\int e^{-i\mathbf{p}'\cdot\mathbf{r}+ipr}\frac{d^3x}{r} \\ &= \sqrt{\frac{4\pi}{v}}\frac{\mathbf{p}'}{\mathbf{p}'^2 - \mathbf{p}^2} \\ &= -\sqrt{\frac{\pi}{v}}\frac{\mathbf{v}'}{\omega};\end{aligned}$$

the integral is calculated by means of (57.6a). The radiation energy is given by (45.8) on multiplying by $d^3p'/(2\pi)^3$ and integrating over the directions of $\mathbf{p}'$ (which is equivalent to multiplying by 4π). The spectral distribution of the emitted energy is then

$$dE_\omega = (2e^2v'^3/3\pi v)\, d\omega.$$

When $\omega \to 0$, the final velocity v' of the electron tends to v, and the formula agrees, as it should, with the non-relativistic limit of the classical result; see *Fields*, §69, Problem. The total emitted energy is (in ordinary units)

$$E = \frac{4}{15\pi}\alpha\left(\frac{v}{c}\right)^2\varepsilon,$$

where $\varepsilon = \frac{1}{2}mv^2$ is the initial energy of the electron.

PROBLEM 4. Determine the energy of the radiation resulting from the reflection of a non-relativistic electron from an infinitely high potential barrier.

SOLUTION. Let the electron be moving perpendicularly to the barrier. Although the photon may be emitted in any direction, in the non-relativistic case the photon momentum is small compared with the electron momentum, and we may therefore suppose that the reflected electron also is moving perpendicularly to the plane of the barrier. Let the barrier be at $x = 0$, and the electron be moving on the side where $x > 0$. The wave functions of the stationary states of one-dimensional motion, normalized by $\delta(p/2\pi)$ $(p = p_x)$, have the form of stationary waves (see *QM*, §21):

$$\psi_i = 2 \sin px, \qquad \psi_f = 2 \sin p'x.$$

The matrix element of the operator $\hat{p} = \hat{p}_x$ is

$$\begin{aligned} p_{fi} &= -4i \int_0^\infty \sin p'x \frac{d}{dx} \sin px \, dx \\ &= -4i \frac{pp'}{p^2 - p'^2}; \end{aligned}$$

integrals of this form are to be understood as the limit, as $\delta \to +0$, of the values obtained by including a factor $e^{-\delta x}$ in the integrand.

The energy radiated in a single reflection of the electron is found from (45.8) by multiplying by $dp' = d\omega/v'$ and dividing by $v/2\pi$ (the flux density of the wave approaching the barrier in the initial function ψ_i):

$$\begin{aligned} dE_\omega &= \frac{4\omega^2 e^2}{3m^2} |p_{fi}|^2 \frac{2\pi \, d\omega}{vv'} \\ &= \frac{8}{3\pi} e^2 vv' \, d\omega. \end{aligned} \tag{1}$$

At low frequencies ($\omega \ll \varepsilon = \frac{1}{2}mv^2$) we have $v' \approx v$, and formula (1) becomes the classical formula (*Fields* (69.5)), which has to be integrated over angles, using the fact that $v = \frac{1}{2}\Delta v$, where Δv is the change in velocity of the electron on reflection; this is as it should be, since the condition for the collision time to be small (*Fields*, (69.1)) is always satisfied in reflection from a barrier. The quantum formula (1), however, also gives the total emitted energy (in ordinary units):

$$E = \int_0^\varepsilon \frac{dE_\omega}{d\omega} d\omega = \frac{16}{9\pi} \alpha\varepsilon \frac{v^2}{c^2}.$$

PROBLEM 5. Determine the bremsstrahlung energy in the scattering of a slow electron by an atom.

SOLUTION. With the condition $pa \ll 1$ (where a denotes the atomic dimensions), the scattering by the atom is isotropic and does not depend on the electron energy; see *QM*, §132. The wave functions of the initial and final states of the electron may be written

$$\psi_i = e^{i\mathbf{p}\cdot\mathbf{r}} + fe^{ipr}/r, \qquad \psi_f = e^{i\mathbf{p}'\cdot\mathbf{r}} + fe^{-ip'r}/r,$$

where f is the constant real scattering amplitude. These expressions pertain to the asymptotic range of distances $r \gg a$, which in the present case is the important range: $r \sim 1/p \gg a$. The matrix element calculated from these functions is

$$\mathbf{p}_{fi} = (2\pi f/\omega)(\mathbf{v} - \mathbf{v}');$$

the integrals are calculated as in Problem 3. Substituting this expression in (92.12), we obtain the cross-section for radiation with scattering of the electron in the direction $\mathbf{p}'$, in ordinary units,

$$d\sigma_{\omega\mathbf{p}'} = \frac{2\alpha p'}{3\pi pc^2} (\mathbf{v} - \mathbf{v}')^2 \, d\sigma_{el} \frac{d\omega}{\omega}, \tag{1}$$

where $d\sigma_{el} = f^2 \, do_{\mathbf{p}'}$ is the differential cross-section for elastic scattering. When $\hbar\omega \ll p^2/2m$, we can take

$p \approx p'$, and this formula then becomes, as it should, the non-relativistic expression for the emission of soft photons; see §98.†

Integrating (1) over the directions of $\mathbf{p}'$, we obtain

$$d\sigma_\omega = \frac{2\alpha p'}{3\pi c^2 p}(v^2 + v'^2)\sigma_{el}\frac{d\omega}{\omega}, \tag{2}$$

where $\sigma_{el} = 4\pi f^2$ is the total elastic scattering cross-section. Lastly, multiplying by $\hbar\omega$ and integrating with respect to ω from 0 to $p^2/2m = \varepsilon$, we obtain the "effective retardation"

$$\kappa_{rad} = \int \hbar\omega\, d\sigma_\omega = \frac{32}{45\pi}\alpha\sigma_{el}\varepsilon(v/c)^2. \tag{3}$$

§93. Electron–nucleus bremsstrahlung. The relativistic case

Let us consider the electron–nucleus bremsstrahlung for the case of relativistic electron velocities.‡ We shall assume that the condition for the Born approximation to be valid is satisfied for both the initial (v) and the final (v') velocity of the electron: $Ze^2/\hbar v \ll 1$, $Ze^2/\hbar v' \ll 1$. The charge on the nucleus must be such that $Z\alpha \ll 1$.

As in §92, we shall neglect the recoil of the nucleus, so that the latter acts only as the source of an external field; the justifiability of this treatment is discussed in §97.

According to (91.4), the cross-section for bremsstrahlung is given in terms of the amplitude by

$$d\sigma = \frac{1}{(2\pi)^5}|M_{fi}|^2 \frac{\omega|\mathbf{p}'|}{|\mathbf{p}|}\, do_{\mathbf{k}}\, do'\, d\omega. \tag{93.1}$$

In the first non-vanishing approximation, the matrix element M_{fi} corresponds to two diagrams:

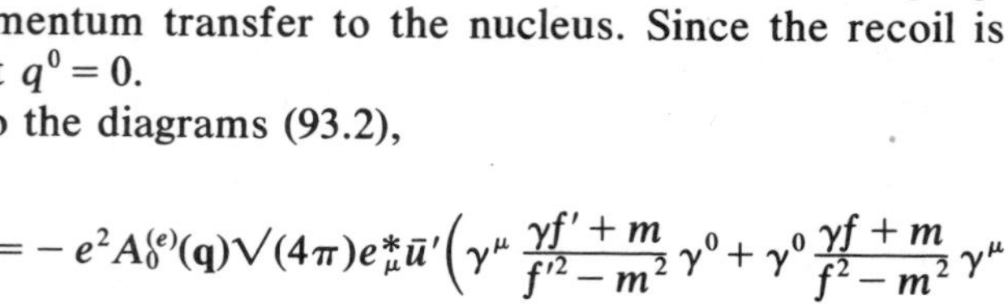

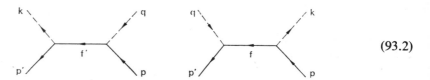

(93.2)

The free end q corresponds to the external field, so that $q = p' - p + k$ is the 4-vector of momentum transfer to the nucleus. Since the recoil is neglected, the time component $q^0 = 0$.

According to the diagrams (93.2),

$$M_{fi} = -e^2 A_0^{(e)}(\mathbf{q})\sqrt{(4\pi)}e_\mu^* \bar{u}'\left(\gamma^\mu \frac{\gamma f' + m}{f'^2 - m^2}\gamma^0 + \gamma^0 \frac{\gamma f + m}{f^2 - m^2}\gamma^\mu\right)u. \tag{93.3}$$

† The fact that "factorization" of the cross-section (the separation of the factor σ_{el}) occurs here for any ω is to some extent accidental, arising because the scattering amplitude is independent of the energy.

‡ The majority of the results given below were first derived by H. A. Bethe and W. Heitler (1934) and independently by F. Sauter (1934).

The intermediate 4-momenta are $f = p - k$, $f' = p' + k$. We shall use the notation

$$f^2 - m^2 = -2kp \equiv -2\kappa\omega, \qquad f'^2 - m^2 = 2kp' \equiv 2\kappa'\omega. \tag{93.4}$$

$A_0^{(e)}$ is the scalar potential of the external field; for a purely Coulomb field,

$$A_0^{(e)}(\mathbf{q}) = 4\pi Ze/\mathbf{q}^2. \tag{93.5}$$

Substitution in (93.1) gives for the cross-section

$$d\sigma = \frac{Z^2e^6}{4\pi^2}\frac{|\mathbf{p}'|\omega}{|\mathbf{p}|\mathbf{q}^4} e_\mu^* e_\nu (\bar{u}'Q^\mu u)(\bar{u}\bar{Q}^\nu u')\, do_\mathbf{k}\, do'\, d\omega, \tag{93.6}$$

where

$$Q^\mu = \gamma^\mu \frac{\gamma f' + m}{2\omega\kappa'}\gamma^0 - \gamma^0 \frac{\gamma f + m}{2\omega\kappa}\gamma^\mu,$$

$$\bar{Q}^\nu = \gamma^0 Q^{\nu+}\gamma^0$$

$$= \gamma^0 \frac{\gamma f' + m}{2\omega\kappa'}\gamma^\nu - \gamma^\nu \frac{\gamma f + m}{2\omega\kappa}\gamma^0.$$

Disregarding polarization effects, we average the cross-section over the directions of spin of the initial electron and sum over the polarizations of the final electron and photon. This is equivalent to the substitution

$$e_\mu^* e_\nu (\bar{u}'Q^\mu u)(\bar{u}\bar{Q}^\nu u') \to -\tfrac{1}{2}\operatorname{tr} Q_\mu(\gamma p + m)\bar{Q}^\mu(\gamma p' + m).$$

The trace is calculated by means of the standard formulae (§22). The calculations are somewhat simplified by using the equation

$$\gamma^0(\gamma p)\gamma^0 = \gamma\tilde{p},$$

where $\tilde{p} = (\varepsilon, -\mathbf{p})$ if $p = (\varepsilon, \mathbf{p})$. Moreover, the number of terms to be calculated can be reduced by using the symmetry with respect to the change $p \leftrightarrow p'$, $k \to -k$, $q \to -q$; this simply interchanges cyclically the factors in the product of matrices, leaving the trace unaltered.

The result is the following expression for the differential cross-section for bremsstrahlung in which a photon of a given frequency is emitted in a given direction and the secondary electron travels in a given direction:†

† Here and in the rest of §93, p, p' and q denote the magnitudes of the three-dimensional vectors: $p = |\mathbf{p}|$, $p' = |\mathbf{p}'|$, $q = |\mathbf{q}|$.

$$d\sigma = \frac{Z^2\alpha r_e^2}{4\pi^2}\frac{p'm^4}{pq^4}\frac{d\omega}{\omega}\,do_{\mathbf{k}}\,do'\times$$
$$\times\left\{\frac{q^2}{\kappa\kappa' m^2}(2\varepsilon^2+2\varepsilon'^2-q^2)+q^2\left(\frac{1}{\kappa}-\frac{1}{\kappa'}\right)^2-4\left(\frac{\varepsilon}{\kappa'}-\frac{\varepsilon'}{\kappa}\right)^2+\right.$$
$$\left.+\frac{2\omega q^2}{m^2}\left(\frac{1}{\kappa'}-\frac{1}{\kappa}\right)-\frac{2\omega^2}{m^2}\left(\frac{\kappa'}{\kappa}+\frac{\kappa}{\kappa'}\right)\right\}, \tag{93.7}$$

where $\kappa = \varepsilon - \mathbf{n}\cdot\mathbf{p}$, $\kappa' = \varepsilon' - \mathbf{n}\cdot\mathbf{p}'$, $\mathbf{n} = \mathbf{k}/\omega$, $\mathbf{q} = \mathbf{p}' + \mathbf{k} - \mathbf{p}$.

By means of simple transformations, this expression can be put in a form somewhat more convenient for analysis:

$$d\sigma = \frac{Z^2\alpha r_e^2}{2\pi}\frac{d\omega}{\omega}\frac{p'm^2}{pq^4}\sin\theta\,d\theta\,\sin\theta'\,d\theta'\,d\phi\times$$
$$\times\left\{\frac{p'^2}{\kappa'^2}(4\varepsilon^2-q^2)\sin^2\theta'+\frac{p^2}{\kappa^2}(4\varepsilon'^2-q^2)\sin^2\theta+\right.$$
$$\left.+\frac{2\omega^2}{\kappa\kappa'}(p^2\sin^2\theta+p'^2\sin^2\theta')-\frac{2pp'}{\kappa\kappa'}(2\varepsilon^2+2\varepsilon'^2-q^2)\sin\theta\sin\theta'\cos\phi\right\}, \tag{93.8}$$

where

$$\kappa = \varepsilon - p\cos\theta, \qquad \kappa' = \varepsilon' - p'\cos\theta',$$
$$q^2 = p^2 + p'^2 + \omega^2 - 2p\omega\cos\theta + 2p'\omega\cos\theta' - 2pp'(\cos\theta\cos\theta' + \sin\theta\sin\theta'\cos\phi);$$

θ, θ' are the angles between $\mathbf{k}$ and $\mathbf{p}$, $\mathbf{p}'$ respectively; ϕ is the angle between the plane of $\mathbf{k}$ and $\mathbf{p}$ and that of $\mathbf{k}$ and $\mathbf{p}'$.

The integration of (93.8) over the directions of the photon and the secondary electron is fairly lengthy. It leads to the following formula for the spectral distribution of the radiation:†

$$d\sigma_\omega = Z^2\alpha r_e^2\frac{d\omega}{\omega}\frac{p'}{p}\left\{\frac{4}{3}-2\varepsilon\varepsilon'\frac{p^2+p'^2}{p^2p'^2}+m^2\left(l\frac{\varepsilon'}{p^3}+l'\frac{\varepsilon}{p'^3}-\frac{ll'}{pp'}\right)+\right.$$
$$\left.+L\left[\frac{8\varepsilon\varepsilon'}{3pp'}+\frac{\omega^2}{p^3p'^3}(\varepsilon^2\varepsilon'^2+p^2p'^2+m^2\varepsilon\varepsilon')+\frac{m^2\omega}{2pp'}\left(l\frac{\varepsilon\varepsilon'+p^2}{p^3}-l'\frac{\varepsilon\varepsilon'+p'^2}{p'^3}\right)\right]\right\}, \tag{93.9}$$

where

$$L = \log\frac{\varepsilon\varepsilon'+pp'-m^2}{\varepsilon\varepsilon'-pp'-m^2},$$
$$l = \log\frac{\varepsilon+p}{\varepsilon-p}, \qquad l' = \log\frac{\varepsilon'+p'}{\varepsilon'-p'}.$$

† The integration over the directions of the secondary electron only can also be completed in an analytical form; see R. L. Gluckstern and M. H. Hull, Jr., *Physical Review* **90**, 1030, 1953.

Reference may also be made to the review paper by H. W. Koch and J. W. Motz, *Reviews of Modern Physics* **31**, 920, 1959, in which the bremsstrahlung formulae are represented graphically.

The permissible values of the frequency in these formulae are limited only by the condition imposed on the final velocity of the electron ($Ze^2/v' \ll 1$): the electron must not lose almost all its energy. As $\omega \to 0$, the emission cross-section diverges as $d\omega/\omega$; this illustrates a general rule which will be discussed in §98.

In the non-relativistic limit ($p \ll m$), the photon momentum is small compared with the electron momentum, since

$$\omega = \frac{p'^2 - p^2}{2m} \ll p.$$

Hence $q^2 \approx (\mathbf{p}' - \mathbf{p})^2$. Putting in (93.8) $\varepsilon = \varepsilon' = m$ and neglecting p, p' and ω in comparison with m, we find

$$d\sigma = \frac{2}{\pi} Z^2 \alpha r_e^2 \frac{d\omega}{\omega} \frac{p' m^2}{p q^4} \sin\theta \, d\theta \sin\theta' \, d\theta' \, d\phi \times$$
$$\times (p^2 \sin^2\theta + p'^2 \sin^2\theta' - 2pp' \sin\theta \sin\theta' \cos\phi),$$

or

$$d\sigma = \frac{Z^2 \alpha^3}{\pi^2} \frac{p'}{p} (\mathbf{n} \times \mathbf{q})^2 \frac{do_{\mathbf{k}} \, do'}{q^4} \frac{d\omega}{\omega}, \tag{93.10}$$

in accordance with the Born-approximation formula derived in §92, Problem 1. Correspondingly, the spectral distribution of the radiation is given by the formula (92.16) already derived.†

In the ultra-relativistic case, when both the initial and the final energies of the electron are large ($\varepsilon, \varepsilon' \gg m$), the angular distribution of photons and secondary electrons is very unusual. For small angles θ, θ', the quantities κ, κ' which appear in the denominators of formula (93.8) are

$$\kappa \approx \tfrac{1}{2}\varepsilon \left(\frac{m^2}{\varepsilon^2} + \theta^2 \right), \qquad \kappa' \approx \tfrac{1}{2}\varepsilon' \left(\frac{m^2}{\varepsilon'^2} + \theta'^2 \right), \tag{93.11}$$

and become very small in the range $\theta \lesssim m/\varepsilon$. In this range the magnitude of the vector $\mathbf{q}$ is also small ($q \sim m$). Thus, in the ultra-relativistic case, the photon and the secondary electron move forwards in a narrow cone with aperture angle $\sim m/\varepsilon$.

A quantitative formula for the angular distribution in the ultra-relativistic case is easily obtained from (93.8) by substituting for κ, κ' from (93.11), replacing p, p' in all other places by ε, ε', and neglecting q^2 in comparison with ε^2. With the convenient notation

$$\delta = \varepsilon\theta/m, \qquad \delta' = \varepsilon'\theta'/m, \tag{93.12}$$

† The derivation of this formula by taking the limit in (93.9) is somewhat laborious, however, because of the cancellation of various terms.

we can put (93.8) in the form

$$d\sigma = \frac{8}{\pi} Z^2\alpha r_e^2 \frac{\varepsilon' m^4}{\varepsilon q^4} \frac{d\omega}{\omega} \delta\, d\delta \cdot \delta'\, d\delta'\, d\phi \times$$

$$\times \left\{\frac{\delta^2}{(1+\delta^2)^2} + \frac{\delta'^2}{(1+\delta'^2)^2} + \frac{\omega^2}{2\varepsilon\varepsilon'} \frac{\delta^2+\delta'^2}{(1+\delta^2)(1+\delta'^2)} - \right.$$

$$\left. - \left(\frac{\varepsilon'}{\varepsilon} + \frac{\varepsilon}{\varepsilon'}\right) \frac{\delta\delta' \cos\phi}{(1+\delta^2)(1+\delta'^2)}\right\}. \qquad (93.13)$$

Putting $\mathbf{q}^2 = (\mathbf{n}\times\mathbf{q})^2 + (\mathbf{n}\cdot\mathbf{q})^2$ ($\mathbf{n} = \mathbf{k}/\omega$), we can easily find that for small angles

$$\frac{q^2}{m^2} = (\delta^2 + \delta'^2 - 2\delta\delta' \cos\phi) + m^2\left(\frac{1+\delta^2}{2\varepsilon} - \frac{1+\delta'^2}{2\varepsilon'}\right)^2. \qquad (93.14)$$

When $\delta \sim \delta' \sim 1$, the second term in (93.14) is small compared with the first. The two terms become comparable at even smaller angles, where $\delta \sim m/\varepsilon$. Although q here becomes particularly small ($q \sim m^2/\varepsilon \ll m$), the integrated contribution from this region to the cross-section is still small compared with that from the whole region $\delta \lesssim 1$ (the ratio of the contributions is easily seen to be m^2/ε^2). But q can also reach values $q \sim m^2/\varepsilon$ when $\delta \sim \delta' \sim 1$ if

$$|\delta - \delta'| \lesssim m/\varepsilon, \qquad \phi \lesssim m/\varepsilon. \qquad (93.15)$$

The contribution from this region is of the same order as the whole integral cross-section, or may even be the principal term in it (see below).

The integration of (93.13) with respect to ϕ and δ' gives the angular distribution of photons (with given frequency), regardless of the direction of the secondary electron:†

$$d\sigma = 8Z^2\alpha r_e^2 \frac{d\omega}{\omega} \frac{\varepsilon'}{\varepsilon} \frac{\delta\, d\delta}{(1+\delta^2)^2} \times$$

$$\times \left\{\left[\frac{\varepsilon}{\varepsilon'} + \frac{\varepsilon'}{\varepsilon} - \frac{4\delta^2}{(1+\delta^2)^2}\right] \log\frac{2\varepsilon\varepsilon'}{m\omega} - \frac{1}{2}\left[\frac{\varepsilon}{\varepsilon'} + \frac{\varepsilon'}{\varepsilon} + 2 - \frac{16\delta^2}{(1+\delta^2)^2}\right]\right\}. \qquad (93.16)$$

Integrating with respect to δ, we find the spectral distribution of the radiation in the ultra-relativistic case:

$$d\sigma_\omega = 4Z^2\alpha r_e^2 \frac{d\omega}{\omega} \frac{\varepsilon'}{\varepsilon} \left(\frac{\varepsilon}{\varepsilon'} + \frac{\varepsilon'}{\varepsilon} - \frac{2}{3}\right)\left(\log\frac{2\varepsilon\varepsilon'}{m\omega} - \frac{1}{2}\right); \qquad (93.17)$$

this formula can also, of course, be obtained directly from (93.9).

The presence of the logarithm of a large quantity (the ratio $\varepsilon\varepsilon'/m\omega \sim \varepsilon'/m \gg 1$

† The integration over ϕ from 0 to 2π is taken first. That over δ' is conveniently replaced by integration over the difference $|\Delta| = |\delta' - \delta|$, dividing the range into parts, from 0 to some Δ_0 and from Δ_0 to ∞, where Δ_0 satisfies the inequalities $m/\varepsilon \ll \Delta_0 \ll 1$. In each region, appropriate approximations are possible in the integrand.

even if $\omega \sim \varepsilon$) should be noted. If this quantity is so large that its logarithm is also large, the logarithmic terms become the principal ones in these formulae. The logarithm arises from integration in the range (93.15).† Thus, in the logarithmic approximation, i.e. when the terms not containing a large logarithm are neglected, the secondary electron moves at an angle $\sim(m/\varepsilon)^2$ to the direction of incidence.

Finally, we shall give the limiting formula for the region near the hard end of the spectrum, when the ultra-relativistic electron radiates almost all its energy: $\omega \approx \varepsilon \gg \varepsilon'$. From (93.9) we easily find

$$d\sigma_\omega = 2Z^2\alpha r_e^2 \frac{d\omega}{\varepsilon}\left\{\frac{\varepsilon'^2}{p'^2}\log\frac{\varepsilon'+p'}{\varepsilon'-p'} - \frac{m^2\varepsilon'}{4p'^3}\log^2\frac{\varepsilon'+p'}{\varepsilon'-p'} - \frac{\varepsilon'}{p'}\right\}. \tag{93.18}$$

Formulae (93.17) and (93.18) together cover the whole range of values of ω for an ultra-relativistic initial electron, and agree for $\omega \approx \varepsilon \gg \varepsilon' \gg m$. If the secondary electron is non-relativistic ($p' \ll m$), then

$$d\sigma_\omega = 2Z^2\alpha r_e^2 \frac{\sqrt{[2m(\varepsilon-\omega)]}}{m}\frac{d\omega}{\varepsilon}. \tag{93.19}$$

Polarization effects

Polarization effects in bremsstrahlung can be studied by the general method described in §65. The choice of the 4-vectors $e^{(1)}$, $e^{(2)}$ is here particularly simple. Since there is only one frame of reference (the rest frame of the nucleus) which is of practical importance, it is sufficient to put $e^{(1)} = (0, \mathbf{e}^{(1)})$, $e^{(2)} = (0, \mathbf{e}^{(2)})$, where $\mathbf{e}^{(1)}$, $\mathbf{e}^{(2)}$ are unit vectors perpendicular to $\mathbf{k}$, one lying in the plane of $\mathbf{k}$ and $\mathbf{p}$ and the other perpendicular to that plane.

We shall not give here either the fairly lengthy calculations or their quantitative results, but merely note some qualitative properties of the polarization effects.‡ These properties can be derived by means of various symmetry relations, as was done for the Compton effect in §87.

The theory under consideration corresponds to the first non-vanishing approximation of perturbation theory. In this approximation the cross-section cannot contain a term proportional only to the polarization vector $\boldsymbol{\zeta}$ of the initial electron or $\boldsymbol{\zeta}'$ of the final electron. The absence of a term $\propto \boldsymbol{\zeta}$ means that the total emission cross-section (summed over the polarizations of the photon and the secondary electron) is independent of the polarization of the incident electron.

Of the terms proportional to only the photon polarization parameters ξ_1', ξ_2', ξ_3',

† This is easily seen by considering the range of integration in which ϕ and $\Delta = \delta' - \delta$ satisfy the conditions $m/\varepsilon \ll \Delta$, $\phi \ll 1$. In this range, $q^2/m^2 \approx \Delta^2 + \phi^2\delta^2$, and the terms in the braces in (93.13) are proportional to ϕ^2 or Δ^2 (becoming zero when $\phi = 0$ and $\Delta = 0$). Integrals of the form

$$\int \frac{\phi^2\, d\phi\, d\Delta}{(\Delta^2+\delta^2\phi^2)^2} \quad \text{or} \quad \int \frac{\Delta^2\, d\phi\, d\Delta}{(\Delta^2+\delta^2\phi^2)^2}$$

diverge logarithmically; they are "cut off" at the limits of the above-mentioned range of the variables.

‡ For a fuller discussion of these effects, see W. H. McMaster, *Reviews of Modern Physics* **33**, 8, 1961, and V. N. Baĭer, V. M. Katkov, and V. S. Fadin, *Radiation from Relativistic Electrons* (*Izluchenie relyativistskikh élektronov*), Atomizdat, Moscow, 1973.

the term $\propto \xi_2'$ is absent. Thus a photon radiated by an unpolarized electron is not circularly polarized. Here, however, there is a difference from the corresponding result for the Compton effect: in the latter case such terms were forbidden by spatial parity because of the impossibility of constructing a pseudoscalar from the only two available independent vectors, **k** and **k**′. For bremsstrahlung, there are three independent momenta **p**, **p**′ and **k**, and these suffice to construct the pseudoscalar $\mathbf{k}\cdot\mathbf{p}\times\mathbf{p}'$. A term of the form $\xi_2'\mathbf{k}\cdot\mathbf{p}\times\mathbf{p}'$ does not violate spatial parity, and therefore, strictly speaking, need not be zero; but it is not invariant under a change of sign of all the momenta (cf. (87.26)), and is consequently absent in the first Born approximation.

The existence of the pseudoscalar $\mathbf{k}\cdot\mathbf{p}\times\mathbf{p}'$ also has the result that, as well as the term proportional to ξ_3', a term proportional to ξ_1' is also allowed in the cross-section, unlike the case of the Compton effect. This term arises as a product of the form

$$S_{\alpha\beta}\nu_\alpha(\mathbf{k}\times\boldsymbol{\nu})_\beta\mathbf{k}\cdot\mathbf{p}\times\mathbf{p}'$$

(where $\boldsymbol{\nu}=\mathbf{k}\times\mathbf{p}$), which is invariant both under spatial inversion and under a change of sign of all the momenta. Thus the emitted photon has linear polarization of both kinds (both along the axes $\mathbf{e}^{(1)}$ and $\mathbf{e}^{(2)}$, and in the "diagonal" directions at 45° to these axes). This refers, however, only to conditions where the direction of motion of the secondary electron is also recorded. On integration over all directions of $\mathbf{p}'$, the term $\propto\xi_1'$ in the cross-section vanishes. This is evident from symmetry, since after the integration the two non-coincident "diagonal" directions become equivalent, and there can therefore be no preferential polarization along one of them, such as occurs when $\xi_1'\neq 0$.

The degree of linear polarization is independent of the state of polarization of the incident electron: the correlation terms of the form $\xi_1'\zeta$ and $\xi_3'\zeta$ in the cross-section are forbidden in the first Born approximation. The term $\xi_2'\zeta$, however, is allowed, so that the photon radiated by a polarized electron is circularly polarized (Ya. B. Zel'dovich, 1952).

Screening

The formulae derived above are for a purely Coulomb field. If radiation in a collision not with a "bare" nucleus but with an atom is considered, allowance must be made for the screening of the nuclear field by the electrons, which reduces the cross-section. For this purpose we must include the atomic form factor $F(q)$ in the potential $A^{(e)}(q)$ of the external field; see *QM*, §139. According to *QM* (139.2), this is done by writing $Z-F(q)$ instead of Z. We shall show under what conditions screening is important.

A given value of q in the form factor corresponds to distances $r\sim 1/q$ in the spatial distribution of the electron charges in the atom. The form factor becomes almost equal to Z (total screening) when $q\lesssim 1/a$, where a is the dimension of the atom.

In the ultra-relativistic case, as we have seen, an important contribution to the emission cross-section comes from the range of values of q near its minimum possible value for given initial and final energies of the electron. In the ultra-relativistic case,

$$
\begin{aligned}
q_{\min} &= p - p' - \omega \\
&= \sqrt{(\varepsilon^2 - m^2)} - \sqrt{(\varepsilon'^2 - m^2)} - (\varepsilon - \varepsilon') \\
&\approx m^2\omega/2\varepsilon\varepsilon'.
\end{aligned}
\tag{93.20}
$$

Screening is important if $q_{\min} \lesssim 1/a$, or

$$\varepsilon\varepsilon'/m\omega \gtrsim am. \tag{93.21}$$

This condition is always satisfied for sufficiently large energies of the incident electron.

If $q_{\min} \ll 1/a$ ("total screening") we can immediately write down, with logarithmic accuracy, the spectral distribution of the radiation. The argument of the logarithm in (93.17) is just the left-hand side of the inequality $\varepsilon\varepsilon'/m\omega \gg am$. When the inequality is satisfied, the integral over q which leads to this logarithm is cut off at a quantity of the order of the right-hand side of the inequality. According to the Thomas–Fermi model $a \sim a_0 Z^{-1/3}$, where $a_0 \sim 1/me^2$ is the Bohr radius (see *QM*, §70); then $am \sim 1/\alpha Z^{1/3}$. Thus, when there is total screening, the logarithm in (93.17) should be replaced by $\log(1/\alpha Z^{1/3})$.

Energy loss

The energy lost as radiation by the electron is expressed by the "effective retardation"

$$\kappa_{\mathrm{rad}} = \int_0^{\varepsilon - m} \omega \, d\sigma_\omega. \tag{93.22}$$

The calculation of the integral, with $d\sigma_\omega$ from (93.17), gives†

$$
\begin{aligned}
\kappa_{\mathrm{rad}} = Z^2\alpha r_e^2 \varepsilon \Bigg\{ &\frac{12\varepsilon^2 + 4m^2}{3\varepsilon p} \log \frac{\varepsilon + p}{m} - \\
&- \frac{(8\varepsilon + 6p)m^2}{3\varepsilon p^2} \log^2 \frac{\varepsilon + p}{m} - \frac{4}{3} + \frac{2m^2}{\varepsilon p} F\left(\frac{2p(\varepsilon + p)}{m^2}\right)\Bigg\},
\end{aligned}
\tag{93.23}
$$

where the function $F(\xi)$ is Spence's function (131.19).

In the non-relativistic case, (93.23) becomes

$$\kappa_{\mathrm{rad}} = 16 Z^2 \alpha r_e^2 m/3, \tag{93.24}$$

† Although formula (93.17) is inapplicable near the upper limit, this fact is unimportant, since the integral converges.

since $F(\xi) \approx \xi$ when $\xi \ll 1$; see (131.23). This formula can, of course, be obtained by direct integration of the non-relativistic Born formula (92.16).

In the ultra-relativistic case,

$$\kappa_{\text{rad}} = 4Z^2\alpha r_e^2 \varepsilon \left(\log \frac{2\varepsilon}{m} - \frac{1}{3}\right); \tag{93.25}$$

when $\xi \gg 1$, $F(\xi) \approx \frac{1}{2}\log^2 \xi$; see (131.20). The two $\log^2$ terms in (93.23) can then be omitted.

The ratio $\kappa_{\text{rad}}/\varepsilon$ is also called the cross-section for energy loss by radiation. It increases logarithmically when ε is large. This increase no longer occurs, however, when screening is taken into account. For total screening, $\kappa_{\text{rad}}/\varepsilon$ tends to a constant limit $\approx 4Z^2\alpha r_e^2 \log(1/\alpha Z^{1/3})$.

For a collision with an atom, it must also be remembered that some radiation originates from the electrons, as well as that from the nucleus. We shall see later (§97) that, in the ultra-relativistic case, the electron–electron emission cross-section differs from the electron–nucleus cross-section only in that the factor Z^2 is absent. Hence the presence of Z atomic electrons can be approximately allowed for by replacing Z^2 by $Z(Z+1)$.

On passing through a medium containing N atoms per unit volume, a fast electron loses its energy, on average, over a distance

$$l_{\text{rad}} \sim \varepsilon/N\kappa_{\text{rad}} \sim [Z^2\alpha N r_e^2 \log(1/\alpha Z^{1/3})]^{-1}, \tag{93.26}$$

called the *radiation length*.

Coherence length

Formula (93.20) may be given a different and more general interpretation. For the expressions derived above to be valid, it is necessary that the external field in which the electron moves should vary only slightly (in the direction of motion) over distances

$$l_{\text{coh}} \sim 1/q_{\text{min}} \sim \varepsilon\varepsilon'/m^2\omega \ (= \varepsilon\varepsilon'/c^3 m^2\omega); \tag{93.27}$$

this is called the *coherence length*.† The value (93.27) obtained in the Born approximation is actually quite generally valid for ultra-relativistic particles: it is easily derived in the opposite case of quasi-classical motion also, since from (90.22)‡ we see at once that the important times for radiation at small angles to the direction of motion are

$$\tau \sim \varepsilon'/\varepsilon\omega(1-v) \sim \varepsilon'\varepsilon/m^2\omega,$$

corresponding to a section of the trajectory with length $c\tau \sim l_{\text{coh}}$.

† The discussion here is due to M. L. Ter-Mikaélyan (1953).

‡ The derivation of (90.22) was based only on the smallness of the curvature of the trajectory, and in that sense does not depend on the fact that a magnetic field was specifically under consideration in §90.

For a given frequency ω, the coherence length increases with the electron energy. However, the formulae obtained for bremsstrahlung at an isolated atom can be valid also for passage through a medium only if there is no secondary photon emission or electron scattering over a distance equal to the coherence length. The condition for no secondary photon emission is $l_{coh} \ll l_{rad}$, but the condition for no electron scattering is violated much sooner: there is repeated scattering of the electron by the atomic nuclei in the medium over a distance $\sim l_{rad}$.

To arrive at a quantitative condition, we return to formula (90.22) before the integration with respect to time in the exponent and write this as

$$-i\frac{\omega\varepsilon}{\varepsilon'}\int_{t_1}^{t_1+\tau}(1-\mathbf{n}\cdot\mathbf{v})\,dt \approx -\frac{i\omega\varepsilon}{\varepsilon'}\left\{(1-v)\tau+\tfrac{1}{2}\int_{t_1}^{t_1+\tau}\theta^2\,dt\right\}, \tag{93.28}$$

where θ is the small angle between $\mathbf{v}$ and $\mathbf{n}$ which results from scattering by nuclei. For Coulomb scattering, θ changes by small jumps, and its time variation is therefore a slow "diffusion in angle". The mean square deflection of the electron over $t - t_1$ is, in order of magnitude,

$$\overline{\theta^2} \sim (t-t_1)/l_{\mathrm{Coul}},$$

where l_{Coul} is the mean free path for Coulomb collisions, given by

$$1/l_{\mathrm{Coul}} \sim (NZ^2e^4/\varepsilon)\log(\chi_{max}/\chi_{min}),$$

where χ_{max} and χ_{min} are the maximum and minimum angles of scattering in one collision for which the process may still be regarded as Rutherford scattering (cf. *PK*, §41).† The value of χ_{min} is determined by the atomic dimensions a, over which the field of the nucleus is screened: $\chi_{min} \sim 1/pa$. Large scattering angles are limited (for an ultra-relativistic electron) by distances of the order of the nuclear radius R: $\chi_{max} \sim 1/pR$. If we put $R \approx 1.5\times 10^{-13}Z^{1/3}\,\mathrm{cm} \sim r_eZ^{1/3}$, we find

$$l_{\mathrm{Coul}} \sim \frac{\varepsilon^2}{NZ^2e^4\log(1/\alpha Z^{1/3})} \sim \frac{\alpha\varepsilon^2}{m^2}\,l_{rad}. \tag{93.29}$$

The second term in (93.28), which covers a time $\tau \sim l_{coh}$, is now estimated as

$$\overline{\theta^2}\tau \sim (\omega\varepsilon/\varepsilon')l_{coh}^2/l_{\mathrm{Coul}} \sim l_{coh}/\alpha l_{rad}.$$

For the bremsstrahlung formulae to be valid that were derived without allowance for multiple scattering, this term must be much less than unity. Hence we find the condition

$$l_{coh} \ll \alpha l_{rad}, \tag{93.30}$$

which is stronger than $l_{coh} \ll l_{rad}$ (L. D. Landau and I. Ya. Pomeranchuk, 1953).

† The mean free path is determined by the transport cross-section $\sigma_t = \int(1-\cos\chi)\,d\sigma(\chi)$. For the scattering of ultra-relativistic electrons by a Coulomb centre, the cross-section $d\sigma(\chi)$ is given by (80.10).

§94. Pair production by a photon in the field of a nucleus

The formation of an electron–positron pair in a collision between a photon and a nucleus ($Z+\gamma \to Z+e^- + e^+$) and electron–nucleus bremsstrahlung ($Z+e^- \to Z+e^- +\gamma$) are two cross-channels of the same reaction. Rules have already been formulated in §91 for the transformation of formulae from the latter case to the former. Applying these rules to (93.8), we find the following expression for the differential cross-section for pair production by an unpolarized photon, averaged over the polarizations of the components of the pair:†

$$d\sigma = -\frac{Z^2\alpha r_e^2}{2\pi}\frac{m^2 p_+ p_- \,d\varepsilon_+}{\omega^3 q^4}\sin\theta_+\,d\theta_+\cdot\sin\theta_-\,d\theta_-\,d\phi\,\times$$
$$\times\left\{\frac{p_+^2}{\kappa_+^2}(4\varepsilon_-^2-q^2)\sin^2\theta_+ + \frac{p_-^2}{\kappa_-^2}(4\varepsilon_+^2-q^2)\sin^2\theta_- -\right.$$
$$\left.-\frac{2\omega^2}{\kappa_+\kappa_-}(p_+^2\sin^2\theta_+ + p_-^2\sin^2\theta_-) - \frac{2p_+p_-}{\kappa_+\kappa_-}(2\varepsilon_+^2+2\varepsilon_-^2-q^2)\sin\theta_+\sin\theta_-\cos\phi\right\}, \tag{94.1}$$

$$\kappa_\pm = \varepsilon_\pm - p_\pm\cos\theta_\pm,\qquad q^2 = (\mathbf{p}_+ + \mathbf{p}_- - \mathbf{k})^2,\qquad \varepsilon_+ + \varepsilon_- = \omega$$

(H. A. Bethe and W. Heitler, 1934).

A similar transformation derives from (93.9) the energy distribution of the components of the pair:

$$d\sigma_{\varepsilon+} = Z^2\alpha r_e^2\frac{p_+p_-}{\omega^3}d\varepsilon_+\left\{-\tfrac{4}{3} - 2\varepsilon_+\varepsilon_-\frac{p_+^2+p_-^2}{p_+^2p_-^2} + \right.$$
$$+m^2\left(l_-\frac{\varepsilon_+}{p_-^3} + l_+\frac{\varepsilon_-}{p_+^3} - l_+l_-\frac{1}{p_+p_-}\right) +$$
$$+L\left[-\frac{8\varepsilon_+\varepsilon_-}{3p_+p_-} + \frac{\omega^2}{p_+^3p_-^3}(\varepsilon_+^2\varepsilon_-^2 + p_+^2p_-^2 - m^2\varepsilon_+\varepsilon_-) -\right.$$
$$\left.\left.-\frac{m^2\omega}{2p_+p_-}\left(l_+\frac{\varepsilon_+\varepsilon_- - p_+^2}{p_+^3} + l_-\frac{\varepsilon_+\varepsilon_- - p_-^2}{p_-^3}\right)\right]\right\},$$
$$L = \log\frac{\varepsilon_+\varepsilon_- + p_+p_- + m^2}{\varepsilon_+\varepsilon_- - p_+p_- + m^2},\qquad l_\pm = \log\frac{\varepsilon_\pm + p_\pm}{\varepsilon_\pm - p_\pm}. \tag{94.2}$$

Since the above formulae are based on the Born approximation, they are valid under the conditions $Ze^2/v_\pm \ll 1$. The symmetry of (94.1) and (94.2) with respect to the electron and the positron is itself a consequence of the Born approximation, and would not occur in higher approximations.

In the ultra-relativistic case ($\varepsilon_\pm \gg m$), the electron and the positron are emitted at angles $\theta_\pm \sim m/\varepsilon_\pm$ relative to the direction of the incident photon. The angular

† Polarization effects in pair production by a photon are discussed in the papers already quoted in §93 in connection with bremsstrahlung.

distribution is given by a formula similar to (93.13):

$$d\sigma = \frac{8}{\pi} Z^2\alpha r_e^2 \frac{m^4\varepsilon_+\varepsilon_-}{\omega^3 q^4} d\varepsilon_+\left\{-\frac{\delta_+^2}{(1+\delta_+^2)^2} - \frac{\delta_-^2}{(1+\delta_-^2)^2} + \frac{\omega^2}{2\varepsilon_+\varepsilon_-}\frac{\delta_+^2+\delta_-^2}{(1+\delta_+^2)(1+\delta_-^2)} + \right.$$

$$\left. + \left(\frac{\varepsilon_+}{\varepsilon_-} + \frac{\varepsilon_-}{\varepsilon_+}\right)\frac{\delta_+\delta_- \cos\phi}{(1+\delta_+^2)(1+\delta_-^2)}\right\} \delta_+\delta_- \cdot d\delta_+\, d\delta_-\, d\phi, \tag{94.3}$$

with

$$\frac{q^2}{m^2} = \delta_+^2 + \delta_-^2 + 2\delta_+\delta_- \cos\phi + m^2\left(\frac{1+\delta_+^2}{2\varepsilon_+} + \frac{1+\delta_-^2}{2\varepsilon_-}\right)^2. \tag{94.4}$$

The energy distribution in this case is

$$d\sigma = 4Z^2\alpha r_e^2 \frac{d\varepsilon_+}{\omega^3}(\varepsilon_+^2 + \varepsilon_-^2 + \tfrac{2}{3}\varepsilon_+\varepsilon_-)\left(\log\frac{2\varepsilon_+\varepsilon_-}{m\omega} - \frac{1}{2}\right) \text{(ultra-relativistic case).} \tag{94.5}$$

Integration of (94.5) over ε_+ from m to ω gives the total cross-section for pair production by a photon having a given energy:†

$$\sigma = \tfrac{28}{9}Z^2\alpha r_e^2\left(\log\frac{2\omega}{m} - \frac{109}{42}\right), \qquad \omega \gg m. \tag{94.6}$$

As with bremsstrahlung, the logarithmic term in the ultra-relativistic cross-section arises from the range of values $q \sim m^2/\varepsilon$. This now corresponds to angles for which

$$|\delta_+ - \delta_-| \lesssim m/\varepsilon, \qquad |\pi - \phi| \lesssim m/\varepsilon,$$

instead of $\phi \lesssim m/\varepsilon$ as in (93.15). Thus, in the logarithmic approximation, the directions of the electron and the positron are at small angles to the direction of the photon and are almost coplanar with the direction of the photon but on opposite sides of it.

Near the reaction threshold ($\omega \to 2m$), the Born approximation is invalid. The derivation of a quantitative formula in this case would require an exact calculation of the Coulomb interaction of the three charged particles (the nucleus and the pair) in the final state. The symmetry with respect to the electron (which is attracted to the nucleus) and the positron (which is repelled from the nucleus) is then, of course, lost.

If

$$Z\alpha \ll \sqrt{\frac{\omega - 2m}{\omega}} \ll 1, \tag{94.7}$$

† Since the integral converges at both limits, the inapplicability of formula (94.5) for small values of $\varepsilon_\pm - m$ is not important.

the Born approximation is still valid. At non-relativistic energies of the pair, $\omega \approx 2m \gg p_\pm$, and therefore $q \approx \omega$. In (94.1) we can everywhere put $\varepsilon_\pm = \kappa_\pm = m$, $\omega = 2m$, and this formula then reduces to

$$d\sigma = \frac{Z^2\alpha r_e^2}{64\pi^2}\frac{p_+p_-}{m^5}(p_+^2 \sin^2\theta_+ + p_-^2 \sin^2\theta_-)\, do_+\, do_-\, d\varepsilon_+. \tag{94.8}$$

After integration over angles,

$$\begin{aligned} d\sigma &= \tfrac{1}{6}Z^2\alpha r_e^2 \frac{p_+p_-(p_+^2+p_-^2)}{m^5}\, d\varepsilon_+ \\ &= \frac{2Z^2\alpha r_e^2}{3m^3}(\omega - 2m)\sqrt{[(\varepsilon_+ - m)(\varepsilon_- - m)]}\, d\varepsilon_+. \end{aligned} \tag{94.9}$$

Finally, integration over ε_+ from m to $\omega - m$ gives the total cross-section

$$\sigma = \frac{\pi}{12} Z^2\alpha r_e^2 \left(\frac{\omega - 2m}{m}\right)^3. \tag{94.10}$$

If the relative velocity (v_0) of the components of the pair formed is small, their Coulomb interaction must be taken into account (A. D. Sakharov, 1948). This interaction becomes important when v_0 is of the order of (or less than) the velocities of the particles in the bound state of the electron and positron (positronium):

$$v_0 \lesssim \alpha. \tag{94.11}$$

Let us consider the process in the centre-of-mass system of the pair. Virtual momenta $\sim m$ are important in the diagrams which represent the process in this system; that is, distances $\sim 1/m$ between the electron and the positron are important. The wave function $\psi(r)$ of their relative motion changes appreciably only over distances $r \sim 1/mv_0 \sim 1/m\alpha$, which are large compared with $1/m$. The allowance for the interaction of the particles therefore amounts to the inclusion of a factor $\psi^*(0)$ in the transition matrix element. The differential cross-section is accordingly multiplied by $|\psi(0)|^2$, i.e. by

$$\frac{2\pi\alpha/v_0}{1 - e^{-2\pi\alpha/v_0}}; \tag{94.12}$$

see *QM*, (136.11). The relative velocity of the two particles is the velocity of one particle in the rest frame of the other. Comparing the values of the invariant $p_{+\mu}p_-^\mu$ here and in the laboratory system (the rest frame of the nucleus), we find

$$\frac{m^2}{\sqrt{(1 - v_0^2)}} = \varepsilon_+\varepsilon_- - \mathbf{p}_+ \cdot \mathbf{p}_-,$$

whence v_0 may be found. If $\mathbf{p}_+$ and $\mathbf{p}_-$ are similar in magnitude and direction, v_0 is given by the approximate formula

$$v_0^2 = p^2\vartheta^2 + (p_+ - p_-)^2/\varepsilon^2 \tag{94.13}$$

valid for $v_0 \ll 1$; here $p = \frac{1}{2}(p_+ + p_-)$, $\varepsilon = \frac{1}{2}(\varepsilon_+ + \varepsilon_-)$, and ϑ is the angle between $\mathbf{p}_+$ and $\mathbf{p}_-$.

The correction to the cross-section according to (94.12) and (94.13) causes an anomaly in the correlation between the momenta of the electron and positron formed: it has a narrow maximum at $\mathbf{p}_+ \approx \mathbf{p}_-$.

§95. Exact theory of pair production in the ultra-relativistic case

In §§93 and 94 we have discussed bremsstrahlung and pair production by a photon in the relativistic case, using the Born approximation, for which the condition $Z\alpha \ll 1$ must always be satisfied. In §§95 and 96 we shall describe a theory of these processes which is not subject to the limitation just mentioned, i.e. is valid even if $Z\alpha \sim 1$ (H. A. Bethe and L. C. Maximon, 1954). We shall assume that both the particles (the initial and final electrons, or the constituents of the pair) are ultra-relativistic, with energy $\varepsilon \gg m$.

We have seen that in the ultra-relativistic case both particles move at small angles (θ, θ' or θ_+, θ_-) to the direction of the photon: $\theta \lesssim m/\varepsilon$. This property is preserved in the exact (with respect to $Z\alpha$) theory, and we shall therefore consider just this range of angles.

The momentum transfer to the nucleus in this range is $q \sim m$. This means that in the wave functions the important values of the impact parameter are $\rho \sim 1/q \sim 1/m$, i.e. "large" distances. At such distances the wave function derived in §39 can be used. The calculations for pair production are as follows.

The pair production cross-section is similar in form to the photoelectric effect cross-section (cf. (56.1), (56.2)):

$$d\sigma = 2\pi \left| e\sqrt{(4\pi)}\frac{1}{\sqrt{(2\omega)}} M_{fi}\right|^2 \delta(\omega - \varepsilon_+ - \varepsilon_-)\frac{d^3p_+\, d^3p_-}{(2\pi)^6}, \tag{95.1}$$

where

$$M_{fi} = \int \psi^{(-)*}_{\varepsilon_-, \mathbf{p}_-}(\boldsymbol{\alpha}\cdot\mathbf{e})\, e^{i\mathbf{k}\cdot\mathbf{r}}\psi^{(+)}_{-\varepsilon_+, -\mathbf{p}_+}\, d^3x. \tag{95.2}$$

Here $\psi^{(-)}_{\varepsilon_-, \mathbf{p}_-}$ is the wave function of the electron, and $\psi^{(+)}_{-\varepsilon_+, -\mathbf{p}_+}$ the wave function with negative energy $-\varepsilon_+$ and momentum $-\mathbf{p}_+$.

The function $\psi^{(-)}_{\varepsilon_-, \mathbf{p}_-}$, which pertains to a particle in the final state, must have an asymptotic form which includes (besides the plane wave) an ingoing spherical wave; this is indicated by the superscript (−). According to (39.10), this wave function is†

$$\left.\begin{aligned}\psi^{(-)}_{\varepsilon_-,\mathbf{p}_-} &= \frac{C^{(-)}}{\sqrt{(2\varepsilon_-)}}\, e^{i\mathbf{p}_-\cdot\mathbf{r}}\left(1-\frac{i\boldsymbol{\alpha}\cdot\nabla}{2\varepsilon_-}\right)F(-i\nu, 1, -i(p_-r+\mathbf{p}_-\cdot\mathbf{r}))u(p_-),\\ C^{(-)} &= e^{\pi\nu/2}\Gamma(1+i\nu), \qquad \nu = Z\alpha.\end{aligned}\right\} \tag{95.3}$$

The function $\psi^{(+)}_{-\varepsilon_+,-\mathbf{p}_+}$ must have an asymptotic form which includes an outgoing spherical wave (indicated by the superscript (+)), since it denotes the wave function of an "initial state with negative energy". The asymptotic form of the wave function of the positron, obtained from $\psi^{(+)*}_{-\varepsilon_+,-\mathbf{p}_+}$, then has an ingoing wave, as is correct for a final particle. According to (39.11), this function is

$$\left.\begin{aligned}\psi^{(+)}_{-\varepsilon_+,-\mathbf{p}_+} &= \frac{C^{(+)}}{\sqrt{(2\varepsilon_+)}}\, e^{-i\mathbf{p}_+\cdot\mathbf{r}}\left(1+\frac{i\boldsymbol{\alpha}\cdot\nabla}{2\varepsilon_+}\right)F(-i\nu, 1, i(p_+r+\mathbf{p}_+\cdot\mathbf{r}))u(-p_+),\\ C^{(+)} &= e^{-\pi\nu/2}\Gamma(1+i\nu).\end{aligned}\right\} \tag{95.4}$$

The terms $\sim 1/\varepsilon$ in (95.3) and (95.4) have to be included because of the matrix structure of M_{fi} (95.2). The matrix element $(\boldsymbol{\alpha})_{fi}$ is a vector whose direction is close to that of $\mathbf{k}$. The leading term in $(\boldsymbol{\alpha}\cdot\mathbf{e})_{fi}$ is therefore small, and the correction terms are of the same order of magnitude as that term.

Substituting (95.3) and (95.4) in (95.2) and neglecting terms $\sim 1/\varepsilon_+\varepsilon_-$, we find

$$M_{fi} = \frac{N}{2\sqrt{(\varepsilon_+\varepsilon_-)}}\, u^*(p_-)\{(\mathbf{e}\cdot\boldsymbol{\alpha})I + (\mathbf{e}\cdot\boldsymbol{\alpha})(\boldsymbol{\alpha}\cdot\mathbf{I}_+) + (\boldsymbol{\alpha}\cdot\mathbf{I}_-)(\mathbf{e}\cdot\boldsymbol{\alpha})\}u(-p_+), \tag{95.5}$$

where

$$N = C^{(+)}C^{(-)} = \pi\nu/\sinh \pi\nu, \tag{95.6}$$

$$\left.\begin{aligned}I &= \int e^{-i\mathbf{q}\cdot\mathbf{r}} F_-^* F_+\, d^3x,\\ \mathbf{I}_+ &= \frac{i}{2\varepsilon_+}\int e^{-i\mathbf{q}\cdot\mathbf{r}} F_-^*\nabla F_+\, d^3x,\\ \mathbf{I}_- &= \frac{i}{2\varepsilon_-}\int e^{-i\mathbf{q}\cdot\mathbf{r}} (\nabla F_-)^* F_+\, d^3x,\\ \mathbf{q} &= \mathbf{p}_+ + \mathbf{p}_- - \mathbf{k};\end{aligned}\right\} \tag{95.7}$$

F_- and F_+ are used for brevity to denote the hypergeometric functions which appear in (95.3) and (95.4). The integrals I, $\mathbf{I}_+$, $\mathbf{I}_-$ satisfy one identical relation: from

$$\int \nabla(e^{-i\mathbf{q}\cdot\mathbf{r}} F_-^* F_+)\, d^3x = 0,$$

† In this section, $p_\pm = |\mathbf{p}_\pm|$, $q = |\mathbf{q}|$.

we have

$$\mathbf{q}I + 2\varepsilon_+\mathbf{I}_+ + 2\varepsilon_-\mathbf{I}_- = 0. \tag{95.8}$$

We average $|M_{fi}|^2$ over polarizations of the incident photon, and sum over directions of the electron and positron spins.† This is done by the tensor substitution

$$e_i e_k^* \to \tfrac{1}{2}(\delta_{ik} - n_i n_k), \quad \mathbf{n} = \mathbf{k}/\omega,$$

and changing the bispinor products according to

$$u_\pm \bar{u}_\pm \to 2\rho_\pm = (\varepsilon_\pm \gamma^0 - \mathbf{p}_\pm \cdot \boldsymbol{\gamma} \mp m).$$

Putting also $\boldsymbol{\alpha} = \gamma^0 \boldsymbol{\gamma}$, we find

$$|M_{fi}|^2 \to (N^2/2\varepsilon_+\varepsilon_-)\{\mathrm{tr}\, \rho_- \mathbf{Q} \cdot \rho_+ \bar{\mathbf{Q}} - \mathrm{tr}\, \rho_-(\mathbf{n} \cdot \mathbf{Q})\rho_+(\mathbf{n} \cdot \bar{\mathbf{Q}})\},$$
$$\mathbf{Q} = \boldsymbol{\gamma} I - \gamma^0 \boldsymbol{\gamma}(\boldsymbol{\gamma} \cdot \mathbf{I}_+) - \gamma^0(\boldsymbol{\gamma} \cdot \mathbf{I}_-)\boldsymbol{\gamma},$$
$$\bar{\mathbf{Q}} = \boldsymbol{\gamma} I^* - \gamma^0(\boldsymbol{\gamma} \cdot \mathbf{I}_+^*)\boldsymbol{\gamma} - \gamma^0 \boldsymbol{\gamma}(\boldsymbol{\gamma} \cdot \mathbf{I}_-^*).$$

The final result, obtained after making the appropriate approximations, for the ultra-relativistic case at small angles

$$\theta_\pm \sim m/\varepsilon \ll 1, \tag{95.9}$$

will be given here. We define the auxiliary vectors

$$\boldsymbol{\delta}_\pm = \frac{1}{m}(\mathbf{p}_\pm)_\perp, \qquad \delta_\pm = \frac{\varepsilon_\pm}{m}\theta_\pm, \tag{95.10}$$

where the suffix $\perp$ denotes the component perpendicular to the direction of $\mathbf{k}$. Then

$$|M_{fi}|^2 \to \tfrac{1}{4}N^2 \left\{ \frac{m^2\omega^2}{2\varepsilon_+^2\varepsilon_-^2}|I|^2 + 2\left| I\frac{m\boldsymbol{\delta}_+}{2\varepsilon_+} + \mathbf{I}_+ \right|^2 + 2\left| I\frac{m\boldsymbol{\delta}_-}{2\varepsilon_-} + \mathbf{I}_- \right|^2 \right\}, \tag{95.11}$$

where we have used the fact that $I \sim \varepsilon I_\pm/q \sim \varepsilon I_\pm/m$ (as is seen from (95.8)), and terms of higher order in m/ε are omitted.

The integrals $\mathbf{I}_\pm$ may be expressed as

$$\mathbf{I}_\pm = i\frac{p_\pm}{2\varepsilon_\pm}\frac{\partial J}{\partial \mathbf{p}_\pm},$$

$$J = \int \frac{e^{-i\mathbf{q}\cdot\mathbf{r}}}{r} F(-i\nu, 1, i(p_+ r + \mathbf{p}_+ \cdot \mathbf{r}))F(i\nu, 1, i(p_- r + \mathbf{p}_- \cdot \mathbf{r}))\, d^3x. \tag{95.12}$$

† Calculations with allowance for the polarizations of all the particles are given by H. Olsen and L. C. Maximon, *Physical Review* **114**, 887, 1959, and in the book by Baĭer *et al.* cited in §93.

The integral J can be written in terms of the complete hypergeometric function:†

$$\left.\begin{aligned} J &= \frac{4\pi}{q^2}\left(\frac{q^2-2\mathbf{p}_+\cdot\mathbf{q}}{q^2-2\mathbf{p}_-\cdot\mathbf{q}}\right)^{i\nu} F(-i\nu, i\nu, 1, z), \\ z &= 2\frac{q^2(p_+p_- - \mathbf{p}_+\cdot\mathbf{p}_-)+2(\mathbf{p}_+\cdot\mathbf{q})(\mathbf{p}_-\cdot\mathbf{q})}{(q^2-2\mathbf{p}_+\cdot\mathbf{q})(q^2-2\mathbf{p}_-\cdot\mathbf{q})}. \end{aligned}\right\} \tag{95.13}$$

The differentiation with respect to $\mathbf{p}_\pm$ must be carried out with $\mathbf{q}$ fixed, only thereafter putting $\mathbf{q}=\mathbf{p}_+ + \mathbf{p}_- - \mathbf{k}$. The result, after making the approximations corresponding to the ultra-relativistic case and the conditions (95.9), is

$$\mathbf{I}_\pm = \frac{4\pi}{q^2}\frac{\varepsilon_\mp}{m^2\omega}\left(\frac{\varepsilon_+\xi_+}{\varepsilon_-\xi_-}\right)^{i\nu}\left\{\pm\nu\mathbf{q}\xi_\mp F(z) + i\frac{q^2}{m^2}F'(z)(\mathbf{q}\xi_\mp - m\boldsymbol{\delta}_\pm)\right\}, \tag{95.14}$$

with, for brevity, the notation

$$\left.\begin{aligned} \xi_\pm &= \frac{1}{1+\delta_\pm^2}, \qquad z = 1-\frac{q^2}{m^2}\xi_+\xi_-, \\ F(z) &= F(-i\nu, i\nu, 1, z), \end{aligned}\right\} \tag{95.15}$$

$F(z)$ being a real function. The integral I is then found immediately from (95.8).

Substituting the values of the integrals in (95.11) and thence in (95.1), we find the required cross-section:

$$\begin{aligned} d\sigma = {} & \frac{4}{\pi}\left(\frac{\pi\nu}{\sinh\pi\nu}\right)^2 Z^2\alpha r_e^2\frac{m^4}{q^4\omega^3}\,\delta_+\,d\delta_+\cdot\delta_-\,d\delta_-\cdot d\phi\,d\varepsilon_+ \times \\ & \times\Big\{F^2(z)[-2\varepsilon_+\varepsilon_-(\delta_+^2\xi_+^2+\delta_-^2\xi_-^2)+\omega^2(\delta_+^2+\delta_-^2)\xi_+\xi_- + \\ & + 2(\varepsilon_+^2+\varepsilon_-^2)\delta_+\delta_-\xi_+\xi_-\cos\phi] + \\ & + \frac{q^4}{m^4}\frac{\xi_+^2\xi_-^2}{\nu^2}F'^2(z)[-2\varepsilon_+\varepsilon_-(\delta_+^2\xi_+^2+\delta_-^2\xi_-^2)+\omega^2(1+\delta_+^2\delta_-^2)\xi_+\xi_- - \\ & - 2(\varepsilon_+^2+\varepsilon_-^2)\delta_+\delta_-\xi_+\xi_-\cos\phi]\Big\}. \end{aligned} \tag{95.16}$$

When $\nu\to 0$,

$$\frac{\pi\nu}{\sinh\pi\nu}\to 1, \qquad F(z)\to 1, \qquad F'(z)\approx\nu^2\to 0.$$

The expression (95.16) then reduces, as it should, to Bethe and Heitler's formula (94.3), which corresponds to the Born approximation. It also reduces to this formula for any ν if the angles of emission of the pair satisfy the conditions

$$|\delta_+ - \delta_-| \ll 1, \qquad |\pi-\phi| \ll 1.$$

† The calculations are given in Nordsieck's paper quoted in §92.

For then $q \ll m$, so that the second term in the braces in (95.16) can be omitted because of the extra factor $(q/m)^4$ as compared with the first term, and in the first term we have (since $1-z \sim q^2/m^2 \ll 1$)†

$$F(z) \to F(1) \equiv F(-i\nu, i\nu, 1, 1) = \frac{1}{\Gamma(1-i\nu)\Gamma(1+i\nu)} = \frac{\sinh \pi\nu}{\pi\nu}, \tag{95.17}$$

so that the similar factor in front of the braces is cancelled.

Let us now consider the integration of the cross-section over the directions of emission of the pair. The integration over angles is divided into two regions I and II, in which we have respectively

$$\text{(I)}\ 1-z > 1-z_1, \qquad \text{(II)}\ 1-z < 1-z_1,$$

where z_1 is a certain value such that $1 \gg 1-z_1 \gg (m/\varepsilon)^2$. Since in region II $1-z \ll 1$, $q^2 \ll m^2$, it follows from the above discussion that in this region $d\sigma \approx d\sigma_B \equiv d\sigma_{\nu=0}$, where $d\sigma_B$ is the cross-section in the Born approximation. The integral over angles is therefore

$$d\sigma_{\varepsilon+} \equiv \int d\sigma = \int_{\text{I}} d\sigma + \int_{\text{II}} d\sigma_{\nu\to 0} = (d\sigma_{\varepsilon+})_B + \int_{\text{I}} (d\sigma - d\sigma_{\nu\to 0}), \tag{95.18}$$

where $(d\sigma_{\varepsilon+})_B$ is the Born cross-section (94.5) integrated over angles.

In region I we have

$$q^2/m^2 \approx \delta_+^2 + \delta_-^2 + 2\delta_+\delta_- \cos\phi.$$

We shall change from the variables δ_+, δ_-, ϕ to ξ_+, ξ_-, z. A direct calculation of the Jacobian for this transformation gives

$$\delta_+\, d\delta_+ \cdot \delta_-\, d\delta_- \cdot d\phi \to \frac{\varepsilon_+\varepsilon_-}{8m^2} \frac{d\xi_+\, d\xi_-\, d\phi}{(\xi_+\xi_-)^3 \sin\phi},$$

where

$$1-z = (q^2/m^2)\xi_+\xi_- = \xi_+ + \xi_- - 2\xi_+\xi_- + 2\sqrt{[\xi_+\xi_-(1-\xi_+)(1-\xi_-)]}\cos\phi.$$

† This value of the function can be obtained from *QM*, (e.7), which relates hypergeometric functions with arguments z and $1-z$.

Expressing $\sin\phi$ and $\cos\phi$ in terms of the other quantities by means of this equation and substituting in (95.16), we obtain after some simple algebra

$$d\sigma = A\, d\varepsilon_+ \frac{2d\xi_+\, d\xi_-\, dz}{[z(1-z)-(1-z)(\xi_+ + \xi_- - 1)^2 - z(\xi_+ - \xi_-)^2]^{1/2}} \times$$
$$\times\left\{\frac{F^2(z)}{(1-z)^2}[(\varepsilon_+^2+\varepsilon_-^2)(1-z)+2\varepsilon_+\varepsilon_-(\xi_+-\xi_-)^2]+\right.$$
$$\left.+\frac{F'^2(z)}{\nu^2}[(\varepsilon_+^2+\varepsilon_-^2)z+2\varepsilon_+\varepsilon_-(\xi_++\xi_--1)^2]\right\},$$
$$A = \left(\frac{\pi\nu}{\sinh \pi\nu}\right)^2 \frac{Z^2\alpha r_e^2}{2\pi\omega^3}.$$

Finally, we replace ξ_+ and ξ_- in terms of new "spherical" variables χ and ψ:

$$\xi_+ + \xi_- - 1 = \sqrt{z}\,\sin\chi\cos\psi,$$
$$\xi_+ - \xi_- = \sqrt{(1-z)}\,\sin\chi\sin\psi,$$
$$0\leqslant\chi\leqslant\tfrac{1}{2}\pi, \qquad 0\leqslant\psi\leqslant 2\pi,$$
$$2\,d\xi_+\,d\xi_- \to \sqrt{[z(1-z)]}\,\sin\chi\cos\chi\,d\chi\,d\psi.$$

These ranges of variation of χ and ψ correspond to the range 0 to 1 for ξ_+ and ξ_-, i.e. to the range 0 to ∞ for δ_+ and δ_- (or, equivalently, θ_+ and θ_-); the rapid convergence of the integral allows the range of variation of the angles to be extended in this way. After the transformation, the root in the denominator becomes $\sqrt{[z(1-z)]}\cos\chi$; the integration over χ and ψ is elementary, and the result is

$$d\sigma = 2A\cdot 2\pi\, dz(\varepsilon_+^2+\varepsilon_-^2+\tfrac{2}{3}\varepsilon_+\varepsilon_-)\left[\frac{F^2(z)}{1-z}+\frac{z}{\nu^2}F'^2(z)\right]d\varepsilon_+.$$

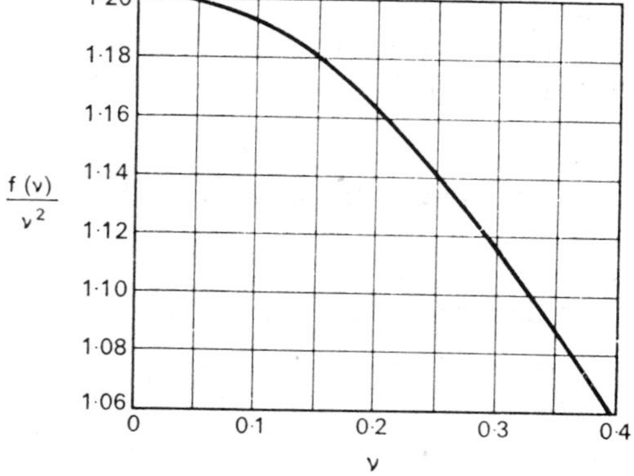

FIG. 18.

An extra factor 2 has been included because the integration over z is to be taken from 0 to z_1, whereas, when the azimuth ϕ varies from 0 to π and from π to 2π, each value of z occurs twice.

The integration over z is effected by means of formula (92.14), which, for $\nu' = -\nu$ (and $F(z)$ accordingly real), becomes

$$\frac{F^2}{1-z^2}+\frac{z}{\nu^2}F'^2=\frac{1}{\nu^2}\frac{d}{dz}(zFF').$$

The integral of this expression is $z_1F(z_1)F'(z_1)/\nu^2$. The value of $z_1F(z_1) \approx F(1)$ is taken from (95.17), and the limit of $F'(z_1 \to 1)$ is given by†

$$\begin{aligned}\frac{1}{\nu^2}F'(z) &= F(1-i\nu, 1+i\nu, 2, z)\\ &\approx -[\log(1-z)+2f(\nu)]\frac{\sinh \pi\nu}{\pi\nu},\end{aligned}$$

where

$$\begin{aligned} f(\nu) &= \tfrac{1}{2}[\Psi(1+i\nu)+\Psi(1-i\nu)-2\Psi(1)]\\ &= \nu^2\sum_{n=1}^{\infty}\frac{1}{n(n^2+\nu^2)},\\ \Psi(z) &= \Gamma'(z)/\Gamma(z).\end{aligned} \tag{95.19}$$

Substituting the above expressions in (95.18), we obtain as the final formula

$$d\sigma_{\varepsilon+} = 4Z^2\alpha r_e^2(\varepsilon_+^2+\varepsilon_-^2+\tfrac{2}{3}\varepsilon_+\varepsilon_-)\left[\log\frac{2\varepsilon_+\varepsilon_-}{m\omega}-\frac{1}{2}-f(\alpha Z)\right]\frac{d\varepsilon_+}{\omega^3}. \tag{95.20}$$

The total cross-section for pair production by a photon with energy ω is

$$\sigma = \tfrac{28}{9}Z^2\alpha r_e^2\left[\log\frac{2\omega}{m}-\frac{109}{42}-f(\alpha Z)\right]. \tag{95.21}$$

We see that the only change in these formulae is that a universal function $f(\alpha Z)$ of the atomic number is subtracted from the logarithm. Figure 18 shows a graph of this function. For $\nu \ll 1$, $f(\nu) \approx 1.2\nu^2$.

§96. Exact theory of bremsstrahlung in the ultra-relativistic case

The matrix element for the bremsstrahlung process is

$$M_{fi} = \int \psi_{\varepsilon'\mathbf{p}'}^{(-)*}(\boldsymbol{\alpha}\cdot\mathbf{e}^*)\, e^{-i\mathbf{k}\cdot\mathbf{r}}\psi_{\varepsilon\mathbf{p}}^{(+)}\, d^3x; \tag{96.1}$$

† The derivation of this formula is in the Appendix to the paper by H. Davies, H. A. Bethe and L. C. Maximon, *Physical Review* **93**, 788, 1954.

the wave functions of the initial electron (ε, $\mathbf{p}$) and the final electron (ε', $\mathbf{p}'$) include respectively outgoing and ingoing spherical waves in their asymptotic forms. The calculation of this integral is similar to that of the matrix element (95.2), but will not be given here. Instead, we shall describe another way of calculating the bremsstrahlung cross-section, based on the fact that the process is quasi-classical, and not using the explicit form of the wave functions of the electron in the field of the nucleus (and in this sense independent of the precise form of the field potential)(V. N. Baĭer and V. M. Katkov, 1968).

In the bremsstrahlung process, the nucleus transfers to the electron and the photon a momentum $\mathbf{q} = \mathbf{p}' + \mathbf{k} - \mathbf{p}$. As in the pair production problem, we must distinguish two ranges of values of the transfer $\mathbf{q}_\perp$ which is transverse relative to $\mathbf{p}$:

$$\text{(I)}\ m \gtrsim q_\perp \gg \omega m^2/\varepsilon^2, \qquad \text{(II)}\ q_\perp \sim \omega m^2/\varepsilon^2 \ll m. \tag{96.2}$$

It is evident that in region I the emission cross-section is equal to the Born value: for these values of $\mathbf{q}_\perp$, the recoil momentum of the nucleus is unimportant, as will be shown in §98 (see the derivation of the condition (98.10)). In region I, the cross-section for the process is therefore the product of the exact cross-section for electron scattering in the field of the nucleus at rest and an emission probability which is independent of the form of the field. But since, according to (80.10), the exact cross-section for scattering at small angles in a Coulomb field is equal to the Born value, so is the cross-section for the whole process in region I.

Thus only region II need be considered. Small momentum transfers correspond to the passage of the electron at large distances from the nucleus: $\rho \sim 1/q_\perp \sim \varepsilon/m^2$. But, at these distances, the motion of the electron is certainly quasi-classical, as is easily seen by direct application of the usual quasi-classical condition, *QM* (46.7), to the ultra-relativistic equation (39.5).

Since the motion is quasi-classical, we can use the method already applied in §90 for synchrotron radiation. The expression (90.7) is in this case the probability of emission when the electron passes the nucleus once.

Formula (90.18) remains valid for the function L used in §90; the only difference is in the form of the quasi-classical electron path $\mathbf{r} = \mathbf{r}(t)$, which is used to calculate the difference $\mathbf{r}_2 - \mathbf{r}_1$.

At large impact parameters, the field of the nucleus may be regarded as weak. In the zero-order approximation, the path is a straight line passing at a distance ρ from the centre. In the next approximation, we have as the equation of motion (cf. *Mechanics*, §20)

$$\frac{d\mathbf{p}}{dt} = -\frac{\boldsymbol{\rho}}{r}\frac{dU}{dr},$$

where $\boldsymbol{\rho}$ is a vector lying in the xy-plane and perpendicular to the initial momentum of the electron, and r on the right-hand side is to be taken as the zero-order function:

$$r \approx \sqrt{(\rho^2 + v^2t^2)} \approx \sqrt{(\rho^2 + t^2)}.$$

Hence

$$\mathbf{p}(t)-\mathbf{p}_1=-\boldsymbol{\rho}\int_{t_1}^{t}\frac{dU}{dt}\frac{dt}{r}. \qquad (96.3)$$

The velocity $\mathbf{v}(t)=\mathbf{p}(t)/\varepsilon$, where the energy ε depends on the magnitude but not the direction of $\mathbf{p}$, may be regarded as constant with sufficient accuracy. A further integration then gives

$$\mathbf{r}(t)-\mathbf{r}_1=\mathbf{v}_1(t-t_1)-\frac{1}{\varepsilon}\int_{t_1}^{t}[\mathbf{p}(t')-\mathbf{p}_1]\,dt'. \qquad (96.4)$$

We shall take $t_1=-\infty$, so that the quantities $\mathbf{p}_1=\mathbf{p}(-\infty)\equiv\mathbf{p}$ and $\mathbf{v}=\mathbf{p}/\varepsilon$ are the initial momentum and velocity of the electron.

We can put the probability (90.7) in the form

$$dw=|a(\boldsymbol{\rho})|^2\,d^3k/(2\pi)^3, \qquad (96.5)$$

where

$$a(\boldsymbol{\rho})=e\sqrt{\frac{2\pi}{\omega}}\int_{-\infty}^{\infty}R(t)\exp\left\{i\frac{\varepsilon}{\varepsilon'}(\omega t-\mathbf{k}\cdot\mathbf{r}(t))\right\}dt,$$

$$R(t)=\frac{u^*_{\varepsilon'\mathbf{p}'}}{\sqrt{(2\varepsilon')}}\boldsymbol{\alpha}\cdot\mathbf{e}^*\frac{u_{\varepsilon\mathbf{p}}}{\sqrt{(2\varepsilon)}}, \qquad (96.6)$$

and $\varepsilon'=\varepsilon-\omega$, $\mathbf{p}'(t)=\mathbf{p}(t)-\mathbf{k}$. The classical function $\mathbf{p}(t)$ is given by (96.3). If $\mathbf{p}$ denotes the initial momentum of the particle, we have for a Coulomb field ($U=-\nu/r$, $\nu=Z\alpha$)

$$\mathbf{p}(t)=\mathbf{p}-\frac{\nu\boldsymbol{\rho}}{\rho^2}\left[\frac{t}{\sqrt{(\rho^2+t^2)}}+1\right],$$

and

$$\mathbf{r}(t)=\frac{\mathbf{p}}{\varepsilon}t-\frac{\boldsymbol{\rho}}{\rho^2}\frac{\nu}{\varepsilon}[\sqrt{(\rho^2+t^2)}+t].$$

In terms of the change of momentum in classical scattering,

$$\boldsymbol{\Delta}=\mathbf{p}(\infty)-\mathbf{p}(-\infty)=-2\boldsymbol{\rho}\nu/\rho^2, \qquad (96.7)$$

we can rewrite these formulae as

$$\left.\begin{aligned}\mathbf{p}(t)&=\mathbf{p}+\tfrac{1}{2}\boldsymbol{\Delta}\left[\frac{t}{\sqrt{(t^2+\rho^2)}}+1\right],\\ \mathbf{r}(t)&=(\mathbf{p}+\tfrac{1}{2}\boldsymbol{\Delta})\frac{t}{\varepsilon}+\frac{\boldsymbol{\Delta}}{2\varepsilon}\sqrt{(t^2+\rho^2)}.\end{aligned}\right\} \qquad (96.8)$$

Now using formula (90.20) for $R(t)$ and the expressions (96.8) for $\mathbf{p}(t)$ and $\mathbf{r}(t)$, we can calculate the integral with respect to time in (96.6). The integration is carried out by replacing the variable t by

$$\xi = -\frac{\varepsilon}{\varepsilon'}(\omega t - \mathbf{k}\cdot\mathbf{r}(t))$$

and using the formula

$$\int_{-\infty}^{\infty} \frac{\xi e^{-i\xi}\, d\xi}{\sqrt{(\chi^2+\xi^2)}} = 2i\chi K_1(\chi),$$

where K_1 is the Macdonald function. There is, however, no need to complete these calculations, since we want the expression $a(\boldsymbol{\rho})$ only for small values of $\boldsymbol{\Delta}$ ($\Delta \ll m$), an independent parameter. Then we obtain

$$a(\boldsymbol{\rho}) = w_f^* \mathbf{D} w_i \cdot \boldsymbol{\Delta} \chi K_1(\chi), \tag{96.9}$$

where

$$\chi = \rho \frac{\omega\varepsilon}{\varepsilon'}(1 - \mathbf{n}\cdot\mathbf{v}),$$

$\mathbf{n} = \mathbf{k}/\omega$, and $\mathbf{D}$ is some function of $\mathbf{p}$, ε and $\mathbf{k}$ (but not of $\boldsymbol{\rho}$), whose precise form is unimportant.† Since, in the ultra-relativistic case, the photon is emitted at a small angle θ to the direction of the electron velocity, we have

$$\chi \approx \rho \frac{\varepsilon}{\varepsilon'}\omega(1 - v + \tfrac{1}{2}\theta^2)$$

or

$$\chi = \rho \frac{\omega m^2}{2\varepsilon\varepsilon'}(1+\delta^2), \qquad \delta = \theta\varepsilon/m. \tag{96.10}$$

It has already been mentioned that (96.5) is the probability of photon emission in a single passage of an electron past a nucleus at impact parameter ρ. The cross-section for the emission of a photon with given frequency and direction is obtained by multiplying by $v^{-1}\, d\rho_x\, d\rho_y \approx d\rho_x\, d\rho_y \equiv d^2\rho$ and integrating with respect to the impact parameter:

$$d\sigma = \frac{d^3k}{(2\pi)^3} \int |a(\boldsymbol{\rho})|^2\, d^2\rho. \tag{96.11}$$

† The spinors w_i and w_f may be taken as constant in the integration, i.e. the change in the electron's polarization in its classical ultra-relativistic motion may be neglected. This can be seen from the equations derived in §41.

However, it should not be thought that this formula without the integration over $d^2\rho$ would also give the directional distribution of the final electrons. The deviation of the electron in motion in a classical path, which is uniquely determined by the external field, is certainly not the same as the indeterminate quantum-mechanical deviation (and the limit $\mathbf{p}'(\infty)$ of the classical function $\mathbf{p}'(t)$ is therefore also different from the actual final momentum of the electron). Consequently, in order to obtain the angular distribution of the electrons, we must re-expand their wave function in plane waves.

It is seen from (96.11) that $a(\boldsymbol{\rho})$ is the amplitude of photon emission in a passage at impact parameter $\boldsymbol{\rho}$. The expressions (96.5) and (96.6), however, define this amplitude only to within a phase factor, which is easily seen to be $e^{-i\mathbf{k}\cdot\boldsymbol{\rho}}$: on account of the time-independent term $\mathbf{r}_\perp(\infty)=\boldsymbol{\rho}$ in $\mathbf{r}(t)$, this constant factor must be present in $V_{fi}(t)$, and may be taken outside the time integral. Since it is not an operator, it is not affected by the process of commutation, and the amplitude for the emission process is thus

$$e^{-i\mathbf{k}\cdot\boldsymbol{\rho}}a(\boldsymbol{\rho}), \tag{96.12}$$

where $a(\boldsymbol{\rho})$ is given by (96.9).

Now let the electron be described, as $z\to-\infty$, by a plane wave with momentum $\mathbf{p}$ along the z-axis. This means that the wave function of the electron as $z\to-\infty$, as a function of x and y, reduces to a constant, which can be taken as unity. Then the wave function of the electron which has passed through the field is, for $z\to\infty$,†

$$\psi(\infty)=S(\boldsymbol{\rho})=\exp\left\{-i\int_{-\infty}^{\infty}U(x,y,z)\,dz\right\}. \tag{96.13}$$

According to the significance of the transition amplitude (96.12), the wave function of an electron which has passed through the field and emitted a photon is

$$e^{-i\mathbf{k}\cdot\boldsymbol{\rho}}a(\boldsymbol{\rho})S(\boldsymbol{\rho}). \tag{96.14}$$

The amplitude for emission in which the electron goes to a state with definite momentum $\mathbf{p}'$ is given by the corresponding Fourier component of (96.14), i.e. by

$$\begin{aligned} a(\mathbf{q}_\perp) &= \int e^{-i\mathbf{p}'\cdot\boldsymbol{\rho}}\,e^{-i\mathbf{k}\cdot\boldsymbol{\rho}}a(\boldsymbol{\rho})S(\boldsymbol{\rho})\,d^2\rho \\ &= \int e^{-i\mathbf{q}_\perp\cdot\boldsymbol{\rho}}a(\boldsymbol{\rho})S(\boldsymbol{\rho})\,d^2\rho, \end{aligned} \tag{96.15}$$

where $\mathbf{q}_\perp$ is the transverse component of the momentum transfer to the nucleus; cf.

† See *QM* (131.4); we have in mind the analogy between equation (39.5) (in which we put $\mathbf{p}^2\approx\varepsilon^2$) and the non-relativistic Schrödinger's equation (39.5a). Bearing in mind the difference in the significance of the coefficients in these equations, it is easily seen that in our case the conditions *QM* (131.1) for the formula *QM* (131.4) to be valid are in fact satisfied. The fact that this formula is not valid for arbitrarily large z is unimportant, for the same reasons as in *QM* §131.

QM (131.7). The cross-section for scattering with a given transfer $\mathbf{q}_\perp$ is

$$d\sigma = |a(\mathbf{q}_\perp)|^2 \frac{d^3k}{(2\pi)^3}\frac{d^2q_\perp}{(2\pi)^2}. \tag{96.16}$$

Let us now calculate $S(\boldsymbol{\rho})$. In the case of a Coulomb field considered here, the integral in the exponent diverges, in accordance with the phase divergence in Coulomb scattering. The integral must therefore be first calculated between finite limits:

$$\begin{aligned}\int_{-R}^{R} U\,dz &= -2\nu\int_0^R \frac{dz}{\sqrt{(\rho^2+z^2)}}\\ &= -2\nu[\log(R+\sqrt{(R^2+\rho^2)})-\log\rho]\\ &\approx -2\nu\log 2R + 2\nu\log\rho\end{aligned}$$

$(R \gg \rho)$. The first term, which is a constant, is unimportant, and therefore

$$S(\boldsymbol{\rho}) = e^{-2i\nu\log\rho} = \rho^{-2i\nu}. \tag{96.17}$$

Substituting (96.9) and (96.17) in (96.15) and integrating over the directions of the vector $\boldsymbol{\rho}$ in the xy-plane, we find

$$a(\mathbf{q}_\perp) \propto \nu\int_0^\infty \rho^{-2i\nu}K_1(\chi)J_1(q_\perp\rho)\rho\,d\rho, \tag{96.18}$$

where J_1 is the Bessel function. The factors not involving $\nu = Z\alpha$ have not been written here.

We see that the dependence of the amplitude $a(\mathbf{q}_\perp)$ (and therefore of the cross-section (96.16)) on ν is contained in a separate factor. On the other hand, when $\nu\to 0$, the cross-section must tend to its value in the Born approximation. It is therefore immediately clear that the cross-section will differ from the Born value only by a factor which is independent of the electron polarization and hence does not influence the polarization effects.

The integral (96.18) can be expressed in terms of the hypergeometric function by means of the formula

$$\int_0^\infty x^{-\lambda}K_1(ax)J_1(bx)x\,dx = \frac{b\Gamma(2-\frac{1}{2}\lambda)\Gamma(1-\frac{1}{2}\lambda)}{2^\lambda a^{3-\lambda}}\left(1+\frac{b^2}{a^2}\right)^{-1+\frac{1}{2}\lambda}F\left(\tfrac{1}{2}\lambda, 1-\tfrac{1}{2}\lambda, 2, \frac{b^2}{a^2+b^2}\right).$$

This gives

$$a(\mathbf{q}_\perp) \propto \nu(1-i\nu)(\tfrac{1}{2}q)^{2i\nu}\Gamma^2(1-i\nu)F(i\nu, 1-i\nu, 2, z), \tag{96.19}$$

where

$$z = 1 - \frac{m^4\omega^2}{4q^2\varepsilon^2\varepsilon'^2}(1+\delta^2)^2, \qquad \delta = \varepsilon\theta/m; \tag{96.20}$$

here we have used the fact that, in region II (see (96.2)), the component of the vector **q** parallel to **p** is

$$q_\parallel^2 \equiv q^2 - \mathbf{q}_\perp^2 \approx \frac{m^4\omega^2}{4\varepsilon^2\varepsilon'^2}(1+\delta^2)^2. \tag{96.21}$$

This is easily proved, since in that region the angles between the momenta **p**, **p**′ and **k** satisfy the conditions (93.15).

The hypergeometric function in (96.19) can be reduced to the function $F(z)$ in (95.15) by means of the formula

$$F(a, b+1, c+1, z) = \frac{c}{c-a}F(a, b, c, z) + \frac{c(1-z)}{b(a-c)}F'(a, b, c, z).$$

The final result is then

$$d\sigma = d\sigma_B \frac{1}{F^2(1)}\left[F^2(z) + \frac{(1-z)^2}{\nu^2}F'^2(z)\right], \tag{96.22}$$

where $d\sigma_B$ is the Born cross-section (93.13) (H. A. Bethe and L. C. Maximon, 1954). When $q \gg m^2/\varepsilon$, we have $z \approx 1$, and the whole coefficient of $d\sigma_B$ tends to unity; in this sense formula (96.22), which has been derived for region II, is automatically satisfied for all $q \lesssim m$. When $q \lesssim m^2/\varepsilon$ and the correction factor in (96.22) is different from unity, the vectors **p**, **p**′ and **k** are almost coplanar, and the quantities δ and δ' are almost equal; this has already been taken into account in (96.22). Thus q^2 in the expression (96.20) for z can be written as

$$\frac{q^2}{m^2} = \delta^2 + \delta'^2 - 2\delta\delta'\cos\phi + \frac{m^2\omega^2}{4\varepsilon^2\varepsilon'^2}(1+\delta^2)^2, \tag{96.23}$$

i.e. we can put $\delta = \delta'$ in the second term in (93.14), but not in the first term, which does not contain a small coefficient ($\sim m^2/\varepsilon^2$).

To find the cross-section integrated over angles, there is no need to repeat the integration: we can proceed as follows (H. Olsen, 1955). Various directions of **p**′ (for a given energy ε') correspond to degeneracy of the final state of the electron. It is evident that the result of summing over states which belong to one degenerate level is independent of how the complete set of these states is chosen. We can therefore use, in summing over directions of **p**′, the set of functions $\psi^{(+)}_{\varepsilon'\mathbf{p}'}$ instead of the $\psi^{(-)}_{\varepsilon'\mathbf{p}'}$ which are needed in calculating the differential cross-section, i.e. we can define the bremsstrahlung matrix element as

$$M_{fi}^{\text{br}} = \int \psi^{(+)*}_{\varepsilon'\mathbf{p}'}(\boldsymbol{\alpha}\cdot\mathbf{e}^*)e^{-i\mathbf{k}\cdot\mathbf{r}}\psi^{(+)}_{\varepsilon\mathbf{p}}\, d^3x.$$

This integral is easily seen to be the same as $(M_{fi}^{pp})^*$ if the parameters of the wave functions in the latter are changed as follows:

$$\mathbf{p}_+, p_+, \varepsilon_+ \to -\mathbf{p}, -p, -\varepsilon; \qquad \mathbf{p}_-, p_-, \varepsilon_- \to \mathbf{p}', p', \varepsilon'; \qquad \mathbf{k} \to -\mathbf{k}$$

and the sign of the integration variables is reversed: $\mathbf{r} \to -\mathbf{r}$.

Hence it is clear that the bremsstrahlung cross-section integrated over angles can be obtained from the integral pair production cross-section (95.20), on multiplication by

$$\frac{\omega^2}{p_+^2}\frac{d\omega}{d\varepsilon_+} \approx \frac{\omega^2}{\varepsilon_+^2}\frac{d\omega}{d\varepsilon_+}$$

(cf. (91.6)) and replacement of ε_+ by $-\varepsilon$, ε_- by ε'. Thus we have

$$d\sigma = 4Z^2\alpha r_e^2 \frac{\varepsilon'}{\varepsilon}\left(\frac{\varepsilon'}{\varepsilon} + \frac{\varepsilon}{\varepsilon'} - \frac{2}{3}\right)\left[\log\frac{2\varepsilon\varepsilon'}{m\omega} - \frac{1}{2} - f(\alpha Z)\right]\frac{d\omega}{\omega}. \qquad (96.24)$$

We see that the corrections to the Born formulae for the integral bremsstrahlung and pair production cross-sections are given by the same function $f(\alpha Z)$.

Formula (96.24), which does not depend on any limitations on the value of $Z\alpha$, allows a passage to the classical limit ($\hbar \to 0$, $Z\alpha \to \infty$). In this limit, we must also put $\varepsilon \approx \varepsilon'$. Bearing in mind the asymptotic expression $\Psi(z) \approx \log z$ as $|z| \to \infty$ and the value $\Psi(1) = -C$ (where C is Euler's constant), we find the effective retardation

$$\hbar\omega\, d\sigma = \frac{16Z^2 r_e^2 e^2}{3c}\left[\log\frac{2\varepsilon^2}{mc\omega Ze^2} - \frac{1}{2} - C\right] d\omega. \qquad (96.25)$$

This expression, which does not contain $\hbar$, is the classical frequency distribution of the bremsstrahlung intensity.

§97. **Electron–electron bremsstrahlung in the ultra-relativistic case**

Electron–electron bremsstrahlung is represented by eight Feynman diagrams: four diagrams

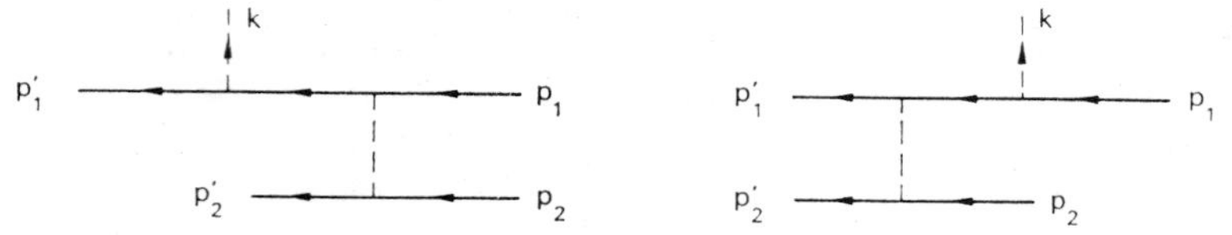

(97.1a)

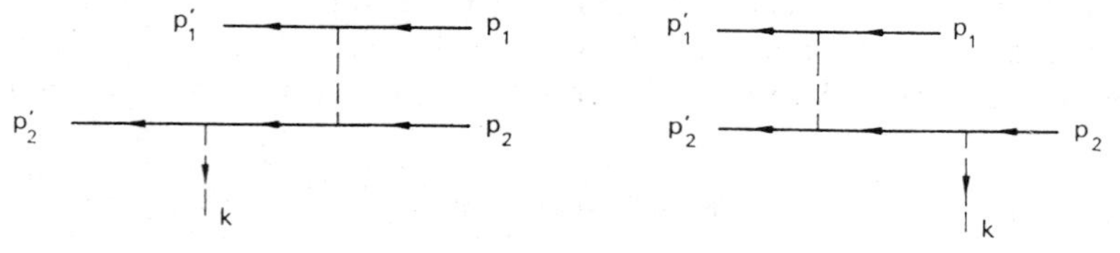

(97.1b)

and four "exchange" diagrams obtained from those shown by interchanging p_1' and p_2'. Here we shall give the results of the calculations for the ultra-relativistic case (G. Altarelli and F. Buccella, 1964; V. N. Baĭer, V. S. Fadin and V. A. Khoze, 1966).†

In the laboratory system (the rest frame of one of the initial electrons, say the second), the emission cross-section integrated over the directions of the photon can be written as a sum $d\sigma = d\sigma^{(1)} + d\sigma^{(2)}$, where

$$d\sigma^{(1)} = 4\alpha r_e^2 \frac{d\omega}{\omega}\frac{\varepsilon-\omega}{\varepsilon}\left(\frac{\varepsilon}{\varepsilon-\omega}+\frac{\varepsilon-\omega}{\varepsilon}-\frac{2}{3}\right)\left(\log\frac{2\varepsilon(\varepsilon-\omega)}{m\omega}-\frac{1}{2}\right); \tag{97.2}$$

$$d\sigma^{(2)} = \tfrac{2}{3}\alpha r_e^2 \frac{m\,d\omega}{\omega^2}\left\{\left(4-\frac{m}{\omega}+\frac{m^2}{4\omega^2}\right)\log\frac{2\varepsilon}{m}-2+\frac{2m}{\omega}-\frac{5m^2}{8\omega^2}\right\} \quad \text{for } \omega \geqslant \tfrac{1}{2}m, \tag{97.3}$$

$$\begin{aligned} d\sigma^{(2)} = \tfrac{2}{3}\alpha r_e^2 \frac{d\omega}{\omega}\Big\{&8\left(1-\frac{\omega}{m}+\frac{\omega^2}{m^2}\right)\log\frac{\varepsilon}{\omega} - \\ &-\left(1-\frac{2\omega}{m}\right)\log\left(1-\frac{2\omega}{m}\right)\left[\frac{m^3}{4\omega^3}-\frac{m^2}{2\omega^2}+\frac{3m}{\omega}-2+\frac{4\omega}{m}\right]- \\ &-\frac{m^2}{2\omega^2}+\frac{3m}{2\omega}-2+\frac{2\omega}{m}-\frac{4\omega^2}{m^2}\Big\} \quad \text{for } \omega \leqslant \tfrac{1}{2}m \end{aligned} \tag{97.4}$$

(ε being the initial energy of the first electron).

These formulae are accurate as far as terms of relative order m/ε. To this accuracy, it is found that the contributions to the cross-section from different diagrams do not interfere, and in this sense $d\sigma^{(1)}$ and $d\sigma^{(2)}$ correspond to emission by each of the two electrons: the fast electron and the recoil electron respectively, diagrams (97.1a) and (97.1b).

The "exchange" diagrams give the same contribution to the cross-section as do the "direct" diagrams. Since the electrons are identical, the total contribution from the direct and exchange diagrams has to be halved, and we may therefore consider only the contribution of the direct diagrams and ignore the identity of the particles. For electron–positron collisions the exchange diagrams are replaced by annihilation diagrams, but their relative contribution is of order m/ε and therefore negligible. Hence the bremsstrahlung cross-sections are the same, to the accuracy indicated, in electron–electron and electron–positron collisions.

For $\omega \gg m$, the ratio

$$\frac{d\sigma^{(2)}}{d\sigma^{(1)}} \sim \frac{m}{\omega} \ll 1,$$

i.e. the emission from the recoil electron is small compared with that from the fast electron; when this ratio becomes of the order of m/ε, formula (97.3) is of course no longer meaningful. When $\omega \ll m$, on the other hand, the two parts of the

† The calculations are given in the book by Baĭer *et al.* cited in §93.

cross-section are almost comparable:

$$\left.\begin{aligned} d\sigma^{(1)} &= \tfrac{16}{3}\alpha r_e^2 \frac{d\omega}{\omega} \log \frac{2\varepsilon^2}{m\omega}, \\ d\sigma^{(2)} &= \tfrac{16}{3}\alpha r_e^2 \frac{d\omega}{\omega} \log \frac{\varepsilon}{\omega}, \end{aligned}\right\} \omega \ll m. \tag{97.5}$$

For formulae (97.2)–(97.5) to be valid, it is necessary that at least one of the electrons should remain ultra-relativistic after emission of radiation, i.e. the photon frequency must be sufficiently far from the hard boundary of the spectrum (the maximum frequency $\omega_{\max}$ that can be emitted). The final energy of the electrons is least, and the photon energy greatest, when both electrons move, after emission, in the direction of the photon at equal speeds. The conservation laws then give

$$\varepsilon + m = \omega_{\max} + 2\varepsilon', \qquad |\mathbf{p}| = \omega_{\max} + 2|\mathbf{p}'|.$$

Hence, eliminating ε' and $\mathbf{p}'$, we have

$$(\varepsilon + m - \omega_{\max})^2 - (|\mathbf{p}| - \omega_{\max})^2 = 4m^2$$

and

$$\omega_{\max} = \frac{m(\varepsilon - m)}{m + \varepsilon - |\mathbf{p}|}. \tag{97.6}$$

When $\varepsilon \gg m$, $\omega_{\max} \approx \varepsilon$. Thus formulae (97.2)–(97.4) are valid if

$$\omega_{\max} - \omega \sim \varepsilon \sim \omega \gg m. \tag{97.7}$$

The cross-section (97.2) for emission by the fast electron is exactly equal to that for electron–nucleus bremsstrahlung when the nucleus has $Z = 1$ (formula (93.17)). This agreement is not fortuitous, and can be explained by considering the significance of recoil in the emission process.

In deriving (93.17) we neglected the recoil of the fixed particle (the nucleus), replacing it by a constant external field. This was equivalent to neglecting the time component of the momentum transfer 4-vector $q = p' - p + k$ (the recoil energy). We shall show that, in the ultra-relativistic case, this treatment is permissible for electron–electron as well as electron–nucleus bremsstrahlung.

We write

$$-q^2 = -(\varepsilon' + \omega - \varepsilon)^2 + (p'_{\parallel} + \omega - p_{\parallel})^2 + (\mathbf{p}'_{\perp} - \mathbf{p}_{\perp})^2, \tag{97.8}$$

where the subscripts indicate the components of the vectors $\mathbf{p}'$ and $\mathbf{p}$ (the initial and final electron momenta) parallel and perpendicular to the direction of the photon $\mathbf{k}$. In the ultra-relativistic case the angles θ, θ' between $\mathbf{k}$ and $\mathbf{p}$, $\mathbf{p}'$ respectively are small: $\theta \lesssim m/\varepsilon$, $\theta' \lesssim m/\varepsilon'$. Hence

$$|\mathbf{p}_\perp| \sim |\mathbf{p}|\theta \sim m,$$
$$p_\parallel \approx |\mathbf{p}| - \frac{\mathbf{p}_\perp^2}{2|\mathbf{p}|} \approx \varepsilon - \frac{m^2}{2\varepsilon} - \frac{\mathbf{p}_\perp^2}{2\varepsilon}, \tag{97.9}$$

and similarly for $\mathbf{p}_\perp'$ and $p_\parallel'$.

Neglecting recoil, we have $\varepsilon' + \omega - \varepsilon = 0$; the term $p_\parallel' + \omega - p_\parallel \sim m^2/\varepsilon$, and so

$$-q^2 \approx (\mathbf{p}_\perp' - \mathbf{p}_\perp)^2 \sim m^2. \tag{97.10}$$

The energy of (electron–electron) recoil is

$$q_0 = \varepsilon' + \omega - \varepsilon \sim \mathbf{q}^2/2m \sim m. \tag{97.11}$$

The change in $\mathbf{p}_\perp'$ due to the change in ε' is negligible. The change in q^2 with allowance for recoil, which we donote by Δq^2, is therefore given by the first two terms in (97.8). Using (97.9), we have

$$\Delta q^2 \approx (\varepsilon' + \omega - \varepsilon)\left(-\frac{m^2}{\varepsilon'} - \frac{\mathbf{p}_\perp'^2}{\varepsilon'} + \frac{m^2}{\varepsilon} + \frac{\mathbf{p}_\perp^2}{\varepsilon}\right)$$
$$\sim m^2 \cdot m/\varepsilon.$$

Comparison with (97.10) shows that $\Delta q^2 \ll |q^2|$, and the neglect of the recoil is therefore justified.†

The fact that the fast particle emits into a narrow cone (with aperture angle $\sim m/\varepsilon$) in the direction of its motion enables us to deduce the cross-section in the centre-of-mass system by a simple conversion of the cross-section (97.2) from the laboratory system.‡

In the centre-of-mass system the two electrons emit in the same manner, each in the direction of its motion. (It may be noted that this gives an intuitive explanation of the absence of interference between the radiation from the two particles.) The energy E of the ultra-relativistic electron in this system is related to its energy ε in the laboratory system by $2E^2 = m\varepsilon$; the respective photon frequencies Ω and ω are related by $\omega/\varepsilon = \Omega/E$. These equations are easily obtained by comparing the values of the invariants (p_1p_2) and (p_1k) in the two systems. The cross-section for emission by each electron in the centre-of-mass system is therefore

$$d\sigma^{(1)} = d\sigma^{(2)}$$
$$= 4\alpha r_e^2 \frac{d\Omega}{\Omega}\frac{E-\Omega}{E}\left(\frac{E}{E-\Omega} + \frac{E-\Omega}{E} - \frac{2}{3}\right)\left(\log\frac{4E^2(E-\Omega)}{m^2\Omega} - \frac{1}{2}\right). \tag{97.12}$$

† This conclusion is, of course, valid *a fortiori* in the case of electron–nucleus bremsstrahlung, for which the recoil energy $q_0 \approx \mathbf{q}^2/2M \sim m^2/M$, where M is the mass of the nucleus.

‡ In general such a conversion is not possible, because the contribution to the spectrum in a given frequency range $d\omega$ comes from photons emitted in quite different directions.

For (97.12) to be valid it is also necessary that the photon frequency should not be close to the boundary of the spectrum. For an ultra-relativistic particle, the above-mentioned transformation gives immediately, when $\omega_{\max} \approx \varepsilon$,

$$\Omega_{\max} \approx \omega_{\max} E/\varepsilon \approx E. \tag{97.13}$$

Thus, in the centre-of-mass system, the electrons can emit only half of their total energy $2E$. A direct calculation of $\Omega_{\max}$ is easily performed by noting that, after the emission of such a photon, the electrons will move (in that system) at equal speeds in the direction opposite to that of the photon. We have

$$2E = 2E' + \Omega_{\max}, \qquad 2|\mathbf{p}'| = \Omega_{\max},$$

whence

$$\Omega_{\max} = \mathbf{p}^2/E = E - m^2/E, \tag{97.14}$$

and in the ultra-relativistic case again (97.13). Thus formula (97.12) is applicable under the condition

$$\Omega_{\max} - \Omega \sim E - \Omega \gg m. \tag{97.15}$$

We shall now give some formulae for emission in the centre-of-mass system in the opposite limiting case, near the boundary of the spectrum, when†

$$\Omega_{\max} - \Omega \ll m. \tag{97.16}$$

Since in this case the recoil is very important, the results differ from those for scattering by a fixed centre and are also different for electron–electron and electron-positron scattering (V. N. Baĭer, V. S. Fadin and V. A. Khoze, 1967).

In electron–electron scattering, besides the squares of the diagrams (97.1), there is also a contribution to the emission cross-section near the boundary of the spectrum from products (interference terms) of the direct and exchange diagrams, in which a given initial particle emits, for example, the product of the second diagram (97.1a) and the diagram

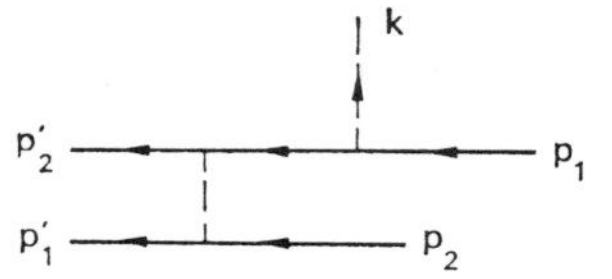

This is because, near the boundary, the final particles have similar momenta and there is no reason for the exchange terms to be small. The final result for the

† The result obtained in the Born approximation is, of course, as usual valid only if the relative velocity of the final electrons is large in comparison with α. If not, the interaction of the particles in the final state has to be taken into account.

cross-section is

$$d\sigma = 2\alpha r_e^2 \frac{[E(\Omega_{\max} - \Omega)]^{\frac{1}{2}}}{m} \frac{d\Omega}{\Omega_{\max}}. \tag{97.17}$$

In electron–positron scattering, a logarithmically large contribution to the emission cross-section comes from the squares of annihilation diagrams, in which there is emission by the initial particles:

(97.18)

The squares of other diagrams are significant when the accuracy is not logarithmic, but the interference terms are small. The final result is

$$d\sigma = 2\alpha r_e^2 \frac{[E(\Omega_{\max} - \Omega)]^{\frac{1}{2}}}{m} \left(\log \frac{2E}{m} + 1\right) \frac{d\Omega}{\Omega_{\max}}. \tag{97.19}$$

Thus the emission in electron–positron scattering is logarithmically large in comparison with that in electron–electron scattering.

§98. Emission of soft photons in collisions

Let $d\sigma_0$ be the cross-section for a given process of scattering of charged particles, which may be accompanied by the emission of a certain number of photons. Together with this process, we shall consider another which differs from it only in that one extra photon is emitted. If the frequency ω of this photon is sufficiently small (the necessary conditions will be formulated below), the cross-section $d\sigma$ for the second process is related in a simple manner to $d\sigma_0$.

When ω is small, we can neglect the influence of the emission of this quantum on the scattering process. The cross-section $d\sigma$ can therefore be represented as a product of two independent factors, the cross-section $d\sigma_0$ and the probability dw of emission of a single photon in the collision. The emission of a soft photon is a quasi-classical process; the probability is therefore the same as the classically calculated number of quanta emitted in the collision, i.e. the same as the classical intensity (total energy) of emission dI, divided by ω ($= \hbar\omega$). Thus

$$d\sigma = d\sigma_0 \, dI/\omega. \tag{98.1}$$

We shall show how this formula can be derived from the general rules of the diagram technique (J. M. Jauch and F. Rohrlich, 1954).

The diagrams for the process involving an additional photon are obtained from those for the original process by adding an external photon line which "branches off" from some (external or internal) electron line, i.e. by replacing

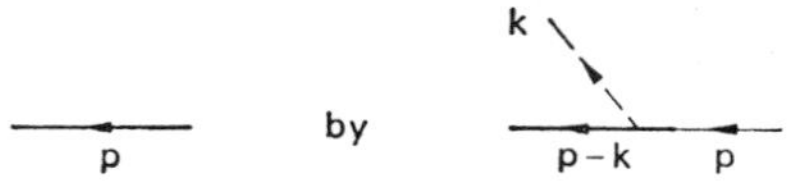

(98.2)

It is easily seen that the most important diagrams will be those in which this change is made in external electron lines. For, if p is the momentum of the external line ($p^2 = m^2$), then for small k we have also $(p-k)^2 \approx m^2$, i.e. the factor $G(p-k)$ added to the diagram is near its pole.

For an initial electron line p the change (98.2) amounts to the following change in the reaction amplitude:

$$\begin{aligned} u(p) &\to e\sqrt{(4\pi)}G(p-k)(\gamma e^*)u(p) \\ &= e\sqrt{(4\pi)}\frac{\gamma p - \gamma k + m}{(p-k)^2 - m^2}(\gamma e^*)u(p) \\ &\approx -e\sqrt{(4\pi)}\frac{\gamma p + m}{2pk}(\gamma e^*)u(p). \end{aligned}$$

Since $(\gamma p)(\gamma e^*) = 2pe^* - (\gamma e^*)(\gamma p)$ and $\gamma p u(p) = m u(p)$, we obtain the following rule:

$$u(p) \to -e\sqrt{(4\pi)}\frac{(pe^*)}{(pk)}u(p). \tag{98.3}$$

Similarly, for a final electron line p', the replacement of

in the diagram implies the change

$$\bar{u}(p') \to e\sqrt{(4\pi)}\bar{u}(p')\frac{(p'e^*)}{(p'k)} \tag{98.4}$$

in the amplitude.

In the rest of the diagram we can everywhere neglect the changes in the momenta of the lines as a result of the emission of the photon k. Here it is assumed that the photon energy ω is always small in comparison with the energies of all the particles participating in the reaction (and in comparison with those of the hard photons, if any, that are emitted).

Let the cross-section $d\sigma_0$ refer, say, to the scattering of an electron by a fixed nucleus (with possible emission of hard photons). The amplitude of this process, which will be conventionally called "elastic", is

$$M_{fi}^{(\mathrm{el})} = \bar{u}(p')Mu(p).$$

Making the successive substitutions (98.3) and (98.4) and adding the results, we obtain the bremsstrahlung amplitude for emission of the same hard photons together with a soft photon k:†

$$M_{fi} = M_{fi}^{(\mathrm{el})} e\sqrt{(4\pi)}\left(\frac{(p'e^*)}{(p'k)} - \frac{(pe^*)}{(pk)}\right). \tag{98.5}$$

Accordingly, the cross-section is

$$d\sigma = d\sigma_{\mathrm{el}} \cdot 4\pi e^2 \left|\frac{(p'e)}{(p'k)} - \frac{(pe)}{(pk)}\right|^2 \frac{d^3k}{(2\pi)^3 2\omega}. \tag{98.6}$$

Summation over polarizations of the photon k gives

$$d\sigma = -e^2\left[\frac{p'}{(p'k)} - \frac{p}{(pk)}\right]^2 \frac{d^3k}{4\pi^2\omega}\, d\sigma_{\mathrm{el}}. \tag{98.7}$$

In terms of three-dimensional quantities, this formula becomes‡

$$d\sigma = \alpha\left(\frac{\mathbf{v}'\times\mathbf{n}}{1-\mathbf{v}'\cdot\mathbf{n}} - \frac{\mathbf{v}\times\mathbf{n}}{1-\mathbf{v}\cdot\mathbf{n}}\right)^2 \frac{d\omega\, do_{\mathbf{k}}}{4\pi^2\omega}\, d\sigma_{\mathrm{el}}, \tag{98.8}$$

where $\mathbf{n} = \mathbf{k}/\omega$, and $\mathbf{v}$ and $\mathbf{v}'$ are the initial and final velocities of the electron. We see that the coefficient of $d\sigma_{\mathrm{el}}$ is in fact the same as the classical intensity of emission (cf. *Fields* (69.4)), divided by ω, as already asserted in formula (98.1).

The condition for the above formulae to be applicable is that not only is ω small compared with ε but also the momentum transfer $\mathbf{q}$ to the nucleus is large compared with the change $\delta\mathbf{q}$ in this quantity due to the emission of the soft photon. We have

$$\begin{aligned}\delta\mathbf{q} &= (\mathbf{p}'-\mathbf{p}-\mathbf{k}) - (\mathbf{p}'-\mathbf{p})_{\omega=0}\\ &= \delta\mathbf{p}' - \mathbf{k},\end{aligned}$$

where $|\delta\mathbf{p}'| \sim \omega\partial|\mathbf{p}'|/\partial\varepsilon \sim \omega/v$ and $|\mathbf{k}| = \omega$. In the non-relativistic case ($v \ll 1$), we therefore obtain the condition

$$\omega/|\mathbf{q}|v \ll 1. \tag{98.9}$$

For scattering by a Coulomb potential (or by any potential that decreases slowly with increasing distance) $|\mathbf{q}| \sim 1/\rho$ (where ρ is the impact parameter), and so this condition can also be written as $\omega\tau \ll 1$, where $\tau \sim \rho/v$ is the characteristic time of the collision.

† It should be noted that the difference term in this formula arises naturally from gauge invariance: the reaction amplitude must be unchanged when the polarization 4-vector e is replaced by e + constant × k.

‡ To derive (98.8) it is convenient to return to (98.6), putting $p = (\varepsilon, \varepsilon\mathbf{v})$, $pk = \varepsilon\omega(1-\mathbf{v}\cdot\mathbf{n}), \ldots, e = (0, \mathbf{e})$, and then summing over polarizations by means of (45.4a).

In the ultra-relativistic case, the photons are emitted chiefly in directions near $\mathbf{v}$ and $\mathbf{v}'$, as is seen from the denominators in (98.8). If the electron scattering angle θ is small, the directions of all three vectors $\mathbf{p}$, $\mathbf{p}'$, $\mathbf{n}$ are close together. Then

$$\begin{aligned}|\delta\mathbf{q}| &= |\delta\mathbf{p}'| - |\mathbf{k}| \\ &= \omega\left(\frac{1}{v} - 1\right) \\ &\sim \omega m^2/\varepsilon^2,\end{aligned}$$

and, since $|\mathbf{q}| \sim \varepsilon\theta$, we obtain the condition

$$\theta \gg \frac{\omega}{\varepsilon}\frac{m^2}{\varepsilon^2}. \tag{98.10}$$

Because the formulae (98.5)–(98.8) are quasi-classical, they are valid for emission by any charged particles, not necessarily electrons as assumed in the derivation. In general, when several such particles take part in the reaction, formula (98.5) must be put in the form

$$M_{fi} = M_{fi}^{(\mathrm{el})} e\sqrt{(4\pi)} \sum Z\left(\frac{p'e^*}{p'k} - \frac{pe^*}{pk}\right), \tag{98.11}$$

where the summation is over all the particles (with charges Ze); formulae (98.6)–(98.8) are changed similarly.

In particular, in the non-relativistic case

$$M_{fi} = M_{fi}^{(\mathrm{el})} \frac{e\sqrt{(4\pi)}}{\omega} \sum Z(\mathbf{v}' - \mathbf{v}) \cdot \mathbf{e}^*. \tag{98.12}$$

For two particles, this formula becomes

$$\left.\begin{aligned} M_{fi} &= M_{fi}^{(\mathrm{el})} \frac{\sqrt{(4\pi)}}{\omega}\left(\frac{Z_1 e}{m_1} - \frac{Z_2 e}{m_2}\right)\mathbf{q} \cdot \mathbf{e}^*, \\ \mathbf{q} &= m(\mathbf{v}' - \mathbf{v}), \qquad m = m_1 m_2/(m_1 + m_2), \end{aligned}\right\} \tag{98.13}$$

where $\mathbf{v}$ and $\mathbf{v}'$ are the relative velocities of the particles before and after the collision. From this, on integrating $|M_{fi}|^2$ over the directions of emission of the photon and summing over the directions of polarization of the photon, we find the nonrelativistic frequency distribution of the radiation:

$$d\sigma_\omega = d\sigma_{\mathrm{el}} \frac{2e^2}{3\pi}\left(\frac{Z_1}{m_1} - \frac{Z_2}{m_2}\right)^2 \mathbf{q}^2 \frac{d\omega}{\omega}.$$

The above results can be generalized to the case of simultaneous emission of several soft photons. For each photon there is an additional factor in M_{fi}, similar to the coefficient of $M_{fi}^{(el)}$ (98.5). This is easily seen directly for the example of two photons, say. The lines of the two emitted photons have to be added on external electron lines, and in two different orders, so that a diagram with external line p is replaced by two diagrams with the lines

respectively. They contain the factors

$$\frac{1}{2(pk_1+pk_2)}\frac{1}{2pk_1} \quad \text{and} \quad \frac{1}{2(pk_1+pk_2)}\frac{1}{2pk_2}$$

(the denominators of the electron propagators) respectively, and their sum is

$$\frac{1}{2pk_1}\frac{1}{2pk_2},$$

i.e. it is the product of two independent factors relating to the first and the second photon. Then, in the sum of all the diagrams, the terms combine (because of gauge invariance) to give the product of differences

$$\left(\frac{p'e_1^*}{p'k_1}-\frac{pe_1^*}{pk_1}\right)\left(\frac{p'e_2^*}{p'k_2}-\frac{pe_2^*}{pk_2}\right).$$

The cross-section for the process separates into factors in accordance with the factorization of the amplitude. The soft photons are therefore emitted independently. The cross-section for emission of n soft photons can be written

$$d\sigma = d\sigma_{el}\, dw_1 \ldots dw_n, \tag{98.14}$$

where $dw_1, dw_2, \ldots$ are the probabilities of individual emission of the photons k_1, $k_2, \ldots$. When this formula is integrated over a finite range of values of the variables (frequencies and directions), the same for all quanta, a factor $1/n!$ must be included in order to take account of the identity of the photons.

If the emission cross-section (98.1) is integrated over frequencies in some finite range from ω_1 to ω_2, the resulting expression is

$$d\sigma \sim \alpha \log(\omega_2/\omega_1)\, d\sigma_{el}; \tag{98.15}$$

cf. (98.8). Here it is assumed that both frequencies are soft, and the possible values of ω_2 are therefore limited by the condition for the method to be applicable. With logarithmic accuracy, however, we can put $\omega_2 \sim \varepsilon$, where ε is the initial energy of the emitting particle. The values of ω_1 have no lower limit, but on letting $\omega_1 \to 0$ we see that the cross-section for emission of all possible soft quanta is infinite. Let us investigate the significance of this "*infra-red catastrophe*" (F. Bloch and A. Nordsieck, 1937).

When

$$\alpha \log(\varepsilon/\omega_1) \gtrsim 1, \tag{98.16}$$

we have $d\sigma \gtrsim d\sigma_{\text{el}}$. This, however, means that perturbation theory is inapplicable, and $d\sigma$ cannot be calculated as a quantity of a higher order of smallness than $d\sigma_{\text{el}}$. Thus in this case the small parameter must be taken as $\alpha \log(\varepsilon/\omega_1)$, not α.

The derivation of formulae (98.5) and (98.6) from perturbation theory is therefore invalid at sufficiently low frequencies. The classical formula for the intensity dI (*Fields* (69.4)), on the other hand, becomes more nearly correct as ω decreases. Hence formula (98.1) remains valid if its meaning is made somewhat more classical. In this formula it has been assumed that one photon is emitted. Then the energy lost by the particle as radiation is equal to ω and the "relative energy loss cross-section" is $\omega\, d\sigma/\varepsilon$ or

$$d\sigma_{\text{el}}\, dI/\varepsilon. \tag{98.17}$$

In reality, for sufficiently small ω, the emission probability is not small, and the probability of emission of two or more photons is greater, not less, than the probability for one photon. Under these conditions, the expression (98.17) remains valid but the classical intensity dI determines, instead of the probability of emission of one photon, the mean number of emitted photons

$$d\bar{n} = dI/\omega, \tag{98.18}$$

or, in a finite range of frequencies,

$$\bar{n} = \int_{\omega=\omega_1}^{\omega_2} dI/\omega. \tag{98.19}$$

Since the soft photons are emitted in a statistically independent manner (this being true in every approximation of perturbation theory), Poisson's formula can be applied to the process of multiple emission: the probability $w(n)$ that n photons are emitted is given in terms of the mean number $\bar{n}$ by

$$w(n) = \bar{n}^n e^{-\bar{n}}/n!. \tag{98.20}$$

The cross-section for a process of scattering with emission of photons may be written

$$d\sigma = d\sigma_{\text{el}} \cdot w(n). \tag{98.21}$$

Since $\Sigma\, w(n) = 1$, $d\sigma_{\text{el}}$ is the total cross-section for scattering accompanied by the emission of any soft radiation. This is evident from a classical treatment; according to perturbation theory, however, $d\sigma_{el}$ is the purely elastic scattering cross-section. But perturbation theory is inapplicable here. Thus we find that $d\sigma_{el}$ calculated by perturbation theory as the elastic scattering cross-section actually includes the emission of any soft photons. The true value of the purely elastic scattering cross-section is zero: as $\omega_1 \to 0$, the mean number $\bar{n} \to \infty$, and according to (98.20) the probability of emission of any finite number of photons vanishes.†

PROBLEMS‡

PROBLEM 1. Find the spectral distribution of soft quanta emitted in ultra-relativistic electron–nucleus bremsstrahlung.

SOLUTION. Integration of (98.8) over $do_{\mathbf{k}}$ gives

$$d\sigma = \alpha F(\xi)(d\omega/\omega)\, d\sigma_{el}, \tag{1}$$

where

$$\left.\begin{aligned} F(\xi) &= \frac{2}{\pi}\left[\frac{2\xi^2+1}{\xi\sqrt{(\xi^2+1)}}\log(\xi+\sqrt{(\xi^2+1)})-1\right],\\ \xi &= \frac{|\mathbf{p}|}{m}\sin\tfrac{1}{2}\theta, \end{aligned}\right\} \tag{2}$$

p being the electron momentum and θ the scattering angle. In the ultra-relativistic case, the most important range of angles is

$$m^2\omega/\varepsilon^3 \ll \theta \ll m/\varepsilon; \tag{3}$$

the lower limit is given by the condition (98.10), and the upper limit is discussed below. Here $\xi \approx \varepsilon\theta/2m \ll 1$, so that

$$F(\xi) \approx (8/3\pi)\xi^2,$$

and the electron–nucleus elastic scattering cross-section is (see (80.10))

$$d\sigma_{el} \approx 4Z^2 r_e^2 \frac{m^2}{\varepsilon^2}\frac{do}{\theta^4}. \tag{4}$$

The integral

$$d\sigma_\omega = \tfrac{16}{3}Z^2\alpha r_e^2 \frac{d\omega}{\omega}\int\frac{d\theta}{\theta}$$

diverges logarithmically; it is cut off below at angles $\theta \sim m^2\omega/\varepsilon^3$ and above at $\xi \sim 1$, i.e. at angles $\theta \sim m/\varepsilon$. When $\xi \to \infty$, $F \sim (4/\pi)\log\xi$ and the integral converges. Thus we have, with logarithmic accuracy,

$$d\sigma_\omega = \tfrac{16}{3}Z^2\alpha r_e^2 \frac{d\omega}{\omega}\log\frac{\varepsilon^2}{m\omega}, \tag{5}$$

which agrees with the logarithmic part of formula (93.17) (where we must put $\varepsilon \approx \varepsilon'$). Non-logarithmic accuracy can be achieved only by going beyond the quasi-classical range.

† We shall return to a more detailed discussion of this in §130, in connection with radiative corrections.

‡ The following applications of formula (98.7) are due to V. N. Baĭer and V. M. Galitskiĭ (1964).

PROBLEM 2. For a collision between two ultra-relativistic electrons, determine (in the centre-of-mass system) the cross-section for simultaneous emission of two soft photons in opposite directions at small angles to the electron momenta.

SOLUTION. Photons moving in opposite directions are emitted by different electrons, each in the direction of its motion. The cross-section for simultaneous emission is

$$\left.\begin{aligned} d\sigma &= d\sigma_{\text{el}} \cdot \alpha F(\xi)\frac{d\omega_1}{\omega_1} \cdot \alpha F(\xi)\frac{d\omega_2}{\omega_2}, \\ \xi &= (\varepsilon/m)\sin\tfrac{1}{2}\theta, \end{aligned}\right\} \tag{6}$$

where ε is the energy of each electron, θ the scattering angle in the centre-of-mass system; θ is the same for each electron. No factor $\frac{1}{2}$ is needed in the cross-section, since the photons are certainly emitted in different directions. The cross-section for elastic scattering of the electrons through small angles in the centre-of-mass system, in the ultra-relativistic case, is the same as (4); cf. (81.11). Unlike (1), the cross-section (6) behaves as $\theta\, d\theta$ when $\theta \to 0$, and the integral therefore converges. On the one hand, this enables us to extend the integration to $\theta = 0$, without any difficulty that the method might cease to be applicable. On the other hand, the main contribution to the integrated cross-section now comes from the region $\theta \sim m/\varepsilon$, not $\theta \ll m/\varepsilon$, and so the exact expression (2) has to be used. The result of integrating the cross-section over scattering angles is

$$\begin{aligned} d\sigma_{\omega_1\omega_2} &= \frac{2}{\pi}[5 + \tfrac{7}{2}\zeta(3)]r_e^2\alpha^2\frac{d\omega_1}{\omega_1}\frac{d\omega_2}{\omega_2} \\ &= 5.9 r_e^2\alpha^2\frac{d\omega_1}{\omega_1}\frac{d\omega_2}{\omega_2}, \end{aligned}$$

the value of the Riemann zeta function being $\zeta(3) = 1.202$.

§99. The method of equivalent photons

Let us compare two processes described by the diagrams

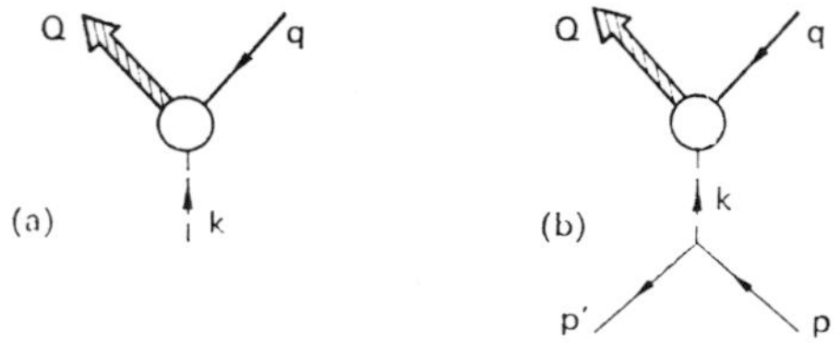

(99.1)

where the circles represent the whole of the internal parts of the diagrams. Diagram (a) represents a collision between a photon k ($k^2 = 0$) and a particle having 4-momentum q (and mass m; $q^2 = m^2$). The system resulting from the collision is a particle or group of particles having total 4-momentum Q. Diagram (b) represents a collision between the same particle q and another particle having 4-momentum p and mass M ($p^2 = M^2$). After the collision, the latter particle has 4-momentum p', and the same system Q is formed. The second process may be regarded as a collision between the particle q and a virtual photon emitted by the particle p and having momentum $k = p - p'$ ($k^2 < 0$). If $|k|^2$ is small, the virtual photon is not greatly different from a real photon. Such a situation is evidently possible in collisions of very fast particles: the electromagnetic field of a charged particle moving with $v \approx 1$ is almost transverse, and therefore has properties similar to those of the field of a light wave. Under these conditions, the cross-section for process (b) can be expressed in terms of that for process (a).†

We shall thus suppose that the particle M is ultra-relativistic, with energy (in the rest frame of the particle m) $\varepsilon \gg M$. If the masses of the colliding particles m and M are different, we shall take the case where $m < M$.

The amplitude of process (a), which involves a real photon, can be written

$$M_{fi}^{(r)} = -e\sqrt{(4\pi)}(e_\mu J^\mu), \tag{99.2}$$

where e_μ is the photon polarization 4-vector and J^μ the transition current corresponding to the vertex (the circle) in the diagram. The amplitude of process (b) is

$$M_{fi} = Ze^2 \frac{4\pi}{k^2}(j_\mu J^\mu), \tag{99.3}$$

where j_μ is the transition current of the particle m (the lower vertex in the diagram), and Ze is the charge on this particle. The current J is a function of $k = Q - q$, and is therefore not the same in the two cases, since $k^2 = 0$ in (99.2) and $k^2 \neq 0$ in (99.3). But if, in the second case,

$$|k^2| \ll m^2, \tag{99.4}$$

we can here also take J for $k^2 = 0$.

The change in the momentum of the particle M when a virtual photon is emitted, $\mathbf{p} - \mathbf{p}' = \mathbf{k}$, is small in comparison with its initial momentum $|\mathbf{p}| \approx \varepsilon$; we can therefore put $\mathbf{p} = \mathbf{p}'$ in the transition current j. That is, the motion of the particle M may be regarded as uniform motion in a straight line. Since such a motion is quasi-classical, the corresponding current is independent of the spin of the particle:‡

$$j^\mu = 2p^\mu. \tag{99.5}$$

The condition for the current to be transverse ($jk = 0$) now gives $\varepsilon\omega - p_x k_x = 0$, the x-axis being taken in the direction of $\mathbf{p}$. Hence

$$\omega = vk_x, \tag{99.6}$$

where $v = p_x/\varepsilon$ is the velocity of the particle M. Since

$$-k^2 = -\omega^2 + k_x^2 + \mathbf{k}_\perp^2 \approx \omega^2(1 - v^2) + \mathbf{k}_\perp^2, \tag{99.7}$$

where $\mathbf{k}_\perp$ is the component of the vector $\mathbf{k}$ transverse to the x-axis, the condition

† The method given below is due to C. F. von Weizsäcker and E. J. Williams (1934); the basic idea had been stated earlier by E. Fermi (1924).

‡ When the wave functions are normalized to one particle in unit volume, the current $j^\mu = (1, \mathbf{v})$, where $\mathbf{v}$ is the velocity. We have, however, decided (§64) to omit the normalization factor $1/\sqrt{(2\varepsilon)}$ in the wave functions. Accordingly, j^μ must include a further factor 2ε, and this gives the expression (99.5).

(99.4) is equivalent to the inequality $|\mathbf{k}_\perp| \ll m$ and to a considerably weaker inequality for ω: $\omega \ll m/\sqrt{(1-v^2)}$.

From the condition for the current J to be transverse ($Jk = 0$) we have, using (99.6),

$$J_0 = \frac{J_x}{v} + \frac{\mathbf{J}_\perp \cdot \mathbf{k}_\perp}{\omega}.$$

We therefore obtain for the scalar product jJ

$$jJ = 2(J_0\varepsilon - J_x p_x)$$
$$\approx 2\frac{\varepsilon}{\omega}\left(\mathbf{k}_\perp \cdot \mathbf{J}_\perp + \frac{\omega M^2}{\varepsilon^2} J_x\right). \quad (99.8)$$

The product Je in (99.2) can be expanded by taking the polarization 4-vector of the real photon in the three-dimensionally transverse gauge: $ek = -\mathbf{e}\cdot\mathbf{k} = 0$, whence $e_x \approx -\mathbf{e}_\perp \cdot \mathbf{k}_\perp/\omega$. Then

$$Je = -\mathbf{e}_\perp \cdot \left(\mathbf{J}_\perp - \frac{\mathbf{k}_\perp}{\omega} J_x\right). \quad (99.9)$$

The expressions (99.8) and (99.9) are proportional if the second terms in the parentheses are negligible. Since the current J pertains to the upper vertex of the diagram (99.1b), it does not depend on the direction of $\mathbf{p}$; hence J_x and $\mathbf{J}_\perp$ must be taken to be quantities of the same order. For the terms in question to be negligible, therefore, we must have $|\mathbf{k}_\perp| \ll \omega$ and $\omega \ll \varepsilon^2|\mathbf{k}_\perp|/M^2$; these conditions are compatible with the previous ones on $\mathbf{k}_\perp$ and ω.

Assuming that in (99.9) the photon is polarized in the plane of x and $\mathbf{k}$ (so that $\mathbf{e}_\perp \| \mathbf{k}_\perp$) and noting that the conditions stated imply that $\mathbf{e}_\perp^2 \approx \mathbf{e}^2 = 1$, we now have

$$M_{fi} = M_{fi}^{(r)} \frac{Ze\sqrt{(4\pi)}}{-k^2} \frac{2\varepsilon}{\omega} |\mathbf{k}_\perp|. \quad (99.10)$$

In accordance with the previous discussion, the following conditions are here assumed satisfied:

$$|\mathbf{k}_\perp| \ll \omega \ll m\gamma, \quad (99.11)$$

$$\omega/\gamma^2 \ll |\mathbf{k}_\perp| \ll m, \quad (99.12)$$

with the notation

$$\gamma = \varepsilon/M = 1/\sqrt{(1-v^2)}.$$

From this we can find the relation between the corresponding cross-sections. According to the general formula (64.18) we have (in the rest frame of the particle m)

$$d\sigma_r = |M_{fi}^{(r)}|^2 (2\pi)^4 \delta^{(4)}(P_f - P_i) \frac{1}{4m\omega}\, d\rho_Q,$$

$$d\sigma = |M_{fi}|^2 (2\pi)^4 \delta^{(4)}(P_f - P_i) \frac{1}{4m\varepsilon} \frac{d^3p'}{2\varepsilon(2\pi)^3}\, d\rho_Q,$$

where $d\rho_Q$ represents the statistical weights of the particles Q. Using (99.10) and (99.7), we find

$$d\sigma = d\sigma_r \cdot n(\mathbf{k}) d^3p', \tag{99.13}$$

where

$$n(\mathbf{k}) = \frac{Z^2e^2}{\pi^2} \frac{\mathbf{k}_\perp^2}{\omega(\mathbf{k}_\perp^2 + \omega^2/\gamma^2)^2}. \tag{99.14}$$

Here $d\sigma_r$ is the cross-section for process (a), resulting from a collision between a real photon and a particle at rest, in which a system of particles Q is formed which have momenta in certain ranges; $d\sigma$ refers to the process (b) of formation of the same system Q when a fast particle (of mass M) collides with the same particle at rest, loses momentum $\mathbf{p} - \mathbf{p}' = \mathbf{k}$, and remains in the range d^3p' of values of $\mathbf{p}'$. The factor $n(\mathbf{k})$ in (99.13) may be interpreted as the number density (in $\mathbf{k}$-space) of the photons equivalent to the electromagnetic field of the fast particle.

The integration over d^3p' is equivalent to one over $d^3k = d\omega\, d^2k_\perp$. On integrating over $d^2k_\perp$, we obtain the cross-section for a process in which the total energy E of the system of particles Q lies in a given range $dE = d\omega$ ($E - m = \varepsilon - \varepsilon' = \omega$, where ε and ε' are the initial and final energies of the particle M). Integration over the directions of $\mathbf{k}_\perp$ signifies averaging over the directions of polarization of the incident photon (and multiplying by 2π). The result is

$$d\sigma = n(\omega)\, d\sigma_r\, d\omega, \tag{99.15}$$

where

$$n(\omega) = \int n(\mathbf{k}) 2\pi k_\perp\, dk_\perp$$

$$= \frac{2Z^2e^2}{\pi\omega} \int \frac{k_\perp^3\, dk_\perp}{(k_\perp^2 + \omega^2/\gamma^2)^2}.$$

The integral over $dk_\perp$ diverges when $k_\perp$ is large, but the divergence is only logarithmic. This enables us (within the range of validity of the method) to obtain a result in the logarithmic approximation: it is assumed that not only the argument of the logarithm but the logarithm itself is large. To this accuracy, it is sufficient to take as the upper limit of integration $k_{\perp\max} \sim m$, the upper limit of the inequality (99.12). Integration then gives for the frequency distribution of equivalent photons

(in ordinary units)

$$n(\omega)\, d\omega = \frac{2}{\pi} Z\alpha \log \frac{\gamma mc^2}{\hbar\omega} \frac{d\omega}{\omega}. \tag{99.16}$$

The approximation used here signifies that the numerical coefficient in the argument of the logarithm remains indeterminate. The inclusion of such a coefficient would mean the addition of a relatively small quantity (~1) to the large logarithm and would be superfluous having regard to the accuracy of the method.

PROBLEMS

PROBLEM 1. From the photon–electron scattering cross-section, find the bremsstrahlung cross-section in a collision between a fast electron and a nucleus.

SOLUTION. In the frame of reference K_1 in which the electron is at rest before the collision, the process may be regarded as the scattering by the electron of the equivalent photons of the field of the nucleus.† According to (86.10) the cross-section for scattering of a photon by an electron in the frame K_1 is

$$d\sigma_{sc}(\omega_1, \omega_1') = \pi r_e^2 \frac{m\, d\omega_1'}{\omega_1^2}\left[\frac{\omega_1}{\omega_1'} + \frac{\omega_1'}{\omega_1} + \left(\frac{m}{\omega_1'} - \frac{m}{\omega_1}\right)^2 - 2m\left(\frac{1}{\omega_1'} - \frac{1}{\omega_1}\right)\right], \tag{1}$$

where ω_1 and ω_1' are the initial and final energies of the photon in this frame. The bremsstrahlung cross-section in the frame K_1 is

$$d\sigma_{br}(\omega_1') = \int d\omega_1 \cdot n(\omega_1)\, d\sigma_{sc}(\omega_1, \omega_1'), \tag{2}$$

where $n(\omega_1)$ is the function (99.16). Since the cross-section is invariant, the change to a frame K in which the nucleus is at rest involves only a change in the frequency ω_1'. The frequencies ω_1' and ω' in the frames K_1 and K are related by the Doppler formula

$$\omega' = \gamma\omega_1'(1 - v\cos\theta_1), \qquad \gamma = 1/\sqrt{(1 - v^2)}, \tag{3}$$

where θ_1 is the scattering angle in the frame K_1. The same angle relates ω_1' and ω_1 according to (86.8):

$$\frac{1}{\omega_1'} - \frac{1}{\omega_1} = \frac{1}{m}(1 - \cos\theta_1). \tag{4}$$

From (3) and (4) we have

$$\omega_1' = \omega_1 \varepsilon'/\varepsilon, \tag{5}$$

where ε $(= m\gamma)$ and ε' are the initial and final energies of the electron in the frame K $(\varepsilon - \varepsilon' = \omega')$. Substituting (5) in (1), we find

$$d\sigma_{sc} = \pi r_e^2 \frac{m\, d\omega'}{\varepsilon\omega_1}\left(\frac{\varepsilon'}{\varepsilon} + \frac{\varepsilon}{\varepsilon'} + \frac{m^2\omega'^2}{\varepsilon'^2\omega_1^2} - \frac{2m\omega'}{\omega_1\varepsilon'}\right).$$

This expression is to be substituted in (2) and the integration over $d\omega_1$ carried out with ω' (i.e. ε') fixed, the range being from $\omega_{1,\min} = m\omega'/2\varepsilon'$ to $\omega_{1,\max} = 2\varepsilon\omega'/m$; these values are given by (3) and (4) with $\theta_1 = 0$ and $\theta_1 = \pi$. Because the integral converges rapidly for large ω_1, the main contribution to it comes

† The scattering of virtual photons by the nucleus (in the rest frame of the nucleus) is excluded by the large mass of the latter: the scattering cross-section tends to zero with increasing mass of the scattering particle.

from the range of ω_1 near the lower limit, i.e. we may put $\omega_{1,\max}\to\infty$. Calculating the integral with logarithmic accuracy†, we have

$$d\sigma_{\rm br} = 4r_e^2\alpha Z \frac{d\omega'}{\omega'}\frac{\varepsilon'}{\varepsilon}\left(\frac{\varepsilon}{\varepsilon'}+\frac{\varepsilon'}{\varepsilon}-\frac{2}{3}\right)\log\frac{\varepsilon\varepsilon'}{m\omega'}.$$

For this result to be valid, besides the condition $\varepsilon \gg m$ (ultra-relativistic electron), the condition (99.11) must also be satisfied: the frequencies $\omega_1 \sim \omega_{1,\min}$ important in the integration must be $\ll\varepsilon$. Hence $\varepsilon-\varepsilon' = \omega' \ll \varepsilon\varepsilon'/m$. Under these conditions the result agrees (to logarithmic accuracy) with (93.17), as it should.

PROBLEM 2. The same as Problem 1, but for electron–electron bremsstrahlung.

SOLUTION. In this case, the virtual photon can be scattered either by the fast electron or by the recoil electron; the photons equivalent to the field of either electron are scattered by the other. The scattering of virtual photons by the fast electron gives the cross-section $d\sigma_{\rm br}^{(1)}$, which is equal to the cross-section for an electron and a nucleus with $Z = 1$.

The scattering of virtual photons by the recoil electron gives a cross-section

$$d\sigma_{\rm br}^{(2)} = \int d\omega\cdot n(\omega)\, d\sigma_{\rm sc}(\omega,\omega'),$$

with $d\sigma_{\rm sc}(\omega,\omega')$ given by (1) with the appropriate change of notation for the frequencies. The range of values of ω for given ω' is (cf. (4))

$$\omega' \leqslant \omega \leqslant \infty \qquad \text{for } \omega' > \tfrac{1}{2}m,$$

$$\omega' \leqslant \omega \leqslant \omega'/(m-2\omega') \qquad \text{for } \omega' < \tfrac{1}{2}m.$$

When $\omega' < \frac{1}{2}m$, integration with respect to ω gives

$$d\sigma_{\rm br}^{(2)} = \tfrac{16}{3}\alpha r_e^2 \frac{d\omega'}{\omega'}\left(1-\frac{\omega'}{m}+\frac{\omega'^2}{m^2}\right)\log\frac{\varepsilon}{\omega'},$$

in agreement with (97.4). But when $\omega' > \frac{1}{2}m$ we must distinguish the cases $\omega' \sim m$ and $\omega' \sim \varepsilon \gg m$. In the former case,

$$d\sigma_{\rm br}^{(2)} = \tfrac{2}{3}\alpha r_e^2 \frac{m\, d\omega'}{\omega'^2}\left(4-\frac{m}{\omega'}+\frac{m^2}{4\omega'^2}\right)\log\frac{\varepsilon}{m},$$

in agreement with (97.3); in the argument of the logarithm we have, with sufficient accuracy, replaced ε/ω' by ε/m. In the case $\omega' \sim \varepsilon$, the method of equivalent photons is not valid for calculating $d\sigma_{\rm br}^{(2)}$. The frequency ω of the virtual photons takes values beginning with ω', and the condition (99.11) is therefore not satisfied when $\omega = \omega' \sim \varepsilon$.

PROBLEM 3. Determine the total pair production cross-section in a photon–nucleus collision from the pair production cross-section in a collision between two photons.

SOLUTION. The energy of the photon in the rest frame K of the nucleus is $\omega \gg m$. If we change to a frame K_0 in which the nucleus moves to meet the photon at a speed v_0 such that $1/\surd(1-v_0^2) = \frac{1}{2}\omega/m$, then in this frame the photon energy is

$$\omega_0 = \omega\frac{1-v_0}{\surd(1-v_0^2)} \approx \tfrac{1}{2}\omega\surd(1-v_0^2) = m.$$

The required cross-section σ is calculated in the frame K_0 as the pair production cross-section in collisions between an incident photon ω_0 and the equivalent photons of the nucleus, whose energy we denote by ω':

$$\sigma = \int \sigma_{\gamma\gamma} n(\omega')\, d\omega',$$

† This means that, by one integration by parts, the term containing the large logarithm is separated and the remaining terms then neglected. This operation is equivalent to taking the logarithm $\log(\varepsilon/\omega_1)$ outside the integral, with $\omega_1 = \omega_{1,\min}$.

where $\sigma_{\gamma\gamma}$ is the cross-section for pair production by two photons and is given by §§88, Problem, formula (1), with

$$v = \sqrt{(1 - m^2/\omega_0\omega')} = \sqrt{(1 - m/\omega')}.$$

Changing to the variable v instead of ω', we have

$$\sigma = 2r_e^2\alpha Z \int_0^1 v \log[\omega(1-v^2)/m]\left\{(3-v^4)\log\frac{1+v}{1-v} - 2v(2-v^2)\right\} dv.$$

Because of the convergence at the upper limit, the integral may be taken over the whole range from the reaction threshold $\omega' = m$ $(v = 0)$ to $\omega' = \infty$ $(v = 1)$ and with logarithmic accuracy (replacing $\log[\omega(1-v^2)/m]$ by its value for $v = 0$ and taking it outside the integral). The result is

$$\sigma = \tfrac{28}{9}\alpha Z^2 r_e^2 \log(\omega/m),$$

in agreement with (94.6); this formula is valid when $\log(\omega/m) \gg 1$.

§ 100. **Pair production in collisions between particles**

Electron–positron pair production in a collision between two charged particles is described by diagrams of two types:

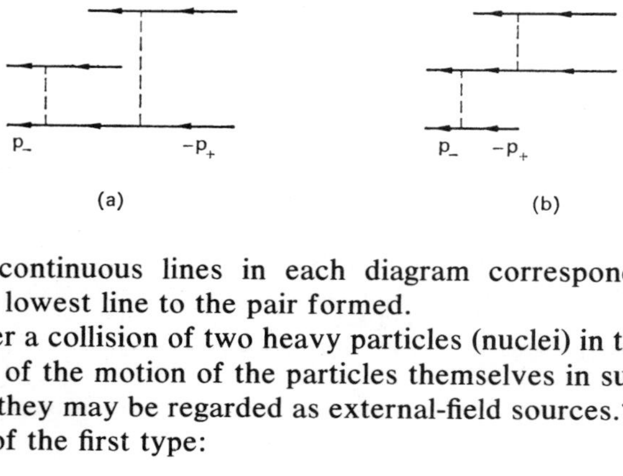

(100.1)

The two upper continuous lines in each diagram correspond to the colliding particles, and the lowest line to the pair formed.

Let us consider a collision of two heavy particles (nuclei) in the ultra-relativistic case. The change of the motion of the particles themselves in such a collision may be neglected, i.e. they may be regarded as external-field sources.† This corresponds to two diagrams of the first type:

(100.2)

where $q^{(1)}$, $q^{(2)}$ are the "momenta" of the Fourier components of the fields of the two particles.

The potential $A^\mu = (A_0, \mathbf{A})$ due to a classical particle moving with a uniform

† The collision of two light particles (electrons), where the change in the motion cannot be neglected, is a considerably more complicated case; see the book by Baĭer *et al.* cited in §93.

velocity **v** satisfies the equations

$$\Box A_0 = -4\pi Ze\delta(\mathbf{r} - \mathbf{v}t - \mathbf{r}_0),$$

$$\Box \mathbf{A} = -4\pi Ze\mathbf{v}\delta(\mathbf{r} - \mathbf{v}t - \mathbf{r}_0).$$

Its Fourier components are

$$A_0(\omega, \mathbf{k}) = -\frac{8\pi^2 Ze}{\omega^2 - \mathbf{k}^2} e^{-i\mathbf{k}\cdot\mathbf{r}_0}\,\delta(\omega - \mathbf{k}\cdot\mathbf{v}),$$

and similarly for $\mathbf{A}(\omega, \mathbf{k})$. In four-dimensional form,

$$A^\mu(q) = -\frac{8\pi^2 Ze}{q^2} e^{iqx_0} U^\mu \delta(Uq),$$

where U is the 4-velocity of the particle, and the 4-vector $x_0 = (0, \mathbf{r}_0)$. If nucleus 1 is at rest at the origin ($\mathbf{r}_0^{(1)} = 0$), then $\boldsymbol{\rho} \equiv \mathbf{r}_0^{(2)}$ is the impact parameter vector (in a plane perpendicular to the direction of motion of nucleus 2). This expression for $A^\mu(q)$ is to be used in writing analytically the diagrams (100.2).

There is, however, no need to use this method for the actual calculations in the present case. The pair production cross-section may be determined by the method of equivalent photons, using the already known photon–nucleus pair production cross-section. The replacement of the field of one particle (the first, say) by a spectrum of equivalent photons implies that in the diagrams (100.2) the lines $q^{(1)}$ are regarded as real-photon lines. The two diagrams then become identical with the diagrams corresponding to pair production by a photon at nucleus 2. When ε_+, $\varepsilon_- \gg m$, the cross-section for the latter process is given by (94.5). Multiplying this cross-section by the spectrum (99.16) of equivalent photons of the first nucleus, we obtain (with logarithmic accuracy) for the differential cross-section for pair production in a collision between particles

$$d\sigma = \frac{8}{\pi} r_e^2 (Z_1 Z_2 \alpha)^2 \frac{d\varepsilon_+\, d\varepsilon_-}{(\varepsilon_+ + \varepsilon_-)^4} (\varepsilon_+^2 + \varepsilon_-^2 + \tfrac{2}{3}\varepsilon_+\varepsilon_-) \log \frac{\varepsilon_+\varepsilon_-}{m(\varepsilon_+ + \varepsilon_-)} \log \frac{m\gamma}{\varepsilon_+ + \varepsilon_-}, \tag{100.3}$$

where $\gamma = 1/\sqrt{(1 - v^2)} \gg 1$.

Here it is assumed that

$$m \ll \varepsilon_+, \varepsilon_- \ll m\gamma; \tag{100.4}$$

the right-hand inequality is the condition for the method of equivalent photons to be applicable. The range defined by the inequalities (100.4) is the same as the electron and positron energy range which is important in the integration of (100.3). On integration over $d\varepsilon_+$ or $d\varepsilon_-$ for a given sum $\varepsilon \equiv \varepsilon_+ + \varepsilon_-$ ($\gg m$), the important range is the one near the upper limit; omitting terms which do not contain the large logarithm, we find

$$d\sigma = \frac{56}{9\pi} r_e^2 (Z_1 Z_2 \alpha)^2 \log\frac{\varepsilon}{m} \log\frac{m\gamma}{\varepsilon}\frac{d\varepsilon}{\varepsilon}.$$

The integral with respect to ε over the range (100.4) diverges as the cube of the logarithm, but only as the square of the logarithm at the boundaries of the range. In the logarithmic approximation ($\log\gamma \gg 1$), therefore, the range (100.4) is in fact the most important one, and the integral can be taken over the range from m to $m\gamma$. Since

$$\int_1^\gamma \log\xi(\log\gamma - \log\xi)\frac{d\xi}{\xi} = \tfrac{1}{6}\log^3\gamma,$$

the total pair production cross-section is

$$\sigma = \frac{28}{27\pi} r_e^2 (Z_1 Z_2 \alpha)^2 \log^3 \frac{1}{\sqrt{(1-v^2)}} \qquad (100.5)$$

(L. D. Landau and E. M. Lifshitz, 1934).

Let us now consider the case of non-relativistic velocities of the colliding nuclei. The change in their motion due to their interaction then becomes important, and the main contribution to the pair production cross-section comes from diagrams of the second type in (100.1). There are four such diagrams: two of them are

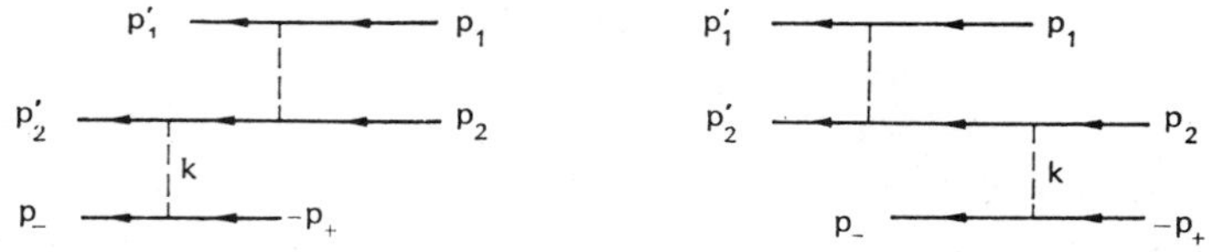

(100.6)

and the other two are similar except that the virtual photon k (which produces the pair) is emitted by the first nucleus and not by the second.†

We shall suppose that the energy of the pair is small compared with the kinetic energy of the relative motion of the nuclei in their centre-of-mass system:

$$\varepsilon_+ + \varepsilon_- \ll \tfrac{1}{2}Mv^2, \qquad (100.7)$$

where $\mathbf{v}$ is the initial relative velocity and $M = M_1 M_2/(M_1 + M_2)$ is the reduced mass of the nuclei. Then the reciprocal effect of pair production on the motion of the nuclei can be neglected. If the electron–positron line in the diagrams (100.6) is omitted, the remainder will represent the emission by the colliding particles of a

† Altogether 36 diagrams correspond to pair production in a collision between two electrons: $2!\times 3! = 12$ diagrams of type a, differing by interchanges of the two initial and three final electrons, and $2\times 2!\times 3! = 24$ diagrams of type b, obtained in a similar way from the two diagrams (100.6).

low-frequency virtual photon ($\omega = \varepsilon_+ + \varepsilon_-$). Thus we return to the situation discussed in §98 for the emission of a real soft photon, and can use the formula (98.13) derived there for the non-relativistic case (except that the amplitude $\sqrt{(4\pi)}e^*$ of the real photon will be replaced by the virtual photon propagator).† Thus the amplitude of the whole process of pair production becomes

$$M_{fi} = M_{fi}^{(\mathrm{el})} \frac{1}{\omega}\left(\frac{Z_1 e}{M_1} - \frac{Z_2 e}{M_2}\right) q^\lambda D_{\lambda\mu}(k)[-ie(\bar{u}_-\gamma^\mu u_+)], \tag{100.8}$$

where $q = (0, \mathbf{q})$, $\mathbf{q} = M(\mathbf{v}' - \mathbf{v})$.

As usual, the photon propagator in the non-relativistic case is to be taken in the gauge (76.14). From the amplitude (100.8) we find the cross-section for the process:

$$d\sigma = d\sigma_{\mathrm{el}} \cdot e^4 \left(\frac{Z_1}{M_1} - \frac{Z_2}{M_2}\right)^2 \frac{d^3p_+\, d^3p_-}{2\varepsilon_+ \cdot 2\varepsilon_-(2\pi)^6 \omega^2(\omega^2 - \mathbf{k}^2)^2} (4\pi)^2 |\bar{u}_- \boldsymbol{\gamma} \cdot \mathbf{Q} u_+|^2, \tag{100.9}$$

where

$$\omega = \varepsilon_+ + \varepsilon_-, \qquad \mathbf{k} = \mathbf{p}_+ + \mathbf{p}_-, \qquad \mathbf{Q} = \mathbf{q} - \frac{1}{\omega^2}\mathbf{k}(\mathbf{q}\cdot\mathbf{k});$$

$d\sigma_{\mathrm{el}}$ is the cross-section for elastic scattering of one nucleus by the other, in their centre-of-mass system, and is given by Rutherford's formula:‡

$$\begin{aligned} d\sigma_{\mathrm{el}} &= 4(Z_1 Z_2 e^2)^2 M^2\, do/\mathbf{q}^4 \\ &\approx 4(Z_1 Z_2 e^2)^2 \frac{dq_y\, dq_z}{v^2 \mathbf{q}^4}; \end{aligned} \tag{100.10}$$

the last equation assumes that the deviation of the nuclei from their original direction of motion (the x-axis) is small. Substituting this expression in (100.9) and summing over polarizations of the pair in the usual manner, we obtain

$$\begin{aligned} d\sigma = {} & (Z_1 Z_2 e^2)^2 \frac{e^4}{v^2}\left(\frac{Z_1}{M_1} - \frac{Z_2}{M_2}\right)^2 \times \\ & \times \operatorname{tr}\{(\gamma p_- + m)(\boldsymbol{\gamma}\cdot\mathbf{Q})(\gamma p_+ - m)(\boldsymbol{\gamma}\cdot\mathbf{Q})\} \frac{d^3p_+\, d^3p_-\, dq_y\, dq_z}{4\pi^4 \varepsilon_+\varepsilon_- \mathbf{q}^4(\omega^2 - \mathbf{k}^2)^2\omega^2}. \end{aligned} \tag{100.11}$$

The remaining calculation is made in the approximation in which all the

† In the non-relativistic case, the photon momentum is small in comparison with the change in momentum of the radiating particles ($|\delta\mathbf{p}| \sim \omega/v$), and can therefore be neglected, in comparison with $\delta\mathbf{p}$, even when the photon energy is not neglected. This applies *a fortiori* here to the virtual photon, for which the four-dimensional square $k^2 = (p_+ + p_-)^2 > 0$, so that $|\mathbf{k}| < \omega$. Under these conditions there is no difference between real and virtual photons, and the use of formula (98.13) is thereby justified.

‡ The diagrams (100.6) are shown on the assumption of the Born approximation for scattering of nuclei. But, since Rutherford's formula is exact (for Coulomb interaction), the validity of the results obtained does not in fact depend on the fulfilment of the condition for the Born approximation to be valid.

logarithms occurring in the integration are assumed large. We shall see that, to this accuracy, pair energies ε_+, $\varepsilon_- \gg m$ and angles θ between $\mathbf{p}_+$ and $\mathbf{p}_-$ such that

$$m/\varepsilon \ll \theta \ll 1 \tag{100.12}$$

are the most important. With the appropriate approximations, the calculation of the trace in (100.11) gives

$$\mathrm{tr}\{\ldots\} = 4\Big[(\varepsilon_+\varepsilon_- - \mathbf{p}_+\cdot\mathbf{p}_-)\Big(\mathbf{q}^2 - \frac{(\mathbf{q}\cdot\mathbf{k})^2}{\omega^2}\Big) +$$

$$+ 2(\mathbf{p}_+\cdot\mathbf{q})(\mathbf{p}_-\cdot\mathbf{q}) + \frac{2\varepsilon_+\varepsilon_-}{\omega^2}(\mathbf{q}\cdot\mathbf{k})^2 - \frac{2\mathbf{q}\cdot\mathbf{k}}{\omega}(\varepsilon_+\mathbf{q}\cdot\mathbf{p}_- + \varepsilon_-\mathbf{q}\cdot\mathbf{p}_+)\Big],$$

where we can also put $|\mathbf{p}_+| = \varepsilon_+$, $|\mathbf{p}_-| = \varepsilon_-$. In the denominator,

$$\omega^2 - \mathbf{k}^2 \approx \varepsilon_+\varepsilon_-\theta^2 + m^2\frac{(\varepsilon_+ + \varepsilon_-)^2}{\varepsilon_+\varepsilon_-}.$$

Integration over the directions of $\mathbf{p}_+$ and $\mathbf{p}_-$, for a given angle between them, gives

$$d\sigma = \frac{8}{3\pi^2}(Z_1Z_2e^2)^2\frac{e^4}{v^2}\Big(\frac{Z_1}{M_1} - \frac{Z_2}{M_2}\Big)^2(\varepsilon_+^2 + \varepsilon_-^2)\,d\varepsilon_+\,d\varepsilon_- \times$$

$$\times \frac{\theta^3\,d\theta}{[\theta^2 + m^2(\varepsilon_+ + \varepsilon_-)^2/\varepsilon_+^2\varepsilon_-^2]^2}\frac{dq_y\,dq_z}{\mathbf{q}^2}. \tag{100.13}$$

The form of the dependence on θ confirms the hypothesis (100.12), and integration with respect to θ gives $\log[\varepsilon_+\varepsilon_-/m(\varepsilon_+ + \varepsilon_-)]$. Integration of the last factor in (100.13) is from $q_y = q_z = 0$ to $\sqrt{(q_y^2 + q_z^2)} \sim 1/R$, where R is a quantity of the order of the radius of the nuclei (corresponding to the smallest impact parameters; see below). This integration gives

$$[\pi \log(q_x^2 + q_y^2 + q_z^2)]_{q_y=q_z=0}^{q_y=q_z=1/R} \approx 2\pi \log\frac{1}{Rq_x}$$

The total energy of the pair, equal to the change in the energy of the nuclei, is

$$\varepsilon \equiv (\varepsilon_+ + \varepsilon_-) = \tfrac{1}{2}M(\mathbf{v}'^2 - \mathbf{v}^2) \approx Mv(v_x' - v_x) = vq_x,$$

whence $q_x = \varepsilon/v$. Thus we find

$$d\sigma = \frac{16}{3\pi}(Z_1Z_2e^2)^2\frac{e^4m^2}{v^2}\Big(\frac{Z_2}{M_2} - \frac{Z_1}{M_1}\Big)^2\frac{\varepsilon_+^2 + \varepsilon_-^2}{\varepsilon^4}\log\frac{v}{R\varepsilon}\log\frac{\varepsilon_+\varepsilon_-}{m\varepsilon}\,d\varepsilon_+\,d\varepsilon_-,$$

and, after integration over $d\varepsilon_+$ or $d\varepsilon_-$ with a given sum ε,

$$d\sigma = \frac{32}{9\pi}(Z_1Z_2e^2)^2\frac{e^4m^2}{v^2}\left(\frac{Z_2}{M_2}-\frac{Z_1}{M_1}\right)^2\log\frac{v}{R\varepsilon}\log\frac{\varepsilon}{m}\frac{d\varepsilon}{\varepsilon}. \tag{100.14}$$

The energy ε may be correlated with the impact parameter $\rho \sim v/\varepsilon$; the pair energy is of the order of the frequency which corresponds to the collision time. Hence the logarithmic divergence on integration over $d\varepsilon$ in (100.14) implies a similar divergence with respect to impact parameters. This means that large values of ρ are important (and this, incidentally, justifies the use of the cross-section (100.10) for scattering in the purely Coulomb field of the nucleus). Accordingly, the important range of energy is given by $m \ll \varepsilon \ll v/R$. Integration of (100.14) gives the total pair production cross-section; the final result is (in ordinary units)

$$\sigma = \frac{16}{27\pi}(Z_1Z_2\alpha)^2r_e^2\left(\frac{c}{v}\right)^2\left(\frac{Z_2m}{M_2}-\frac{Z_1m}{M_1}\right)^2\log^3\frac{\hbar v}{mc^2R} \tag{100.15}$$

(E. M. Lifshitz, 1935).†

§ 101. Emission of a photon by an electron in the field of a strong electromagnetic wave

The application of perturbation theory to processes of interaction between an electron and a radiation field requires not only that the interaction constant α should be small but also that the field should be sufficiently weak. If a is the amplitude of the classical 4-potential of an electromagnetic wave field, the characteristic quantity in this respect is the dimensionless invariant ratio

$$\xi = e\sqrt{(-a^2)}/m. \tag{101.1}$$

In this section we shall consider emission processes occurring in the interaction of an electron with a field of a strong electromagnetic wave, for which ξ can have any value. The method used is based on an exact treatment of this interaction; the interaction of the electron with the newly emitted photons can, as before, be regarded as a small perturbation (A. I. Nikishov and V. I. Ritus, 1964).

Let us consider a monochromatic plane wave, say a circularly polarized one. Its 4-potential may be written in the form

$$A = a_1\cos\phi + a_2\sin\phi,\ \phi = kx, \tag{101.2}$$

where $k^\mu = (\omega, \mathbf{k})$ is the wave 4-vector ($k^2 = 0$), and the 4-amplitudes a_1 and a_2 are equal in magnitude and orthogonal:

$$a_1^2 = a_2^2 \equiv a^2, \qquad a_1a_2 = 0.$$

† A numerical error was corrected by L. B. Okun' (1953).

We shall assume that the Lorentz gauge condition is applied to the potential, so that $a_1k = a_2k = 0$.

The exact wave function for an electron in the field of an arbitrary plane electromagnetic wave has been derived in §40, and is given by formulae (40.7) and (40.8). We shall, however, change the normalization by making ψ_p correspond to unit mean spatial number density of particles, in the same way as the wave functions of free particles are normalized to "one particle in unit volume". Since the mean density for the function (40.7) is $\bar{j}_0 = q_0/p_0$, in order to obtain the required normalization this function must be multiplied by $\sqrt{(p_0/q_0)}$, i.e. the factor $1/\sqrt{(2p_0)}$ in (40.7) must be replaced by $1/\sqrt{(2q_0)}$. For a wave with the 4-potential (101.2), we find

$$\psi_p = \left[1 + \frac{e}{2(kp)}\{(\gamma k)(\gamma a_1)\cos\phi + (\gamma k)(\gamma a_2)\sin\phi\}\right]\frac{u(p)}{\sqrt{(2q_0)}}\times$$

$$\times\exp\left\{-ie\frac{(a_1p)}{(kp)}\sin\phi + ie\frac{(a_2p)}{(kp)}\cos\phi - iqx\right\}, \tag{101.3}$$

where

$$q^\mu = p^\mu - e^2\frac{a^2}{2(kp)}k^\mu. \tag{101.4}$$

According to (40.14), the 4-vector q is the mean 4-momentum of the electron; we shall call it the *quasi-momentum*.

The S-matrix element for a transition of the electron from the state ψ_p to the state $\psi_{p'}$ with emission of a photon having 4-momentum $k^{\mu\prime} = (\omega', \mathbf{k}')$ and polarization 4-vector e' is

$$S_{fi} = -ie\int \bar{\psi}_{p'}(\gamma e'^*)\psi_p\frac{e^{ik'x}}{\sqrt{(2\omega')}}d^4x. \tag{101.5}$$

The integrand in (101.5) is a linear combination of the quantities

$$e^{-i\alpha_1\sin\phi + i\alpha_2\cos\phi},$$

$$\cos\phi\cdot e^{-i\alpha_1\sin\phi + i\alpha_2\cos\phi},$$

$$\sin\phi\cdot e^{-i\alpha_1\sin\phi + i\alpha_2\cos\phi},$$

where

$$\alpha_1 = e\left(\frac{a_1p}{kp} - \frac{a_1p'}{kp'}\right), \qquad \alpha_2 = e\left(\frac{a_2p}{kp} - \frac{a_2p'}{kp'}\right). \tag{101.6}$$

These quantities, together with the factor $\exp[i(k'+p'-p)x]$, give the whole dependence of the integrand on x. We expand them in Fourier series, denoting the expansion coefficients by B_s, B_{1s}, B_{2s} respectively; for example,

$$e^{-i\alpha_1 \sin\phi + i\alpha_2 \cos\phi} = e^{-i\sqrt{(\alpha_1^2+\alpha_2^2)}\sin(\phi-\phi_0)}$$

$$= \sum_{s=-\infty}^{\infty} B_s\, e^{-is\phi}.$$

These coefficients can be expressed in terms of Bessel functions by the formulae

$$\left.\begin{aligned} B_s &= J_s(z)\, e^{is\phi_0}, \\ B_{1s} &= \tfrac{1}{2}[J_{s+1}(z)\, e^{i(s+1)\phi_0} + J_{s-1}(z)\, e^{i(s-1)\phi_0}], \\ B_{2s} &= \frac{1}{2i}[J_{s+1}(z)\, e^{i(s+1)\phi_0} - J_{s-1}(z)\, e^{i(s-1)\phi_0}], \end{aligned}\right\} \tag{101.7}$$

where

$$z = \sqrt{(\alpha_1^2 + \alpha_2^2)}, \qquad \cos\phi_0 = \alpha_1/z, \qquad \sin\phi_0 = \alpha_2/z.$$

The functions B_s, B_{1s}, B_{2s} are related by

$$\alpha_1 B_{1s} + \alpha_2 B_{2s} = sB_s, \tag{101.8}$$

which follows from the familiar relation

$$J_{s-1}(z) + J_{s+1}(z) = 2sJ_s(z)/z$$

between the Bessel functions.

The matrix element (101.5) then becomes

$$S_{fi} = \frac{1}{(2\omega' \cdot 2q_0 \cdot 2q_0')^{1/2}} \sum_s M_{fi}^{(s)} (2\pi)^4 i\delta^{(4)}(sk + q - q' - k'); \tag{101.9}$$

we shall not give here the fairly complicated expressions for the amplitudes $M_{fi}^{(s)}$.

Thus S_{fi} is an infinite sum of terms, each corresponding to a conservation law

$$sk + q = q' + k'. \tag{101.10}$$

Since

$$q^2 = q'^2 = m^2(1+\xi^2) \equiv m_*^2 \tag{101.11}$$

(cf. (40.15)), and $k^2 = k'^2 = 0$, the equation (101.10) can be satisfied only if $s \geqslant 1$. The sth term of the sum describes the emission of a photon k' by the absorption from the wave of s photons with 4-momenta k. The form (101.10) shows that all the kinematic relationships which occur for the Compton effect will apply to the processes considered here if the electron momenta are replaced by the quasi-momenta q and the incident photon momentum by the 4-vector sk. In particular, the frequency of the emitted photon in the frame of reference where the electron is at rest on average ($\mathbf{q}=0$, $q_0=m_*$) is

$$\omega' = \frac{s\omega}{1+(s\omega/m_*)(1-\cos\theta)}, \tag{101.12}$$

where θ is the angle between $\mathbf{k}$ and $\mathbf{k}'$; cf. (86.8). We may say that the frequencies ω' are harmonics of ω.

In the notation previously used (§64), the amplitude of the process of emission of the sth harmonic is $M_{fi}^{(s)}$, and the expression

$$dW_s = |M_{fi}^{(s)}|^2 \frac{d^3k'\, d^3q'}{(2\pi)^6\cdot 2\omega'\cdot 2q_0\cdot 2q_0'}(2\pi)^4\delta^{(4)}(sk+q-q'-k') \tag{101.13}$$

gives the corresponding differential probability per unit volume and unit time.†

The amplitudes $M_{fi}^{(s)}$ have a structure similar to that of the scattering amplitudes with plane waves, $\bar{u}(p')\ldots u(p)$; the operations of summation over polarizations of the particles are therefore carried out in the usual manner. After summation over the polarizations of the final electrons and the photon and averaging over the polarizations of the initial electron, we have

$$dW_s = \frac{e^2m^2}{4\pi}\frac{d^3k'\,d^3q'}{q_0q_0'\omega'}\delta^{(4)}(sk+q-q'-k')\times$$

$$\times\left\{-2J_s^2(z)+\xi^2\left(1+\frac{(kk')^2}{2(kp)(kp')}\right)(J_{s+1}^2+J_{s-1}^2-2J_s^2)\right\}. \tag{101.14}$$

In order to integrate this expression, we note that, owing to the axial symmetry of the field of a circularly polarized wave, the differential probability is independent of the azimuthal angle ϕ around the direction of $\mathbf{k}$. This fact, together with the presence of the delta function, enables us to integrate over all variables except one, which we take to be the invariant $u=(kk')/(kp')$. Then, after integration over $d^3k\,d\phi\,d(q_0'+\omega')$, we find

$$\delta^{(4)}(sk+q-q'-k')\frac{d^3q'\,d^3k'}{q_0'\omega'}\to\frac{2\pi\,du}{(1+u)^2}.$$

For, in the centre-of-mass system (in which $s\mathbf{k}+\mathbf{q}=\mathbf{q}+\mathbf{k}'=0$), this integration gives $2\pi|\mathbf{q}'|d\cos\theta/E_s$, where $E_s=s\omega+q_0=\omega'+q_0'$ and θ is the angle between $\mathbf{k}$ and $\mathbf{q}'$; cf. the transformation (64.12). In the same system, moreover,

$$u=\frac{E_s}{q_0'-|\mathbf{q}'|\cos\theta}-1,\qquad d\cos\theta=\frac{E_s\,du}{|\mathbf{q}'|(1+u)^2}.$$

† It should be noted that the normalization of the functions ψ_p to unit density corresponds to normalization by the delta function "on the $\mathbf{q}/2\pi$ scale"; cf. (40.17), where the factor q_0/p_0 on the right will now be absent. It is for this reason that the number of final states of the electron must be measured by the element $d^3q'/(2\pi)^3$.

The range $-1 \leqslant \cos\theta \leqslant 1$ corresponds to

$$0 \leqslant u \leqslant u_s \equiv E_s^2/m_*^2 - 1$$
$$= 2s(kp)/m_*^2;$$

in making the transformations it must be remembered that $kp = kq$.

Thus the total probability of emission from unit volume in unit time is

$$W = \sum_{s=1}^{\infty} W_s = \frac{e^2 m^2}{4q_0} \sum_{s=1}^{\infty} \int_0^{u_s} \frac{du}{(1+u)^2} \left\{ -4J_s^2(z) + \xi^2 \left(2 + \frac{u^2}{1+u}\right)(J_{s+1}^2 + J_{s-1}^2 - 2J_s^2)\right\}, \qquad (101.15)$$

where†

$$\left.\begin{aligned} u &= (kk')/(kp'), \qquad u_s = 2s(kp)/m_*^2, \\ z &= 2sm^2 \frac{\xi}{\sqrt{(1+\xi^2)}} \sqrt{\left[\frac{u}{u_s}\left(1 - \frac{u}{u_s}\right)\right]}. \end{aligned}\right\} \qquad (101.16)$$

When $\xi \ll 1$ (the condition for perturbation theory to be valid), the integrand in (101.15) can be expanded in powers of ξ. For example, the first term in the expansion of W_1 is

$$W_1 = \frac{e^2 m^2}{4p_0} \xi^2 \int_0^{u_1} \left[2 + \frac{u^2}{1+u} - 4\frac{u}{u_1}\left(1 - \frac{u}{u_1}\right)\right] du$$
$$= \frac{e^2 m^2}{4p_0} \xi^2 \left[\left(1 - \frac{4}{u_1} - \frac{8}{u_1^2}\right) \log(1+u_1) + \frac{1}{2} + \frac{8}{u_1} - \frac{1}{2(1+u_1)^2}\right], \qquad (101.17)$$

with $u_1 \approx 2(kp)/m^2$. This result agrees, as it should, with the Klein–Nishina formula for the scattering of a photon by an electron: putting in (101.17) $-a^2 = 4\pi/\omega$, $\xi^2 = 4\pi e^2/m^2\omega$, and dividing by the incident flux density (64.14), we return to (86.16) (the integrated scattering cross-section is independent of the initial polarization of the photon).‡

The expression for the probability of emission of the second harmonic (the first term in the expansion of W_2 for $\xi \ll 1$) is

† To calculate z, we first note that

$$z^2 = (a_1 Q)^2 + (a_2 Q)^2 = a^2 Q^2,$$

where $Q = q/(kq) - q'/(kq')$. This is easily shown by choosing a frame of reference in which $(a_1)_0 = (a_2)_0 = 0$ and the vectors $\mathbf{a}_1$, $\mathbf{a}_2$, $\mathbf{k}$ are along the axes x^1, x^2, x^3, and noting that $Q_0 = Q_3$ because $kQ = 0$.

‡ This value of a^2 corresponds to normalization of the 4-potential to "one particle in unit volume". To determine it, ω must be equated to the energy of a classical field with the (real) 4-potential (101.2).

$$W_2 = \frac{e^2m^2\xi^4}{p_0}\int_0^{u_2}\frac{du}{(1+u)^2}\frac{u}{u_2}\left(1-\frac{u}{u_2}\right)\left[2+\frac{u^2}{1+u}-4\frac{u}{u_2}\left(1-\frac{u}{u_2}\right)\right]$$

$$= \frac{e^2m^2\xi^4}{p_0}\left[\frac{1}{2}+\frac{1}{3u_1}-\frac{4}{u_1^2}-\frac{2}{u_1^3}-\frac{1}{2(1+2u_1)}-\right.$$

$$\left.-\left(\frac{1}{2u_1}-\frac{3}{2u_1^2}-\frac{3}{u_1^3}-\frac{1}{u_1^4}\right)\log(1+2u_1)\right]. \qquad (101.18)$$

The leading term in W_s for fairly small s is proportional to ξ^{2s}.

Let us now consider the opposite case ($\xi \gg 1$). The parameter ξ can be made large, for instance, by decreasing the frequency ω with a fixed field strength; evidently $\xi = eF/m\omega$, where F is the amplitude of the field strength. It is therefore clear that the case $\xi \gg 1$ essentially refers to processes in a constant and uniform field where **E** and **H** are orthogonal and equal in magnitude; this will be called a *crossed* field. The probability of emission in this field can be found by taking the limit $\xi \to \infty$, but it is simpler to assume a constant field in the calculations, taking the 4-potential in the form

$$A^\mu = a^\mu\phi, \qquad \phi = kx, \qquad ak = 0 \qquad (101.19)$$

(so that $F_{\mu\nu} = k_\mu a_\nu - k_\nu a_\mu =$ constant). The exact wave function of the electron in this field is obtained by substituting (101.19) in (40.7), (40.8):

$$\psi_p = \left[1+e\frac{(\gamma k)(\gamma a)}{2(kp)}\phi\right]\frac{u(p)}{\sqrt{(2p_0)}}\exp\left\{-ie\frac{(ap)}{2(kp)}\phi^2+ie^2\frac{a^2}{6(kp)}\phi^3-ipx\right\}. \qquad (101.20)$$

The result given by using this function is exact for emission by an electron with any energy in a crossed field. However, in the ultra-relativistic case this result (when put in the appropriate form; see below) applies to emission by an electron not only in a crossed field but in any constant and uniform electromagnetic field, including a constant magnetic field as discussed in §90.

To formulate this assertion we note that the state of a particle in any constant and uniform field is defined by as many quantum numbers as the state of a free particle, and these may always be so chosen as to become, when the field is removed, those of a free particle, i.e. its 4-momentum p^μ ($p^2 = m^2$). Thus the state of a particle in a constant field is described by a constant 4-vector p.

The total intensity of emission, being an invariant, depends only on the invariants which can be constructed from the constant 4-tensor $F_{\mu\nu}$ and the constant 4-vector p^μ. Since $F_{\mu\nu}$ can appear in the intensity only in combination with the charge e, we obtain three dimensionless invariants:

$$\left.\begin{aligned} \chi^2 &= -\frac{e^2}{m^6}(F_{\mu\nu}p^\nu)^2 = -\frac{e^2}{m^6}a^2(kp)^2, \\ f &= e^2(F_{\mu\nu})^2/m^4, \\ g &= \frac{e^2}{m^4}e_{\lambda\mu\nu\rho}F^{\lambda\mu}F^{\nu\rho}. \end{aligned}\right\} \qquad (101.21)$$

In a crossed field $f = g \equiv 0$, whereas in general all three invariants are non-zero. If the electron is ultra-relativistic ($p_0 \gg m$), however, and the vector **p** makes angles $\theta \gg m/p_0$ with the fields **E** and **H**, then $\chi^2 \gg f, g$ (that is, for an ultra-relativistic particle any constant field appears to be a crossed field for almost all directions **p**). If also the fields $|\mathbf{E}|, |\mathbf{H}| \ll m^2/e$ $(= m^2c^3/e\hbar)$, then $|f|, |g| \ll 1$.† Under these conditions the intensity calculated for a crossed field and expressed in terms of the invariant χ will apply also to the emission in any constant field.

The invariant χ is given in terms of the fields **E** and **H** by

$$\chi^2 = \frac{e^2}{m^6}\{(\mathbf{p} \times \mathbf{H} + p_0\mathbf{E})^2 - (\mathbf{p} \cdot \mathbf{E})^2\}.$$

For a constant magnetic field, χ is equal to the quantity (90.3), and the above arguments are therefore another means of deriving the results in §90.‡

† And p in χ may be regarded, with the same accuracy, as being the ordinary 4-momentum of the particle.

‡ A detailed account of the theory of various processes in strong fields is given in the review papers by A. I. Nikishov and V. I. Ritus in *Proceedings (Trudy) of the P. N. Lebedev Physics Institute*, Vol. **111**, pp. 5 and 152.

CHAPTER XI

EXACT PROPAGATORS AND VERTEX PARTS

§ 102. Field operators in the Heisenberg representation

HITHERTO, in considering various specific processes in electrodynamics, we have used only the first non-vanishing approximation of perturbation theory. We shall now go on to discuss the effects which occur in higher approximations. These are called *radiative corrections.*

A better understanding of the structure of the higher approximations can be obtained by first examining some general properties of exact scattering amplitudes (i.e. those which have not been expanded in powers of e^2). We have seen in §72 that the successive terms of the series in perturbation theory can be expressed in terms of the field operators in the interaction representation, whose time dependence is determined by the Hamiltonian $\hat{H}_0$ of a system of free particles. The exact scattering amplitudes, however, are more conveniently expressed in terms of the field operators in the Heisenberg representation, where the time dependence is determined by the exact Hamiltonian $\hat{H} = \hat{H}_0 + \hat{V}$ of a system of interacting particles.

The general rule for constructing the Heisenberg operators gives

$$\hat{\psi}(x) \equiv \hat{\psi}(t, \mathbf{r}) = e^{i\hat{H}t}\hat{\psi}(\mathbf{r})e^{-i\hat{H}t}, \tag{102.1}$$

and similarly for $\hat{\bar{\psi}}(x)$ and $\hat{A}(x)$, $\hat{\psi}(\mathbf{r})$, etc., being time-independent (Schrödinger) operators.† It may be noted immediately that the Heisenberg operators for a given time obey the same commutation rules as the operators in the Schrödinger representation or the interaction representation: for example,

$$\{\hat{\psi}_i(t, \mathbf{r})\hat{\bar{\psi}}_k(t, \mathbf{r}')\}_t = e^{i\hat{H}t}\{\hat{\psi}_i(\mathbf{r}), \hat{\bar{\psi}}_k(\mathbf{r}')\}_+ e^{-i\hat{H}t} = \gamma^0_{ik}\,\delta(\mathbf{r}-\mathbf{r}'); \tag{102.2}$$

cf. (75.6). Similarly, the operators $\hat{\psi}(t, \mathbf{r})$ and $\hat{A}(t, \mathbf{r}')$ commute:

$$\{\hat{\psi}_i(t, \mathbf{r}), \hat{A}(t, \mathbf{r}')\}_- = 0,$$

but this does *not* hold good for operators pertaining to different times.

The "equation of motion" satisfied by the Heisenberg ψ-operator can be derived from the general formula *QM*, (13.7):

$$-i\,\frac{\partial\hat{\psi}(x)}{\partial t} = \hat{H}\hat{\psi}(x) - \hat{\psi}(x)\hat{H}. \tag{102.3}$$

† In this chapter, operators with a time argument belong to the Heisenberg representation; those in the interaction representation will be given the suffix int.

The Schrödinger and Heisenberg representations are the same as regards the Hamiltonian, which is expressed in the same way in terms of the field operators. Here, to calculate the right-hand side of (102.3), we may omit from the Hamiltonian the part which depends only on the operator $\hat{A}(x)$ (the Hamiltonian of the free electromagnetic field), since this part commutes with $\hat{\psi}(x)$. According to (21.13) and (43.3),

$$\hat{H} = \int \hat{\psi}^*(t, \mathbf{r})(\boldsymbol{\alpha}\cdot\hat{\mathbf{p}} + \beta m)\hat{\psi}(t, \mathbf{r})\, d^3x + e\int \hat{\bar{\psi}}(t, \mathbf{r})(\gamma\hat{A}(t, \mathbf{r}))\hat{\psi}(t, \mathbf{r})\, d^3x$$

$$= \int \hat{\bar{\psi}}(t, \mathbf{r})\{\gamma\hat{p} + m + e(\gamma\hat{A}(t, \mathbf{r}))\}\hat{\psi}(t, \mathbf{r})\, d^3x. \tag{102.4}$$

When the commutator $\{\hat{H}, \hat{\psi}(t, \mathbf{r})\}_-$ is calculated from (102.2) and the delta function is eliminated by integration over d^3x, we get

$$(\gamma\hat{p} - e\gamma\hat{A} - m)\hat{\psi}(t, \mathbf{r}) = 0. \tag{102.5}$$

As we should expect, the operator $\hat{\psi}(t, \mathbf{r})$ satisfies an equation which is formally the same as Dirac's equation.

The equation for the electromagnetic field operator $\hat{A}(t, \mathbf{r})$ is obvious from the correlation with the classical case. When that case applies, i.e. when the occupation numbers are large (cf. §5), the operator equation must become the classical Maxwell's equation for the potentials, *Fields* (30.2), after averaging over the state of the field. It is therefore clear that the equation for the operator is simply the same as Maxwell's equation, so that we have (for an arbitrary gauge)

$$\partial^\nu\partial_\mu\hat{A}^\mu(x) - \partial^\mu\partial_\mu\hat{A}^\nu(x) = -4\pi e\hat{j}^\nu(x), \tag{102.6}$$

where $\hat{j}^\nu(x) = \hat{\bar{\psi}}(x)\,\gamma^\nu\,\hat{\psi}(x)$ is the current operator, satisfying identically the equation of continuity

$$\partial_\nu\hat{j}^\nu(x) = 0. \tag{102.7}$$

It is important to note that the equations (102.6) are linear in $\hat{A}^\mu$ and $\hat{j}^\mu$, and the question of the sequence of these operators therefore does not arise.

Like the similar equations for wave functions, the operator equations (102.6) and (102.7) are invariant under the gauge transformation

$$\left.\begin{aligned} \hat{A}_\mu(x) &\to \hat{A}_\mu(x) - \partial_\mu\hat{\chi}(x), \\ \hat{\psi}(x) &\to \hat{\psi}(x)e^{ie\hat{\chi}}, \\ \hat{\bar{\psi}}(x) &\to e^{-ie\hat{\chi}}\hat{\bar{\psi}}(x), \end{aligned}\right\} \tag{102.8}$$

where $\hat{\chi}(x)$ is any Hermitian operator which commutes (at a particular time) with $\hat{\psi}$.†

† This refers specifically to the Heisenberg ψ-operators. In the interaction representation, the gauge transformation of the electromagnetic potentials does not affect the ψ-operators.

Let us now ascertain the relationship between the operators in the Heisenberg representation and those in the interaction representation. To simplify the discussion, it is convenient to make the formal assumption (which will not affect the final result) that the interaction $\hat{V}(t)$ is adiabatically "switched on" from $t = -\infty$ to finite times. Then the Heisenberg and interaction representations are the same for $t \to -\infty$, and the wave functions of the system, Φ and Φ_{int}, are the same:

$$\Phi_{\text{int}}(t = -\infty) = \Phi. \tag{102.9}$$

But the wave function in the Heisenberg representation is independent of time (since the whole of the time dependence is in the operators); in the interaction representation, the time dependence of the wave function is given by (72.7):

$$\Phi_{\text{int}}(t) = \hat{S}(t, -\infty)\Phi_{\text{int}}(-\infty), \tag{102.10}$$

where

$$\hat{S}(t_2, t_1) = \text{T} \exp\left\{-i \int_{t_1}^{t_2} \hat{V}(t')\, dt'\right\}, \tag{102.11}$$

and the following properties of $\hat{S}$ are obvious:

$$\left.\begin{aligned} \hat{S}(t, t_1)\hat{S}(t_1, t_0) &= \hat{S}(t, t_0), \\ \hat{S}^{-1}(t, t_1) &= \hat{S}(t_1, t). \end{aligned}\right\} \tag{102.12}$$

Comparison of (102.10) and (102.9) gives

$$\Phi_{\text{int}}(t) = \hat{S}(t, -\infty)\Phi \tag{102.13}$$

as the relation between the wave functions in the two representations. The operator transformation formula is similarly

$$\begin{aligned} \hat{\psi}(t, \mathbf{r}) &= \hat{S}^{-1}(t, -\infty)\hat{\psi}_{\text{int}}(t, \mathbf{r})\hat{S}(t, -\infty) \\ &= \hat{S}(-\infty, t)\hat{\psi}_{\text{int}}(t, \mathbf{r})\hat{S}(t, -\infty), \end{aligned} \tag{102.14}$$

and likewise for $\hat{\bar{\psi}}$ and $\hat{A}$.

One further general remark may be added. It has already been mentioned more than once that, in relativistic quantum theory, the physical significance of the field operators is very limited because the zero-point fluctuations are infinite. This is even more true of operators in the Heisenberg representation, which contain also divergences due to the interaction. In this chapter, §§102–109 deal with the formal theory, which ignores the question of eliminating these singularities and which treats all quantities as if they were finite. The results thus obtained have mainly heuristic value: they lead to a fuller understanding of the significance of the expansions given by perturbation theory, and they may also remain valid in some form in a future theory which is free from the present difficulties.

§ 103. The exact photon propagator

The concepts of exact propagators play a central role in the formalism of the exact theory (i.e. without expansion in powers of e^2).†

The *exact photon propagator* (denoted by the script letter $\mathscr{D}$) is defined by

$$\mathscr{D}_{\mu\nu}(x-x') = i\langle 0|\mathrm{T}A_\mu(x)A_\nu(x')|0\rangle, \tag{103.1}$$

where $\hat{A}_\mu(x)$ are Heisenberg operators, in contrast to the definition (76.1):

$$D_{\mu\nu}(x-x') = i\langle 0|\mathrm{T}A_\mu^{\mathrm{int}}(x)A_\nu^{\mathrm{int}}(x')|0\rangle, \tag{103.2}$$

in which the operators in the interaction representation were used. The function (103.2) may be called the *free* (or *bare*)-*photon propagator* to distinguish it from the exact propagator (103.1).

Since the mean value in (103.1) cannot be exactly calculated, it is impossible to obtain an exact analytical expression for $\mathscr{D}_{\mu\nu}$, although the definition does lead to some general properties of this function, as will be discussed in §111; here we shall consider the calculation of $\mathscr{D}_{\mu\nu}$ by perturbation theory, using the diagram technique. For this purpose, we must express $\mathscr{D}_{\mu\nu}$ in terms of the operators in the interaction representation.
in the interaction representation.

First, let $t > t'$. Using the relationship between $\hat{A}(x)$ and $\hat{A}_{\mathrm{int}}(x)$ (cf. 102.14)), we can write

$$\begin{aligned}\mathscr{D}_{\mu\nu}(x-x') &= i\langle 0|A_\mu(x)A_\nu(x')|0\rangle \\ &= i\langle 0|S(-\infty, t)A_\mu^{\mathrm{int}}(x)S(t, -\infty)S(-\infty, t')\times \\ &\quad\times A_\nu^{\mathrm{int}}(x')S(t', -\infty)|0\rangle.\end{aligned}$$

According to (102.12) we can make the substitutions

$$\hat{S}(t, -\infty)\hat{S}(-\infty, t') = \hat{S}(t, t'),$$
$$\hat{S}(-\infty, t) = \hat{S}(-\infty, +\infty)\hat{S}(\infty, t).$$

Then

$$\mathscr{D}_{\mu\nu}(x-x') = i\langle 0|S^{-1}[S(\infty, t)A_\mu^{\mathrm{int}}(x)S(t, t')A_\nu^{\mathrm{int}}(x')S(t', -\infty)]|0\rangle, \tag{103.3}$$

with

$$\hat{S} \equiv \hat{S}(+\infty, -\infty). \tag{103.4}$$

Since, according to the definition (102.11), $\hat{S}(t_2, t_1)$ includes only operators for times

† These concepts were introduced by F. J. Dyson (1949), who also developed essentially the whole of the treatment given in this chapter.

between t_1 and t_2 arranged in chronological sequence, it is evident that all the operator factors in the brackets in (103.3) are in order of decreasing time from left to right. If the time-ordering symbol T is placed before the bracket, we can rearrange the factors in any manner, since the operator T will automatically put them in the necessary order. Then we write the bracket as

$$[\cdots] = \mathrm{T}[\hat{A}_\mu^{\mathrm{int}}(x)\hat{A}_\nu^{\mathrm{int}}(x')\hat{S}(\infty, t)\hat{S}(t, t')\hat{S}(t', -\infty)]$$
$$= \mathrm{T}[\hat{A}_\mu^{\mathrm{int}}(x)\hat{A}_\nu^{\mathrm{int}}(x')\hat{S}].$$

Thus

$$\mathscr{D}_{\mu\nu}(x - x') = i\langle 0|S^{-1}\mathrm{T}[A_\mu^{\mathrm{int}}(x)A_\nu^{\mathrm{int}}(x')S]|0\rangle. \tag{103.5}$$

It is easily shown by a similar argument that this formula is also valid if $t < t'$.

We shall now prove that the factor $\hat{S}^{-1}$ can be taken outside the averaging over the vacuum to form a phase factor. To do so, we recall that the Heisenberg vacuum wave function Φ is the same as the value $\Phi_{\mathrm{int}}(-\infty)$ of the wave function of the same state in the interaction representation (see (102.9)). From (72.8),

$$\hat{S}\Phi_{\mathrm{int}}(-\infty) \equiv \hat{S}(+\infty, -\infty)\Phi_{\mathrm{int}}(-\infty) = \Phi_{\mathrm{int}}(+\infty).$$

The vacuum is a strictly stationary state, in which no spontaneous processes of particle generation can occur. In other words, in the course of time the vacuum remains the vacuum; this means that $\Phi_{\mathrm{int}}(+\infty)$ can differ from $\Phi_{\mathrm{int}}(-\infty)$ only by a phase factor $e^{i\alpha}$. Hence

$$\hat{S}\Phi_{\mathrm{int}}(-\infty) = e^{i\alpha}\Phi_{\mathrm{int}}(-\infty) = \langle 0|S|0\rangle\Phi_{\mathrm{int}}(-\infty), \tag{103.6}$$

or, taking the complex conjugate and using the unitarity of the operator $\hat{S}$,

$$\Phi_{\mathrm{int}}^*(-\infty)\hat{S}^{-1} = \langle 0|S|0\rangle^{-1}\Phi_{\mathrm{int}}^*(-\infty).$$

Hence it is clear that (103.5) can be written

$$\mathscr{D}_{\mu\nu}(x - x') = i\,\frac{\langle 0|\mathrm{T}A_\mu^{\mathrm{int}}(x)A_\nu^{\mathrm{int}}(x')S|0\rangle}{\langle 0|S|0\rangle}. \tag{103.7}$$

Substituting in the numerator and the denominator the expansion (72.10) for $\hat{S}$ and averaging by means of Wick's theorem (§77), we get an expansion of $\mathscr{D}_{\mu\nu}$ in powers of e^2.

In the numerator of (103.7), the quantities to be averaged differ from the matrix elements of the type (77.1) only in that the "external" photon creation and annihilation operators are replaced by $\hat{A}_\mu^{\mathrm{int}}(x)$ and $\hat{A}_\nu^{\mathrm{int}}(x')$. Since all the factors in the products to be averaged are preceded by the time-ordering symbol, the pairwise contractions of these operators with the "internal" operators $\hat{A}^{\mathrm{int}}(x_1)$, $\hat{A}^{\mathrm{int}}(x_2)$, ... will give the photon propagators $D_{\mu\nu}$. Thus the results of the averaging are expressed by sets of diagrams with two free ends, constructed in accordance with the rules in §77, except that propagators $D_{\mu\nu}$, not the amplitudes e of real photons, correspond to external (and internal) photon lines. In the zero-order approximation, with $\hat{S} = 1$, the numerator of (103.7) is simply $D_{\mu\nu}(x - x')$. The next non-zero terms will be of the order of e^2. They are represented by a set of diagrams having two free ends and two vertices:

a b (103.8)

The second of these diagrams consists of two disconnected parts: a broken line (corresponding to $-iD_{\mu\nu}$) and a closed loop. The separation of the parts of the diagram signifies that the corresponding analytical expression separates into two independent factors. On adding to the diagrams (103.8) the zero-order approximation diagram (a single broken line) and "taking it outside the brackets", we find that the numerator in (103.7) is, as far as second-order terms,

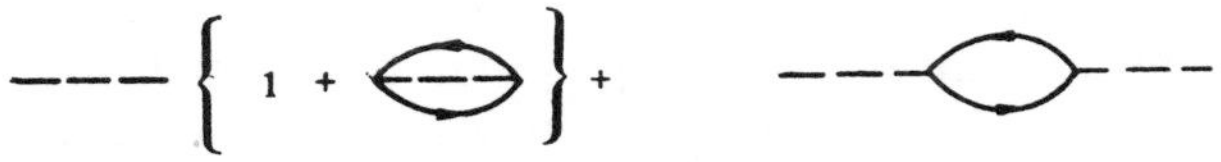

The expression $\langle 0|S|0\rangle$ in the denominator of (103.7) is the amplitude of the "transition" from the vacuum to the vacuum. Its expansion therefore contains only diagrams without free ends. In the zero-order approximation, $\langle 0|S|0\rangle = 1$, and as far as second-order terms we have

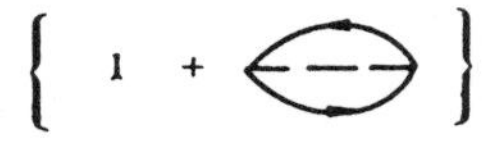

When the numerator is divided by the denominator we get, to the same order, the expression

Thus the diagram with the detached loop does not occur in the result. This is a general theorem. Having regard to the way in which the diagrams are constructed which correspond to the numerator and denominator in (103.7), we can easily see that the role of the denominator $\langle 0|S|0\rangle$ is simply to ensure that in all orders of perturbation theory the exact propagator $\mathscr{D}_{\mu\nu}$ will be represented only by diagrams which do not contain separated parts.

The diagrams with no free ends, forming closed loops, have no physical significance and need not be taken into account, quite apart from the fact that they disappear when the propagator $\mathscr{D}$ is formed. Such loops represent radiative corrections to the diagonal element of the S-matrix for a vacuum–vacuum transition; but, according to (103.6), the sum of all these loops, together with the unity

QE4 - EE

given by the zero-order approximation, gives only an unimportant phase factor, which cannot affect any physical results.

The change from the coordinate representation to the momentum representation is made in the usual way. For example, in the second-order approximation of perturbation theory, the propagator $-i\mathscr{D}_{\mu\nu}(k)$, which will be shown by a thick broken line, is the sum

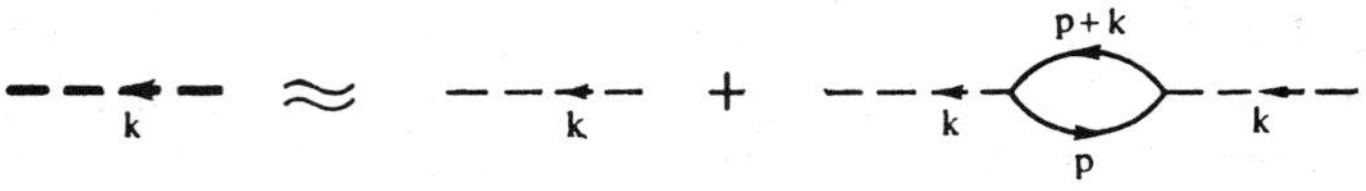

(103.9)

in which all the diagrams are calculated by the general rules given in §77 except that factors $-iD_{\mu\nu}(k)$ are assigned to the external as well as the internal photon lines. In analytical form, we therefore have†

$$\mathscr{D}_{\mu\nu}(k) \approx D_{\mu\nu}(k) + ie^2 D_{\mu\lambda}(k) \int \operatorname{tr} \gamma^\lambda G(p+k)\gamma^\rho G(p) \frac{d^4p}{(2\pi)^4} D_{\rho\nu}(k); \quad (103.10)$$

the bispinor indices of the matrices γ and G are, as usual, omitted.

The terms in subsequent approximations are constructed in a similar manner, and are represented by sets of diagrams having two external photon lines and the appropriate number of vertices. For example, the terms in e^4 correspond to the following four-vertex diagrams:

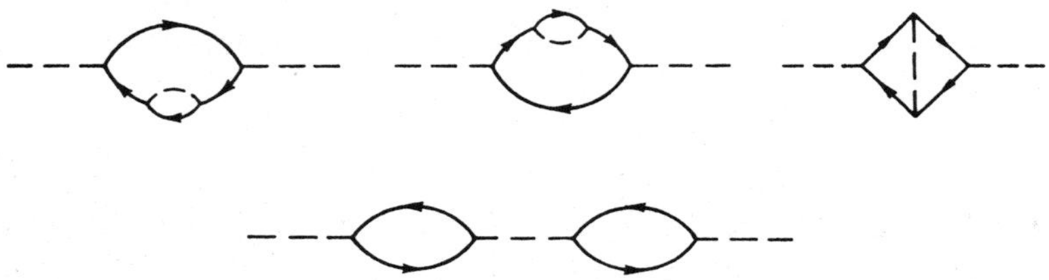

(103.11)

The diagram

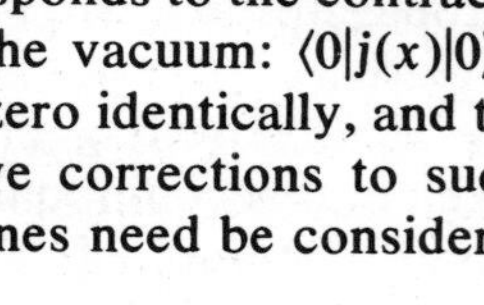

also has four vertices; its upper part is a loop formed by a single "self-closed" electron line. Such a loop corresponds to the contraction $\hat{\bar{\psi}}(x)\gamma\hat{\psi}(x)$, i.e. to the value of the current averaged over the vacuum: $\langle 0|j(x)|0\rangle$. But, by the definition of the vacuum, this quantity must be zero identically, and the identity cannot of course be altered by any further radiative corrections to such a loop.‡ Thus no diagrams having "self-closed" electron lines need be considered in any approximation.

† The factor -1 from the closed electron loop must be taken into account when deriving the signs.
‡ Although a direct calculation from the diagrams would lead to divergent integrals.

The part of a diagram which lies between two (external or internal) photon lines is called a *photon self-energy part*. In the general case, it can itself be divided into parts joined in pairs by a single photon line, i.e. it has a structure of the form

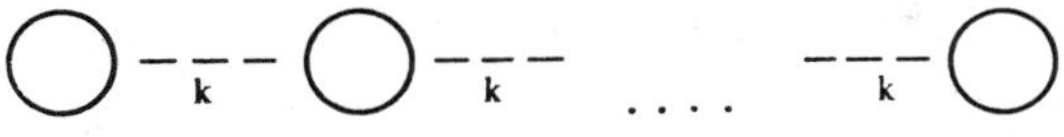

where the circles denote parts which cannot be further subdivided in the same manner; such parts are said to be *compact* or *proper*. For example, the first three of the four fourth-order self-energy parts (103.11) are compact.

Let $i\mathscr{P}_{\mu\nu}/4\pi$ denote the sum of the infinity of compact self-energy parts. The function $\mathscr{P}_{\mu\nu}(k)$ is called the *polarization operator*. When the diagrams are classified by the number of compact parts which they contain, the exact propagator $\mathscr{D}_{\mu\nu}$ can be put in the form of a series

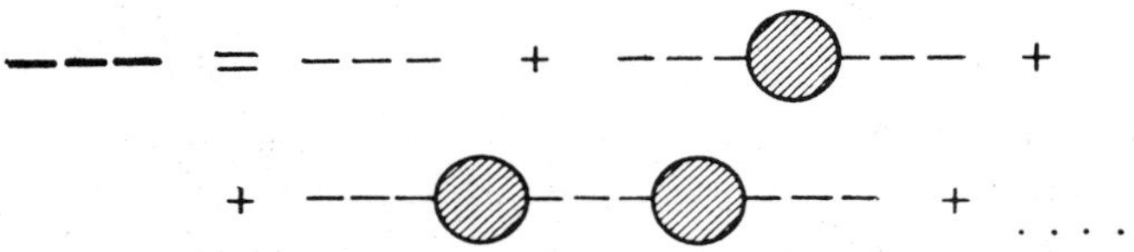

where $i\mathscr{P}_{\mu\nu}/4\pi$ corresponds to each shaded circle. The analytical form of this series is

$$\begin{aligned}\mathscr{D} &= D + D\frac{\mathscr{P}}{4\pi}D + D\frac{\mathscr{P}}{4\pi}D\frac{\mathscr{P}}{4\pi}D + \cdots \\ &= D\left\{1 + \frac{\mathscr{P}}{4\pi}\left[D + D\frac{\mathscr{P}}{4\pi}D + \cdots\right]\right\},\end{aligned} \tag{103.12}$$

where the indices are omitted, for brevity. The series in the brackets is again $\mathscr{D}$. Hence

$$\mathscr{D}_{\mu\nu}(k) = D_{\mu\nu}(k) + D_{\mu\lambda}(k)\frac{\mathscr{P}^{\lambda\rho}(k)}{4\pi}\mathscr{D}_{\rho\nu}(k). \tag{103.13}$$

Multiplying this equation on the left by the inverse tensor $(D^{-1})^{\tau\mu}$ and on the right by $(\mathscr{D}^{-1})^{\nu\sigma}$, and renaming the indices, we get the equivalent form

$$\mathscr{D}^{-1}{}_{\mu\nu} = D^{-1}{}_{\mu\nu} - \frac{1}{4\pi}\mathscr{P}_{\mu\nu}. \tag{103.14}$$

It must be emphasized that writing $\mathscr{D}$ in the form (103.12) assumes that the diagrams can be broken down into simpler parts calculated by the general rules of the diagram technique, and that the combination of such parts gives the correct expressions for the entire diagrams. The admissibility of this breakdown of the diagrams is an important and by no means trivial feature of the diagram technique, which arises from the fact that the overall numerical factor in the diagram does not depend on the order of the diagram.

The same property enables us to use the function $\mathscr{D}$ (assumed known) to simplify the calculations of the radiative corrections to the amplitudes of various scattering processes: instead of treating afresh each time the diagrams with different corrections to the internal photon lines, we can simply make these lines thick, i.e. assign to them the propagators $\mathscr{D}$ (instead of D) in the appropriate approximation.

If the photon line corresponds to a real and not a virtual photon, i.e. if it is a free end of the whole diagram, the application to it of all the self-energy corrections gives what is called an *effective external line.* It corresponds to the expression obtained from (103.13) by replacing the factor D by the polarization amplitude of the real photon:

$$e_\mu + \mathscr{D}_{\mu\rho}(k)\frac{\mathscr{P}^{\rho\lambda}(k)}{4\pi}e_\lambda. \qquad (103.15)$$

For an external-field line, e_μ in this expression is to be replaced by $A_\mu^{(e)}$.

The discussion in §76 of the tensor structure and the gauge non-uniqueness of the approximate propagator $D_{\mu\nu}$ applies to the exact function $\mathscr{D}_{\mu\nu}$ also. Considering only the relativistically invariant representations of this function, we can write it in the general form

$$\mathscr{D}_{\mu\nu}(k) = \mathscr{D}(k^2)\left(g_{\mu\nu} - \frac{k_\mu k_\nu}{k^2}\right) + \mathscr{D}^{(l)}(k^2)\frac{k_\mu k_\nu}{k^2}; \qquad (103.16)$$

the first term corresponds to the Landau gauge, and in the second term $\mathscr{D}^{(l)}$ is a gauge-arbitrary function. The corresponding form of the approximate propagator† is

$$D_{\mu\nu}(k) = D(k^2)\left(g_{\mu\nu} - \frac{k_\mu k_\nu}{k^2}\right) + D^{(l)}(k^2)\frac{k_\mu k_\nu}{k^2}. \qquad (103.17)$$

The longitudinal part $\mathscr{D}^{(l)}$ of the propagator is related to the longitudinal part of the potential 4-vector, which has no physical significance. It is therefore not concerned in the interaction and is unaffected by the latter, so that

$$\mathscr{D}^{(l)}(k^2) = D^{(l)}(k^2). \qquad (103.18)$$

The inverse tensors must, by definition, satisfy the equations

$$\mathscr{D}^{-1}{}_{\mu\nu}\mathscr{D}^{\lambda\nu} = \delta_\mu^\lambda, \qquad D^{-1}{}_{\mu\nu}D^{\lambda\nu} = \delta_\mu^\lambda.$$

When the original tensors have the form (103.16) or (103.17), the inverse tensors are, from (103.18),

$$\left.\begin{aligned} \mathscr{D}^{-1}{}_{\mu\nu} &= \frac{1}{\mathscr{D}}\left(g_{\mu\nu} - \frac{k_\mu k_\nu}{k^2}\right) + \frac{1}{D^{(l)}}\frac{k_\mu k_\nu}{k^2}, \\ D^{-1}{}_{\mu\nu} &= \frac{1}{D}\left(g_{\mu\nu} - \frac{k_\mu k_\nu}{k^2}\right) + \frac{1}{D^{(l)}}\frac{k_\mu k_\nu}{k^2}. \end{aligned}\right\} \qquad (103.19)$$

† In this formula $D^{(l)}$ is not the same as in (76.3).

From these, it follows that the polarization operator $\mathcal{P}_{\mu\nu}$ is a transverse tensor:

$$\mathcal{P}_{\mu\nu} = \mathcal{P}(k^2)\left(g_{\mu\nu} - \frac{k_\mu k_\nu}{k^2}\right), \tag{103.20}$$

where $\mathcal{P} = k^2 - 4\pi/\mathcal{D}$, or

$$\mathcal{D}(k^2) = \frac{4\pi}{k^2[1 - \mathcal{P}(k^2)/k^2]}. \tag{103.21}$$

Thus the polarization operator, unlike the photon propagator itself, is a gauge-invariant quantity.

§ 104. The self-energy function of the photon

In order to examine further the analytical properties of the photon propagator, it is useful to define, as well as the polarization operator, another auxiliary function $\Pi_{\mu\nu}(k)$, called the *self-energy function of the photon*: $i\Pi_{\mu\nu}/4\pi$ is defined as the sum of all self-energy photon parts (not only the compact ones). If this sum is represented in the diagram by a square, we can write the exact propagator as the sum

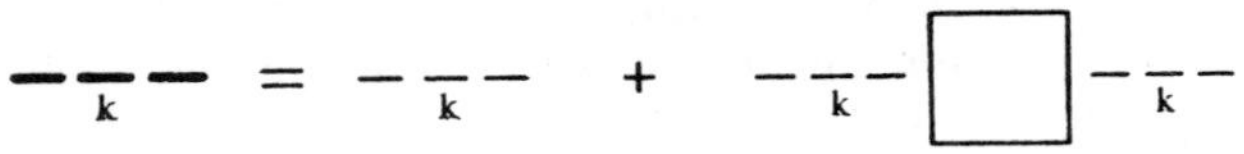

i.e.

$$\mathcal{D}_{\mu\nu} = D_{\mu\nu} + D_{\mu\lambda}\frac{\Pi^{\lambda\rho}}{4\pi}D_{\rho\nu}. \tag{104.1}$$

Hence, expressing $\Pi_{\mu\nu}$ as

$$\frac{1}{4\pi}\Pi_{\mu\nu} = D^{-1}{}_{\mu\lambda}\mathcal{D}^{\lambda\rho}D^{-1}{}_{\rho\nu} - D^{-1}{}_{\mu\nu}$$

and substituting (103.16) and (103.19) followed by (103.21), we get

$$\Pi_{\mu\nu} = \Pi(k^2)\left(g_{\mu\nu} - \frac{k_\mu k_\nu}{k^2}\right), \qquad \Pi = \frac{\mathcal{P}}{1 - \mathcal{P}/k^2}. \tag{104.2}$$

Thus $\Pi_{\mu\nu}$, like $\mathcal{P}_{\mu\nu}$, is a gauge-invariant tensor.

The usefulness of $\Pi_{\mu\nu}$ arises from the expression for it in the coordinate representation. This is easily found by noting that the equation

$$\frac{1}{4\pi}\Pi_{\mu\nu}(k) = D^{-1}{}_{\mu\lambda}D^{-1}{}_{\rho\nu}\{\mathcal{D}^{\lambda\rho}(k) - D^{\lambda\rho}(k)\},$$

in which the tensor $\mathscr{D}^{\lambda\rho} - D^{\lambda\rho}$ is transverse by (103.18), can be written in the coordinate representation

$$\Pi_{\mu\nu}(x-x') = \frac{1}{4\pi}(\partial_\mu\partial_\lambda - g_{\mu\lambda}\partial_\sigma\partial^\sigma)(\partial'_\nu\partial'_\rho - g_{\nu\rho}\partial'_\sigma\partial'^\sigma)\{\mathscr{D}^{\lambda\rho}(x-x') - D^{\lambda\rho}(x-x')\}.$$

In order to carry out the differentiation, we must substitute

$$\mathscr{D}^{\lambda\rho}(x-x') - D^{\lambda\rho}(x-x') = i\langle 0|\mathrm{T}A^\lambda(x)A^\rho(x') - \mathrm{T}A^\lambda_{\mathrm{int}}(x)A^\rho_{\mathrm{int}}(x')|0\rangle. \quad (104.3)$$

In §75 we have seen that the differentiation of a T product generally demands caution, because the product has discontinuities. But the difference that is to be averaged in (104.3) is continuous, and so are its first derivatives, since the commutation rules are the same for the components of the operators $\hat{A}^\lambda(x)$ and $\hat{A}^\lambda_{\mathrm{int}}(x)$ for a given time, and the corresponding discontinuities cancel out (cf. §75). The difference in (104.3) may therefore be differentiated under the symbol T. According to (102.6) and the corresponding equation with zero on the right for the free electromagnetic field operators $\hat{A}^\mu_{\mathrm{int}}(x)$, the result is

$$\Pi_{\mu\nu}(x-x') = 4\pi i e^2\langle 0|\mathrm{T}j_\mu(x)j_\nu(x')|0\rangle. \quad (104.4)$$

This shows explicitly the gauge-invariance of $\Pi_{\mu\nu}$, since the current operators are gauge-invariant.

From (104.4) we can derive an important integral form of this function. According to (104.2), it is sufficient to consider the scalar function $\Pi = \frac{1}{3}\Pi^\mu_\mu$. In the coordinate representation,

$$\Pi(x-x') = \frac{4\pi}{3} ie^2\langle 0|\mathrm{T}j_\mu(x)j^\mu(x')|0\rangle$$

$$= \frac{4\pi}{3} ie^2 \begin{cases} \sum_n \langle 0|j_\mu(x)|n\rangle\langle n|j^\mu(x')|0\rangle & \text{for} \quad t>t', \\ \sum_n \langle 0|j_\mu(x')|n\rangle\langle n|j^\mu(x)|0\rangle & \text{for} \quad t<t', \end{cases} \quad (104.5)$$

where n labels the states of the system electromagnetic field + electron–positron field.† Since the current operator $\hat{j}(x)$ depends on $x^\mu = (t, \mathbf{r})$, its matrix elements also depend on x. The relationship can be found explicitly by taking as the states $|n\rangle$ states which have definite values of the total 4-momentum.

The time dependence of the current matrix elements, like that of any Heisenberg operator, is given by

$$\langle n|j^\mu(t,\mathbf{r})|m\rangle = \langle n|j^\mu(\mathbf{r})|m\rangle e^{-i(E_m-E_n)t},$$

where E_n and E_m are the energies of the states $|n\rangle$ and $|m\rangle$, and $\hat{j}(\mathbf{r})$ is the Schrödinger operator.

† The current operator conserves charge; hence the states $|n\rangle$ in (104.5) can contain only the same numbers of electrons and positrons.

To determine the coordinate dependence of the matrix elements, we consider the operator $\hat{j}(\mathbf{r})$ as being the result of transforming the operator $\hat{j}(0)$ by a parallel translation over the distance $\mathbf{r}$. The operator of this translation is $\exp(i\mathbf{r}\cdot\hat{\mathbf{P}})$, where $\hat{\mathbf{P}}$ is the total momentum operator of the system (see *QM*, (15.15)). Using the general rule for the transformation of the matrix elements (see *QM*, (12.7)), we therefore have

$$\begin{aligned}\langle n|j^{\mu}(\mathbf{r})|m\rangle &= \langle n|e^{-i\mathbf{r}\cdot\mathbf{P}}j^{\mu}(0)e^{i\mathbf{r}\cdot\mathbf{P}}|m\rangle \\ &= \langle n|j^{\mu}(0)|m\rangle e^{i(\mathbf{P}_m-\mathbf{P}_n)\cdot\mathbf{r}}.\end{aligned}$$

Together with the previous formula, this gives finally

$$\langle n|j^{\mu}(t,\mathbf{r})|m\rangle = \langle n|j^{\mu}(0)|m\rangle e^{-i(P_m-P_n)x}. \tag{104.6}$$

The matrix $\langle n|j^{\mu}(0)|m\rangle$ is Hermitian, like the matrix (104.6) of the entire operator $\hat{j}^{\mu}(t,\mathbf{r})$, and according to the equation of continuity (102.7) it satisfies the transversality condition

$$(P_n-P_m)^{\mu}\langle n|j_{\mu}(0)|m\rangle = 0. \tag{104.7}$$

Let us now calculate the function $\Pi(x-x')$. Substitution of (104.6) in (104.5) gives

$$\Pi(\xi) = \frac{4\pi i e^2}{3}\sum_n \langle 0|j_{\mu}(0)|n\rangle\langle n|j^{\mu}(0)|0\rangle e^{\mp iP_n\xi} \quad \text{for} \quad \tau \gtrless 0, \tag{104.8}$$

where $x-x' = \xi = (\tau, \boldsymbol{\xi})$. We use the notation

$$\rho(k^2) = -\frac{4\pi e^2}{3}(2\pi)^3\sum_n \langle 0|j_{\mu}(0)|n\rangle\langle 0|j^{\mu}(0)|n\rangle^*\delta^{(4)}(k-P_n). \tag{104.9}$$

The sum is taken over all systems of real electron–positron pairs and photons that can be generated by a virtual photon having 4-momentum $k = (\omega, \mathbf{k})$ $(\omega > 0)$, and for each such system there is summation over the internal variables (the polarizations and momenta of the particles in the centre-of-mass system).† After this summation, the function ρ can depend only on k, and since it is a scalar it can depend only on k^2. In particular, it does not depend on the direction of $\mathbf{k}$. Using these properties of ρ, we can rewrite (104.8) as

$$\begin{aligned}\Pi(\xi) &= -i\int_0^{\infty} d\omega \int \frac{d^3k}{(2\pi)^3}\rho(k^2)e^{i\mathbf{k}\cdot\boldsymbol{\xi}-i\omega|\tau|} \\ &= -i\int \frac{d^3k}{(2\pi)^3}\int_0^{\infty}\int_0^{\infty} d\omega\, d(\mu^2)\delta(\mu^2-k^2)\rho(\mu^2)e^{i\mathbf{k}\cdot\boldsymbol{\xi}-i\omega|\tau|}.\end{aligned}$$

† This definition of the states $|n\rangle$ is evidently identical with their definition as states for which the matrix elements $\langle 0|j|n\rangle$ of a charge-odd operator are non-zero.

The momentum representation is obtained by substituting

$$e^{-i\omega|\tau|} = 2i\omega \int_{-\infty}^{\infty} e^{-ik_0\tau} \frac{1}{k_0^2 - \omega^2 + i0} \frac{dk_0}{2\pi} \tag{104.10}$$

(see §76); the result is

$$\Pi(k^2) = \int_0^\infty d(\mu^2) \int_0^\infty d(\omega^2)\delta(\mu^2 + \mathbf{k}^2 - \omega^2) \frac{\rho(\mu^2)}{k_0^2 - \omega^2 + i0},$$

or, finally,†

$$\Pi(k^2) = \int_0^\infty \frac{\rho(\mu^2)\, d\mu^2}{k^2 - \mu^2 + i0}. \tag{104.11}$$

The coefficient ρ in this integral form is called the *spectral density* of the function $\Pi(k^2)$, and has the properties

$$\left.\begin{array}{lll} \rho(k^2) = 0 & \text{for} & k^2 < 0, \\ \rho(k^2) > 0 & \text{for} & k^2 > 0, \end{array}\right\} \tag{104.12}$$

since the 4-momentum k of a virtual photon which can generate a system of real particles must necessarily be time-like; k^2 is equal to the square of the total energy of the particles in their centre-of-mass system. The transversality condition (104.7) gives

$$P_n^\mu \langle 0|j_\mu(0)|n\rangle = 0.$$

The 4-vector $\langle 0|j|n\rangle$ is orthogonal to the time-like 4-vector P_n and must be space-like:

$$\langle 0|j_\mu(0)|n\rangle\langle 0|j^\mu(0)|n\rangle^* < 0;$$

thus, from the definition (104.9), $\rho > 0$.

§ 105. **The exact electron propagator**

The exact electron propagator, similarly to that of the photon, is defined by

$$\mathcal{G}_{ik}(x - x') = -i\langle 0|\mathrm{T}\psi_i(x)\bar{\psi}_k(x')|0\rangle \tag{105.1}$$

† The formal calculations analogous to those given above require caution, on account of the presence of the divergences previously mentioned. These give rise, in particular, to the occurrence on the right of (104.11) of further divergent terms which do not have an explicitly relativistically invariant form, called *Schwinger terms*. They will not be written out here, since they in any case disappear on renormalization (§110) and do not affect the subsequent results.

where i and k are bispinor indices, which differs from the definition (75.1) of the free-particle propagator

$$G_{ik}(x-x') = -i\langle 0|\mathrm{T}\psi_i^{\mathrm{int}}(x)\bar{\psi}_k^{\mathrm{int}}(x')|0\rangle \qquad (105.2)$$

in that the ψ-operators in the interaction representation are replaced by Heisenberg operators.

The same arguments as were used to derive (103.7) lead to

$$\mathcal{G}_{ik}(x-x') = -i\,\frac{\langle 0|\mathrm{T}\psi_i^{\mathrm{int}}(x)\bar{\psi}_k^{\mathrm{int}}(x')S|0\rangle}{\langle 0|S|0\rangle}. \qquad (105.3)$$

The expansion of this expression in powers of e^2 puts the $\mathcal{G}$ function in the form of a set of diagrams with two external electron lines and various numbers of vertices. The denominator in (105.3) again has the function of retaining only the diagrams which do not have detached "vacuum loops". For example, as far as the terms in e^4, the graphical representation of the propagator $\mathcal{G}$ (denoted by a thick continuous line) is†

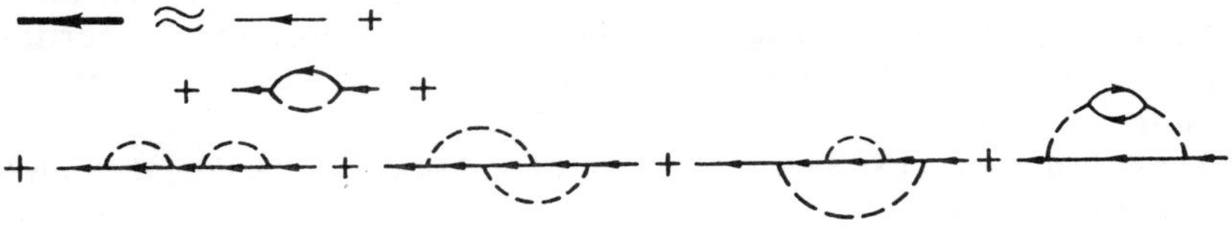

(105.4)

The thick continuous line corresponds to the function $i\mathcal{G}(p)$ in the momentum representation, and the sets of continuous and broken lines in the diagrams on the right of the equation correspond to the free-particle propagators iG and $-iD$ respectively.

The section between two electron lines is called an *electron self-energy part.* As with the photon, it is said to be *compact* if it cannot be further subdivided into two self-energy parts by cutting a single electron line. The sum of all possible compact parts will be denoted by $-i\mathcal{M}_{ik}$; the function $\mathcal{M}_{ik}(p)$ is called the *mass operator.* For example, as far as the terms in e^4,

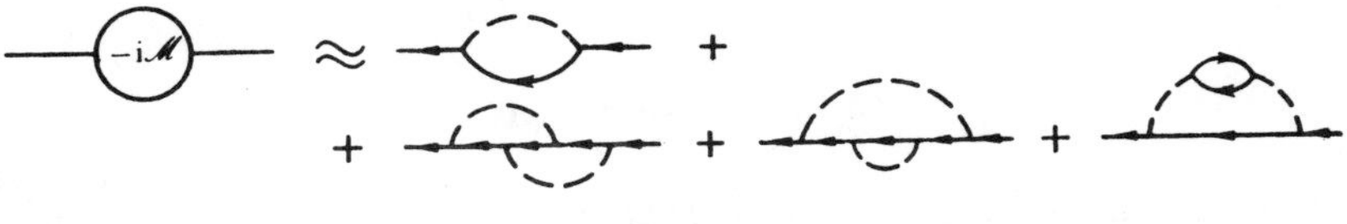

(105.5)

† It has already been shown in §103 that there is also no need to take account of diagrams which contain "self-closed" lines; these would here appear in the second order:

By a summation exactly similar to the derivation of (103.13), we find

$$\mathcal{G}(p) = G(p) + G(p)\mathcal{M}(p)\mathcal{G}(p) \tag{105.6}$$

(omitting the bispinor indices) or, for the inverse matrices,

$$\begin{aligned}\mathcal{G}^{-1}(p) &= G^{-1}(p) - \mathcal{M}(p)\\ &= \gamma p - m - \mathcal{M}(p).\end{aligned} \tag{105.7}$$

It has already been noted in §102 that the Heisenberg ψ-operators (unlike those in the interaction representation) are altered by a gauge transformation of the electromagnetic potentials. The exact electron propagator $\mathcal{G}$ is therefore also not gauge-invariant. Its gauge transformation behaviour may be derived as follows (L. D. Landau and I. M. Khalatnikov, 1952).

The change in $\mathcal{G}$ under the gauge transformation must evidently be expressed in terms of the same quantity $D^{(l)}$ as is added to the photon propagator by this transformation. This is clear, since in the calculation of $\mathcal{G}$ by the perturbation-theory diagrams each term of the series is expressed in terms of the functions D, and no other electromagnetic quantities are involved. The analysis can therefore be simplified: any special assumptions can be made regarding the properties of the arbitrary operator $\hat{\chi}$ in the transfromation (102.8), provided that the result is expressed in terms of $D^{(l)}$.

The transformation (102.8) brings the propagators $\mathcal{D}$ (103.1) and $\mathcal{G}$ (105.1) into the following forms:

$$\left.\begin{aligned}\mathcal{D}_{\mu\nu} &\to i\langle 0|T[A_\mu(x) - \partial_\mu\chi(x)][A_\nu(x') - \partial'_\nu\chi(x')]|0\rangle,\\ \mathcal{G}_{ik} &\to -i\langle 0|T\psi_i(x)e^{ie\chi(x)}e^{-ie\chi(x')}\bar{\psi}_k(x')|0\rangle.\end{aligned}\right\} \tag{105.8}$$

We shall now suppose that the operators $\hat{\chi}$ are averaged independently of all the remaining operators in the T product. This is a reasonable assumption, since the "field" $\hat{\chi}$ takes no part in the interaction, because of the gauge invariance. We also assume that the mean value, over the vacuum, of the operator $\hat{\chi}$ is zero: $\langle 0|\chi|0\rangle = 0$. Then the terms in $\hat{\chi}$ in (105.8) can be separated, and the result is

$$\mathcal{D}_{\mu\nu} \to \mathcal{D}_{\mu\nu} + i\langle 0|T\partial_\mu\chi(x)\cdot\partial'_\nu\chi(x')|0\rangle, \tag{105.9}$$

$$\mathcal{G}_{ik} \to \mathcal{G}_{ik}\langle 0|Te^{ie\chi(x)}e^{-ie\chi(x')}|0\rangle. \tag{105.10}$$

The rest of the derivation will be given for the case of an infinitesimal transformation, and we shall emphasize this by writing $\delta\hat{\chi}$ in place of $\hat{\chi}$.

The transformation (105.9) may be written† (independently of the smallness of

† Formula (105.11) can be derived from (105.9) if the function $d^{(l)}$ and its derivative with respect to t are continuous at $t = t'$; if they are discontinuous, the right-hand sides of these expressions differ by delta-function terms (cf. the derivation of (75.2)). In the momentum representation, this condition is equivalent to assuming that $d^{(l)}(q)$ decreases more rapidly than $1/q^2$ as $|q^2| \to \infty$.

$\delta\hat{\chi}$) as

$$\mathscr{D}_{\mu\nu} \to \mathscr{D}_{\mu\nu} + \delta\mathscr{D}_{\mu\nu}, \quad \delta\mathscr{D}_{\mu\nu} = \partial_\mu \partial'_\nu d^{(l)}(x - x'), \tag{105.11}$$

where

$$d^{(l)}(x - x') = i\langle 0|\mathrm{T}\delta\chi(x)\delta\chi(x')|0\rangle. \tag{105.12}$$

Hence it is clear that $d^{(l)}$ determines the change caused by the gauge transformation in the longitudinal part $\mathscr{D}^{(l)}$ of the photon propagator. The assumption that $d^{(l)}$ depends only on $x - x'$ implies, of course, a certain limitation on the properties of the operator $\delta\hat{\chi}$; in the general case of a completely arbitrary gauge transformation, the propagator may cease to be homogeneous in space and time.

In the transformation (105.10), we expand the exponential factors in powers of $\delta\hat{\chi}$ as far as the quadratic terms:

$$\begin{aligned} &\langle 0|\mathrm{T}e^{ie\delta\chi(x)}e^{-ie\delta\chi(x')}|0\rangle \\ &\quad \approx -\tfrac{1}{2}e^2\langle 0|\delta\chi^2(x) + \delta\chi^2(x') - 2\mathrm{T}\delta\chi(x)\delta\chi(x')|0\rangle. \end{aligned}$$

Using the definition (105.12), we thus find the following transformation rule for the electron propagator:

$$\mathscr{G} \to \mathscr{G} + \delta\mathscr{G}, \; \delta\mathscr{G} = ie^2\mathscr{G}(x - x')[d^{(l)}(0) - d^{(l)}(x - x')]. \tag{105.13}$$

In the momentum representation,† we have

$$\delta\mathscr{G}(p) = ie^2 \int d^{(l)}(q)[\mathscr{G}(p) - \mathscr{G}(p - q)]\frac{d^4q}{(2\pi)^4}; \tag{105.14}$$

$d^{(l)}(q)$ is related to the change in the function $\mathscr{D}^{(l)}$ by

$$\delta\mathscr{D}^{(l)}(q) = q^2 d^{(l)}(q). \tag{105.15}$$

† If the function $f(x) = f_1(x)f_2(x)$, its Fourier components are

$$\begin{aligned} f(p) &= \int f(x)e^{ipx}\,d^4x \\ &= \iiint d^4x \frac{d^4q_1\,d^4q_2}{(2\pi)^8} e^{ix(p-q_1-q_2)}f_1(q_1)f_2(q_2) \\ &= \iint \frac{d^4q_1\,d^4q_2}{(2\pi)^4}\delta^{(4)}(p - q_1 - q_2)f_1(q_1)f_2(q_2) \\ &= \int \frac{d^4q}{(2\pi)^4} f_1(q)f_2(p - q). \end{aligned}$$

In deriving (105.14) from (105.13), we also use the result

$$f(x = 0) = \int f(q)\frac{d^4q}{(2\pi)^4}.$$

An integral expression analogous to (104.11) could be derived for the electron propagator, using the expressions

$$\psi_{nm}(x) = \psi_{nm}(0)e^{-i(P_m-P_n)x} \tag{105.16}$$

for the matrix elements of the ψ-operator, similarly to the expressions (104.6) for the current matrix elements. Unlike the current, however, the ψ-operators are not gauge-invariant. The coordinate dependence (105.16) is therefore not general, but applies only to some particular gauge. The same is true as regards the integral representation based on (105.16). The deeper physical reason for this situation is that the zero photon mass leads to the infra-red catastrophe (§98). In consequence, the electron emits an infinite number of soft quanta during the interaction, and this means that the "single-particle" propagator (105.1) loses much of its direct significance.

§ 106. **Vertex parts**

In complicated diagrams it is possible to distinguish both self-energy parts and sections of another type which are not equivalent to them. An important class of such sections is found by considering the function

$$K^{\mu}_{ik}(x_1, x_2, x_3) = \langle 0|\mathrm{T}A^{\mu}(x_1)\psi_i(x_2)\bar{\psi}_k(x_3)|0\rangle \tag{106.1}$$

which has one 4-vector index and two bispinor indices; since space–time is homogeneous, this function depends only on the differences of the arguments x_1, x_2, x_3. When expressed in terms of the operators in the interaction representation, the function K has the form

$$K^{\mu}_{ik}(x_1, x_2, x_3) = \frac{\langle 0|\mathrm{T}A^{\mu}_{\mathrm{int}}(x_1)\psi^{\mathrm{int}}_i(x_2)\bar{\psi}^{\mathrm{int}}_k(x_3)S|0\rangle}{\langle 0|S|0\rangle}. \tag{106.2}$$

The momentum representation is obtained by using the formula

$$(2\pi)^4\delta^{(4)}(p_1+k-p_2)K^{\mu}_{ik}(p_2, p_1; k) = \iiint K^{\mu}_{ik}(x_1, x_2, x_3)e^{-ikx_1+ip_2x_2-ip_1x_3}\, d^4x_1\, d^4x_2\, d^4x_3. \tag{106.3}$$

In the diagram technique, the functions K^{μ}_{ik} correspond to three-ended (one photon and two electron) sections of the form

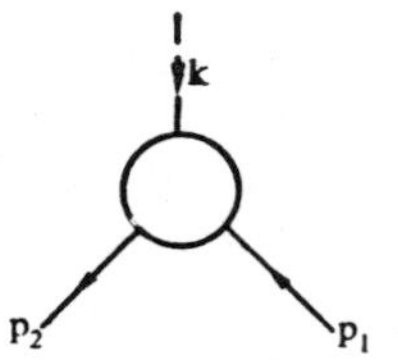

(106.4)

where the momenta are related by the conservation law

$$p_1 + k = p_2. \tag{106.5}$$

The zero-order term in the expansion of this function is zero; the first-order term is

$$K^{\mu}(x_1, x_2, x_3) = e \int G(x_2 - x)\gamma_\nu G(x - x_3) \cdot D^{\nu\mu}(x_1 - x)\, d^4x$$

in the coordinate representation, and

$$K^{\mu}(p_2, p_1; k) = eG(p_2)\gamma_\nu G(p_1) \cdot D^{\nu\mu}(k) \tag{106.6}$$

in the momentum representation (omitting the bispinor indices); the corresponding diagram is

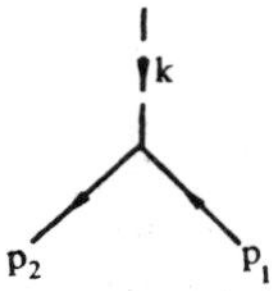

(106.7)

In the subsequent approximations, the diagrams are complicated by the addition of new vertices, but not all such diagrams provide essentially new information. For instance, in the third order we have the diagrams

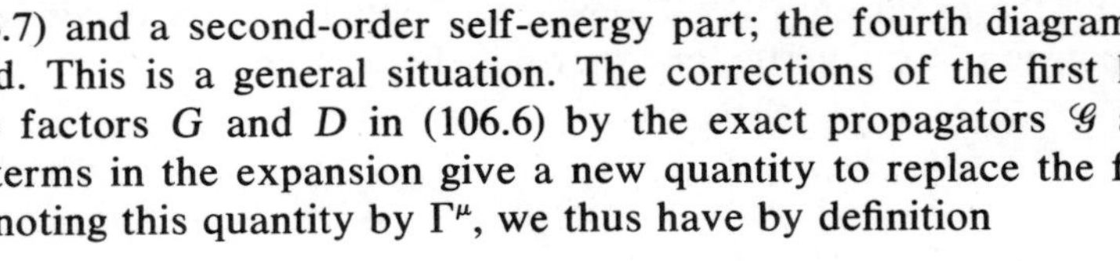

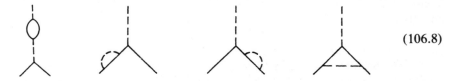

(106.8)

The first three of these can be cut (across one photon or electron line) into a simple vertex (106.7) and a second-order self-energy part; the fourth diagram cannot be thus treated. This is a general situation. The corrections of the first kind simply replace the factors G and D in (106.6) by the exact propagators $\mathcal{G}$ and $\mathcal{D}$. The remaining terms in the expansion give a new quantity to replace the factor γ^μ in (106.6). Denoting this quantity by Γ^μ, we thus have by definition

$$K^{\mu}(p_2, p_1; k) = \{i\mathcal{G}(p_2)[-ie\Gamma_\nu(p_2, p_1; k)]i\mathcal{G}(p_1)\}[-i\mathcal{D}^{\nu\mu}(k)]. \tag{106.9}$$

A section joined to other parts of the diagram by one photon line and two electron lines is called a *vertex part* if it cannot be divided into parts joined by only one (electron or photon) line. The quantity Γ^μ is the sum of an infinity of vertex parts, including the simple vertex γ^μ, and is called a *vertex operator* or *vertex function*.

The following are all the vertex-operator diagrams as far as the fifth-order quantities:

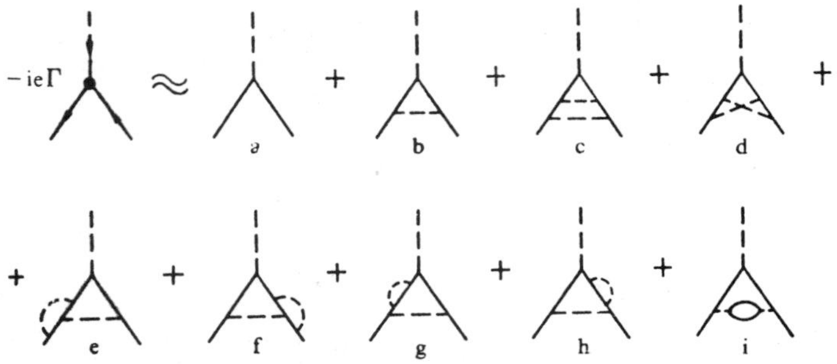

(106.10)

the black dot denoting the exact vertex operator $-ie\Gamma$.

The operator Γ (like the operator γ of the simple vertex) has two matrix (bispinor) indices and one 4-vector index; it is a function of two electron momenta (p_1, p_2) and one photon momentum (k). The three momenta cannot all relate to real particles simultaneously: the diagram (106.4) in itself (not as part of a larger diagram) would correspond to the absorption of a photon by a free electron, but this process is incompatible with the conservation of the 4-momentum of real particles. Hence at least one of the three free ends of the diagram must pertain to a virtual particle (or to an external field).

The vertex parts may also be classified as *reducible* and *irreducible.* The irreducible ones are those which do not contain self-energy corrections to internal lines and in which it is not possible to separate parts which constitute (lower-order) corrections to internal vertices. For example, of the diagrams in (106.10), the only irreducible ones are (b) and (d) (apart from the simple vertex (a)). Diagrams (g), (h) and (i) contain self-energy parts; in diagram (c) the upper broken horizontal line may be regarded as a correction to the upper vertex, and in diagrams (e) and (f) the lateral broken lines may be regarded as corrections to the lateral vertices.

When the internal lines in irreducible diagrams are replaced by corresponding thick lines, and the vertices by black dots, i.e. when the approximate propagators D and G are replaced by the exact propagators $\mathscr{D}$ and $\mathscr{G}$, and the approximate vertex operators γ by the exact ones Γ,† we evidently obtain the set of all vertex parts. Thus the expansion of the vertex operator may be written

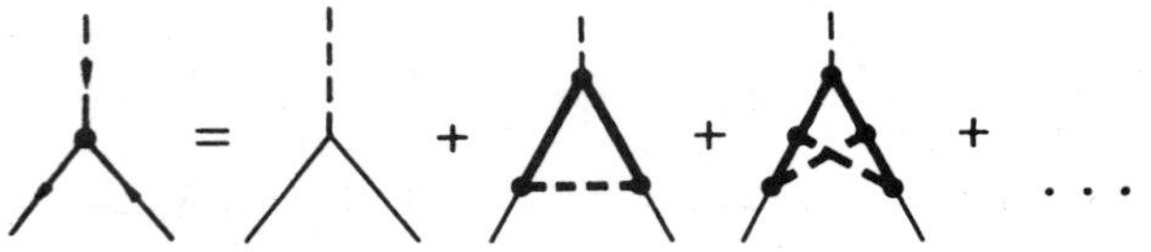

(106.11)

This equation is an integral equation for Γ, with an infinity of terms on the right.

From the above discussion we can easily derive the general principle of construction of the exact expressions for sections having any number of ends.

† The resulting diagrams are called *skeleton diagrams.*

They are obtained as vacuum mean values of T products of Heisenberg operators, with one operator $\hat{\bar{\psi}}(x)$ for each initial electron, one $\hat{\psi}(x)$ for each final electron, and one $\hat{A}(x)$ for each photon.

A further example is given by diagrams of the form

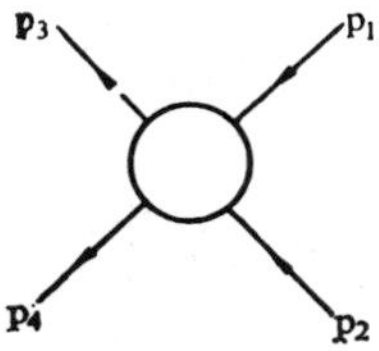

(106.12)

with four external electron lines. These are obtained from the function

$$K_{ik,\,lm}(x_1, x_2; x_3, x_4) = \langle 0|T\psi_i(x_1)\psi_k(x_2)\bar{\psi}_l(x_3)\bar{\psi}_m(x_4)|0\rangle, \tag{106.13}$$

which, of course, depends only on the differences of the four arguments. Its Fourier components may be written

$$\int K_{ik,\,lm}(x_1, x_2; x_3, x_4)e^{i(p_3x_1+p_4x_2-p_1x_3-p_2x_4)}\, d^4x_1\, d^4x_2\, d^4x_3\, d^4x_4$$
$$= (2\pi)^4\delta^{(4)}(p_1+p_2-p_3-p_4)K_{ik,\,lm}(p_3, p_4; p_1, p_2), \tag{106.14}$$

with

$$K_{ik,\,lm}(p_3, p_4; p_1, p_2)$$
$$= (2\pi)^4\delta^{(4)}(p_1-p_3)\mathcal{G}_{il}(p_1)\mathcal{G}_{km}(p_2) - (2\pi)^4\delta^{(4)}(p_2-p_3)\mathcal{G}_{im}(p_1)\mathcal{G}_{kl}(p_2) +$$
$$+ \mathcal{G}_{in}(p_3)\mathcal{G}_{kr}(p_4)[-i\Gamma_{nr,\,st}(p_3, p_4; p_1, p_2)]\mathcal{G}_{sl}(p_1)\mathcal{G}_{tm}(p_2). \tag{106.15}$$

In the latter expression, the first two terms exclude from the definition of the function $\Gamma(p_3, p_4; p_1, p_2)$ diagrams which fall into two disconnected parts, each having two free ends:

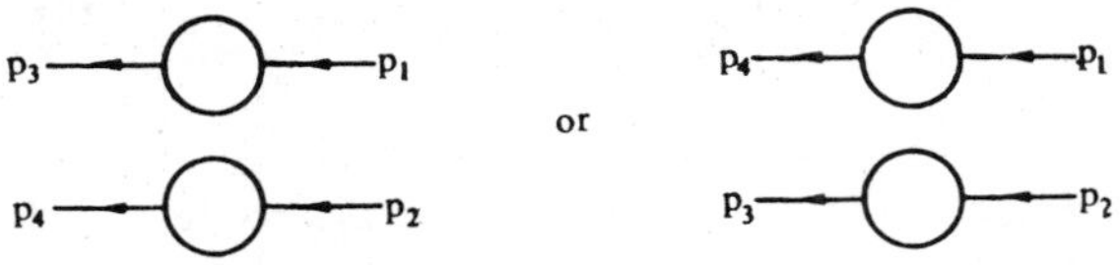

In the third term the $\mathcal{G}$ factors exclude from the definition of Γ those parts of diagrams which are corrections to external electron lines.

From the properties of the T product of the Fermi ψ-operators, it follows that the functions $\Gamma(p_3, p_4; p_1, p_2)$ are antisymmetric:

$$\Gamma_{ik,\,lm}(p_3, p_4; p_1, p_2) = -\Gamma_{ki,\,lm}(p_4, p_3; p_1, p_2)$$
$$= -\Gamma_{ik,\,ml}(p_3, p_4; p_2, p_1). \tag{106.16}$$

If the momenta p_1, p_2, p_3, p_4 correspond to real particles, the non-separating (i.e.

connected) diagrams (106.12) represent the scattering of two electrons. The scattering amplitude is found by assigning the wave amplitudes of the particles (instead of the propagators $\mathscr{G}$) to the free ends of the diagram:†

$$iM_{fi} = \bar{u}_i(p_3)\bar{u}_k(p_4)[-ie\Gamma_{ik,\,lm}(p_3, p_4; p_1, p_2)]u_l(p_1)u_m(p_2). \tag{106.17}$$

According to (106.16) this amplitude must have the appropriate antisymmetry with respect to interchanges of electrons.

§ 107. **Dyson's equations**

The exact propagators and the vertex part satisfy certain integral relations, the origin of which is particularly clear if the diagram technique is used.

The concepts of reducibility and irreducibility defined in §106 can be applied not only to vertex parts but also to any other diagrams or parts thereof. Let us consider from this aspect the compact self-energy electron diagrams.

It is easily seen that only one diagram out of this infinity is irreducible, namely the second-order diagram

Any complication of this diagram can be regarded as the application of further corrections to its internal (electron or photon) lines or to one of its vertices. Here it is important to note that, owing to the obvious symmetry of the diagram, any vertex correction need by assigned only to either one or the other vertex.‡

Since, therefore, only one of the compact self-energy electron parts is irreducible, the ensemble of all such parts (i.e. the mass operator $\mathscr{M}$) is represented by only one skeleton diagram:

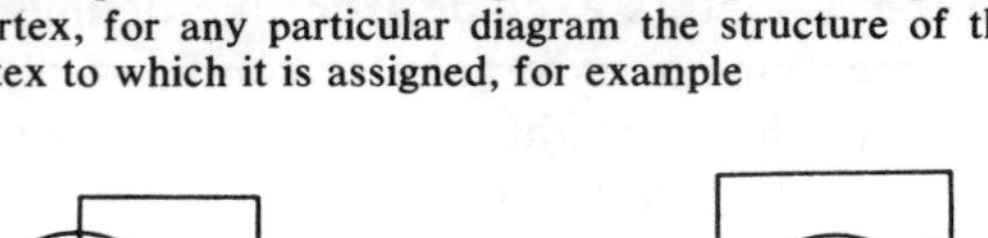

(107.1)

† It will be seen later (§110) that the self-energy parts in the free ends can be ignored in deriving the amplitudes of real processes.

‡ For clarity, it should be emphasized that, although all the required diagrams are found by applying corrections to only one vertex, for any particular diagram the structure of the correction section in general depends on the vertex to which it is assigned, for example

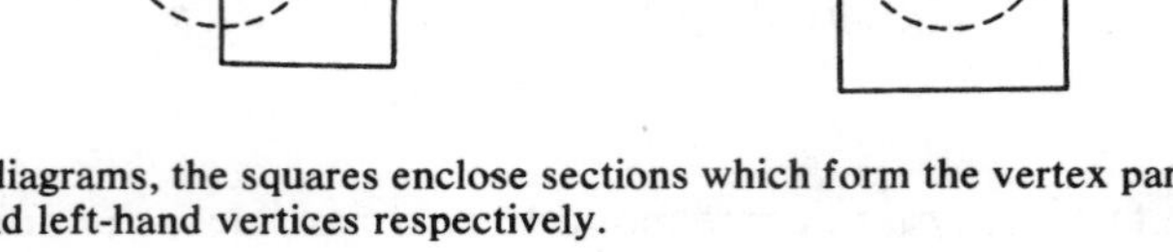

where, in identical diagrams, the squares enclose sections which form the vertex part when it is assigned to the right-hand and left-hand vertices respectively.

In analytical form, this graphical equation becomes†

$$\mathcal{M}(p) = G^{-1}(p) - \mathcal{G}^{-1}(p)$$

$$= -ie^2 \int \gamma^\nu \mathcal{G}(p+k)\Gamma^\mu(p+k, p; k) \cdot \mathcal{D}_{\mu\nu}(k) \frac{d^4k}{(2\pi)^4}. \qquad (107.2)$$

A similar expression can be derived for the polarization operator $\mathcal{P}$. Again only one of the compact self-energy photon parts is irreducible, and $\mathcal{P}$ is therefore represented by a single skeleton diagram:

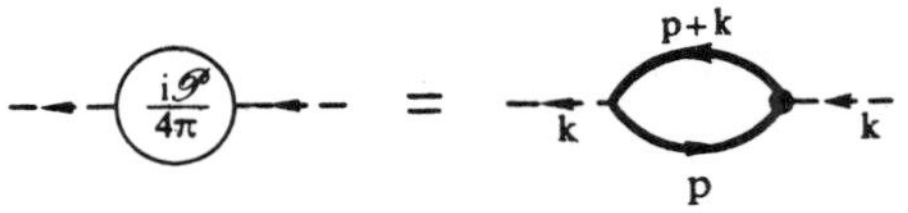

(107.3)

The corresponding analytical equation is

$$\frac{\mathcal{P}_{\mu\nu}(k)}{4\pi} = D^{-1}{}_{\mu\nu}(k) - \mathcal{D}^{-1}{}_{\mu\nu}(k)$$

$$= ie^2 \operatorname{tr} \int \gamma_\mu \mathcal{G}(p+k)\Gamma_\nu(p+k, p; k)\mathcal{G}(p) \frac{d^4p}{(2\pi)^4}; \qquad (107.4)$$

the bispinor indices are omitted from (107.2) and (107.4).

The relations (107.2) and (107.4) are called *Dyson's equations*; they can also be obtained by direct calculation. For example, to derive (107.2) we consider the quantity

$$(\gamma\hat{p} - m)_{il}\mathcal{G}_{lk}(x - x') = -i(\gamma\hat{p} - m)_{il}\langle 0|\mathrm{T}\psi_l(x)\bar{\psi}_k(x')|0\rangle,$$

where $\hat{p} = i\partial$ is the operator of differentiation with respect to x, which is found from (102.5) in exactly the same way as was done when deriving (75.7) for the free-particle propagator. The result is

$$(\gamma\hat{p} - m)_{il}\mathcal{G}_{lk}(x - x')$$

$$= -ie\gamma^\nu_{il}\langle 0|\mathrm{T}A_\nu(x)\psi_l(x)\bar{\psi}_k(x')|0\rangle + \delta_{ik}\delta^{(4)}(x - x');$$

the delta-function term on the right is the same as in (75.7), since the commutation properties at $t = t'$ are the same for ψ-operators in the Heisenberg and interaction representations. The first term is $-ie\gamma_\nu K^\nu_{lk}(x, x, x')$, and we can thus write (again omitting the bispinor indices)

$$(\gamma\hat{p} - m)\mathcal{G}(x - x') = -ie\gamma^\mu K_\mu(x, x, x') + \delta^{(4)}(x - x'). \qquad (107.5)$$

To obtain the Fourier components we note that, if the definition (106.3) is

† If the exact vertex part is assigned to the left-hand vertex in (107.1), the factors γ and Γ are interchanged in (107.2). The two forms of the equation are, of course, essentially equivalent.

integrated over $d^4k d^4p_2/(2\pi)^8$, the result is

$$\int K^\mu(p+k, p; k)\frac{d^4k}{(2\pi)^4} = \int K^\mu(0, 0, x_3)e^{-ipx_3}\, d^4x_3$$
$$= \int K^\mu(x, x, x')e^{ip(x-x')}\, d^4(x-x'), \qquad (107.6)$$

from which it is seen that the integral on the left is the Fourier component of $K^\mu(x, x, x')$. Thus, by taking the Fourier components of both sides of (107.5), and using the definition (106.9) and the formula $\gamma p - m = G^{-1}(p)$, we find

$$G^{-1}(p)\mathcal{G}(p) = 1 - ie^2 \int \gamma^\nu \mathcal{G}(p+k)\Gamma^\mu(p+k, p; k)\mathcal{G}(p)D_{\mu\nu}(k)\frac{d^4k}{(2\pi)^4}.$$

Finally, multiplying this on the right by $\mathcal{G}^{-1}(p)$, we obtain (107.2).

§ 108. **Ward's identity**

Another relationship between the photon propagator and the vertex part, simpler than Dyson's equation, follows from gauge invariance. To derive, it, we apply the gauge transformation (102.8), assuming that $\chi(x) \equiv \delta\chi(x)$ is an infinitesimal non-operator function of the 4-coordinates x. Then the change in the electron propagator is

$$\delta\mathcal{G}(x, x') = ie\mathcal{G}(x-x')[\delta\chi(x) - \delta\chi(x')]. \qquad (108.1)$$

Note that this gauge transformation violates the homogeneity of space–time, and the function $\delta\mathcal{G}$ depends on the arguments x and x' separately, not only on the difference $x - x'$. Its Fourier expansion must therefore be made in the variables x and x' separately. Thus, in the momentum representation $\delta\mathcal{G}$ is a function of two 4-momenta:

$$\delta\mathcal{G}(p_2, p_1) = \int\int \delta\mathcal{G}(x, x')e^{ip_2x - ip_1x'}\, d^4x\, d^4x'.$$

Substituting (108.1) and integrating over $d^4x\, d^4\xi$ or $d^4\xi\, d^4x'$ ($\xi = x - x'$), we get

$$\delta\mathcal{G}(p+q, p) = ie\delta\chi(q)[\mathcal{G}(p) - \mathcal{G}(p+q)]. \qquad (108.2)$$

With the same gauge transformation, the operator $\hat{A}_\mu(x)$ is augmented by the function

$$\delta A_\mu^{(e)}(x) = -\frac{\partial}{\partial x^\mu}\delta\chi, \qquad (108.3)$$

which may be regarded as an infinitesimal external field. In the momentum representation,

$$\delta A_\mu^{(e)}(q) = iq_\mu \delta\chi(q). \tag{108.4}$$

The quantity $\delta\mathcal{G}$ can also be calculated as the change in the propagator under the action of this field. As far as quantities of the first order in $\delta\chi$, this change can evidently be represented by a single skeleton diagram:

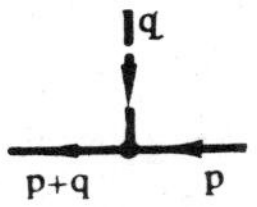

The thick broken line is the effective external-field line, corresponding to the factor (see (103.15))

$$\delta A_\mu^{(e)}(q) + \delta A_\lambda^{(e)}(q)\frac{\mathcal{P}^{\lambda\nu}(q)}{4\pi}\mathcal{D}_{\nu\mu}(q).$$

The 4-vector $\delta A_\lambda^{(e)}(q)$ is longitudinal (with respect to q) and the tensor $\mathcal{P}^{\lambda\nu}$ is transverse. The second term is therefore zero, leaving

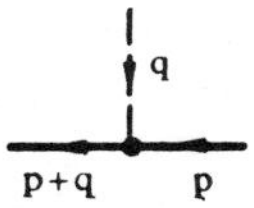

(108.5)

where the thin broken line corresponds, in the usual manner, to the field $\delta A^{(e)}$ simply. In the analytical form,

$$\delta\mathcal{G} = e\mathcal{G}(p+q)\Gamma^\mu(p+q, p; q)\mathcal{G}(p)\cdot\delta A_\mu^{(e)}. \tag{108.6}$$

Substituting (108.4) and comparing with (108.2), we get

$$\mathcal{G}(p+q) - \mathcal{G}(p) = -\mathcal{G}(p+q)\Gamma^\mu(p+q, p; q)\mathcal{G}(p)\cdot q_\mu$$

or, in terms of the inverse matrices,

$$\mathcal{G}^{-1}(p+q) - \mathcal{G}^{-1}(p) = q_\mu\Gamma^\mu(p+q, p; q) \tag{108.7}$$

(H. S. Green, 1953).

Taking the limit of this equation as $q\to 0$ and equating coefficients when q_μ is infinitesimal, we get

$$\frac{\partial}{\partial p_\mu}\mathcal{G}^{-1}(p) = \Gamma^\mu(p, p; 0). \tag{108.8}$$

This is *Ward's identity* (J. C. Ward, 1950). We see that the momentum derivative of

$\mathcal{G}^{-1}(p)$ is equal to the vertex operator with zero momentum transfer.† The derivative of the function $\mathcal{G}(p)$ itself is

$$-\frac{\partial}{\partial p_\mu} i\mathcal{G}(p) = i\mathcal{G}(p)[-i\Gamma^\mu(p, p; 0)]i\mathcal{G}(p). \qquad (108.9)$$

The higher derivatives could be found similarly by continuing the calculations to higher orders in $\delta\chi$, but we shall not need these expressions.

Let us now consider the derivative $\partial\mathcal{P}(k)/\partial k_\mu$ of the polarization operator. Unlike $\mathcal{G}(p)$, $\mathcal{P}(k)$ is gauge-invariant and is unchanged by the application of the fictitious external field (108.4). Its derivative therefore cannot be calculated in the same way, but a diagram expression can be obtained for this derivative too. To do so, we consider the first diagram in the definition of $\mathcal{P}$: the second-order diagram

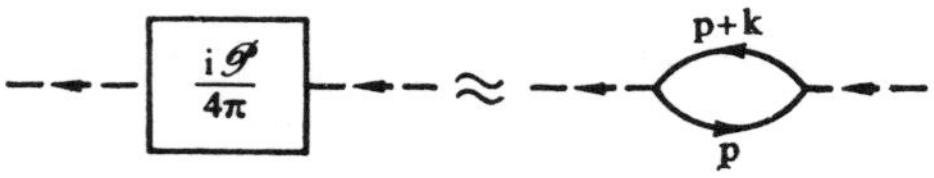

(108.10)

The continuous lines correspond to the factors $iG(p)$ and $iG(p+k)$. Differentiation with respect to k replaces the second factor by $i\partial G(p+k)/\partial k$, and according to the identity (108.9) this change is equivalent to adding a further vertex on the electron line:

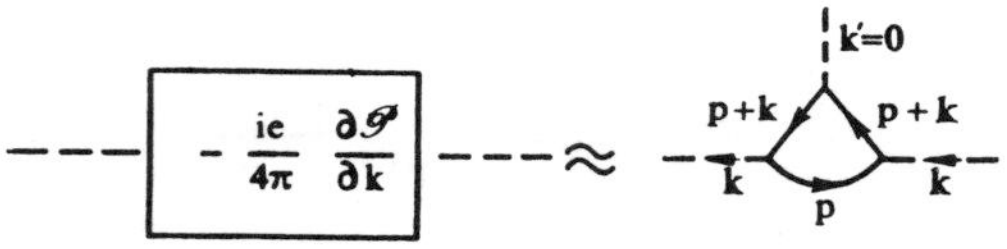

(108.11)

We see that, in the first non-vanishing order, the required derivative has been expressed in terms of a diagram having three photon ends. It must be stressed immediately that this diagram does not itself give the amplitude for the transformation of one photon into two. The amplitude of this process is the sum of (108.11) and a similar diagram in which the loop is traversed in the other direction, and the sum is zero by Furry's theorem. The diagram (108.11) is not itself zero.

In a similar manner, we can differentiate more complicated diagrams by successively adding vertices with $k' = 0$ on all the electron lines which depend on k. There are, however, diagrams in which the dependence on k occurs in the internal photon lines also, for instance the diagram on the left in the next equation:

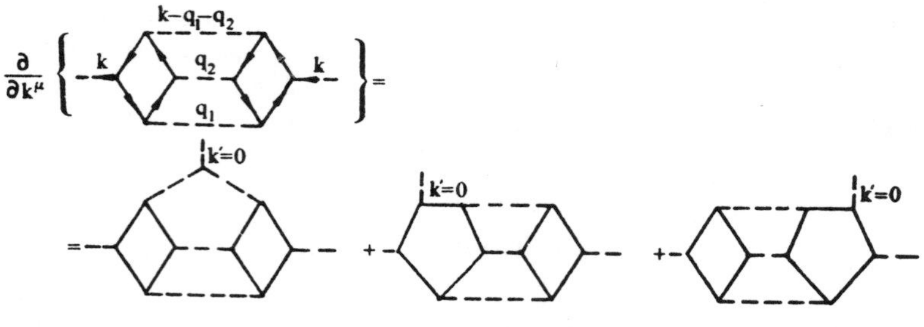

† In the zero-order approximation, i.e. for the free-particle propagator, this identity is obvious: $G^{-1}(p) = \gamma p - m$, and therefore $\partial G^{-1}/\partial p_\mu = \gamma^\mu$.

The derivative of the diagram in the braces is shown here in diagram form by means of a new graphical symbolism, a fictitious three-particle photon vertex, i.e. a point where three broken lines meet, corresponding to the quantity

$$4\pi i \frac{\partial D^{-1}}{\partial k^{\mu}} = 2ik_{\mu} \equiv v_{\mu}. \tag{108.12}$$

We can now differentiate any diagram by adding, to the lines depending on k, vertices v_{μ} or γ_{μ} and continuing in accordance with the general rules. Summation of these higher-order corrections gives

$$-\frac{1}{4\pi}\frac{\partial \mathcal{P}_{\mu\nu}}{\partial k^{\lambda}} = \mathcal{V}_{\mu\lambda\nu}, \tag{108.13}$$

where $ie\mathcal{V}_{\mu\lambda\nu}$ is the sum of the internal parts of all the diagrams with three photon ends thus obtained.

We shall also need the second derivative of the polarization operator. Differentiating the equation (108.13) once more in a similar manner, we have

$$\frac{1}{4\pi}\frac{\partial^2 \mathcal{P}_{\mu\nu}}{\partial k^{\rho}\partial k^{\sigma}} = \mathcal{S}_{\mu\rho\sigma\nu} + \mathcal{S}_{\mu\sigma\rho\nu}, \tag{108.14}$$

where $ie^2\mathcal{S}$ is the sum of the internal parts of all the diagrams with four photon ends such as

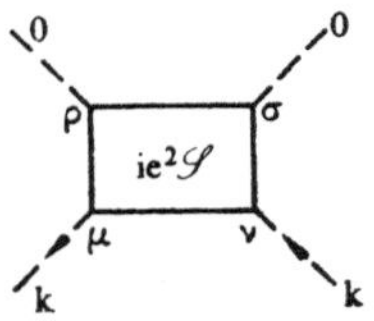

(108.15)

including, of course, those containing the fictitious three-particle vertices (108.12).

§ 109. Electron propagators in an external field

If a system is in a given external field $A^{(e)}(x)$, the exact electron propagator is expressed by the same formula (105.1), but in the Hamiltonian $\hat{H} = \hat{H}_0 + \hat{V}$ which converts to the Heisenberg representation of operators we have also the interaction between the electrons and the external field:

$$\hat{V} = e\int \hat{A}_{\mu}\hat{j}^{\mu}\, d^3x + e\int A^{(e)}_{\mu}\hat{j}^{\mu}\, d^3x. \tag{109.1}$$

Since the external field makes space and time no longer homogeneous, the propagator $\mathcal{G}(x, x')$ will now depend on the two arguments x and x' separately and not only on the difference $x - x'$.

If we proceed in the usual manner to the interaction representation, the ordinary diagram technique is obtained, with external-field lines as well as virtual photon lines. This technique is, however, unsuitable when the external field cannot be regarded as a small perturbation, in particular when the particles may be in bound states in the field. Now the electron propagator in an external field is in fact required principally for the analysis of the properties of bound states, and in particular for determining the energy levels with allowance for radiative corrections. In order to derive such a propagator, we have to start from a representation of operators where the external field is exactly taken into account even in the "zero-order" approximation with respect to the electron–photon interaction (W. H. Furry, 1951).

We shall henceforward assume the external field to be independent of time. The desired representation of the ψ-operators is given by the formulae (32.9) for second quantization in an external field:

$$\left.\begin{aligned}\hat{\psi}^{(e)}(t,\mathbf{r}) &= \sum_n \{\hat{a}_n\psi_n^{(+)}(\mathbf{r})e^{-i\varepsilon_n^{(+)}t} + \hat{b}_n^+\psi_n^{(-)}(\mathbf{r})e^{i\varepsilon_n^{(-)}t}\},\\ \hat{\bar{\psi}}^{(e)}(t,\mathbf{r}) &= \sum_n \{\hat{a}_n^+\bar{\psi}_n^{(+)}(\mathbf{r})e^{i\varepsilon_n^{(+)}t} + \hat{b}_n\bar{\psi}_n^{(-)}(\mathbf{r})e^{-i\varepsilon_n^{(-)}t}\},\end{aligned}\right\} \tag{109.2}$$

where $\psi_n^{(\pm)}(\mathbf{r})$ and $\varepsilon_n^{(\pm)}$ are the wave functions and energy levels of the electron and the positron respectively, which are solutions of the "single-particle" problem, i.e. of Dirac's equation for a particle in a field. It is easily seen that the operators (109.2) are ψ-operators in a certain representation (the *Furry representation*) which is, as it were, intermediate between the Heisenberg and interaction representations. They may be written

$$\left.\begin{aligned}\hat{\psi}^{(e)}(t,\mathbf{r}) &= e^{i\hat{H}_1t}\hat{\psi}(\mathbf{r})e^{-i\hat{H}_1t},\\ \hat{\bar{\psi}}^{(e)}(t,\mathbf{r}) &= e^{i\hat{H}_1t}\hat{\bar{\psi}}(\mathbf{r})e^{-i\hat{H}_1t},\end{aligned}\right\} \tag{109.3}$$

where

$$\hat{H}_1 = \hat{H}_0 + e\int A_\mu^{(e)}(x)\hat{j}^\mu(x)\,d^3x.$$

The electromagnetic-field operator $\hat{A}_\mu$ of course commutes with the second term in $\hat{H}_1$, and so the Furry representation is the same as the interaction representation for this operator.

The electron propagator in the zero-order approximation, in the new representation, is defined as

$$G_{ik}^{(e)}(x,x') = -i\langle 0|\mathrm{T}\psi_i^{(e)}(x)\bar{\psi}_k^{(e)}(x')|0\rangle. \tag{109.4}$$

The operator $\hat{\psi}^{(e)}(t,\mathbf{r})$ satisfies Dirac's equation in the external field:

$$[\gamma\hat{p} - e\gamma A^{(e)}(x) - m]\hat{\psi}^{(e)}(t,\mathbf{r}) = 0, \tag{109.5}$$

and the function $G^{(e)}$ correspondingly satisfies the equation

$$[\gamma\hat{p} - e\gamma A^{(e)}(x) - m]G^{(e)}(x, x') = \delta^{(4)}(x - x'); \tag{109.6}$$

cf. the derivation of (107.5).

The diagram technique, which expresses the exact propagator $\mathcal{G}$ as a series in powers of e^2, is obtained by changing from the Heisenberg to the Furry representation, in exactly the same way as the earlier change to the interaction representation. The resulting diagrams are of the same form, with the continuous lines now corresponding to factors $iG^{(e)}$ instead of iG.

One slight difference in the rules for writing the analytical expressions for the diagrams arises because in the coordinate representation $G^{(e)}$ is not a function of the difference $x - x'$ only. In a constant external field, however, the homogeneity of time is preserved, and so the times t and t' will again appear only as the difference $t - t' \equiv \tau$:

$$G^{(e)} = G^{(e)}(\tau, \mathbf{r}, \mathbf{r}').$$

The momentum representation is obtained by a Fourier expansion with respect to each of the arguments of the function:

$$G^{(e)}(\tau, \mathbf{r}, \mathbf{r}') = \iiint e^{i(\mathbf{p}_2\cdot\mathbf{r} - \mathbf{p}_1\cdot\mathbf{r}' - \varepsilon\tau)} G(\varepsilon, \mathbf{p}_2, \mathbf{p}_1)\frac{d\varepsilon}{2\pi}\frac{d^3p_1}{(2\pi)^3}\frac{d^3p_2}{(2\pi)^3}. \tag{109.7}$$

Each line corresponding to the factor $iG^{(e)}(\varepsilon, \mathbf{p}_2, \mathbf{p}_1)$ must now be assigned one value of the virtual energy ε and two values of the momentum, the initial value $\mathbf{p}_1$ and the final value $\mathbf{p}_2$:

$$iG^{(e)}(\varepsilon, \mathbf{p}_2, \mathbf{p}_1) = \underline{\mathbf{p}_2\ \varepsilon\ \mathbf{p}_1}. \tag{109.8}$$

This leads to the rule for writing the analytical expressions, in which the integration over $d\varepsilon/2\pi$ is normal, but those over $d^3p_1/(2\pi)^3$ and $d^3p_2/(2\pi)^3$ are independent, the conservation of momentum at each vertex being taken into account. For example,

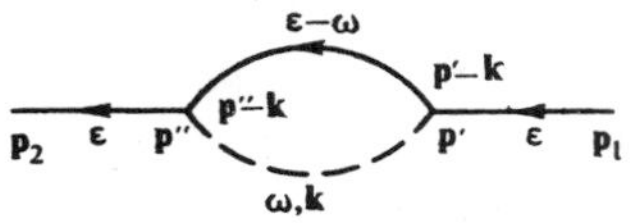

=

$$= e^2 \iiint G^{(e)}(\varepsilon, \mathbf{p}_2, \mathbf{p}'')\gamma^\mu G^{(e)}(\varepsilon - \omega, \mathbf{p}'' - \mathbf{k}, \mathbf{p}' - \mathbf{k}) \times$$

$$\times\, \gamma^\nu G^{(e)}(\varepsilon, \mathbf{p}', \mathbf{p}_1) D_{\mu\nu}(\omega, \mathbf{k})\frac{d^4k}{(2\pi)^4}\frac{d^3p'}{(2\pi)^3}\frac{d^3p''}{(2\pi)^3}. \tag{109.9}$$

It is important to note that in this technique one must also take account of diagrams with "self-closed" electron lines, which in the ordinary technique are rejected as being associated with a "vacuum current". When an external field is present, this current need not be zero, because of the "vacuum polarization" caused by the field. For instance, in the diagram

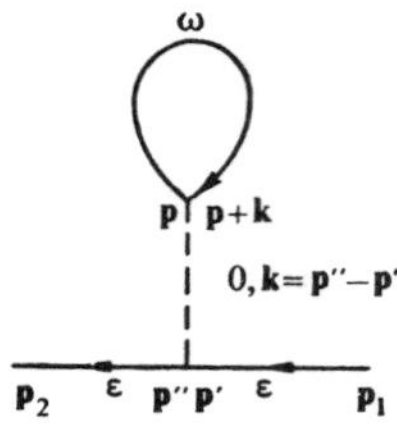

(109.10)

the loop at the top corresponds to the factor

$$i\iint G^{(e)}(\omega, \mathbf{p}+\mathbf{k}, \mathbf{p})\frac{d^3p}{(2\pi)^3}\frac{d\omega}{2\pi}. \tag{109.11}$$

Here, however, we must still specify the meaning of the integral over $d\omega$. This is because the integration of the Fourier component of the function $G^{(e)}(\tau)$ with respect to ω amounts to taking the value of that function at $\tau = 0$, and $G^{(e)}(\tau)$ is discontinuous at $\tau = 0$; we must therefore indicate which of its two limiting values is to be taken. To resolve this question, we need only note that the integral (109.11) arises from the contraction of ψ-operators in the same current operator:

$$\hat{j}^\mu = \hat{\bar{\psi}}^{(e)}(t, \mathbf{r})\gamma^\mu\hat{\psi}^{(e)}(t, \mathbf{r}),$$

where $\hat{\bar{\psi}}^{(e)}$ is to the left of $\hat{\psi}^{(e)}$. According to the definition of the propagator (109.4), this order of factors for $t = t'$ is obtained if t' is taken as $t + 0$, i.e. if the limiting value of the function $G^{(e)}(t - t')$ as $t - t' \to -0$ is taken. In other words, the integral over $d\omega/2\pi$ in (109.11) is to be taken as

$$\int \cdots e^{-i\omega\tau}\frac{d\omega}{2\pi} \quad \text{for} \quad \tau \to -0. \tag{109.12}$$

The mass operator in the external field is defined as in §105: $-i\mathcal{M}$ is the sum of all the compact self-energy parts. It is now a function of the energy ε and the momenta $\mathbf{p}_1$ and $\mathbf{p}_2$ at the ends of the external lines where they respectively enter and leave the part in question:

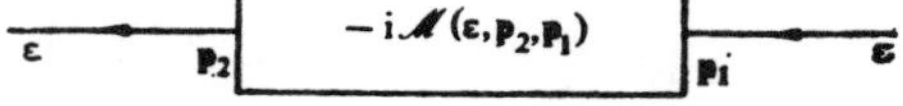

(109.13)

Proceeding exactly as in the derivation of (105.6), we get the equation

$$\mathcal{G}(\varepsilon, \mathbf{p}_2, \mathbf{p}_1) - G^{(e)}(\varepsilon, \mathbf{p}_2, \mathbf{p}_1)$$
$$= \iint G^{(e)}(\varepsilon, \mathbf{p}_2, \mathbf{p}'')\mathcal{M}(\varepsilon, \mathbf{p}'', \mathbf{p}')\mathcal{G}(\varepsilon, \mathbf{p}', \mathbf{p}_1)\frac{d^3p'}{(2\pi)^3}\frac{d^3p''}{(2\pi)^3}. \tag{109.14}$$

This can be put in a more natural form by returning to the coordinate representation in terms of the spatial variables, using the function

$$\mathscr{G}(\varepsilon, \mathbf{r}, \mathbf{r}') = \iint \mathscr{G}(\varepsilon, \mathbf{p}_2, \mathbf{p}_1) e^{i(\mathbf{p}_2\cdot\mathbf{r} - \mathbf{p}_1\cdot\mathbf{r}')} \frac{d^3p_1\, d^3p_2}{(2\pi)^6}, \tag{109.15}$$

and similarly for the other quantities. Taking the inverse Fourier transform of (109.14), we obtain

$$\mathscr{G}(\varepsilon, \mathbf{r}, \mathbf{r}') - G^{(e)}(\varepsilon, \mathbf{r}, \mathbf{r}') = \iint G^{(e)}(\varepsilon, \mathbf{r}, \mathbf{r}_2)\mathscr{M}(\varepsilon, \mathbf{r}_2, \mathbf{r}_1)\mathscr{G}(\varepsilon, \mathbf{r}_1, \mathbf{r}')\, d^3x_1\, d^3x_2.$$

Next we apply to both sides the operator

$$\gamma^0\varepsilon - \boldsymbol{\gamma}\cdot\hat{\mathbf{p}} - e\gamma^\mu A_\mu^{(e)}(x),$$

where ε is a number, and $\hat{\mathbf{p}} = -i\nabla$ is the operator of differentiation with respect to the coordinates $\mathbf{r}$. Here it must be noted that, by (109.6),

$$[\gamma^0\varepsilon - \boldsymbol{\gamma}\cdot\hat{\mathbf{p}} - e\gamma A^{(e)}(x)]G^{(e)}(\varepsilon, \mathbf{r}, \mathbf{r}') = \delta(\mathbf{r} - \mathbf{r}'). \tag{109.16}$$

The resulting equation is

$$[\gamma^0\varepsilon - \boldsymbol{\gamma}\cdot\hat{\mathbf{p}} - e\gamma A^{(e)}(x)]\mathscr{G}(\varepsilon, \mathbf{r}, \mathbf{r}') - \int \mathscr{M}(\varepsilon, \mathbf{r}, \mathbf{r}_1)\mathscr{G}(\varepsilon, \mathbf{r}_1, \mathbf{r}')\, d^3x_1 = \delta(\mathbf{r} - \mathbf{r}'). \tag{109.17}$$

The function $\mathscr{G}(\varepsilon, \mathbf{r}, \mathbf{r}')$ has the especially valuable property that its poles determine the energy levels of the electron in the external field. We shall prove this first for the approximate function $G^{(e)}(\varepsilon, \mathbf{r}, \mathbf{r}')$. Substituting the operators (109.2) in the definition of the propagator (109.4), we obtain, in exact analogy to formulae (75.12) for the free-particle propagator,

$$G_{ik}^{(e)}(t - t', \mathbf{r}, \mathbf{r}') = \begin{cases} -i \sum_n \psi_{ni}^{(+)}(\mathbf{r})\bar{\psi}_{nk}^{(+)}(\mathbf{r}') \exp\{-i\varepsilon_n^{(+)}(t - t')\}, & t > t', \\ i \sum_n \psi_{ni}^{(-)}(\mathbf{r})\bar{\psi}_{nk}^{(-)}(\mathbf{r}') \exp\{i\varepsilon_n^{(-)}(t - t')\}, & t < t', \end{cases} \tag{109.18}$$

and the Fourier time component is

$$G_{ik}^{(e)}(\varepsilon, \mathbf{r}, \mathbf{r}') = \sum_n \left\{ \frac{\psi_{ni}^{(+)}(\mathbf{r})\bar{\psi}_{nk}^{(+)}(\mathbf{r}')}{\varepsilon - \varepsilon_n^{(+)} + i0} + \frac{\psi_{ni}^{(-)}(\mathbf{r})\bar{\psi}_{nk}^{(-)}(\mathbf{r}')}{\varepsilon + \varepsilon_n^{(-)} - i0} \right\}. \tag{109.19}$$

We see that $G^{(e)}(\varepsilon, \mathbf{r}, \mathbf{r}')$, as an analytic function of ε, has poles on the positive real axis which coincide with the electron energy levels, and poles on the negative real axis which coincide with the positron energy levels. The values $\varepsilon_n^{(\pm)} > m$ form a continuous spectrum,† and the corresponding poles form two cuts in the ε-plane, from $-\infty$ to $-m$ and from m to ∞. The segment $|\varepsilon| < m$ contains poles which give the discrete energy levels.

For the exact propagator $\mathscr{G}(\varepsilon, \mathbf{r}, \mathbf{r}')$ we can obtain a similar expansion by expressing it in terms of the matrix elements of Schrödinger operators; the matrix elements of Heisenberg ψ-operators are related to these by

$$\langle m|\psi(t, \mathbf{r})|n\rangle = \langle m|\psi(\mathbf{r})|n\rangle e^{-i(E_n - E_m)t}. \tag{109.20}$$

Here the E_n are the exact energy levels (i.e. with all radiative corrections) of the system in the external field. The operator $\hat{\psi}$ increases the charge of the system by 1 (i.e. by $+|e|$), and $\hat{\bar{\psi}}$ decreases it by one. This means that in the matrix elements $\langle n|\psi|0\rangle$ and $\langle 0|\bar{\psi}|n\rangle$ the states $|n\rangle$ must correspond to a charge of the system of $+1$, i.e. they can contain, besides a single positron, only a certain number of electron–positron pairs and a certain number of photons; the energies of these states will be denoted by $E_n^{(-)}$. Similarly, in the matrix elements $\langle 0|\psi|n\rangle$ and $\langle n|\bar{\psi}|0\rangle$ the states $|n\rangle$ contain one electron and some pairs and photons (energy $E_n^{(+)}$). Instead of (109.18) we now have

$$\mathscr{G}_{ik}(t - t', \mathbf{r}, \mathbf{r}') = \begin{cases} -i \sum_n \langle 0|\psi_i(\mathbf{r})|n\rangle\langle n|\bar{\psi}_k(\mathbf{r}')|0\rangle \exp\{-iE_n^{(+)}(t - t')\}, & t > t', \\ i \sum_n \langle 0|\bar{\psi}_k(\mathbf{r}')|n\rangle\langle n|\psi_i(\mathbf{r})|0\rangle \exp\{iE_n^{(-)}(t - t')\}, & t < t', \end{cases} \tag{109.21}$$

and hence

$$\mathscr{G}_{ik}(\varepsilon, \mathbf{r}, \mathbf{r}') = \sum_n \left\{\frac{\langle 0|\psi_i(\mathbf{r})|n\rangle\langle n|\bar{\psi}_k(\mathbf{r}')|0\rangle}{\varepsilon - E_n^{(+)} + i0} + \frac{\langle 0|\bar{\psi}_k(\mathbf{r}')|n\rangle\langle n|\psi_i(\mathbf{r})|0\rangle}{\varepsilon + E_n^{(-)} - i0}\right\}. \tag{109.22}$$

Let ε be close to one of the discrete energy levels $E_n^{(+)}$ (or $-E_n^{(-)}$). Then only the corresponding pole term need be retained in the sum (109.22). Substitution in (109.17) shows that the factors which depend on the second argument $\mathbf{r}'$ (when $\mathbf{r} \neq \mathbf{r}'$) do not appear in the equation. The result is a homogeneous integro-differential equation for the function $\langle 0|\psi(\mathbf{r})|n\rangle$ (or $\langle n|\psi(\mathbf{r})|0\rangle$), which we denote for brevity by $\Psi_n(\mathbf{r})$.‡ Omitting the subscript n, we have

$$[\gamma^0\varepsilon + i\boldsymbol{\gamma}\cdot\nabla - e\gamma A^{(e)}(\mathbf{r})]_{ik}\Psi_k(\mathbf{r}) - \int \mathscr{M}_{ik}(\varepsilon, \mathbf{r}, \mathbf{r}_1)\Psi_k(\mathbf{r}_1)\, d^3x_1 = 0 \tag{109.23}$$

(J. Schwinger, 1951). The discrete energy levels E_n now appear as the eigenvalues of this equation. Thus (109.23) becomes the regular basis for determining these levels.

† We assume that the external field is zero at infinity.

‡ When radiative corrections are neglected, the $\Psi_n(\mathbf{r})$ are the same (for states with one electron or positron) as the wave functions $\psi_n^{(+)}$ or $\psi_n^{(-)}$ which are solutions of Dirac's equation.

For example, it can be used to determine the correction, in the first order with respect to $\mathcal{M}$, to the discrete electron energy level ε_n given by solving Dirac's equation:

$$[\gamma^0\varepsilon_n + i\boldsymbol{\gamma}\cdot\nabla - e\gamma A^{(e)}(\mathbf{r})]\psi_n(\mathbf{r}) = 0; \tag{109.24}$$

let the wave function $\psi_n(\mathbf{r})$ be normalized by the condition

$$\int \psi_n^*\psi_n\, d^3x = 1. \tag{109.25}$$

The eigenfunction of equation (109.23) may be written

$$\Psi_n(\mathbf{r}) = \psi_n(\mathbf{r}) + \psi_n^{(1)}(\mathbf{r}), \tag{109.26}$$

where $\psi_n^{(1)}$ is the correction to $\psi_n(\mathbf{r})$. Substituting (109.26) in (109.23), multiplying on the left by $\bar\psi_n(\mathbf{r})$ and integrating† over d^3x, we get the required expression

$$E_n - \varepsilon_n \approx \iint \bar\psi_{ni}(\mathbf{r})\mathcal{M}_{ik}(\varepsilon_n, \mathbf{r}, \mathbf{r}_1)\psi_{nk}(\mathbf{r}_1)\, d^3x\, d^3x_1. \tag{109.27}$$

§ 110. **Physical conditions for renormalization**

The theory discussed so far in this chapter has been largely formal. We have treated all quantities as if they were finite, and have deliberately passed over any infinities which occur in the theory. In the practical calculation of the functions $\mathcal{D}$, $\mathcal{G}$ and Γ by perturbation theory, however, divergent integrals arise, which cannot be assigned any definite values without further consideration. These divergences are a manifestation of the logical incompleteness of the existing quantum electrodynamics. It will be seen below, nevertheless, that in this theory it is possible to establish certain rules which allow an unambiguous "subtraction of infinities", and thus to obtain finite values for all quantities which have a direct physical meaning. These rules are based on obvious physical requirements that the photon mass is zero and the electron charge and mass are equal to their observed values.

Let us first ascertain the conditions to be imposed on the photon propagator, and consider a scattering process which can occur through one-particle intermediate states having one virtual photon. The amplitude of such a process must have a pole when the square of the total 4-momentum P of the initial particles is equal to the squared mass of the real photon, i.e. when $P^2 = 0$; we have seen in §79 that this requirement follows from the general condition of unitarity. The pole term

† In the integration we use the fact that the differential operator in (109.24) is self-conjugate, and thus transfer its action from $\psi_n^{(1)}$ to $\bar\psi_n$.

in the amplitude arises from a diagram of the form (79.1):

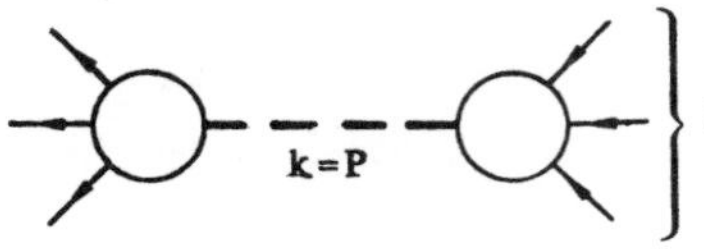

(110.1)

and when radiative corrections are taken into account the two parts of the diagram must be joined by a thick broken line (the exact photon propagator). This means that the function $\mathscr{D}(k^2)$ must have a pole at $k^2 = 0$, i.e. must be such that

$$\mathscr{D} \to 4\pi Z/k^2 \quad \text{when} \quad k^2 \to 0, \tag{110.2}$$

Z being a constant. Hence, for the polarization operator $\mathscr{P}(k^2)$, (103.21) gives

$$\mathscr{P}(0) = 0. \tag{110.3}$$

The coefficient in (110.2) is given by

$$\frac{1}{Z} = \left[1 - \frac{\mathscr{P}(k^2)}{k^2}\right]_{k^2 \to 0}.$$

Further restrictions on the function $\mathscr{P}(k^2)$ can be derived from an analysis of the physical definition of the particle's electric charge: two classical (i.e. infinitely heavy) particles at rest at a large distance apart must interact in accordance with Coulomb's law, $U = e^2/r$. (These are distances much greater than $1/m$, where m is the electron mass.) This interaction can also be represented by the diagram

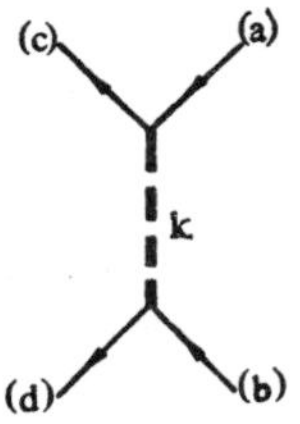

(110.4)

in which the upper and lower lines correspond to classical particles. The photon self-energy corrections are taken into account in the virtual photon line. All other corrections, affecting the heavy-particle lines, would make the diagram equal to zero: the addition of any further internal lines in the diagram (110.4), for example a photon line joining the lines a and c or a and b, would produce lines of virtual heavy particles, with corresponding propagators. But the propagator of a particle has its mass M in the denominator, and tends to zero as $M \to \infty$.

The form of the diagram (110.4) makes it clear (cf. §83) that the factor $e^2\mathscr{D}(k^2)$ in it must be (apart from the sign) the Fourier transform of the particle interaction potential. Since the interaction is steady, the virtual-photon frequency $\omega = 0$, and large distances correspond to small wave vectors **k**. The Fourier transform of the

Coulomb potential is $4\pi e^2/\mathbf{k}^2$. Since $\mathscr{D}$ depends only on $k^2 = \omega^2 - \mathbf{k}^2$, we finally arrive at the condition

$$\mathscr{D} \to 4\pi/k^2 \quad \text{when} \quad k^2 \to 0, \tag{110.5}$$

i.e. the coefficient in (110.2) must be $Z = 1$; the sign is obvious, since $\mathscr{D}(k^2)$ tends to the free-photon propagator $D(k^2)$. The polarization operator $\mathscr{P}(k^2)$ must therefore satisfy

$$\mathscr{P}(k^2)/k^2 \to 0 \quad \text{when} \quad k^2 \to 0. \tag{110.6}$$

This leads not only to the condition (110.3) given previously but also to the result

$$\mathscr{P}'(0) = 0. \tag{110.7}$$

It has been noted in §103 that an effective external real-photon line corresponds to the factor (103.15) or, using (103.16) and (103.20),

$$\left[1 + \frac{1}{4\pi}\mathscr{P}(0)\mathscr{D}(0)\right]e_\mu.$$

We now see from (110.5) and (110.6) that the correction term is zero. Thus we have the important result that radiative corrections need not be considered in external photon lines.

The natural physical requirements therefore lead to the establishment of definite values (namely zero) for the quantities $\mathscr{P}(0)$ and $\mathscr{P}'(0)$. The calculation of these quantities from the perturbation-theory diagrams would lead to divergent integrals, and we see that the way to eliminate such infinities is to assign fixed values a priori to the divergent expressions, these values being determined by physical requirements. This procedure is called *renormalization* of the quantities concerned.†

The procedure can also be formulated in a somewhat different manner. For instance, in renormalizing the particle charge one can define a non-physical intrinsic (bare or unrenormalized) charge e_c as a parameter which appears in the expression for the original electromagnetic interaction operator in formal perturbation theory. The renormalization condition then becomes $e_c^2\mathscr{D}(k^2) \to 4\pi e^2/k^2$ (when $k^2 \to 0$), e being the actual physical charge. Hence we have the relation $e_c^2 Z = e^2$, which is used to eliminate the non-physical quantity e_c from formulae which concern observable effects. By putting immediately $Z = 1$, the renormalization is effected "en route", and there is no need to use fictitious quantities even in the intermediate steps.

Let us now investigate the conditions for renormalization of the electron propagator. To do so, we now consider a scattering process which can take place

† The idea of this approach was first put forward by H. A. Kramers (1947); the systematic application of the renormalization method in quantum electrodynamics is due to Dyson, Tomonaga, Feynman, and Schwinger.

through a one-particle intermediate state with one virtual electron. The amplitude of such a process must have a pole when the square of the total 4-momentum P_i of the initial particles is equal to the squared mass of the real electron, i.e. when $P_i^2 = m^2$. The pole term in the amplitude arises from a diagram of the form

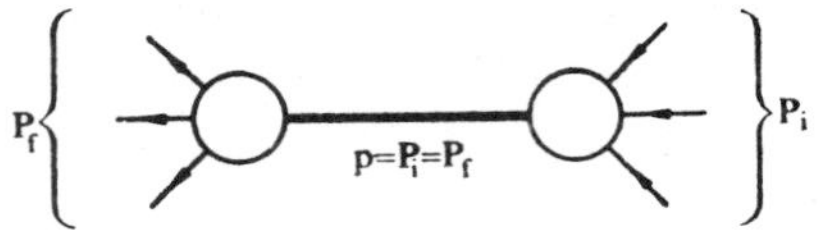

(110.8)

and when radiative corrections are taken into account the thick line is the exact electron propagator. This means that the function $\mathcal{G}(p)$ must have a pole at $p^2 = m^2$, i.e. its limiting form there must be

$$\mathcal{G}(p) \approx Z_1 \frac{\gamma p + m}{p^2 - m^2 + i0} + g(p) \quad \text{when} \quad p^2 \to m^2, \tag{110.9}$$

Z_1 being a scalar constant and $g(p)$ remaining finite as $p^2 \to m^2$. The matrix structure of the pole term in (110.9) (proportional to $\gamma p + m$) is a consequence of the same unitarity condition that causes the existence of the pole. We shall prove this statement and at the same time elucidate the important question of the renormalization conditions for the external electron lines.

If $\mathcal{G}(p)$ has the limiting form (110.9), the inverse matrix is

$$\mathcal{G}^{-1}(p) \approx \frac{1}{Z_1}(\gamma p - m) - (\gamma p - m) g (\gamma p - m) \quad \text{when} \quad p^2 \to m^2. \tag{110.10}$$

The mass operator is

$$\mathcal{M} = G^{-1} - \mathcal{G}^{-1} \approx \left(1 - \frac{1}{Z_1}\right)(\gamma p - m) + (\gamma p - m) g (\gamma p - m) \quad \text{when} \quad p^2 \to m^2. \tag{110.11}$$

The effective external (say incoming) electron line corresponds (cf. (103.15)) to a factor

$$\mathcal{U}(p) = u(p) + \mathcal{G}(p)\mathcal{M}(p)u(p), \tag{110.12}$$

where $u(p)$ is the ordinary amplitude of the electron wave function, which satisfies Dirac's equation $(\gamma p - m)u = 0$. Because of the requirements of relativistic invariance ($\mathcal{U}$, like u, is a bispinor), the limiting value of $\mathcal{U}(p)$ for $p^2 \to m^2$ can differ from that of $u(p)$ only by a constant scalar factor:

$$\mathcal{U}(p) = Z' u(p). \tag{110.13}$$

This factor Z' is related in a definite manner to the factor Z_1, but the relation cannot be determined simply by substituting (110.10) and (110.11) in (110.12), because there is an indeterminacy; the result depends on the order in which the limits of the various factors in (110.12) are taken.

It is, however, possible to avoid the problem of the correct method of taking the limit, by using instead the unitarity condition for the reaction shown by the diagram (110.8). The unitarity relation generally applies to amplitudes of processes as a whole, not to individual diagrams. But when $p^2 \to m^2$ the pole diagram (110.8) gives the main contribution to the corresponding amplitude M_{fi}, so that the other diagrams which pertain to the same reaction can be ignored.

As has been shown in §79, the unitary conditions require that a one-particle intermediate state should produce in the reaction amplitude an imaginary part with a delta function:

$$i\pi\delta(p^2 - m^2) \sum_{\text{polar.}} M_{fn} M_{in}^*, \tag{110.14}$$

where the subscript n refers to a state having one real electron, and the summation is over the latter's polarizations; to avoid additional complications we assume, as in §79, that both sides of the unitarity relation are symmetrized with respect to the helicities of the initial and final particles, so that $M_{fi} = M_{if}$. The amplitude M_{fn} corresponds to a process represented by the diagram

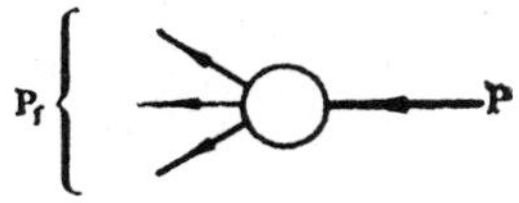

and is

$$M_{fn} = (M'_{fn}\mathcal{U}) = Z'(M'_{fn}u),$$

where M'_{fn} is a factor with one free bispinor index.† Similarly, the structure of the amplitude M_{in}^* is

$$M_{in}^* = (\bar{\mathcal{U}} M_{in}'^*) = Z'(\bar{u} M_{in}'^*).$$

Summation over the polarizations of the electron replaces the product $(M'_{fn}u) \times (\bar{u}M_{in}'^*)$ by $M'_{fn}(\gamma p + m)M_{in}$, and so the term (110.14) in the amplitude M_{fi} becomes

$$Z'^2 i\pi\delta(p^2 - m^2)\{M'_{fn}(\gamma p + m)M_{in}'^*\}.$$

Using this term in the imaginary part, we can reconstruct the entire pole term in the scattering amplitude; from (79.5),

$$M_{fi} = -\frac{Z'^2\{M'_{fn}(\gamma p + m)M_{in}'^*\}}{p^2 - m^2 + i0}, \quad p^2 \to m^2.$$

† One point should be clarified here. The electron, a stable particle, cannot really be transformed into an assembly of real particles, but we may formally take as the latter certain fictitious particles whose masses are such as to allow this transformation. The resulting relationship is then to be taken as an analytical continuation to real masses.

Calculation of the same amplitude directly from the diagram (110.8) gives

$$iM_{fi} = iM'_{fn} \cdot i\mathscr{G}(p) \cdot iM'^{*}_{in}.$$

A comparison of the two formulae confirms the limiting expression written above for $\mathscr{G}(p)$ (the first term in (110.9)), and shows that

$$Z' = \sqrt{Z_1}. \tag{110.15}$$

We shall now show that, when the limiting form of the electron propagator is known, there is no need to establish any further conditions for the vertex operator. Let us consider the diagram

(110.16)

which represents the scattering of an electron in an external field $A^{(e)}(k)$, in the first order with respect to the field, and taking account of all radiative corrections. In the limit $k \to 0$, $p_2 \to p_1 \equiv p$, the self-energy corrections to the external-field line are zero (since they vanish for any $k^2 = 0$). Then the diagram corresponds to the amplitude

$$M_{fi} = -e\bar{\mathscr{U}}(p)\Gamma(p, p; 0)\mathscr{U}(p) \cdot A^{(e)} \quad (k \to 0), \tag{110.17}$$

i.e. the product of the potential $A^{(e)}$ and the electron transition current $\bar{\mathscr{U}}\Gamma\mathscr{U}$. But when $k \to 0$ the potential $A^{(e)}(x)$ reduces to a constant independent of coordinates and time. No physical field corresponds to this potential (a particular case of gauge invariance), which therefore can cause no change in the electron current. Thus, in the limit considered, the transition current $\bar{\mathscr{U}}\Gamma\mathscr{U}$ must be simply the free current $\bar{u}\gamma u$:

$$\begin{aligned}\bar{\mathscr{U}}(p)\Gamma^{\mu}(p, p; 0)\mathscr{U}(p) &= Z_1\bar{u}(p)\Gamma^{\mu}u(p) \\ &= \bar{u}(p)\gamma^{\mu}u(p).\end{aligned} \tag{110.18}$$

This is essentially also a definition of the physical charge on the electron. It is easily seen to be necessarily satisfied, whatever the value of Z_1: substituting $\mathscr{G}^{-1}(p)$ from (110.10) in Ward's identity (108.8), we find

$$\Gamma^{\mu}(p, p; 0) = \frac{1}{Z_1}\gamma^{\mu} - \gamma^{\mu}g(p)(\gamma p - m) - (\gamma p - m)g(p)\gamma^{\mu},$$

and (110.18) is satisfied, since $(\gamma p - m)u = 0$, $\bar{u}(\gamma p - m) = 0$.

We see that, when the amplitude of the physical process is calculated, the "renormalization constant" Z_1 disappears. Moreover, by using the indeterminacy that arises from divergences in the calculation of Γ, we can simply require that

$$\bar{u}(p)\Gamma^{\mu}(p,p;0)u(p)=\bar{u}(p)\gamma^{\mu}u(p) \quad \text{when} \quad p^2=m^2, \tag{110.19}$$

i.e. put $Z_1=1$.

The convenience of this definition lies in the fact that there is no need to apply corrections to the external electron lines: we have simply

$$\mathcal{U}(p)=u(p).$$

This can also be deduced directly by noticing that for $Z_1=1$ the mass operator (110.11) is

$$\mathcal{M}=(\gamma p-m)g(\gamma p-m) \tag{110.20}$$

and the second term in (110.12) obviously vanishes. Thus there is no need to "renormalize" the external lines of any real particles, either photons or electrons.†

§ 111. Analytical properties of photon propagators

It is convenient to begin the study of the analytical properties of the photon propagator with the function $\Pi(k^2)$. The reason is that the direct use of the definition (103.1) for this purpose is difficult because the operators $\hat{A}^{\mu}(x)$ are gauge-ambiguous and their properties are therefore indeterminate.

The integral representation of the function $\Pi(k^2)$ (104.11) was derived from the expression for the photon self-energy function in terms of the matrix elements of the gauge-invariant current operator. Denoting the variable k^2 by‡ t, let us consider the properties of the function $\Pi(t)$ in the complex t-plane.

From the integral representation

$$\Pi(t)=\int_0^{\infty}\frac{\rho(t')\,dt}{t-t'+i0} \tag{111.1}$$

we see that $\Pi(t)$ is real on the negative real axis, and elsewhere in the plane satisfies the symmetry relation

$$\Pi(t^*)=\Pi^*(t). \tag{111.2}$$

The function $\Pi(t)$ can have singularities only at singular points of $\rho(t)$. These are at values of $t=k^2$ which are threshold values for the creation of various groups

† In renormalizing the photon propagator, the condition $Z=1$ arose as a necessary physical condition, after which the corrections to the external photon lines disappear automatically. Formally, however, the situation is similar for both photon and electron external lines: when $Z\neq 1$, the wave amplitude e_{μ} of a real photon, with corrections, would be multiplied by $\sqrt{Z}$.

‡ Not to be confused with the symbol for the time.

of real particles by a virtual photon. At these values, new types of intermediate states "come into play" in the sum (104.9). Their contribution is zero below the threshold, but not zero above the threshold, and this causes the singularity at the threshold itself. These threshold values are, of course, real and non-negative.† The singularities of $\Pi(t)$ therefore also lie on the positive real t-axis. If a cut is made along this axis, the function $\Pi(t)$ is analytic throughout the cut plane.

The term $+i0$ in the denominator of the integrand in (111.1) shows that we must pass below the pole $t' = t$. This means that the value of $\Pi(t)$ for real t must be taken as its value on the upper edge of the cut. Using the rule (75.18):

$$\frac{1}{x \pm i0} = P\frac{1}{x} \mp i\pi\delta(x), \tag{111.3}$$

we find that for real t

$$\operatorname{im} \Pi(t) \equiv \operatorname{im} \Pi(t + i0) = -\pi\rho(t). \tag{111.4}$$

On the lower edge of the cut, im Π has the opposite sign; re Π is the same on both edges. Thus the discontinuity of $\Pi(t)$ at the cut is

$$\Pi(t + i0) - \Pi(t - i0) = -2\pi i\rho(t). \tag{111.5}$$

The integral representation (111.1) itself can in this way be regarded simply as Cauchy's formula for the analytic function $\Pi(t)$: applying this formula

$$\Pi(t) = \frac{1}{2\pi i}\int_C \frac{\Pi(t')\,dt'}{t' - t} \tag{111.6}$$

to the contour

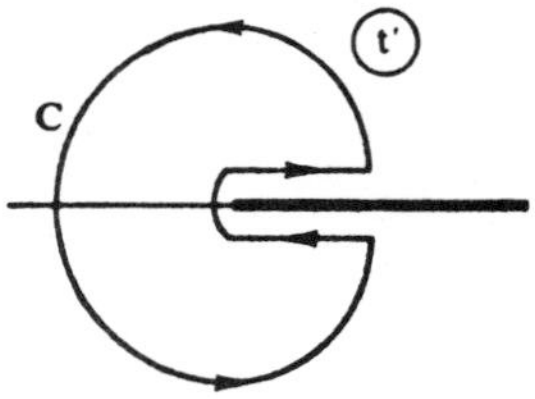

(111.7)

which passes around the cut, and assuming that $\Pi(t)$ decreases sufficiently rapidly at infinity, we find that the integral along the large circle is zero, and those along the edges of the cut give the following *dispersion relation* between $\Pi(t)$ and its

† For example, the point $k^2 = 0$ is a threshold for the production of three (or a higher odd number of) real photons; $k^2 = 4m^2$ is a threshold for electron–positron pair production, and so on.

imaginary part:

$$\Pi(t) = \frac{1}{\pi} \int_0^\infty \frac{\operatorname{im} \Pi(t' + i0)}{t' - t} dt'$$

$$= \frac{1}{\pi} \int_0^\infty \frac{\operatorname{im} \Pi(t')}{t' - t - i0} dt'. \qquad (111.8)$$

Substitution of (111.4) then gives (111.1).†

The analytical properties of the functions $\mathscr{P}(t)$ and $\mathscr{D}(t)$ are the same as those of $\Pi(t)$, in terms of which they are expressed by the simple formulae (104.2) and (103.21). For $\mathscr{D}(t)$ we have

$$\mathscr{D}(t) = \frac{4\pi}{t}\left(1 + \frac{\Pi(t)}{t}\right). \qquad (111.9)$$

On the positive real t-axis, we must take t as $t + i0$, as shown above. The imaginary part of $\mathscr{D}(t)$ can then be calculated from (111.3) and (111.4), bearing in mind that $\Pi(t)/t \to 0$ when $t \to 0$, by (110.6). We thus find

$$\operatorname{im} \mathscr{D}(t) = -4\pi^2 \delta(t) + (4\pi/t^2) \operatorname{im} \Pi(t)$$

$$= -4\pi^2 \delta(t) - (4\pi^2/t^2)\rho(t). \qquad (111.10)$$

Now, applying to $\mathscr{D}(t)$ a dispersion relation of the form (111.8), we obtain the integral representation

$$\mathscr{D}(t) = \frac{4\pi}{t + i0} + 4\pi \int_0^\infty \frac{\rho(t')}{t'^2} \frac{dt'}{t - t' + i0}, \qquad (111.11)$$

called the *Källén–Lehmann expansion* (G. Källén, 1952; H. Lehmann, 1954).

There is a close relationship between the position of the cut for the function $\mathscr{D}(t)$ (and therefore its imaginary part on the cut) and the unitarity condition for the amplitude of the process $a + b \to c + d$ represented by the diagram (110.4); this reaction is, of course, a purely imagined one, but it does not violate the conservation laws, and the unitarity condition is formally valid for it.

In the initial state i of this process there are two "classical" particles a and b, and in the final state there are two other such particles c and d. The unitary condition is (71.2)‡

$$T_{fi} - T_{if}^* = i(2\pi)^4 \sum_n T_{fn} T_{in}^* \delta^{(4)}(P_f - P_i); \qquad (111.12)$$

† Dispersion relations were first used in quantum field theory by M. Gell-Mann, M. L. Goldberger and W. E. Thirring (1954).

‡ The amplitudes T_{fi} differ from the M_{fi} only by the factors shown in (64.10).

the summation on the right is taken over all the physical "intermediate" states n. In the present case these states are evidently those of the systems of real pairs and photons which can be created by the virtual photon k, i.e. the states which occur in the matrix elements in the definition (104.9) of the function $\rho(k^2)$. The amplitudes M_{fi} and M^*_{if} include the factors $\mathcal{D}(k^2)$ and $\mathcal{D}^*(k^2)$ respectively, and their difference contains im $\mathcal{D}(k^2)$. We see, therefore, that the relation given by (111.4) between an imaginary part of $\mathcal{D}$ and the existence of these intermediate states is a consequence of the necessary requirements of unitarity.

We shall see later that it is convenient, in practical calculations of $\mathcal{D}(t)$ (or, equivalently, $\mathcal{P}(t)$) by perturbation theory, to begin by finding the imaginary part of $\mathcal{P}$, which does not involve divergent expressions. But if $\mathcal{P}(t)$ is then calculated from a dispersion relation of the type (111.8), the integral diverges and further subtraction operations are necessary in order to satisfy the conditions $\mathcal{P}(0)=0$ and $\mathcal{P}'(0)=0$. This subtraction can, however, be effected without the explicit use of divergent integrals. To do so, we need only apply the dispersion relation (111.8) not to $\mathcal{P}(t)$ itself but to $\mathcal{P}(t)/t^2$. Then we have

$$\mathcal{P}(t)=\frac{t^2}{\pi}\int_0^\infty \frac{\operatorname{im}\mathcal{P}(t')}{t'^2(t'-t-i0)}\,dt'. \tag{111.13}$$

This integral is convergent, and the function $\mathcal{P}(t)$ thus obtained must necessarily satisfy the required conditions. A relationship such as (111.13) is called a "double-subtraction" dispersion relation. The significance of the change to $\mathcal{P}(t)/t^2$ becomes especially clear if (111.13) is written in the form

$$\mathcal{P}(t)=\frac{1}{\pi}\int_0^\infty \frac{\operatorname{im}\mathcal{P}(t')}{t'-t-i0}\,dt'-\frac{1}{\pi}\int_0^\infty \frac{\operatorname{im}\mathcal{P}(t')}{t'}\,dt'-\frac{t}{\pi}\int_0^\infty \frac{\operatorname{im}\mathcal{P}(t')}{t'^2}\,dt'.$$

If the first ("non-regularized") integral is denoted by $\bar{\mathcal{P}}(t)$, the right-hand side is $\bar{\mathcal{P}}(t)-\bar{\mathcal{P}}(0)-t\bar{\mathcal{P}}'(0)$.

§ 112. **Regularization of Feynman integrals**

The physical conditions of renormalization discussed in §110 enable us, in principle, to derive a unique finite value for the amplitude of any electrodynamic process in any approximation of perturbation theory.

Let us first of all ascertain the nature of the divergences that occur in the integrals derived directly from Feynman diagrams. This is considerably facilitated by counting the powers of the virtual 4-momenta which appear in the integrands.

Let us consider a diagram of order n (i.e. one containing n vertices), with N_e external electron lines and N_γ external photon lines; N_e is even, and the electron lines form $\frac{1}{2}N_e$ continuous sequences, each beginning and ending at a free end. The number of internal electron lines in each such sequence is one less than the number of vertices in it; the total number of internal electron lines in the diagram is therefore $n - \frac{1}{2}N_e$. One photon line comes to each vertex: at N_γ vertices the photon line is external, and at the other $n - N_\gamma$ it is internal. Since each internal photon line joins two vertices, the total number of such lines is $\frac{1}{2}(n - N_\gamma)$.

Each internal photon line is associated with a factor $D(k)$, which contains k to the power -2. Each internal electron line is associated with a factor $G(p)$, which contains p to the power -1 (when $p^2 \gg m^2$). Thus the total power of the 4-momenta in the denominator of the diagram is $2n - \frac{1}{2}N_e - N_\gamma$.

The number of integrations (over d^4p or d^4k) in the diagram is equal to the number of internal lines minus the number $(n-1)$ of additional conditions imposed on the virtual momenta (of the n conservation laws at the vertices, one relates the momenta at the free ends of the diagram). Multiplying by 4, we obtain the number of integrations over all the 4-momentum components, $2(n - N_e - N_\gamma + 2)$.

Lastly, the difference r between the number of integrations and the power of the momenta in the denominator of the integrand is

$$r = 4 - \tfrac{3}{2}N_e - N_\gamma, \tag{112.1}$$

and is independent of the order n of the diagram.

The condition $r < 0$ for the diagram as a whole is not in general sufficient for convergence of the integral: the corresponding numbers r' for the internal sections which can be taken from the diagram must also be negative. The existence of sections having $r' > 0$ would make them divergent although the other integrals in the diagram would converge "with something to spare". The condition $r < 0$ is, however, sufficient for the convergence of the simplest diagrams, in which $n = N_e + N_\gamma$ and there is only one integration over d^4p.

If $r \geq 0$, the integral always diverges. The order of divergence is at least r if r is even, and at least $r-1$ if r is odd; the decrease by one in the latter case is due to the vanishing of the integrals over all 4-space of products of an odd number of 4-vectors. The order of divergence may be higher if there are internal sections with $r' > 0$.

Since N_e and N_γ are positive integers, we see from (112.1) that there exist only a few pairs of values of N_e and N_γ for which $r \geq 0$. The simplest diagrams of each such type may be enumerated, omitting the cases $N_e = N_\gamma = 0$ (vacuum loops) and $N_e = 0$, $N_\gamma = 1$ (mean value of the vacuum current), since they have no physical meaning and the corresponding diagrams must be rejected, as already shown in §103. The remaining cases are:

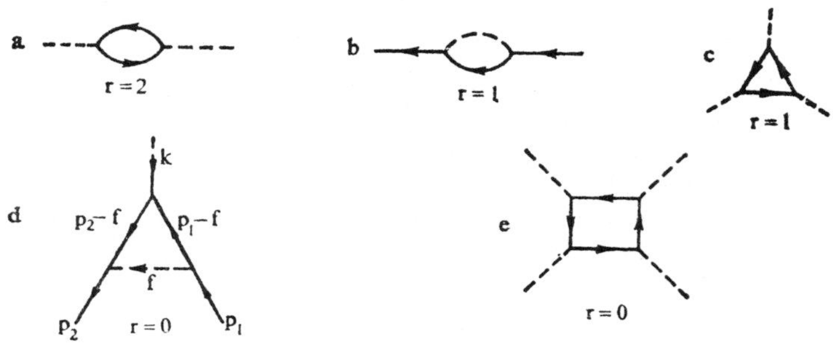

(112.2)

In (a) the divergence is quadratic; in all the others ($r=0$ and $r=1$) it is logarithmic.

The diagram (112.2d) is the first correction to the vertex operator. It must satisfy the condition (110.19), which we shall here write as

$$\bar{u}(p)\Lambda^{\mu}(p, p; 0)u(p)=0 \quad \text{when} \quad p^2=m^2, \tag{112.3}$$

with

$$\Lambda^{\mu}=\Gamma^{\mu}-\gamma^{\mu}. \tag{112.4}$$

Let $\bar{\Lambda}^{\mu}(p_2, p_1; k)$ be the Feynman integral as derived directly from the diagram. This integral is logarithmically divergent, and does not itself satisfy the condition (112.3), but we can obtain a quantity which does satisfy this condition by taking the difference

$$\Lambda^{\mu}(p_2, p_1; k)=\bar{\Lambda}^{\mu}(p_2, p_1; k)-[\bar{\Lambda}^{\mu}(p_1, p_1; 0)]_{p_1^2=m^2}. \tag{112.5}$$

The leading divergent term in $\bar{\Lambda}^{\mu}(p_2, p_1; k)$ is obtained by taking the virtual photon 4-momentum f in the integrand to be arbitrarily large. It is†

$$-4\pi ie^2\int\frac{\gamma^{\nu}(\gamma f)\gamma^{\mu}(\gamma f)\gamma_{\nu}}{f^2\cdot f^2\cdot f^2}\frac{d^4f}{(2\pi)^4}$$

and is independent of the 4-momenta of the external lines. In the difference (112.5) the divergence therefore cancels, leaving a finite quantity. This operation of removing the divergence by subtraction is called *regularization* of the integral.

It must be emphasized that the integral $\bar{\Lambda}^{\mu}(p_2, p_1; k)$ can be regularized by one subtraction because here the divergence is only logarithmic, i.e. as weak as it can be. If the integral involved divergences of various orders, a single subtraction with $k=0$ might be insufficient to eliminate all the divergent terms.

When the first correction in Γ^{μ} (i.e. the first term in the expansion of Λ^{μ}) has been determined, the first correction in the electron propagator (the diagram (112.2a)) can be calculated from Ward's identity (108.8), which may also be written in the form

$$\partial\mathcal{M}(p)/\partial p_{\mu}=\Lambda^{\mu}(p, p; 0), \tag{112.6}$$

with the mass operator $\mathcal{M}$ in place of $\mathcal{G}$, and Λ^{μ} in place of Γ^{μ}. This equation is to be integrated with the boundary condition

$$\bar{u}(p)\mathcal{M}(p)u(p)=0 \quad \text{for} \quad p^2=m^2, \tag{112.7}$$

which follows from (110.20).

Finally, to calculate the first term in the expansion of the polarization operator, we use the identity (108.14), which after contraction with respect to two pairs of

† The complete expression for the integral is given in (117.2).

indices gives

$$\frac{3}{4\pi}\frac{\partial^2 \mathcal{P}}{\partial k_\sigma \partial k^\sigma} = 2\mathcal{S},$$

a relation between the scalar functions $\mathcal{P} = \frac{1}{3}\mathcal{P}^\mu_\mu$ and $\mathcal{S} = \mathcal{S}^{\rho\mu}_{\mu\rho}$. Both these functions depend only on the scalar variable k^2, and so we have

$$2k^2\mathcal{P}''(k^2) + \mathcal{P}'(k^2) = \frac{4\pi}{3}\mathcal{S}(k^2), \tag{112.8}$$

the primes denoting differentiation with respect to k^2. With the condition $\mathcal{P}'(0) = 0$, this equation shows that

$$\mathcal{S}(0) = 0. \tag{112.9}$$

In the first approximation of perturbation theory, $\mathcal{S}(k^2)$ is given by the diagram (112.2e), with the free ends having 4-momenta k, k, 0, 0. The corresponding Feynman integral $\bar{\mathcal{S}}(k^2)$ diverges logarithmically, and can be regularized by a single subtraction, using the condition (112.9):

$$\mathcal{S}(k^2) = \bar{\mathcal{S}}(k^2) - \bar{\mathcal{S}}(0).$$

Then $\mathcal{P}(k^2)$ is found by solving equation (112.8) with the boundary conditions $\mathcal{P}(0) = 0$, $\mathcal{P}'(0) = 0$.

In the next approximation of perturbation theory, the correction to the vertex operator $\Lambda_\mu^{(2)}$ is determined by the diagrams (106.10, (c)–(i)). The irreducible diagrams (d)–(f) are calculated by a similar regularization of the integrals, using a single subtraction as in (112.5), in the same way as in calculating the first-approximation correction $\Lambda_\mu^{(1)}$. In the reducible diagrams, the internal self-energy and vertex parts of lower order are immediately replaced by the already known (regularized) first-approximation quantities ($\mathcal{P}^{(1)}$, $\mathcal{M}^{(1)}$, $\Lambda_\mu^{(1)}$), after which the integrals obtained are again regularized in accordance with (112.5).† The corrections $\mathcal{P}^{(2)}$ and $\mathcal{M}^{(2)}$ can then be calculated from (112.6) and (112.8).

This systematic procedure will in principle enable us to derive finite values for $\mathcal{P}$, $\mathcal{M}$ and Λ_μ in any approximation of perturbation theory. It thus becomes possible to calculate the amplitudes of physical scattering processes that are described by diagrams containing $\mathcal{P}$, $\mathcal{M}$ and Λ_μ as constituent sections.

The physical conditions derived in §111 are therefore sufficient for an unambiguous regularization of all the Feynman diagrams occurring in the theory. This *renormalizability* is a far from trivial property of quantum electrodynamics.‡

In a practical calculation of radiative corrections, the above procedure may, however, not be the simplest and most rational. In Chapter XII we shall see, in

† In diagrams for still higher approximations, it may be necessary to replace the four-ended sections $\mathcal{S}$ also by already regularized values.

‡ A different approach to renormalization theory in quantum electrodynamics is given by N. N. Bogoliubov and D. V. Shirkov, *Introduction to the Theory of Quantized Fields*, Wiley, New York, 1980.

particular, that a convenient treatment may start by calculating the imaginary parts of the corresponding quantities; these are given by integrals which do not involve divergences. The quantity itself is then found by analytical continuation, using the dispersion relations. It thus becomes possible to avoid the lengthy calculations which are needed in a direct regularization by subtractions.

CHAPTER XII

RADIATIVE CORRECTIONS

§ 113. Calculation of the polarization operator

LET us now go on to the actual calculation of the radiative corrections, and begin with that of the polarization operator (J. Schwinger, 1949; R. P. Feynman, 1949). In the first approximation of perturbation theory, this is given by the loop in the diagram

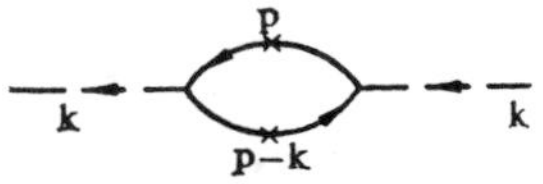

(113.1)

As already mentioned, the problem becomes easier if we first calculate the imaginary part of the required function. This in turn is most conveniently done by using the unitarity relation. The virtual-photon lines are regarded as corresponding to a fictitious "real" particle, a vector boson with mass $M^2 = k^2$ which interacts with an electron in the same way as a photon does. Then (113.1) becomes the diagram of a "real" process, and the unitarity condition can be justifiably applied to it.

Thus we regard (113.1) as a diagram giving the amplitude of the transition of the boson into itself (a diagonal element of the S-matrix) via a decay into an electron–positron pair. The crosses in the diagram show where it is to be cut into two parts so as to show the intermediate state that figures in the application of the unitarity relation. This state contains an electron with 4-momentum $p_- = p$ and a positron with $p_+ = -(p-k)$.

The unitarity relation with a two-particle intermediate state (71.4), for coincident initial and final states, gives

$$2 \operatorname{im} M_{ii} = \frac{|\mathbf{p}|}{(4\pi)^2 \varepsilon} \sum_{\text{polar.}} \int |M_{ni}|^2 do. \tag{113.2}$$

Here the amplitude M_{ii}, from (113.1), is

$$iM_{ii} = \sqrt{(4\pi)} e_\mu^* \cdot \sqrt{(4\pi)} e_\nu \frac{i\mathscr{P}^{\mu\nu}}{4\pi}, \tag{113.3}$$

where e_μ is the boson polarization 4-vector, which according to (14.13) satisfies the equation

$$e_\mu k^\mu = 0.$$

The amplitude M_{ni} corresponds to the diagram for the decay of the boson into an electron–positron pair:

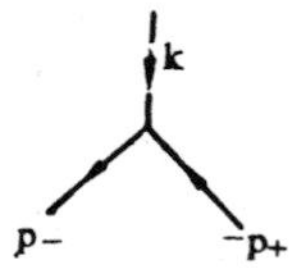

The corresponding expression is

$$M_{ni} = -e\sqrt{(4\pi)}e_\mu j^\mu, \quad j^\mu = \bar{u}(p_-)\gamma^\mu u(-p_+). \tag{113.4}$$

Substitution of (113.3) and (113.4) in (113.2) gives

$$2e_\mu^* e_\nu \operatorname{im} \mathcal{P}^{\mu\nu} = \frac{e^2}{4\pi}\frac{|\mathbf{p}|}{\varepsilon}\sum_{\text{polar}}\int j^{\mu *} j^\nu e_\mu^* e_\nu \, do. \tag{113.5}$$

Here $\mathbf{p} = \mathbf{p}_- = -\mathbf{p}_+$ and $\varepsilon = \varepsilon_+ + \varepsilon_- = 2\varepsilon_+$ are the momenta and total energy of the pair in its centre-of-mass system; the integration is over the directions of $\mathbf{p}$, and the summation is over the polarizations of the two particles.

Let us now average both sides of (113.5) over the polarizations of the boson. This is done by means of the formula (14.15):

$$\overline{e_\mu^* e_\nu} = -\frac{1}{3}\left(g_{\mu\nu} - \frac{k_\mu k_\nu}{k^2}\right).$$

Since the tensor $\mathcal{P}^{\mu\nu}$ and the vector j^μ are transverse ($\mathcal{P}^{\mu\nu}k_\nu = 0$, $j^\mu k_\mu = 0$), we have as the result

$$2 \operatorname{im} \mathcal{P} = \frac{1}{12\pi}\frac{|\mathbf{p}|}{\varepsilon}\sum_{\text{polar.}}\int (jj^*)\, do, \tag{113.6}$$

where $\mathcal{P} = \frac{1}{3}\mathcal{P}_\mu^\mu$.

The summation over the polarizations is effected in the usual manner, the integration over do reduces to a multiplication by 4π, and so

$$\begin{aligned} 2 \operatorname{im} \mathcal{P} &= e^2 \frac{|\mathbf{p}|}{3\varepsilon} \operatorname{tr} \gamma_\mu(\gamma p_- + m)\gamma^\mu(\gamma p_+ - m) \\ &= -e^2 \frac{8|\mathbf{p}|}{3\varepsilon}(p_+ p_- + 2m^2). \end{aligned}$$

In terms of the variable

$$\begin{aligned} t &= k^2 = (p_+ + p_-)^2 \\ &= 2(m^2 + p_+ p_-), \end{aligned} \tag{113.7}$$

we have

$$\varepsilon^2 = t, \ \ \mathbf{p}^2 = \tfrac{1}{4}t - m^2,$$

and hence finally

$$\operatorname{im} \mathscr{P}(t) = -\frac{1}{3}\alpha\sqrt{\frac{t-4m^2}{t}}(t+2m^2), \ \ t \geqslant 4m^2. \tag{113.8}$$

The value $t = 4m^2$ is the threshold value for the production of one electron–positron pair by a virtual photon (cf. the second footnote to §111); in the approximation considered ($\sim e^2$), the state with a single pair is the only one that can appear as an intermediate state in the unitarity condition (113.2). In this approximation, therefore, the right-hand side of (113.2) is zero for $t < 4m^2$, and hence

$$\operatorname{im} \mathscr{P}(t) = 0, \ \ t < 4m^2. \tag{113.9}$$

For the same reason, in this approximation the cut in the complex t-plane for the function $\mathscr{P}(t)$ extends only from $t = 4m^2$ on the real axis, and this point must be the lower limit in the dispersion integral (111.13). Thus

$$\mathscr{P}(t) = -\frac{\alpha}{3\pi}t^2 \int_{4m^2}^{\infty} \frac{dt'}{t'-t-i0}\sqrt{\left(\frac{t'-4m^2}{t'}\right)}\frac{t'+2m^2}{t'^2}. \tag{113.10}$$

It is convenient for the expression of the result to replace t by a variable ξ, defined as follows:

$$t/m^2 = -(1-\xi)^2/\xi. \tag{113.11}$$

This transformation maps the upper half-plane of t on a semicircle of unit radius in the upper half-plane of ξ, as shown in Fig. 19, where corresponding line segments in the two planes are indicated by similar markings. The semicircle $\xi = e^{i\phi}, 0 \leqslant \phi \leqslant \pi$,

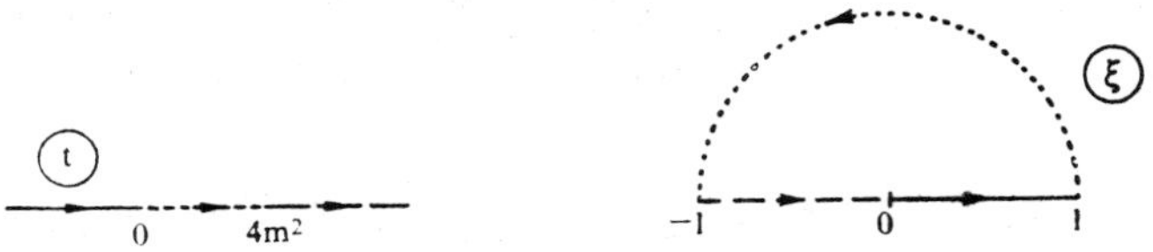

FIG. 19.

corresponds to the non-physical region $0 \leqslant t/m^2 \leqslant 4$. The right-hand and left-hand radii on the real axis correspond to the physical regions $t < 0$ and $t/m^2 > 4$.

The integral in (113.10) is most simply calculated by means of the substitution

$$\frac{t'}{4m^2} = \frac{1}{1-x^2},$$

and we first take the case $t<0$ (so that the denominator does not vanish in the range of integration and the imaginary term $i0$ may be omitted). The result of the integration, in terms of the variable ξ, is

$$\mathcal{P}(\xi)=\frac{\alpha m^2}{3\pi}\left\{-\frac{22}{3}+\frac{5}{3}\left(\xi+\frac{1}{\xi}\right)+\left(\xi+\frac{1}{\xi}-4\right)\frac{1+\xi}{1-\xi}\log\xi\right\}. \tag{113.12}$$

The analytical continuation of this formula gives the function $\mathcal{P}(t)$ in the range $t>4m^2$ also; this is done by putting $\xi=|\xi|e^{i\pi}$ (and the logarithm gives a contribution to the imaginary part: $\log\xi=\log|\xi|+i\pi$).† For the non-physical region, we must put $\xi=e^{i\phi}$, and then

$$\mathcal{P}(t)=\frac{2\alpha m^2}{3\pi}\{-\tfrac{10}{3}\sin^2\tfrac{1}{2}\phi-2+(1+2\sin^2\tfrac{1}{2}\phi)\phi\cot\tfrac{1}{2}\phi\}, \tag{113.13}$$

$$t/4m^2=\sin^2\tfrac{1}{2}\phi.$$

In the limit of small $|t|$ (i.e. $\xi\to 1$), these formulae become

$$\mathcal{P}(t)=-\frac{\alpha}{15\pi}\frac{t^2}{m^2},\quad |t|\ll 4m^2. \tag{113.14}$$

In the opposite case of large $|t|$ (i.e. $\xi\to 0$), we have

$$\left.\begin{aligned}\mathcal{P}(t)&=-\frac{\alpha}{3\pi}|t|\log\frac{|t|}{m^2}, & -t&\gg 4m^2,\\ \mathcal{P}(t)&=\frac{\alpha}{3\pi}t\left(\log\frac{t}{m^2}-i\pi\right), & t&\gg 4m^2.\end{aligned}\right\} \tag{113.15}$$

In accordance with the significance of perturbation theory, these formulae are valid if $\mathcal{P}/4\pi\ll D^{-1}=t/4\pi$. The condition for (113.15) to be applicable is thus

$$\frac{\alpha}{3\pi}\log\frac{|t|}{m^2}\ll 1. \tag{113.16}$$

The radiative corrections which involve $\alpha\log(|t|/m)$ are called *logarithmic corrections*.

§ 114. **Radiative corrections to Coulomb's law**

Let us apply the formulae derived above to the problem of the radiative corrections to Coulomb's law. These corrections may be intuitively described as resulting from the polarization of the vacuum around a point charge.

† The analytical continuation thus obtained is, as it should be, a continuation on the upper edge of the cut, since the semicircle in the ξ-plane corresponds to the upper half-plane of t.

If corrections are neglected, the field of a fixed centre (with charge e_1) is given by the Coulomb scalar potential $\Phi \equiv A_0^{(e)} = e_1/r$. The components of its three-dimensional Fourier expansion are

$$\Phi(\mathbf{k}) \equiv A_0^{(e)}(\mathbf{k}) = 4\pi e_1/\mathbf{k}^2.$$

When the radiative corrections are included, this field is replaced by the "effective field"

$$\mathscr{A}_0^{(e)} = A_0^{(e)} + \mathscr{D}_{0\rho}\frac{\mathscr{P}^{\rho\lambda}}{4\pi}A_\lambda^{(e)} = A_0^{(e)} + \frac{1}{4\pi}\mathscr{P}\mathscr{D}A_0^{(e)}; \tag{114.1}$$

cf. (103.15). The second term gives the required change in the scalar potential. In the first approximation of perturbation theory for $\mathscr{P}(k^2)$, we must take the expression derived in §113 and replace $\mathscr{D}(k^2)$ by the zero-order approximation:

$$\mathscr{D}(k^2) \approx D(k^2) = -4\pi/\mathbf{k}^2.$$

Thus the radiative correction to the field potential is

$$\delta\Phi(\mathbf{k}) = -\frac{4\pi e_1}{(\mathbf{k}^2)^2}\mathscr{P}(-\mathbf{k}^2). \tag{114.2}$$

To determine the form of this correction in the coordinate representation, we must take the inverse Fourier transform:

$$\delta\Phi(\mathbf{r}) = \int e^{i\mathbf{k}\cdot\mathbf{r}}\delta\Phi(\mathbf{k})\, d^3k/(2\pi)^3. \tag{114.3}$$

Since $\delta\Phi(\mathbf{k})$ is a function of $t = -\mathbf{k}^2$ only, integration over angles gives

$$\delta\Phi(r) = \frac{1}{4\pi^2}\int_0^\infty \delta\Phi(t)\frac{\sin(r\sqrt{-t})}{r}\,d(-t)$$

$$= \frac{1}{4\pi^2 r}\operatorname{im}\int_{-\infty}^{\infty}\delta\Phi(-y^2)\,e^{iry}y\,dy;$$

in the last transformation, we use the fact that the integrand is an even function of $y = \sqrt{-t}$. The contour of integration can now be moved into the upper half-plane of y, and made to coincide with the cut of the function $\mathscr{P}(-y^2)$ (Fig. 20). This cut

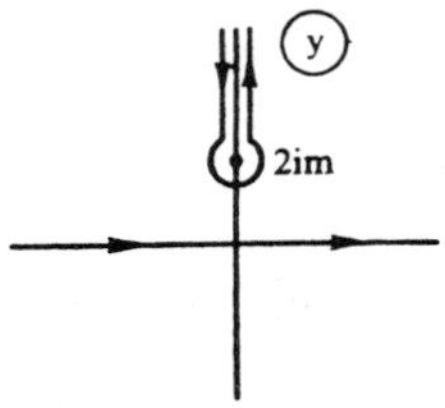

FIG. 20.

extends upwards along the imaginary axis from the point $2im$, the physical sheet corresponding to the left side of the cut. Replacing y by a new variable, $y = ix$, we get

$$\delta\Phi(r) = \frac{1}{2\pi^2 r} \int_{2m}^{\infty} \operatorname{im} \delta\Phi(x^2)\, e^{-rx} x\, dx.$$

Finally, on returning to the integration over $t = x^2$, we have

$$\delta\Phi(r) = \frac{1}{4\pi^2 r} \int_{4m^2}^{\infty} \operatorname{im} \delta\Phi(t)\, e^{-r\sqrt{t}}\, dt. \qquad (114.4)$$

The imaginary part

$$\operatorname{im} \delta\Phi(t) = -\frac{4\pi e}{t^2} \operatorname{im} \mathcal{P}(t)$$

is taken from (113.8), and after an obvious change of variable we have

$$\Phi(r) = \frac{e_1}{r} + \delta\Phi(r) = \frac{e_1}{r}\left\{1 + \frac{2\alpha}{3\pi} \int_1^{\infty} e^{-2mr\zeta}\left(1 + \frac{1}{2\zeta^2}\right) \frac{\sqrt{(\zeta^2 - 1)}}{\zeta^2}\, d\zeta\right\} \qquad (114.5)$$

(E. A. Uehling and R. Serber, 1935).

The integral can be evaluated in two limiting cases. Let us first take that of small r ($mr \ll 1$), and divide the integral of the first term in the parenthesis into two parts:

$$I = \int_1^{\infty} e^{-2mr\zeta} \frac{\sqrt{(\zeta^2 - 1)}}{\zeta^2}\, d\zeta$$

$$= \int_1^{\zeta_1} \cdots d\zeta + \int_{\zeta_1}^{\infty} \cdots d\zeta$$

$$\equiv I_1 + I_2,$$

with ζ_1 chosen so that $1/mr \gg \zeta_1 \gg 1$. Consequently, we can take $r = 0$ in the first integral, so that

$$I_1 \approx \int_1^{\zeta_1} \frac{\sqrt{(\zeta^2 - 1)}}{\zeta^2}\, d\zeta$$

$$\approx \log 2\zeta_1 - 1.$$

In I_2, on the other hand, unity may be omitted from $\zeta^2 - 1$:

$$I_2 \approx \int_{\zeta_1}^{\infty} e^{-2mr\zeta} \frac{d\zeta}{\zeta}$$

$$= -\log \zeta_1 \cdot e^{-2mr\zeta_1} + 2mr \int_{\zeta_1}^{\infty} e^{-2mr\zeta} \log \zeta \, d\zeta.$$

In the exponential and the lower limit, it is permissible to put $\zeta_1 = 0$. Then, with the change of variable $2mr\zeta = x$, we have

$$I_2 = -\log 2\zeta_1 + \log \frac{1}{mr} + \int_0^{\infty} e^{-x} \log x \, dx$$

$$= -\log 2\zeta_1 + \log \frac{1}{mr} - C,$$

where $C = 0.577\ldots$ is Euler's constant.

In the integral of the second term in (114.5) we can immediately put $r = 0$:

$$I_3 \approx \frac{1}{2} \int_1^{\infty} \frac{\sqrt{(\zeta^2 - 1)}}{\zeta^4} \, d\zeta = \frac{1}{6}.$$

When the three integrals are added, ζ_1 disappears, leaving

$$\Phi(r) = \frac{e_1}{r} \left[1 + \frac{2\alpha}{3\pi} \left(\log \frac{1}{mr} - C - \frac{5}{6}\right)\right], \quad r \ll 1/m. \tag{114.6}$$

When $mr \gg 1$, the range $\zeta - 1 \sim 1/mr \ll 1$ is important in the integral. The change of variable $\zeta = 1 + \xi$ and appropriate transformations reduce it to

$$e^{-2mr} \int_0^{\infty} e^{-2mr\xi} \frac{3}{2} \sqrt{(2\xi)} \, d\xi = \frac{3}{8(mr)^{3/2}} \sqrt{\pi} \, e^{-2mr}.$$

In this case, therefore,†

$$\Phi(r) = \frac{e_1}{r} \left(1 + \frac{\alpha}{4\sqrt{\pi}} \frac{e^{-2mr}}{(mr)^{3/2}}\right), \quad r \gg 1/m. \tag{114.7}$$

We see that the polarization of the vacuum alters the Coulomb field of a point

† The origin of the factor e^{-2mr} in $\delta\Phi(r)$ is evident from the form of the initial integral (114.4): when r is large, the important values of t are those near the lower limit. Thus the exponent is determined by the position of the first singularity of the function $\delta\Phi(t)$.

charge in a region $r \sim 1/m(=\hbar/mc)$, where m is the electron mass. Outside this region, the change in the field decreases exponentially.

One further general comment may be made. Hitherto we have implicitly assumed that the radiative corrections arise from the interaction between the photon field and the electron–positron field. Thus, by associating the internal closed loops in the photon self-energy diagrams with the electrons, we have taken into account the interaction of the photon with the "electron vacuum". But the photon also interacts with the fields of other particles; the interaction with the "vacua" of these fields is described by similar self-energy diagrams, in which the internal loops are associated with the appropriate particles. The contributions of these diagrams differ in order of magnitude from those of the electron diagrams by several factors of m_e/m, where m is the mass of the particle concerned and m_e the electron mass.

The particles whose mass is closest to that of the electron are the muons and the pions. The numerical ratios m_e/m_μ and m_e/m_π are close to α. The radiative corrections from these particles would therefore have to be taken into account together with the electron corrections of higher orders. For muons, the radiative corrections can in principle be calculated by means of the existing theory, but for pions (which are strongly interacting particles) they cannot.

This places a fundamental limitation on exact calculations of specific effects in present-day quantum electrodynamics. The use of arbitrarily high-order corrections from the photon–electron interaction alone would be an invalid exaggeration of the attainable accuracy.

The radiative corrections to Coulomb's law discussed in this section are, as we have seen, valid even at distances $r \lessdot 1/m_e$. We can now add that the formulae obtained cease to be valid at distances $r < 1/m_\mu$ (or $1/m_\pi$), where polarization of the vacua of other particles becomes significant.

§ 115. Calculation of the imaginary part of the polarization operator from the Feynman integral

In a direct calculation from the diagram (the loop in (113.1)), the polarization operator in the first approximation of perturbation theory would be given by the integral

$$\frac{i\mathcal{P}^{\mu\nu}}{4\pi} \to -e^2 \int \mathrm{tr}\, \gamma^\mu G(p) \gamma^\nu G(p-k) \frac{d^4 p}{(2\pi)^4}. \tag{115.1}$$

This integral, however, taken over all four-dimensional p-space, is quadratically divergent, and in order to obtain a finite result it would be necessary to regularize the integral by the procedure described in §112.

Here, we shall not give the complete derivation, but show how one can use the integral (115.1) to calculate the imaginary part of the polarization operator (which has been determined in §113 by means of the unitarity condition); this derivation includes a number of instructive points.

The imaginary part of the integral (115.1) does not diverge and therefore does not need regularization. For the scalar function $\operatorname{im} \mathscr{P} = \frac{1}{3} \operatorname{im} \mathscr{P}^{\mu}_{\mu}$ we have

$$\operatorname{im} \mathscr{P} = \operatorname{im} \left\{ i \frac{4\pi e^2}{3(2\pi)^4} \int \frac{\operatorname{tr} \gamma^{\mu}(\gamma p + m)\gamma_{\mu}(\gamma p + \gamma k + m)}{(p^2 - m^2 + i0)[(p-k)^2 - m^2 + i0]} d^4 p \right\}.$$

After calculating the trace, we get

$$\left.\begin{aligned} \operatorname{im} \mathscr{P}(k^2) &= \operatorname{im} \int \frac{i\phi(p)\, d^4 p}{(p^2 - m^2 + i0)[(p-k)^2 - m^2 + i0]}, \\ \phi(p) &= \frac{2e^2}{3\pi^3}(2m^2 + pk - p^2). \end{aligned}\right\} \tag{115.2}$$

Let $k^2 > 0$. We shall use a frame of reference in which $k = (k_0, 0)$, and

$$(p - k)^2 = (p_0 - k_0)^2 - \mathbf{p}^2.$$

Using also the notation $\varepsilon = \sqrt{(\mathbf{p} + m^2)}$ (this is not the "energy" of the virtual electron p_0), we can write (115.2) in the form

$$\left.\begin{aligned} \operatorname{im} \mathscr{P}(k^2) &= \operatorname{im} \int d^3 p \int_{-\infty}^{\infty} dp_0 \frac{i\phi(p_0, \mathbf{p})}{(p_0^2 - \varepsilon^2 + i0)[(p_0 - k_0)^2 - \varepsilon^2 + i0]}, \\ \phi(p_0, \mathbf{p}) &= \frac{2e^2}{3\pi^3}(m^2 + \varepsilon^2 + p_0 k_0 - p_0^2). \end{aligned}\right\} \tag{115.3}$$

The integrand has poles at four values of p_0:

(a) $p_0 = \varepsilon - i0$, (a′) $p_0 = -\varepsilon + i0$,
(b) $p_0 = k_0 - \varepsilon + i0$, (b′) $p_0 = k_0 + \varepsilon - i0$.

Figure 21 shows the configuration of these poles; we shall take the specific case $k_0 > 0$, but the final answer depends only on k_0^2, and not on the sign of k_0. We can

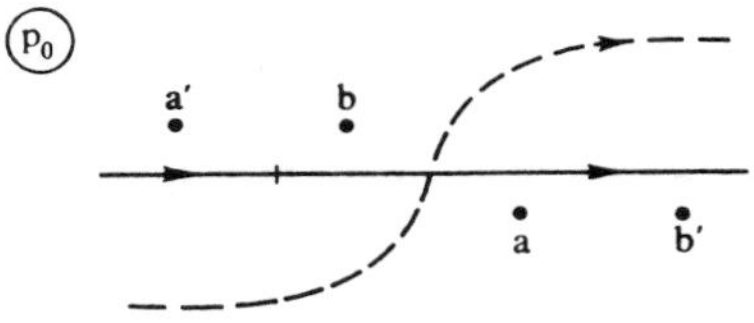

FIG. 21.

calculate the discontinuity of the function $\mathscr{P}(t)$ at the cut in the complex plane of $t = k^2 = k_0^2$ or, equivalently, on the real axis in the k_0-plane. The real part of $\mathscr{P}(t)$ is continuous at the cut, and the discontinuity is therefore

$$\Delta \mathscr{P}(t) = 2i \operatorname{im} \mathscr{P}(t). \tag{115.4}$$

We shall first show how the position of the cut may be established from the form of the integral. Let $I(\mathbf{p}, k_0)$ denote the inner integral (over dp_0) in (115.3). So long as the upper and lower poles in Fig. 21 are at finite distances apart, the path of integration over p_0 can be taken far from the poles, as shown by the broken line. It is therefore evident that in this case the integral $I(\mathbf{p}, k_0)$ is unaltered by an infinitesimal upward or downward movement of the poles b and b' respectively away from the real axis, i.e. by the change $k_0 \to k_0 \pm i\delta$, $\delta \to 0$. Thus, where k_0 tends to its real value from above or from below, the value of $I(\mathbf{p}, k_0)$ is the same, and therefore makes no contribution to the discontinuity $\Delta\mathcal{P}$. The situation is different only if two poles (which may be a and b when $k_0 > 0$) are exactly one beneath the other, so that the contour of integration is "trapped" between them and cannot be moved to a great distance. Thus the discontinuity $\Delta\mathcal{P} \neq 0$ only if the condition $k_0 - \varepsilon = \varepsilon$, i.e. $k_0 = 2\varepsilon = 2\sqrt{(\mathbf{p}^2 + m^2)}$, can be satisfied somewhere in the region of integration over d^3p. For this to be so, we must evidently have $k_0 \geqslant 2m$, i.e. $t \geqslant 4m^2$.†

The integral $I(\mathbf{p}, k_0)$ can be written in the form

$$I(\mathbf{p}, k_0) = \int\limits_C \frac{i\phi(p_0, \mathbf{p})\, dp_0}{(p_0^2 - \varepsilon^2)[(p_0 - k_0)^2 - \varepsilon^2]}, \tag{115.5}$$

with the terms $i0$ omitted from the denominator and the integration contour C correspondingly modified, as shown in Fig. 22. We see that the discontinuity $\Delta\mathcal{P}(t)$ is due to the impossibility of bringing the contour away from the pole a when it is trapped between a and b. The contour C is therefore replaced by C', which passes

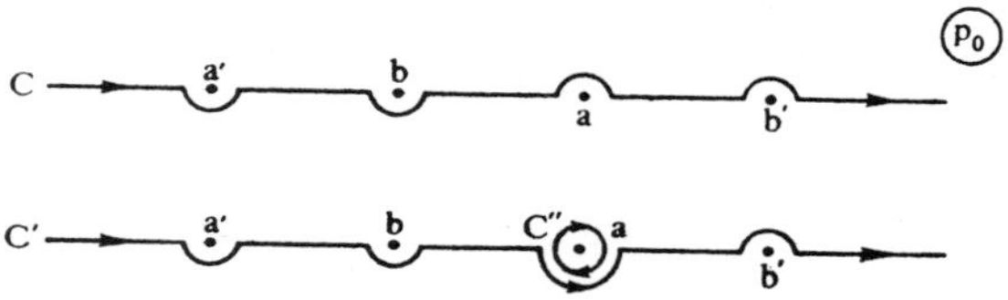

FIG. 22.

beneath a, and the integral around a small circle C'' centred at a is added. Then C' can always be moved away from the poles without any difficulty, and the integration along it therefore contributes only to the regular part of $\mathcal{P}(t)$. To determine the required discontinuity, we need only consider the integral along the circle C'', which is done by calculating the residue at the pole a. This calculation can be made by substituting in the integrand

$$\frac{1}{p_0^2 - \varepsilon^2} \to -2\pi i \delta(p_0^2 - \varepsilon^2); \tag{115.6}$$

† It can be shown similarly that there is no cut when $t = k^2 < 0$. Taking in this case a frame of reference where $k = (0, \mathbf{k})$, we find that the poles of the integrand are at

$$p_0 = \pm(\varepsilon - i0), \quad p_0 = \pm(\sqrt{[(\mathbf{p} - \mathbf{k})^2 + m^2]} - i0).$$

The two lower poles are always in the right half-plane of p_0, and the two upper ones in the left half-plane, so that no pair can be vertically aligned.

the minus sign is used because the circle around the pole is traversed in the negative direction. In the argument of the delta function, only the zero $p_0 = +\varepsilon$ is to be used (the pole a, not a'); this is automatically done if we agree to integrate over only half of momentum 4-space ($p_0 > 0$).

After the substitution (115.6), the discontinuity of the integral $I(\mathbf{p}, k_0)$ can be calculated immediately:

$$\Delta I = \{I(\mathbf{p}, k_0 + i\delta) - I(\mathbf{p}, k_0 - i\delta)\}_{\delta\to+0}$$

$$= -2\pi i \int_0^\infty \delta(p_0^2 - \varepsilon^2) i\phi(p_0, \mathbf{p}) \times$$

$$\times \left[\frac{1}{(k_0 - p_0)^2 - \varepsilon^2 + i\delta} - \frac{1}{(k_0 - p_0)^2 - \varepsilon^2 - i\delta}\right] dp_0.$$

With the equation

$$\frac{1}{(k_0 - p_0)^2 - \varepsilon^2 \pm i\delta} = P\frac{1}{(k_0 - p_0)^2 - \varepsilon^2} \mp i\pi\delta[(k_0 - p_0)^2 - \varepsilon^2]$$

(see (111.3)), we have

$$\Delta I = i(2\pi i)^2 \int_0^\infty \delta(p_0^2 - \varepsilon^2)\delta[(k_0 - p_0)^2 - \varepsilon^2]\phi(p_0, \mathbf{p})\, dp_0.$$

The arguments of the delta functions can be rewritten in an invariant form by adding and subtracting $\mathbf{p}^2$:

$$p_0^2 - \varepsilon^2 = p^2 - m^2, \quad (k_0 - p_0)^2 - \varepsilon^2 = (k - p)^2 - m^2.$$

We then obtain finally

$$\Delta\mathscr{P}(k^2) = i(2\pi i)^2 \int_{p_0>0} d^4p \cdot \phi(p)\delta(p^2 - m^2)\delta[(p - k)^2 - m^2]. \tag{115.7}$$

Because of the delta functions, the integration is actually taken only over the region of intersection of the hypersurfaces

$$p^2 = m^2, \quad (p - k)^2 = m^2. \tag{115.8}$$

Since in this region all the 4-vectors p are time-like, the condition of integration over $p_0 > 0$ is invariant (the upper interior region of the cone $p^2 = m^2$).

Let us compare (115.7) with the original formula (115.2). We see that the discontinuity of the function $\mathscr{P}(t)$ at the cut in the t-plane can be found by applying in the original Feynman integral the substitution

$$\frac{1}{p^2-m^2+i0}\to -2\pi i\delta(p^2-m^2) \tag{115.9}$$

in the propagators which correspond to the loop lines intersected in the diagram (113.1) (S. Mandelstam, 1958; R. E. Cutkosky, 1960).

The conditions (115.8) select the region of momentum space in which the lines of virtual particles in the diagram correspond to real particles (the 4-momenta p and $p-k$ are then said to lie on the *mass surface*). Here we see clearly the relationship to the unitarity-relation method, where these lines are replaced by lines of intermediate-state real particles.

We also observe the mathematical reason for the absence of divergence in the imaginary part of the diagram: the integration is over a finite region of the mass surface, not over the whole of infinite momentum 4-space as in the original Feynman integral.

In order to derive from (115.7) the formula obtained in §113, we return to the frame of reference in which $\mathbf{k}=0$ and integrate over

$$d^4p = |\mathbf{p}|\varepsilon\, d\varepsilon\, dp_0\, do.$$

The integration amounts to removing the delta functions, with

$$\delta(p^2-m^2)\,dp_0 = \delta(p_0^2-\varepsilon^2)\,dp_0 \to \frac{1}{2\varepsilon}\,\delta(p_0-\varepsilon)\,dp_0,$$

and then

$$\begin{aligned}\delta[(p-k)^2-m^2]\,d\varepsilon &= \delta[(p_0-k_0)^2-\varepsilon^2]\,d\varepsilon\\ &= \delta(-2\varepsilon k_0+k_0^2)\,d\varepsilon\\ &\to \frac{1}{2k_0}\,\delta(\varepsilon-\tfrac{1}{2}k_0)\,d\varepsilon.\end{aligned}$$

The result is

$$\Delta\mathscr{P}(t) = -\frac{1}{2}i\pi^2\int\sqrt{\left(\frac{t-4m^2}{t}\right)}\phi(\varepsilon,\mathbf{p})\,do, \tag{115.10}$$

where $t=k^2=k_0^2$; the value of ϕ is taken for

$$p_0=\varepsilon=\tfrac{1}{2}k_0,\quad \mathbf{p}^2=\varepsilon^2-m^2=\tfrac{1}{4}k_0^2-m^2,$$

i.e. it is

$$\phi(\varepsilon,\mathbf{p}) = \frac{e^2}{3\pi^3}(2m^2+t)$$

and is independent of the angle. The integration over do reduces to multiplication by 4π, and we come back to (113.8).

The only vital point in the foregoing derivation is that the diagram is divided into two parts by cutting no more than two lines. The rule stated therefore remains valid for diagrams comprising any two sections joined by two (electron or photon) lines. The integral calculated by the substitution (115.9) then gives the contribution to the imaginary part of the diagram that arises (in the unitarity-relation method) from the corresponding two-particle intermediate state.

§ 116. **Electromagnetic form factors of the electron**

Let us consider the vertex operator $\Gamma^\mu = \Gamma^\mu(p_2, p_1; k)$ in the case where the two electron lines are external lines and the photon line is internal. The external electron lines correspond to factors $u_1 = u(p_1)$ and $\bar{u}_2 = \bar{u}(p_2)$, and Γ therefore appears in the expression for the diagram as the product

$$j_{fi}^\mu = \bar{u}_2 \Gamma^\mu u_1. \tag{116.1}$$

As already noted in §111, this is the electron transition current, including radiative corrections. The conditions of relativistic and gauge invariance enable us to establish the general matrix structure of this current.

The electromagnetic interaction operator $\hat{V} = e(\hat{j}\hat{A})$ is a true scalar (not a pseudoscalar), in accordance with the conservation of spatial parity in these interactions. The transition current j_{fi} is therefore a true 4-vector (not a pseudo-vector), and hence can be expressed only in terms of other true 4-vectors formed from the two available 4-vectors p_1 and p_2 ($k = p_2 - p_1$ is a third) and the bispinors u_1 and u_2. There are three independent 4-vectors of this kind bilinear in $\bar{u}_2$ and u_1:

$$\bar{u}_2 \gamma u_1, (\bar{u}_2 u_1) p_1, (\bar{u}_2 u_1) p_2,$$

or, equivalently,

$$\bar{u}_2 \gamma u_1, (\bar{u}_2 u_1) P, (\bar{u}_2 u_1) k, \tag{116.2}$$

where $P = p_1 + p_2$. The condition of gauge invariance requires that the transition current should be transverse to the photon 4-momentum k:

$$j_{fi} k = 0. \tag{116.3}$$

This is satisfied by the first two 4-vectors (116.2), respectively because of Dirac's equations

$$(\gamma p_1 - m) u_1 = 0, \quad \bar{u}_2(\gamma p_2 - m) = 0 \tag{116.4}$$

and because $Pk = 0$. The current j_{fi} is given by a linear combination of these two

4-vectors:

$$j^\mu_{fi} = f_1(\bar{u}_2 u_1)P^\mu + f_2(\bar{u}_2\gamma^\mu u_1),$$

where f_1 and f_2 are invariant functions, called the *electromagnetic form factors* of the electron.

Since the 4-momenta p_1 and p_2 relate to the free electron, $p_1^2 = p_2^2 = m^2$, and from the three 4-vectors p_1, p_2 and k (which are related by $k = p_2 - p_1$) we can construct only one independent scalar variable, which we take as k^2. Then the form factors are functions of k^2.

The expression for the current can also be put in other forms with different choices of the two independent terms. Using the equations (116.4) and the commutation rules for the matrices γ, we can easily show that

$$(\bar{u}_2\sigma^{\mu\nu}u_1)k_\nu = -2m(\bar{u}_2\gamma^\mu u_1) + (\bar{u}_2 u_1)P^\mu, \tag{116.5}$$

where $\sigma^{\mu\nu} = \frac{1}{2}(\gamma^\mu\gamma^\nu - \gamma^\nu\gamma^\mu)$. The coefficient of this term will later be seen to have an important physical significance, and we therefore write

$$\Gamma^\mu = \gamma^\mu f(k^2) - \frac{1}{2m} g(k^2)\sigma^{\mu\nu}k_\nu, \tag{116.6}$$

where f and g are two other form factors; the reason for writing the factor $1/2m$ separately will be explained below.† For brevity, we shall always use the vertex operator instead of the current, the $\bar{u}_2$ and u_1 on either side being understood.

In order to determine the properties of the form factors, let us consider the diagram (110.16) for the interaction of an electron with an external field. The corresponding scattering amplitude is

$$M_{fi} = -ej^\mu_{fi}\mathscr{A}^{(e)}_\mu(k), \tag{116.7}$$

where $\mathscr{A}^{(e)}_\mu$ is the effective external field (taking account of the polarization of the vacuum).

The amplitude (116.7) describes two reaction channels. In the scattering channel, the invariant t is such that

$$t = k^2 = (p_2 - p_1)^2 \leqslant 0.$$

Putting p_- for p_2 and $-p_+$ for p_1, we change to the annihilation channel, which corresponds to pair production with 4-momenta p_- and p_+. In this channel,

$$t = (p_- + p_+)^2 \geqslant 4m^2.$$

The range $0 < t < 4m^2$ is non-physical.

† To avoid misunderstanding, it may be mentioned that in the definition (116.6) k is assumed to be the 4-momentum of the photon line coming to the vertex; for the outgoing line, the sign of the second term would be reversed.

Let us now consider the unitarity condition (111.12). In the scattering channel ($t<0$) there are no physical intermediate states in this case: one free electron cannot change its momentum or give rise to any other particles. There are also, of course, no intermediate states in the non-physical region. Hence, when $t<4m^2$, the right-hand side of (111.12) is zero, so that the matrix T_{fi} (or, equivalently, M_{fi}) is Hermitian:

$$M_{fi} = M^*_{if}.$$

The interchange of the initial and final states corresponds to the interchange of p_2 and p_1, and therefore to a reversal of the sign of k. Putting M_{fi} in the form (116.7), we therefore have

$$j^{\mu}_{fi}\mathscr{A}^{(e)}_{\mu}(k) = j^{\mu *}_{if}\mathscr{A}^{(e)*}_{\mu}(-k).$$

Since $\mathscr{A}^{(e)}(-k) = \mathscr{A}^{(e)*}(k)$, it follows that the transition-current matrix is also Hermitian:

$$j_{fi} = j^*_{if} \quad \text{when} \quad t<4m^2. \tag{116.8}$$

Using the properties of the matrices γ (21.7), we can easily prove that

$$(\bar{u}_2\gamma^{\mu}u_1) = (\bar{u}_1\gamma^{\mu}u_2)^*,$$
$$(\bar{u}_2\sigma^{\mu\nu}u_1) = -(\bar{u}_1\sigma^{\mu\nu}u_2)^*.$$

Thus j^*_{if} differs from j_{fi} only in that the functions $f(t)$ and $g(t)$ are replaced by their complex conjugates, and it then follows from (116.8) that these functions are real. Thus

$$\operatorname{im} f(t) = \operatorname{im} g(t) = 0 \quad \text{when} \quad t<4m^2. \tag{116.9}$$

In the annihilation channel ($t>4m^2$), the f state is a pair which can be transformed into another pair with different momenta (elastic scattering) or into a more complex system. The right-hand side of the unitarity condition is therefore not zero, the matrix M_{fi} (and hence j_{fi}) is not Hermitian, and so the form factors are complex.

The analytical properties of the functions $f(t)$ and $g(t)$ are exactly similar to those of the function $\mathscr{P}(t)$ discussed in §111, although it is difficult to prove them so directly. These functions are analytic in the complex t-plane cut along the positive real axis $t>4m^2$, with

$$f^*(t) = f(t^*), \quad g^*(t) = g(t^*).$$

The renormalization condition (110.19) applied to the vertex operator (116.6) leads to the requirement

$$f(0) = 1. \tag{116.10}$$

In order to include this condition automatically (when calculating $f(t)$ from its imaginary part), we must apply a dispersion relation of the form (111.8) not to the function $f(t)$ itself but to $(f-1)/t$. Then we get the dispersion relation "with one subtraction":

$$f(t)-1=\frac{t}{\pi}\int_{4m^2}^{\infty}\frac{\operatorname{im} f(t')}{t'(t'-t-i0)}\,dt'. \tag{116.11}$$

No values for the form factor $g(t)$ are prescribed by physical conditions. Its dispersion relation will therefore be written "without subtractions":

$$g(t)=\frac{1}{\pi}\int_{4m^2}^{\infty}\frac{\operatorname{im} g(t')}{t'-t-i0}\,dt'. \tag{116.12}$$

The value of $g(0)$ has an important physical significance, as it specifies the correction to the magnetic moment of the electron. In order to see this, let us consider the scattering of a non-relativistic electron in a constant magnetic field which is almost uniform in space.

The term in the scattering amplitude (116.7) which depends on the form factor $g(k^2)$ is

$$\delta M_{fi}=\frac{e}{2m}\,g(k^2)(\bar{u}_2\sigma^{\mu\nu}u_1)k_\nu A_\mu^{(e)}(k). \tag{116.13}$$

For a purely magnetic field, $A^{(e)\mu}=(0,\mathbf{A})$; since the field is constant in time, the 4-vector $k^\mu=(0,\mathbf{k})$, and since it varies only slowly in space, $\mathbf{k}$ is small. With a view to subsequently taking the limit $\mathbf{k}\to 0$, we have already replaced the effective $\mathscr{A}^{(e)}$ by $A^{(e)}$ in (116.13). Expanding (116.13) and expressing it in terms of three-dimensional quantities, we find

$$\delta M_{fi}=\frac{e}{2m}\,g(-\mathbf{k}^2)(\bar{u}_2\boldsymbol{\Sigma}u_1)i\mathbf{k}\times\mathbf{A}_\mathbf{k},$$

where $\boldsymbol{\Sigma}$ is the matrix (21.21). The product $i\mathbf{k}\times\mathbf{A}_\mathbf{k}$ is replaced by the magnetic field $\mathbf{H}_\mathbf{k}$, and we can then take the limit $\mathbf{k}\to 0$. Finally, with the non-relativistic spinor amplitudes w_1, w_2 given by (23.12):

$$\bar{u}_2=\sqrt{(2m)}(w_2^*0),$$

$$u_1=\sqrt{(2m)}\begin{pmatrix}w_1\\0\end{pmatrix},$$

we have

$$\delta M_{fi}=\frac{e}{2m}\,g(0)\,\mathbf{H}_\mathbf{k}\cdot 2m(w_2^*\boldsymbol{\sigma}w_1). \tag{116.14}$$

This expression may be compared with the scattering amplitude in a constant electric field having the scalar potential $\Phi_{\mathbf{k}}$:

$$M_{fi} = -e(\bar{u}_2\gamma^0 u_1)\Phi_{\mathbf{k}} \approx -e\Phi_{\mathbf{k}} \cdot 2m(w_2^* w_1).$$

We see that an electron in a magnetic field can be regarded as having an additional potential energy

$$-\frac{e}{2m} g(0)\boldsymbol{\sigma} \cdot \mathbf{H}_{\mathbf{k}}.$$

This means that the electron has an "anomalous" magnetic moment (in ordinary units)

$$\mu' = (e\hbar/2mc)g(0) \tag{116.15}$$

in addition to its "normal" Dirac magnetic moment $e\hbar/2mc$.

§ 117. Calculation of electron form factors

Let us now go on to the actual calculation of the electron form factors (J. Schwinger, 1949). In the zero-order approximation of perturbation theory, the vertex operator $\Gamma^\mu = \gamma^\mu$, i.e. the electron form factors are $f = 1$, $g = 0$. The first radiative correction to the form factors is given by the vertex diagram

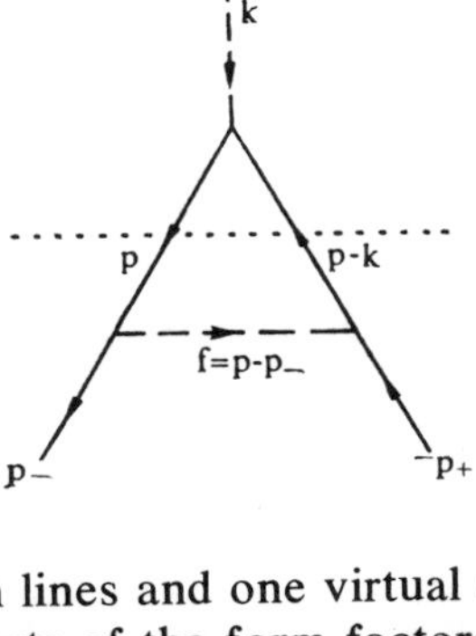

(117.1)

with two real external electron lines and one virtual external photon line. We shall first calculate the imaginary parts of the form factors. As has been shown in §116, these differ from zero only in the annihilation channel ($k^2 > 4m^2$); accordingly, the 4-momenta of the external electron lines in the diagram (117.1) correspond to the production of an electron and a positron, and are denoted by p_- and $-p_+$. The analytical expression of the diagram (117.1) is

$$-ie\bar{u}(p_-)\Gamma^\mu u(-p_+) = (-ie)^3\bar{u}(p_-)\gamma^\nu i \int G(p)\gamma^\mu G(p-k)\gamma^\nu D_{\mu\nu}(f)\frac{d^4p}{(2\pi)^4} u(-p_+) \tag{117.2}$$

or, in expanded form,

$$\gamma^\mu f(k^2) - \frac{1}{2m} g(k^2)\sigma^{\mu\nu}k_\nu = \int \frac{i\phi^\mu(p)\, d^4p}{(p^2 - m^2)[(p-k)^2 - m^2]}, \tag{117.3}$$

with the notation

$$\phi^\mu(p) = -e^2 \frac{\gamma^\nu(\gamma p + m)\gamma^\mu(\gamma p - \gamma k + m)\gamma_\nu}{4\pi^3(p_- - p)^2} \tag{117.4}$$

and omitting for brevity the factors $\bar{u}(p_-)$, $u(-p_+)$; it is understood that both sides of the equations below are placed between these factors.

The horizontal dotted line divides the diagram (117.1) into two parts, in such a way as to indicate the intermediate state which would figure in a calculation of the imaginary part of the form factor by means of the unitarity condition, namely the state of an electron–positron pair with momenta differing from p_-, p_+. The intersection also shows where the pole factors are to be replaced in the integral (117.2) in order to use the rule (115.9) in the calculation (in (117.3), these factors have been separated in the integrand).

The integral in (117.3) has the same form as that in (115.2), and we can therefore immediately write the result of the transformation as in (115.10):

$$2\gamma^\mu \operatorname{im} f(t) - \frac{2}{2m}\sigma^{\mu\nu}k_\nu \operatorname{im} g(t) = -\tfrac{1}{2}\pi^2\sqrt{\left(\frac{t-4m^2}{t}\right)} \int \phi^\mu(p)\, do_{\mathbf{p}}, \tag{117.5}$$

where $t = k^2$, the integration is over the directions of the vector $\mathbf{p}$, and the 4-vectors $p_-' \equiv p$ and $p_+' \equiv k - p$ in the definition of the function $\phi^\mu(p)$ (117.4) become the 4-momenta of real (not virtual) particles. The expression (117.5) relates to a frame of reference in which $\mathbf{k} = 0$, i.e. the centre-of-mass system of the pair p_-, p_+ (and hence of the "intermediate" pair p_-', p_+'). In this frame, therefore, $k = (k_0, 0)$, $p_- = (\frac{1}{2}k_0, \mathbf{p}_-)$, $p_+ = (\frac{1}{2}k_0, -\mathbf{p}_-)$, $p = (\frac{1}{2}k_0, \mathbf{p})$, and it is easy to verify that

$$f^2 = (p - p_-)^2 = -2\mathbf{p}^2(1 - \cos\theta) = -\tfrac{1}{2}(t - 4m^2)(1 - \cos\theta), \tag{117.6}$$

where θ is the angle between $\mathbf{p}$ and $\mathbf{p}_-$ (and $\mathbf{p}^2 = \mathbf{p}_-^2$). Not substituting (117.4) in (117.5) and eliminating the matrices $\gamma^\nu \ldots \gamma_\nu$ in the integrand by means of (22.6), we have

$$\begin{aligned}
&\gamma^\mu \operatorname{im} f(t) - \frac{1}{2m}\sigma^{\mu\nu}k_\nu \operatorname{im} g(t) \\
&= -\frac{e^2}{4\sqrt{[t(t-4m^2)]}} \int \frac{do_{\mathbf{p}}}{2\pi(1-\cos\theta)} \gamma^\nu(\gamma p + m)\gamma^\mu(\gamma p - \gamma k + m)\gamma_\nu \\
&= -\frac{e^2}{4\sqrt{[t(t-4m^2)]}} \int \frac{do_{\mathbf{f}}}{2\pi(1-\cos\theta)} [-2m^2\gamma^\mu + 4m(P^\mu + 2f^\mu) + \\
&\quad + 2(\gamma p_+ - \gamma f)\gamma^\mu(\gamma p_- + \gamma f)],
\end{aligned} \tag{117.7}$$

with the 4-vectors

$$f = p - p_- = (0, \mathbf{f}), \quad P = p_- - p_+ = (0, 2\mathbf{p}_-). \tag{117.8}$$

The integration now amounts to the calculation of the three integrals with different numerators

$$I, I^\mu, I^{\mu\nu} = \int \frac{1, f^\mu, f^\mu f^\nu}{1 - \cos\theta} \frac{do_f}{2\pi}. \tag{117.9}$$

The integral I is logarithmically divergent when $\theta \to 0$. If it is written as

$$I = \int_0^{t-4m^2} d(\mathbf{f}^2)/\mathbf{f}^2 = \int_0^{-(t-4m^2)} d(f^2)/f^2,$$

we see that the divergence corresponds to small "masses" of the virtual photon, i.e. it is an "infra-red" divergence, which will be further discussed in §122; here we shall merely note that the divergence is fictitious, in the sense that, when all physical effects are correctly taken into account, the divergences of this kind cancel out. We can therefore cut the integral off at any lower limit, and then, in the subsequent calculation of real physical phenomena, make this limit tend to zero.

Here it will be simplest to apply the cut-off in a relativistically invariant manner. To do so, we assign to the virtual photon f a small but finite mass λ ($\ll m$), i.e. make the change

$$f^2 \to f^2 - \lambda^2 \tag{117.10}$$

in the photon propagator $D(f^2)$ in (117.2). Then

$$I = \int_0^{-(t-4m^2)} \frac{d(f^2)}{f^2 - \lambda^2} = \log\frac{t - 4m^2}{\lambda^2}. \tag{117.11}$$

The integral I^μ, in which f^μ is a space-like 4-vector, must be expressed in terms of the 4-vector P^μ, which (unlike k^μ, the only other available 4-vector) is space-like for all p_+ and p_-. Hence $I^\mu = AP^\mu$. If this equation is multiplied by P_μ and the integral $P_\mu I^\mu$ is calculated in the centre-of-mass system of the pair (with the components of the 4-vectors f and P given by (117.8)), the result is

$$A = \frac{1}{2\mathbf{p}^2} \int_{-1}^{1} \frac{\mathbf{f}\cdot\mathbf{p}\, d\cos\theta}{1 - \cos\theta}$$

$$= -\tfrac{1}{2} \int_{-1}^{1} d\cos\theta = -1.$$

Thus

$$I^{\mu} = -P^{\mu}. \tag{117.12}$$

We can similarly calculate the integral

$$I^{\mu\nu} = \tfrac{1}{4}P^2\left(g^{\mu\nu} - \frac{P^{\mu}P^{\nu}}{P^2}\right) + \tfrac{1}{4}P^{\mu}P^{\nu}; \tag{117.13}$$

to determine the coefficients in this expression, we have only to evaluate the integrals I^{μ}_{μ} and $I^{\mu\nu}P_{\mu}P_{\nu}$.

The calculation continues as follows: (117.11)–(117.13) are substituted in (117.7) to obtain a series of terms between $\bar{u}(p_-)$ and $u(-p_+)$. In each term we use the commutation rules for the matrices γ^{μ} to "propel" the factor γp_+ to the right and γp_- to the left; then we can make the replacements $\gamma p_- \to m$, $\gamma p_+ \to -m$, since

$$\bar{u}(p_-)\gamma p_- = m\bar{u}(p_-), \quad \gamma p_+ u(-p_+) = -mu(-p_+).$$

In the resulting sum

$$-4(p_+p_-)I\gamma^{\mu} + 2mP^{\mu} - 3P^2\gamma^{\mu},$$

we can then replace P^{μ} by $2m\gamma^{\mu} + \sigma^{\mu\nu}k_{\nu}$, which is equivalent to it when between the $\bar{u}$ and u factors; cf. (116.5). Finally, all quantities are expressed in terms of the invariant $t = k^2$ $(2p_+p_- = t - 2m^2,\ P^2 = 4m^2 - t)$ and the two sides of (117.7) are compared; this yields the following expressions for the imaginary parts of the form factors:

$$\operatorname{im} g(t) = \frac{\alpha m^2}{\sqrt{[t(t-4m^2)]}}, \tag{117.14}$$

$$\operatorname{im} f(t) = \frac{\alpha}{4\sqrt{[t(t-4m^2)]}}\left[-3t + 8m^2 + 2(t-2m^2)\log\frac{t-4m^2}{\lambda^2}\right]. \tag{117.15}$$

The infra-red divergence occurs only in $\operatorname{im} f(t)$.

The functions $f(t)$ and $g(t)$ themselves are obtained from their imaginary parts by means of (116.11) and (116.12), in which the integrations are effected by the same substitutions as were used in §113 to calculate $\mathcal{P}(t)$. The form factors are given in terms of the variable ξ (113.11) *by*

$$g(\xi) = \frac{\alpha}{\pi}\frac{\xi\log\xi}{\xi^2-1}, \tag{117.16}$$

$$f(\xi) - 1 = \frac{\alpha}{2\pi}\left\{2\left(1 + \frac{1+\xi^2}{1-\xi^2}\log\xi\right)\log\frac{m}{\lambda} - \frac{3(1+\xi^2)+2\xi}{2(1-\xi^2)}\log\xi + \right.$$
$$\left. + \frac{1+\xi^2}{1-\xi^2}\left[\tfrac{1}{6}\pi^2 - \tfrac{1}{2}\log^2\xi - 2F(\xi) + 2\log\xi\log(1+\xi)\right]\right\}, \tag{117.17}$$

where $F(\xi)$ is Spence's function (131.19).

In the non-physical region ($0 < t/m^2 < 4$), we must put $\xi = e^{i\phi}$. The expressions for the form factors can then be written

$$f(\phi) = \frac{\alpha}{\pi}\left\{\left(1 - \frac{\phi}{\tan\phi}\right)\log\frac{m}{\lambda} + \frac{3\cos\phi + 1}{2\sin\phi}\phi + \frac{2}{\tan\phi}\int\limits_0^{\phi/2} x\tan x\,dx\right\},$$
(117.18)

$$g(\phi) = \frac{\alpha}{2\pi}\frac{\phi}{\sin\phi}. \qquad (117.19)$$

Finally, the following are the limiting expressions: for small $|t|$,

$$f(t) - 1 = \frac{\alpha t}{3\pi m^2}\left(\log\frac{m}{\lambda} - \frac{3}{8}\right), \quad |t| \ll 4m^2, \quad g(t) = \alpha/2\pi, \qquad (117.20)$$

and for large $|t|$,

$$f(t) - 1 = -\frac{\alpha}{2\pi}\left(\frac{1}{2}\log^2\frac{|t|}{m^2} + 2\log\frac{m}{\lambda}\log\frac{|t|}{m^2}\right) + \begin{cases}\frac{1}{2}i\alpha\log(t/\lambda^2), & t \gg 4m^2,\\ 0, & -t \gg 4m^2,\end{cases}$$
(117.21)

$$g(t) = -\frac{\alpha m^2}{\pi t}\log\frac{|t|}{m^2} + \begin{cases}i\alpha m^2/t, & t \gg 4m^2,\\ 0, & -t \gg 4m^2.\end{cases} \qquad (117.22)$$

Formula (117.21) is valid (as regards re f) with what is called double-logarithmic accuracy, i.e. as far as the squares of the large logarithms.†

§ 118. Anomalous magnetic moment of the electron

As has been shown in §116, the value of $g(0)$ determines the radiative correction to the magnetic moment of the electron. If we seek to calculate only this quantity, there is of course no need to find the whole function $g(t)$. From (117.14) and (116.12),

$$g(0) = \frac{1}{\pi}\int\limits_{4m^2}^{\infty}\frac{\operatorname{im} g(t')}{t'}dt' = \frac{\alpha}{4\pi}\int\limits_1^{\infty}\frac{dx}{x^{3/2}\sqrt{(x-1)}} = \frac{\alpha}{2\pi}. \qquad (118.1)$$

With this correction, the magnetic moment of the electron is

$$\mu = \frac{e\hbar}{2mc}\left(1 + \frac{\alpha}{2\pi}\right), \qquad (118.2)$$

a formula first derived by Schwinger (1949).

† The expression for the vertex operator in the case of one virtual and one real external electron line and a real external photon line is given by A. I. Akhiezer and V. B. Berestetskii, *Quantum Electrodynamics*, Interscience, New York, 1965, §47.5.

In the next approximation (with terms in α^2), the radiative corrections in the form factors are represented by the seven diagrams (106.10, (c)–(i)). Even to find the value of $g(0)$ in this approximation demands very lengthy calculations. Details of these may be found in the original papers; here we shall give only the final value† for the correction in the second approximation:

$$g^{(2)}(0) = \left(\frac{\alpha}{\pi}\right)^2\left(\frac{197}{144}+\frac{\pi^2}{12}-\frac{1}{2}\pi^2\log 2+\frac{3}{4}\zeta(3)\right)$$
$$= -0.328\,\alpha^2/\pi^2. \tag{118.3}$$

The magnetic moment of the electron is therefore

$$\mu = \frac{e\hbar}{2mc}\left(1+\frac{\alpha}{2\pi}-0.328\frac{\alpha^2}{\pi^2}\right) \tag{118.4}$$

(C. M. Sommerfield, 1957; A. Petermann, 1957).

Particular consideration may be given to the contribution of the vacuum polarization to the correction $g^{(2)}(0)$, namely the diagram

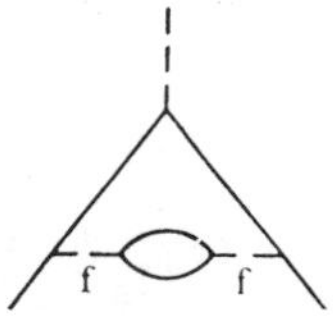

(118.5)

which contains a photon self-energy part. It differs from the first-approximation diagram (117.1) only by having instead of the photon propagator $D(f^2) = 4\pi/f^2$ the product

$$D(f^2)\frac{\mathcal{P}(f^2)}{4\pi}D(f^2) = \frac{4\pi}{f^2}\frac{\mathcal{P}(f^2)}{f^2},$$

where $\mathcal{P}(f^2)$ is the polarization operator in the first approximation (terms in α), calculated in §113. Repeating, with this difference, some of the calculations in §117, we find as the "polarization part" of the correction

$$\text{im}\, g^{(2)}_{\text{polar.}}(t) = \frac{\alpha m^2}{\sqrt{[t(t-4m^2)]}}\int_{-1}^{1}\frac{\mathcal{P}(f^2)}{f^2}\frac{1+3\cos\theta}{2}\,d\cos\theta, \tag{118.6}$$

with

$$f^2 = -\tfrac{1}{2}(t-4m^2)(1-\cos\theta); \tag{118.7}$$

† The calculation by the unitarity method is given by M. V. Terent'ev, *Soviet Physics JETP* **16**, 444, 1963.

see (117.6). When this integral and then

$$g^{(2)}_{\text{polar.}}(0) = \frac{1}{\pi} \int\limits_{4m^2}^{\infty} \text{im}\, g^{(2)}_{\text{polar.}}(t') \frac{dt'}{t'} \tag{118.8}$$

are calculated, the result is

$$g^{(2)}_{\text{polar.}}(0) = \frac{\alpha^2}{\pi^2}\left(\frac{119}{36} - \frac{1}{3}\pi^2\right) = 0.016\frac{\alpha^2}{\pi^2}, \tag{118.9}$$

which is about 5% of the total quantity (118.3).

It has already been noted at the end of §114 that vacuum polarization for other particles may also make a contribution to the radiative corrections. The contribution of the muon vacuum to the anomalous magnetic moment of the electron is obtained from the same formulae (118.6)–(118.8), in which m is again the electron mass m_e (this applies also to the definition of the variable f^2) but the parameter m in the expression for $\mathcal{P}(f^2)$ must be the muon mass m_μ. The function $\mathcal{P}(f^2)/f^2$ depends only on the ratio f^2/m_μ^2. In the integral (118.8) the important values of t (and therefore of f^2) are those comparable with m_e^2; hence the ratio $f^2/m_\mu^2 \sim (m_e/m_\mu)^2 \ll 1$, and in evaluating the integrals we can use the limiting formula (113.14), according to which

$$\frac{\mathcal{P}(f^2)}{f^2} = -\frac{\alpha}{15\pi}\frac{f^2}{m_\mu^2}.$$

From this we see that the contribution to $g^2(0)$ from the muon vacuum polarization contains the extra small factor $(m_e/m_\mu)^2$.

The opposite situation, however, occurs for the corrections to the magnetic moment of the muon. Since the particle mass does not appear in (118.3), this value of $g^2(0)$ is valid for the muon also, and it takes into account the contribution of the muon vacuum polarization. But the contribution from the vacuum polarization of other particles, namely the electrons, is here considerably greater. It can be calculated from formulae (118.6)–(118.8), with m_μ for m and the electron polarization operator for $\mathcal{P}(t)$. Unlike the previous case, the important range of values is now given by $f^2/m_e^2 \sim (m_\mu/m_e)^2 \gg 1$, and $\mathcal{P}(f^2)$ must be taken as the limiting expression (113.15):

$$\frac{\mathcal{P}(f^2)}{f^2} = \frac{\alpha}{3\pi}\log\frac{|f^2|}{m_e^2}.$$

The calculation of the integrals gives

$$\begin{aligned}[g^2(0)]_{\text{electr. polar.}} &= \left(\frac{\alpha}{\pi}\right)^2\left(\frac{1}{3}\log\frac{m_\mu}{m_e} - \frac{25}{36}\right)\\ &= 1.09\,\alpha^2/\pi^2 \end{aligned} \tag{118.10}$$

(H. Suura and E. H. Wichmann, 1957; A. Petermann, 1957).

Adding (118.10) and (118.3), we find as the magnetic moment of the muon

$$\mu_{\text{muon}} = \frac{e\hbar}{2m_\mu c}\left(1+\frac{\alpha}{2\pi}+0.76\frac{\alpha^2}{\pi^2}\right). \tag{118.11}$$

The contribution of the muon vacuum polarization (118.9) is here about 2% of the total value of $g^{(2)}(0)$. The pion vacuum polarization would give a contribution of the same order (because of the similarity of the masses), but this cannot be calculated exactly, and so there would be no point in finding the corrections $\sim\alpha^3$ in the magnetic moment of the muon.

§ 119. **Calculation of the mass operator**

The method of direct regularization of the Feynman integrals may be demonstrated by a calculation of the mass operator. In the first non-vanishing approximation, the mass operator is represented by the loop in the diagram

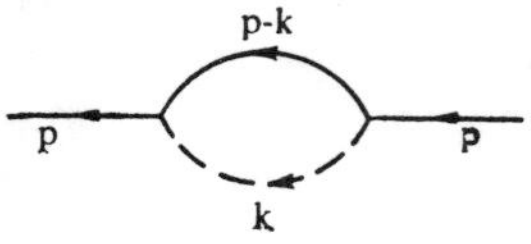

(119.1)

corresponding to the integral

$$-i\mathscr{M}(p) = (-ie)^2\int \gamma^\mu G(p-k)\gamma^\nu D_{\mu\nu}(k)\frac{d^4k}{(2\pi)^4};$$

on substituting the propagators and combining the factors $\gamma^\mu\ldots\gamma_\mu$ by means of formulae (22.6), we get

$$\bar{\mathscr{M}}(p) = -\frac{8\pi i}{(2\pi)^4}e^2\int\frac{2m-\gamma p+\gamma k}{[(p-k)^2-m^2](k^2-\lambda^2)}d^4k \tag{119.2}$$

(the bar over $\mathscr{M}$ denotes the non-regularized value of the integral). The fictitious "photon mass" λ is used in the photon propagator in order to eliminate the infra-red divergence (as in §117).

The integral may be transformed by means of formula (131.4), with a_1 and a_2 the two factors in the denominator of (119.2). A simple rearrangement of the terms in the denominator of the new integral gives

$$\bar{\mathscr{M}}(p) = -\frac{8\pi i}{(2\pi)^4}e^2\int d^4k\int_0^1 dx\,\frac{2m-\gamma p+\gamma k}{[(k-px)^2-a^2]^2}, \tag{119.3}$$

with

$$a^2 = m^2x^2-(p^2-m^2)x(1-x)+\lambda^2(1-x). \tag{119.4}$$

The change of variable $k \to k + px$ brings the integrand in (119.3) to a form in which its denominator depends only on k^2; according to (131.17), (131.18) a constant is then added to the integral:

$$\bar{\mathscr{M}}(p) = -\frac{8\pi i}{(2\pi)^4} e^2 \left\{ \int d^4k \int_0^1 dx \frac{2m - \gamma p(1-x)}{(k^2 - a^2)^2} - \frac{1}{4} i\pi^2 \gamma p \right\}. \tag{119.5}$$

(The term in γk in the numerator is here omitted, since it gives zero on integration over the directions of the 4-vector k; cf. (131.8).)

The regularization of this integral involves subtractions such as to reduce it to the form (110.20). The latter expression gives zero on multiplication by the wave amplitude $u(p)$, if p is the 4-momentum of a real electron. Without bringing in $u(p)$ explicitly, we can formulate this condition as stating that $\mathscr{M}(p)$ vanishes when we substitute

$$\gamma p \to m, \quad p^2 \to m^2. \tag{119.6}$$

The form of the integral (119.5) is convenient in that the 4-vector p appears in it only as γp and p^2, but not as kp.

Subtracting from (119.5) a similar expression after the substitution (119.6), we get

$$-\frac{8\pi i}{(2\pi)^4} \left\{ \int d^4k \int_0^1 dx [2m - \gamma p(1-x)] \left[\frac{1}{(k^2 - a^2)^2} - \frac{1}{(k^2 - a_0^2)^2} \right] - \right.$$

$$\left. - \int d^4k \int_0^1 dx \frac{1-x}{(k^2 - a_0^2)^2} (\gamma p - m) - \tfrac{1}{4} i\pi^2 (\gamma p - m) \right\}, \tag{119.7}$$

with $a_0^2 = m^2x^2 + \lambda^2(1-x)$.

To complete the regularization, however, a further subtraction is needed: according to (110.20), the substitution (119.6) should reduce to zero not only $\mathscr{M}(p)$ itself but also $\mathscr{M}(p)$ without the factor $\gamma p - m$. A corresponding subtraction removes entirely the second and third terms in the braces in (119.7).† The first integral is transformed by incorporating a further integration (using (131.5)), taking $n = 2$, $a = k^2 - a^2$, and $b = k^2 - a_0^2$. Then (119.7) becomes

$$(\gamma p - m) \frac{16\pi i}{(2\pi)^4} e^2 \int d^4k \int_0^1 dx \int_0^1 dz \frac{(\gamma p + m)[2m - \gamma p(1-x)]x(1-x)}{[k^2 - a_0^2 + (p^2 - m^2)x(1-x)z]^3},$$

† Thus, in the process of renormalization "en route" (§110) we omit corrections to the renormalization constant Z_1. The corresponding integrals are logarithmically divergent. If we use a "cut-off parameter" $\Lambda^2 \gg m^2$, p^2, and limit the region of integration over d^4k by the condition $k^2 \leqslant \Lambda^2$, the correction can be explicitly calculated; the result is

$$Z_1 = 1 + Z_1^{(1)}, \quad Z_1^{(1)} = -\frac{\alpha}{2\pi} \left[\frac{1}{2} \log \frac{\Lambda^2}{m^2} + \log \frac{\lambda^2}{m^2} + \frac{9}{4} \right]. \tag{119.7a}$$

where we have also used the identity $p^2 - m^2 = (\gamma p - m)(\gamma p + m)$. The integration over d^4k is immediate: assuming that $p^2 - m^2 < 0$ and using (131.14), we have

$$(\gamma p - m)\frac{e^2}{2\pi}\int_0^1 dx \int_0^1 dz \frac{(\gamma p + m)[2m - \gamma p(1-x)]x(1-x)}{m^2x^2 + \lambda^2(1-x) + (m^2 - p^2)x(1-x)z}.$$

We now need to subtract a similar integral with the substitution (119.6), omitting temporarily the factor $\gamma p - m$; after some simple calculations, we get

$$\mathcal{M}(p) = (\gamma p - m)^2 \frac{e^2}{2\pi}\int_0^1 dx \int_0^1 dz \times$$
$$\times \frac{m(1-x^2) - (\gamma p + m)(1-x)^2\left[1 - \dfrac{2x(1+x)z}{x^2 + (\lambda/m)^2}\right]}{m^2x + (m^2 - p^2)(1-x)z} \tag{119.8}$$

(the term in λ^2 is omitted from the common denominator, since this causes no divergence in the present case; elsewhere, $\lambda^2(1-x)$ is replaced by λ^2, since the infra-red divergence will correspond to the divergence as $x \to 0$).

The integration in (119.8) (first over dz and then over dx) is fairly lengthy but elementary, and the result is

$$\mathcal{M}(p) = \frac{\alpha}{2\pi m}(\gamma p - m)^2\left\{\frac{1}{2(1-\rho)}\left(1 - \frac{2-3\rho}{1-\rho}\log\rho\right) - \right.$$
$$\left. - \frac{\gamma p + m}{m\rho}\left[\frac{1}{2(1-\rho)}\left(2 - \rho + \frac{\rho^2 + 4\rho - 4}{1-\rho}\log\rho\right) + 1 + 2\log\frac{\lambda}{m}\right]\right\}, \tag{119.9}$$

with

$$\rho = (m^2 - p^2)/m^2$$

(R. Karplus and N. M. Kroll, 1950). The integral has been calculated on the assumption that $\rho > 0$ and $\rho \gg \lambda/m$. In accordance with the rule for passing round poles, in the analytical continuation of (119.9) into the region $\rho < 0$ the phase of the logarithm is found by making $m \to m - i0$; then $\rho \to \rho - i0$, and $\log\rho$ must be taken for $\rho < 0$ as

$$\log\rho = \log|\rho| - i\pi, \quad \rho < 0. \tag{119.10}$$

Let us now consider how the mass operator behaves when $p^2 \gg m^2$. Then $-\rho \approx p^2/m^2 \gg 1$, and with logarithmic accuracy we have

$$\mathcal{M}(p) = -[\mathcal{G}^{-1}(p) - G^{-1}(p)]$$
$$\approx \frac{\alpha}{4\pi}(\gamma p)\log\frac{p^2}{m^2}. \tag{119.11}$$

As in the case of the photon propagator (cf. formulae (113.15) and (113.16) for the polarization operator), the correction to G^{-1} is small only when the energy is so small that

$$\frac{\alpha}{4\pi}\log\frac{p^2}{m^2}\ll 1.$$

In the present case, however, the logarithmic increase is in a certain sense fictitious, and can be eliminated by a suitable choice of gauge, i.e. of the function $D^{(l)}$ in the photon propagator (L. D. Landau, A. A. Abrikosov and I. M. Khalatnikov, 1954). To achieve this, we put (in the notation of §103)

$$D^{(l)} = 0, \tag{119.12}$$

whereas formula (119.9) has been derived with the gauge

$$D^{(l)} = D. \tag{119.13}$$

This property of the gauge (119.12) makes it especially suitable for investigating the theory when $p^2 \gg m^2$, as we shall do in §132.

To prove the result stated, we note that, if only terms in e^2 are concerned, the transformation from the gauge (119.13) to (119.12) can be regarded as infinitesimal. Accordingly, we can apply immediately the formula (105.14), with

$$d^{(l)}(q) = -D/q^2 = -4\pi/(q^2)^2,$$

and also replace $\mathcal{G}$ in the integrand by G with the necessary accuracy. In the integral over d^4q, the important range is $q \gg p$, for which $G(p-q)$ in the integrand is much less than $G(p)$ and can be neglected. Then

$$\begin{aligned}\delta\mathcal{G}^{-1} &= -G^{-2}(p)\delta\mathcal{G}(p)\\ &= -ie^2G^{-1}(p)\int d^{(l)}(q)\, d^4q/(2\pi)^4.\end{aligned}$$

Finally, with the transformation (131.11) and (131.12), we have

$$\begin{aligned}\delta\mathcal{G}^{-1}(p) &= -\frac{e^2}{4\pi}G^{-1}(p)\int\frac{d(-q^2)}{-q^2}\\ &\approx -\frac{e^2}{4\pi}\gamma p\log\frac{\Lambda^2}{p^2},\end{aligned}$$

where Λ is an upper limit, at which the divergence is removed by renormalization; the renormalization consists in subtracting the same expression with $p^2 \approx m^2$, giving finally

$$\delta\mathcal{G}^{-1} = \frac{e^2}{4\pi}\gamma p\log\frac{p^2}{m^2}.$$

This just cancels the difference $\mathcal{G}^{-1} - G^{-1}$ in (119.11).

Lastly, let us consider why it is necessary to use a finite "photon mass" λ in regularizing the integral (119.2), which is closely related to its behaviour when $p^2 \to m^2$. Firstly, this integral itself is finite when $p^2 = m^2$ and $\lambda = 0$; to exclude the divergence for large k, which is here unimportant, we assume that the integral is taken over a large but finite region of k-space. The need to use λ arises in the subtraction of the renormalization integral, which would otherwise diverge at $p^2 = m^2$. Let us therefore ascertain how the non-regularized mass operator would behave as $p^2 \to m^2$. Since this behaviour depends essentially on the choice of gauge, we shall consider the general case of an arbitrary gauge, whereas the integral (119.2) has been written for the specific gauge (119.13).

We again apply the transformation (105.14). Writing

$$\begin{aligned} d^{(l)}(q) &= \delta D^{(l)}/q^2 \\ &= \frac{4\pi}{(q^2)^2}\,\delta a(q^2), \end{aligned} \tag{119.14}$$

we assume that δa is the variation of a function $a(q^2)$ which changes appreciably only over intervals $q^2 \sim m^2$ and is finite when $q^2 \approx m^2$. In the integrand on the right of (105.14), the two terms in the difference $\mathcal{G}(p) - \mathcal{G}(p-q)$ are almost equal when q is small, and the integral converges. For small q,

$$\mathcal{G}(p-q) \sim \frac{1}{p^2 - m^2 - 2pq},$$

and so $\mathcal{G}(p-q)$ may be neglected in comparison with $\mathcal{G}(p)$ when $q \gg (p^2 - m^2)/m$. The integral

$$\begin{aligned} \delta\mathcal{G}(p) &= ie^2 \mathcal{G}(p) \int d^{(l)}(q) \frac{d^4 q}{(2\pi)^4} \\ &= -\frac{e^2}{4\pi}\mathcal{G}(p) \int \delta a(q^2) \frac{d(-q^2)}{-q^2} \end{aligned}$$

diverges logarithmically in the range

$$\frac{(p^2 - m^2)^2}{m^2} \ll q^2 \ll m^2.$$

We therefore have, with logarithmic accuracy,

$$\frac{\delta\mathcal{G}}{\mathcal{G}} = -\frac{e^2}{2\pi}\,\delta a(m^2) \log \frac{m^2}{p^2 - m^2}.$$

This can be integrated as follows. When $\alpha \equiv e^2 \to 0$, the exact propagator $\mathcal{G}$ must be the same as the free-particle propagator G, and therefore

$$\mathcal{G}(p) = \frac{1}{\gamma p - m}\left(\frac{m^2}{p^2 - m^2}\right)^{\alpha(C - a_0)/2\pi}, \tag{119.15}$$

where $a_0 = a(m^2)$ and C is a constant. To determine C, we compare the expression

$$\mathcal{G}^{-1}(p) = (\gamma p - m)\left[1 + \frac{\alpha}{2\pi}(C - a_0)\log\rho\right], \tag{119.16}$$

which is obtained from (119.15) in the first approximation with respect to α, and the corresponding expression given by the integral (119.2) when $\lambda = 0$:†

$$\mathcal{G}^{-1}(p) = (\gamma p - m)\left[1 + \frac{\alpha}{\pi}\log\rho\right]. \tag{119.17}$$

According to the definition (119.14), the function $a(q^2)$ is equal to the ratio $D^{(l)}/D$. The gauge (119.13) to which (119.17) belongs therefore corresponds to $a = a_0 = 1$. Equating (119.16) and (119.17) for this value of a_0, we find $C = 3$.

Thus we have finally as the limiting (*infra-red asymptotic*) expression for the unrenormalized electron propagator with $p^2 \to m^2$

$$\mathcal{G}(p) = \frac{\gamma p + m}{p^2 - m^2}\left(\frac{m^2}{p^2 - m^2}\right)^{\alpha(3-a_0)/2\pi} \tag{119.18}$$

(A. A. Abrikosov, 1955). The validity of this formula depends only on the inequalities $\alpha \ll 1$, $|\log\rho| \gg 1$, whereas the formulae of perturbation theory would require also that $\alpha|\log\rho|/2\pi \ll 1$. The sign of the difference $p^2 - m^2$ is also unimportant here, since the imaginary part of (119.18) is in any case beyond the limits of its accuracy.

The renormalized propagator must have a simple pole when $p^2 = m^2$. We see that (119.18) satisfies this condition only in the gauge where

$$D^{(l)} = 3D \tag{119.19}$$

(so that $a_0 = 3$). Then the regularization of the Feynman integral (in order to prevent its divergence at the upper limits) will not require the use of a finite "photon mass". In other gauges, the zero mass of the photon produces a branch point instead of a simple pole at $p^2 = m^2$, and the finite parameter λ is needed in order to remove this "defect".

§ 120. Emission of soft photons with non-zero mass

In calculating the electron form factors in §117, we encountered a divergence of the integrals when the frequencies of the virtual photons are small. This divergence is closely related to the infra-red catastrophe discussed in §98, where it was pointed out that the cross-section for any process involving charged particles (including the

† It is not necessary to repeat the calculations in order to derive (119.17). The term in $\log\rho$ in (119.9) is obtained on the assumption that $\rho \gg \lambda$, which allows the limit $\lambda \to 0$ to be taken. The term in $\log(\lambda/m)$ arises from the subtraction of the renormalization integral, and is not present in the original integral (119.2). The subtraction is easily seen to have no effect on the $\log\rho$ terms.

scattering of electrons by an external field, represented by a diagram such as (117.1)) has no significance in itself, but only when the simultaneous emission of any number of soft photons is taken into account. It will be shown in §122 that all the divergences cancel in the total cross-section, which includes the emission of soft quanta. Here, of course, in order to obtain the correct result it is necessary for the initial "cut-off" of the divergent integrals to be taken in the same manner in all the cross-sections in the sum.

In §117, this cut-off was applied by means of a fictitious finite mass λ of the virtual photon. We must therefore now modify the formulae of §98 in such a way that they describe the emission of soft "photons" with non-zero mass.

Formally, such a photon is a "vector" particle with spin 1, whose free field has been discussed in §14. Such particles are described by a 4-vector ψ-operator

$$\hat{\psi}_\mu = \sqrt{(4\pi)} \sum_{\mathbf{k},\alpha} \frac{1}{\sqrt{(2\omega)}} (\hat{c}_{\mathbf{k}\alpha} e_\mu^{(\alpha)} e^{-ikx} + \hat{c}^+_{\mathbf{k}\alpha} e_\mu^{(\alpha)*} e^{ikx}), \quad \alpha = 1, 2, 3; \tag{120.1}$$

the notation and normalization differ from that in (14.16), in order to bring them into line with the photon case.

The interaction of the "photons" (120.1) with electrons is to be described by a Lagrangian of the same form as for true photons:

$$-e\hat{j}^\mu \hat{\psi}_\mu, \tag{120.2}$$

the potential $\hat{A}_\mu$ being replaced by $\hat{\psi}_\mu$. Then the amplitudes for the processes of emission of photons with finite mass will be given by the usual rules of the diagram technique, the only differences being that

$$k^2 = \lambda^2 \tag{120.3}$$

and that the summation over the polarizations of the emitted photon must be taken over three independent polarizations (two transverse and one longitudinal) instead of two as for the ordinary photon.This is equivalent to averaging with respect to the density matrix of unpolarized particles

$$\rho_{\mu\nu} = -\tfrac{1}{3}\left(g_{\mu\nu} - \frac{k_\mu k_\nu}{\lambda^2}\right) \tag{120.4}$$

(cf. (14.15)), followed by multiplication by 3.

The propagator for "photons with non-zero mass" is

$$D_{\mu\nu} = \frac{4\pi}{k^2 - \lambda^2}\left(g_{\mu\nu} - \frac{k_\mu k_\nu}{\lambda^2}\right)$$

(cf. (76.18)), but in the case of gauge invariance the amplitudes of real scattering processes do not depend on the longitudinal part of the photon propagator, and this property is not the result of the specific form of its transverse part. The second term in the parentheses can therefore be omitted, leaving an expression of the same type as for ordinary photons:

$$D_{\mu\nu} = \frac{4\pi}{k^2 - \lambda^2} g_{\mu\nu}, \qquad (120.5)$$

as has been used in §§117 and 119.

Let us now consider the emission of soft photons (in the sense explained in §98). The derivation of (98.5) and (98.6) can be applied to the present case, the only difference being that the term $k^2 = \lambda^2$ is added in expanding the squares $(p \pm k)^2$ in the denominators of the electron propagators. We thus have, instead of (98.6),

$$d\sigma = d\sigma_{\text{el}} \cdot e^2 \left| \frac{p'e}{p'k + \lambda^2/2} - \frac{pe}{pk - \lambda^2/2} \right|^2 \frac{d^3k}{4\pi^2\omega},$$

where $d\sigma_{\text{el}}$ is the cross-section for the same process without emission of a soft quantum, which we shall conventionally call an "elastic process". In the integrations over d^3k, the important range is $|\mathbf{k}| \sim \lambda$. Then $p'k \sim pk \gg \lambda^2$, so that the terms in λ^2 in the denominators may be neglected. The summation over polarizations of the photon is carried out by means of (120.4), as already described. When the approximation stated is used, the second term in (120.4) makes no contribution to the cross-section, and there remains†

$$d\sigma = - d\sigma_{\text{el}} \cdot e^2 \left(\frac{p'}{(p'k)} - \frac{p}{(pk)} \right)^2 \frac{d^3k}{4\pi^2\omega}. \qquad (120.6)$$

We thus recover formula (98.7), but ω must now be taken as

$$\omega = \sqrt{(\mathbf{k}^2 + \lambda^2)}. \qquad (120.7)$$

Formula (120.6) is completely general: it is applicable to both elastic and inelastic scattering and even when the type of particle changes. The result of the further integration over d^3k depends on the 4-vectors p and p', i.e. on the nature of the basic scattering process.

Let us take the case of elastic scattering, for which

$$|\mathbf{p}| = |\mathbf{p}'|, \quad \varepsilon = \varepsilon',$$

and determine the total probability of photon emission with a frequency less than some $\omega_{\max}$, with the assumption that

$$\lambda \ll \omega_{\max} \qquad (120.8)$$

and that $\omega_{\max}$ is subject to an upper limit governed by the conditions (98.9) and (98.10) for the theory of soft-photon emission to be valid. We first calculate the

† There may be some doubt at first as to the validity of neglecting λ^2 before averaging, since it occurs in the denominator of the second term in (120.4), but we can easily show directly that this term gives, on averaging, a contribution $\sim \lambda^4 \cdot 1/\lambda^2$, which is negligible.

integral over d^3k in the non-relativistic limit. For $|\mathbf{p}| = |\mathbf{p}'| \ll m$,

$$\left(\frac{p'}{(p'k)} - \frac{p}{(pk)}\right)^2 \approx \frac{(\mathbf{q}\cdot\mathbf{k})^2}{m^2\omega^4} - \frac{\mathbf{q}^2}{m^2\omega^2},$$

where $\mathbf{q} = \mathbf{p}' - \mathbf{p}$. Integration over the directions of $\mathbf{k}$ gives

$$\frac{4\pi\mathbf{q}^2}{m^2\omega^2}\left(\frac{\mathbf{k}^2}{3\omega^2} - 1\right).$$

Then, from (120.6),

$$d\sigma = d\sigma_{\text{el}} \cdot \frac{e^2\mathbf{q}^2}{\pi m^2} \int_0^{\omega=\omega_{\max}} \left[1 - \frac{\mathbf{k}^2}{3(\mathbf{k}^2 + \lambda^2)}\right] \frac{\mathbf{k}^2\, d|\mathbf{k}|}{(\mathbf{k}^2 + \lambda^2)^{3/2}},$$

or, after integration with the assumption that $\omega_{\max}/\lambda \gg 1$,

$$d\sigma = d\sigma_{\text{el}} \cdot \frac{2\alpha}{3\pi}\frac{\mathbf{q}^2}{m^2}\left(\log\frac{2\omega_{\max}}{\lambda} - \frac{5}{6}\right), \quad \mathbf{q}^2 \ll m^2. \tag{120.9}$$

In the general relativistic case, the integral is calculated by means of (131.4). The angle integral is then

$$I = \int \frac{do_{\mathbf{k}}}{(pk)(p'k)} = \int_0^1 dx \int \frac{do_{\mathbf{k}}}{[(pk)x + (p'k)(1-x)]^2}$$

or, expanding the scalar products with $p = (\varepsilon, \mathbf{p})$, $p' = (\varepsilon, \mathbf{p}')$,

$$I = \int_0^1 dx \int \frac{do_{\mathbf{k}}}{\{\varepsilon\omega - \mathbf{k}\cdot[\mathbf{p}x + \mathbf{p}'(1-x)]\}^2}.$$

The inner integral is now easily calculated in spherical coordinates with the polar axis along the vector $\mathbf{p}x + \mathbf{p}'(1-x)$, giving

$$I = \int_0^1 \frac{4\pi\, dx}{(\varepsilon\omega)^2 - [\mathbf{p}x + \mathbf{p}'(1-x)]^2\mathbf{k}^2}$$

$$= \int_0^1 \frac{4\pi\, dx}{[m^2 + \mathbf{q}^2 x(1-x)]\mathbf{k}^2 + \varepsilon^2\lambda^2}.$$

The other two integrals, with $(pk)^2$ and $(p'k)^2$ in the denominators, are derived from this by putting $\mathbf{q}=0$. Using also the formula

$$pp' = \varepsilon^2 - \mathbf{p}\cdot\mathbf{p}' = m^2 + \tfrac{1}{2}\mathbf{q}^2,$$

we get

$$d\sigma = \frac{2e^2}{\pi}\int_0^1 dx \int_0^{\omega_{\max}} \frac{\mathbf{k}^2\, d|\mathbf{k}|}{\sqrt{(\mathbf{k}^2+\lambda^2)}}\left\{\frac{m^2+\frac{1}{2}\mathbf{q}^2}{[m^2+\mathbf{q}^2x(1-x)]\mathbf{k}^2+\varepsilon^2\lambda^2} - \frac{m^2}{m^2\mathbf{k}^2+\varepsilon^2\lambda^2}\right\}. \tag{120.10}$$

The integration over $d|\mathbf{k}|$ calls for the calculation of integrals having the form

$$\int_0^{\omega_{\max}} \frac{\mathbf{k}^2\, d|\mathbf{k}|}{(a\mathbf{k}^2+\lambda^2)\sqrt{(\mathbf{k}^2+\lambda^2)}}$$

$$= \frac{1}{a}\int_0^{\omega_{\max}} \frac{d|\mathbf{k}|}{\sqrt{(\mathbf{k}^2+\lambda^2)}} - \frac{\lambda^2}{a}\int_0^{\omega_{\max}} \frac{d|\mathbf{k}|}{(a\mathbf{k}^2+\lambda^2)\sqrt{(\mathbf{k}^2+\lambda^2)}}$$

$$\approx \frac{1}{a}\log\frac{2\omega_{\max}}{\lambda} - \frac{1}{a}\int_0^{\infty}\frac{dz}{(az^2+1)\sqrt{(z^2+1)}}.$$

In the second integral we have put λz for $|\mathbf{k}|$ and replaced the upper limit $\omega_{\max}/\lambda$ by infinity; this is permissible, since the integral converges.

The integrals over dx which then occur in (120.10) cannot be expressed entirely in terms of elementary functions. The result may be written in the form

$$d\sigma = \alpha\left[F\left(\frac{|\mathbf{q}|}{2m}\right)\log\frac{2\omega_{\max}}{\lambda} + F_1\right] d\sigma_{\text{el}}, \tag{120.11}$$

where†

$$F(\xi) = \frac{2}{\pi}\left[\frac{2\xi^2+1}{\xi\sqrt{(\xi^2+1)}}\log\left(\xi+\sqrt{(\xi^2+1)}\right) - 1\right], \tag{120.12}$$

$$F_1 = \frac{2\varepsilon}{\pi|\mathbf{p}|}\log\frac{\varepsilon+|\mathbf{p}|}{m} - \frac{2m^2+\mathbf{q}^2}{\pi\varepsilon^2}\int_0^1 \frac{dx}{a\sqrt{(1-a)}}\log\frac{1+\sqrt{(1-a)}}{\sqrt{a}}, \tag{120.13}$$

$$a = \frac{1}{\varepsilon^2}[m^2+\mathbf{q}^2x(1-x)].$$

† The function $F(\xi)$ has already occurred in §98, Problems. This is not surprising, since (120.11) can be derived with logarithmic accuracy by integrating the cross-section (98.8) for the emission of zero-mass photons over $d\omega$ from λ to $\omega_{\max}$.

If ξ is replaced by a variable θ such that $\xi = \sinh\frac{1}{2}\theta$, then

$$F(\theta) = (2/\pi)(\theta\coth\theta - 1). \tag{120.12a}$$

An asymptotic expression for the cross-section in the ultra-relativistic case can be obtained, assuming not only that $\varepsilon \gg m$ but also that $|\mathbf{q}| \gg m$, i.e. that the scattering angle is not very small. Then the important range of values of x in the integral in (120.13) is that for which $a \ll 1$; the appropriate approximations give

$$F_1 \approx \frac{\mathbf{q}^2}{2\pi\varepsilon^2} \int_0^1 \frac{\log a}{a}\, dx$$

$$\approx \frac{1}{2\pi} \int \frac{\log(\mathbf{q}^2/\varepsilon^2) + \log x + \log(1-x)}{x(1-x)}\, dx.$$

The integral is to be cut off at $a \sim 1$, i.e. at $x \sim m^2/\mathbf{q}^2$ at the lower limit and $1-x \sim m^2/\mathbf{q}^2$ at the upper limit. Then

$$F_1 \approx \frac{1}{2\pi}\left[2\log\frac{\mathbf{q}^2}{\varepsilon^2}\log\frac{\mathbf{q}^2}{m^2} - \log^2\frac{\mathbf{q}^2}{m^2}\right]$$

$$= \frac{1}{2\pi}\left[\log^2\frac{\mathbf{q}^2}{m^2} - 4\log\frac{\varepsilon}{m}\log\frac{\mathbf{q}^2}{m^2}\right].$$

This formula is valid as far as the squares of the logarithms (with double-logarithmic accuracy). To the same accuracy, it is sufficient to put in the first term in (120.11)

$$F(\xi) \approx (4/\pi)\log\xi \quad (\xi \gg 1).$$

The final result is

$$d\sigma = \frac{2\alpha}{\pi}\left[\log\frac{\mathbf{q}^2}{m^2}\log\frac{\omega_{\max}}{\lambda} - \log\frac{\varepsilon}{m}\log\frac{\mathbf{q}^2}{m^2} + \frac{1}{4}\log^2\frac{\mathbf{q}^2}{m^2}\right] d\sigma_{\text{el}}, \quad \mathbf{q}^2 \gg m^2. \tag{120.14}$$

§ 121. Electron scattering in an external field in the second Born approximation

In the first two approximations with respect to the external field, the scattering of an electron is represented by the diagrams

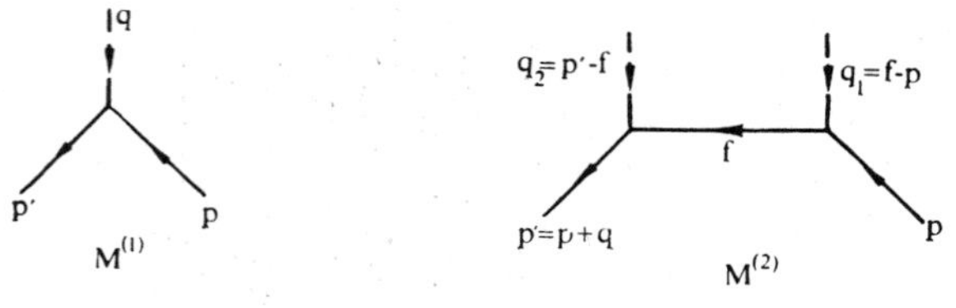

(121.1)

The first of these corresponds to the amplitude $M^{(1)} \sim Ze^2$ considered in §80. The amplitude of the second approximation is $M^{(2)} \sim (Ze^2)^2$.

It is easily seen that terms of the same order as this arise from the radiative corrections. In the third order of perturbation theory, the radiative corrections to the scattering amplitude are represented by the diagrams

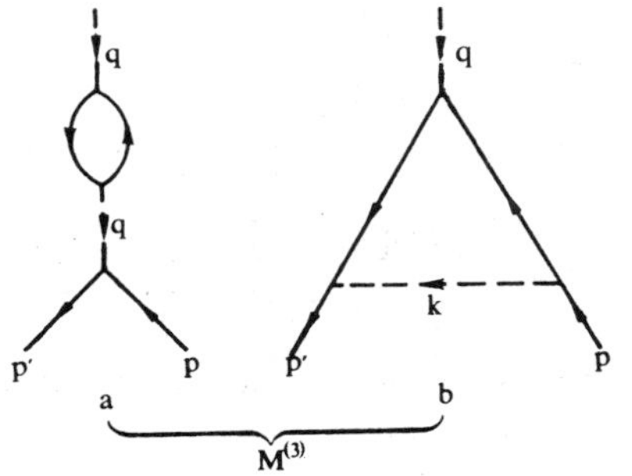

(121.2)

Here $M^{(3)} \sim Ze^2 \cdot e^2$, and $M^{(3)} \sim M^{(2)}$ (if $Z \sim 1$).

According to (64.26), the scattering cross-section is

$$d\sigma = |M_{fi}^{(1)} + M_{fi}^{(2)} + M_{fi}^{(3)}|^2 \, do'/16\pi^2. \tag{121.3}$$

In the squared amplitude we can retain not only $|M_{fi}^{(1)}|^2$ but also the interference terms between $M_{fi}^{(1)}$ and $M_{fi}^{(2)}$ and between $M_{fi}^{(1)}$ and $M_{fi}^{(3)}$. Thus the cross-section is given, as far as the terms in e^6, by the sum

$$d\sigma = d\sigma^{(1)} + d\sigma^{(2)} + d\sigma_{\text{rad}}, \tag{121.4}$$

where $d\sigma^{(1)}$ is the cross-section in the first Born approximation (§80), and the corrections are

$$\left.\begin{aligned} d\sigma^{(2)} &= 2 \,\text{re}\, M_{fi}^{(1)} M_{fi}^{(2)*} \, do'/16\pi^2, \\ d\sigma_{\text{rad}} &= 2 \,\text{re}\, M_{fi}^{(1)} M_{fi}^{(3)*} \, do'/16\pi^2. \end{aligned}\right\} \tag{121.5}$$

From §80,

$$M_{fi}^{(1)} = |e|(\bar{u}'\gamma^0 u)\Phi(\mathbf{q}), \tag{121.6}$$

where $\Phi(\mathbf{q})$ is a Fourier component of the scalar potential of the constant external field ($\Phi \equiv A_0^{(e)}$), and we have used the fact that the electron charge $e = -|e|$.

The two expressions (121.5) can evidently be calculated independently. The first will now be discussed, and the second in §122.

The second-approximation amplitude is given by the diagram (121.1) as the integral†

$$M_{fi}^{(2)} = -e^2 \int \left\{ \bar{u}(p')\gamma^0 \frac{\gamma f + m}{f^2 - m^2 + i0} \gamma^0 u(p) \right\} \Phi(\mathbf{p}' - \mathbf{f})\Phi(\mathbf{f} - \mathbf{p}) \frac{d^3 f}{(2\pi)^3}. \tag{121.7}$$

The "4-momenta" $q_1 = f - p$ and $q_2 = p' - f$ of the constant external field have no

† Here it is necessary to apply the diagram-technique rule concerning a constant external field; see rule 8 in §77.

time components. Hence

$$f_0 = \varepsilon = \varepsilon', \tag{121.8}$$

where ε and ε' are the initial and final electron energies, which in elastic scattering are the same.

In the purely Coulomb field of a stationary charge $Z|e|$,

$$\Phi(\mathbf{q}) = 4\pi Z|e|/\mathbf{q}^2.$$

For this potential the integral (121.7) is logarithmically divergent (when $\mathbf{f} \approx \mathbf{p}$ and $\mathbf{f} \approx \mathbf{p}'$). This divergence is specific to the Coulomb field, and arises from the slowness with which the field decreases at large distances. Its origin is most easily shown for the non-relativistic case. According to *QM* (135.8), the coefficient of the spherical wave $e^{i|\mathbf{p}|r}/r$ in the asymptotic expression for the wave function of an electron in a Coulomb field is

$$f(\theta) \exp\left(-i\frac{Z\alpha m}{|\mathbf{p}|}\log|\mathbf{p}|\,r\right).$$

This coefficient is also the electron scattering amplitude in the field, and we see that its phase includes a term which diverges as $r \to \infty$. When the scattering amplitude is expanded in powers of $Z\alpha$, this term causes divergence of all the terms in the expansion from the second term onwards (since the function $f(\theta)$ is proportional to $Z\alpha$). In the relativistic case there is, of course, a similar situation.

These arguments also show that the divergent terms must cancel when we calculate the scattering cross-section, in which the phase of the amplitude is unimportant. The simplest procedure for a correct calculation is to consider first the scattering in a screened Coulomb field, putting

$$\Phi(q) = 4\pi Z|e|/(\mathbf{q}^2 + \delta^2) \tag{121.9}$$

with a small screening constant $\delta \ll |\mathbf{p}|$. This eliminates the divergence in the scattering amplitude, and we can then put $\delta = 0$ in the final formula for the cross-section.

Substituting (121.9) in (121.7), we have

$$M_{fi}^{(2)} = -\frac{2}{\pi}Z^2\alpha^2\bar{u}(p')[(\gamma^0\varepsilon + m)J_1 + \boldsymbol{\gamma}\cdot\mathbf{J}]\,u(p),$$

with the notation

$$\left.\begin{aligned} J_1 &= \int\frac{d^3f}{[(\mathbf{p}'-\mathbf{f})^2+\delta^2][(\mathbf{f}-\mathbf{p})^2+\delta^2][\mathbf{p}^2-\mathbf{f}^2+i0]}, \\ \mathbf{J} &= \int\frac{\mathbf{f}\,d^3f}{[(\mathbf{p}'-\mathbf{f})^2+\delta^2][(\mathbf{f}-\mathbf{p})^2+\delta^2][\mathbf{p}^2-\mathbf{f}^2+i0]} \equiv \tfrac{1}{2}(\mathbf{p}+\mathbf{p}')\,J_2; \end{aligned}\right\} \tag{121.10}$$

$\mathbf{p}^2 = \varepsilon^2 - m^2 = \mathbf{p}'^2$, and the integral $\mathbf{J}$ is symmetrical in $\mathbf{p}$ and $\mathbf{p}'$; from considerations of vector symmetry, it is immediately obvious that the vector $\mathbf{J}$ must be parallel to $\mathbf{p}+\mathbf{p}'$. Now, eliminating the matrices $\boldsymbol{\gamma}$ by means of the equations

$$\boldsymbol{\gamma}\cdot\mathbf{p}u = (\gamma^0\varepsilon - m)u,$$
$$\bar{u}'\boldsymbol{\gamma}\cdot\mathbf{p}' = \bar{u}'(\gamma^0\varepsilon - m),$$

we obtain

$$M_{fi}^{(2)} = -\frac{2}{\pi}Z^2\alpha^2\bar{u}(p')[\gamma^0\varepsilon(J_1+J_2)+m(J_1-J_2)]\,u(p). \tag{121.11}$$

In order to continue the calculations, we change (as in §80) from the bispinor amplitudes u and u' to the three-dimensional spinors w and w' which correspond to them in accordance with (23.9) and (23.11). A direct multiplication gives

$$\bar{u}'u = w'^*\{(\varepsilon+m)-(\varepsilon-m)\cos\theta + i\,\boldsymbol{\nu}\cdot\boldsymbol{\sigma}(\varepsilon-m)\sin\theta\}\,w,$$

$$\bar{u}'\gamma^0 u = w'^*\{(\varepsilon+m)+(\varepsilon-m)\cos\theta - i\,\boldsymbol{\nu}\cdot\boldsymbol{\sigma}(\varepsilon-m)\sin\theta\}\,w,$$

where $\boldsymbol{\nu} = \mathbf{n}\times\mathbf{n}'/\sin\theta$, $\mathbf{n} = \mathbf{p}/|\mathbf{p}|$, $\mathbf{n}' = \mathbf{p}'/|\mathbf{p}'|$, $\cos\theta = \mathbf{n}\cdot\mathbf{n}'$. Then the amplitude (121.11) may be written†

$$\left.\begin{aligned}
M_{fi}^{(2)} &= 4\pi w'^*(A^{(2)} + B^{(2)}\,\boldsymbol{\nu}\cdot\boldsymbol{\sigma})\,w,\\
A^{(2)} &= -\frac{1}{2\pi^2}Z^2\alpha^2\{[(\varepsilon+m)+(\varepsilon-m)\cos\theta]\,\varepsilon(J_1+J_2)+\\
&\quad +[(\varepsilon+m)-(\varepsilon-m)\cos\theta]\,m(J_1-J_2)\},\\
B^{(2)} &= \frac{i}{2\pi^2}Z^2\alpha^2(\varepsilon-m)\sin\theta[\varepsilon(J_1+J_2)-m(J_1-J_2)].
\end{aligned}\right\} \tag{121.12}$$

The first-approximation scattering amplitude is, in corresponding notation,

$$\left.\begin{aligned}
M_{fi}^{(1)} &= 4\pi w'^*(A^{(1)} + B^{(1)}\,\boldsymbol{\nu}\cdot\boldsymbol{\sigma})\,w,\\
A^{(1)} &= \frac{Z\alpha}{\mathbf{q}^2}[(\varepsilon+m)+(\varepsilon-m)\cos\theta],\\
B^{(1)} &= -i\frac{Z\alpha}{\mathbf{q}^2}(\varepsilon-m)\sin\theta,
\end{aligned}\right\} \tag{121.13}$$

where $\mathbf{q} = \mathbf{p}'-\mathbf{p}$.

The scattering cross-section and the polarization effects are expressed in terms of the quantities $A = A^{(1)}+A^{(2)}$ and $B = B^{(1)}+B^{(2)}$ by the formulae derived in *QM*,

† The definition of A and B is as in §37 and in *QM*, §140, and differs by a factor from that in §80.

§140. For example, the scattering cross-section for unpolarized electrons is

$$d\sigma = (|A|^2 + |B|^2)\, do'$$
$$\approx d\sigma^{(1)} + 2(A^{(1)} \operatorname{re} A^{(2)} - iB^{(1)} \operatorname{im} B^{(2)})\, do'.$$

Substitution of (121.12) and (121.13) and straightforward calculation gives

$$d\sigma^{(2)} = -\,do' \frac{Z^3\alpha^3\varepsilon^3}{\pi^2\mathbf{p}^2 \sin^2\frac{1}{2}\theta}[(1 - \mathbf{v}^2 \sin^2\tfrac{1}{2}\theta) \operatorname{re}(J_1 + J_2) + (m^2/\varepsilon^2) \operatorname{re}(J_1 - J_2)], \tag{121.14}$$

where $\mathbf{v} = \mathbf{p}/\varepsilon$ is the electron velocity and θ the scattering angle. The electrons are polarized by scattering, and the polarization vector of the final electrons is

$$\boldsymbol{\zeta}' = \frac{2 \operatorname{re}(AB^*)}{|A|^2 + |B|^2}\boldsymbol{\nu}$$
$$\approx \frac{2(A^{(1)} \operatorname{re} B^{(2)} - iB^{(1)} \operatorname{im} A^{(2)})}{|A^{(1)}|^2 + |B^{(1)}|^2}\boldsymbol{\nu}$$

or, substituting (121.12) and (121.13),

$$\boldsymbol{\zeta}' = \frac{4Z\alpha m \mathbf{p}^4}{\pi^2\varepsilon^2}\frac{\sin^3\frac{1}{2}\theta \cos\frac{1}{2}\theta}{1 - v^2 \sin^2\frac{1}{2}\theta} \operatorname{im}(J_1 - J_2)\, \boldsymbol{\nu}. \tag{121.15}$$

Let us now calculate the integrals J_1 and J_2. This is more easily done by using the parametrization method (131.2). The integral J_1 then becomes

$$J_1 = -2\int_0^1\int_0^1\int_0^1\int \frac{d^3f\, d\xi_1\, d\xi_2\, d\xi_3\, \delta(1 - \xi_1 - \xi_2 - \xi_3)}{\{[(\mathbf{p}' - \mathbf{f})^2 + \delta^2]\xi_1 + [(\mathbf{p} - \mathbf{f})^2 + \delta^2]\xi_2 + [\mathbf{f}^2 - \mathbf{p}^2 - i0]\xi_3\}^3}.$$

The integration over $d\xi_3$ eliminates the delta function, and reduction of the denominator then gives

$$J_1 = -2\int_0^1\int_0^{1-\xi_2}\int \frac{d^3f\, d\xi_1\, d\xi_2}{\{\delta^2(\xi_1 + \xi_2) + \mathbf{p}^2(2\xi_1 + 2\xi_2 - 1) - 2\mathbf{f}\cdot(\xi_1\mathbf{p}' + \xi_2\mathbf{p}) + \mathbf{f}^2 - i0\}^3}.$$

Using in place of $\mathbf{f}$ a new variable $\mathbf{k} = \mathbf{f} - \xi_1\mathbf{p}' - \xi_2\mathbf{p}$, we can reduce the integration over d^3f to the form

$$\int \frac{d^3k}{(\mathbf{k}^2 - a^2 - i0)^3} = i\frac{\pi^2}{4a^3},$$

so that

$$J_1 = -\tfrac{1}{2}i\pi^2\int_0^1\int_0^{1-\xi_2} \frac{d\xi_1\, d\xi_2}{\{\mathbf{p}^2(\xi_1^2 + \xi_2^2 - 2\xi_1 - 2\xi_2 + 1) + 2\xi_1\xi_2\mathbf{p}\cdot\mathbf{p}' - \delta^2(\xi_1 + \xi_2) - i0\}^{3/2}}.$$

Instead of ξ_1 and ξ_2 we use the symmetrical combinations $x = \xi_1 + \xi_2$, $y = \xi_1 - \xi_2$. The integration over dy from 0 to x is elementary, and gives

$$J_1 = -\frac{i\pi^2}{2|\mathbf{p}|^3}\int_0^1 \frac{x\,dx}{[bx^2 - 2x + 1 - (\delta^2/\mathbf{p}^2)x - i0][(1-x)^2 - (\delta^2/\mathbf{p}^2)x - i0]^{1/2}},$$

where

$$\begin{aligned} b &= (\mathbf{p}^2 + \mathbf{p}\cdot\mathbf{p}')/2\mathbf{p}^2 \\ &= \cos^2 \tfrac{1}{2}\theta. \end{aligned}$$

To calculate the integral over dx as $\delta \to 0$, we divide the range of integration into two parts:

$$\int_0^1 \cdots dx = \int_0^{1-\delta_1} \cdots dx + \int_{1-\delta_1}^1 \cdots dx, \quad 1 \gg \delta_1 \gg \delta/|\mathbf{p}|.$$

In the first integral we can put $\delta = 0$; then†

$$\begin{aligned} \int_0^{1-\delta_1} \cdots dx &= \frac{1}{2(1-b)}\left[\log \frac{(1-x)^2}{bx^2 - 2x + 1 - i0}\right]_0^{1-\delta_1} \\ &= \frac{1}{2(1-b)}\left[\log \frac{\delta_1^2}{1-b} + i\pi\right]. \end{aligned}$$

In the second integral we can put $x = 1$ everywhere except in the term $(1-x)^2$ and $\delta = 0$ in the first bracket in the denominator. Then‡

$$\begin{aligned} \int_{1-\delta_1}^1 \cdots dx &= -\frac{1}{1-b}\int_0^{\delta_1} \frac{dx'}{[x'^2 - (\delta^2/\mathbf{p}^2) - i0]^{1/2}} \\ &= -\frac{1}{1-b}\left[\int_{\delta/|\mathbf{p}|}^{\delta_1} \frac{dx'}{[x'^2 - (\delta^2/\mathbf{p}^2)]^{1/2}} + \right. \\ &\qquad \left. + i\int_0^{\delta/|\mathbf{p}|} \frac{dx'}{[(\delta^2/\mathbf{p}^2) - x'^2]^{1/2}}\right] \\ &= -\frac{1}{1-b}\left[\log \frac{2|\mathbf{p}|\,\delta_1}{\delta} + \tfrac{1}{2}i\pi\right]. \end{aligned}$$

† The term $i0$ arises from the rule for avoiding the singularity, which gives the change in the argument of the logarithm between 0 and $1 - \delta_1$, namely from 0 to $-\pi$ as we pass below the branch point.

‡ Here again the singularity avoidance rule gives the sign of the square root as we go from positive to negative values of the radicand.

On adding the two integrals, δ_1 disappears, as it should, leaving

$$J_1 = \frac{i\pi^2}{2|\mathbf{p}|^3 \sin^2\frac{1}{2}\theta} \log\left(\frac{2|\mathbf{p}|}{\delta}\sin\tfrac{1}{2}\theta\right). \tag{121.16}$$

The integral J_2 is calculated similarly:

$$J_2 = J_1 - \frac{\pi^3(1-\sin\frac{1}{2}\theta)}{4|\mathbf{p}|^3\cos^2\frac{1}{2}\theta\sin\frac{1}{2}\theta} - \frac{i\pi^2}{2|\mathbf{p}|^3\cos^2\frac{1}{2}\theta}\log\sin\tfrac{1}{2}\theta. \tag{121.17}$$

We now have only to substitute these expressions in (121.14) and (121.15), obtaining as the final results

$$d\sigma^{(2)} = \frac{\pi(Z\alpha)^3\varepsilon}{4|\mathbf{p}|^3\sin^3\frac{1}{2}\theta}(1-\sin\tfrac{1}{2}\theta)\,do', \tag{121.18}$$

$$\boldsymbol{\zeta}' = \frac{2Z\alpha m|\mathbf{p}|}{\varepsilon^2}\frac{\sin^3\frac{1}{2}\theta\log\sin\frac{1}{2}\theta}{(1-v^2\sin^2\frac{1}{2}\theta)\cos\frac{1}{2}\theta}\boldsymbol{\nu} \tag{121.19}$$

(W. A. McKinley and H. Feshbach, 1948; R. H. Dalitz, 1950).

In the first Born approximation, the electron and positron scattering cross-sections are the same (in the same external field); in the second approximation, this symmetry does not occur. In the scattering of a positron (charge $+|e|$) the amplitude of the first approximation (121.6) has the opposite sign, but the sign of $M_{fi}^{(2)}$ is unchanged. The cross-section $d\sigma^{(2)}$, which is the interference term between $M_{fi}^{(1)}$ and $M_{fi}^{(2)}$, therefore changes sign. The same occurs for the expression (121.19) for the polarization vector. The formulae for electron scattering are all converted to those for positron scattering by the formal change $Z \to -Z$.

§ 122. Radiative corrections to electron scattering in an external field

Let us now calculate the radiative corrections to electron scattering in an external field (J. Schwinger, 1949). The corresponding part of the scattering amplitude is represented by the two diagrams (121.2). The contribution from the first of these to the amplitude is

$$-(\bar{u}'\gamma^0 u)\frac{\mathcal{P}(-\mathbf{q}^2)}{4\pi}D(-\mathbf{q}^2)\cdot e\Phi(\mathbf{q}),$$

where $\mathcal{P}(-\mathbf{q}^2)$ is the polarization operator corresponding to the loop in the diagram. The contribution from the second diagram is

$$-(\bar{u}'\Lambda^0 u)e\Phi(\mathbf{q}),$$

where Λ^0 is the correction term in the vertex operator ($\Gamma^\mu = \gamma^\mu + \Lambda^\mu$); according to

(116.6),

$$\Lambda^0 = \gamma^0[f(-\mathbf{q}^2) - 1] - \frac{1}{2m}\sigma^{0\nu} q_\nu g(-\mathbf{q}^2).$$

Adding the two contributions, we have†

$$\left.\begin{aligned} M_{fi}^{(3)} &= -(\bar{u}'\gamma^0 Q_{\text{rad}}\, u)e\Phi(\mathbf{q}), \\ Q_{\text{rad}}(\mathbf{q}) &= f(-\mathbf{q}^2) - 1 - \frac{1}{\mathbf{q}^2}\mathcal{P}(-\mathbf{q}^2) + \frac{1}{2m}g(-\mathbf{q}^2)\mathbf{q}\cdot\boldsymbol{\gamma}. \end{aligned}\right\} \tag{122.1}$$

Let us first consider the infra-red divergence in the form factor $f(-\mathbf{q}^2)$ and therefore in the scattering amplitude (122.1). It has already been mentioned in §98 that the exact value of the purely elastic scattering amplitude is zero, i.e. it has no meaning. The only physically significant thing is the amplitude of scattering defined as a process in which any number of soft photons can be emitted, each having an energy less than a specified value $\omega_{\max}$ satisfying the conditions for soft-photon emission theory to be valid. That is, only the sum

$$d\sigma = d\sigma_{\text{el}} + d\sigma_{\text{el}} \int\limits_0^{\omega_{\max}} dw_\omega + d\sigma_{\text{el}}\cdot\frac{1}{2!}\int\limits_0^{\omega_{\max}} dw_{\omega_1}\int\limits_0^{\omega_{\max}} dw_{\omega_2} + \cdots \tag{122.2}$$

is meaningful, where $d\sigma_{\text{el}}$ is the cross-section for scattering without emission of photons, and dw_ω the differential probability for the emission by the electron of a photon with frequency ω. Here it is assumed that $d\sigma_{\text{el}}$ itself is calculated as a perturbation-theory series, i.e. as an expansion in powers of α.‡ Then, on bringing together the terms of each order in α in (122.2), we obtain $d\sigma$ as an expansion in powers of α in which each term is finite.

In the first Born approximation, $d\sigma_{\text{el}} \sim \alpha^2$. This term has, of course, an independent significance. If, however, the next correction ($\sim\alpha^3$) to $d\sigma_{\text{el}}$ is to be taken into account, we must also include the second term in the sum (122.2): since $dw_\omega \sim \alpha$, multiplication by $d\sigma_{\text{el}} \sim \alpha^2$ likewise gives a quantity $\sim\alpha^3$. We shall show that the infra-red divergence disappears when these two quantities are added.

The divergent term in the form factor f (117.17) is§

$$-\tfrac{1}{2}\alpha F(|\mathbf{q}|/2m)\log(m/\lambda).$$

The corresponding term in the amplitude (122.1) is

$$\tfrac{1}{2}\alpha F\log(m/\lambda)(\bar{u}'\gamma^0 u)\, e\Phi(\mathbf{q}),$$

† Note that $q_\mu = (0, -\mathbf{q})$ if $q^\mu = (0, \mathbf{q})$, and therefore $\sigma^{0\nu} q_\nu = -\gamma^0\mathbf{q}\cdot\boldsymbol{\gamma}$.

‡ The need to take account of radiative corrections in the probability dw_ω is governed by the value of $\omega_{\max}$; the limit $\omega \to 0$ corresponds to the classical case where the radiative corrections are zero, and so the latter can always be made small by taking a sufficiently small $\omega_{\max}$.

§ This expression is easily verified by using the relation

$$|\mathbf{q}|/m = (1-\xi)/\sqrt{\xi}$$

between $|\mathbf{q}|$ and the variable ξ in terms of which (117.17) is written.

and in the cross-section (121.5)

$$d\sigma^{\text{infra}} = -\alpha F \log(m/\lambda)|\bar{u}'\gamma^0 u|^2|e\Phi(\mathbf{q})|^2\, do'/16\pi^2.$$

Comparing this with the Born cross-section

$$d\sigma^{(1)} = |\bar{u}'\gamma^0 u|^2|e\Phi(\mathbf{q})|^2\, do'/16\pi^2,$$

we find that

$$d\sigma^{\text{infra}} = -\alpha F \log(m/\lambda)\, d\sigma^{(1)}. \tag{122.3}$$

The second term in (122.2), with $\int dw_\omega$ from (120.11), gives

$$d\sigma_{\text{el}}\int_0^{\omega_{\max}} dw_\omega = \alpha F \log(2\omega_{\max}/\lambda)\, d\sigma^{(1)}. \tag{122.4}$$

Finally, adding (122.3) and (122.4), we obtain

$$-d\sigma^{(1)}\alpha F(|\mathbf{q}|/2m)\log(m/2\omega_{\max}). \tag{122.5}$$

We see that the divergent contribution from soft ($|\mathbf{k}| \sim \lambda$) virtual photons does in fact cancel with that from the emission of real photons of the same kind. A similar result occurs in any other scattering process.

There is also a dependence of the scattering cross-section on $\omega_{\max}$, resulting from the fact that $\omega_{\max}$ appears in the definition of scattering as a process in which any number of soft photons can be emitted. The cross-section for such a process will of course decrease with the upper frequency limit $\omega_{\max}$ for photons whose emission we regard as belonging to the scattering process in question.

Let us now determine the complete radiative correction to the scattering cross-section. Proceeding in accordance with the standard rules (see (65.7)), we find as the cross-section averaged over the polarizations of the initial electron and summed over the polarizations of the final electron

$$\begin{aligned} d\sigma &= d\sigma^{(1)} + d\sigma_{\text{rad}} \\ &= |e\Phi(\mathbf{q})|^2 \operatorname{tr}\{(\gamma p' + m)(\gamma^0 + \gamma^0 Q_{\text{rad}})(\gamma p + m)(\gamma^0 + \gamma^0 \bar{Q}_{\text{rad}})\}\, do'/32\pi^2. \end{aligned} \tag{122.6}$$

According to (122.1),

$$Q_{\text{rad}} = a + b\,\boldsymbol{\gamma}\cdot\mathbf{q}, \quad \bar{Q}_{\text{rad}} = \gamma^0 Q_{\text{rad}}^{+}\gamma^0 = a - b\,\boldsymbol{\gamma}\cdot\mathbf{q},$$

$$a = f(-\mathbf{q}^2) - 1 - \frac{1}{\mathbf{q}^2}\mathscr{P}(-\mathbf{q}^2), \quad b = \frac{1}{2m}g(-\mathbf{q}^2).$$

As far as the terms linear in a and b, the trace in (122.6) is given by

$$\tfrac{1}{4}\operatorname{tr}\{\cdots\} = 2(\varepsilon^2 - \tfrac{1}{4}\mathbf{q}^2)(1 + 2a) - 2bm\mathbf{q}^2.$$

Hence

$$d\sigma_{\text{rad}} = 2\left\{f_\lambda(-\mathbf{q}^2) - 1 - \frac{1}{\mathbf{q}^2}\mathcal{P}(-\mathbf{q}^2) - \frac{\mathbf{q}^2}{4\varepsilon^2 - \mathbf{q}^2}g(-\mathbf{q}^2)\right\} d\sigma^{(1)}, \qquad (122.7)$$

where $d\sigma^{(1)}$ is the Born cross-section (80.5) for the scattering of unpolarized electrons, and a subscript λ is added to the form factor f in order to show explicitly that it is cut off at photon mass λ.

We now have only to add to (122.7) the cross-section for the emission of soft photons. If we write f_λ in the form

$$f_\lambda(-\mathbf{q}^2) = 1 - \tfrac{1}{2}\alpha F(|\mathbf{q}|/2m)\log(m/\lambda) + \alpha F_2, \qquad (122.8)$$

then from (120.11) this addition simply means replacing f_λ in (122.7) by

$$f_{\omega_{\max}} = 1 - \tfrac{1}{2}\alpha F(|\mathbf{q}|/2m)\log(m/2\omega_{\max}) + \tfrac{1}{2}\alpha F_1 + \alpha F_2. \qquad (122.9)$$

With this change, (122.7) gives the final answer.

In the non-relativistic limit we have†

$$f_{\omega_{\max}} = 1 - \frac{\alpha\mathbf{q}^2}{3\pi m^2}\left(\log\frac{m}{2\omega_{\max}} + \frac{11}{24}\right), \quad \mathbf{q}^2 \ll m^2. \qquad (122.10)$$

The particular form of the external field appears in the radiative correction to the cross-section only through $d\sigma^{(1)}$; the factor in the braces in (122.7) is universal. In the non-relativistic approximation,

$$d\sigma_{\text{rad}} = -d\sigma^{(1)}\cdot\frac{2\alpha}{3\pi}\frac{\mathbf{q}^2}{m^2}\left(\log\frac{m}{2\omega_{\max}} + \frac{19}{30}\right), \quad \mathbf{q}^2 \ll m^2, \qquad (122.11)$$

which includes contributions from all the terms in (122.7). In the opposite (ultra-relativistic) limit, the main contribution comes only from the term in $f_{\omega_{\max}} - 1$:

$$d\sigma_{\text{rad}} = -d\sigma^{(1)}\cdot\frac{2\alpha}{\pi}\log\frac{\mathbf{q}^2}{m^2}\log\frac{\varepsilon}{\omega_{\max}}, \quad \mathbf{q}^2 \gg m^2. \qquad (122.12)$$

Finally, it may be noted that the radiative corrections considered here do not cause any additional polarization effects that are not present in the first Born approximation (unlike the corrections of the second Born approximation, discussed in §121). The reason is that the particular features of the first Born approximation are ultimately due to the fact that the S-matrix is Hermitian. This property is maintained even when the radiative corrections described above are taken into

† This differs from the non-relativistic formula (117.20) by the change

$$\log\lambda \to \log 2\omega_{\max} - \tfrac{5}{6}.$$

account, since in this approximation there are no real intermediate states in the scattering channel (and so the right-hand side of the unitarity relation is zero).†

§ 123. **Radiative shift of atomic levels**

The radiative corrections cause a shift of the energy levels of bound states of an electron in an external field, called the *Lamb shift*. The most interesting case of this kind is that of a hydrogen atom (or hydrogen-like ion).‡

A consistent method of finding the energy-level corrections is based on the use of the exact electron propagator in an external field (§109). But, if

$$Z\alpha \ll 1, \tag{123.1}$$

it is possible to use a simpler procedure in which the external field is regarded as a perturbation.

In the first approximation with respect to the external field, the radiative correction in the interaction between an electron and a constant electric field is described by the two diagrams (121.2) already used in connection with the problem of electron scattering in such a field; the change from one problem to the other needs no more than a simple reformulation (see below).

However, it is easily seen that this treatment can give only the part of the level shift that is due to the interaction with virtual photons of sufficiently high frequency. Let us consider, for example, the next radiative correction (as regards order with respect to the external field) to the electron scattering amplitude:

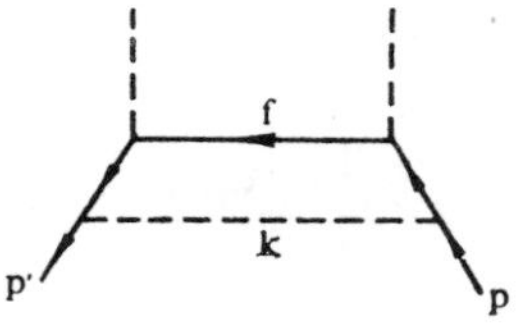

(123.2)

(unlike (121.2b), this diagram contains two external-field vertices). In the range of integration over d^4k where k_0 is sufficiently large, this correction involves an extra power of $Z\alpha$, and is therefore unimportant. But the addition of a second external-field vertex to the diagram also brings in a further electron propagator $G(f)$. When

† The calculation of the radiative corrections for processes which appear only in the second approximation of perturbation theory is considerably more laborious, and will not be given here. We shall simply list some references: L. M. Brown and R. P. Feynman, *Physical Review* **85**, 231, 1952 (radiative corrections to photon scattering by an electron); I. Harris and L. M. Brown, *ibid*, **105**, 1656, 1957 (r.c. to a two-photon pair-annihilation); M. L. G. Redhead, *Proceedings of the Royal Society* A**220**, 219, 1953, and R. V. Polovin, *Soviet Physics JETP* **4**, 385, 1957 (r.c. to electron scattering by an electron or a positron); P. I. Fomin, *ibid.* **8**, 491, 1959 (r.c. to bremsstrahlung).

‡ The shift of the hydrogen levels was first calculated by H. A. Bethe (1947) with logarithmic accuracy, using a non-relativistic treatment; this work provided the initial stimulus for the whole subsequent development of quantum electrodynamics. The difference between the $2s_{1/2}$ and $2p_{1/2}$ levels (in the first non-vanishing approximation of perturbation theory) was exactly calculated by N. M. Kroll and W. E. Lamb (1949); the complete formula for the level shift is due to V. F. Weisskopf and J. B. French (1949).

k is small, and the free ends p and p' are non-relativistic, the important values of the virtual-electron momenta f are those close to the pole of the propagator $G(f)$. The small denominator which thus occurs cancels the extra small factor $Z\alpha$. The same evidently applies to the corrections of all orders with respect to the external field. Thus, at low frequencies of the virtual photons, the external field must be taken into account exactly.

We can divide the required level shift† δE_s into two parts:

$$\delta E_s = \delta E_s^{(I)} + \delta E_s^{(II)}, \tag{123.3}$$

which originate from the interaction with virtual photons having frequencies in the ranges (I) $k_0 > \kappa$ and (II) $k_0 < \kappa$; κ is chosen so that

$$(Z\alpha)^2 m \ll \kappa \ll m, \tag{123.4}$$

where $Z^2\alpha^2 m$ is of the same order as the binding energy of the electron in the atom. Then, in region I, it is sufficient to take account of the nuclear field in the first approximation. In region II, the nuclear field must be treated exactly, but on the other hand, since $\kappa \ll m$, we can solve the problem in the non-relativistic approximation—not only as regards the electron itself, but for all the intermediate states. With the condition (123.4), the ranges of validity of the two methods of calculation overlap, and it is therefore possible to make an exact "joining" of the two parts of the level correction.

The high-frequency part of the shift

Let us first consider region I. Here it is possible to use the correction (122.1) to the scattering amplitude, after removing the contribution of the virtual photons which pertain to region II. These make only a small contribution to the form factor g, which therefore can be left unaltered. The low-frequency virtual photons make a large contribution to f, however, because of the infra-red divergence. Thus f in (122.1) must be taken as a function f_κ from which the region $k_0 < \kappa$ has been excluded.

This could be done directly by subtracting from f the integral over the region $k_0 < \kappa$, but the required result can be obtained without fresh calculations by using the results of §122. To do so, we note that the exclusion of frequencies $k_0 < \kappa$ can be regarded as one possible type of infra-red cut-off. The result for the correction to the scattering cross-section must, of course, be independent of the cut-off used, provided that the real soft photon emission probability is cut off in the same way, i.e. the concept of "elastic" scattering includes the emission only of photons with frequencies from κ to the specified ω_{max}. If we take $\omega_{max} = \kappa$, there is no need to take explicit account of the photon emission. Hence we see that f_κ is obtained from the $f_{\omega_{max}}$ determined in §122 by simply replacing ω_{max} by κ. In particular, in the

† In this section, E_s denotes the energy of an electron in an atom, not including its rest energy. The suffix s stands for all the quantum numbers which define the state of the atom.

non-relativistic case

$$f_\kappa - 1 = -\frac{\alpha \mathbf{q}^2}{3\pi m^2}\left(\log\frac{m}{2\kappa} + \frac{11}{24}\right). \tag{123.5}$$

Let us now transform the correction (122.1) to the scattering amplitude by representing it as the result of a corresponding correction to the effective potential energy of the electron in the field. Comparing the amplitude (122.1)

$$-e(u'^* Q_{\rm rad}\Phi u)$$

with the Born scattering amplitude (121.6)

$$-e(u'^*\Phi u),$$

we see that the correction is given (in the momentum representation) by the function

$$e\delta\Phi(\mathbf{q}) = eQ_{\rm rad}(\mathbf{q})\Phi(\mathbf{q}). \tag{123.6}$$

In the non-relativistic case, taking $\mathcal{P}$ and g from (113.14) and (117.20), and substituting f_κ from (123.5) for f, we get

$$\delta\Phi(\mathbf{q}) = \left\{-\frac{\alpha \mathbf{q}^2}{3\pi m^2}\left(\log\frac{m}{2\kappa} + \frac{11}{24} - \frac{1}{5}\right) + \frac{\alpha}{4\pi m}\mathbf{q}\cdot\boldsymbol{\gamma}\right\}\Phi(\mathbf{q}). \tag{123.7}$$

The corresponding function $\delta\Phi(\mathbf{r})$ in the coordinate representation is†

$$\delta\Phi(\mathbf{r}) = \frac{\alpha}{3\pi m^2}\left(\log\frac{m}{2\kappa} + \frac{11}{24} - \frac{1}{5}\right)\triangle\Phi(\mathbf{r}) - i\frac{\alpha}{4\pi m}\boldsymbol{\gamma}\cdot\nabla\Phi(\mathbf{r}). \tag{123.8}$$

The level shift $\delta E_s^{(\rm I)}$ is found by averaging $e\delta\Phi(\mathbf{r})$ over the wave function of the unperturbed state of the electron in the atom, i.e. as the corresponding diagonal matrix element:‡

$$\delta E_s^{(\rm I)} = \frac{e\alpha}{3\pi m^2}\left(\log\frac{m}{2\kappa} + \frac{11}{24} - \frac{1}{5}\right)\langle s|\triangle\Phi|s\rangle - $$

$$- i\frac{e\alpha}{4\pi m}\langle s|\boldsymbol{\gamma}\cdot\nabla\Phi|s\rangle. \tag{123.9}$$

In the first term, the non-relativistic electron function suffices for the averaging.

† Note that this correction to the potential is not the same as the one discussed in §114, which included only the effect of the vacuum polarization (diagram (121.2a)) on the Coulomb field as such. The correction (123.8) relates to the interaction of the field with the electron, and includes also the effect of a change in the motion of the electron (diagram (121.2b)).

‡ Strictly speaking, the form factors determined in §117 related to the vertex operator with two external electron lines ($p^2 = p'^2 = m^2$). For an electron in an atom, the energy E_s is a level which is unrelated to $\mathbf{p}$. The distinction may, however, be neglected in region I.

In the second term, this approximation is insufficient: the zero-order approximation with respect to the non-relativistic functions is zero on account of the absence of diagonal elements in the matrices $\boldsymbol{\gamma}$. Here, therefore, we must use the approximate relativistic function

$$\psi = \begin{pmatrix} \phi \\ \chi \end{pmatrix}$$

derived in §33, retaining the components χ which are small (in the standard representation). We have

$$\psi^* \boldsymbol{\gamma} \phi = \phi^* \boldsymbol{\sigma} \chi - \chi^* \boldsymbol{\sigma} \phi$$

and, substituting from (33.4)

$$\chi = \frac{1}{2m} \boldsymbol{\sigma} \cdot \hat{\mathbf{p}} \phi = -\frac{i}{2m} \boldsymbol{\sigma} \cdot \nabla \phi,$$

we get, using the identity (33.5) and integrating by parts,

$$\begin{aligned} \langle s | \boldsymbol{\gamma} \cdot \nabla \Phi | s \rangle &= -\frac{i}{2m} \int \{ \phi^* (\boldsymbol{\sigma} \cdot \nabla \Phi)(\boldsymbol{\sigma} \cdot \nabla \phi) + (\nabla \phi^* \cdot \boldsymbol{\sigma})(\boldsymbol{\sigma} \cdot \nabla \Phi) \phi \} \, d^3x \\ &= \frac{i}{2m} \int \{ \phi^* \triangle \Phi \cdot \phi - 2i \boldsymbol{\sigma} \cdot \phi^* [\nabla \Phi \times \nabla \phi] \} \, d^3x. \end{aligned}$$

Since $\Phi = \Phi(r)$,

$$\nabla \Phi = \frac{\mathbf{r}}{r} \frac{d\Phi}{dr},$$

and hence

$$-i \boldsymbol{\sigma} \cdot [\nabla \Phi \times \nabla] = \frac{1}{r} \frac{d\Phi}{dr} \boldsymbol{\sigma} \cdot \hat{\mathbf{l}},$$

where $\hat{\mathbf{l}} = -i \mathbf{r} \times \nabla$ is the orbital angular momentum operator. Finally, bringing together the expressions obtained and substituting in (123.9), we have

$$\delta E_s^{(\mathrm{I})} = \frac{e^3}{3\pi m^2} \left(\log \frac{m}{2\kappa} + \frac{19}{30} \right) \langle s | \triangle \Phi | s \rangle + \frac{e^3}{4\pi m^2} \left\langle s \left| \boldsymbol{\sigma} \cdot \mathbf{l} \frac{1}{r} \frac{d\Phi}{dr} \right| s \right\rangle, \qquad (123.10)$$

in which the averaging is over the non-relativistic wave function in both terms.

The low-frequency part of the shift

In order to calculate the second part of the level shift, we use a technique based ultimately on the unitarity condition.

Since a photon can be emitted, the excited state of the atom is not strictly stationary, but only quasi-stationary. A complex energy value can be assigned to such a state, its imaginary part being $-\frac{1}{2}w$ if w is the decay probability of the state, in this case the total photon emission probability (see *QM*, §134). In the non-relativistic approximation there is dipole radiation, and from (45.7)

$$\operatorname{im} \delta E_s = -\tfrac{1}{2}w_s = -\tfrac{2}{3}\sum_{s'} |\mathbf{d}_{ss'}|^2 (E_s - E_{s'})^3,$$

where the summation is over all the lower levels ($E_{s'} < E_s$), or, equivalently,

$$\operatorname{im} \delta E_s = -\tfrac{2}{3}\int_0^\infty d\omega \cdot \sum_{s'} |\mathbf{d}_{ss'}|^2 (E_s - E_{s'})^3 \delta(E_s - E_{s'} - \omega). \tag{123.11}$$

In order to find the real part of δE_s, we must regard E_s as a complex variable and use analytical continuation. This may be done by treating the delta functions as originating from poles. The rule for the avoidance of poles is, as usual, specified by adding a negative imaginary part to the masses of the virtual particles—in this case, to the masses $m_{s'}$ of the electron in the intermediate states of the atom. These are $m_{s'} = m + E_{s'}$, and so we must put

$$E_{s'} \to E_{s'} - i0,$$

whence

$$\delta(E_s - E_{s'} - \omega) = -\frac{1}{\pi}\operatorname{im}\frac{1}{E_s - E_{s'} - \omega + i0}; \tag{123.12}$$

cf. (111.3).

Thus, substituting (123.12) in (123.11), we find

$$\operatorname{im} \delta E_s = \operatorname{im}\frac{2}{3\pi}\int_0^\infty d\omega \cdot \sum_{s'} |\mathbf{d}_{ss'}|^2 \frac{(E_s - E_{s'})^3}{E_s - E_{s'} - \omega + i0}.$$

The required analytical continuation is now obtained by simply omitting the symbol im, but we must take from δE_s only the part due to the contribution from frequencies in region II ($\omega < \kappa$). To do so, we need only replace the upper limit of integration by κ. The result of the integration is

$$\delta E_s^{(\mathrm{II})} = \frac{2}{3\pi}\sum_{s'} |\mathbf{d}_{ss'}|^2 (E_{s'} - E_s)^3 \log\frac{\kappa}{E_{s'} - E_s + i0}; \tag{123.13}$$

because of the inequality (123.4), the difference $E_s - E_{s'}$ is neglected in comparison with κ at the upper limit. We shall henceforward be concerned only with the real part of the level, which is obtained by using $\kappa/|E_{s'} - E_s|$ as the argument of the logarithm in (123.13).

In the expression (123.13), the term in log κ can be transformed by replacing the matrix elements of the dipole moment $\mathbf{d} = e\mathbf{r}$ by those of the momentum $\mathbf{p} = m\mathbf{v}$ and its derivative $\dot{\mathbf{p}}$:

$$\sum_{s'} |\mathbf{d}_{ss'}|^2(E_{s'} - E_s)^3 = -\frac{e^2}{m^2}\sum_{s'} |\mathbf{p}_{ss'}|^2(E_{s'} - E_s)$$
$$= \frac{ie^2}{2m^2}\sum_{s'} \{(\dot{\mathbf{p}})_{ss'} \cdot \mathbf{p}_{s's} - \mathbf{p}_{ss'} \cdot (\dot{\mathbf{p}})_{s's}\}.$$

Now replacing $\hat{\dot{\mathbf{p}}}$ in accordance with the operator equation of motion of the electron, $\hat{\dot{\mathbf{p}}} = -e\nabla\Phi$, we get

$$\sum_{s'} |\mathbf{d}_{ss'}|^3(E_{s'} - E_s)^3 = -\frac{ie^3}{2m^2}\sum_{s'} \{(\nabla\Phi)_{ss'} \cdot \mathbf{p}_{s's} - \mathbf{p}_{ss'} \cdot (\nabla\Phi)_{s's}\}$$
$$= \frac{ie^3}{2m^2}\langle s|\dot{\mathbf{p}} \cdot \nabla\Phi - \nabla\Phi \cdot \dot{\mathbf{p}}|s\rangle$$
$$= \frac{e^3}{2m^2}\langle s|\triangle\Phi|s\rangle. \tag{123.14}$$

We can therefore write instead of (123.13)

$$\delta E_s^{(\mathrm{II})} = \frac{e^3}{3\pi m^2}\langle s|\triangle\Phi|s\rangle \log\frac{2\kappa}{m} +$$
$$+\frac{2e^2}{3\pi}\sum_{s'} |\mathbf{r}_{ss'}|^2(E_{s'} - E_s)^3 \log\frac{m}{2|E_s - E_{s'}|}. \tag{123.15}$$

THE TOTAL SHIFT

Finally, adding the two parts, we have the following formula for the level shift:

$$\delta E_s = \frac{2e^2}{3\pi}\sum_{s'} |\mathbf{r}_{ss'}|^2(E_{s'} - E_s)^3 \log\frac{m}{2|E_s - E_{s'}|} +$$
$$+\frac{e^3}{3\pi m^2}\cdot\frac{19}{30}\langle s|\triangle\Phi|s\rangle = \frac{e^3}{4\pi m^2}\left\langle s\left|\boldsymbol{\sigma}\cdot\mathbf{l}\frac{1}{r}\frac{d\Phi}{dr}\right|s\right\rangle; \tag{123.16}$$

as was to be expected, the auxiliary quantity κ does not appear.†

All the matrix elements in (123.16) are taken with respect to the non-relativistic wave functions of the electron in the atom. For a hydrogen atom or a hydrogen-like ion, these functions depend only on three quantum numbers: the principal quantum number n, the orbital angular momentum l and its component m, but not on the

† The determination of the next-order corrections in the level shift involves very complicated calculations. The most complete tabulation and a systematic derivation of the corrections, together with further references, are given by G. W. Erickson and D. R. Yennie, *Annals of Physics* **35**, 271, 447, 1965.

total angular momentum j; the corresponding energy levels depend on n only. We shall use the notation‡

$$L_{nl} = \frac{n^3}{2m(Ze^2)^4} \sum_{n'l'm'} |\langle n'l'm'|\mathbf{r}|nlm\rangle|^2 (E_{n'} - E_n)^3 \cdot \log \frac{m(Ze^2)^2}{2|E_{n'} - E_n|}. \tag{123.17}$$

The energy levels are proportional to $(Ze^2)^2$, and the characteristic dimension of the atom is proportional to Ze^2, so that the L_{nl} defined by (123.17) are independent of Z. They can be calculated numerically.

We shall take separately the cases $l = 0$ and $l \neq 0$. When $l = 0$, the last term in (123.16) is zero. In the second term, we use the equation

$$e\triangle\Phi = 4\pi Ze^2\delta(\mathbf{r}),$$

which is satisfied by the potential of the Coulomb field of the nucleus. Hence

$$\langle nlm|\triangle\Phi|nlm\rangle = 4\pi Ze^2|\psi_{nlm}(0)|^2 = \begin{cases} 4m^3(Ze^2)^4 n^{-3} & (l = 0) \\ 0 & (l \neq 0) \end{cases}$$

(cf. (34.3)). In the first term, with the notation (123.17) and again using (123.14),

$$\sum_{n'l'm'} |\langle n'l'm'|\mathbf{r}|n00\rangle|^2 (E_{n'} - E_n)^3 = \frac{e}{2m^2}\langle n00|\triangle\Phi|n00\rangle = 2m(Ze^2)^4/n^3.$$

This gives the following expression for the shift of the s terms (in ordinary units):

$$\delta E_{n0} = \frac{4mc^2Z^4\alpha^5}{3\pi n^3}\left[\log\frac{1}{(Z\alpha)^2} + L_{n0} + \frac{19}{30}\right]. \tag{123.18}$$

The numerical values of some of the L_{n0} are:

n	1	2	3	4	∞
L_{n0}	-2.984	-2.812	-2.768	-2.750	-2.721

The unperturbed levels are $E_n = -mc^2(Z\alpha)^2/2n^2$, and so the relative magnitude of the radiative shift is

$$|\delta E_{n0}/E_{n0}| \sim Z^2\alpha^3 \log(1/Z\alpha). \tag{123.19}$$

When $l \neq 0$, the second term in (123.16) is zero. The third term can be calculated

‡ The matrix elements of $\mathbf{r}$ are diagonal in j and independent of j; the summation over s in (123.16) therefore reduces to summation over n, l and m. Because of the isotropy of space, the sum (123.17) is also, of course, independent of m.

by means of the formulae in §34, and leads to a dependence of the level shift on the number j also. The result is

$$\delta E_{nlj} = \frac{4mc^2Z^4\alpha^5}{3\pi n^3}\left[L_{nl} + \frac{3}{8}\frac{j(j+1) - l(l+1) - \frac{3}{4}}{l(l+1)(2l+1)}\right], \quad l \neq 0. \tag{123.20}$$

Thus the radiative shift removes the last degeneracy which remains after the spin–orbit interaction has been taken into account, namely the degeneracy of levels having the same n and j but different $l = j \pm \frac{1}{2}$. For example, the numerical value of L_{21} is $+0.030$, and formulae (123.18)–(123.20) give as the difference of the $2s_{1/2}$ and $2p_{1/2}$ levels of the hydrogen atom

$$E_{20(1/2)} - E_{21(1/2)} = 0.41mc^2\alpha^5,$$

corresponding to a frequency of 1050 MHz.

§ 124. **Radiative shift of mesic-atom levels**

At the end of §118 the electron vacuum polarization has been shown to play an important role in the (second-approximation) radiative correction to the magnetic moment of the muon. This still more true (and even in the first approximation) as regards the radiative level shift in μ-mesic hydrogen, a hydrogen-like system consisting of a proton and a muon (A. D. Galanin and I. Ya. Pomeranchuk, 1952).

In calculating the level shift for an ordinary atom in §123, we took account, in particular, of the electron vacuum polarization effect (the electron loop in the diagram (121.2a)). If the muon vacuum polarization effect is similarly treated in the mesic atom, the entire calculation can be applied to this case, simply replacing the electron mass $m = m_e$ by the muon mass m_μ. Since the relative shift (123.19) of the levels does not depend on the electron mass, the same result is obtained for mesic hydrogen.

It is easily seen that the electron vacuum polarization has a much stronger effect on the level shift in the mesic atom, because the replacement of the muon loop in the diagram by an electron loop implies the replacement of the muon polarization operator by the electron polarization operator; and the polarization operator $\mathcal{P}(q^2)$ is inversely proportional to the square of the particle mass for non-relativistic values of q^2. Hence the change mentioned must increase the effect by a factor $(m_\mu/m_e)^2$, and it is this contribution which determines the order of magnitude of the level shift:

$$\delta E/|E| \sim \alpha^3(m_\mu/m_e)^2,$$

or four orders of magnitude greater than in ordinary hydrogen.† The origin of this effect can be more clearly seen by noting that the distortion of the Coulomb

† For a similar reason, the contribution of the muon vacuum polarization to the level shift in the ordinary hydrogen atom is, conversely, negligible.

potential by the electron vacuum polarization extends to distances $\sim 1/m_e$ (§114). In the ordinary hydrogen atom the electron is at distances from the nucleus that are of the order of $1/m_e\alpha$, i.e. outside the main region of field distortion, but in mesic hydrogen the muon is at distances $\sim 1/m_\mu\alpha$ which are *in* this region.

To calculate precisely the level shift in the mesic atom, however, it is not possible to use the approximate non-relativistic expression for the polarization operator, as was done when using (123.7) to find the level shift in the ordinary atom. The reason is that the characteristic momenta of the muon in the mesic hydrogen atom are $|\mathbf{p}_\mu| \sim \alpha m_\mu$. For the muon these momenta are non-relativistic, but for the electron they are relativistic.

We must therefore use the full relativistic formula (114.5) for the effective potential of the nuclear field as modified by the electron vacuum polarization. The level shift is found by averaging over the wave function of the muon in the atom:

$$\delta E_{nl} = -|e| \int |\psi_{nl}|^2 \delta\Phi(r)\, d^3x$$

$$= -|e| \int_0^\infty R_{nl}^2(r)\delta\Phi(r) r^2\, dr, \tag{124.1}$$

where R_{nl} is the radial part of the (non-relativistic) Coulomb wave function. For a hydrogen-like ion with nuclear charge $Z|e|$, the functions $R_{nl}(r)$ depend on r only through the dimensionless combination $\rho = Z\alpha m_\mu r$ (the distance in Coulomb units). Using this fact and substituting $\delta\Phi(r)$ from (114.5) (with the charge $Z|e|$ in place of e_1), we can bring the integral (124.1) to the form

$$\delta E_{nl} = -\frac{2}{3\pi} Z\alpha^3 m_\mu Q_{nl}(m_e/Z\alpha m_\mu), \tag{124.2}$$

where

$$Q_{nl}(x) = \int_0^\infty \rho\, d\rho \int_1^\infty R_{nl}^2(\rho) e^{-2x\rho\zeta}\left(1+\frac{1}{2\zeta^2}\right)\frac{\sqrt{(\zeta^2-1)}}{\zeta^2}\, d\zeta.$$

The first few levels of mesic hydrogen are shown by numerical evaluation to have the following relative shifts:

$$\delta E_{10}/|E_{10}| = -6.4\times 10^{-3},$$

$$\delta E_{20}/|E_{20}| = -2.8\times 10^{-4},$$

$$\delta E_{21}/|E_{21}| = -2.0\times 10^{-5},$$

§ 125. **The relativistic equation for bound states**

The method used in the preceding sections to calculate the radiative shift of the atomic levels is not valid for solving a problem such as that of determining the corrections to the levels of positronium, a system consisting of two particles of equal rank, neither of which can be regarded as the source of the external field acting on the other.

The systematic procedure for solving this problem is based on the fact that the energy levels of the bound states are poles of the exact amplitude of mutual scattering of these two particles, as a function of their total energy in the centre-of-mass system. In any of its discrete states, positronium may be regarded as an "intermediate particle" having a definite mass, which can be formed as a stage in the electron–positron scattering process, and a pole of the scattering amplitude corresponds to each "one-particle" intermediate state; these poles of course lie in the non-physical region of 4-momenta of the particles undergoing scattering.

According to (106.17), the exact scattering amplitude comprises the exact four-ended vertex part $\Gamma_{ik,lm}$ and the polarization amplitudes u of the particles. The latter are clearly unconnected with the pole singularities, and it is therefore more convenient to ignore them, referring instead to the poles of the vertex part itself, i.e. of the function

$$\Gamma_{ik,lm}(p'_-, -p_+; p_-, -p'_+), \tag{125.1}$$

where the notation for the 4-momenta of the external lines of the diagram (106.12) corresponds to the scattering of a positron by an electron.

It should be stressed that the assertion that poles are present refers to the exact scattering amplitude or the exact vertex part; there is no pole in any separate term of the perturbation-theory series, as can be seen from the fact that the Feynman diagrams in each approximation include only electron (and photon) lines, not lines belonging to the composite particle positronium as a whole. Hence it follows in turn that the calculation of the scattering amplitude near its poles involves summation of an infinite series of diagrams. The diagrams concerned can be determined as follows.

In the first non-vanishing approximation of perturbation theory (the first approximation with respect to α), the vertex part (125.1) corresponds to two second-order diagrams:

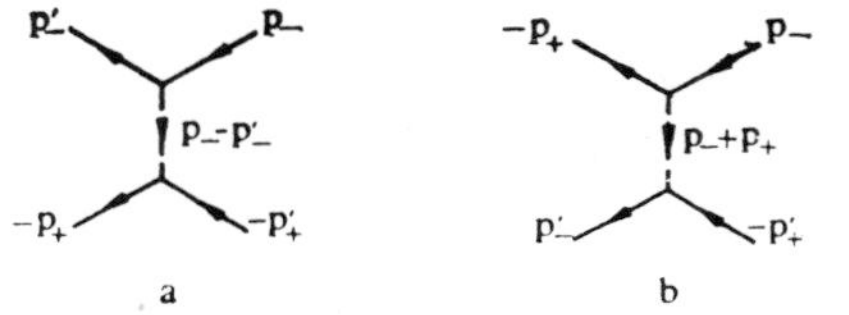

(125.2)

or, in analytical form,

$$\Gamma_{ik,lm} = -e^2\gamma^{\mu}_{il}\gamma^{\nu}_{km}D_{\mu\nu}(p_- - p'_-) + e^2\gamma^{\mu}_{im}\gamma^{\nu}_{kl}D_{\mu\nu}(p_- + p_+). \tag{125.3}$$

In the next approximation (the second with respect to α) there are ten fourth-order diagrams:

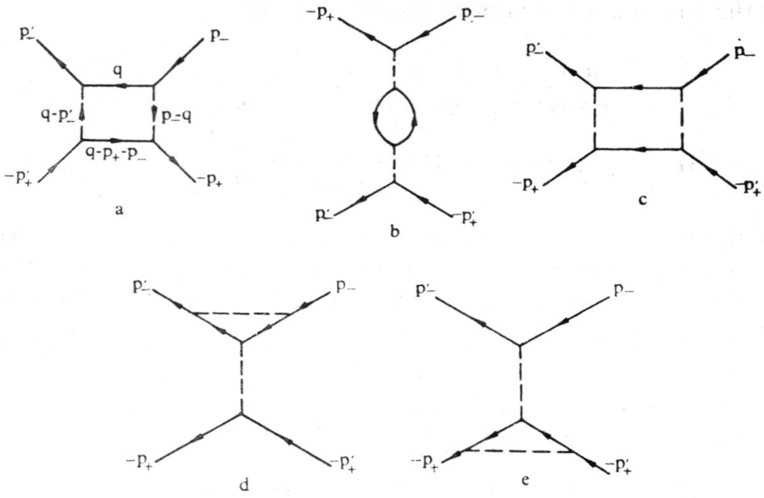

(125.4)

and a further five diagrams obtained from (125.4) by interchanging p_- and $-p'_+$. All these include an extra power of $e^2 = \alpha$ in comparison with (125.2), but we shall show that in diagram (a) this extra order of smallness is cancelled by a denominator which is also small when the electron and positron momenta are small.

All quantities will be taken in the centre-of-mass system, but, since the 4-momenta of the external lines in the diagrams are not assumed to be physical (i.e. $p^2 \neq m^2$), $\varepsilon_+ \neq \varepsilon_-$ in this system, although $\mathbf{p}_+ = -\mathbf{p}_-$. Thus these 4-momenta are

$$\left.\begin{aligned} p_- &= (\varepsilon_-, \mathbf{p}), \quad p_+ = (\varepsilon_+, -\mathbf{p}), \\ p'_- &= (\varepsilon'_-, \mathbf{p}') \quad p'_+ = (\varepsilon'_+, -\mathbf{p}'), \\ \varepsilon_- &+ \varepsilon_+ = \varepsilon'_- + \varepsilon'_+. \end{aligned}\right\} \tag{125.5}$$

The binding energy of the electron and the positron in positronium is $\sim m\alpha^2$. Thus, in the neighbourhood of the scattering-amplitude poles, with which we are concerned,

$$\begin{gathered} |\mathbf{p}| \sim |\mathbf{p}'| \sim m\alpha \ll m, \\ |\varepsilon_- - m| \sim |\varepsilon_+ - m| \sim \mathbf{p}^2/m \sim m\alpha^2, \ldots \end{gathered} \tag{125.6}$$

The contribution to the vertex part from the diagram (125.4a) is

$$\Gamma^{(4a)}_{ik,lm} = -ie^4 \int (\gamma^\lambda G(q)\gamma^\mu)_{il}(\gamma^\nu G(q-p_--p_+)\gamma^\rho)_{km} \times \\ \times D_{\lambda\rho}(q-p'_-)D_{\mu\nu}(p_--q)\, d^4q/(2\pi)^4. \tag{125.7}$$

The important range of values of $q^\mu = (q_0, \mathbf{q})$ in the integral (125.7) is that which is close to poles of both functions G simultaneously. In this range, $|\mathbf{q}|$ and $|q_0 - m|$ are small, and the electron propagators are

$$\left.\begin{aligned} G(q) &= \frac{\gamma^0 q_0 - \boldsymbol{\gamma}\cdot\mathbf{q} + m}{(q_0+m)(q_0-m) - \mathbf{q}^2 + i0} \approx \tfrac{1}{2}(\gamma^0+1)\frac{1}{q_0 - m - (\mathbf{q}^2/2m) + i0}, \\ G(q - p_- - p_+) &\approx \tfrac{1}{2}(\gamma^0 - 1)\frac{1}{q_0 - \varepsilon_- - \varepsilon_+ + m + (\mathbf{q}^2/2m) - i0}. \end{aligned}\right\} \tag{125.8}$$

The poles of these two expressions are on opposite sides of the real axis of the complex variable q_0; closing the path of integration along this axis by a contour in the upper half-plane (say), we can calculate the integral over dq_0 from the residue at the corresponding pole.† The result is

$$\Gamma^{(4a)} \sim e^4 \int \frac{d^3q}{(q - p_-')^2 (p_- - q)^2 (2m - \varepsilon_- - \varepsilon_+ + \mathbf{q}^2/m)},$$

and so, using (125.6), we have in order of magnitude

$$\Gamma^{(4a)} \sim \alpha^2 \frac{(m\alpha)^3}{(m\alpha)^4\, m\alpha^2} = \frac{1}{m^2\alpha}.$$

The contribution to Γ from the second-order diagram (125.2a) (the first term in (125.3)) is of the same order, and this proves the statement made above about the order of smallness of the diagram (125.4a). A similar situation occurs in all higher approximations of perturbation theory.

Thus the calculation of the relevant vertex part near its poles calls for the summation of an infinite succession of "anomalously large" diagrams with intermediate states resembling the internal lines of (125.4a). A typical property of these diagrams is that they can be cut between the ends p_-, $-p_+$ and p_-', $-p_+'$ into parts joined only by two electron lines.‡ The set of all diagrams which do *not* satisfy this condition will be called a "compact" vertex part and denoted by $\tilde{\Gamma}_{ik,lm}$; since it does not include the anomalously large diagrams, such quantities can be calculated by ordinary perturbation theory. For example, in the first approximation $\tilde{\Gamma}$ is given by the two second-order diagrams (125.2), and in the second approximation by the eight fourth-order diagrams in (125.4), excluding diagrams (a) and (b).

If the non-compact vertex parts are classified according to the number of "double bonds", we can represent the total Γ as an infinite series:

(125.9)

where the continuous thick internal lines are exact propagators $\mathcal{G}$; this is often

† For the diagram (125.4c), which differs from (125.4a) only as regards the relative direction of the electron lines, both poles would be on the same side of the real axis, and so the integral would be zero in the approximation considered.

‡ This definition includes all the anomalously large diagrams and also some "normal" diagrams such as (125.4b).

called a "ladder" series. To sum the series, we "multiply" it on the left by a further $\tilde{\Gamma}$:†

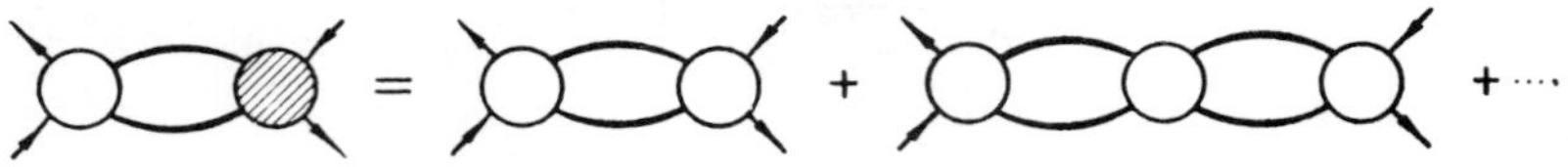

Comparison with the original series (125.9) shows that

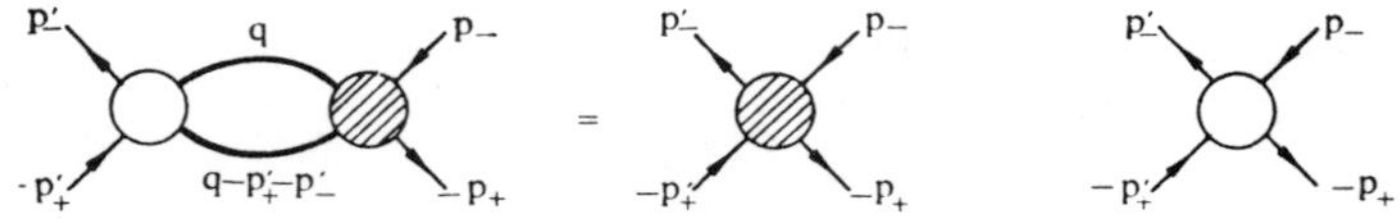

(125.10)

This graphical equation corresponds to the integral equation

$$i\Gamma_{ik,lm}(p'_-, -p_+; p_-, -p'_+) = i\tilde{\Gamma}_{ik.lm}(p'_-, -p_+; p_-, -p'_+) +$$
$$+\int \tilde{\Gamma}_{ir,sm}(p'_-, q-p'_+-p'_-; q, -p'_+)\mathscr{G}_{st}(q)\mathscr{G}_{nr}(q-p'_+-p'_-)\times$$
$$\times\Gamma_{tk,ln}(q, -p_+; p_-, q-p'_+-p'_-)\, d^4q/(2\pi)^4. \qquad (125.11)$$

The functions $\tilde{\Gamma}$ and $\mathscr{G}$ are calculated by perturbation theory, and equation (125.11) then allows, in principle, the determination of Γ with any desired accuracy.

To find the energy levels, we need to know only the positions of the poles of Γ. Near the poles, $\Gamma \gg \tilde{\Gamma}$, and so the first term on the right of (125.11) (the second diagram from the right in (125.10)) may be neglected, the equation then becoming homogeneous in Γ. The variables p_+, p_- and the indices k and l become parameters, the dependence on which is arbitrary and is not defined by the equation itself. Omitting these parameters and also the primes in the remaining variables p'_+, p'_-, we have

$$i\Gamma_{i,m}(p_-; -p_+) = \int \tilde{\Gamma}_{ir,sm}(p_-, q-p_+-p_-; q, -p_+)\mathscr{G}_{st}(q)\times$$
$$\times\mathscr{G}_{nr}(q-p_+-p_-)\Gamma_{t,n}(q; q-p_+-p_-)\, d^4q/(2\pi)^4 \qquad (125.12)$$

(E. E. Salpeter and H. A. Bethe, 1951).

Equation (125.12), written in the centre-of-mass system ($\mathbf{p}_+ + \mathbf{p}_- = 0$), has solutions only for certain values of $\varepsilon_+ + \varepsilon_-$, and these give the positronium energy levels. The function $\Gamma_{i,m}$ plays only an auxiliary role. Another function is more convenient in practice:

$$\chi_{sr}(p_1, p_2) = \mathscr{G}_{st}(p_1)\Gamma_{t,n}(p_1; p_2)\mathscr{G}_{nr}(p_2). \qquad (125.13)$$

† That is, we multiply each term in the series by $\tilde{\Gamma}$ and two $\mathscr{G}$, and integrate appropriately over the 4-momenta of the new internal bonds.

Then equation (125.12) becomes

$$i[\mathcal{G}^{-1}(p_-)\chi(p_-, -p_+)\mathcal{G}^{-1}(-p_+)]_{im}$$
$$= \int \tilde{\Gamma}_{ir,sm}(p_-, q-p_+-p_-; q, -p_+)\chi_{sr}(q, q-p_+-p_-)\, d^4q/(2\pi)^4, \qquad (125.14)$$

where $\tilde{\Gamma}$ appears as the kernel of an integral operator. As already mentioned, $\tilde{\Gamma}$ may be calculated by perturbation theory, and the same is of course true of $\mathcal{G}^{-1}$.

We shall show that, in the first approximation of perturbation theory (with respect to α), (125.14) reduces, as we should expect, to the non-relativistic Schrödinger's equation for positronium.

In the first non-relativistic approximation, $\tilde{\Gamma}$ is determined by the diagram (125.2a) alone; the annihilation-type diagram (125.2b) is zero in this approximation.† For a similar reason to that in §83, it is convenient to take the photon propagator in the Coulomb gauge (76.12), (76.13), and only D_{00} need be retained in it. Then

$$\tilde{\Gamma}_{ir,sm}(p_-, q-p_+-p_-; q, -p_+) = -e^2\gamma^0_{is}\gamma^0_{rm}D_{00}(q-p_-)$$
$$= -U(\mathbf{q}-\mathbf{p}_-)\gamma^0_{is}\gamma^0_{rm},$$

where

$$U(\mathbf{q}) = -4\pi e^2/\mathbf{q}^2$$

is the Fourier component of the potential energy of the Coulomb interaction between the positron and the electron. Equation (125.14) becomes

$$i\chi_{im}(p_-, -p_+)$$
$$= \left[G(p_-)\gamma^0 \int U(\mathbf{q}-\mathbf{p}_-)\chi(q, q-p_+-p_-)\frac{d^4q}{(2\pi)^4}\cdot \gamma^0 G(-p_+)\right]_{im}, \qquad (125.15)$$

where we have also replaced the exact propagators $\mathcal{G}$ by the free-electron propagators G. The latter are given by the approximate expression (cf. (125.8))

$$G(p_-) \approx \tfrac{1}{2}(1+\gamma^0)g(p_-), \quad G(-p_+) \approx \tfrac{1}{2}(1-\gamma^0)g(p_+),$$

where the matrix factors have been separated, and $g(p)$ is the scalar function

$$g(p) = \frac{1}{\varepsilon - m - (\mathbf{p}^2/2m) + i0}. \qquad (125.16)$$

In substituting these expressions in (125.15), we note that all the non-zero matrix

† The particle velocities in positronium are such that $v/c \sim \alpha$. In this sense the expansions in powers of α and of $1/c$ are interrelated.

elements

$$[\tfrac{1}{2}(1+\gamma^0)\gamma^0\chi\gamma^0\cdot\tfrac{1}{2}(1-\gamma^0)]_{im} = [\tfrac{1}{2}(\gamma^0+1)\chi\cdot\tfrac{1}{2}(\gamma^0-1)]_{im}$$

are equal to the elements $-\chi_{im}$. The matrix equation (125.15) is therefore equivalent to one for the scalar function

$$i\chi(p_-,-p_+) = -g(p_-)g(p_+)\int U(\mathbf{q}-\mathbf{p}_-)\chi(q, q-p_+-p_-)\, d^4q/(2\pi)^4. \qquad (125.17)$$

We now replace p_+ and p_- by the variables

$$p \equiv (\varepsilon, \mathbf{p}) = \tfrac{1}{2}(p_- - p_+), \quad P = p_- + p_+;$$

these are the 4-momentum of the relative motion of the particles and that of the positronium as a whole. In the centre-of-mass system, $P = (E+2m, 0)$, where $E+2m$ is the total energy, and E therefore the energy level relative to the rest mass. In terms of these variables, (125.17) becomes

$$\begin{aligned} i\chi(p, P) &= -g(p+\tfrac{1}{2}P)g(-p+\tfrac{1}{2}P)\int U(\mathbf{q}-\mathbf{p}_-)\chi(q-\tfrac{1}{2}P, P)\, d^4q/(2\pi)^4 \\ &= -g(p+\tfrac{1}{2}P)g(-p+\tfrac{1}{2}P)\int U(\mathbf{q}'-\mathbf{p})\chi(q', P)\, d^4q'/(2\pi)^4. \end{aligned}$$

In this equation, P occurs only as a parameter, and χ figures on the right-hand side only as the integral

$$\psi(\mathbf{q}) = \int_{-\infty}^{\infty} \chi(q, P)\, dq_0.$$

Integrating both sides of the equation over $d\varepsilon$, we get an equation for ψ in a closed form:

$$\psi(\mathbf{p}) = -\frac{1}{2\pi i}\int_{-\infty}^{\infty} g(p+\tfrac{1}{2}P)g(-p+\tfrac{1}{2}P)\, d\varepsilon \int U(\mathbf{q}-\mathbf{p})\psi(\mathbf{q})\frac{d^3q}{(2\pi)^3},$$

where

$$g(\pm p+\tfrac{1}{2}P) = \frac{1}{\pm\varepsilon+\tfrac{1}{2}E-(\mathbf{p}^2/2m)+i0}.$$

If the path of integration over $d\varepsilon$ is closed by a contour in the upper half-plane (say) of the complex variable ε, we can evaluate the integral from the residue at the corresponding pole, obtaining

$$\left(\frac{\mathbf{p}^2}{m}-E\right)\psi(\mathbf{p}) + \int U(\mathbf{p}-\mathbf{q})\psi(\mathbf{q})\frac{d^3q}{(2\pi)^3} = 0. \qquad (125.18)$$

This is Schrödinger's equation for positronium in the momentum representation; see *QM*, (130.4).

If only the diagrams (125.2) were used in $\tilde{\Gamma}$, but including in them (and in $\mathcal{G}$) the next terms in the expansion in powers of $1/c$, we should arrive at Breit's equation (§83). The inclusion of the diagrams (125.4) (together with the further terms in the expansion in $1/c$) gives the radiative corrections to the positronium levels, but the calculations become very complicated.

The following is the difference between the ground levels of ortho- and para-positronium, including the above-mentioned corrections:†

$$E(^3S_1) - E(^1S_0) = \alpha^2 \frac{me^4}{2\hbar^2}\left\{\frac{7}{6} - \left(\frac{16}{9} + \log 2\right)\frac{\alpha}{\pi} - \tfrac{1}{2}i\alpha\right\}; \qquad (125.19)$$

the first term in the braces is the fine splitting (see §84, Problem 2). The second term is the radiative correction to the difference between the levels. The imaginary part of the difference arises from the parapositronium annihilation probability (see (89.4)), i.e. from the fact that the level $^1S_0$ is complex; for parapositronium, the level width is found to be of the same order as the radiative correction to the real part of the level.

§ 126. The double dispersion relation

After the vertex part with three external lines, the next in order of complexity is a section with four external lines. In quantum electrodynamics, three basic diagrams of this type are possible:

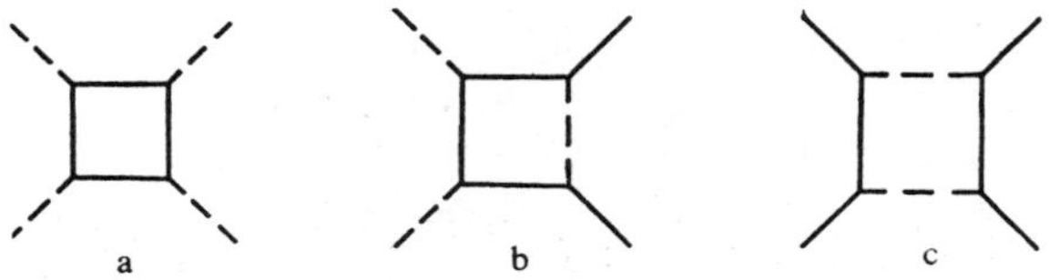

(126.1)

The first describes the scattering of a photon by a photon; the others are individual terms in the radiative corrections to the scattering of (b) a photon and (c) an electron, by an electron.

This §126 deals with some general properties of such diagrams, but to be simple and specific we shall refer only to (126.1a).

The momenta of the lines in such a diagram will be denoted as follows:

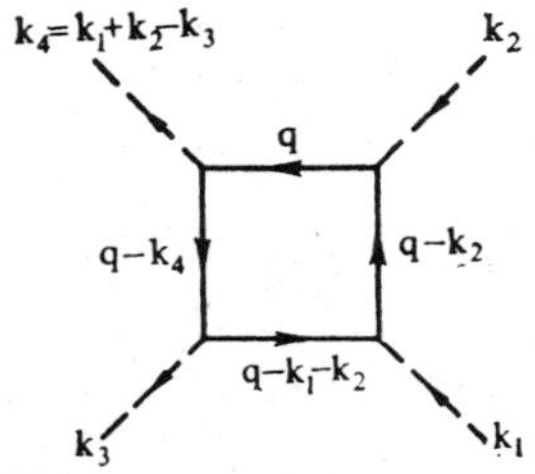

(126.2)

† R. Karplus and A. Klein, *Physical Review* **87**, 848, 1952.

The 4-momenta k_1, k_2, k_3, k_4 correspond to real photons, and their squares are therefore zero.

If the dependence on the photon polarizations is written separately, the amplitude M_{fi} which corresponds to the diagram (126.2) can be expressed in terms of various scalar functions of the photon 4-momenta. These are the invariant amplitudes discussed in §70; they will be derived in §127 for the specific case of photon–photon scattering. Being scalars, they depend only on scalar variables, which may be taken as, for example, any two of the quantities

$$s = (k_1 + k_2)^2, \quad t = (k_1 - k_3)^2, \quad u = (k_1 - k_4)^2, \quad s + t + u = 0; \tag{126.3}$$

in what follows we shall take s and t as independent variables.

Each of the invariant amplitudes, which will be denoted here by the same letter M, can be written as an integral:

$$M = \int \frac{iB\,d^4q}{[q^2 - m^2][(q-k_4)^2 - m^2][(q-k_1-k_2)^2 - m^2][(q-k_2)^2 - m^2]}, \quad m^2 \to m^2 - i0, \tag{126.4}$$

where B is some function of all the 4-momenta; the factors in the denominator arise from the propagators of the four virtual electrons.

When s and t are sufficiently small, the amplitudes M are real (more precisely, they can be made real by a suitable choice of the phase factor), since if s is small the photons cannot generate real particles (an electron–positron pair) in the s channel, and if t is small the same applies to the t channel.† Thus neither channel has real intermediate states which could, according to the unitarity condition, lead to an imaginary part of the amplitude.

Now let s increase while t remains at a fixed small value. When $s \geq 4m^2$, the amplitude M has an imaginary part due to the possibility of pair production by two photons in the s channel. We can therefore write for M a dispersion relation in the variable s:

$$M(s,t) = \frac{1}{\pi} \int_{4m^2}^{\infty} \frac{A_{1s}(s',t)}{s'-s-i0}\,ds', \tag{126.5}$$

where $A_{1s}(s,t)$ denotes the imaginary part of $M(s,t)$.

As in any diagram having the form

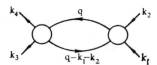

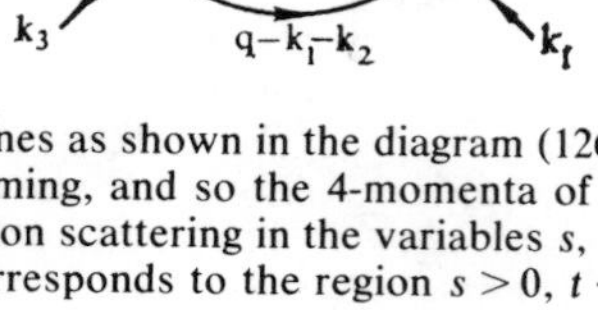

† The directions of the external lines as shown in the diagram (126.2) correspond to the s channel. In the t channel, lines 1 and 3 are incoming, and so the 4-momenta of the initial photons are k_1 and $-k_3$. The physical regions for photon–photon scattering in the variables s, t, u are the shaded sectors in Fig. 8 (§67). For example, the s channel corresponds to the region $s > 0$, $t < 0$, $u < 0$.

$A_{1s}(s, t)$ is calculated by the rule (115.9), replacing the pole factors in the integral (126.4) by delta functions:

$$2iA_{1s}(s, t) = (2\pi i)^2 \int \frac{iB\delta(q^2 - m^2)\delta[(q - k_1 - k_2)^2 - m^2]}{[(q - k_4)^2 - m^2][(q - k_2)^2 - m^2]} d^4p; \qquad (126.6)$$

the integration is taken over the half of q-space in which $q^0 > 0$.

An important further step can be taken by noting that the integral (126.6) has a structure (of pole factors) similar to that of the amplitude for a reaction represented by a diagram having the form

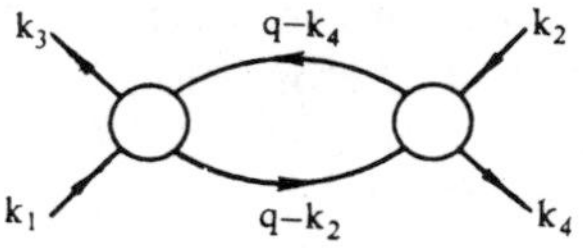

The analytical properties of $A_{1s}(s, t)$ as a function of t are therefore similar to the analytical properties of this amplitude. In particular, the function $A_{1s}(s, t)$ can acquire an imaginary part (as t increases) only if both factors in the denominator become zero simultaneously. This will not, however, occur as soon as t reaches the value $4m^2$ which is the threshold for pair production in the t channel. The reason is that the presence of the delta functions in the integrand restricts the region of integration in q-space, which may be incompatible with the value $t = 4m^2$. The extent of the region of integration depends on s (the arguments of the delta functions contain k_1 and k_2), and therefore so does the limiting value $t = t_c(s)$ beyond which $A_{1s}(s, t)$ becomes complex.

In the same way as $M(s, t)$ is expressed in terms of its imaginary part $A_{1s}(s, t)$ by (126.5), the function $A_{1s}(s, t)$ is in turn expressed in terms of $A_2(s, t) = \text{im}\, A_{1s}(s, t)$ by a dispersion relation in the variable t:

$$A_{1s}(s, t) = \frac{1}{\pi} \int_{t_c(s)}^{\infty} \frac{A_2(s, t')}{t' - t - i0} dt'. \qquad (126.7)$$

If we now substitute (126.7) in (126.5), we get the *double dispersion relation* or *Mandelstam representation* for the amplitude $M(s, t)$:

$$M(s, t) = \frac{1}{\pi^2} \int_{4m^2}^{\infty} \int_{t_c(s)}^{\infty} \frac{A_2(s', t')}{(s' - s - i0)(t' - t - i0)} dt'\, ds' \qquad (126.8)$$

(S. Mandelstam, 1958).

The function $A_2(s, t)$ is called the *double spectral density* of $M(s, t)$. It can be obtained from the integral (126.6) by twice applying the substitution rule (115.9). Putting for brevity

$$l_1 = q, \quad l_2 = q - k_4, \quad l_3 = q - k_2, \quad l_4 = q - k_1 - k_2, \qquad (126.9)$$

we have

$$(2i)^2 A_2(s,t) = (2\pi i)^4 \int iB\delta(l_1^2 - m^2)\delta(l_2^2 - m^2)\delta(l_3^2 - m^2)\delta(l_4^2 - m^2)\, d^4q, \tag{126.10}$$

the integration being taken over the region $q^0 > 0$.

It should be noted, however, that formula (126.10) is purely symbolic, since the region $s > 0$, $t > 0$ is non-physical, and accordingly $l_1, l_2, \ldots$ are in general complex in this region when q is real; and the delta function is not fully defined for a complex argument. It would be more accurate to refer immediately to the taking of residues at the corresponding poles of the original integral (126.4). In our case this is, however, of no importance. The condition for the four expressions in the denominator in (126.4), or the four arguments of the delta functions, to be zero, entirely determines the components of the 4-vector q. On changing to integration with respect to $l_1^2, l_2^2, \ldots$ (see below) and formally applying the usual rules to (126.10), we obtain (apart from the sign) the expression for A_2.

To continue the calculations, we use the centre-of-mass system (in the s channel). Then

$$k_1 = (\omega, \mathbf{k}), \quad k_2 = (\omega, -\mathbf{k}), \quad k_3 = (\omega, \mathbf{k}'), \quad k_4 = (\omega, -\mathbf{k}'), \tag{126.11}$$

$$\left.\begin{aligned} s &= 4\omega^2, \quad t = -(\mathbf{k}-\mathbf{k}')^2 = -4\omega^2 \sin^2 \tfrac{1}{2}\theta, \\ u &= -(\mathbf{k}+\mathbf{k}')^2 = -4\omega^2 \cos^2 \tfrac{1}{2}\theta, \end{aligned}\right\} \tag{126.12}$$

where θ is the angle between $\mathbf{k}$ and $\mathbf{k}'$ (the scattering angle). The x-axis of spatial Cartesian coordinates is taken along the vector $\mathbf{k}+\mathbf{k}'$, and the y-axis along $\mathbf{k}-\mathbf{k}'$.†

We shall now transform the integral (126.10) by taking $l_1^2, l_2^2, \ldots$ as new variables of integration in place of the four components of q. Then

$$\partial(l_1^2)/\partial q^\mu = 2 l_{1\mu}, \ldots,$$

and the Jacobian of the transformation is therefore

$$\frac{\partial(l_1^2, l_2^2, l_3^2, l_4^2)}{\partial(q^0, q_x, q_y, q_z)} = 16D,$$

where D is the determinant formed by the sixteen components of the four 4-vectors $l_1, l_2, \ldots$. The integration in (126.10) amounts simply to replacing the functions B and D in the integrand by their values‡ when

$$l_1^2 = l_2^2 = l_3^2 = l_4^2 = m^2. \tag{126.13}$$

† When $t > 0$, $(\mathbf{k}-\mathbf{k}')^2 < 0$, i.e. the vector $\mathbf{k}-\mathbf{k}'$ is imaginary. This difficulty is, however, easily circumvented by expanding all vector expressions with $t < 0$ and using analytical continuation to $t > 0$.

‡ This method of integration automatically takes account of only one zero of each argument of the delta functions.

From the conditions $l_1^2 = l_4^2 = m^2$ we have, as in §115,

$$q^0 = \omega, \quad \mathbf{q}^2 = \omega^2 - m^2. \tag{126.14}$$

The other two conditions give

$$(q-k_4)^2 - m^2 = -2qk_4 = -2\omega^2 - 2\mathbf{q}\cdot\mathbf{k}' = 0,$$
$$(q-k_2)^2 - m^2 = -2\omega^2 - 2\mathbf{q}\cdot\mathbf{k} = 0,$$

and hence

$$\mathbf{q}\cdot\mathbf{k} = \mathbf{q}\cdot\mathbf{k}' = -\tfrac{1}{4}s,$$

or, in components,

$$q^0 = \omega, \quad q_x = -s/2(s+t), \quad q_y = 0,$$
$$q_z = \pm\sqrt{(\omega^2 - m^2 - q_x^2)}$$
$$= \pm\left[\frac{st - 4m^2(s+t)}{4(s+t)}\right]^{1/2}. \tag{126.15}$$

Thus the integral (126.10) is

$$A_2(s,t) = \frac{\pi^4}{4D}\sum(-iB), \tag{126.16}$$

where the summation is over the two values of $\mathbf{q}$ given by (126.15).

The determinant D can be written in terms of the antisymmetric unit tensor:

$$D = e_{\mu\nu\rho\sigma} l_1^\mu l_2^\nu l_3^\rho l_4^\sigma$$
$$= -e_{\mu\nu\rho\sigma} q^\mu k_4^\nu k_2^\rho k_1^\sigma$$
$$= -e_{\mu\nu\rho\sigma}(q-k_1)^\mu (k_4-k_1)^\nu (k_2-k_1)^\rho k_1^\sigma,$$

where the antisymmetry of $e_{\mu\nu\rho\sigma}$ has been used. Since only k_1 among the four factors has a time component, we deduce that

$$D = -\omega\mathbf{q}\cdot(\mathbf{k}+\mathbf{k}')\times(\mathbf{k}-\mathbf{k}').$$

Expanding this expression with $t<0$ and then continuing to $t>0$, we find

$$D = -\omega q_z \sqrt{(s+t)}\sqrt{(-t)} \to \pm\tfrac{1}{4}i\{st[st-4m^2(s+t)]\}^{1/2}. \tag{126.17}$$

The choice of sign needed here can be made as follows. For simplicity, let $B=1$. Then $A_{1s}(s,t)<0$ in the physical region ($s>0, t<0$), since the two factors in the denominator in (126.6) have the same (negative) sign:

$$(q-k_4)^2 - m^2 = -2\omega^2 - 2\mathbf{q}\cdot\mathbf{k}' < -2\omega(\omega - |\mathbf{q}|) < 0,$$
$$(q-k_2)^2 - m^2 = -2\omega^2 - 2\mathbf{q}\cdot\mathbf{k} < -2\omega(\omega - |\mathbf{q}|) < 0$$

(here we use the results (126.14) which follow from the presence of the two delta functions in the numerator, and which show that $|\mathbf{q}| < \omega$).† From (126.7) it is then seen that $A_2(s, t)$ also must be negative when $s > 0$ and $t > 0$ (since, as is evident from (126.16), $A_2(s, t)$ does not change sign). This means that the upper sign must be taken in (126.17), giving finally

$$A_2 = -\pi^4 \frac{\sum B}{\{st[st - 4m^2(s+t)]\}^{1/2}}. \tag{126.18}$$

Since, from its significance, $A_2(s, t)$ must be real, there is a further condition: as well as s and t, the expression in brackets in the denominator must be positive:

$$\left.\begin{aligned} st - 4m^2(s+t) &\geq 0, \\ s > 0, \quad t &> 0. \end{aligned}\right\} \tag{126.19}$$

These inequalities define the region (shaded in Fig. 23) over which the integration is to be taken in the double dispersion integral (126.8). The region is bounded by the curve

$$st - 4m^2(s+t) = 0,$$

with asymptotes $s = 4m^2$ and $t = 4m^2$.

The dispersion relations in the form (126.5) and (126.8) do not yet take account of the renormalization conditions; if they were applied as they stand, the integrals would be divergent and would need to be regularized. The renormalization condition for the amplitudes $M(s, t)$ is

$$M(0, 0) = 0: \tag{126.20}$$

the photon–photon scattering amplitude must be zero when $k_1 = k_2 = k_3 = k_4 = 0$ (and therefore $s = t = 0$), since $k = 0$ implies a potential constant in time and space, corresponding to no physical field; this condition will be further discussed in §127.

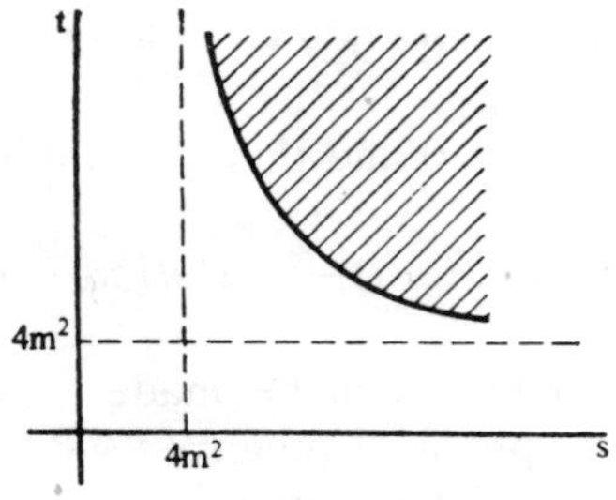

FIG. 23.

† This is, of course, not fortuitous: A_{1s} is negative, in fact, because of the unitarity condition, as is especially clear when $t = 0$ and A_{1s} determines the total cross-section.

To include this condition automatically, we must write the dispersion relation "with subtraction" (as in deriving (111.13) from (111.8)). The required relation is obtained in a natural manner by first using an identical transformation:

$$\frac{1}{(s'-s)(t'-t)} \equiv \frac{st}{(s'-s)(t'-t)s't'} + \frac{s}{(s'-s)s't'} + \frac{t}{(t'-t)s't'} + \frac{1}{s't'}.$$

Substitution of this in the integrand (126.8) gives

$$M(s,t) = \frac{st}{\pi^2}\int\int \frac{A_2(s',t')\,ds'\,dt'}{(s'-s)(t'-t)s't'} + \frac{s}{\pi}\int \frac{f(s')\,ds'}{(s'-s)s'} + \frac{t}{\pi}\int \frac{g(t')\,dt'}{(t'-t)t'} + C,$$

where

$$f(s) = \frac{1}{\pi}\int \frac{A_2(s,t')}{t'}\,dt', \quad g(t) = \frac{1}{\pi}\int \frac{A_2(s',t')}{s'}\,ds',$$

$$C = \frac{1}{\pi^2}\int\int \frac{A_2(s',t')}{s't'}\,ds'\,dt'.$$

These equations would, however, be meaningful only if all the integrals converged. If not, the functions $f(s)$, $g(t)$ and the constant C must be assigned specified values in accordance with the renormalization condition, putting

$$C = 0, \quad f(s) = A_{1s}(s,0), \quad g(t) = A_{1t}(0,t),$$

where A_{1t} is the imaginary part of $M(s,t)$ which appears as t increases for a given small s (just as A_{1s} is the imaginary part which appears as s increases for a given small t). The first of these equations is obvious: $C = M(0,0) = 0$. The second (and similarly the third) follows on comparing the equation

$$M(s,0) = \frac{s}{\pi}\int \frac{f(s')\,ds'}{(s'-s)s'}$$

with the single dispersion relation (126.5) written "with subtraction" according to (126.20):

$$M(s,t) = \frac{s}{\pi}\int \frac{A_{1s}(s',t)}{(s'-s)s'}\,ds'. \qquad (126.21)$$

Thus the double dispersion relation "with subtraction" is finally

$$M(s,t) = \frac{st}{\pi^2}\int\int \frac{A_2(s',t')}{(s'-s)(t'-t)\,s't'}\,ds'\,dt' + \\ + \frac{s}{\pi}\int \frac{A_{1s}(s',0)}{(s'-s)s'}\,ds' + \frac{t}{\pi}\int \frac{A_{1t}(0,t')}{(t'-t)t'}\,dt'. \qquad (126.22)$$

If s and t are themselves within the region of integration, the integrals (126.21),

(126.22) must as usual be taken in the sense

$$s \to s + i0, \quad t \to t + i0. \tag{126.23}$$

§ 127. **Photon–photon scattering**

The scattering of light by light (in a vacuum) is a specifically quantum-electrodynamic process; in classical electrodynamics it does not occur, owing to the fact that Maxwell's equations are linear.†

In quantum electrodynamics, photon–photon scattering is described as the result of the production of a virtual electron–positron pair by the two initial photons, followed by the annihilation of the pair into the final photons. The amplitude of this process (in the first non-vanishing approximation) is represented by six "square" diagrams with every possible relative position of the four external ends. These include the diagrams

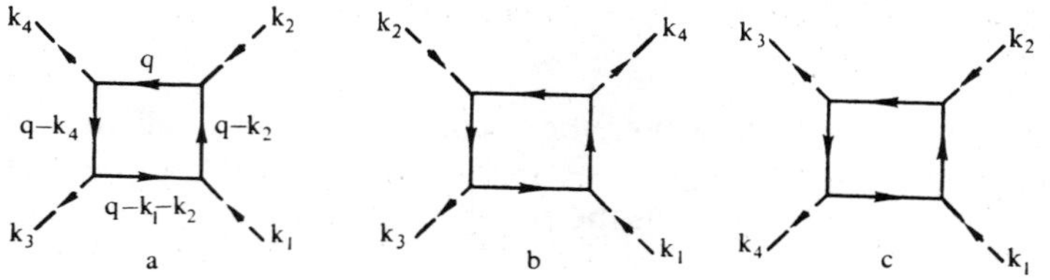

(127.1)

and another three which differ from these only in that the internal electron loop is traversed in the opposite direction. The contribution of these three diagrams is the same as that of the diagrams (127.1), and the total scattering amplitude is therefore

$$M_{fi} = 2(M^{(a)} + M^{(b)} + M^{(c)}), \tag{127.2}$$

where $M^{(a)}$, $M^{(b)}$ and $M^{(c)}$ are the contributions of diagrams (a), (b) and (c).

According to (64.19) the scattering cross-section is

$$d\sigma = \frac{1}{64\pi^2}|M_{fi}|^2 \frac{do'}{(2\omega)^2}, \tag{127.3}$$

where do' is the solid-angle element for the direction $\mathbf{k}'$ in the centre-of-mass system. The scattering angle in that system is denoted by θ.

Invariant amplitudes

Writing separately the polarization factors of the four photons, we have M_{fi} in the form

$$M_{fi} = e_1^\lambda e_2^\mu e_3^{\nu *} e_4^{\rho *} M_{\lambda\mu\nu\rho}; \tag{127.4}$$

† In the limit of low frequencies, this process was first discussed by H. Euler (1936), and in the ultra-relativistic case by A. I. Akhiezer (1937). The complete solution is due to R. Karplus and M. Neumann (1951).

the 4-tensor $M_{\lambda\mu\nu\rho}$ (called the *photon–photon scattering tensor*) is a function of the 4-momenta of all the photons. If the arguments of functions are written with the signs which correspond to like directions of the external lines in the diagram, it is evident from the symmetry of the group of diagrams (127.1) that

$$M_{\lambda\mu\nu\rho}(k_1, k_2, -k_3, -k_4)$$

is symmetrical with respect to any interchange of the four arguments together with a simultaneous corresponding interchange of the four suffixes. Because of the gauge invariance, the amplitude (127.4) is unchanged when e is replaced by $e + \text{constant} \cdot k$. Thus we must have

$$k_1^\lambda M_{\lambda\mu\rho\sigma} = k_2^\mu M_{\lambda\mu\rho\sigma} = \cdots = 0. \tag{127.5}$$

It is easily deduced from this that, in particular, the expansion of the scattering tensor in powers of the 4-momenta $k_1, k_2, \ldots$ must begin with terms containing quaternary products of the components, and certainly

$$M_{\lambda\mu\nu\rho}(0, 0, 0, 0) = 0. \tag{127.6}$$

To determine the actual invariant amplitudes, however, it is desirable to take from the start a particular gauge of the polarization 4-vectors e, in which

$$e_1^\mu = (0, \mathbf{e}_1), \quad e_2^\mu = (0, \mathbf{e}_2), \ldots. \tag{127.7}$$

Then

$$M_{fi} = M_{iklm} e_{1i} e_{2k} e_{3l}^* e_{4m}^*, \tag{127.8}$$

where M_{iklm} is a three-dimensional tensor.

We take as the two independent polarizations for each photon the circular polarizations with opposite directions of rotation, i.e. two helical states with helicities $\lambda = \pm 1$.

The tensor M_{iklm} can then be written

$$M_{iklm} = \sum_{\lambda_1\lambda_2\lambda_3\lambda_4} M_{\lambda_1\lambda_2\lambda_3\lambda_4} e_{1i}^{(\lambda_1)*} e_{2k}^{(\lambda_2)*} e_{3l}^{(\lambda_3)} e_{4m}^{(\lambda_4)}; \tag{127.9}$$

the sixteen quantities $M_{\lambda_1\lambda_2\lambda_3\lambda_4}$ are functions of s, t and u, and act as invariant amplitudes, but they are not all independent.

The quantities $M_{\lambda_1\lambda_2\lambda_3\lambda_4}$ are three-dimensional scalars. Spatial inversion changes the sign of the helicities, while the invariant quantities s, t and u remain unaltered. The condition of P invariance therefore gives the relations

$$M_{\lambda_1\lambda_2\lambda_3\lambda_4}(s, t, u) = M_{-\lambda_1,-\lambda_2,-\lambda_3,-\lambda_4}(s, t, u). \tag{127.10}$$

Time reversal interchanges the initial and final photons without affecting their helicities; s, t and u again remain unaltered. The condition of T invariance therefore gives the equation

$$M_{\lambda_1\lambda_2\lambda_3\lambda_4}(s, t, u) = M_{\lambda_3\lambda_4\lambda_1\lambda_2}(s, t, u). \tag{127.11}$$

Lastly, one further relation follows from the invariance of the amplitude M_{fi} under the interchange of the two initial or the two final photons. If both interchanges are made ($k_1 \leftrightarrow k_2$, $k_3 \leftrightarrow k_4$), the variables s, t, u are unaltered, and the interchange in the polarization indices leads to

$$M_{\lambda_1\lambda_2\lambda_3\lambda_4}(s, t, u) = M_{\lambda_2\lambda_1\lambda_4\lambda_3}(s, t, u). \tag{127.12}$$

It is easy to see that, because of the symmetry properties (127.10)–(127.12), the number of independent invariant amplitudes is only five, which may be chosen, for example, as

$$M_{++++}, \quad M_{++--}, \quad M_{+-+-}, \quad M_{+--+}, \quad M_{+++-}$$

(the suffixes + and − denoting, for brevity, helicity values +1 and −1).

If one of the amplitudes $M_{\lambda_1\lambda_2\lambda_3\lambda_4}$ is substituted for M_{fi} in (127.3), the result is the cross-section for scattering with specified polarizations of the initial and final photons. The cross-section summed over the final polarizations and averaged over the initial polarizations is obtained by the substitution

$$|M_{fi}|^2 \to \tfrac{1}{4}\{2|M_{++++}|^2 + 2|M_{++--}|^2 + 2|M_{+-+-}|^2 + 2|M_{+--+}|^2 + 8|M_{+++-}|^2\}. \tag{127.13}$$

The symmetry relations (127.10)–(127.12) connect different invariant amplitudes as functions of the same variables. Further functional relations are obtained from crossing invariance (§78), since the amplitude M_{fi} describes the same reaction (photon–photon scattering) in every channel, and therefore must be the same for every channel.

The s channel (corresponding to the arrow directions as shown in the diagrams (127.1)) is converted to the t channel by interchanging the 4-momenta k_2 and $-k_3$ (i.e. by changing the variables $s \leftrightarrow t$) and interchanging the helicity suffixes $\lambda_2 \leftrightarrow -\lambda_3$. Similarly, it is converted to the u channel by interchanging k_2 and $-k_4$ ($s \leftrightarrow u$) and $\lambda_2 \leftrightarrow -\lambda_4$. This leads to the relations

$$\left.\begin{aligned} M_{+-+-}(s, t, u) &= M_{++++}(u, t, s), \\ M_{+--+}(s, t, u) &= M_{++++}(t, s, u), \\ M_{++++}(s, t, u) &= M_{++++}(s, u, t); \end{aligned}\right\} \tag{127.14}$$

M_{++--} and M_{+++-} are completely symmetrical in s, t and u.† It is therefore sufficient

† Here we have also used the symmetry with respect to the two final photons. Since the three variables s, t and u are not independent, it would be sufficient to write two arguments (for example, the first two), but we have retained all three, simply in order to clarify the symmetry of the interchanges.

to calculate only three of the sixteen amplitudes, for instance M_{++++}, M_{++--} and M_{+++-}.

The relations (127.10)–(127.12) and (127.14) apply to the total amplitudes, i.e. the sums of the contributions of all three diagrams (127.1). But these contributions themselves are related in a manner which is obvious on comparing the diagrams. For example, diagram (b) is obtained from diagram (a) by the substitutions $k_2 \leftrightarrow -k_4$, $e_2 \leftrightarrow e_4^*$, and so their contributions to the invariant amplitudes are obtained from each other by interchanging the variables $s \leftrightarrow u$ and the suffixes $\lambda_2 \leftrightarrow -\lambda_4$; similarly, the contribution of diagram (c) is obtained from that of diagram (a) by the changes $t \leftrightarrow u$, $\lambda_3 \leftrightarrow -\lambda_4$.

Calculation of the amplitudes

The integral $M_{fi}^{(a)}$ corresponding to the diagram (127.1a) has the form (126.4), with

$$B^{(a)} = \frac{e^4}{\pi^2} \operatorname{tr} \{\gamma e_1(\gamma q - \gamma k_2 + m)(\gamma e_2) \times \\ \times (\gamma q + m)(\gamma e_4^*)(\gamma q - \gamma k_4 + m)(\gamma e_3^*)(\gamma q - \gamma k_1 - \gamma k_2 + m)\}. \quad (127.15)$$

The integrals (126.4) are logarithmically divergent. In accordance with the condition (127.6), they are regularized by subtracting the value when $k_1 = k_2 = \cdots = 0$.† The calculation of the regularized integrals is, however, exceedingly laborious.

The most straightforward way to calculate the photon–photon scattering amplitudes is based on the use of the double dispersion relation (B. De Tollis, 1964). This method makes the most complete allowance for the symmetry of the diagrams, and almost entirely eliminates the difficulties of the integrations.

The function $A_{1s}^{(a)}(s, t)$ (and similarly $A_{1t}^{(a)}$) for any given set of helicities λ_1, λ_2, λ_3, λ_4 is calculated in accordance with (126.6); owing to the presence of two delta functions in the integrand, the value of $B^{(a)}$ is needed only for

$$l_1^2 \equiv q^2 = m^2, \quad l_4^2 \equiv (q - k_1 - k_2)^2 = m^2. \quad (127.16)$$

These equations can be utilized in calculating the trace (127.15). For substitution in

† In the summation of the contributions from all the diagrams, the divergent parts of the integrals cancel, as is easily seen by noting the asymptotic form of the integral as $q \to \infty$:

$$M_{\lambda\mu\nu\rho}^{(a)} \propto \int \operatorname{tr}\{\gamma_\lambda(\gamma q)\gamma_\mu(\gamma q)\gamma_\rho(\gamma q)\gamma_\nu(\gamma q)\}\, d^4q/(q^2)^4.$$

After averaging over the directions of q (cf. (131.10)), the trace is easily calculated, giving

$$M_{\lambda\mu\nu\rho}^{(a)} \propto (g_{\lambda\mu}g_{\nu\rho} + g_{\lambda\nu}g_{\mu\rho} - 2g_{\lambda\rho}g_{\mu\nu}) \int d^4q/(q^2)^2.$$

The summation over the diagrams is equivalent to symmetrizing this expression with respect to the suffixes λ, μ, ν and ρ, the result of which is zero. However, this is in a certain sense fortuitous, and does not remove the need for regularization, even though the latter amounts only to the subtraction of a finite quantity.

(126.22) we need the value of $A_{1s}^{(a)}$ only for $t = 0$. This implies that $\mathbf{k} = \mathbf{k}'$ and $k_2 = k_4$. Then the integral (126.6) becomes

$$A_{1s}^{(a)}(s, 0) = -\tfrac{1}{4}\pi^2\sqrt{\frac{s-4m^2}{s}}\int\frac{B^{(a)}\,do_{\mathbf{q}}}{[(q-k_2)^2-m^2]^2}; \tag{127.17}$$

cf. the derivation of (115.10). With the angle ϑ between $\mathbf{q}$ and $\mathbf{k}$, we have

$$\begin{aligned}(q-k_2)^2 - m^2 &= -2\omega(1-|\mathbf{q}|\cos\vartheta)\\ &= -\sqrt{s}[1-\tfrac{1}{2}\sqrt{(s-4m^2)}\cos\vartheta].\end{aligned}$$

The integrals (127.17) can in fact be expressed in terms of elementary functions. The calculation of $A_2^{(a)}(s, t)$ from its definition (126.18) involves no integration; here the expression for $B^{(a)}$ is to be taken for the values of q given by (126.15), which satisfy not only (127.16) but also the conditions $(q-k_2)^2 = m^2$, $(q-k_4)^2 = m^2$.

When the functions A_{1s}, A_{1t} and A_2 have been calculated, the dispersion relation (126.22) gives the amplitude directly as single and double definite integrals. We shall give the final result for the three invariant amplitudes which, according to the preceding discussion, are sufficient to determine all the other amplitudes:†

$$\begin{aligned}\frac{1}{8\alpha^2}M_{++++} = &-1-\left(2+\frac{4t}{s}\right)B(t)-\left(2+\frac{4u}{s}\right)B(u)-\\ &-\left[\frac{2(t^2+u^2)}{s^2}-\frac{8}{s}\right][T(t)+T(u)]+\frac{4}{t}\left(1-\frac{2}{s}\right)I(s,t)+\\ &+\frac{4}{u}\left(1-\frac{2}{s}\right)I(s,u)+\left[\frac{2(t^2+u^2)}{s^2}-\frac{16}{s}-\frac{4}{t}-\frac{4}{u}-\frac{8}{tu}\right]I(t,u),\end{aligned} \tag{127.18}$$

$$\begin{aligned}\frac{1}{8\alpha^2}M_{+++-} = &\;1+4\left(\frac{1}{s}+\frac{1}{t}+\frac{1}{u}\right)[T(s)+T(t)+T(u)]-\\ &-4\left(\frac{1}{u}+\frac{2}{st}\right)I(s,t)-4\left(\frac{1}{t}+\frac{2}{su}\right)I(s,u)-4\left(\frac{1}{s}+\frac{2}{tu}\right)I(t,u),\end{aligned}$$

$$\frac{1}{8\alpha^2}M_{++--} = 1-\frac{8}{st}I(s,t)-\frac{8}{su}I(s,u)-\frac{8}{tu}I(t,u).$$

Here, $B(s)$, $T(s)$ and $I(s, t)$ denote the functions

$$B(s) = \sqrt{(1-4/s)}\sinh^{-1}\tfrac{1}{2}\sqrt{-s}-1, \quad s<0,$$

$$T(s) = (\sinh^{-1}\tfrac{1}{2}\sqrt{-s})^2, \quad s<0,$$

$$I(s,t) = \tfrac{1}{4}\int_0^1\frac{dy}{y(1-y)-(s+t)/st}\{\log[1-i0-sy(1-y)]+\log[1-i0-ty(1-y)]\}, \tag{127.19}$$

† Some further details of the transformations of the integrals, various representations of the transcendental functions B, T and I, and some limiting forms are given by B. De Tollis, *Nuovo Cimento* [10] **32**, 757, 1964; **35**, 1182, 1965; V. Costantini, B. De Tollis and G. Pistoni, *ibid.* [11] **2A**, 733, 1971.

and the expressions in the ranges $0 < s < 4$ and $s > 4$ are obtained from (127.19) by analytical continuation with the rule $s \to s + i0$, i.e. through the upper half-plane of these variables. To simplify the notation, s and t denote s/m^2 and t/m^2, in (127.18) and (127.19) only.

Scattering cross-section

The limiting case of low frequencies ($\omega \ll m$) corresponds to small values of the variables s, t and u. The first terms in the expansion of the invariant amplitudes in powers of these variables are

$$\left.\begin{aligned} M_{++++} \approx 11e^4 s^2/45m^4, \quad M_{+--+} \approx 11e^4 t^2/45m^4, \\ M_{+-+-} \approx 11e^4 u^2/45m^4, \quad M_{++--} \approx -e^4(s^2+t^2+u^2)/15m^4, \\ M_{+++-} \approx 0. \end{aligned}\right\} \tag{127.20}$$

Substituting these values in (127.3), we find the cross-sections for the scattering of polarized photons. The differential scattering cross-section for unpolarized photons, calculated from (127.13), is (in ordinary units)

$$d\sigma = \frac{139}{4\pi^2(90)^2}\alpha^2 r_e^2\left(\frac{\hbar\omega}{mc^2}\right)^6(3+\cos^2\theta)\ do' \tag{127.21}$$

and the total cross-section is†

$$\begin{aligned} \sigma &= \frac{973}{10125\pi}\alpha^2 r_e^2\left(\frac{\hbar\omega}{mc^2}\right)^6 \\ &= 0.031\alpha^2 r_e^2\left(\frac{\hbar\omega}{mc^2}\right)^6, \quad \hbar\omega \ll mc^2. \end{aligned} \tag{127.22}$$

In the opposite (ultra-relativistic) case, the total scattering cross-section for unpolarized photons is‡

$$\sigma = 4.7\alpha^4(c/\omega)^2, \quad \hbar\omega \gg mc^2. \tag{127.23}$$

Finally, the differential cross-section for small-angle scattering in the ultra-relativistic case is

$$d\sigma = \frac{\alpha^4 c^2}{\pi^2\omega^2}\log^4\frac{1}{\theta}\ do, \quad mc^2/\hbar\omega \ll \theta \ll 1. \tag{127.24}$$

This expression is valid with logarithmic accuracy (the next term in the expansion contains a power of the large logarithm lower by one unit). In the limit $\theta = 0$

† In going from $d\sigma$ to σ, a factor $\frac{1}{2}$ has to be included to take account of the identity of the two final photons.

‡ The origin of this dependence of σ on ω will be further discussed at the end of §134.

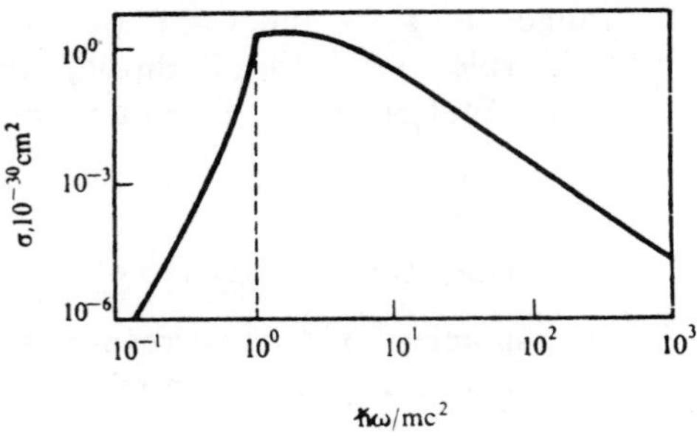

FIG. 24.

(forward scattering), (127.24) is invalid, and is replaced by

$$d\sigma = \frac{\alpha^4 c^2}{\pi^2\omega^2}\log^4\frac{\hbar\omega}{mc^2}\,do, \quad \theta \ll mc^2/\hbar\omega. \tag{127.25}$$

This expression is easily derived from the general formulae (127.18), putting $t = 0$ and noting that, for $s \gg 1$, the highest power (the square) of the large logarithm is present only in the function

$$T(s/m^2) \approx \tfrac{1}{4}\log^2(s/m^2) \approx \log^2(\omega/m).$$

To this accuracy, the only non-zero amplitudes are

$$M_{++++} = M_{----} = M_{+-+-} = -16e^4\log^2(\omega/m).$$

In particular, therefore, the photon polarization is in this case unchanged on scattering.

Figure 24 shows the total scattering cross-section as a function of the frequency, plotted on a double logarithmic scale. The cross-section decreases towards both low and high frequencies, reaching a maximum when $\hbar\omega \approx 1.5mc^2$. The break in the curve at $\hbar\omega = mc^2$ corresponds to the change in the nature of the process when the production of a real electron–positron pair becomes possible.

LOW FREQUENCIES

For low frequencies ($\omega \ll m$), the photon–photon scattering amplitude can also be derived by a totally different method, based on the correction terms in the Lagrangian of a weak electromagnetic field (§129).

The small correction $\hat{V}'$ to the interaction Hamiltonian differs only in sign from that to the Lagrangian. From (129.21),

$$\hat{V}' = -\frac{e^4}{45\times 8\pi^2 m^4}\int \{(\hat{\mathbf{E}}^2 - \hat{\mathbf{H}}^2)^2 + 7(\hat{\mathbf{E}}\cdot\hat{\mathbf{H}})^2\}\,d^3x. \tag{127.26}$$

Since this operator is of the fourth order in the field, it has matrix elements for the relevant transition, even in the first approximation.

For the calculation, we substitute in (127.26)

$$\hat{\mathbf{E}} = -\,\partial \hat{\mathbf{A}}/\partial t, \quad \hat{\mathbf{H}} = \operatorname{curl} \hat{\mathbf{A}},$$

$$\hat{\mathbf{A}} = \sqrt{(4\pi)} \sum_{\mathbf{k},\lambda} (\hat{c}_{\mathbf{k}\lambda}\, \mathbf{e}_{\mathbf{k}\lambda}\, e^{-ikx} + \hat{c}^{+}_{\mathbf{k}\lambda}\, \mathbf{e}^{*}_{\mathbf{k}\lambda}\, e^{ikx}), \tag{127.27}$$

where λ numbers the polarization; the S-matrix element is then given by

$$\begin{aligned} S_{fi} &= -\,i\langle f|\int V'\,dt|i\rangle \\ &= -\,i\langle 0|c_{\mathbf{k}_3\lambda_3}\, c_{\mathbf{k}_4\lambda_4} \int V'\,dt\, c^{+}_{\mathbf{k}_1\lambda_1}\, c^{+}_{\mathbf{k}_2\lambda_2}|0\rangle \end{aligned} \tag{127.28}$$

(cf. §§72 and 77). When $\hat{\mathbf{A}}$ is normalized as in (127.27), the scattering amplitude M_{fi} is found immediately from S_{fi}:

$$S_{fi} = i(2\pi)^4 \delta^{(4)}(k_3 + k_4 - k_1 - k_2) M_{fi} \tag{127.29}$$

(cf. §64). The mean value in (127.28) is calculated by means of Wick's theorem, using (77.3), with contraction of only the "external" operators $\hat{c}_{\mathbf{k}\lambda}$, $\hat{c}^{+}_{\mathbf{k}\lambda}$ with the internal operators $\hat{\mathbf{A}}$.

§ 128. Coherent scattering of a photon in the field of a nucleus

Other effects which are non-linear, like photon–photon scattering, and are described by square diagrams of the form (127.1), are the disintegration of one photon into two in an external field (and the reverse process of combination of two photons into one), and photon scattering in an external field. The former corresponds to diagrams in which one of the four external photon lines is replaced by an external field line; the latter process corresponds to diagrams with two external lines of real photons and two of virtual photons.

This class includes, in particular, coherent (elastic) scattering of a photon in the constant electric field of a stationary nucleus. In general, the calculations lead to very lengthy formulae involving multiple quadratures.† Here, only some estimates will be given.

Because of the requirements of gauge invariance, the scattering amplitude as $\omega \to 0$ must contain products of the components of the 4-momenta of the initial photon (k) and the final photon (k'), just as the expansion of the photon–photon scattering amplitude begins with the quaternary products of the components of the 4-momenta of all the photons. Thus the scattering amplitude for a low-frequency photon is proportional to ω^2. Since also this amplitude involves the external field

† See V. Costantini, B. De Tollis and G. Pistoni, *Nuovo Cimento* [11] 2A, 733, 1971; B. De Tollis, M. Lusignoli and G. Pistoni, *ibid.* 32A, 227, 1976.

(the field of the nucleus with charge Ze) in the second order, we conclude that the scattering cross-section is

$$d\sigma \sim Z^4\alpha^4 r_e^2(\omega/m)^4\, do \quad (\omega \ll m). \tag{128.1}$$

The frequency dependence is, of course, in agreement with the general results of §59.

The coefficient in (128.1) cannot be calculated from the Lagrangian for a uniform electromagnetic field (as was done for photon–photon scattering). The reason is that, in the process here considered, distances from the nucleus $r \sim 1/m$ at which its field cannot be regarded as uniform are important.

The result of the exact calculation is

$$\left.\begin{aligned} d\sigma_{++} &= d\sigma_{--} = 1.004\times 10^{-3}(Z\alpha)^4 r_e^2(\omega/m)^4 \cos^4 \tfrac{1}{2}\theta\, do,\\ d\sigma_{+-} &= d\sigma_{-+} = 3.81\times 10^{-4}(Z\alpha)^4 r_e^2(\omega/m)^4 \sin^4 \tfrac{1}{2}\theta\, do. \end{aligned}\right\} \tag{128.2}$$

Here, as in §127, the suffixes + and − denote the helicities +1 and −1 of the final and initial photons; θ is the scattering angle in the rest frame of the nucleus (V. Costantini, B. De Tollis and G. Pistoni, 1971).

To estimate the cross-section at high frequencies, we use the optical theorem (§71). The intermediate state which appears on the right-hand side of the unitarity relation is here a state of the electron–positron pair (corresponding to the division of the diagrams at two internal electron lines between external photon lines). The optical theorem therefore relates the amplitude for elastic scattering of a photon through an angle of zero and the total cross-section σ_{pair} for photon pair production in the field of the nucleus. If the amplitude $f(\omega, \theta)$ for scattering through an angle θ is so defined that the scattering cross-section is $d\sigma = |f|^2\, do$ (cf. (71.5)), we have

$$\operatorname{im} f(\omega, 0) = \omega\sigma_{\text{pair}}/4\pi.$$

The cross-section σ_{pair} is, of course, zero unless $\omega > 2m$. In the ultra-relativistic case, taking σ_{pair} from (94.6), we get

$$f''(\omega) \equiv \operatorname{im} f(\omega, 0) = \frac{7}{9\pi}(Z\alpha)^2 r_e \frac{\omega}{m}\left[\log\frac{2\omega}{m} - \frac{109}{42}\right] \quad (\omega \gg m). \tag{128.3}$$

The real part of the scattering amplitude is determined by the imaginary part, through the dispersion relation. The latter must here be written "with one subtraction", i.e. for the function f/t (where $t = \omega^2$), since as $\omega \to 0$ the amplitude $f \propto \omega^2$; compare the dispersion relation "with two subtractions" (111.13). Separating the real part of the dispersion integral (for which it is sufficient to take the integral as a principal value), and changing from integration with respect to $t' = \omega'^2$ to that with respect to ω', we have

$$f'(\omega) \equiv \operatorname{re} f(\omega, 0) = \frac{2\omega^2}{\pi} P \int_{2m}^{\infty} \frac{f''(\omega')\, d\omega'}{\omega'(\omega'^2 - \omega^2)}. \tag{128.4}$$

When $\omega \gg m$, the important values in the integral are $\omega' \sim \omega \gg m$, so that we can use the expression (128.3) for $f''(\omega')$; the lower limit of integration may then be replaced by zero. The principal value of the integral can be represented as half the sum of the integrals along paths on the upper and lower edges of the positive real axis in the complex ω'-plane; these paths may then in turn be rotated in the ω'-plane to lie along the positive and negative imaginary axes respectively. Then

$$f'(\omega) = -\frac{\omega^2}{\pi} \int_0^\infty \frac{f''(i\xi) + f''(-i\xi)}{\xi(\xi^2+\omega^2)} d\xi$$

$$= \frac{7}{9\pi}(Z\alpha)^2 \frac{r_e}{m} \omega^2 \int_0^\infty \frac{d\xi}{\xi^2+\omega^2}$$

and the final result is

$$\text{re}\, f(\omega, 0) = \frac{7}{18}(Z\alpha)^2 r_e \omega / m. \qquad (128.5)$$

Note that the real part of the amplitude, unlike the imaginary part, does not contain a large logarithm.

The sum of the squares of (128.3) and (128.5) gives the cross-section for scattering through an angle of zero as

$$d\sigma_{\theta=0} = \frac{49}{81\pi^2}(Z\alpha)^4 r_e^2 \left(\frac{\omega}{m}\right)^2 \left\{\log^2\left(\frac{0.15\omega}{m}\right) + \tfrac{1}{4}\pi^2\right\} do \qquad (128.6)$$

(F. Rohrlich and R. L. Gluckstern, 1952).

The result (128.6) derived for scattering exactly forwards is valid also over a certain range of small angles. The condition for its validity can be shown to be $\theta \ll (m/\omega)^2$. This range, however, makes only a small contribution to the total scattering cross-section. The main contribution to the latter comes from angles $\theta \lesssim m/\omega$, as is easily seen from the general (not only for angle zero) unitarity relation between the amplitudes for photon–photon scattering and photon pair production. In that range, however, the logarithmic term is absent, and the total scattering cross-section is thus

$$\sigma \sim (Z\alpha)^4 r_e^2 (\omega/m)^2 \theta^2 \sim (Z\alpha)^4 r_e^2 \qquad (128.7)$$

(H. A. Bethe and F. Rohrlich, 1952). For large ω, therefore, the coherent scattering cross-section tends to a constant limit.

§ 129. Radiative corrections to the electromagnetic field equations

In the quantization of the electron–positron field (§25) it has been shown that the expression for the vacuum energy contains an infinite constant, which may be written†

$$\mathscr{E}_0 = -\sum_{\mathbf{p},\sigma} \varepsilon^{(-)}_{\mathbf{p}\sigma}, \tag{129.1}$$

where $-\varepsilon^{(-)}_{\mathbf{p}\sigma}$ are the negative frequencies of the solutions of Dirac's equation. This constant itself has no physical meaning, since the vacuum energy is, by definition, zero. When an electromagnetic field is present, however, the energy levels $\varepsilon^{(-)}_{\mathbf{p}\sigma}$ will change. The changes are finite and are physically significant. They describe the field dependence of the properties of space, and alter the equations of the electromagnetic field in a vacuum.

The changes in the field equations correspond to the change in the field Lagrangian. The density L of the Lagrangian is a relativistic invariant, and therefore can depend only on the invariants $\mathbf{E}^2 - \mathbf{H}^2$ and $\mathbf{E}\cdot\mathbf{H}$. The usual expression

$$L_0 = (\mathbf{E}^2 - \mathbf{H}^2)/8\pi \tag{129.2}$$

is the first term in the expansion of the general expression in powers of the invariants.

Let us derive the Lagrangian for the case where the fields $\mathbf{E}$ and $\mathbf{H}$ vary so slowly in space and time that they can be regarded as uniform and constant. Then L may be assumed not to involve the derivatives of the fields. The necessary conditions for this will be discussed at the end of the section.

However, if the problem stated is to be meaningful, we must also assume the electric field to be sufficiently weak. The reason is that a uniform electric field can generate pairs from the vacuum. The field itself can be treated as a closed system only if the pair production probability is sufficiently small:

$$|\mathbf{E}| \ll m^2/|e| \quad (= m^2c^3/|e|\hbar), \tag{129.3}$$

i.e. the change in the energy of a charge e over a distance $\hbar/mc$ must be small in comparison with mc^2. We shall see below (cf. also Problem 2) that the pair production probability is then exponentially small.

If there is a magnetic field as well as an electric field, it is in general possible to choose a frame of reference in which $\mathbf{E}$ and $\mathbf{H}$ are parallel. Then the magnetic field does not influence the motion of the charge in the direction of $\mathbf{E}$. The condition (129.3) is to be satisfied in this frame, which will be the one used in the subsequent calculations.

The calculation of the Lagrangian begins with that of the change W' in the vacuum energy. This is given by the change in the "zero energy" (129.1) due to the field. From this, however, we must subtract the mean values of the potential energy of the electrons in the "states" of negative energy. The subtraction simply makes the total charge of the vacuum zero by definition.

The zero energy in the presence of the field is

$$\mathscr{E}_0 = -\sum_{\mathbf{p},\sigma} \varepsilon^{(-)}_{\mathbf{p}\sigma} = \sum_{\mathbf{p},\sigma} \int \psi^{(-)*}_{\mathbf{p}\sigma} \cdot i\frac{\partial}{\partial t}\psi^{(-)}_{\mathbf{p}\sigma}\, d^3x, \tag{129.4}$$

† Here we shall write $\mathscr{E}$ in place of E, to avoid confusion with the electric field.

where $\psi^{(-)}_{\mathbf{p}\sigma}$ are the negative-frequency solutions of Dirac's equation in the field concerned. We shall assume that the integration is over a unit volume, and that the wave functions are normalized to unity in that volume; then $\mathscr{E}_0$ is the energy per unit volume. According to the preceding discussion, we have to subtract from $\mathscr{E}_0$ the quantity

$$U_0 = \sum_{\mathbf{p},\sigma} \int \psi^{(-)*}_{\mathbf{p}\sigma}\, e\phi \psi^{(-)}_{\mathbf{p}\sigma}\, d^3x,$$

where $\phi = -\mathbf{E}\cdot\mathbf{r}$ is the potential of the uniform field. According to the theorem on the differentiation of an operator with respect to a parameter (see *QM*, (11.16)),

$$\begin{aligned} U_0 &\equiv \mathbf{E}\cdot \sum_{\mathbf{p},\sigma} \psi^{(-)*}_{\mathbf{p}\sigma} \frac{\partial \hat{H}}{\partial \mathbf{E}} \psi^{(-)}_{\mathbf{p}\sigma}\, d^3x \\ &= -\mathbf{E}\cdot \sum_{\mathbf{p},\sigma} \partial \varepsilon^{(-)}_{\mathbf{p}\sigma}/\partial \mathbf{E} \\ &= \mathbf{E}\cdot \partial\mathscr{E}_0/\partial\mathbf{E}. \end{aligned}$$

Thus the total change in the vacuum energy density is

$$W' = (\mathscr{E}_0 - \mathbf{E}\cdot\partial\mathscr{E}_0/\partial\mathbf{E}) - (\mathscr{E}_0 - \mathbf{E}\cdot\partial\mathscr{E}_0/\partial\mathbf{E})_{\mathbf{E}=\mathbf{H}=0}. \tag{129.5}$$

We can relate W' to the change L' in the Lagrangian density ($L = L_0 + L'$) by using the general formula

$$W = \sum \dot{q}\, \partial L/\partial \dot{q} - L,$$

where q represents the "generalized coordinates" of the field (see *Fields*, §32). For an electromagnetic field, the quantities q are the potentials $\mathbf{A}$ and ϕ. Since

$$\mathbf{E} = -\dot{\mathbf{A}} - \nabla\phi, \quad \mathbf{H} = \text{curl}\,\mathbf{A}, \tag{129.6}$$

$\dot{\mathbf{A}}$ is the only "velocity" $\dot{q}$ which appears in L, and the differentiation with respect to $\dot{\mathbf{A}}$ is equivalent to one with respect to $\mathbf{E}$; hence

$$W' = \mathbf{E}\cdot\partial L'/\partial\mathbf{E} - L'. \tag{129.7}$$

Comparison of (129.5) and (129.7) gives

$$L' = -[\mathscr{E}_0 - (\mathscr{E}_0)_{\mathbf{E}=\mathbf{H}=0}]. \tag{129.8}$$

Thus L' can be calculated by means of the sum (129.1).

Let us first take the case where there is only a magnetic field. The "negative" energy levels of the electron (charge $e = -|e|$) in a constant uniform field $H_z = H$

are

$$-\varepsilon_{\mathbf{p}}^{(-)} = -\sqrt{[m^2 + |e|H(2n - 1 + \sigma) + p_z^2]}, \tag{129.9}$$

$$n = 0, 1, 2, \ldots; \sigma = \pm 1$$

(see §32, Problem). To find the sum, we note that the number of states in the interval dp_z is

$$\frac{|e|H}{2\pi} \frac{dp_z}{2\pi}$$

(see *QM*, §112); the first factor is the number of states with various values of p_x, which do not affect the energy. Moreover, all the levels except $n = 0$, $\sigma = -1$ are doubly degenerate, the levels n, $\sigma = +1$ and $n + 1$, $\sigma = -1$ coinciding. Hence

$$-\mathscr{E}_0 = \frac{|e|H}{(2\pi)^2} \int_{-\infty}^{\infty} \left\{ \sqrt{(m^2 + p_z^2)} + 2 \sum_{n=1}^{\infty} \sqrt{(m^2 + 2|e|Hn + p_z^2)} \right\} dp_z. \tag{129.10}$$

The divergence of the integrals in (129.10) is eliminated in the calculation of L' (129.8) by subtracting the value of the sum when $H = 0$. To carry out this "renormalization", it is convenient to calculate first the convergent expression

$$\begin{aligned}
\Phi &\equiv -\frac{\partial^2 \mathscr{E}_0}{(\partial m^2)^2} \\
&= -\frac{|e|H}{2(2\pi)^2} \int_0^{\infty} \left\{ (m^2 + p_z^2)^{-3/2} + 2 \sum_{n=1}^{\infty} (m^2 + 2|e|Hn + p_z^2)^{-3/2} \right\} dp_z \\
&= -\frac{|e|H}{8\pi^2} \left\{ \frac{1}{m^2} + 2 \sum_{n=1}^{\infty} \frac{1}{m^2 + 2|e|Hn} \right\}.
\end{aligned}$$

The sum in the braces can be reduced to that of a geometrical progression, as follows:

$$\begin{aligned}
\Phi &= -\frac{|e|H}{8\pi^2} \int_0^{\infty} e^{-m^2\eta} \left[2 \sum_{n=0}^{\infty} e^{-2|e|Hn\eta} - 1 \right] d\eta \\
&= -\frac{|e|H}{8\pi^2} \int_0^{\infty} e^{-m^2\eta} \left[\frac{2}{1 - e^{-2|e|H\eta}} - 1 \right] d\eta \\
&= -\frac{|e|H}{8\pi^2} \int_0^{\infty} e^{-m^2\eta} \coth(|e|H\eta)\, d\eta.
\end{aligned} \tag{129.11}$$

To find L', we must now integrate Φ twice with respect to m^2 and then subtract the value of the resulting quantity when $H = 0$. This gives

$$L' = -\frac{1}{8\pi^2}\int_0^\infty \frac{e^{-m^2\eta}}{\eta^3}\{\eta|e|H \coth(\eta|e|H) - 1\}\, d\eta + c_1 + c_2 m^2, \qquad (129.12)$$

where c_1 and c_2 depend on H but not on m^2.

From considerations of dimensions and of parity with respect to $\mathbf{H}$, it is evident that L' as a function of H and m must have the form

$$L' = m^4 f(H^2/m^4).$$

Hence there can be no terms in L' which are odd in m^2, and so $c_2 = 0$. The coefficient c_1 is given by the condition that the expansion of L' in powers of H^2 begins with a term in H^4: a term in H^2 would simply alter the coefficient in the original Lagrangian $L_0 = -H^2/8\pi$, and this would essentially signify a changed definition of the field and therefore of the charge. The elimination of the H^2 terms thus corresponds to a renormalization of charge. It is easily verified that this is achieved by putting

$$c_1 = \frac{H^2 e^2}{3\times 8\pi^2}\int_0^\infty \frac{e^{-\eta}}{\eta}\, d\eta.$$

Finally, making the change of variable $m^2\eta \to \eta$ in (129.12), we have

$$L'(H; E = 0) = \frac{m^4}{8\pi^2}\int_0^\infty \{-\eta b \coth b\eta + 1 + \tfrac{1}{3}b^2\eta^2\}\, e^{-\eta}\, \frac{d\eta}{\eta^3}, \qquad (129.13)$$

where $b = |e|H/m^2$.

Let us now go to the general case where there is not only a magnetic field but also an electric field $\mathbf{E}$ parallel to it, satisfying the condition (129.3).

To find L' in this case it is not necessary to determine afresh the energy levels $\varepsilon_{\mathbf{p}}^{(-)}$ of the electron in the field; we need only note that, if the wave function (the solution of the second-order equation (32.7)) is sought as a product

$$\psi = \psi_E(z)\, e^{ip_x x}\chi_{n\sigma}(y),$$

where $\chi_{n\sigma}(y)$ is the wave function in the magnetic field when $\mathbf{E} = 0$ and $p_z = 0$, then the mass m and the field H appear in the equation for $\psi_E(z)$ only in the combination

$$m^2 + |e|H(2n + 1 + \sigma).$$

If now the factor $|e|H/2\pi$ is again taken from the summation over p_x (the energy levels being independent of p_x), the dimensional argument shows that

$$\begin{aligned}\Phi(H, E) &\equiv \partial^2 L'/(\partial m^2)^2 \\ &= -\frac{|e|H}{8\pi^2}\sum_{n=0}^{\infty}\sum_{\sigma=\pm 1}\frac{F([m^2+|e|H(2n+1+\sigma)]/|e|H)}{m^2+|e|H(2n+1+\sigma)} \\ &= -\frac{b}{8\pi^2}\left\{F\left(\frac{1}{a}\right)+2\sum_{n=1}^{\infty}\frac{F(1+2bn/a)}{1+2bn}\right\},\end{aligned} \qquad (129.14)$$

$$a = |e|E/m^2;$$

each term in the sum is $-d^2\varepsilon_{\mathbf{p}}^{(-)}/(dm^2)^2$ summed over all the quantum numbers except n. Here F is a function as yet unknown, which will be derived from considerations of relativistic invariance.

Φ must be a function of the scalars $H^2 - E^2$ and $(EH)^2 = (\mathbf{E}\cdot\mathbf{H})^2$:

$$\Phi(H, E) = f(H^2 - E^2, (EH)^2).$$

Hence

$$\Phi(0, E) = f(-E^2, 0) = \Phi(iE, 0).$$

The function $\Phi(iE, 0)$ is obtained from (129.11) by putting $H \to iE$; after a change of notation for the variable of integration, this gives

$$\Phi(iE, 0) = \frac{1}{8\pi^2}\int_0^{\infty} e^{-\eta/a}\cot\eta\, d\eta. \qquad (129.15)$$

The function F can be found by comparing this expression with the limit $\Phi(H \to 0, E)$ given by (129.14). The passage to the limit $H \to 0$ in (129.14) can be effected by replacing the summation over n by integration over $dn = dx/2b$:

$$\begin{aligned}\Phi(0, E) &= -\frac{1}{8\pi^2}\int_0^{\infty} F\left(\frac{1+x}{\alpha}\right)\frac{dx}{1+x} \\ &= -\frac{1}{8\pi^2}\int_{1/a}^{\infty}\frac{F(y)}{y}\,dy.\end{aligned} \qquad (129.16)$$

Equating the expressions (129.15) and (129.16) and differentiating with respect to $1/a \equiv z$, we find

$$F(z)/z = -\int_0^{\infty} e^{-\eta z}\,\eta\cot\eta\, d\eta.$$

The summation in (129.14) then reduces again to the summation of a geometric progression, and the subsequent calculations are similar to those given previously: we express Φ in terms of m^2, E and H, integrate twice with respect to m^2, subtract the value for $E = H = 0$, and determine the constants of integration as in the derivation of (129.13). The final result is†

$$\left.\begin{aligned} L' &= \frac{m^4}{8\pi^2}\int_0^\infty \frac{e^{-\eta}}{\eta^3}\{-(\eta a \cot \eta a)(\eta b \coth \eta b) + 1 - \tfrac{1}{3}\eta^2(a^2 - b^2)\}\, d\eta, \\ a &= |e|E/m^2\ (= |e|\hbar E/m^2c^3), \qquad b = |e|H/m^2\ (= |e|\hbar H/m^2c^3). \end{aligned}\right\} \quad (129.17)$$

The parameters a and b may be written in the invariant form

$$\left.\begin{aligned} a &= -\frac{i|e|}{\sqrt{2}m^2}\{(\mathcal{F} + i\mathcal{G})^{1/2} - (\mathcal{F} - i\mathcal{G})^{1/2}\}, \\ b &= \frac{|e|}{\sqrt{2}m^2}\{(\mathcal{F} + i\mathcal{G})^{1/2} + (\mathcal{F} - i\mathcal{G})^{1/2}\}, \end{aligned}\right\} \quad (129.18)$$

where $\mathcal{F}$ and $\mathcal{G}$ denote the invariants

$$\mathcal{F} = \tfrac{1}{2}(\mathbf{H}^2 - \mathbf{E}^2), \quad \mathcal{G} = \mathbf{E}\cdot\mathbf{H}, \quad \mathcal{F} \pm i\mathcal{G} = \tfrac{1}{2}(\mathbf{H} \pm i\mathbf{E})^2. \quad (129.19)$$

When (129.17) is expressed in terms of the invariants $\mathcal{F}$ and $\mathcal{G}$, it becomes applicable in any frame of reference (not only in that where $\mathbf{E}\|\mathbf{H}$).

The formula (129.17) is written in a somewhat arbitrary manner. It is valid only if the electric field is small: $a \ll 1$ (129.3); this condition is not shown explicitly in (129.17), but can be seen from the fact that the integrand in (129.17) has poles at $\eta = n\pi/a$ ($n = 1, 2, \ldots$), and the integral as written above has, strictly speaking, no meaning. Hence (129.17) can essentially be used only to derive the terms of the asymptotic series (see below) in powers of a by a formal expansion of $\cot a$.

The integral (129.17) can be given a mathematical meaning by passing round the poles in the complex η-plane. Then L', and therefore the energy density W', have an imaginary part. Since the energy is complex, there are quasi-stationary states.‡ In the present case, the stationarity condition is violated by pair production, and $-2\,\mathrm{im}\,W'$ is the probability w of pair production per unit volume and time; since the small increments of W and L differ only in sign, the probability w, expressed in terms of E and H, is simply

$$w = 2\,\mathrm{im}\,L'. \quad (129.20)$$

This is clearly proportional to $e^{-\pi/a}$ (see (129.22) below). Because $\mathrm{im}\,W'$ is exponentially small when $a \ll 1$, an asymptotic series in powers of a, retaining any finite number of terms, is meaningful.

† This was first derived by W. Heisenberg and H. Euler (1935). The analysis given above makes use also of the principles of a proof suggested by V. F. Weisskopf (1936).

‡ The direction of passage round the poles must be chosen so that $\mathrm{im}\,W' < 0$; this corresponds to the usual rule $m^2 \to m^2 - i0$ (i.e. here $a \to a + i0$).

Let us consider the limiting cases of formula (129.17). In weak fields ($a \ll 1$, $b \ll 1$), the leading terms of the expansion are

$$L' = \frac{m^4}{8\pi^2}\frac{(a^2-b^2)^2+7(ab)^2}{45} = \frac{e^4}{45\times 8\pi^2 m^4}(4\mathcal{F}^2+7\mathcal{G}^2). \tag{129.21}$$

In particular, when $b = 0$ the relative correction is

$$L'/L_0 = \alpha a^2/45\pi.$$

The imaginary part of L' for $a \ll 1$ is obtained from the integral (129.17) by taking half the residue at the pole of the cotangent nearest to the origin, i.e. at $\eta a = \pi - i0$. From (129.20), this gives the probability of pair production by a weak electric field:

$$w = \frac{m^4}{4\pi^3}a^2 e^{-\pi/a}$$

or, in ordinary units,

$$w = \frac{1}{4\pi^3}\left(\frac{eE\hbar}{m^2c^3}\right)^2\frac{mc^2}{\hbar}\left(\frac{mc}{\hbar}\right)^3\exp\left(-\frac{\pi m^2c^3}{|e|\hbar E}\right). \tag{129.22}$$

In a strong magnetic field ($a = 0$, $b \gg 1$), we start from (129.13), written (with $b\eta \to \eta$) as

$$L' = \frac{m^4b^2}{8\pi^2}\int_0^\infty \frac{e^{-\eta/b}}{\eta}\left[\frac{1}{3}-\frac{\eta\coth\eta-1}{\eta^2}\right]d\eta.$$

When $b \gg 1$, the important range in this integral is $1 \ll \eta \ll b$, in which $e^{-\eta/b} \approx 1$ and we can neglect the second term in the brackets, terminating the range of integration (with logarithmic accuracy) at $\eta \approx 1$ and $\eta \approx b$. Then

$$L' = (m^4b^2/24\pi^2)\log b; \tag{129.23}$$

in a more exact result, $\log b$ becomes $\log b - 2.29$. The ratio L'/L_0 is here

$$L'/L_0 \approx (\alpha/3\pi)\log b,$$

from which we see that the radiative corrections to the field equations may become of relative order unity only in exponentially strong fields:

$$H \sim (m^2/|e|)e^{3\pi/\alpha}. \tag{129.24}$$

The corrections calculated above are, nevertheless, meaningful: they remove the linearity of Maxwell's equations, and thus lead to effects which are in principle observable (e.g. scattering of light by light or in an external field).

The relation between the fields **E** and **H** and the potentials **A** and ϕ remains, by definition, as before (129.6), and there is therefore also no change in the first pair of Maxwell's equations:

$$\operatorname{div} \mathbf{H} = 0, \quad \operatorname{curl} \mathbf{E} = -\frac{\partial \mathbf{H}}{\partial t}. \tag{129.25}$$

The second pair of equations are obtained by varying the action

$$S = \int (L_0 + L')\, d^4x$$

with respect to **A** and ϕ, and can be written

$$\operatorname{curl} (\mathbf{H} - 4\pi \mathbf{M}) = \frac{\partial}{\partial t} (\mathbf{E} + 4\pi \mathbf{P}), \tag{129.26}$$

$$\operatorname{div} (\mathbf{E} + 4\pi \mathbf{P}) = 0, \tag{129.27}$$

with the notation

$$\mathbf{P} = \partial L'/\partial \mathbf{E}, \quad \mathbf{M} = \partial L'/\partial \mathbf{H}. \tag{129.28}$$

Equations (129.25)–(129.27) agree in form with the macroscopic Maxwell's equations for a field in matter.† Hence we see that **P** and **M** signify the electric and magnetic polarization vectors of the vacuum.

Note that **P** and **M** are zero for the field of a plane wave, where both invariants $\mathbf{E}^2 - \mathbf{H}^2$ and $\mathbf{E} \cdot \mathbf{H}$ are zero. For a plane wave, therefore, the non-linear corrections are zero in a vacuum.

Lastly, let us consider the conditions for the above formulae to be valid. If the fields are to be regarded as constant, their relative changes over distances or times of the order of $1/m$ must be small; this ensures that the corrections to L_0 arising from the derivatives are small in comparison with L_0 itself. For instance, if the field is only time-dependent, this gives the obvious condition

$$\omega \ll m. \tag{129.29}$$

For a weak field, however, there is also a more stringent condition. This occurs because the fourth-order term (129.21) must be much larger than the correction to L_0 quadratic in the derivatives; otherwise, the fourth-order term would have no meaning. For example, in an electric field depending only on the time, this leads to the condition

$$\omega \ll m|e|E/m^2, \tag{129.30}$$

which is more stringent than (129.29).

† In making the comparison, it must be remembered that in macroscopic electrodynamics the mean value of the magnetic field is denoted by **B**, not by **H** as here.

The condition (129.30) does not arise, however, in the problem of photon–photon scattering considered in the last part of §128. There, we are concerned from the start with a four-photon process only, described by the fourth-order terms in the Lagrangian, and the relative magnitude of the other terms in L' is irrelevant. It is therefore sufficient if the condition (129.29) is satisfied.

PROBLEMS

PROBLEM 1. Determine the correction to the field of a small stationary charge e_1 due to the non-linearity of Maxwell's equations.

SOLUTION. For $\mathbf{H} = 0$, (129.21) gives

$$\mathbf{P} = \partial L'/\partial \mathbf{E}$$
$$= (\alpha^2/90\pi^2 m^4)\, E^2\mathbf{E}. \tag{1}$$

In the case of central symmetry, (129.27) gives

$$(E + 4\pi P)r^2 = \text{constant} = e_1, \tag{2}$$

the value of the constant being obtained from the condition that as $r \to \infty$ the field is the Coulomb field of the charge e_1. An approximate solution of (2) is

$$E = (e_1/r^2)(1 - 2\alpha^2 e_1^2/45\pi m^4 r^4),$$

or

$$\Phi = (e_1/r)(1 - 2\alpha^2 e_1^2/225\pi m^4 r^4). \tag{3}$$

The correction in (3) that is non-linear in e_1 is to be distinguished from the linear correction in (114.6), due ultimately to the non-uniformity of the Coulomb field. The correction (3) is of a higher order in α, but decreases more slowly with increasing distance and increases more rapidly with e_1.

PROBLEM 2. Estimate directly the probability of pair production in a weak uniform constant electric field in the quasi-classical approximation, to exponential accuracy (F. Sauter, 1931).

SOLUTION. The motion is quasi-classical in a weak field $\mathbf{E}$ (which has a slowly varying potential $\phi = -\mathbf{E}\cdot\mathbf{r} = -Ez$). Since the reaction amplitude contains the wave function of the final positron as the initial "negative-frequency" function, pair production may be regarded as a transition of an electron from a "negative-frequency" to a "positive-frequency" state. In the former state, with the field present, the quasi-classical momentum is determined by the equation

$$\varepsilon = -\sqrt{[p^2(z) + m^2]} + |e|\, Ez, \tag{1}$$

and in the latter state by

$$\varepsilon = +\sqrt{[p^2(z) + m^2]} + |e|\, Ez. \tag{2}$$

The change from the first to the second state implies a passage through a potential barrier (the region of imaginary $p(z)$), which separates the regions where the functions (1) and (2) apply with real $p(z)$ for a given ε. The boundaries z_1 and z_2 of this barrier occur at $p(z) = 0$, i.e.

$$\varepsilon = -m + |e|\, Ez_1, \quad \varepsilon = +m + |e|\, Ez_2.$$

The probability of passage through a quasi-classical barrier is

$$w \propto \exp\left(-2\int_{z_2}^{z_1} |p(z)|\, dz\right)$$
$$= \exp\left(-4\frac{m^2}{eE}\int_0^1 \sqrt{(1-\xi^2)}\, d\xi\right),$$

whence

$$w \propto \exp(-\pi m^2/|e|E),$$

in agreement with (129.22).

§ 130. Photon splitting in a magnetic field

The non-linear corrections in the electromagnetic field equations give rise to a number of specific effects in photon propagation in external fields.

In order to put these equations in a more familiar form (cf. the last footnote), we shall denote the electric and magnetic fields in this section by **E** and **B**; **D** and **H** will denote the quantities

$$\mathbf{D} = \mathbf{E} + 4\pi\mathbf{P}, \quad \mathbf{H} = \mathbf{B} - 4\pi\mathbf{M}, \quad \mathbf{P} = \partial L'/\partial\mathbf{E}, \quad \mathbf{M} = \partial L'/\partial\mathbf{B}.$$

Equations (129.25)–(129.27) then become

$$\left.\begin{aligned} &\operatorname{div}\mathbf{B} = 0, \quad \operatorname{curl}\mathbf{E} = -\partial\mathbf{B}/\partial t, \\ &\operatorname{div}\mathbf{D} = 0, \quad \operatorname{curl}\mathbf{H} = \partial\mathbf{D}/\partial t. \end{aligned}\right\} \tag{130.1}$$

Let us consider photon propagation in a constant uniform magnetic field $\mathbf{B}_0$. Denoting by a prime the quantities which relate to the weak field of the electromagnetic wave, we have for these the equations

$$\left.\begin{aligned} &\mathbf{k}\times\mathbf{H}' = -\omega\mathbf{D}', \quad \mathbf{k}\times\mathbf{E}' = \omega\mathbf{B}', \\ &\mathbf{k}\cdot\mathbf{B}' = 0, \quad \mathbf{k}\cdot\mathbf{D}' = 0, \end{aligned}\right\} \tag{130.2}$$

with

$$D'_i = \varepsilon_{ik}E'_k, \quad B'_i = \mu_{ik}H'_k; \tag{130.3}$$

the vacuum permittivity and permeability tensors are functions of the external field $\mathbf{B}_0$. Assuming this field to be so weak that $|e|B_0/m^2 \ll 1$, we find from the Lagrangian (129.21)

$$\left.\begin{aligned} \varepsilon_{ik} &= \delta_{ik} + \frac{2e^4}{45m^4}B_0^2\left(-\delta_{ik} + \tfrac{7}{2}b_ib_k\right), \\ \mu_{ik} &= \delta_{ik} + \frac{2e^4}{45m^4}B_0^2\left(\delta_{ik} + 2b_ib_k\right), \end{aligned}\right\} \tag{130.4}$$

where $\mathbf{b} = \mathbf{B}_0/B_0$.

The photon frequency is assumed so small that $\omega \ll m$ (129.29). However, the structure of the tensors ε_{ik} and μ_{ik} does not depend on this assumption; it follows from the invariance of quantum electrodynamics under spatial inversion and charge conjugation. The first of these prevents the occurrence in $\mathbf{D}'$ of terms having the form constant $\times\,\mathbf{B}'$ or constant $\times\,\mathbf{B}_0(\mathbf{B}_0\cdot\mathbf{B}')$ (since inversion changes the sign of **E** and **D** but leaves **H** and **B** unchanged); the second prevents the occurrence in ε_{ik} and μ_{ik} of terms antisymmetric and odd in $\mathbf{B}_0$, of the form $e_{ikl}B_{0l}$ (since charge conjugation changes the sign of all fields).

Since the problem under consideration has a distinctive plane, namely the **kb**-plane, it is reasonable to take the linear polarizations in and normal to this plane as the two independent polarizations of the photon. The subscripts $\perp$ and $\|$ will denote polarizations in which the vector $\mathbf{B}'$ is respectively perpendicular to the **kb**-plane and in that plane.

For perpendicular polarization, the vector $\mathbf{H}'$ is, like $\mathbf{B}'$, at right angles to the **kb**-plane:

$$\mathbf{B}' = \left(1 + \frac{2e^4}{45m^4} B_0^2\right)\mathbf{H}'.$$

The vectors $\mathbf{E}'$ and $\mathbf{D}'$ are in that plane. Then, from equations (130.2), we obtain the photon dispersion relation $k = n_\perp \omega$, with the "refractive index" (in ordinary units)

$$n_\perp = 1 + \frac{7e^4\hbar}{90m^4c^7} B_0^2 \sin^2\theta, \tag{130.5}$$

where θ is the angle between **k** and $\mathbf{B}_0$.†

In the second case, $\mathbf{B}'$ and $\mathbf{H}'$ are in the **kb**-plane, $\mathbf{E}'$ and $\mathbf{D}'$ perpendicular to it. The refractive index is found to be

$$n_\| = 1 + \frac{2e^4\hbar}{45m^4c^7} B_0^2 \sin^2\theta. \tag{130.6}$$

Note that $n_\perp \geqslant n_\|$. The equality occurs when $\theta = 0$, $n_\perp = n_\| = 1$.

The most interesting manifestation of the non-linearity of Maxwell's equations with radiative corrections is the splitting of a photon into two in an external magnetic field (S. L. Adler, J. N. Bahcall, C. G. Callan and M. N. Rosenbluth, 1970).

In a constant uniform field, this process occurs with conservation of energy and momentum.‡ In the decay of a photon **k** into photons $\mathbf{k}_1$ and $\mathbf{k}_2$, we have

$$\omega(\mathbf{k}) = \omega(\mathbf{k}_1) + \omega(\mathbf{k}_2), \quad \mathbf{k}_1 + \mathbf{k}_2 = \mathbf{k}. \tag{130.7}$$

For photons in a vacuum, in the absence of external fields, $\omega = k$ and the equations (130.7) can be satisfied only for three photons moving in the same direction. In that case, however, the decay is rigorously forbidden by the invariance under charge conjugation: Furry's theorem (§79) shows that the sum of diagrams with three photon free ends is zero.

† Expressing $\mathbf{B}'$ in terms of $\mathbf{H}'$ in the second equation (130.2), we substitute $\mathbf{H}'$ from there in the first equation, and then take the projection of the latter on the direction of **b**. The product $\mathbf{k}\cdot\mathbf{E}'$ is expressed in terms of $\mathbf{b}\cdot\mathbf{E}'$ by means of the equation $\mathbf{k}\cdot\mathbf{D}' = 0$.

‡ The conservation of momentum is due to the spatial uniformity of the field, but of course occurs only for processes involving uncharged particles. The Lagrangian for charged particles contains not only the fields but also the field potentials, which depend on the coordinates even in a uniform field.

The presence of an external field makes the decay of the photon possible; this decay is represented by diagrams with three photon ends and one or more external-field lines. The possibility is, however, dependent on the nature of the photon polarization. The dependence may be deduced from the conservation laws (130.7) and the change in the photon dispersion relation in a magnetic field.

The dispersion relation may be written

$$\omega = k + \beta(\mathbf{k}), \tag{130.8}$$

where $\beta(\mathbf{k})$ is an increment that is small (in a weak field). Its presence makes possible, in principle, the fulfilment of equations (130.7) for momenta $\mathbf{k}_1$ and $\mathbf{k}_2$ lying in a certain narrow cone near the direction of $\mathbf{k}$. Since the directions of all three vectors $\mathbf{k}$, $\mathbf{k}_1$, $\mathbf{k}_2$ are close together, they can all be regarded as parallel to $\mathbf{k}$ in the small terms $\beta(\mathbf{k})$, and we can take $k_1 + k_2 = k$. The law of conservation of energy then becomes

$$\beta(\boldsymbol{\kappa}k) - \beta_1(\boldsymbol{\kappa}k_1) - \beta_2(\boldsymbol{\kappa}k - \boldsymbol{\kappa}k_1) = k_1 + |\mathbf{k} - \mathbf{k}_1| - k$$

(where $\boldsymbol{\kappa} = \mathbf{k}/k$); since the dispersion relation depends on the polarization of the photon, the functions β, β_1, β_2 may be different. Since

$$|\mathbf{k} - \mathbf{k}_1| = [(k - k_1)^2 + 2kk_1(1 - \cos\vartheta)]^{1/2} \approx k - k_1 + \frac{kk_1}{2(k - k_1)}\vartheta^2,$$

where ϑ is the small angle between $\mathbf{k}$ and $\mathbf{k}_1$, we have

$$\beta(\boldsymbol{\kappa}k) - \beta_1(\boldsymbol{\kappa}k_1) - \beta_2(\boldsymbol{\kappa}k_1) = kk_1\vartheta^2/2(k - k_1) > 0. \tag{130.9}$$

This inequality specifies the properties of the dispersion relation that are necessary for decay.

For frequencies $\omega \ll m$, the dispersion relation is given by (130.5) and (130.6), so that $\beta(\mathbf{k}) \approx -k[n(\boldsymbol{\kappa}) - 1]$, where the function $n(\boldsymbol{\kappa})$ depends on the direction of the vector $\mathbf{k}$ but not on its magnitude. Then we must have

$$k_1 n_1(\boldsymbol{\kappa}) + (k - k_1)n_2(\boldsymbol{\kappa}) - kn(\boldsymbol{\kappa}) > 0. \tag{130.10}$$

Since $n_\perp > n_\parallel$, this condition immediately excludes the decays

$$\gamma_\perp \to \gamma_\parallel + \gamma_\parallel, \quad \gamma_\perp \to \gamma_\parallel + \gamma_\perp,$$

where γ denotes a photon, and $\perp$ and $\parallel$ correspond to the two polarizations defined above.†

For the decays

$$\gamma_\perp \to \gamma_\perp + \gamma_\perp, \quad \gamma_\parallel \to \gamma_\parallel + \gamma_\parallel,$$

† Numerical calculations show that $n_\perp > n_\parallel$ is true not only when $\omega \ll m$ and (130.5) and (130.6) are valid, but for all $\omega < 2m$, the threshold for pair production by the photon.

the left-hand side of (130.10) is zero, since the functions n, n_1, n_2 are the same. To solve the problem in this case, we have to take into account the dependence of the refractive index on k which appears as ω increases. The required inequality is

$$k_1 n(\boldsymbol{\kappa}, k_1) + (k - k_1) n(\boldsymbol{\kappa}, k - k_1) - kn(\boldsymbol{\kappa}, k) > 0.$$

It can be shown by general arguments that $n(\boldsymbol{\kappa}, k)$ is an increasing function of k, and so this inequality cannot be satisfied, so that the above decays also are impossible: replacing $n(k - k_1)$ and $n(k_1)$ by $n(k)$ will certainly increase the sum, and the result of these changes is a sum equal to zero. This conclusion applies to any transparent media, and follows from Kramers and Kronig's formula for the refractive index (see *ECM*, §64). In the present case, the external field is a "transparent medium" for photons of all frequencies $\omega < 2m$, up to the pair production threshold, i.e. the photon absorption threshold.

Thus the only decay processes allowed are

$$\gamma_{\parallel} \to \gamma_{\perp} + \gamma_{\perp}, \tag{130.11}$$

$$\gamma_{\parallel} \to \gamma_{\parallel} + \gamma_{\perp}. \tag{130.12}$$

It has already been noted that the momenta $\mathbf{k}_1$ and $\mathbf{k}_2$ are at small angles ϑ to the initial photon momentum $\mathbf{k}$. If these angles are neglected, i.e. if the momenta of all the photons are assumed to be parallel (the collinear approximation), then the decay (130.12) is impossible, as may be shown in the following way.

Similarly to (127.4), we express the decay amplitude as

$$M_{fi} = M_{\lambda\mu\nu} e^{\lambda} e_1^{\mu *} e_2^{\nu *},$$

where e, e_1, e_2 are the photon polarization 4-vectors, defined as usual by their 4-potentials A. With the three-dimensional potential gauge, $e = (0, \mathbf{e})$, we can rewrite this as

$$M_{fi} = M_{ikl} e_i e_{1k}^* e_{2l}^*.$$

Two independent polarizations are defined by the unit vectors†

$$\mathbf{e}_{\parallel} \| \mathbf{k} \times \mathbf{b}, \quad \mathbf{e}_{\perp} \| \mathbf{k} \times (\mathbf{k} \times \mathbf{b}). \tag{130.13}$$

It is easy to see that, in the expansion

$$M_{ikl} = \sum_{\lambda, \lambda_1, \lambda_2} M_{\lambda\lambda_1\lambda_2} e_i^{(\lambda)*} e_k^{(\lambda_1)} e_l^{(\lambda_2)},$$

where λ, λ_1, λ_2 take the values $\perp$ and $\parallel$ (cf. (127.9)), the vectors $\mathbf{e}_{\perp}$ must occur an

† The suffixes $\parallel$ and $\perp$ refer to the polarizations defined above. It must be remembered that the unit vectors $\mathbf{e}$ determine the direction of the vector potential $\mathbf{A}$ (and therefore of the field $\mathbf{E}'$) and are perpendicular to $\mathbf{B}'$.

even number of times (0 or 2) in each term. The amplitude M_{fi} is invariant under the CP transformation, and since the potentials $\mathbf{A}$ (and therefore $\mathbf{e}$) are CP-invariant, so must be the tensor M_{ikl}. Under the CP transformation $\mathbf{e}_\parallel \to \mathbf{e}_\parallel$, $\mathbf{e}_\perp \to -\mathbf{e}_\perp$; charge conjugation changes the sign of $\mathbf{b}$, inversion changes that of $\mathbf{k}$ while leaving the axial vector $\mathbf{b}$ unaltered. Hence, if the vector $\mathbf{e}_\perp$ occurs once in any term of the expansion, the corresponding scalar $M_{\lambda\lambda_1\lambda_2}$ must be CP-odd. But it is impossible to construct a CP-odd scalar from only two vectors (in the collinear approximation) $\mathbf{k} = \mathbf{k}_1 = \mathbf{k}_2$ and $\mathbf{b}$, both of which change sign under the CP transformation. This proves the above statement.

The decay (130.12) is therefore forbidden in the collinear approximation. A more detailed analysis shows that the ratio of the amplitude of this process to that of the decay (130.11) allowed in the collinear approximation,

$$\frac{M_{\parallel\perp,\parallel}}{M_{\perp\perp,\parallel}} \sim \vartheta^2 \sim \alpha(B_0/B_{\mathrm{cr}})^2, \tag{130.14}$$

where

$$B_{\mathrm{cr}} = m^2/|e| \quad (= m^2c^3/|e|\hbar = 4.4\times 10^{13}\,\mathrm{G});$$

the angles ϑ are estimated from (130.9) as $\vartheta^2 \sim n_\perp - n_\parallel$.

The fact that the only possible decay (in the principal approximation) is $\gamma_\parallel \to \gamma_\perp + \gamma_\perp$ implies that $\perp$ polarization is eventually established in an unpolarized photon beam propagating in a magnetic field.

Let us now calculate the decay amplitude $M_{fi} \equiv M_{\perp\perp,\parallel}$ by perturbation theory, assuming that $B_0 \ll B_{\mathrm{cr}}$.

The first non-vanishing Feynman diagrams (with respect to α, and with respect to the external field) are of the form

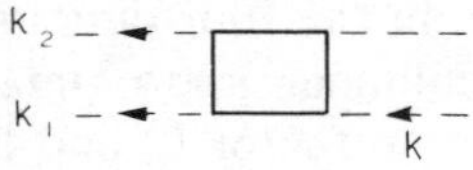

(130.15)

with all possible permutations of the ends, three ends corresponding to photons and one to the external field. In the collinear approximation, however, the amplitude corresponding to these diagrams is zero. For, as a result of gauge invariance, the external field can appear in the amplitude of the process only as a 4-tensor of its field strengths $F_{\mu\nu}$, and the photon polarization 4-vectors only in the antisymmetric combinations

$$f_{\mu\nu} = k_\mu e_\nu - k_\nu e_\mu$$

with the wave 4-vectors. The final expression for the amplitude is constructed from the external field tensor $F_{\mu\nu}$, the tensors $f_{\mu\nu}$, $f_{1\mu\nu}$ and $f_{2\mu\nu}$ of the three photons and their wave 4-vectors k_μ, $k_{1\mu}$, $k_{2\mu}$; it must be linear in each of the tensors $f_{\mu\nu}$, and for the diagrams (130.15) it must be linear in $F_{\mu\nu}$ also. In the collinear approximation the 4-vectors k_1 and k_2 reduce to k: $k_1 = k\omega_1/\omega$, $k_2 = k\omega_2/\omega$. Under these conditions, any scalar product formed as stated above is identically zero: we can easily show that such a product will contain at least one zero factor k^2 or ke.

In the collinear approximation, therefore, the first non-zero contribution to the decay amplitude comes from hexagonal diagrams of the form

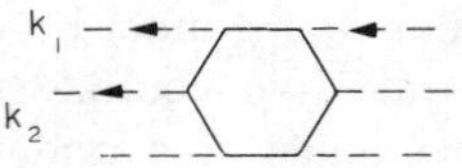

(130.16)

with three external-field lines.† The amplitude corresponding to such diagrams involves three factors $F_{\mu\nu}$. Scalar products of this kind need not be zero, but all non-zero products contain the photon wave vectors only through the tensors $f_{\mu\nu}$: it is easy to see that the addition of further factors k will lead to the presence of zero factors k^2 or ke in the products. The components of the tensor $f_{\mu\nu}$ are the same as those of the photon fields $\mathbf{E}'$ and $\mathbf{B}'$. This means that, if the decay amplitude corresponding to the diagrams (130.16) is represented as the matrix element of an operator, then that operator, expressed in terms of the photon field operators, is independent of the photon frequencies. Hence it follows in turn that the calculation of the scattering amplitude corresponding to the diagram (130.16), by means of the Lagrangian (129.17), gives the correct answer without restriction to the case $\omega \ll m$.

It has been shown at the end of §127 how the interaction Hamiltonian is obtained from the Lagrangian L found in §129. We now have a process involving three photons, and the corresponding interaction operator is found from the terms in the expansion of L which contain products of three photon fields $\mathbf{E}'$ and $\mathbf{B}'$. Here we need consider only the term

$$(\mathbf{B}' \cdot \mathbf{B}_0)(\mathbf{E}' \cdot \mathbf{B}_0)^2, \qquad (130.17)$$

in which each of the vectors $\mathbf{B}'$ and $\mathbf{E}'$ appears as a scalar product with $\mathbf{B}_0$: the products $\mathbf{E}'^2$, $\mathbf{B}'^2$ and $\mathbf{E}' \cdot \mathbf{B}'$ arise, in the four-dimensional notation, from scalars of the form $f_{\mu\nu}f^{\mu\nu}$, which in the collinear approximation are identically zero. The selection of the term containing one factor $\mathbf{B}'$ and two factors $\mathbf{E}'$ is made because the process under consideration involves one $\parallel$ photon and two $\perp$ photons; for the former, the field $\mathbf{B}'$ has a component along $\mathbf{B}_0$, and for the latter the field $\mathbf{E}'$ has such a component.

The Lagrangian L is expressed in terms of the invariants $\mathcal{F} = \frac{1}{2}(\mathbf{B}^2 - \mathbf{E}^2)$ and $\mathcal{G} = \mathbf{E} \cdot \mathbf{B}$. The required term in the expansion comes from a term proportional to $\mathcal{F}\mathcal{G}^2$. A calculation using (129.17) gives for this term

$$-\frac{13e^6}{630\pi^2 m^8}\mathcal{F}\mathcal{G}^2.$$

Putting $\mathbf{B} = \mathbf{B}_0 + \mathbf{B}'$, $\mathbf{E} = \mathbf{E}'$, and taking the term $\mathbf{B}_0 \cdot \mathbf{B}'$ from $\mathcal{F}$ and $\mathbf{B}_0 \cdot \mathbf{E}'$ from $\mathcal{G}$, we get the required expansion term having the form (130.17). Thus the three-photon

† The corrections arising from the inclusion of non-collinearity in the diagrams (130.15) would give a contribution to the amplitude that is of the next higher order in α relative to that from (130.16).

interaction operator for the decay $\gamma_{\parallel} \to \gamma_{1\perp} + \gamma_{2\perp}$ is

$$\hat{V}^{(3)} = \frac{13e^6}{315\pi^2 m^8} \int (\mathbf{B}_0 \cdot \hat{\mathbf{E}}_1')(\mathbf{B}_0 \cdot \hat{\mathbf{E}}_2')(\mathbf{B}_0 \cdot \hat{\mathbf{B}}')\, d^3x, \tag{130.18}$$

where

$$\hat{\mathbf{B}}' = i\sqrt{(4\pi)}\mathbf{k} \times \mathbf{e}_{\parallel} e^{i(\mathbf{k}\cdot\mathbf{r}-\omega t)} \hat{c}_{\mathbf{k}\parallel},$$

$$\hat{\mathbf{E}}_1' = -i\sqrt{(4\pi)}\omega_1 \mathbf{e}_{\perp}\, e^{i(\mathbf{k}_1\cdot\mathbf{r}-\omega_1 t)} \hat{c}^+_{\mathbf{k}_1\perp},$$

and similarly for $\hat{\mathbf{E}}_2'$; cf. (127.26), (127.27).†

According to the rules given in §64, the decay amplitude M_{fi} is calculated from the definition

$$S_{fi} = \left\langle f \right| \int \hat{V}^{(3)}\, dt \left| i \right\rangle = -i(2\pi)^4 \delta^{(4)}(k - k_1 - k_2) M_{fi},$$

and is

$$M_{fi} = -i \frac{13e^6}{315\pi^2 m^8} (4\pi)^{3/2} \omega\omega_1\omega_2 B_0^3 \sin^3\theta,$$

θ being the angle between $\mathbf{k}$ and $\mathbf{B}_0$. The decay probability per unit time is (see (64.11))

$$dw = (2\pi)^4 \delta(\mathbf{k} - \mathbf{k}_1 - \mathbf{k}_2)\delta(\omega - \omega_1 - \omega_2)|M_{fi}|^2 \frac{d^3k_1 d^3k_2}{2\cdot 2\omega\cdot 2\omega_1 \cdot 2\omega_2 \cdot (2\pi)^6};$$

the extra factor $\frac{1}{2}$ takes account of the decrease in the phase volume due to the identity of the two final photons. The first delta function is eliminated by integration over d^3k_2. To eliminate the second we note that, if dispersion is neglected,

$$\begin{aligned}\omega - \omega_1 - \omega_2 &= k - k_1 - |\mathbf{k} - \mathbf{k}_1| \\ &\approx -\frac{kk_1}{k - k_1}(1 - \cos\vartheta_1),\end{aligned}$$

and therefore‡

$$\int_0^{\omega}\int_0^1 \omega\omega_1\omega_2 \delta(\omega - \omega_1 - \omega_2)\, d\cos\vartheta_1 \cdot 2\pi\omega_1^2\, d\omega_1$$

$$= 2\pi \int_0^{\omega} \omega_1^2(\omega - \omega_1)^2 d\omega_1 = \pi\omega^5/15.$$

† The coefficient in (130.18) is doubled because $\mathbf{E}_1'$ and $\mathbf{E}_2'$ can be taken from either of the two factors $\mathbf{E}'$ in L.

‡ Here it is assumed that, with dispersion taken into account, the argument of the delta function in fact has a zero at some $\cos\vartheta_1 < 1$. Thus a dispersion of this type is necessary for decay to be possible, but the decay probability itself does not depend on the amount of the small dispersion.

We finally have for the total photon decay probability per unit time (in ordinary units)

$$w = \frac{\alpha^3}{15\pi^2}\left(\frac{13}{315}\right)^2 \frac{mc^2}{\hbar}\left(\frac{\hbar\omega}{mc^2}\right)^5\left(\frac{B_0 \sin\theta}{B_{\mathrm{cr}}}\right)^6$$

$$= 0.18\alpha^6 \frac{mc^2}{\hbar}\left(\frac{B_0^2 \sin^2\theta}{8\pi mc^2}\right)^3\left(\frac{\hbar}{mc}\right)^9\left(\frac{\hbar\omega}{mc^2}\right)^5. \qquad (130.19)$$

As already mentioned, the condition $\omega \ll m$ is not necessary for this formula to be valid. Its validity is restricted only by the condition that the terms corresponding to eighth-order diagrams be small. To obtain an estimate, we note that the eighth-order matrix element may contain, for example, a term differing from the sixth-order terms by a dimensionless invariant factor of the form $(eF^{\mu\nu}k_\nu/m^3)^2$. The condition for this to be small is a very weak one:

$$\omega \ll m(m^2/|e|B_0).$$

§ 131. **Calculation of integrals over four-dimensional regions**

We shall now give some rules and formulae that are useful in the calculation of integrals arising in the theory of radiative corrections.

The typical form of integral corresponding to a Feynman diagram is

$$\int \frac{f(k)\, d^4k}{a_1 a_2 \ldots a_n}, \qquad (131.1)$$

where $a_1, a_2, \ldots$ are second-degree polynomials in the 4-vector k, $f(k)$ is a polynomial of some degree n', and the integration is over the whole of four-dimensional k-space.

A convenient method for calculating such integrals, due to Feynman (1949), is based on an initial transformation or parametrization of the integrand by the use of further integrations with respect to auxiliary variables $\xi_1, \xi_2, \ldots$:

$$\frac{1}{a_1 a_2 \ldots a_n} = (n-1)! \int_0^1 d\xi_1 \ldots \int_0^1 d\xi_n \frac{\delta(\xi_1 + \xi_2 + \cdots + \xi_n - 1)}{(a_1\xi_1 + a_2\xi_2 + \cdots + a_n\xi_n)^n}. \qquad (131.2)$$

This transformation replaces the n different quadratics in the denominator by the nth power of a single quadratic polynomial.

Eliminating the delta function by integrating over $d\xi_n$ and using new variables defined by

$$\xi_1 = x_{n-1}, \quad \xi_2 = x_{n-2} - x_{n-1}, \ldots, \xi_{n-1} = x_1 - x_2,$$
$$\xi_1 + \xi_2 + \cdots + \xi_{n-1} = x_1,$$

we get as an equivalent form of (131.2)

$$\frac{1}{a_1 a_2 \dots a_n} = (n-1)! \int_0^1 dx_1 \int_0^{x_1} dx_2 \dots \int_0^{x_{n-2}} dx_{n-1} \times$$

$$\times \frac{1}{[a_1 x_{n-1} + a_2(x_{n-2} - x_{n-1}) + \dots + a_n(1 - x_1)]^n}. \tag{131.3}$$

When $n = 2$, this formula becomes

$$\frac{1}{a_1 a_2} = \int_0^1 \frac{dx}{[a_1 x + a_2(1-x)]^2}, \tag{131.4}$$

as can be easily verified. For any value of n, the formula can be proved by induction: on carrying out the integration over dx_{n-1} in (131.3), we get on the right-hand side the difference of two $(n-2)$-fold integrals of the same form. If the formula is assumed valid for these, we have

$$\frac{1}{a_1 - a_2}\left[\frac{1}{a_2 a_3 \dots a_n} - \frac{1}{a_1 a_3 \dots a_n}\right],$$

which is equal to the left-hand side of (131.3).

By differentiating (131.3) with respect to a_1, a_2, etc., we can derive similar formulae which can be used for the parametrization of integrals whose denominators contain any of the polynomials in powers above the first.

The divergent integrals are regularized by subtracting from them other integrals of similar form. To determine the difference, it may be convenient first to transform the difference of the integrands (each of which is already transformed by means of (131.2)) as follows:

$$\frac{1}{a^n} - \frac{1}{b^n} = -\int_0^1 \frac{n(a-b)\, dz}{[(a-b)z + b]^{n+1}}. \tag{131.5}$$

After the application of (131.3), the four-dimensional integration in (131.1) becomes

$$\int \frac{f(k)\, d^4k}{[(k-l)^2 - \alpha^2]^n}, \tag{131.6}$$

where l is a 4-vector and α^2 is a scalar, both of these depending on the parameters $x_1, \dots, x_{n-1}$; the scalar α^2 will be assumed positive.

If the integral (131.6) converges, we can make the change of variables $k - l \to k$ (a shift of the origin), which gives (with a different function $f(k)$)

$$\int \frac{f(k)\, d^4k}{(k^2 - \alpha^2)^n}, \tag{131.7}$$

the denominator now involving only the square k^2. In the numerator we need only consider scalar functions $f = F(k^2)$: for integrals having numerators of any other form,

$$\int \frac{k^\mu F(k^2)\, d^4k}{(k^2-\alpha^2)^n} = 0, \tag{131.8}$$

$$\int \frac{k^\mu k^\nu F(k^2)\, d^4k}{(k^2-\alpha^2)^n} = \frac{1}{4} g^{\mu\nu} \int \frac{k^2 F(k^2)\, d^4k}{(k^2-\alpha^2)^n}, \tag{131.9}$$

$$\int \frac{k^\mu k^\nu k^\rho k^\sigma F(k^2)\, d^4k}{(k^2-\alpha^2)^n}$$

$$= \frac{1}{24}(g^{\mu\nu}g^{\rho\sigma} + g^{\mu\rho}g^{\nu\sigma} + g^{\mu\sigma}g^{\nu\rho}) \int \frac{(k^2)^2 F(k^2)\, d^4k}{(k^2-\alpha^2)^n} \tag{131.10}$$

and so on, as is evident from symmetry on integration over all directions of k.

In the original integral (131.1), each of the factors $a_1, a_2, \ldots$ in the denominator has (as a function of k_0) two zeros, which are avoided in the integration over dk_0 according to the general rule (§75). After the transformation to the form (131.7), the $2n$ simple poles of the integrand are replaced by two poles of order n, which are avoided according to the same rule (path C in Fig. 25). By moving the contour of integration as shown by the arrows, it can be converted to the imaginary axis in the k_0 plane (path C'). Thus the variable k_0 is replaced by ik_0' with k_0' real. Then, writing also $\mathbf{k}'$ for $\mathbf{k}$, we have

$$k^2 = k_0^2 - \mathbf{k}^2 \to -(k_0'^2 + \mathbf{k}'^2) = -k'^2, \tag{131.11}$$

where k' is a 4-vector in the Euclidean metric; and

$$d^4k \to id^4k' = ik'^2 d(\tfrac{1}{2}k'^2)\, d\Omega,$$

where $d\Omega$ is an element of four-dimensional solid angle. The integration over $d\Omega$ gives $2\pi^2$ (see *Fields*, §111), and so

$$d^4k \to i\pi^2 k'^2 d(k'^2). \tag{131.12}$$

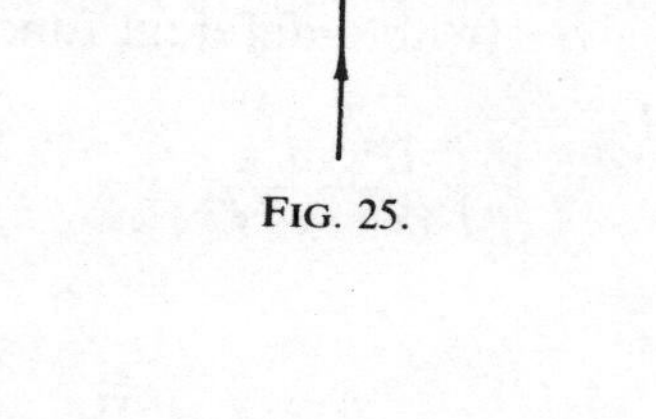

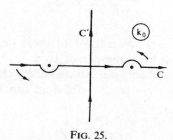

FIG. 25.

Putting $k'^2 = z$, we have finally

$$\int \frac{F(k^2)\, d^4k}{(k^2-\alpha^2)^n} = (-1)^n i\pi^2 \int_0^\infty \frac{F(-z) z\, dz}{(z+\alpha^2)^n}. \tag{131.13}$$

In particular,

$$\int \frac{d^4k}{(k^2-\alpha^2)^n} = \frac{(-1)^n i\pi^2}{\alpha^{2(n-2)}(n-1)(n-2)}. \tag{131.14}$$

The logarithmically divergent part of the integrals (131.7) can be separated as

$$\int \frac{d^4k}{[(k-l)^2-\alpha^2]^2}. \tag{131.15}$$

It is easily seen that in an integral of this type we can replace k by $k+l$: the difference is

$$\int \left\{\frac{1}{[(k-l)^2-\alpha^2]^2} - \frac{1}{(k^2-\alpha^2)^2}\right\} d^4k$$

and is a convergent integral, so that the change $k \to k+l$ is certainly permissible in this integral. On doing so and also changing the sign of k, we get the same quantity with the opposite sign, and it must therefore be zero.

A linearly divergent integral would have to have the form

$$\int \frac{k^\mu\, d^4k}{[(k-l)^2-\alpha^2]^2}, \tag{131.16}$$

but in fact such an integral is only logarithmically divergent: the integrand tends asymptotically to $k^\mu/(k^2)^2$ as $k \to \infty$ and gives zero on averaging over directions. The change of origin, however, adds a constant to the integral (131.16). This can be shown for the infinitesimal change $k \to k + \delta l$, by calculating the difference

$$\Delta^\mu = \int \left\{\frac{k^\mu}{[(k-\delta l)^2-\alpha^2]^2} - \frac{k^\mu+\delta l^\mu}{(k^2-\alpha^2)^2}\right\} d^4k. \tag{131.17}$$

As far as the first order in δl,

$$\Delta^\mu = \int \left\{\frac{4k^\mu(k\,\delta l)}{(k^2-\alpha^2)^3} - \frac{\delta l^\mu}{(k^2-\alpha^2)^2}\right\} d^4k.$$

In the first term, averaging over directions replaces the numerator by $k^2 \delta l^\mu$ (cf. (131.9)), and hence†

$$\Delta^\mu = \alpha^2\, \delta l^\mu \int \frac{d^4k}{(k^2-\alpha^2)^3} = -\tfrac{1}{2} i\pi^2\, \delta l^\mu. \tag{131.18}$$

† The corresponding result for finite l can be obtained by more laborious calculations.

In the final expressions for the radiative corrections there is frequently a transcendental function defined by the integral

$$F(\xi) = \int_0^\xi \frac{\log(1+x)}{x}\, dx, \tag{131.19}$$

sometimes called Spence's function. Some of its properties (given here for reference) are

$$F(\xi) + F(1/\xi) = \tfrac{1}{6}\pi^2 + \tfrac{1}{2}\log^2 \xi, \tag{131.20}$$

$$F(-\xi) + F(-1+\xi) = -\tfrac{1}{6}\pi^2 + \log \xi \log(1-\xi), \tag{131.21}$$

$$F(1) = \tfrac{1}{12}\pi^2, \quad F(-1) = -\tfrac{1}{6}\pi^2. \tag{131.22}$$

The expansion for small ξ is

$$F(\xi) = \xi - \tfrac{1}{4}\xi^2 + \tfrac{1}{9}\xi^3 - \tfrac{1}{16}\xi^4 + \cdots. \tag{131.23}$$

CHAPTER XIII

ASYMPTOTIC FORMULAE OF QUANTUM ELECTRODYNAMICS

§ 132. Asymptotic form of the photon propagator for large momenta

THE first-order term (with respect to α) in the expansion of the polarization operator $\mathscr{P}(k^2)$ has been calculated in §113, and it was found that for $|k^2| \gg m^2$ this term is, with logarithmic accuracy,

$$\mathscr{P}(k^2) = \frac{\alpha}{3\pi} k^2 \log \frac{|k^2|}{m^2}. \tag{132.1}$$

It was also mentioned that the derivation of this formula (as a first-approximation correction to the propagator $4\pi D^{-1} = k^2$) assumed the condition

$$\frac{\alpha}{3\pi} \log \frac{|k^2|}{m^2} \ll 1, \tag{132.2}$$

which placed an upper limit on the permissible values of $|k^2|$. We shall now show that the expression (132.1) is in fact valid also with the much weaker condition

$$\frac{\alpha}{3\pi} \log \frac{|k^2|}{m^2} \lesssim 1. \tag{132.3}$$

The proof is as follows.† We first note that, although in principle the condition (132.3) allows contributions to $\mathscr{P}(k^2)$ from the terms of all orders (in α) in the perturbation-theory series, in every order n only the terms $\sim \alpha^n \log^n(|k|^2/m^2)$ need be considered, which contain the large logarithm in the same power as α; the terms containing lower powers of the logarithm are certainly small, since $\alpha \ll 1$.

The analysis of the perturbation-theory series for $\mathscr{P}$ can be reduced to that of the series $\mathscr{G}$ and Γ^μ, using Dyson's equation

$$\mathscr{P}(k^2) = i \cdot \frac{4\pi\alpha}{3} \operatorname{tr} \int \gamma_\mu \mathscr{G}(p+k)\Gamma^\mu(p+k, p; k)\mathscr{G}(p) \frac{d^4p}{(2\pi)^4}; \tag{132.4}$$

see (107.4). Since the function $\mathscr{P}(k^2)$ is gauge-invariant, it can be calculated with any gauge for $\mathscr{G}$ and Γ. The most convenient here is the Landau gauge, in which the

† This formulation and conclusions are due to L. D. Landau, A. A. Abrikosov and I. M. Khalatnikov (1954).

free-photon propagator has the form (76.11):

$$D_{\mu\nu}(k) = \frac{4\pi}{k^2}\left(g_{\mu\nu} - \frac{k_\mu k_\nu}{k^2}\right) \tag{132.5}$$

($D^{(l)} = 0$ in (103.17)). It is found that in this gauge the perturbation-theory series for $\mathcal{G}$ and Γ^μ do not contain any terms with zero power of the logarithms. In (132.4) it is therefore sufficient to substitute the zero-order approximations $\mathcal{G} = G, \Gamma^\mu = \gamma^\mu$, obtaining

$$\mathcal{P}(k^2) = i \cdot \frac{4\pi\alpha}{3} \operatorname{tr} \int \gamma_\mu G(p+k)\gamma^\mu G(p) \frac{d^4p}{(2\pi)^4}. \tag{132.6}$$

This is the Feynman integral corresponding to the diagram (113.1) of the first approximation (with respect to α), and leads (after appropriate renormalization) to (132.1).

In order to prove the foregoing statements, let us first examine the origin of the logarithm in the integral (132.6). It is easily seen that the logarithmic term comes from the range of integration

$$p^2 \gg |k^2| \quad \text{when} \quad |k^2| \gg m^2: \tag{132.7}$$

formally expanding G in powers of $1/\gamma p$, we, have

$$\begin{aligned} G(p) &\approx 1/\gamma p = \gamma p/p^2, \\ G(p-k) &\approx 1/(\gamma p - \gamma k) \\ &\approx \frac{1}{\gamma p} + \frac{1}{\gamma p}\gamma k\frac{1}{\gamma p} + \frac{1}{\gamma p}\gamma k\frac{1}{\gamma p}\gamma k\frac{1}{\gamma p} \\ &= \frac{\gamma p}{p^2} + \frac{(\gamma p)(\gamma k)(\gamma p)}{(p^2)^2} + \\ &\quad + \frac{(\gamma p)(\gamma k)(\gamma p)(\gamma k)(\gamma p)}{(p^2)^3}. \end{aligned}$$

On substitution in (132.6) the first term, which is independent of k, is removed by regularization (in accordance with the condition $\mathcal{P}/k^2 \to 0$ when $k^2 \to 0$); the second term gives zero on integration over the directions of p; the third integral is logarithmically divergent with respect to p^2, and on taking it from $p^2 \sim |k^2|$ (the lower limit of the range (132.7)) to some "cut-off parameter" Λ^2 we get

$$-\frac{\alpha}{3\pi} k^2 \log \frac{\Lambda^2}{|k^2|}. \tag{132.8}$$

In regularizing, we have to subtract from $\mathcal{P}/k^2$ its value for $k^2 = 0$. But, since logarithmic accuracy presupposes the condition $|k^2| \gg m^2$, in calculation to this accuracy the regularization is achieved by subtracting the value for $|k^2| \sim m^2$, and Λ^2 in the argument of the logarithm is replaced by m^2, giving (132.1).

Since the desired corrections to $\mathcal{G}$ and Γ^μ are logarithmic, their inclusion makes these quantities differ from G and γ^μ by slowly varying logarithmic factors. In the exact integral (132.4), therefore, the important range will be the same one (132.7) as in the approximate integral (132.6). However, we cannot simply put $k = 0$ in $\Gamma^\mu(p+k, p; k)$: since the integral is quadratically divergent, its regularization involves also the next two terms in the expansion of $\Gamma^\mu(p+k, p; k)$ in powers of k. Here we shall consider only the corrections to $\Gamma^\mu(p, p; 0)$, which show sufficiently clearly the importance of the choice of gauge and the difference in the nature of the integrals arising from diagrams of different types. It may also be noted that a similar analysis of $\mathcal{G}$ is not needed, since the corrections fo Γ and $\mathcal{G}$ are related by Ward's identity (108.8).

The first-order correction (with respect to α) to $\Gamma(p, p; 0)$ corresponds to the diagram

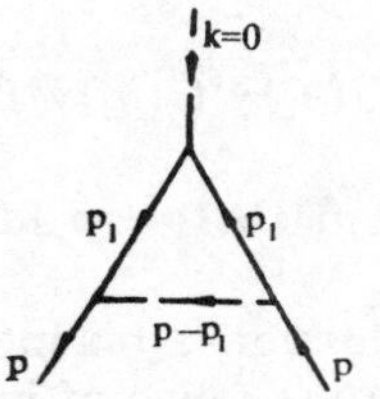

and hence to the integral†

$$\Gamma^{\mu(1)} = -i\alpha \int \gamma^\lambda G(p_1)\gamma^\mu G(p_1)\gamma^\nu D_{\lambda\nu}(p-p_1)d^4p_1/(2\pi)^4. \tag{132.9}$$

In the ordinary gauge,

$$D_{\lambda\nu}(p-p_1) = g_{\lambda\nu}\cdot 4\pi/(p-p_1)^2,$$

and the important range in the integral is $p_1^2 \gg p^2$, where it is logarithmically divergent. On calculating the integral

$$\Gamma^{\mu(1)} \approx -4\pi\alpha i \int \frac{\gamma^\lambda(\gamma p_1)\gamma^\mu(\gamma p_1)\gamma_\lambda}{(p_1^2)^3}\frac{d^4p_1}{(2\pi)^4} \tag{132.10}$$

and regularizing the logarithm, we get

$$\Gamma^{\mu(1)} \approx -\frac{\alpha}{4\pi}\gamma^\mu \log\frac{p^2}{m^2}.$$

In the Landau gauge, (132.10) is replaced by

$$\Gamma^{\mu(1)} \approx -4\pi\alpha i \int \{\gamma^\lambda(\gamma p_1)\gamma^\mu(\gamma p_1)\gamma_\lambda - p_1^2\gamma^\mu\}\, d^4p_1/(p_1^2)^3(2\pi)^4.$$

† To avoid misunderstandings in a comparison with the results of §117, we may note that in §117 the two electron ends of the diagram were assumed to be physical, whereas here we assume that $p^2 \gg |k^2| \gg m^2$, and these two lines are therefore certainly non-physical.

Averaging over the directions of p_1 and reducing the matrices γ, we find that this integral is zero and the logarithmic term in $\Gamma^{\mu(1)}$ disappears.†

In the second-order corrections (with respect to α) we take the diagram

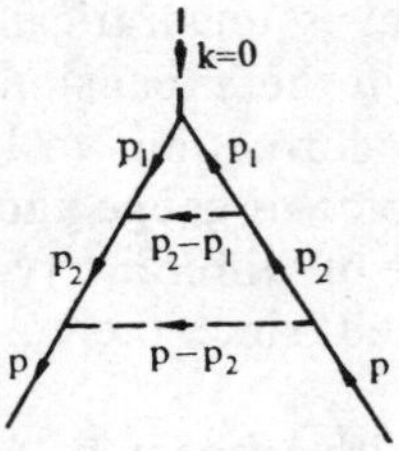

The corresponding integral is

$$\Gamma^{\mu(2)} = -\alpha^2 \int \gamma^\lambda G(p_2)\gamma^\nu G(p_1)\gamma^\mu G(p_1)\gamma^\rho G(p_2)\gamma^\sigma \times$$
$$\times D_{\nu\rho}(p_2 - p_1) D_{\lambda\sigma}(p - p_2) d^4p_1 d^4p_2/(2\pi)^8.$$

In the ordinary gauge of the D functions, this integral includes a term in which the logarithm is squared, arising from the range of integration

$$p_1^2 \gg p_2^2 \gg p^2: \tag{132.11}$$

when p_2 is neglected in the argument of $D_{\nu\rho}(p_2 - p_1)$, the integration over d^4p_1 becomes the same as in (132.9) and gives $\log p_2^2$, and the subsequent integration over d^4p_2 is likewise logarithmic, giving $\log^2(p_2^2/m^2)$. When the Landau gauge is used for the D functions, however, the logarithmic terms disappear from both integrations.

A similar situation occurs for all the other diagrams in the skeleton diagram

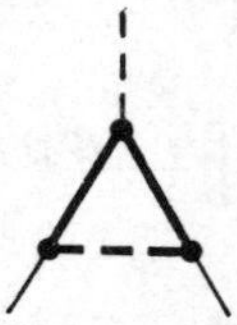

(132.12)

Diagrams of other types, with intersecting photon lines, for example those in the skeleton diagram

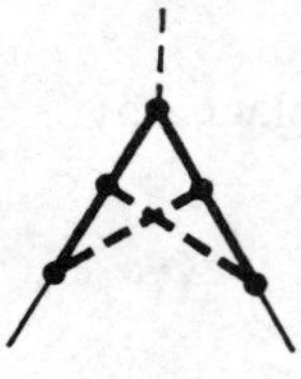

(132.13)

† The corrections to G^{-1} in both gauges, as derived from the correction $\Gamma^{(1)}$ by means of the identity (108.8), are of course in agreement with the results of §119.

(cf. (106.11)), do not in any gauge contain terms involving the necessary power of the logarithm; there is no range of the variables in which the integral reduces to several successive logarithmic integrations.

These arguments, and similar ones for the subsequent terms in the expansion of Γ in powers of k, confirm that in the Landau gauge there are no corrections to $\mathcal{G}$ and Γ involving the necessary powers of the logarithm; thus the expression (132.1) is in fact valid even if the condition (132.3) applies.

The function $\mathcal{D}(k^2)$ which corresponds to the polarization operator (132.1) is

$$\mathcal{D}(k^2) = \frac{4\pi}{k^2}\frac{1}{1-(\alpha/3\pi)\log(|k^2|/m^2)}. \tag{132.14}$$

Because of the condition (132.3), there is no need to expand this expression in powers of α.

§ 133. The relation between unrenormalized and actual charges

The applicability of (132.14) is, however, limited on the side of large $|k^2|$, because the denominator decreases. The derivation of that formula was based on neglecting the diagram (132.13), and others with even more intersecting thick photon lines, in comparison with (132.12). But each such added line brings into the diagram a factor $e^2\mathcal{D}$, with $\mathcal{D}$ the exact propagator. The small parameter is then not $\alpha = e^2$ but

$$\frac{\alpha}{1-(\alpha/3\pi)\log(|k^2|/m^2)} \ll 1. \tag{133.1}$$

When $|k^2|$ increases and this quantity becomes comparable with unity, the small parameter essentially disappears from the theory.

The resulting situation can be more clearly understood if the renormalization in the derivation of (132.14) is made not "en route" but through the use of an "intrinsic" electron charge e_c, which will here be chosen so as to give the correct value of the observed physical charge e (§110). If the integral is "cut off", as was done previously, at some upper limit Λ^2, the intrinsic charge is a function $e_c(\Lambda^2)$ and the limit $\Lambda \to \infty$ must finally be taken.

With this treatment, the polarization operator is

$$\mathcal{P}(k^2) = \frac{e_c^2}{3\pi}k^2 \log\frac{\Lambda^2}{|k^2|}$$

(the expression (132.8) with e_c in place of e), and therefore

$$\mathcal{D}(k^2) = \frac{4\pi}{k^2}\frac{1}{1+(e_c^2/3\pi)\log(\Lambda^2/|k^2|)}. \tag{133.2}$$

Now determining the physical charge e from the condition

$$e_c^2\mathcal{D}(k^2) \to 4\pi e^2/k^2 \quad \text{when} \quad k^2 \to \sim m^2,$$

we have

$$e^2 = \frac{e_c^2}{1+(e_c^2/3\pi)\log(\Lambda^2/m^2)}, \tag{133.3}$$

or

$$e_c^2 = \frac{e^2}{1-(e^2/3\pi)\log(\Lambda^2/m^2)}. \tag{133.4}$$

If we formally take the limit $\Lambda\to\infty$ in (133.3), then $e^2\to 0$ whatever the form of the function $e_c^2(\Lambda)$. This "zeroizing" of the charge means, of course, that no rigorous renormalization is possible. The limit cannot be taken, however, without violating the assumptions made in deriving (133.3). It is seen from (133.4) that, as Λ increases (for a given value of e^2), e_c^2 also increases; and the formulae cease to be valid even when $e_c^2 \sim 1$, since their derivation is based on the assumption that

$$e_c^2 \ll 1 \tag{133.5}$$

as the condition for perturbation theory to be applicable to the "intrinsic" interaction. The failure of the inequality (133.5) to be satisfied as Λ increases is of fundamental significance. It shows that quantum electrodynamics is logically incomplete as a theory that is based on weak interaction. This essentially implies that the existing theory as a whole is logically incomplete, since its entire formalism is based on the possibility of treating the electromagnetic interaction as a weak perturbation. All the quantities calculated by the theory are found as series in powers of e_c^2, which are in fact asymptotic series. In order for them to have a definite meaning when e_c^2 is not small, further arguments would be needed which do not follow from the general principles of the existing theory.

It must also be emphasized, however, that in quantum electrodynamics these difficulties cannot be more than theoretical ones. They arise at enormous energies that are of no practical significance.† We may expect that in reality the electromagnetic interactions will very much sooner be "merged" with the weak and strong interactions, so that pure electrodynamics is no longer meaningful.‡

To conclude this section, we shall show how formulae (133.3) and (133.4) can be derived by simple arguments based on the significance of renormalization and on dimensional reasoning (M. Gell-Mann and F. E. Low, 1954).

Let us consider the square of the unrenormalized charge as a function of the cut-off parameter, $e_c^2(\Lambda^2)$, and define a function d as the ratio of the values of e_c^2 for two different arguments: $e_c^2(\Lambda_2^2) = e_c^2(\Lambda_1^2)\, d$. When $\Lambda_1^2, \Lambda_2^2 \gg m$, the function d is independent of m and, being dimensionless, can depend only on the likewise

† For example, the equation $(\alpha/\pi)\log(\varepsilon^2/m^2) = 1$ is satisfied when $\varepsilon \sim 10^{93}m$.

‡ The opposite situation occurs in theories where the interaction between particles is mediated not by the electromagnetic field but by Yang–Mills fields. The relation between the renormalized and unrenormalized charges in such theories is given by an expression of the type (133.4), but with the opposite sign in the denominator, so that, for a given value of e^2, the unrenormalized charge e_c^2 decreases with increasing Λ. This is called *asymptotic freedom* of the theory. Such a theory is, of course, fundamentally different from the theory with zeroized charge.

dimensionless quantities $e_c^2(\Lambda_1^2)$ and Λ_2^2/Λ_1^2:

$$e_c^2(\Lambda_2^2) = e_c^2(\Lambda_1^2)\, d[e_c^2(\Lambda_1^2),\ \Lambda_2^2/\Lambda_1^2]. \tag{133.6}$$

From this functional relation, we can derive a differential equation, by writing it for infinitesimally close values of Λ_1^2 and Λ_2^2. Denoting Λ_1^2 by ξ and putting $\Lambda_2^2 = \xi + d\xi$, we obtain for $\alpha_c(\xi) \equiv e_c^2(\Lambda_1^2)$ the differential equation

$$d\alpha_c = \phi(\alpha_c) d\xi/\xi, \tag{133.7}$$

with

$$\phi(\alpha_c) = \alpha_c[\partial d(\alpha_c, x)/\partial x]_{x=1}; \tag{133.8}$$

we have used the fact that $d(\alpha_c, 1) \equiv 1$, from the definition (133.6). Integration of (133.7) from $\xi = \Lambda_1^2$ to $\xi = \Lambda_2^2$ gives

$$\log(\Lambda_2^2/\Lambda_1^2) = \int_{e_c^2(\Lambda_1^2)}^{e_c^2(\Lambda_2^2)} d\alpha/\phi(\alpha). \tag{133.9}$$

Throughout the range of integration, e_c^2 is small. We can therefore use for $\phi(\alpha)$ the expression corresponding to the first approximation of perturbation theory. The correction to the unrenormalized charge e_c^2 is $e_c^2 k^2 \mathscr{P}(k^2)$. Taking the first approximation (132.1) for the polarization operator, we find

$$d(\alpha_c, \Lambda_2^2/\Lambda_1^2) = 1 + (\alpha_c/3\pi)\log(\Lambda_2^2/\Lambda_1^2),\ \phi(\alpha_c) = \alpha_c^2/3\pi,$$

and the integration in (133.9) then gives

$$\frac{1}{3\pi}\log\frac{\Lambda_2^2}{\Lambda_1^2} = \frac{1}{e_c^2(\Lambda_1^2)} - \frac{1}{e_c^2(\Lambda_2^2)}. \tag{133.10}$$

As $\Lambda_1^2 \to \sim m^2$, the unrenormalized charge $e_c(\Lambda_1^2)$ tends to the actual charge e, and (133.10) then agrees with (133.3) and (133.4).†

§ 134. Asymptotic form of the scattering amplitudes at high energies

Let us consider the asymptotic form (at high energies) of the amplitudes and cross-sections for two-particle scattering processes $(1+2 \to 3+4)$. For the basic electrodynamic processes in the first non-vanishing approximation (with respect to α), this problem can be solved by means of the specific formulae derived in the preceding chapters, which are valid at all energies. Here, however, we shall discuss

† The *renormalization group method* based on the functional properties of propagators and vertex parts is systematically developed in the book by Bogoliubov and Shirkov cited at the end of § 112.

the question from a more general standpoint, which enables such asymptotic forms to be derived directly.

As in §66, we use the invariants

$$s = (p_1 + p_2)^2,\ t = (p_1 - p_3)^2,\ u = (p_1 - p_4)^2, \tag{134.1}$$

with $p_1 + p_2 = p_3 + p_4$; the notation corresponds to reactions in the s channel, which will be considered here. In the ultra-relativistic case, when the energies are much greater than the particle masses, the energies of the two particles in the centre-of-mass system are approximately equal. We denote by ε the sum of the energies of the colliding particles, and then have, in the centre-of-mass system, $p_1 = (\frac{1}{2}\varepsilon, \mathbf{p}_1)$, $p_2 = (\frac{1}{2}\varepsilon, -\mathbf{p}_1)$, $p_3 = (\frac{1}{2}\varepsilon, \mathbf{p}_3)$, $p_4 = (\frac{1}{2}\varepsilon, -\mathbf{p}_3)$, $\mathbf{p}_1^2 = \mathbf{p}_3^2 = \frac{1}{4}\varepsilon^2$, with

$$s = \varepsilon^2,\ t = -\tfrac{1}{2}s(1 - \cos\theta),\ u = -\tfrac{1}{2}s(1 + \cos\theta), \tag{134.2}$$

θ being the angle between $\mathbf{p}_1$ and $\mathbf{p}_3$.

Let us first consider the asymptotic form of the reaction cross-section for a fixed value of the scattering angle θ. Then all three variables s, t and u are proportional, and tend to infinity together. In the ultra-relativistic case, the particle masses cannot appear in the result, and the only quantity having the dimensions of length is $1/\varepsilon$ $(=\hbar c/\varepsilon)$. Hence it follows from dimensional arguments that the differential cross-section for two-particle reactions decreases with increasing energy in the asymptotic form

$$d\sigma/do \propto 1/s \quad \text{as} \quad s, |t|, |u| \to \infty. \tag{134.3}$$

If the cross-section is related not to the solid-angle element do but to the differential dt, we have, since $do \propto dt/s$,

$$d\sigma/dt \propto 1/s^2. \tag{134.4}$$

The cross-section is expressed in terms of the scattering amplitude (in the ultra-relativistic case) as $d\sigma/do \propto |M_{fi}|^2/s$; see (64.22), (64.23). The law (134.3) therefore means that in the asymptotic limit the scattering amplitude is independent of s:

$$M_{fi} = \text{constant}. \tag{134.5}$$

It is clear from the manner of the derivation that these results apply not only to the first non-vanishing approximation of perturbation theory, but also to higher approximations (those taking account of radiative corrections), if logarithmic factors (of the form $\log s/m^2$) are ignored; the dependence on dimensionless logarithms cannot, of course, be determined from dimensional arguments.†

A different situation arises if s increases with t fixed, i.e. with a fixed square of the momentum transfer. This is scattering through small angles, which decrease as

† The summation of series containing logarithmic corrections may lead to an exponential dependence on the logarithms, which changes the exponent in the power law. This change is, however, small if α is small.

the energy increases:

$$s \to \infty, |t| \sim s\theta^2 = \text{constant}, \theta \sim (|t|/s)^{\frac{1}{2}}. \tag{134.6}$$

In such a case, dimensional arguments allow us to establish only that the combined power of $1/s$ and $1/t$ in $d\sigma/dt$ is 2, and in the amplitude M_{fi} zero.† Hence, to find the part of the cross-section that decreases least rapidly with increasing s, we have to separate the factor that has the greatest power of $1/t$. Such factors arise only if the Feynman diagram can be divided into two parts between the ends 1, 3 and 2, 4 by cutting the lines of virtual particles. The total 4-momentum of such lines is $p_1 - p_3$, and this leads to the factor that depends on $t = (p_1 - p_3)^2$. Thus the asymptotic form of the diagram in the range (134.6) depends on the nature of the possible cuttings of diagrams in the t channel.

Similarly, the asymptotic behaviour in the range

$$s \to \infty, \quad |u| \sim s(\pi - \theta)^2 = \text{constant}, \quad |\pi - \theta| \sim (|u|/s)^{\frac{1}{2}}, \tag{134.7}$$

corresponding to scattering through angles close to π, is governed by the nature of the possible cuttings of diagrams in the u channel, i.e. between the ends 1, 4 and 2, 3.

The simplest example is electron–electron scattering, described by the two diagrams (73.13) and (73.14). The first of these allows cutting in the t channel at the virtual photon line, and it determines the asymptotic form of the scattering amplitude in the range (134.6). The virtual photon line corresponds to a D function that is proportional to $1/t$. The asymptotic forms of the amplitude and the differential cross-section are

$$M_{fi} \propto s/t, \quad d\sigma \propto dt/t^2. \tag{134.8}$$

In the limit (134.7), close to the backward direction, the asymptotic form is determined by the "exchange" diagram (73.14); in that limit,

$$M_{fi} \propto s/u, \quad d\sigma \propto du/u^2.$$

For mutual scattering of different particles (electron and muon), there is no exchange diagram, and so the cross-section for scattering through angles $\theta \approx \pi$ decreases in accordance with (134.3) and (134.4).‡

We shall show that these results for the asymptotic behaviour of electron–electron scattering are unaffected by the inclusion of radiative corrections. To do so, let us consider the corrections of various orders to the diagram (73.13).

It has already been shown that the diagrams which are corrections to the internal D function (see (113.11)) or to the vertex parts (see (117.1)) lead only to logarithmic corrections in the amplitude; they do not alter the power law (134.8). We shall show that the same is true of the diagram allowing cutting at two (not one)

† These arguments assume a constant $|t| \gg m^2$. The results thus obtained remain valid, as regards the dependence on s (i.e. on the energy), even when $|t| \sim m^2$.

‡ All these statements are, of course, in accordance with the results of §81; see (81.11) and Problem 6.

internal photon lines:

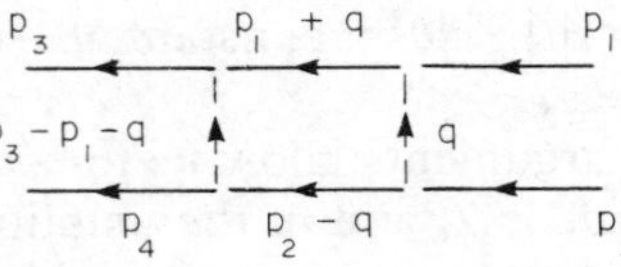

(134.9)

The scattering amplitude corresponding to this diagram differs from that corresponding to (73.13), in that the factor $1/t$ is replaced by

$$\frac{(\gamma(p_1+q))(\gamma(p_2-q))}{(p_1+q)^2(p_2-q)^2q^2(p_3-p_1-q)^2}\,d^4q,$$

followed by integration over d^4q. The important range of integration is the one which gives rise to the lowest power of $1/s$. For this, q must always be small in comparison with p_1 and p_2. Rejecting the terms which are then small (and also the terms $p_1^2=p_2^2=m^2$), we can rewrite this expression as

$$\frac{(\gamma p_1)(\gamma p_2)}{(p_1q)(p_2q)q^2(p_3-p_1-q)^2}\,d^4q.$$

The denominator does not contain s if q_0 and q_x (with the x-axis along $\mathbf{p}_1=-\mathbf{p}_2$) are $\propto 1/\sqrt{s}$; q_y and q_z may be $\propto\sqrt{|t|}$; then the range of integration $\propto 1/s$. The order of magnitude of the numerator is $p_1p_2\propto s$. Thus the replacement of one internal photon line in the diagram by two does not affect the dependence of the diagram on s (for a given t).† That is, the contribution of the diagram (134.9) to the scattering amplitude has the same asymptotic behaviour (134.8) as that of the principal diagram. The position is unaffected by adding other parallel internal photon lines in the diagram, and also by including corrections to the internal electron lines.

This is a general result: any diagram which can be cut in the t or u channel into two parts across any number of internal photon lines corresponds to an amplitude contribution which has the asymptotic form $M_{fi}\propto s/t$ with t constant or s/u with u constant (V. G. Gorshkov, V. N. Gribov, L. N. Lipatov and G. V. Frolov, 1967; H. Cheng and T. T. Wu, 1969).

As a second example, let us consider Compton scattering described by the two diagrams (74.12). These do not allow cutting in the t channel, but the second diagram can be cut in the u channel at an internal electron line. In the notation of the present section, it is

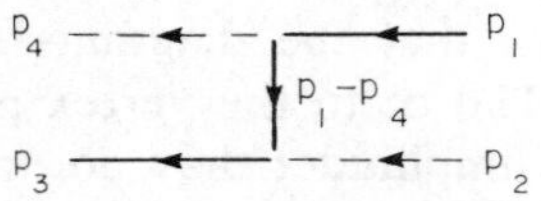

(134.10)

This means that the scattering is largely concentrated near the backward direction, as

† Let us mention again that only power-law asymptotic forms are under consideration, and so we need not take account of logarithmic divergences in the integrations. Diagrams of the form (134.9) will be further studied in §137.

already noted at the end of §86; see (86.20). To find the asymptotic behaviour in this range, we note that the factor G corresponding to the internal line in (134.10) is in order of magnitude $1/\gamma(p_1 - p_4) \propto 1/\sqrt{|u|}$. Hence the scattering amplitude $M_{fi} \propto \alpha(s/|u|)^{\frac{1}{2}}$; this includes a factor α because the diagram (134.10) is of the second order. The differential cross-section is therefore $d\sigma/du \propto \alpha^2/|u|s$. The integral of this expression with respect to $|u|$ is governed by the range $|u| \ll s$. The total cross-section then decreases with increasing energy as $\sigma \propto \alpha^2/s$, or more exactly $\sigma \propto (\alpha^2/s)\log(s/m^2)$; cf. (86.20).†

For this process, however, radiative corrections alter the asymptotic behaviour. The change is due to sixth-order diagrams of the type

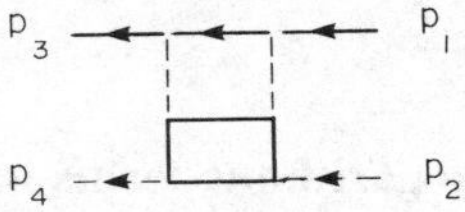

(134.11)

In the t channel, these allow a cut across two internal photon lines, and therefore contribute to the amplitude, with asymptotic form $M_{fi} \propto \alpha^3 s/t$; the factor α^3 corresponds to a sixth-order diagram. When s is sufficiently large, this part of the amplitude becomes the principal one, and the differential cross-section is then

$$d\sigma/dt \propto \alpha^6/t^2.$$

The integral of this expression with respect to t is governed by the range of small $|t| \sim m^2$, i.e. by the range of scattering angles $\theta \sim m\sqrt{s}$; note that the scattering is now mainly forward, not backward. The total cross-section then no longer decreases with increasing energy:

$$\sigma \propto \alpha^6/m^2 = \alpha^4 r_e^2. \tag{134.12}$$

The decreasing part of the cross-section becomes comparable with this constant part when $\varepsilon = \sqrt{s} \propto m/\alpha^2$.

A similar situation occurs for photon–photon scattering. In the first non-vanishing approximation, this is described by the "square" diagrams (127.1), which can be cut across two internal electron lines. Integration is carried out with respect to the 4-momentum of these lines in the diagram; momenta $\sim\sqrt{s}$ are important, and small values of t or u are not especially significant. The asymptotic form of these diagrams for any constant t or u is given by (134.5): $M_{fi} = \text{constant} \propto \alpha^2$. The total cross-section decreases with increasing energy: $\sigma \propto \alpha^4/s$ (cf. (127.23)); angles close to zero or π have no special significance here. In the eighth order, however, there are diagrams which can be cut (in the t or u channel) across two internal

† The exact form of the dependence of the cross-section on $|u|$ or $|t|$ when these are $\lesssim m^2$ cannot, of course, be ascertained from the arguments given here. It is assumed that the integral with respect to $|u|$ or $|t|$ converges at values $\sim m^2$. This is in fact so for all processes except elastic scattering of charged particles.

photon lines, for example

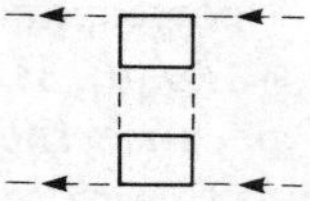

(134.13)

These diagrams give an asymptotically constant cross-section: $\sigma \propto \alpha^8/m^2$ when $\sqrt{s} \gg m/\alpha^2$.†

The asymptotic constancy of the total cross-section is a characteristic property of scattering processes whose diagrams can be cut (in the t or u channel) across internal photon lines. It occurs even when more than two particles are present in the final state of the reaction.

§135. Separation of the double-logarithmic terms in the vertex operator

The corrections having the form $(\alpha L)^n$ (where L is the large logarithm) can become important only at enormous energies, as already mentioned at the end of §133, and therefore are of purely theoretical significance. But there are also much larger corrections, of the form $(\alpha L^2)^n$, in the amplitudes of actual scattering processes. Such terms, containing the square of the logarithm to the same power as α, are called *double-logarithmic terms.*

The characteristic expansion parameter in the double-logarithmic corrections is

$$(\alpha/\pi)\log^2(\varepsilon^2/m^2), \tag{135.1}$$

where ε denotes the energies occurring in the problem (for example, the total energy of the colliding particles in the centre-of-mass system). The condition for perturbation theory to be valid is that this quantity be small; it ceases to be satisfied at energies

$$\varepsilon \sim m \exp[\tfrac{1}{2}\sqrt{(\pi/\alpha)}] \sim 3\times 10^4 m. \tag{135.2}$$

Let us now try to avoid this limitation and derive formulae valid when

$$(\alpha/\pi)\log^2(\varepsilon^2/m^2) \lesssim 1. \tag{135.3}$$

This clearly requires the summation of an infinite series of corrections in all powers $(\alpha L^2)^n$.

The double-logarithmic corrections occur in cases of two kinds. One includes scattering through a fixed finite angle; as has been shown in §134, the

† The cross-section for coherent scattering of a photon in the field of a nucleus is asymptotically constant even in the first non-vanishing approximation, described by "square" diagrams in which two ends are external-field lines; see (128.7). In reality, however, these diagrams would have to be represented in the form (134.11), the upper continuous line being the nucleus line. The external-field lines then become internal lines in the diagram, and the reason for the asymptotically constant form becomes evident.

cross-sections always decrease in the asymptotic high-energy range. In such cases, the double-logarithmic corrections are closely associated with the infra-red divergence. These cases include, in particular, elastic scattering of electrons in an external Coulomb field; the first double-logarithmic correction to the cross-section has been found in §122. The present section and §136 deal with the complete determination of these corrections, under the condition (135.3).

The other class of cases includes reaction cross-sections which decrease with increasing energy for a given square of the momentum transfer, i.e. for scattering angles which asymptotically approach zero or π; as shown in §134, this occurs for processes whose diagrams cannot be cut across internal photon lines in the t or u channel. Here, the double-logarithmic corrections are not connected with the infra-red divergence. As an example of this, we shall discuss in §137 the problem of backward (u = constant) electron–muon scattering.

First of all, with the condition (135.3) the single-logarithmic corrections are $\sim(\alpha/\pi)\log(\varepsilon^2/m^2) \lesssim \sqrt{(\alpha/\pi)} \ll 1$ and can therefore be omitted. Since the double-logarithmic corrections do not appear in $\mathcal{G}$ and $\mathcal{D}$, the latter functions can now be taken as simply equal to their unperturbed values G and D.

The calculation of the vertex operator Γ involves the summation of the double-logarithmic terms which arise from an infinite sequence of diagrams. This problem will be analysed in §136, but first a method will be described for separating the double-logarithmic terms in the various Feynman integrals before actually performing the integrations over all the variables in them (V. V. Sudakov, 1956).

Let us consider the first-order correction (with respect to α) to the vertex operator, represented by the diagram (117.1), which is here conveniently taken (with a renaming of the variables) in the form

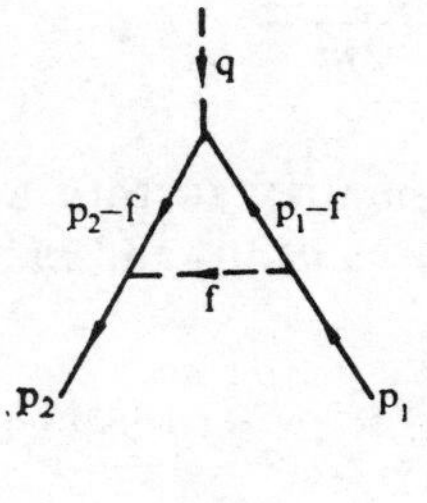

(135.4)

or, analytically,

$$\Gamma^{\mu(1)}(p_2, p_1; q) = -\frac{ie^2}{4\pi^3}\int \frac{\gamma^\nu(\gamma p_2 - \gamma f + m)\gamma^\mu(\gamma p_1 - \gamma f + m)\gamma_\nu d^4 f}{[(p_2-f)^2 - m^2 + i0][(p_1-f)^2 - m^2 + i0][f^2 + i0]}. \tag{135.5}$$

We shall assume that

$$|q^2| \gg p_1^2, p_2^2, m^2, \tag{135.6}$$

and that the ends p_1, p_2 may be either physical or virtual. From (135.6) it follows that

$$|p_1 p_2| \approx \tfrac{1}{2}|q^2| \gg p_1^2, p_2^2, m^2, \tag{135.7}$$

i.e. the 4-vectors p_1 and p_2 have large components but small squares; this is possible because the four-dimensional metric is pseudo-Euclidean. The double-logarithmic terms in fact occur when the conditions (135.6) are satisfied.

We shall see below that relatively small values of f are important in the integration over d^4f. We can therefore neglect f in the numerator of the integrand, and $\Gamma^{(1)}$ becomes

$$\Gamma^{\mu(1)} = -\frac{ie^2}{4\pi^3}\gamma^\nu(\gamma p_2 + m)\gamma^\mu(\gamma p_1 + m)\gamma_\nu I_1, \tag{135.8}$$

where

$$I_1 = \int \frac{d^4f}{[(p_2-f)^2 - m^2 + i0][(p_1-f)^2 - m^2 + i0][f^2 + i0]}. \tag{135.9}$$

The matrix factor in (135.8) can be simplified by using the fact that when Γ appears in diagrams it is always, in effect, multiplied by the matrices $(\gamma p_2 + m)$ and $(\gamma p_1 + m)$:

$$(\gamma p_2 + m)\Gamma(\gamma p_1 + m). \tag{135.10}$$

For, if the lines p_1 and p_2 are virtual, these factors come from $G(p_1)$ and $G(p_2)$; if the lines correspond to real electrons, Γ is multiplied by $\bar{u}_2$ and u_1, and Dirac's equations show that

$$\bar{u}_2 = \bar{u}_2\frac{\gamma p_2 + m}{2m}, \qquad u_1 = \frac{\gamma p_1 + m}{2m}u_1.$$

Interchanging the order of the matrix factors and neglecting at each stage, in accordance with (135.7), the squares p_1^2, p_2^2 and m^2 in comparison with $p_1 p_2$, we get

$$(\gamma p_2 + m)\Gamma^{\mu(1)}(\gamma p_1 + m) \approx -\frac{ie^2}{\pi^3}(p_1 p_2)(\gamma p_2 + m)\gamma^\mu(\gamma p_1 + m)I_1.$$

We can therefore put $\Gamma^{(1)}$ in the final form

$$\Gamma^{\mu(1)} = (ie^2/2\pi^3)\gamma^\mu t I_1, \tag{135.11}$$

where

$$t = q^2 \approx -2(p_1 p_2). \tag{135.12}$$

The integral I_1 converges when f is large, and therefore need not be regularized.

The main point of the subsequent calculations is the introduction of new and more convenient variables of integration. Let f be resolved into components tangential and normal to the p_1p_2 plane:

$$f = up_1 + vp_2 + f_\perp$$
$$\equiv f_\parallel + f_\perp, \tag{135.13}$$

$$f_\perp p_1 = f_\perp p_2 = 0. \tag{135.14}$$

As new variables, we take the coefficients u and v and

$$\rho = -f_\perp^2. \tag{135.15}$$

It is evident from the conditions (135.7) that the metric in the p_1p_2 plane is pseudo-Euclidean. The time axis can therefore be taken in this plane, so that $f_\perp$ is a space-like 4-vector and $\rho > 0$.

Let the indices 0 and x temporarily denote the components of 4-vectors in the p_1p_2 plane; y and z those in the normal plane. To transform the 4-volume element

$$d^4f = d^2f_\perp d^2f_\parallel$$

to the new variables, we write

$$d^2f_\perp = |\mathbf{f}_\perp|\, d|\mathbf{f}_\perp|\, d\phi$$
$$= \tfrac{1}{2}\, d\rho\, d\phi \to \pi\, d\rho$$

(since the integrand in (135.9) is independent of the angle ϕ). Also

$$d^2f_\parallel = \left|\frac{\partial(f_0, f_x)}{\partial(u, v)}\right| du\, dv$$
$$= |p_{10}p_{2x} - p_{20}p_{1x}|\, du\, dv$$
$$\approx \tfrac{1}{2}|q^2|\, du\, dv,$$

since p_2^2 is small, so that $p_{2x}^2 \approx p_{20}^2$ and

$$(p_{10}p_{2x} - p_{20}p_{1x})^2 \approx (p_{10}p_{20} - p_{2x}p_{1x})^2$$
$$= (p_1p_2)^2 = (\tfrac{1}{2}q^2)^2.$$

Thus

$$d^4f = \tfrac{1}{2}|t|\, du\, dv\, d^2f_\perp$$
$$\to \tfrac{1}{2}\pi|t|\, du\, dv\, d\rho. \tag{135.16}$$

The calculations now depend on the relation between p_1^2, p_2^2 and m^2; two cases will be considered.

VIRTUAL ELECTRON LINES

Let the momenta p_1 and p_2 correspond to virtual electrons, with

$$|p_1^2|, |p_2^2| \gg m^2. \tag{135.17}$$

We shall see that the most important range of integration, which leads to a double-logarithmic expression, is in this case given by the inequalities

$$\begin{gathered} 0 < \rho \ll |tu|, |tv|, \\ |p_1^2/t| \ll |v| \ll 1, |p_2^2/t| \ll |u| \ll 1. \end{gathered} \tag{135.18}$$

Accordingly, in the denominator of the integrand in (135.9) we can neglect m^2, p_1^2, p_2^2 and f^2 in comparison with $(p_1 f)$ and $(p_2 f)$:

$$I_1 = \int \frac{d^4 f}{2(p_2 f) \cdot 2(p_1 f)(f^2 + i0)}. \tag{135.19}$$

The quantities $p_1 f$, $p_2 f$ and f^2 are given by

$$\begin{aligned} f^2 &= (up_1 + vp_2)^2 - \rho \approx -tuv - \rho, \\ 2(p_1 f) &= 2p_1(up_1 + vp_2) \approx -tv, \\ 2(p_2 f) &\approx -tu. \end{aligned}$$

Then

$$I_1 = -\frac{\pi}{2|t|} \int \frac{d\rho}{\rho + tuv - i0} \frac{du}{u} \frac{dv}{v}. \tag{135.20}$$

In accordance with the conditions (135.18), the integration over $d\rho$ is taken from 0 to the smaller of $|tv|$ and $|tu|$; the result is

$$\int\limits_0^{\min(|tu|, |tv|)} \frac{d\rho}{\rho + tuv - i0} = \log \min\left\{\frac{1}{|u|}, \frac{1}{|v|}\right\} + \begin{cases} i\pi \text{ when } tuv < 0, \\ 0 \text{ when } tuv > 0. \end{cases} \tag{135.21}$$

The logarithmic integration over dv is taken from -1 to $-|p_1^2/t|$ and from $|p_1^2/t|$ to 1 (and similarly over du). When (135.21) is substituted in (135.20), the integral of the first term over $du\,dv$ is zero, because the integrand is an odd function. The integration of the second term is carried over ranges of u and v having the same sign when $t < 0$ and opposite signs when $t > 0$. In either case the ranges $v > 0$ and $v < 0$ give the same contribution after integration over du, and the result is

$$I_1 = \frac{i\pi^2}{2t} \cdot 2 \int\limits_{|p_1^2/t|}^{1} \frac{du}{u} \int\limits_{|p_2^2/t|}^{1} \frac{dv}{v} = \frac{i\pi^2}{t} \log\left|\frac{t}{p_1^2}\right| \log\left|\frac{t}{p_2^2}\right|, \tag{135.22}$$

the sign being the same as that of t.

Finally, substituting in (135.11), we get

$$\Gamma^{\mu(1)}(p_2, p_1; q) = -\frac{\alpha}{2\pi}\gamma^\mu \log\left|\frac{q^2}{p_1^2}\right| \log\left|\frac{q^2}{p_2^2}\right|, \tag{135.23}$$
$$|q^2| \gg |p_1^2|, |p_2^2| \gg m^2.$$

Physical external electron lines

Now let the momenta p_1 and p_2 correspond to real electrons, so that

$$p_1^2 = p_2^2 = m^2. \tag{135.24}$$

Then the important range of integration is

$$0 < \rho \ll |tu|, |tv|,$$
$$0 < |v|, |u| \ll 1. \tag{135.25}$$

Since $p_1^2 - m^2 = p_2^2 - m^2 = 0$, we can neglect p_1^2 and p_2^2 in comparison with $p_1 f$ and $p_2 f$, and again bring the integral (135.9) to the form (135.19). To eliminate the infra-red divergence which then occurs, however, we must apply a finite photon mass $\lambda \ll m$ in the photon propagator (cf. §117):

$$I_1 = \int \frac{d^4 f}{2(p_1 f)\cdot 2(p_2 f)(f^2 - \lambda^2 + i0)}. \tag{135.26}$$

In this case

$$f^2 \approx -tuv - \rho,$$
$$2p_1 f \approx -tv + 2m^2 u,$$
$$2p_2 f \approx -tu + 2m^2 v,$$

and hence

$$I_1 = -\frac{\pi}{2|t|}\int \frac{d\rho}{\rho + tuv + \lambda^2 - i0}\frac{du}{u - \tau v}\frac{dv}{v - \tau u}, \tag{135.27}$$

where $\tau = 2m^2/t \ll 1$.

After the integration over $d\rho$, similarly to (135.21), we obtain

$$I_1 = -\frac{i\pi^2}{2|t|}\int\int \frac{du}{u - \tau v}\frac{dv}{v - \tau u},$$

the integration being subject to the condition $tuv + \lambda^2 < 0$. The ranges $v > 0$ and

$v < 0$ again make equal contributions, and the result of the integration over du is

$$I_1 = \frac{i\pi^2}{t}\int_0^1 dv \int_{\delta/v}^1 \frac{du}{(u-\tau v)(v-\tau u)}$$

$$= \frac{i\pi^2}{t}\int_0^1 \log\frac{\tau\delta - v^2}{(\delta-\tau v^2)(\tau - v)}\frac{dv}{v}, \tag{135.28}$$

where $\delta = \lambda^2 t$, $|\delta| \ll |\tau|$, and we have used the inequality $|\tau| \ll 1$.

In the integral (135.28), three ranges of v lead to double-logarithmic expressions: (I) $|\tau| \ll v \ll 1$, (II) $\sqrt{(\delta/\tau)} \ll v \ll |\tau|$, (III) $\sqrt{(\tau\delta)} \ll v \ll \sqrt{(\delta/\tau)}$. (We take the specific case $\sqrt{(\delta/\tau)} \ll |\tau|$; the result does not depend on this assumption.) With the appropriate approximations in each range, we find

$$I_1 = \frac{i\pi^2}{2t}\left(\log^2\frac{|t|}{m^2} + 4\log\frac{|t|}{m^2}\log\frac{m}{\lambda}\right). \tag{135.29}$$

Finally, substituting in (135.11), we have

$$\Gamma^{\mu(1)}(p_2, p_1; q) = -\frac{\alpha}{4\pi}\gamma^\mu\left(\log^2\frac{|q^2|}{m^2} + 4\log\frac{|q^2|}{m^2}\log\frac{m}{\lambda}\right), \tag{135.30}$$

$$|q^2| \gg p_1^2 = p_2^2 = m^2,$$

in agreement with (117.21).

§ 136. Double-logarithmic asymptotic form of the vertex operator

When the corrections $\Gamma^{(1)}$ calculated in §135 become of the order of unity, the vertex operator has to be found by summation of the infinite sequence of double-logarithmic terms of all orders in α. This problem can be solved because such terms arise only from diagrams of a particular type, and the contributions from diagrams of different orders are related in a simple manner. As we shall see below, the double-logarithmic terms arise from all the diagrams that have the form

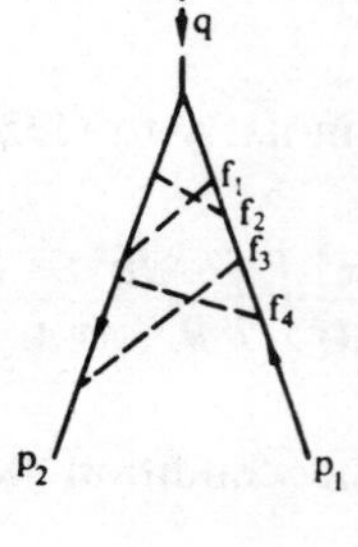

(136.1)

etc., in which each photon line joins the two electron lines, and the photon lines themselves may intersect in any manner.

Let the photon momenta $f_1, f_2, \ldots$ be numbered in the sequence of, say, the right-hand ends of their lines. Then the various diagrams of a given order will differ in the sequence of the left-hand ends of the photon lines. In each Feynman integral, we neglect terms in the numerator and denominator as in (135.5), and then treat the numerator in the same way as in the derivation of (135.11). Then the sum of all the diagrams having n photon lines, giving the term $\propto \alpha^n$ in Γ, is

$$\Gamma^{\mu(n)} = \gamma^\mu (i\alpha t/2\pi^3)^n I_n, \tag{136.2}$$

$$I_n = \sum_{\text{int } 2} \int \frac{d^4f_1 \ldots d^4f_n}{2(p_1f_1)\cdot 2(p_1f_1+p_1f_2)\ldots 2(p_1f_1+\cdots+p_1f_n)\cdot 2(p_2f_1)\ldots 2(p_2f_1+\cdots+p_2f_n)f_1^2f_2^2\ldots f_n^2}, \tag{136.3}$$

where the sum is taken over all interchanges (permutations) of the subscripts k in the products p_2f_k; the terms $i0$ and λ^2 in the denominators are omitted for brevity.

It is clear that, if the subscripts k in the products p_1f_k in the sum (136.3) are interchanged in any manner, this is simply equivalent to renaming the momenta and so does not affect the value of I_n. We can therefore extend the summation in (136.3) to all interchanges of the factors f_k in both p_2f_k and p_1f_k, and divide the result by $n!$.

We now make use of the important formula

$$\sum_{\text{int}} \frac{1}{a_1(a_1+a_2)\ldots(a_1+a_2+\cdots+a_n)} = \frac{1}{a_1}\cdot\frac{1}{a_2}\cdots\frac{1}{a_n}, \tag{136.4}$$

where the sum is taken over interchanges of the subscripts $1, 2, \ldots, n$.† When this formula is twice applied to the sum of integrals, we obtain a product of n identical integrals of the form (135.19) (or (135.26)), so that

$$I_n = \frac{1}{n!} I_1^n. \tag{136.5}$$

Substitution in (136.2) and summation of $\Gamma^{(n)}$ over all $n = 0, 1, 2, \ldots$ gives finally

$$\Gamma^\mu(p_2, p_1; q) = \gamma^\mu \exp(ie^2tI_1/2\pi^3). \tag{136.6}$$

In particular, substitution of I_1 from (135.22) gives the double-logarithmic asymptotic form of the vertex operator with virtual external electron lines:

$$\Gamma^\mu(p_2, p_1; q) = \gamma^\mu \exp\left\{-\frac{\alpha}{2\pi}\log\left|\frac{q^2}{p_1^2}\right|\log\left|\frac{q^2}{p_2^2}\right|\right\}, \tag{136.7}$$

$$|q^2| \gg |p_1^2|, |p_2^2| \gg m^2$$

(V. V. Sudakov, 1956).

† The formula is obviously true for $n = 2$, and can easily be proved by induction.

Substitution of I_1 from (135.29) gives the asymptotic form of the vertex operator for real external electron lines:

$$\Gamma^{\mu}(p_2, p_1; q) = \gamma^{\mu} \exp\left\{ -\frac{\alpha}{4\pi}\left(\log^2\frac{|q^2|}{m^2} + 4\log\frac{|q^2|}{m^2}\log\frac{m}{\lambda}\right)\right\}, \qquad (136.8)$$

$$|q^2| \gg p_1^2 = p_2^2 = m^2.$$

The factor which distinguishes this Γ^{μ} from its unperturbed value γ^{μ} defines also the difference between the amplitude for electron scattering in the external field and its Born value. The scattering cross-section is therefore

$$d\sigma = d\sigma_B \exp\left\{ -\frac{\alpha}{2\pi}\left(\log^2\frac{|q^2|}{m^2} + 4\log\frac{|q^2|}{m^2}\log\frac{m}{\lambda}\right)\right\}. \qquad (136.9)$$

To eliminate the infra-red divergence, we still have to multiply this expression by the sum of the probabilities for the emission of various numbers of soft photons with energy not exceeding some small value $\omega_{\max}$, i.e. by the quantity (see (122.2))

$$1 + \int_0^{\omega_{\max}} dw_{\omega} + \frac{1}{2!}\int_0^{\omega_{\max}} dw_{\omega_1}\int_0^{\omega_{\max}} dw_{\omega_2} + \cdots = \exp\left\{\int_0^{\omega_{\max}} dw_{\omega}\right\}. \qquad (136.10)$$

The integral in the exponential is given by (120.14) (as the expression which there multiplies $d\sigma_{\text{el}}$), and the final result is the following asymptotic formula for the cross-section for scattering of an electron with energy ε at a high momentum transfer:

$$d\sigma = d\sigma_B \exp\left\{ -\frac{2\alpha}{\pi}\log\frac{|q^2|}{m^2}\log\frac{\varepsilon}{\omega_{\max}}\right\}, \qquad (136.11)$$

$$|q^2| \gg m^2, \quad (\alpha/2\pi)\log^2(\varepsilon/m) \sim 1$$

(A. A. Abrikosov, 1956). The first-order term (with respect to α) in the expansion of this expression is, of course, (122.12).

Note that, if we put $\omega_{\max} \sim \varepsilon$, one of the logarithms in (136.11) becomes of the order of unity; that is, the double-logarithmic corrections cancel in the cross-section for simultaneous emission of photons of any energy.† In the approximation used, the exponential factor in (136.11) then becomes unity, and the cross-section has its Born value, in agreement with the general statement at the end of §98.

§137. Double-logarithmic asymptotic form of the electron–muon scattering amplitude

As an example of the second kind, let us consider the scattering of an electron by a negative muon, taking only the case of scattering exactly backwards through an

† For scattering through a finite angle, the condition formulated in §98 for the photon to be soft requires only that $\omega_{\max} \ll \varepsilon$, and so the formulae derived here can be used with logarithmic accuracy even when $\omega_{\max} \sim \varepsilon$.

angle $\theta = \pi$ (V. G. Gorshkov, V. N. Gribov, L. N. Lipatov and G. V. Frolov, 1967). This is a simple process in two respects. Firstly, exchange diagrams do not appear, because the two particles are not identical. Secondly, in back-scattering there is very little emission of soft photons, and therefore no infra-red divergence: according to (98.8) the soft-photon emission cross-section is

$$d\sigma = \alpha\left[\left(\frac{\mathbf{v}'_e}{1-\mathbf{v}'_e\cdot\mathbf{n}} + \frac{\mathbf{v}'_\mu}{1-\mathbf{v}'_\mu\cdot\mathbf{n}} - \frac{\mathbf{v}_e}{1-\mathbf{v}_e\cdot\mathbf{n}} - \frac{\mathbf{v}_\mu}{1-\mathbf{v}_\mu\cdot\mathbf{n}}\right)\times\mathbf{n}\right]^2 \cdot \frac{d\omega\, do_{\mathbf{n}}}{4\pi^2\omega}\, d\sigma_{\mathrm{el}}, \tag{137.1}$$

where $\mathbf{v}_e$, $\mathbf{v}_\mu$ and $\mathbf{v}'_e$, $\mathbf{v}'_\mu$ are the particle velocities before and after the collision; in the ultra-relativistic case equality of momenta implies equality of velocities, and to this accuracy, in the centre-of-mass system, for back-scattering, $\mathbf{v}_e = -\mathbf{v}_\mu = -\mathbf{v}'_e = \mathbf{v}'_\mu$, so that (137.1) is zero.

If the scattering process considered corresponds to the s channel of the reaction, in the t channel it becomes the conversion of an electron–positron pair into a $\mu^+\mu^-$ pair. In this channel the condition $\theta = \pi$ signifies that the directions of motion of e^- and μ^- (and of e^+ and μ^+) coincide. The elimination of the bremsstrahlung in this channel is particularly clear, since the direction of motion of the charge of each sign is unchanged.

The cancelling of the leading terms in the emission cross-section has the result that its asymptotic form does not contain double-logarithmic corrections. Correspondingly, there is (with the same double-logarithmic accuracy) no infra-red divergence even on integration over the momenta of the virtual photons in the scattering amplitude.

If the process is described by means of the invariants

$$s = (p_e + p_\mu)^2, \quad t = (p_e - p'_e)^2, \quad u = (p_e - p'_\mu)^2,$$

the values corresponding to back-scattering in the ultra-relativistic case are

$$s = -t \gg m_\mu^2, \quad u = 0. \tag{137.2}$$

In the first approximation (with respect to α) of perturbation theory, electron–muon scattering is represented by the diagram

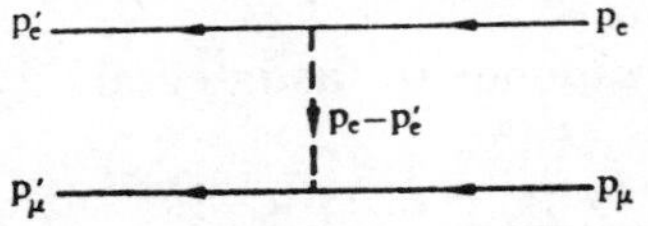

(137.3)

The corresponding amplitude is

$$M_{fi}^{(1)} = \frac{4\pi\alpha}{t}(\bar{u}^{(\mu)\prime}\gamma^\nu u^{(\mu)})(u^{(e)\prime}\gamma_\nu u^{(e)}). \tag{137.4}$$

The value of this expression in the limit (137.2) is obtained by replacing the matrix

4-vector γ^ν by its "projection" $\gamma_\perp^\nu$ on a plane perpendicular to the $p_e p_e'$ plane (or, equivalently, the $p_\mu p_\mu'$ plane, since in ultra-relativistic back-scattering $p_e \approx p_\mu'$ and $p_e' \approx p_\mu$): the components parallel to the $p_e p_e'$ plane are the matrices

$$\frac{1}{\sqrt{s}}(\gamma p_e + \gamma p_e'), \qquad \frac{1}{\sqrt{s}}(\gamma p_e - \gamma p_e')$$

(the first being equal to γ^0 and the second to $\mathbf{n}_e \cdot \boldsymbol{\gamma}$, with $\mathbf{n}_e$ a unit vector along $\mathbf{p}_e$), and on using Dirac's equations for the bispinors $u^{(e)}$ and $u^{(\mu)}$, we have

$$(\bar{u}^{(\mu)\prime}\gamma_\parallel^\nu u^{(\mu)})(\bar{u}^{(e)\prime}\gamma_{\nu\parallel}u^{(e)}) \sim 1/s,$$

so that these terms may be omitted.

In the next approximation we add the diagram

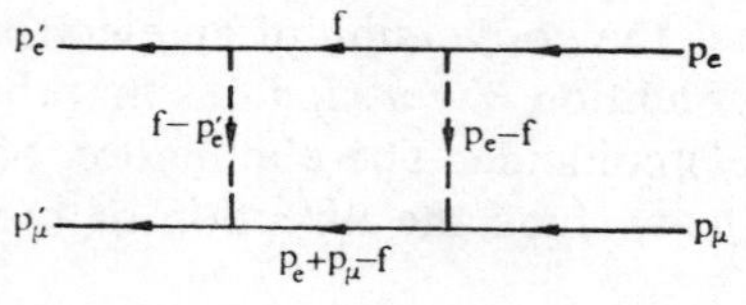

(137.5)

and a diagram with the photon lines "crossed", which is conveniently taken in a form which differs from (137.5) only as regards the direction of one of the continuous lines:

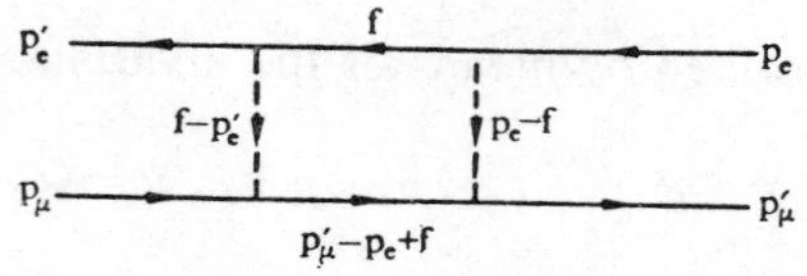

(137.6)

An analysis of the corresponding integrals shows that in both diagrams we have double-logarithmic contributions from the regions of "soft" virtual photons: $|(f - p_e)^2| \ll m_e^2$ or $|(f - p_e')^2| \ll m_e^2$. These contributions arise from the infra-red divergences of the integrals and, according to the foregoing discussion, must certainly cancel here. In the diagram (137.6), however, there is a double-logarithmic contribution also from large momenta: $|f^2| \gg m_\mu^2$, and this contribution must be calculated.

The diagram (137.6) corresponds to the integral

$$M_{fi}^{(2)} = -\frac{i\alpha^2}{\pi^2}\int \frac{[\bar{u}^{(e)\prime}\gamma^\nu(\gamma f + m_e)\gamma_\lambda u^{(e)}][\bar{u}^{(\mu)\prime}\gamma^\lambda(\gamma f + m_\mu)\gamma_\nu u^{(\mu)}]}{(p_e' - f)^2(f^2 - m_e^2)(f^2 - m_\mu^2)(p_e - f)^2}\, d^4 f, \qquad (137.7)$$

where we we have already used the fact that $p_e \approx p_\mu'$. We again put

$$f = u p_e + v p_e' + f_\perp \qquad (137.8)$$

(cf. (135.13)). The double-logarithmic contribution comes from the range defined by the inequalities

$$\left.\begin{aligned} |su|, |sv| \gg \rho \gg m_\mu^2, \\ m_\mu^2/s \ll |u|, |v| \ll 1, \end{aligned}\right\} \tag{137.9}$$

where $\rho = -f_\perp^2$. The 4-vector $f_\perp$ is defined so that $f_\perp p_e = f_\perp p'_e = 0$; in the present case of back-scattering, it follows that $f_\perp^0 = 0$ in the centre-of-mass system, and so $\rho = \mathbf{f}_\perp^2$.

In the numerator in (137.7) we can neglect m_e and m_μ, as well as all terms containing u or v; the factors u and v there would cancel the corresponding poles in the denominator (see below), and so the required squares of logarithms would not occur. Since

$$(p'_e - f)^2 \approx tu \approx -su, \quad (p_e - f)^2 \approx -sv, \quad f^2 \approx suv - \rho,$$

we can transform the element of integration d^4f in accordance with (135.16) and rewrite (137.7) as

$$M_{fi}^{(2)} = -\frac{i\alpha^2}{2\pi^2}\int \frac{[\bar{u}^{(e)\prime}\gamma^\nu(\gamma f_\perp)\gamma_\lambda u^{(e)}][\bar{u}^{(\mu)\prime}\gamma^\lambda(\gamma f_\perp)\gamma_\nu u^{(\mu)}]}{su\cdot sv(suv - \rho + i0)^2}\, s\, du\, dv\, d^2 f_\perp.$$

The numerator in the integrand is further transformed by averaging over the directions of $\mathbf{f}_\perp$ and replacing γ^ν and γ^λ by $\gamma_\perp^\nu$ and $\gamma_\perp^\lambda$ (on the same principle as in (137.4)). A simple calculation gives

$$\left.\begin{aligned} M_{fi}^{(2)} &= M_{fi}^{(1)} J^{(1)}, \\ J^{(1)} &= -\frac{i\alpha}{4\pi^2}\int \frac{\rho\, du\, dv\, d\rho}{uv\,(suv - \rho + i0)^2}. \end{aligned}\right\} \tag{137.10}$$

Finally, using in the numerator the identity $\rho \equiv (\rho - suv) + suv$, we can omit the second term, which would cancel simple poles and therefore give no double-logarithmic contribution. Thus

$$J^{(1)} = -\frac{i\alpha}{4\pi^2}\int \frac{du\, dv\, d\rho}{uv\,(\rho - suv - i0)}. \tag{137.11}$$

This integral has the same form as (135.20), and the integration over $d\rho$ can therefore be performed in the same manner, but since now $\rho \gg m_\mu^2$ we have the condition $suv \gg m_\mu^2$ instead of $suv > 0$. The result is

$$J^{(1)} = \frac{\alpha}{2\pi}\int \frac{du\, dv}{uv}, \tag{137.12}$$

the range of integration being defined by the inequalities

$$m_\mu^2/s < u, v < 1, \quad suv > m_\mu^2;$$

in the calculation with logarithmic accuracy, the strong inequalities $\gg$ are replaced by > simply. A straightforward calculation gives

$$J^{(1)} = \frac{\alpha}{4\pi}\log^2\frac{s}{m_\mu^2}. \qquad (137.13)$$

In the higher approximations of perturbation theory, the desired terms $\sim\alpha^n \log^{2n}s$ arise from "ladder" diagrams similar to (137.6) but with a larger number of "rungs". The complete double-logarithmic asymptotic form of the scattering amplitude is therefore given by the infinite sum

$iM_n =$ 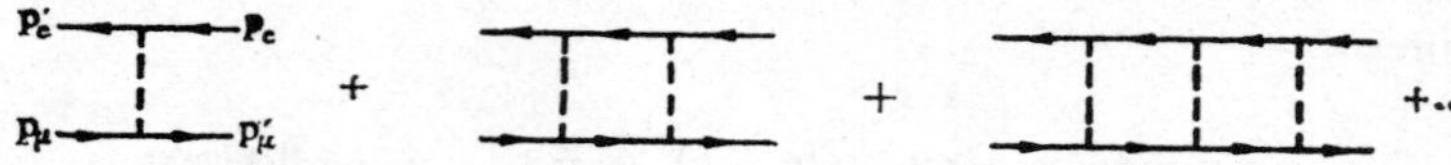

(137.14)

To determine the general form of the terms in this sum, let us consider the diagram for the third approximation (the third term in the series (137.14)). The corresponding integral may be written

$$\left.\begin{aligned} M_{fi}^{(3)} &= M_{fi}^{(1)} J^{(2)}, \\ J^{(2)} &= \left(\frac{\alpha}{2\pi}\right)^2 \int \frac{du_1\,dv_1\,du_2\,dv_2}{u_1 v_1 (u_1+u_2)(v_1+v_2)}, \end{aligned}\right\} \qquad (137.15)$$

with the range of integration

$$m_\mu^2 s < u_{1,2},\ v_{1,2} < 1, \quad s u_1 v_1,\ s u_2 v_2 > m_\mu^2.$$

The double-logarithmic term in this integral can be separated by applying to the variables of integration the further conditions

$$v_2 \gg v_1, \quad u_2 \gg u_1. \qquad (137.16)$$

Then

$$\begin{aligned} J^{(2)} &= \left(\frac{\alpha}{2\pi}\right)^2 \int \frac{du_1 dv_1 du_2 dv_2}{u_1 u_2 v_1 v_2} \\ &= \left(\frac{\alpha}{2\pi}\right)^2 \int d\xi_1\, d\eta_1\, d\xi_2\, d\eta_2, \end{aligned}$$

where $\xi_i = \log(su_i/m_\mu^2)$, $\eta_i = -\log v_i$, and the range of integration is defined by the inequalities

$$\xi_1 > \eta_1, \quad \xi_2 > \eta_2, \quad \sigma > \xi_2,\ \eta_2 > 0, \quad \sigma = \log(s/m_\mu^2).$$

Similarly, the nth term of the series can be written as $M_{fi}^{(n)} = M_{fi}^{(1)} J^{(n)}$, where

$$J^{(n)}(\sigma) = \left(\frac{\alpha}{2\pi}\right)^n \int d\xi_1 d\eta_1 \ldots d\xi_n d\eta_n, \qquad (137.17)$$

with the range of integration

$$\xi_i > \eta_i \ (i = 1, 2, \ldots, n), \quad \sigma > \xi_n, \eta_n > 0. \tag{137.18}$$

The total scattering amplitude is

$$M_{fi} = M_{fi}^{(1)} \left[1 + \sum_{n=1}^{\infty} J^{(n)}(\sigma) \right]. \tag{137.19}$$

To calculate this sum, we use auxiliary functions $A^{(n)}(\xi, \eta)$, given by the same integrals (137.17) but with ranges of integration

$$\xi_i > \eta_i \ (i = 1, 2, \ldots, n), \quad \xi > \xi_n > 0, \quad \eta > \eta_n > 0 \tag{137.20}$$

(i.e. with different limits of integration with respect to ξ_n and η_n, instead of uniform limits as in (137.18)). It is evident that $M_{fi} = M_{fi}^{(1)} A(\sigma, \sigma)$, where

$$A(\xi, \eta) = \sum_{n=0}^{\infty} A^{(n)}(\xi, \eta), \quad A^{(0)} = 1. \tag{137.21}$$

The definition of the functions $A^{(n)}(\xi, \eta)$ shows that they satisfy the recurrence relations

$$A^{(n)}(\xi, \eta) = \frac{\alpha}{2\pi} \int d\xi_1 d\eta_1 A^{(n-1)}(\xi_1, \eta_1),$$

and summation of this equation with respect to n from 1 to ∞ gives an integral equation for the function $A(\xi, \eta)$:

$$\left. \begin{aligned} & A(\xi, \eta) = 1 + \frac{\alpha}{2\pi} \int A(\xi_1, \eta_1)\, d\xi_1 d\eta_1, \\ & \xi_1 > \eta_1, \quad \xi > \xi_1 > 0, \quad \eta > \eta_1 > 0. \end{aligned} \right\} \tag{137.22}$$

For the subsequent analysis it will be sufficient to consider $A(\xi, \eta)$ in the range $\xi > \eta$. Then equation (137.22) can be written

$$A(\xi, \eta) = 1 + \frac{\alpha}{2\pi} \int_0^{\eta} \int_{\eta_1}^{\xi} A(\xi_1, \eta_1)\, d\xi_1\, d\eta_1. \tag{137.23}$$

Differentiating this with respect to η, we have

$$\frac{\partial A(\xi, \eta)}{\partial \eta} = \frac{\alpha}{2\pi} \int_{\eta}^{\xi} A(\xi_1, \eta)\, d\xi_1, \tag{137.24}$$

and a further differentiation with respect to ξ gives the differential equation

$$\frac{\partial^2 A}{\partial\eta\,\partial\xi} - \frac{\alpha}{2\pi} A = 0. \tag{137.25}$$

This has to be solved with the boundary conditions

$$A(\xi, 0) = 1, \quad [\partial A/\partial\eta]_{\xi=\eta} = 0, \tag{137.26}$$

which follow immediately from (137.23) and (137.24).

The solution can be found by means of a Laplace transformation with respect to ξ:

$$A(\xi, \eta) = \frac{1}{2\pi i}\int\limits_C e^{p\xi} Q(p, \eta)\, dp, \tag{137.27}$$

where the contour C in the complex p-plane is a closed curve around the point $p = 0$. Substituting (137.27) in (137.25) and equating the integrand to zero, we have

$$p\frac{\partial Q}{\partial\eta} = \frac{\alpha}{2\pi} Q, \quad Q = \phi(p) e^{\alpha\eta/2\pi p},$$

with $\phi(p)$ an arbitrary function. The first boundary condition (137.26) now gives

$$\phi(p) = \frac{1}{p} + \psi(p),$$

with $\psi(p)$ an analytic function, having no singularity within C. The second condition (137.26) can be satisfied by putting $\psi(p) = -2\pi p/\alpha$: then

$$\left[\frac{\partial A}{\partial\eta}\right]_{\xi=\eta} = -\frac{1}{2\pi i\xi}\int\limits_C \frac{d}{dp} e^{\xi(p+\alpha/2\pi p)}\, dp = 0.$$

Combination of these expressions with $\xi = \eta = \sigma$ gives

$$A(\sigma, \sigma) = -\frac{1}{2\pi i}\frac{2\pi}{\alpha\sigma}\int\limits_C p\frac{d}{dp} e^{\sigma(p+\alpha/2\pi p)}\, dp.$$

Finally, integrating by parts and using the familiar formula

$$I_1(z) = \frac{1}{2\pi i}\int\limits_C e^{\frac{1}{2}z(p+1/p)}\, dp$$

(where $I_1(z) = -iJ_1(iz)$ is the Bessel function with imaginary argument), we find the scattering amplitude

$$M_{fi} = M_{fi}^{(1)} \sqrt{\frac{2\pi}{\alpha\sigma^2}}\, I_1\left(\sqrt{\frac{2\alpha}{\pi}}\,\sigma\right). \tag{137.28}$$

The cross-section for scattering through $\theta = \pi$ is correspondingly

$$\left.\begin{aligned} d\sigma &= d\sigma^{(1)} \frac{2\pi}{\alpha \log^2(s/m_\mu^2)} I_1^2\left(\sqrt{\frac{2\alpha}{\pi}} \log\frac{s}{m_\mu^2}\right), \\ d\sigma^{(1)} &= (2\pi\alpha^2/s^2)\, dt \end{aligned}\right\} \tag{137.29}$$

being the cross-section in the Born approximation in the ultra-relativistic case (see §81, Problem 6).†

† References to further literature on double-logarithmic asymptotic forms are given in the review article by V. G. Gorshkov, *Soviet Physics Uspekhi* **16**, 322. 1973.

CHAPTER XIV

ELECTRODYNAMICS OF HADRONS

§ 138. Electromagnetic form factors of hadrons

So far, we have been discussing in this book the quantum electrodynamics of particles not capable of strong interactions (electrons, positrons and muons). There are also many particles known as *hadrons*,† which take part in strong interactions. These include, for example, protons and neutrons with spin $\frac{1}{2}$, pions with spin zero, and other particles. Atomic nuclei, which consist of protons and neutrons, are of course hadrons also.

Present-day theory does not enable us to derive a complete electrodynamics of hadrons. It is clearly impossible to set up equations which determine the electromagnetic interactions of hadrons without taking account of the considerably more powerful strong interactions. In particular, the latter must be included in order to obtain the explicit form of the hadron current, in order to describe the interactions in quantum electrodynamics. The hadron current will therefore be introduced as a phenomenological quantity whose structure is determined only by the general kinematic requirements independent of any assumption about the dynamics of the interactions.‡

The electromagnetic interaction operator will again have the form

$$e(\hat{J}\hat{A}), \tag{138.1}$$

where the current is now denoted by the capital letter J to distinguish it from the electron current j. Since the order of magnitude of this interaction is specified by the same elementary charge e, we can again use the methods of perturbation theory.§

Let us first establish the form of the transition current between two states of a hadron in free motion (without any transformation of the hadron itself). This current occurs in the three-ended diagram

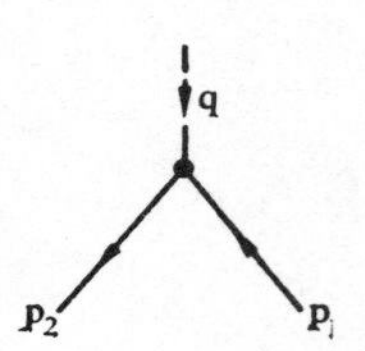

(138.2)

† From the Greek *hadros*, "large, massive".

‡ Topics of hadron electrodynamics which involve the quark model will not be discussed in this book.

§ In this chapter, e (>0) denotes the unit charge.

which itself may be part of a more complex diagram (for example, that for elastic electron scattering by a hadron). The broken line in the diagram (138.2) represents a virtual photon; it cannot correspond to a real photon, since a free particle cannot absorb (or emit) such a photon; and

$$q^2 = (p_2 - p_1)^2 < 0.$$

If we take first a hadron with spin zero, let u_1 and u_2 be the wave amplitudes of the initial and final states of the hadron, in which its 4-momenta are p_1 and p_2; for a spin-zero particle these amplitudes are scalars (or pseudoscalars).† The hadron transition current J_{fi} between these two states must be bilinear in u_1 and u_2^*. We can write it

$$J_{fi} = u_2^* \Gamma u_1, \tag{138.3}$$

where the 4-vector Γ is an unknown vertex operator (the circle in the diagram (138.2)). If we put $u_1 = u_2 = 1$, then $J_{fi} = \Gamma$.

The conservation of current is a universal property in electrodynamics, due to the gauge invariance of the theory. In the momentum representation, it is expressed by the orthogonality of the transition current and the photon 4-momentum $q = p_2 - p_1$:

$$qJ_{fi} = 0. \tag{138.4}$$

Here, this means that Γ must have the form

$$\Gamma = PF(q^2), \tag{138.5}$$

where $P = p_1 + p_2$ and $F(q^2)$ is a scalar function of q^2, which is the only invariant independent variable. Since the type of hadron is unchanged by the transition, $p_1^2 = p_2^2 = M^2$, where M is the mass of the hadron, and hence $Pq = 0$.

The matrix elements (138.3) with Γ given by (138.5), and therefore the operator $\hat{J}$, are true 4-vectors. The interaction operator (138.1) is consequently a true scalar. Thus the electromagnetic interaction of spin-zero hadrons is necessarily P-invariant. It is also T-invariant. Time reversal interchanges the initial and final 4-momenta, leaving the sum $P = p_1 + p_2$ unaltered, and changes the sign of the space components of the 4-momenta but not the time components. This is also the way in which the components of the 4-potential A are transformed, so that the product $\hat{J}\hat{A}$ is unaltered.

The invariant function $F(q^2)$ is called the *electromagnetic form factor* of the hadron. The explicit form of this quantity cannot, of course, be established in a phenomenological theory, but it is real (in the region $q^2 < 0$ at present under consideration), as follows from the same arguments as were used in §116 for the

† The plane wave is written in the form $\psi = [u/\sqrt{(2\varepsilon)}]e^{-ipx}$. The normalization to one particle per unit volume corresponds (for spin-zero particles) to the normalization of the scalar by $u^*u = 1$, and we can take simply $u = 1$ (§10). In the following we define the transition current with respect to the amplitudes u_1, u_2, in accordance with the notation used in §64.

electron form factors: when $q^2 < 0$, there are no intermediate states which could appear on the right-hand side of the unitarity relation, and so the matrix M_{fi} (and therefore J_{fi}) is Hermitian.

If $q = 0$, the initial and final states are the same, and J_{fi} becomes a diagonal matrix element. In particular, $e(J^0)_{ii}/2\varepsilon_i = eF(0)$ is the charge density, which is equal to the total charge Ze of the particle because of the normalization to one particle per unit volume.

For an electrically neutral particle, $F(0) = 0$, but it must be emphasized that this does not imply a strictly neutral particle. If a particle is strictly neutral and has a definite charge parity, $F(q^2) \equiv 0$ for all q^2; since the current operator is charge-odd (§13), its matrix elements between two states of the same hadron are zero.†

Let us now go on to hadrons with spin $\frac{1}{2}$. In this case the wave amplitudes u_1, u_2 are bispinors, and the hadron current is

$$J_{fi} = \bar{u}_2 \Gamma u_1. \tag{138.6}$$

From the bilinear combinations of $\bar{u}_2$ and u_1 and the 4-vectors p_1 and p_2, both true 4-vectors and pseudovectors (satisfying the condition (138.4)) can be constructed. Hence the condition of P invariance of the interaction is not necessarily satisfied, and must be imposed separately.‡ As has been shown in §116, under this condition the vertex operator contains two independent and (if $q^2 < 0$) real form factors. We shall now write it as

$$\begin{aligned}\Gamma^\mu &= 2M(F_e - F_m)\frac{P^\mu}{P^2} + F_m\gamma^\mu \\ &= 2M\left(F_e - \frac{q^2}{4M^2}F_m\right)\frac{P^\mu}{P^2} - \frac{F_m}{2M}\sigma^{\mu\nu}q_\nu \\ &= (4M^2F_e - q^2F_m)\frac{\gamma^\mu}{P^2} + \frac{2M}{P^2}(F_e - F_m)\sigma^{\mu\nu}q_\nu,\end{aligned} \tag{138.7}$$

where $F_e(q^2)$ and $F_m(q^2)$ are invariant form factors (M being the mass of the hadron). It is easily seen from the equation $P^2 + q^2 = 4M^2$ and (116.5) that the three expressions in (138.7) are equivalent.§

† This does not, of course, mean that such a hadron has no interaction with an electromagnetic field. The product of two current operations, $\hat{J}(x)\hat{J}(x')$, is charge-even, and its matrix elements are non-zero for transitions between states having the same charge parity. Thus a strictly neutral hadron can scatter a photon or simultaneously emit two photons, i.e. can take part in processes of a higher order in α.

‡ We shall not consider possible violations of parity conservation in electromagnetic interactions in consequence of virtual weak interactions.

§ The convenience of defining the form factors as in (138.7) (R. Sachs, 1962) will be shown below. In the literature, use is also made of form factors F_1 and F_2 defined similarly to f and g in (116.6):

$$\Gamma^\mu = F_1\gamma^\mu - \frac{F_2}{2M}\sigma^{\mu\nu}q_\nu.$$

These are related to F_e and F_m by

$$F_e = F_1 + F_2q^2/4M^2, \qquad F_m = F_1 + F_2.$$

The electromagnetic form factors are among the invariant amplitudes defined in §70. They may be regarded as the amplitudes of a "reaction" which is (in its annihilation channel) the decay of a virtual photon into a hadron and an antihadron. The virtual photon is a "particle" with spin 1. The fact that its decay into two particles with spin $\frac{1}{2}$ must be described by two independent amplitudes is easily seen by calculating the corresponding helicity amplitudes $\langle\lambda_b\lambda_c|S^J|\lambda_a\rangle$ (see §69): the P invariance means that the four non-zero elements of the S-matrix must be equal in pairs:

$$\langle\tfrac{1}{2}\tfrac{1}{2}|S^1|1\rangle = \langle-\tfrac{1}{2}-\tfrac{1}{2}|S^1|-1\rangle,$$
$$\langle\tfrac{1}{2}-\tfrac{1}{2}|S^1|0\rangle = \langle-\tfrac{1}{2}\tfrac{1}{2}|S^1|0\rangle.$$

The requirement of T invariance (or C invariance in the annihilation channel) imposes no further relations between these elements. This is connected with the fact that the interaction described by the vertex operator (138.7) is necessarily T-invariant also (but this does not apply for particles with higher spins).

When $q \to 0$, the terms of zero and first order (in q) in (138.7) are

$$\Gamma^\mu = F_e(0)\gamma^\mu - \frac{1}{2M}[F_m(0) - F_e(0)]\sigma^{\mu\nu}q_\nu. \qquad (138.8)$$

Hence it follows (see §116) that $F_e(0) \equiv Z$ is the electric charge of the particle (in units of e), and $F_m(0) - F_e(0)$ is its anomalous magnetic moment (in units of $e/2M$).†

So far we have used only form factors in momentum space. This is, of course, sufficient to describe the observed phenomena. Purely as an illustration, however, we shall give a somewhat more intuitive interpretation of the form factors, regarding them as the Fourier transforms of certain functions of the coordinates. To do this, it is convenient to take a frame of reference in which $\mathbf{P} = \mathbf{p}_1 + \mathbf{p}_2 = 0$ (called the *Breit frame*); this is always possible, since $P^2 > 4M^2 > 0$. In this frame, $\varepsilon_1 = \varepsilon_2 \equiv \varepsilon$, so that $P^0 = 2\varepsilon$, and the components of the 4-vector q are $q^0 = 0$, $\mathbf{q} = 2\mathbf{p}_2 = -2\mathbf{p}_1$.

For a spin-zero hadron, the transition current in the Breit frame has a particularly simple form:

$$J^0_{fi}/2\varepsilon = F(-\mathbf{q}^2), \qquad \mathbf{J} = 0.$$

From this we see that $F(-\mathbf{q}^2)$ may be interpreted as the Fourier transform of a static distribution of charges with density

$$e\rho(\mathbf{r}) = e\frac{1}{(2\pi)^3}\int F(-\mathbf{q}^2)\, e^{i\mathbf{q}\cdot\mathbf{r}}\, d^3q. \qquad (138.9)$$

In this sense, the particle is said to have a spatial electromagnetic structure: when $F = \text{constant} = Z$, we have $\rho(\mathbf{r}) = Z\delta(\mathbf{r})$, and the dependence of the form factor on $\mathbf{q}$

† For example, the proton has $F_e(0) = 1$, $F_m(0) - F_e(0) = 1.79$; the neutron has $F_e(0) = 0$, $F_m(0) = -1.91$ (the magnetic moment being entirely "anomalous").

is interpreted as the difference between the charge distribution and a point charge. It must be stressed, however, that this interpretation is not to be taken literally. The function $\rho(\mathbf{r})$ does not relate to any particular frame of reference, since there is a different frame for each value of $\mathbf{q}$.

The Breit frame is the same as the rest frame of the particle, and independent of $\mathbf{q}$, only in the non-relativistic limit of small $\mathbf{q}^2 \ll M^2$, when the change in the particle energy in scattering is negligible. The initial and final states of the particle are the same in this approximation, and so the transition current becomes a diagonal matrix element, with $\rho(\mathbf{r})$ the actual spatial distribution of charges. For the elementary particles, however, the values of $|\mathbf{q}|$ for which the form factors vary considerably are only slightly less than M. In the non-relativistic limit for these particles, we can therefore replace $F(-\mathbf{q}^2)$ by $F(0)$, i.e. regard the particle as a point. The situation is different for nuclei. The mass M of a nucleus is proportional to the number A of nucleons in it, and the typical value of $|\mathbf{q}| \sim 1/R$, i.e. is proportional to $A^{-1/3}$ (R being the radius of the nucleus). Hence, for sufficiently heavy nuclei, the typical $\mathbf{q}^2 \ll M^2$, and so the non-relativistic treatment is permissible throughout the significant range. Thus the concept of the electromagnetic structure of the nucleus becomes a quite definite one.

For a spin-$\frac{1}{2}$ particle, (138.7) gives in the Breit frame

$$\begin{aligned} J^0_{fi} &= (F_e - F_m)\frac{M}{\varepsilon}(\bar{u}_2 u_1) + F_m(\bar{u}_2\gamma^0 u_1) \\ &= F_e(\bar{u}_2\gamma^0 u_1), \end{aligned} \tag{138.10}$$

$$\mathbf{J}_{fi} = \frac{1}{2M} F_m i\mathbf{q}\times(\bar{u}_2 \mathbf{\Sigma} u_1), \tag{138.11}$$

where $\mathbf{\Sigma}$ is the three-dimensional spin operator (matrix) (21.21), and in (138.10) the equation

$$\varepsilon(\bar{u}_2\gamma^0 u_1) = M(\bar{u}_2 u_1)$$

has been used; this is easily verified by means of Dirac's equations for u_1 and $\bar{u}_2$ with $\mathbf{p}_1 = -\mathbf{p}_2$.

The time component of the transition current (138.10) differs from the expression for a "point particle" (an electron) by a factor $F_e(-\mathbf{q}^2)$. We can therefore say that the form factor F_e (called the *electric form factor*) describes the "spatial distribution of charge" in accordance with (138.9).

Similarly, the three-dimensional vector (138.11) can be correlated with the "spatial distribution" of the current density $e\mathbf{j}(\mathbf{r}) = \operatorname{curl} \boldsymbol{\mu}(\mathbf{r})$, where

$$\boldsymbol{\mu}(\mathbf{r}) = \frac{e}{2M}\mathbf{\Sigma}\int F_m(-\mathbf{q}^2)e^{i\mathbf{q}\cdot\mathbf{r}}\, d^3q$$

is the "magnetic moment density". Thus the *magnetic form factor* F_m may be interpreted as the magnetic moment spatial distribution density, of course with the same reservations as were expressed above with regard to the charge distribution.

F_m includes both the "normal" Dirac magnetic moment and the "anomalous" magnetic moment specific to the hadron, the "density" of the latter corresponding to the difference $F_m - F_e$.

It is reasonable to suppose that the singular points of the hadron electromagnetic form factors, like those of the electron form factors, occur at real positive values of the argument $t = q^2 = -\mathbf{q}^2$. From this we can derive certain conclusions as to the asymptotic behaviour of the distribution $\rho(r)$ (and $\boldsymbol{\mu}(r)$) as $r \to \infty$. The same transformation of the integral (138.9) as was applied in §114 to derive (114.4) from (114.3) gives the following result for large r:

$$\rho(r) \propto e^{-\kappa_0 r},$$

where κ_0^2 is the abscissa of the first singularity of the form factor $F(q^2)$; cf. the footnote to §114. If the nearest singularity is given by the threshold for the production of a pair of hadrons (each with mass M_0) by the virtual photon, then $\kappa_0 = 2M_0$.

§ 139. Electron–hadron scattering

The formulae derived in §138 can be applied to the elastic scattering of an electron by a hadron. Let the initial and final 4-momenta of the hadron be p_h and p'_h, and those of the electron p_e and p'_e; then

$$p_e + p_h = p'_e + p'_h. \tag{139.1}$$

The process is represented by the diagram

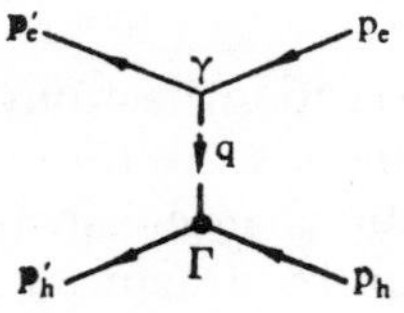

(139.2)

The emission of a virtual photon by the electron corresponds to the ordinary vertex operator γ, and its absorption by the hadron corresponds to the operator Γ.

Let us take the most interesting case, that of a hadron with spin $\frac{1}{2}$ (for example, the scattering of an electron by a proton or neutron). The diagram (139.2) corresponds to the scattering amplitude

$$M_{fi} = -4\pi e^2 \frac{1}{q^2} (\bar{u}'_e \gamma^\mu u_e)(\bar{u}'_h \Gamma_\mu u_h); \tag{139.3}$$

in this chapter, the electron charge is $-e$. The calculation of the cross-section from this amplitude is essentially the same as the calculations in §81; the operator Γ is conveniently taken in the form of the first expression (138.7).

For the scattering of unpolarized particles, the result is

$$d\sigma = \frac{\pi\alpha^2 dt}{[s-(M+m)^2][s-(M-m)^2]t^2(1-t/4M^2)}\times$$
$$\times\left\{F_e^2[(s-u)^2+(4M^2-t)t]-\frac{t}{4M^2}F_m^2[(s-u)^2-(4M^2-t)(4m^2+t)]\right\}, \quad (139.4)$$

where M is the hadron mass and m the electron mass,

$$s=(p_e+p_h)^2, \qquad t=q^2=(p_e-p_e')^2, \qquad u=(p_e-p_h')^2,$$
$$s+t+u=2m^2+2M^2.$$

The following are some limiting cases.

For the scattering of an electron by a heavy nucleus, an important case is that in which the momentum transfer $|\mathbf{q}|$ from the electron to the nucleus is small compared with the mass of the nucleus, but not small compared with $1/R$ (where R is the radius of the nucleus), so that the nucleus cannot be regarded as a point. In this case, the centre-of-mass system approximately coincides with the rest frame of the nucleus, the recoil of the nucleus may be neglected, and the electron energy is unchanged. Then

$$-t=\mathbf{q}^2\ll M^2, \qquad \pi|dt|=\mathbf{p}_e^2\,do_e',$$
$$s-M^2\approx M^2-u\approx 2M\varepsilon_e,$$

and formula (139.4) becomes

$$d\sigma=\frac{\alpha^2 do_e'}{\mathbf{q}^4}(4\varepsilon_e^2-q^2)F_e^2(-\mathbf{q}^2). \quad (139.5)$$

In this approximation, the cross-section has only the term containing the electric form factor, and (139.5) corresponds to formula (80.5), which applies to the scattering of an electron by a static charge distribution.

In the scattering of an electron by a neutron at rest, in the same limiting case $\varepsilon_e\ll M$ (where M is the mass of the neutron), the form factors can be replaced by their values for $\mathbf{q}=0$, since, as already mentioned, for a single nucleon the characteristic "radius" of the charge distribution is comparable with $1/M$.† Since the neutron is electrically neutral, $F_e(0)=0$, and the cross-section becomes

$$d\sigma=\alpha\mu^2\left[\frac{4(\varepsilon_e^2-m^2)}{\mathbf{q}^2}+1\right]do_e'$$
$$=\alpha\mu^2(\operatorname{cosec}^2\tfrac{1}{2}\vartheta+1)\,do_e', \quad (139.6)$$

where $\mu=(e/2M)F_m(0)$ is the magnetic moment of the neutron and ϑ is the

† The empirical value of the r.m.s. "radius" of the nucleon is about $3.5/M\approx 1/2m_\pi$ (where m_π is the pion mass).

scattering angle. This formula corresponds to the scattering of an electron by a point magnetic moment at rest.

Finally, we shall give the cross-section for the scattering of an ultra-relativistic electron by a nucleon, with $|\mathbf{q}| \gg m$. As before, $\mathbf{q}^2$ denotes the square of the momentum transfer in the centre-of-mass system, and hence the invariant $t = -\mathbf{q}^2$. In the rest frame of the original nucleon (the laboratory system), we have

$$-t \approx 2(p_e p'_e) = 2\varepsilon_e \varepsilon'_e (1 - \cos\vartheta),$$

where ε_e and ε'_e are the initial and final energies of the electron, and ϑ is the scattering angle in this system. In the ultra-relativistic case, ε'_e is related to ϑ by the same formula as in the scattering of a photon (cf. (86.8)):

$$\frac{1}{\varepsilon'_e} - \frac{1}{\varepsilon_e} = \frac{1}{M}(1 - \cos\vartheta).$$

Hence

$$-t = \frac{4\varepsilon_e^2 \sin^2 \frac{1}{2}\vartheta}{1 + (2\varepsilon_e/M)\sin^2 \frac{1}{2}\vartheta}, \qquad (139.7)$$

$$\pi d|t| = \frac{\varepsilon_e^2 \, do'_e}{[1 + (2\varepsilon_e/M)\sin^2 \frac{1}{2}\vartheta]^2}, \qquad (139.8)$$

where $do'_e = 2\pi \sin\vartheta \, d\vartheta$. In formula (139.4), we can everywhere omit the electron mass m; expressing all quantities in terms of t and $s - M^2 = 2M\varepsilon_e$, we have

$$d\sigma = \frac{\pi\alpha^2 d|t|}{\varepsilon_e^2 t^2}\left\{F_e^2(t)\left[\frac{(4M\varepsilon_e + t)^2}{4M^2 - t} + t\right] - \right.$$

$$\left. - \frac{t}{4M^2} F_m^2(t)\left[\frac{(4M\varepsilon_e + t)^2}{4M^2 - t} - t\right]\right\}, \qquad (139.9)$$

or, using (139.7) and (139.8),

$$d\sigma = do'_e \frac{\alpha^2 \cos^2 \frac{1}{2}\vartheta}{4\varepsilon_e^2 \sin^4 \frac{1}{2}\vartheta} \frac{1}{1 + (2\varepsilon_e/M)\sin^2 \frac{1}{2}\vartheta} \times$$

$$\times \left\{\frac{F_e^2 - (t/4M^2)F_m^2}{1 - t/4M^2} - \frac{t}{2M^2} F_m^2 \tan^2 \frac{1}{2}\vartheta\right\} \qquad (139.10)$$

(M. N. Rosenbluth, 1950).

Note that the form factors F_e and F_m contribute independently to the cross-section, and there are no interference terms between them. This shows that the form factors have been appropriately chosen.

PROBLEM

Find the cross-section for scattering of an electron by a hadron with spin zero.

SOLUTION. Using (138.5), we have instead of (139.3)

$$M_{fi} = -\frac{4\pi e^2}{q^2}(\bar{u}'_e(\gamma P_h)u_e)F(q^2).$$

The cross-section is found to be

$$d\sigma = \frac{\pi\alpha^2\,dt[(s-u)^2+(4M^2-t)t]}{[s-(M+m)^2][s-(M-m)^2]t^2}F^2(t),$$

with the same notation as in (139.4). When $|t| \gg m^2$,

$$d\sigma = do'_e\frac{\alpha^2}{4\varepsilon_e^2}\frac{\cos^2\frac{1}{2}\vartheta}{\sin^4\frac{1}{2}\vartheta}\frac{F^2(t)}{1+(2\varepsilon_e/M)\sin^2\frac{1}{2}\vartheta}$$

with the same notation as in (139.10).

§ 140. The low-energy theorem for bremsstrahlung

In §98 we have investigated the emission of a photon in a collision of particles, in the limit of zero photon frequency, and found that the amplitude of the process is inversely proportional to ω and can be simply expressed in terms of the amplitude for the same collision without emission of a soft photon; the latter will again be conventionally referred to below as the amplitude for "elastic" scattering and denoted by $M_{fi}^{(\mathrm{el})}$. In the next approximation with respect to ω,

$$M_{fi} = M_{fi}^{(-1)} + M_{fi}^{(0)}, \tag{140.1}$$

where a correction term independent of ω ($\propto \omega^0$) has been added to the principal term ($\propto \omega^{-1}$). We shall see that this correction term also, like the principal term, can be expressed in terms of $M_{fi}^{(\mathrm{el})}$, and that this is true whatever the detailed electromagnetic structure of the hadron. This result is called the *low-energy theorem* for bremsstrahlung (F. E. Low, 1958).

We have seen in §98 that the main contribution to the amplitude for emission of a soft photon (corresponding to the first term in (140.1)) arises from diagrams in which the photon is emitted by the initial or final particle. These are diagrams of the form

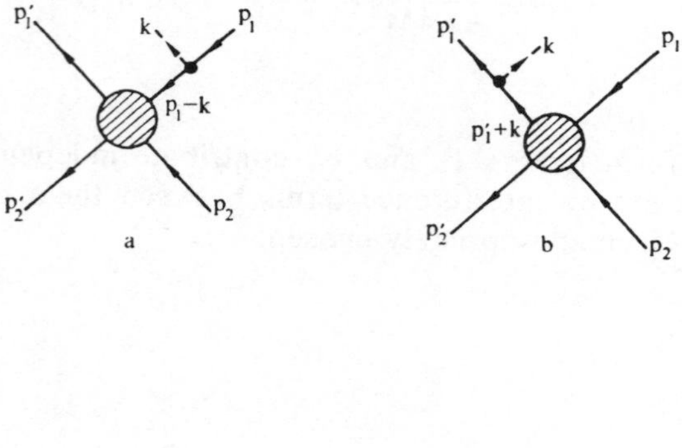

(140.2)

in contrast to those of the form

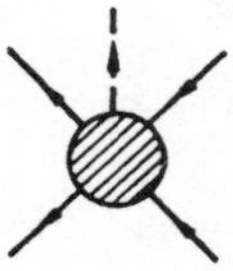

(140.3)

where the photon line comes from the internal parts of the diagram. A characteristic feature of the diagrams (140.2) is that they can be divided into two parts by cutting only one (initial or final) virtual-hadron line. Thus they illustrate an important property: there exists a one-particle intermediate state with one hadron. We have seen in §79 that, because of the unitarity conditions, this property necessarily causes a pole singularity of the amplitude.

Let us assume for simplicity that only one (denoted by the subscript 1) of two colliding hadrons has an electric charge and therefore can radiate, and that neither hadron has any spin. The wave amplitudes u of such hadrons are scalars, which will be taken as unity. Then the contribution of the pole part of the diagram (140.2a) to the amplitude is

$$iM_{fi}^{(a)} = \sqrt{(4\pi)}\, e_\mu^*(2p_1^\mu - k^\mu)\, eF \frac{1}{(p_1 - k)^2 - M^2}\, i\Gamma. \tag{140.4}$$

The first factor corresponds to the photon k (e_μ being its polarization 4-vector). The second factor corresponds to the electromagnetic hadron vertex (the black dot in the diagram), and is written in the form (138.5), with F the hadron form factor. The third factor is the propagator of the virtual hadron $p_1 - k$ (M being its mass). Finally, the factor $i\Gamma$ denotes the whole remaining section. This differs from the amplitude of the elastic process

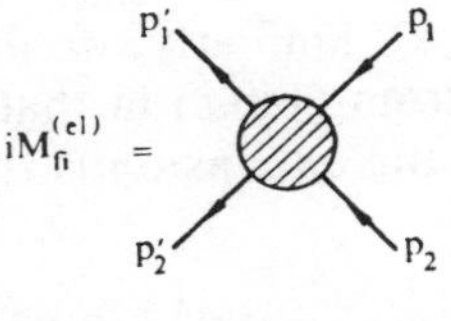

(140.5)

in that the real hadron p_1 is replaced by the virtual hadron $p_1 - k$.

The first few terms in the expansion of (140.4) in powers of ω include (1) terms inversely proportional to ω, (2) terms independent of ω but depending on the direction of $\mathbf{k}$, (3) terms independent of both ω and $\mathbf{k}$. The terms of the third kind only are given also by non-singular diagrams of the type (140.3), which do not have a pole singularity, and by the non-pole parts of the diagrams (140.2). We shall see that all such terms of this sort are jointly and unambiguously given by the terms of the first and second kinds when the condition of gauge invariance is applied, so that no separate calculation of these terms is necessary.

The amplitude of the elastic process (140.5) depends only on two invariants:

$$\left.\begin{aligned} s &= (p_1 + p_2)^2 = (p_1' + p_2')^2, \\ t &= (p_2' - p_2)^2. \end{aligned}\right\} \tag{140.6}$$

The replacement of p_1 by $p_1 - k$ not only changes s into $(p_1 - k + p_2)^2$ but also brings in a dependence on a further variable,

$$(p_1 - k)^2 - M^2 = -2p_1 k,$$

which represents the "non-physicalness" of the momentum $p_1 - k$. But the first term in the expansion in powers of this new variable (a small quantity) already eliminates the singularity in the amplitude (140.4), and therefore can yield in this amplitude only terms independent of k, which according to the foregoing discussion are not yet relevant. Thus we reach the important conclusion that Γ in (140.4) can be replaced by the physical amplitude $M_{fi}^{(el)}(s, t)$ with the change

$$s \to (p_1 + p_2 - k^2) = s - 2k(p_1 + p_2). \tag{140.7}$$

The first terms in its expansion are given by

$$\Gamma \to M_{fi}^{(el)}(s, t) - 2(kp_1 + kp_2)(\partial M_{fi}^{(el)}/\partial s)_t.$$

For a similar reason, it is unimportant that the electromagnetic form factor F here relates to a vertex at which only one of the two external hadron lines (p_1 and $p_1 - k$) is physical. The form factor can therefore be replaced by the one described in §138, for a vertex with two physical external lines; since the photon k is then a real photon, we have $F(k^2) = F(0) = Z_1$, where eZ_1 is the hadron charge.

Thus (140.4) gives

$$M_{fi}^{(a)} = Z_1 e\sqrt{(4\pi)}\frac{2(e^*p_1)}{-2(kp_1)} - Z_1 e\sqrt{(4\pi)}\,2(e^*p_1)\frac{1}{-2(kp_1)}2(p_2k)\frac{\partial M_{fi}^{(el)}}{\partial s} + \cdots, \tag{140.8}$$

where the dots represent terms independent of $\mathbf{k}$ (whereas the second term in (140.8) depends on the direction of $\mathbf{k}$). Similarly, we find that the contribution to M_{fi} from the diagram (140.2b) differs from (140.8) in that p_1, p_2 and k are replaced by p_1', p_2' and $-k$. The leading term in the expansion is the already familiar expression (cf. (98.5))

$$M_{fi}^{(-1)} = Z_1 e\sqrt{(4\pi)}\left(\frac{p_1'e^*}{p_1'k} - \frac{p_1e^*}{p_1k}\right)M_{fi}^{(el)}. \tag{140.9}$$

The terms independent of k can be found from the condition for the amplitude as a whole to be gauge-invariant: it must be unaffected by the change $e^* \to e^* +$ constant $\times\, k$, i.e. must have the form $M_{fi} = e_\mu^* J^\mu$, with $k_\mu J^\mu = 0$. It is easy to see that, for this to be so, we must add to (140.8) the term independent of k

$$-2Z_1 e\sqrt{(4\pi)}(p_2e^*),$$

and similarly for the diagram (140.2b). The final result is

$$M_{fi}^{(0)} = 2Z_1 e\sqrt{(4\pi)}e_\mu^*\left[p_1^\mu\frac{(p_2k)}{(p_1k)} - p_2^\mu + p_1'^\mu\frac{(p_2'k)}{(p_1'k)} - p_2'^\mu\right]\frac{\partial M_{fi}^{(el)}}{\partial s}. \tag{140.10}$$

The problem can be solved by means of this formula, which may be more compactly written by using the identity

$$2p_{2\nu}\left(\frac{\partial}{\partial s}\right)_t \equiv \left(\frac{\partial}{\partial p_1^\nu}\right)_{p_1', p_2, p_2'}$$

and similarly for $\partial/\partial p_1'$, and the differential operators

$$\hat{d}_{1\mu} = \frac{p_{1\mu}}{(p_1 k)}k^\nu \frac{\partial}{\partial p_1^\nu} - \frac{\partial}{\partial p_1^\mu} \tag{140.11}$$

and similarly for $\hat{d}_{1\mu}'$. Then

$$M_{fi}^{(0)} = Z_1 e\sqrt{(4\pi)}\, e_\mu^*(\hat{d}_1^\mu + \hat{d}_1'^\mu)\, M_{fi}^{(\mathrm{el})}. \tag{140.12}$$

The cross-section is given by $|M_{fi}|^2$; to the appropriate accuracy,

$$|M_{fi}|^2 = |M_{fi}^{(-1)}|^2 + 2\,\mathrm{re}\,(M_{fi}^{(-1)}M_{fi}^{(0)*}). \tag{140.13}$$

The second term gives the required correction to the emission cross-section. Summation over the polarizations of the photon gives as the value of this correction

$$-4\pi(Z_1 e)^2\left(\frac{p'}{(p'k)} - \frac{p}{(pk)}\right)^\mu (\hat{d}_1' + \hat{d}_1)_\mu\, |M_{fi}^{(\mathrm{el})}|^2. \tag{140.14}$$

Thus the correction to the emission cross-section is expressed in terms of the cross-section for the elastic process and its derivative with respect to s.

If the charged hadron has spin $\frac{1}{2}$, the calculations are unchanged in principle; only the specific form of the vertices and propagators is altered. It is found that formula (140.14) remains valid after averaging over the polarizations of the hadrons and the photon (T. H. Burnett and N. M. Kroll, 1968).

§141. The low-energy theorem for photon–hadron scattering

In the limit of low frequencies, the cross-section for the scattering of a photon by any charged particle at rest tends to its classical value given by Thomson's formula. This limit corresponds to an amplitude independent of the photon frequency ω, which we denote by $M_{fi}^{(0)}$. It is found, however, that not only the first term in the expansion of the amplitude in powers of ω,

$$M_{fi} = M_{fi}^{(0)} + M_{fi}^{(1)}. \tag{141.1}$$

but also the next term ($M^{(1)} \sim \omega$), are independent of the specific electromagnetic structure of the hadron for photon scattering, as well as for the bremsstrahlung discussed in §140 (F. E. Low, 1954; M. Gell-Mann and M. L. Goldberger, 1954).

This process is represented by diagrams of three types:

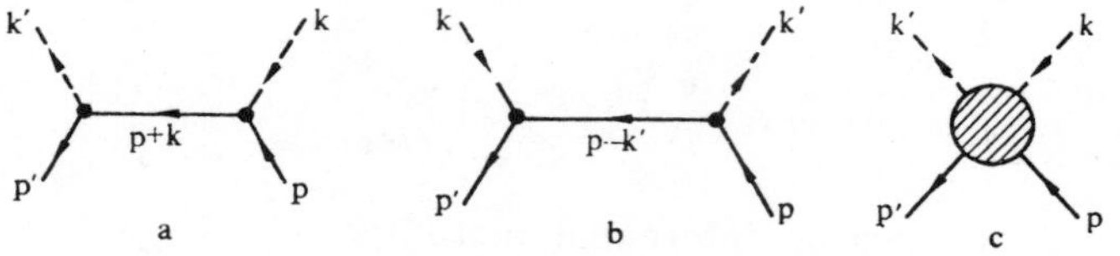

(141.2)

of which the first two again have a one-particle intermediate state, and therefore a pole singularity.

The analysis and the principle of the calculations are the same as in §140. In practice, we need only determine the contribution from the pole parts of the diagrams (141.2a) and (141.2b), expressing their electromagnetic vertices in terms of the static form factors (the charge Ze and the anomalous magnetic moment μ_{an}), as in (140.15).

Unlike the bremsstrahlung case, however, the corrections to the Compton-effect cross-section are important only for particles that have spin. This is because, for bremsstrahlung, as well as the spin-dependent corrections, there are corrections arising from the energy dependence of the amplitude of the "elastic" process. In photon scattering, this amplitude is replaced by form factors which, for "physical external lines", are constants independent of the energy, and therefore the cor-reactions arise only from the magnetic moment, which is zero for spinless particles. We shall discuss the scattering of a photon by a spin-$\frac{1}{2}$ hadron.

If M_{fi} denotes the contribution of the pole diagrams to the scattering amplitude, then (cf. (86.3), (86.4))

$$M_{fi} = -4\pi(Ze)^2 e'^{*}_{\mu} e_{\nu} (\bar{u} Q^{\mu\nu} u), \tag{141.3}$$

where

$$\left.\begin{aligned} Q^{\mu\nu} = (\gamma^\mu + S'^\mu)\frac{\gamma p + \gamma k + M}{s - M^2}(\gamma^\nu - S^\nu) + (\gamma^\nu - S^\nu)\frac{\gamma p - \gamma k' + M}{u - M^2}(\gamma^\mu + S'^\mu),\\ s = (p+k)^2 = (p'+k')^2, \quad u = (p-k')^2 = (p'-k)^2, \end{aligned}\right\} \tag{141.4}$$

and for brevity we have put

$$\mu_{\text{an}}\sigma^{\mu\lambda}k_\lambda = ZeS^\mu, \qquad \mu_{\text{an}}\sigma^{\mu\lambda}k'_\lambda = ZeS'^\mu. \tag{141.5}$$

By interchanging the operators $\gamma p + M$ and using the equations $\bar{u}'(\gamma p' - M) = (\gamma p - M)u = 0$, we can transform (141.4) to

$$\begin{aligned} Q^{\mu\nu} = \left[(\gamma^\mu + S'^\mu)\frac{(\gamma k)\gamma^\nu + 2p^\nu}{2pk} + \frac{\gamma^\nu(\gamma k) - 2p'^\nu}{2p'k}(\gamma^\mu + S'^\mu)\right] - \\ - \left[\frac{\gamma^\mu(\gamma k') + 2p'^\mu}{2p'k'}S^\nu + S^\nu\frac{\gamma^\mu(\gamma k') - 2p^\mu}{2pk'}\right] - \\ - \left[S'^\mu\frac{\gamma p + \gamma k + M}{2pk}S^\nu - S^\nu\frac{\gamma p - \gamma k' + M}{2pk'}S'^\mu\right]. \end{aligned} \tag{141.6}$$

This form, and the corresponding one with k and k' interchanged, clearly show that (141.3) is gauge-invariant; the relevant condition is

$$k'_\mu(\bar{u}' Q^{\mu\nu} u) = (\bar{u}' Q^{\mu\nu} u) k_\nu = 0, \tag{141.7}$$

in verifying which it must be remembered that $(\gamma k)(\gamma k) = 0$, $kS = k'S' = 0$.

Since the pole part of the scattering amplitude is thus gauge-invariant by itself, so must be the regular part of the amplitude (which includes the contribution of the diagram (141.2c)). Hence in turn it follows that the expansion of this part in powers of k and k' must begin with quadratic terms; cf. the similar comment relating to the condition (127.5). That is, the regular part of the amplitude includes only terms starting with those proportional to $\omega\omega' \sim \omega^2$, and makes no contribution to the terms concerned here, which are proportional to ω^0 and ω^1. These are therefore included in (141.3).

To calculate the terms in question, we use the laboratory system, in which the initial hadron is at rest. For the photons, we take a three-dimensionally transverse gauge, in which $e_0 = e'_0 = 0$. Then $pe = 0$, $p'e'^* \sim |\mathbf{p}'| \sim \omega$, and from (141.6) it is obvious that the leading terms in the expansion of M_{fi} will be proportional to ω^0, and that the terms in μ_{an} will contribute only to the ω^1 terms.

The wave amplitudes of the initial and final hadrons in the laboratory system are, with the necessary accuracy,

$$u = \sqrt{(2M)} \begin{pmatrix} w \\ 0 \end{pmatrix}, \qquad \bar{u}' = \sqrt{(2M)} \left[w'^*, -\frac{w'^*}{2M} (\mathbf{k} - \mathbf{k}') \cdot \boldsymbol{\sigma} \right],$$

where w and w' are three-dimensional spinors.

A straightforward calculation gives the result

$$M_{fi}^{(0)} = -8\pi (Ze)^2 (\mathbf{e}'^* \cdot \mathbf{e})(w'^* w), \tag{141.8}$$

$$\begin{aligned} M_{fi}^{(1)} = &- 16\pi i M \mu_{\text{an}}^2 \omega (w'^* \boldsymbol{\sigma} w) \cdot (\mathbf{n}' \times \mathbf{e}'^*) \times (\mathbf{n} \times \mathbf{e}) - \\ &- 4\pi i Z e \mu_{\text{an}} \omega (w'^* \boldsymbol{\sigma} w) \cdot \{\mathbf{n}(\mathbf{n} \times \mathbf{e} \cdot \mathbf{e}'^*) + \\ &+ (\mathbf{n} \times \mathbf{e}) \mathbf{n} \cdot \mathbf{e}'^* - \mathbf{n}'(\mathbf{n}' \times \mathbf{e}'^* \cdot \mathbf{e}) - \\ &- (\mathbf{n}' \times \mathbf{e}'^*) \mathbf{n} \cdot \mathbf{e} - 2\mathbf{e}'^* \times \mathbf{e}\}, \end{aligned} \tag{141.9}$$

where $\mathbf{n} = \mathbf{k}/\omega$, $\mathbf{n}' = \mathbf{k}'/\omega'$.

The scattering cross-section is

$$d\sigma = \frac{1}{64\pi^2} |M_{fi}|^2 \frac{\omega'^2}{M^2 \omega^2} do'; \tag{141.10}$$

see (64.19). For scattering by a charged particle, both $M_{fi}^{(1)}$ and $M_{fi}^{(0)}$ are non-zero. The accuracy used allows us to retain in $|M_{fi}|^2$ the terms $|M_{fi}^{(0)}|^2$ and $\text{re}(M_{fi}^{(0)} M_{fi}^{(1)*})$. The first of these gives the Thomson scattering. The second becomes zero on averaging over the polarizations of the photons and hadrons. In scattering by a charged hadron, therefore, the corrections under consideration occur only in the polarization effects.

For scattering by an electrically neutral hadron, $M_{fi}^{(0)} = 0$ and the cross-section is determined by $|M_{fi}^{(1)}|^2$. After averaging over the polarizations of the final particles and summing over those of the initial particles, it is (in ordinary units)

$$d\sigma = \frac{2\mu^4\omega^2}{\hbar^2 c^4}(2 + \sin^2\vartheta)\, do', \tag{141.11}$$

where ϑ is the photon scattering angle and the anomalous magnetic moment is equal to the total moment μ. The angle dependence of this cross-section is the same as for antisymmetric scattering (see §60, Problem 2).

§ 142. Multipole moments of hadrons

Let us now consider the transition current corresponding to a diagram of the same kind as (138.2):

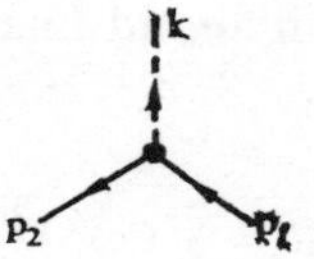

(142.1)

but with the lines p_1 and p_2 pertaining to different particles (masses M_1 and M_2); the photon line $k = p_1 - p_2$ will be more conveniently represented here as leaving the vertex. The photon may be either virtual or real, the only necessary condition being $k^2 < (M_1 - M_2)^2$, so that the value $k^2 = 0$ is permissible. Thus the applications of this diagram include, in particular, processes of photon emission in transformations of nuclei as well as other particles (for nuclei, the initial and final particles are the same nucleus in different states).

The most interesting case here is that in which the wavelength of the photon is large compared with the characteristic "dimensions" of the particle (i.e. those which appear in its form factors, equal to the "radius" in the case of a nucleus). Then the transition current can be expanded in powers of k.†

Note first of all that we must have

$$J_{fi} = 0 \text{ when } k = 0, \tag{142.2}$$

since the limit $k \to 0$ corresponds to a potential constant in space and time, but such a potential has no physical significance and cannot give rise to real processes. The same conclusion can be reached by a more formal argument: the currents discussed in §138 were non-zero for $k = 0$ on account of the terms proportional to the 4-vector $P = p_1 + p_2$, but when $M_1 \neq M_2$ the product $Pk \neq 0$, and such terms are therefore forbidden by the condition for the current to be transverse.

† The following treatment is due to V. B. Berestetskiĭ (1948).

This condition for the current $J_{fi} = (\rho_{fi}, \mathbf{J}_{fi})$ is, in three-dimensional form,

$$\mathbf{k}\cdot\mathbf{J}_{fi} = \omega\rho_{fi}, \tag{142.3}$$

and can be satisfied in two ways:

$$\mathbf{J}_{fi} = \omega\mathbf{v}(\mathbf{k}, \omega), \qquad \rho_{fi} = \mathbf{k}\cdot\mathbf{v}(\mathbf{k}, \omega) \tag{142.4}$$

or

$$\mathbf{J}_{fi} = \mathbf{k}\times\mathbf{a}(\mathbf{k}, \omega), \qquad \rho_{fi} = 0. \tag{142.5}$$

Here $\mathbf{v}$ is a polar vector and $\mathbf{a}$ an axial vector. The current is said to be of the electric and magnetic type respectively. According to (142.2), $\mathbf{v}$ and $\mathbf{a}$ are finite or zero when $\mathbf{k}, \omega \to 0$.

Let the photon energy $\omega \ll M_1$. Then the recoil may be neglected, and the final particle M_2 also may be regarded as being at rest (in the rest frame of M_1); $\omega = M_1 - M_2$ is given quantity. The states of the particles M_1 and M_2 at rest are specified by three-dimensional spinors w_1 and w_2 of ranks $2s_1$ and $2s_2$, where s_1 and s_2 are the spins of the particles. The transition current must be a bilinear combination of w_1 and w_2^*. From the products of the components of these spinors, we can form irreducible tensors with ranks $l = s_1 + s_2, \ldots, |s_1 - s_2|$; for a given l, they are true tensors or pseudotensors according to the internal parities of the particles M_1 and M_2. Apart from these tensors, we have available only the vector $\mathbf{k}$. In order to obtain the first term in the expansion of the current in powers of $\mathbf{k}$, we must form from these quantities a vector of the lowest possible power of $\mathbf{k}$. This is done by taking the tensor of lowest rank and contracting it $l-1$ times with the vector $\mathbf{k}$. This will give the polar vector $\mathbf{v}$ or the axial vector $\mathbf{a}$.

Let Q_{lm} be the spherical components of the tensor formed from the wave amplitudes of the particles. The spherical components of the tensor of rank $l-1$ formed from the components of $\mathbf{k}$ are $|\mathbf{k}|^{l-1}\, Y_{l-1,m}(\mathbf{n})$, where $\mathbf{n} = \mathbf{k}/\omega$. From the general rule for the addition of spherical tensors (see *QM*, (107.3)), the spherical components of the vector $\mathbf{v}$ may be written

$$\mathbf{v}_\lambda = (-1)^{\lambda+1} i^l \frac{\sqrt{(4\pi)}}{(2l-1)!!}\sqrt{\frac{2l+1}{l}}\,|\mathbf{k}|^{l-1}\sum_m \begin{pmatrix} l-1 & 1 & l \\ \lambda+m & -\lambda & -m\end{pmatrix} Q_{l,-m} Y_{l-1,\lambda+m}(\mathbf{n}),$$

where λ takes the values 0 and ± 1; the choice of the common factor is explained below. Using formulae (7.16), we can express $\mathbf{v}$ in terms of spherical harmonic vectors:

$$\mathbf{v} = i^l \frac{\sqrt{(4\pi)}|\mathbf{k}|^{l-1}}{(2l-1)!!\sqrt{[l(2l+1)]}}\sum_m (-1)^{l-m} Q_{l,-m}[\sqrt{(l+1)}\mathbf{Y}^{(e)}_{lm}(\mathbf{n}) + \sqrt{l}\,\mathbf{Y}^{(l)}_{lm}(\mathbf{n})]. \tag{142.6}$$

Substitution in (142.4) gives the El transition current:

$$\mathbf{J}_{fi} = i^l \frac{\sqrt{(4\pi)}\omega|\mathbf{k}|^{l-1}}{(2l-1)!!\sqrt{[l(2l+1)]}}\sum_m (-1)^{l-m} Q^{(e)}_{l,-m}[\sqrt{(l+1)}\mathbf{Y}^{(e)}_{lm}(\mathbf{n}) + \sqrt{l}\,\mathbf{Y}^{(l)}_{lm}(\mathbf{n})], \tag{142.7}$$

$$\rho_{fi} = i^l \frac{\sqrt{(4\pi)}|\mathbf{k}|^{l}}{(2l-1)!!\sqrt{(2l+1)}}\sum_m (-1)^{l-m} Q^{(e)}_{l,-m}\, Y_{lm}(\mathbf{n}); \tag{142.8}$$

$|\mathbf{k}|$ and ω are distinguished in each formula, with a view to possible applications to real photons and also virtual photons, for which the two quantities are not equal.

In (142.7) and (142.8) it is assumed that the spherical tensor Q_{lm} (here denoted by $Q_{lm}^{(e)}$) is a true tensor. If it is a pseudotensor (denoted by $Q_{lm}^{(m)}$), then (142.6) defines the pseudovector **a**, and substitution in (142.5) gives the Ml transition current:

$$\left.\begin{aligned} \mathbf{J}_{fi} &= i^l \frac{\sqrt{(4\pi)}}{(2l-1)!!}\sqrt{\frac{l+1}{l(2l+1)}}|\mathbf{k}|^l \sum_m (-1)^{l-m} Q_{l,-m}^{(m)} \mathbf{Y}_{lm}^{(m)}(\mathbf{n}), \\ \rho_{fi} &= 0. \end{aligned}\right\} \tag{142.9}$$

The quantities $Q_{lm}^{(e)}$ and $Q_{lm}^{(m)}$ are the hadron electric and magnetic multipole transition moments. Their role in hadron electrodynamics is exactly analogous to that of the corresponding quantities in electron electrodynamics. However, for electron systems these moments can in principle be calculated from the wave functions (as the matrix elements of the corresponding operators), whereas in hadron electrodynamics they occur as phenomenological quantities whose values are determined from experiment.

The normalization of these quantities in (142.7)–(142.9) is chosen so as to agree with their definition in §46. This can be verified by regarding the currents (142.7)–(142.9) as Fourier components of the transition current in the coordinate representation. For example, expanding the factor $e^{-i\mathbf{k}\cdot\mathbf{r}}$ in the integral

$$\rho_{fi}(\mathbf{k}) = \int \rho_{fi}(\mathbf{r}) e^{-i\mathbf{k}\cdot\mathbf{r}}\, d^3x \tag{142.10}$$

by means of (46.3), we get

$$\rho_{fi}(\mathbf{k}) = 4\pi i^l \sum_{l,m} Y_{lm}(\mathbf{n}) \int \rho_{fi}(\mathbf{r}) Y_{lm}^*(\mathbf{r}/r) g_l(|\mathbf{k}|r)\, d^3x.$$

Retaining the term with the smallest value of l such that the integral is non-zero, and replacing $g_l\,(|\mathbf{k}|\,r)$ for $|\mathbf{k}|r \ll 1$ by the first term in its expansion (46.5), we return to (142.9), with

$$Q_{lm}^{(e)} = \sqrt{\frac{4\pi}{2l+1}} \int r^l \rho_{fi}(\mathbf{r}) Y_{lm}(\mathbf{r}/r) d^3x, \tag{142.11}$$

in agreement with the definition (46.7).

It can also be shown that, when applied to the emission of a real photon, the formulae derived above give results already known. The amplitude of a transition with emission of a photon having momentum $\mathbf{k} = \omega\mathbf{n}$ and polarization $e = (0, \mathbf{e})$ is

$$M_{fi} = -e\sqrt{(4\pi)}\mathbf{e}^* \cdot \mathbf{J}_{fi}. \tag{142.12}$$

If the nucleus has definite values of the angular-momentum component M_i and M_f in the initial and final states, only one term remains in each sum over m in (142.7)–(142.9), namely that with $m = M_i - M_f$. Since, from (16.23), the products

$\mathbf{Y}^{(e)}_{lm}\cdot\mathbf{e}^{(\lambda)*}$ and $\mathbf{Y}^{(m)}_{lm}\cdot\mathbf{e}^{(\lambda)*}$ (where $\lambda=\pm 1$ is the helicity of the photon, and $\mathbf{e}^{(\lambda)}\perp\mathbf{n}$) are proportional to $D^{l}_{\lambda m}$, we obtain again the formulae given in §48.

The differential emission probability is†

$$dw = 2\pi\delta[\omega-(E_i-E_f)]|M_{fi}|^2 d^3k/2\omega(2\pi)^3, \tag{142.13}$$

where E_i and E_f are the initial and final energies of the nucleus. The total probability is found by summation over the polarizations and integration over d^3k. Substituting (142.7) or (142.9) in (142.12) and thence in (142.13), and performing the operations just mentioned, we again obtain (46.9) or (47.2).

Formulae (142.7)–(142.9) include all possible cases of the emission of a real photon. For virtual photons, there is another possible case which they do not include (R. H. Fowler, 1930).

If the spins and parities of the initial and final states of the nucleus are the same, we can obtain from their wave amplitudes a scalar Q_0, and from this a transition current of the form

$$\rho_{fi} = Q_0\mathbf{k}^2, \qquad \mathbf{J}_{fi} = Q_0\omega\mathbf{k}. \tag{142.14}$$

Q_0 is called the *monopole* (E0) transition moment. The corresponding transition amplitude for the emission of a real photon is zero, since $\mathbf{e}^*\cdot\mathbf{k}=0$. The monopole current, however, may give rise to transitions involving the emission of a virtual photon. It is, moreover, the only such source when $s_1=s_2=0$ and all the multipole moments are zero.

The monopole current (142.14) is analogous to the electric quadrupole current as regards its dependence on ω and $\mathbf{k}$. Accordingly, the moment Q_0 also is a quantity of the same order as the quadrupole moment. The same conclusion can be reached by regarding (142.14) as the Fourier components of the current in the coordinate representation. Using in (142.10) the expansion of $e^{-i\mathbf{k}\cdot\mathbf{r}}$ in powers of $\mathbf{k}\cdot\mathbf{r}$ and assuming that $\rho_{fi}(\mathbf{r})$ is spherically symmetrical, we get

$$\rho_{fi}(\mathbf{k}) = -\tfrac{1}{6}\mathbf{k}^2\int\rho_{fi}(r)r^2\,d^3x.$$

Comparison with (142.14) shows that

$$Q_0 = -\tfrac{1}{6}\int\rho_{fi}(r)r^2\,d^3x. \tag{142.15}$$

The similarity to the quadrupole moment is obvious.

PROBLEMS

PROBLEM 1. Find the probability of ionization of an atom from the K shell because of the excitation energy ω of the nucleus (called internal conversion of γ-rays), in an Ml nuclear transition,

† The factor $2\pi\delta$ in this formula, replacing $(2\pi)^4\delta^{(4)}$ in (64.11), arises because momentum is not conserved when the recoil of the nucleus is neglected, and so only the conservation of energy remains.

neglecting the binding energy of the electron in the atom and the influence of the nuclear field on the wave functions of the nucleus.†

SOLUTION. The process is described by the diagram

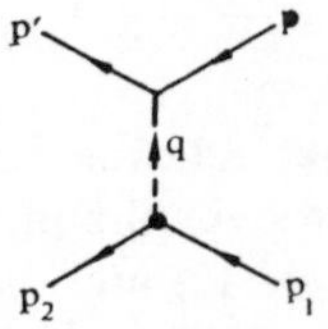

(1)

where p_1 and p_2 pertain to the nucleus at rest in different states, and $p = (m, 0)$ and $p' = (m + \omega, \mathbf{p}')$ are the 4-momenta of the initial and final electrons. This diagram corresponds to the amplitude

$$M_{fi} = -e^2 \frac{4\pi}{q^2} \bar{u}(p')(\gamma J_{fi})u(p),$$

where J_{fi} is the transition current of the nucleus. After summation over the final polarizations of the electron, and averaging over its initial polarizations, we get

$$\tfrac{1}{2} \sum_{\text{polar.}} |M_{fi}|^2 = e^4 \frac{16\pi^2}{(q^2)^2} \{q^2(J_{fi}J^*_{fi}) + 4(J_{fi}p)(J^*_{fi}p)\},$$

using the fact that $J_{fi}q = 0$ and therefore $J_{fi}p = J_{fi}p'$. The conversion probability is calculated from

$$dw_{\text{conv}} = 2|\psi_i(0)|^2 \left(\frac{|\mathbf{p}|}{m}\, d\sigma\right)_{\mathbf{p}\to 0},$$

where $d\sigma$ is the cross-section for the scattering process represented by the diagram (1) with $p = (\varepsilon, \mathbf{p})$, and ψ_i is the wave function of the atomic electron; for a K electron $|\psi_i(0)|^2 = (Z\alpha m)^3/\pi$. The factor 2 takes account of the two electrons in the K shell of the atom. The cross-section $d\sigma$ is

$$d\sigma = 2\pi\delta(\varepsilon + \omega - \varepsilon')|M_{fi}|^2 \frac{d^3p'}{2|\mathbf{p}| \cdot 2\varepsilon'(2\pi)^3};$$

cf. the last footnote.

For Ml transitions, the current J_{fi} must be taken from (142.9). The integration of dw_{conv} over $d\varepsilon'$ removes the delta function, and the integration over do' replaces $|\mathbf{Y}^{(m)}_{lm}|^2$ by unity. The conversion probability is thus expressed in terms of $|Q^{(m)}_{l,-m}|^2$. But, according to (46.9), the probability w_γ for the spontaneous emission of a photon in the same nuclear transition can be expressed in terms of the same quantity. The final result is

$$\frac{w_{\text{conv}}}{w_\gamma} = 2\alpha(Z\alpha)^3 \frac{m}{\omega} \left(1 + \frac{2m}{\omega}\right)^{l+\frac{1}{2}},$$

this ratio being called the conversion coefficient.

PROBLEM 2. The same as Problem 1, but for an El nuclear transition.

SOLUTION. By the same method, with the transition current given by (142.7) and (142.8), we obtain

$$\frac{w_{\text{conv}}}{w_\gamma} = 2\alpha(Z\alpha)^3 \left(1 + \frac{l}{l+1} \frac{m^2}{\omega^2}\right) \left(1 + \frac{2m}{\omega}\right)^{l-\frac{1}{2}}.$$

† This approximation implies that the nuclear charge is small and the excitation energies ω are sufficiently large (but $1/\omega$ is assumed large compared with the dimensions of the nucleus). In practice, the approximation is somewhat unsatisfactory, and a more precise calculation has to take into account the Coulomb field of the nucleus.

PROBLEM 3. The same as Problem 1, but for a monopole nuclear transition.

SOLUTION. With the transition current given by (142.14), the result is

$$w_{\text{conv}} = 16\alpha^2(Z\alpha)^3 m^3 \omega^2 \left(1 + \frac{2m}{\omega}\right)^{3/2} |Q_0|^2.$$

Since monopole emission of a photon is impossible, $|Q_0|^2$ cannot be eliminated.

§ 143. Inelastic electron–hadron scattering

Elastic electron–hadron scattering has been discussed in §139. The problem of inelastic scattering may be formulated similarly. The only difference is that the final hadron state now corresponds to another hadron or several hadrons. The law of conservation of momentum (139.1) remains valid if p'_h denotes the 4-momentum of the final hadron or the total 4-momentum of the group of hadrons formed in the scattering process. We thus have now $p'^2_h \neq p^2_h = M^2$, where M is the mass of the initial hadron.

With this difference, the inelastic scattering process is described by the same diagram (139.2). The lower vertex of the diagram is denoted by J_{fi} as in §138. However, in contrast to (138.3) or (138.6), we shall not express the transition current in terms of the vertex operator and the amplitudes of the states, in order not to specify in advance the nature of the final hadron state.

We can now write the scattering amplitude in a form analogous to (139.3):

$$M_{fi} = -\frac{4\pi e^2}{(p_e - p'_e)^2}(\bar{u}'_e \gamma_\mu u_e) J^\mu_{fi}; \tag{143.1}$$

a similar amplitude has already been used in §142, Problem 1, where energy transfer to an electron was considered, and the amplitude has a similar structure in the problem of the excitation of nuclei by electrons.

We shall assume that the initial electron energy is so high that many hadrons can be formed in the final state, and consider the "inclusive" cross-section, for which only the electron momentum in the final state is fixed, with summation over all the hadron states. This differential cross-section is written as follows, in accordance with the formulae of §64:

$$d\sigma = \frac{d^3p'_e}{4I(2\pi)^3 2\varepsilon'_e} \sum_f (2\pi)^4 \delta^{(4)}(p_h + p'_h - p_e - p'_e)|M_{fi}|^2. \tag{143.2}$$

The inclusive cross-section can depend only on three kinematic invariants, which may be determined by measurements on electrons only. The three invariants are

$$t = q^2 \equiv (p_e - p'_e)^2, \qquad s = (p_e + p_h)^2, \tag{143.3}$$

and p'^2_h. The need to include the third invariant arises because, in contrast to the case of elastic scattering, $p_h'^2$, the "mass" of the final hadron state, is now unspecified. Instead of it, however, another invariant is more convenient, namely

$$\nu = qp_h. \tag{143.4}$$

The relation between ν and $p_h'^2$ follows from $p_h' = p_h + q$:

$$p_h'^2 = M^2 + t + 2\nu. \tag{143.5}$$

If the initial hadron is stable (for instance, a proton), the rest energy of the final state exceeds M, i.e. $p_h'^2 \geqslant M^2$, and from (143.5), since $t < 0$, we have

$$\nu \geqslant \tfrac{1}{2}|t|, \tag{143.6}$$

the equality occurring for elastic scattering.

The kinematic invariants can be expressed in terms of the electron energies ε_e and ε_e' in the initial and final states and the scattering angle θ. We shall assume the electron to be ultra-relativistic ($\varepsilon_e \gg m$, $\varepsilon_e' \gg m$), and neglect its mass. Then, in the rest frame of the initial hadron (the laboratory system),

$$t = -4\varepsilon_e\varepsilon_e' \sin^2 \tfrac{1}{2}\theta, \quad \nu = M(\varepsilon_e - \varepsilon_e'), \quad s - M^2 = 2M\varepsilon. \tag{143.7}$$

Substituting (143.1) in (143.2) and summing as usual over the electron polarizations, we obtain the scattering cross-section for unpolarized electrons, which we write as

$$d\sigma = \frac{\alpha^2}{(q^2)^2} \frac{d^3p_e'}{(2\pi)^3 \cdot 8M\varepsilon_e\varepsilon_e'} w_{\mu\nu}W^{\mu\nu}, \tag{143.8}$$

or

$$d\sigma = \frac{\alpha^2}{(q^2)^2} \frac{dt\, d\nu}{4(p_hp_e)^2} w_{\mu\nu}W^{\mu\nu}, \tag{143.9}$$

where

$$w_{\mu\nu} = 4p_{e\mu}p_{e\nu}' - 2(p_{e\mu}q_\nu + p_{e\nu}q_\mu) + q^2 g_{\mu\nu}, \tag{143.10}$$

$$W^{\mu\nu} = \sum_f (2\pi)^4 \delta^{(4)}(p_h' - p_h - q) J_{fi}^\mu J_{fi}^{\nu *}. \tag{143.11}$$

The tensor $W^{\mu\nu}$ of course depends essentially on the properties of the hadron currents, and in general we can only pose the problem of its phenomenological structure, similarly to the problem of hadron form factors. We first use the fact that the tensor structure of $W^{\mu\nu}$ must be determined only by the 4-vectors relating to the lower vertex of the diagram (139.2), i.e. p_h and q. From these (and the metric tensor $g_{\mu\nu}$) five independent tensors can be constructed. The requirement of invariance under time reversal means that the tensor must be symmetrical, and there are four such. Lastly, the condition of current conservation, i.e.

$$W^{\mu\nu}q_\nu = 0, \qquad W^{\mu\nu}q_\mu = 0,$$

reduces the number of independent tensors to two. These may be taken as

$$\tau^{(1)}_{\mu\nu} = \frac{q_\mu q_\nu}{q^2} - g_{\mu\nu}, \quad \tau^{(2)}_{\mu\nu} = (p_{h\mu} - \nu q_\mu/t)(p_{h\nu} - \nu q_\nu/t), \tag{143.12}$$

and $W_{\mu\nu}$ be written as

$$W_{\mu\nu} = 4\pi M W_1 \tau^{(1)}_{\mu\nu} + (4\pi/M) W_2 \tau^{(2)}_{\mu\nu}. \tag{143.13}$$

Substituting (143.10) and (143.13) in (143.8), we put the cross-section in the form

$$d\sigma = (W_2 + 2W_1 \tan^2 \tfrac{1}{2}\theta)\, d\varepsilon'_e\, d\sigma_{\text{el}}, \tag{143.14}$$

where

$$d\sigma_{\text{el}} = \frac{\alpha^2 \cos^2 \frac{1}{2}\theta}{4\varepsilon_e^2 \sin^4 \frac{1}{2}\theta}\, do'$$

is the cross-section for scattering of an ultra-relativistic electron in a Coulomb field; cf. (80.7).

We see that the cross-section is determined by two structure functions which depend on the two invariants t and ν. If the physics of hadrons at high energies does not contain characteristic quantities having the dimensions of mass (the hypothesis of scale invariance), we may expect that the structure functions depend, at high energies, on the only dimensionless parameter, t/ν. Then the functions W_1, W_2 must be functions of one variable:

$$W_1 = (M/\nu)F_1(t/\nu), \qquad W_2 = (M/\nu)F_2(t/\nu); \tag{143.15}$$

the ratio M/ν is independent of M.

§ 144. Hadron formation from an electron–positron pair

Let us now consider the transformation of an electron–positron pair into hadrons. We denote the 4-momenta of the electron and positron by p_- and p_+, and the (total) 4-momentum of the hadrons formed by p_h, with $p_- + p_+ = p_h$. The process is represented by the diagram

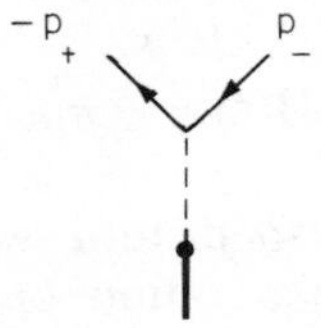

(144.1)

The lower vertex here corresponds to the transition current from the vacuum to a hadron state $|n\rangle$, which we denote by $\langle n|J|0\rangle$ as in §104.

The diagram (144.1) corresponds to the scattering amplitude

$$M_n = -(4\pi\alpha/q^2)\bar{u}(-p_+)\gamma_\mu u(p_-)n\langle n|J^\mu|0\rangle. \tag{144.2}$$

We shall consider the total cross-section σ_h for annihilation into hadrons, i.e. sum over all final states $|n\rangle$. Then, in accordance with (64.18),

$$\sigma_h = \frac{1}{4I}\sum_n |M_n|^2 (2\pi)^4\delta^{(4)}(p_h - q), \tag{144.3}$$

where $q = p_- + p_+$. The mass of the electron will henceforward be neglected; then $q^2 = 2(p_-p_+)$, $I = \frac{1}{2}q^2$.

As in §143, we write the cross-section in the form

$$\sigma_h = (4\pi)^2 w^{\mu\nu}W_{\mu\nu}/2t^3, \tag{144.4}$$

where

$$w^{\mu\nu} = \alpha(p_-^\mu q^\nu + p_-^\nu q^\mu - 2p_-^\mu p_-^\nu - \tfrac{1}{2}q^2 g^{\mu\nu}), \tag{144.5}$$

$$W_{\mu\nu} = \alpha\sum_n (2\pi)^4\delta^{(4)}(p_h - q)\langle 0|J_\nu|n\rangle\langle n|J_\mu|0\rangle \tag{144.6}$$

and $t = q^2 > 0$.

Note that t is the only kinematic invariant in this problem, with the three-ended diagram (144.1), and q is the only 4-vector on which $W_{\mu\nu}$ can depend. Hence, by the requirement of current conservation, the tensor $W_{\mu\nu}$ may be written

$$W_{\mu\nu} = \tfrac{1}{2}\rho_h(t)\left(\frac{q_\mu q_\nu}{q^2} - g_{\mu\nu}\right), \tag{144.7}$$

where $\rho_h(t)$ is the only invariant function, depending on the properties of the hadron current and determining the annihilation cross-section. Substitution of (144.5)–(144.7) in (144.4) gives

$$\sigma_h = (4\pi^2\alpha/t^2)\rho_h(t). \tag{144.8}$$

Note that the function $\rho_h(t) = -\frac{2}{3}W_\mu^\mu$ is exactly the same as $\rho(t)$ defined in (104.9) if the currents in the latter equation are taken to be hadron currents. Moreover, $\rho(t)$ is the spectral density of the self-energy function $\Pi(t)$:

$$\operatorname{im}\Pi(t) = -\pi\rho(t).$$

In the lowest approximation with respect to α, which is being considered here, the function Π is the same as the polarization operator $\mathscr{P}$. In this approximation, therefore, $\rho_h(t)$ is also the spectral density of the hadron contribution to the polarization operator:

$$\text{im}\,\mathcal{P}_h(t) = -\pi\rho_h(t). \tag{144.9}$$

Using the dispersion relation (111.13) and expressing ρ_h in terms of σ_h by (144.8), we get

$$\mathcal{P}_h(t) = -\frac{t^2}{4\pi^2\alpha}\int_0^\infty \frac{\sigma_h(t')\,dt'}{t'-t-i0}, \tag{144.10}$$

which expresses the hadron contribution to the polarization of the vacuum in terms of the measured cross-section for annihilation into hadrons.

Note that we could, in exactly the same way, solve the problem of electron–positron pair annihilation to form a muon pair (in the first approximation with respect to α, only one such pair is formed). Corresponding to the formula (144.8), the result is

$$\sigma_\mu = (4\pi^2\alpha/t^2)\rho_\mu(t), \tag{144.11}$$

where $\rho_\mu(t)$ is the spectral density of the muon polarization of the vacuum. It differs from the electron polarization only in that the electron mass m is replaced by the muon mass μ, and, from (113.8), it is

$$\rho_\mu(t) = (\alpha/3\pi)(t+2\mu^2)\sqrt{(1-4\mu^2/t)}.$$

Substitution in (144.11) brings us back to the result already derived in §81, Problem 8.

INDEX